Neutrosophic Operational Research

Florentin Smarandache • Mohamed Abdel-Basset
Editors

Neutrosophic Operational Research

Methods and Applications

 Springer

Editors
Florentin Smarandache
Department of Mathematics and Science
University of New Mexico
Gallup, NM, USA

Mohamed Abdel-Basset
Faculty of Computers and Informatics
Zagazig University
Zagazig, Sharqiyah, Egypt

ISBN 978-3-030-57199-3   ISBN 978-3-030-57197-9   (eBook)
https://doi.org/10.1007/978-3-030-57197-9

This Springer imprint is published by the registered company Springer Nature Switzerland AG
The registered company address is: Gewerbestrasse 11, 6330 Cham, Switzerland

Contents

Neutrosophical Plant Hybridization in Decision-Making Problems

M. Arockia Dasan, E. Bementa, Florentin Smarandache, and X. Tubax

1 Introduction

The world life of every human being has some problems of uncertainty, imprecise, incomplete, and inconsistent information. Fuzzy set theory is one of the wide frameworks for uncertainty. In the year 1965, A. Zadeh [1] initiated this theory to analyze imprecise, incomplete mathematical information, and this set is a generalization of a crisp set that considers the membership degree of each element from the crisp set. Adlassnig [2] applied the fuzzy logic to the computerized diagnosis system and analyzed the medical relationships. This theory [3–8] has been scientifically used in various fields such as control systems, medical diagnosis, and engineering, respectively, by M. Sugeno [3], P. R. Innocent [4], and T. J. Roos [5]. To handle the fuzzy problems, different types of similarity measures are introduced by the researchers [6–8]. By considering the degree of nonmembership of an element along with the degree of membership, K. Atanassov [9] introduced intuitionistic

M. A. Dasan (✉)
Department of Mathematics, St. Jude's College, Kanyakumari, India

Manonmaniam Sundaranar University, Tirunelveli, Tamil Nadu, India
e-mail: dassfredy@gmail.com

E. Bementa
PG and Research Department of Physics, Arulanandar College, Madurai, India
e-mail: bementa88@gmail.com

F. Smarandache
Department of Mathematics and Science, University of New Mexico, Gallup, NM, USA
e-mail: fsmarandache@gmail.com

X. Tubax
School of Information Technology, Cyryx College, Male, Maldives
e-mail: tubax_x@yahoo.co.in

© The Author(s), under exclusive license to Springer Nature Switzerland AG 2021
F. Smarandache, M. Abdel-Basset (eds.), *Neutrosophic Operational Research*,
https://doi.org/10.1007/978-3-030-57197-9_1

fuzzy sets as a generalization of a fuzzy set. S. De et al. [10] and Szmidt and Kacprzyk [11] applied the intuitionistic fuzzy sets in medical diagnosis. Biswas et al. [12] defined intuitionistic fuzzy cosine similarity measure to study professionals' health problems. Khatibi and Montazer [13] compared the relations of intuitionistic fuzzy sets, fuzzy sets with the application in medical pattern recognition. Hung and Tuan [14] reported that the approach of [10] has some questionable results on the false diagnosis of the patient's symptoms. It generally recognized the available information about the patients and medical relationships are inherently uncertain. There may be indeterminacy components for data mining in real-life problems. The neutrosophic logic can be used in this situation, which is a generalization of fuzzy, intuitionistic, Boolean, paraconsistent logics, etc. Florentin Smarandache [15, 16] developed the concept of neutrosophic logic and set which deals with three various components such as truth membership, indeterminacy membership, and falsity membership whose values are real standard or a nonstandard subset of unit interval $]0^-, 1^+[$. The single-valued neutrosophic set was first initiated by Wang et al. [17] in 1998, which is a neutrosophic set, and can be used in real-life engineering and scientific applications. Majumdar and Samanta [18] defined some similarity measures of single-valued neutrosophic sets in decision-making problems.

Neutrosophic set is an effective and useful tool to describe problems with uncertainty, imprecise, incomplete, and inconsistent information. In this regard, Smarandache and Pramanik [19] widely founded the solutions of neutrosophic decision-making problems. The multi-attributed decision-making (MADM) and multi-criteria decision-making (MCDM) problems have a wide scope in the research area of neutrosophic sets utilization. J. Ye and S. Ye [20–22] solved the multi-attributed decision-making problems by using single-valued similarity measure, tangent similarity measure, and distance-based similarity measures in neutrosophic environments. Biswas et al. [23] proposed the cosine similarity measure for solving multiple-attribute decision problems in neutrosophic single-valued sets. S. Broumi and F. Smarandache [24] introduced some kinds of similarity measures of the neutrosophic sets. In the neutrosophic environment, the cotangent similarity measure is introduced by Pramanik and Mondal [25]. V. Ulucay et al. [26] defined a hybrid distance-based similarity measure for refined neutrosophic sets with its application in medical diagnosis. N. Nabeeh et al. [27, 28] introduced a new neutrosophic technique and integrated neutrosophic-TOPSIS approach for personal selections in multi-criteria decision-making problems. Following this, M. Abdel-Basset et al. [29–31] solved the supplier selection and smart medical device selection by using the TOPSIS approach, which includes many conflicting criteria in the MCDM problems. Gaussian single-valued neutrosophic numbers and its application are invented in the MADM problem by F. Karaaslan [32]. B. C. Giri et al. [33] introduced the TOPSIS method for the MADM problem based on interval trapezoidal neutrosophic numbers. S. Aal et al. [34] formulated the two ranking methods based on single-valued triangular neutrosophic numbers to evaluate the quality of the systems. Recently the concept of the difference of two neutrosophic sets is defined by G. Jayaparthasarathy et al. [35] with the real-life application by using single-valued neutrosophic score function. Furthermore, Hossein with his collaborators

[36] introduced an ELECTRE approach to finding the best alternative for multi-attributed decision-making problems in a refined neutrosophic environment. Under the neutrosophic environment, every abovementioned researcher founded the best alternatives or the alternatives with the alternatives or the alternatives with the attributes for the entire region (problem). We may here ask some questions "Is it possible to find the best alternatives not only in the entire region but in each region?" and "Can we formulate a new method to identify the alternatives in each and entire region?"

The hybridization would cross different plants such as grasses rice, maize, cotton, and wheat for the new hybrid plant which give different increased yield and improved grain quality in both cross- and self-pollinated crops. In the eighteenth century, Mendel [37], an Augustinian who is the author of *Experiments on Plant Hybridization*, produced the F_1 hybrid by cross-breeding pea plants. The present chapter formulates a new method to answer the above questions for multi-criteria decision-making problems under neutrosophic environments. The beauty of this chapter is the proposed method uses the neutrosophic set-theoretical operation such as complements, intersections, unions, and single-valued neutrosophic score functions in both direct and reverse direction to identify the best alternatives in each and the entire region. The plant hybridization problem is solved under neutrosophic environments as a real-life application to demonstrate the effectiveness of the proposed method because the hybrid grain yields are varying place to place.

2 Preliminaries

This section presents some of the basic properties of neutrosophic sets and operations on neutrosophic sets which are used for further study.

Definition 1 [15] Let X be a non-empty set. A neutrosophic set A having the form $A = \{(x, \mu_A(x), \sigma_A(x), \gamma_A(x)) : x \in X\}$, where $\mu_A(x)$, $\sigma_A(x)$ and $\gamma_A(x) \in]0^-, 1^+[$ represent the degree of membership (namely, $\mu_A(x)$), the degree of indeterminacy (namely, $\sigma_A(x)$), and the degree of nonmembership (namely, $\gamma_A(x)$), respectively, for each $x \in X$ to the set A such that $0^- \leq \mu_A(x) + \sigma_A(x) + \gamma_A(x) \leq 3^+$ for all $x \in X$. For a non-empty set X, $N(X)$ denotes the collection of all neutrosophic sets of X.

Definition 2 [16] The following statements are true for neutrosophic sets A and B on X:

i. $\mu_A(x) \leq \mu_B(x)$, $\sigma_A(x) \leq \sigma_B(x)$, and $\gamma_A(x) \geq \gamma_B(x)$ for all $x \in X$ if and only if $A \subseteq B$.
ii. $A \subseteq B$ and $B \subseteq A$ if and only if $A = B$.
iii. $A \cap B = \{(x, min\{\mu_A(x), \mu_B(x)\}, min\{\sigma_A(x), \sigma_B(x)\}, max\{\gamma_A(x), \gamma_B(x)\}) : x \in X\}$.
iv. $A \cup B = \{(x, max\{\mu_A(x), \mu_B(x)\}, max\{\sigma_A(x), \sigma_B(x)\}, min\{\gamma_A(x), \gamma_B(x)\}) : x \in X\}$.

More generally, the intersection and the union of a collection of neutrosophic sets $\{A_i\}_{i \in \Lambda}$ are defined by $\bigcap_{i \in \Lambda} A_i = \{(x, \inf_{i \in \Lambda} \{\mu_{A_i}(x)\}, \inf_{i \in \Lambda} \{\sigma_{A_i}(x)\},$ $\sup_{i \in \Lambda} \{\gamma_{A_i}(x)\}) : x \in X\}$ and $\bigcup_{i \in \Lambda} A_i = \{(x, \sup_{i \in \Lambda} \{\mu_{A_i}(x)\}, \sup_{i \in \Lambda} \sigma_{A_i} \{(x)\},$ $\inf_{i \in \Lambda} \{\gamma_{A_i}(x)\}) : x \in X\}$.

Corollary 1 [16] The following statements are true for the neutrosophic sets A, B, C, and D on X:

  i. $A \cap C \subseteq B \cap D$ and $A \cup C \subseteq B \cup D$, if $A \subseteq B$ and $C \subseteq D$.
 ii. $A \subseteq B \cap C$, if $A \subseteq B$ and $A \subseteq C$. $A \cup B \subseteq C$, if $A \subseteq C$ and $B \subseteq C$.
iii. $A \subseteq C$, if $A \subseteq B$ and $B \subseteq C$.

Definition 3 [35] The difference of neutrosophic sets A and B on X is a neutrosophic set on X, defined as $A \setminus B = \{(x, |\mu_A(x) - \mu_B(x)|, |\sigma_A(x) - \sigma_B(x)|,$ $1 - |\gamma_A(x) - \gamma_B(x)|) x \in X\}$. Clearly $(1_n)^c = 1_n \setminus 1_n = (0, 0, 1) = 0_n$ and $(0_n)^c = 1_n \backslash 0_n = (1, 1, 0) = 1_n$; here the neutrosophic empty set is $0_n = \{(x, 0, 0, 1) : x \in X\}$ and the neutrosophic whole set is $1_n = \{(x, 1, 1, 0) : x \in X\}$.

Definition 4 [8] A neutrosophic set $A = \{(x, \mu_A(x), \sigma_A(x), \gamma_A(x)) : x \in X\}$ is called a single-valued neutrosophic set on a non-empty set X, if $\mu_A(x)$, $\sigma_A(x)$ and $\gamma_A(x) \in [0, 1]$ and $0 \leq \mu_A(x) + \sigma_A(x) + \gamma_A(x) \leq 3$ for all $x \in X$ to the set A. For each attribute, the single-valued neutrosophic score function (shortly SVNSF) of A is defined as $SVNSF(A) = 1/3m \left[\sum_{i=1}^{m} [2 + \mu_i - \sigma_i - \gamma_i]\right]$. A single-valued neutrosophic number is a neutrosophic set that is symbolized by $<T, I, F>$ such that $T, I, F \in [0, 1]$ and $0 \leq T + I + F \leq 3$.

3   Neutrosophic Methodologies in Multi-criteria Decision-Making Problems

This section systematically develops a new methodological approach in multi-criteria decision making (MCDM) problems with single-valued neutrosophic information for neutrosophic sets structure. The following methodological approach gives the necessary steps to select the best alternatives in each division and the best alternative overall regions in the MCDM situations.

Step 1: Problem Selection Consider the multi-criteria decision-making problem shown in Table 1, with m alternatives A_1, A_2, ..., A_m and p attributes B_1, B_2, ..., B_p for n regions D_1, D_2, ..., D_n to identify the best alternatives of each region and the best alternative of the entire regions.

Step 2: Problem Division Divide the selected problem into n subproblems for n regions.

Step 3: Direct Direction

Step 3(a): Subproblem Selection Select first subproblem for the corresponding region.

Table 1 Problem selection

D_1	Attributes			
alternatives	B_1	B_2	$\ldots$	B_p
A_1	$(a_{11})_1$	$(a_{12})_1$	$\ldots$	$(a_{1p})_1$
A_2	$(a_{21})_1$	$(a_{22})_1$	$\ldots$	$(a_{2p})_1$
.	.	.	$\ldots$	.
.	.	.	$\ldots$	.
.	.	.	$\ldots$	.
A_m	$(a_{m1})_1$	$(a_{m2})_1$	$\ldots$	$(a_{mp})_1$
D_2 alternatives	B_1	B_2	$\ldots$	B_p
A_1	$(a_{11})_2$	$(a_{12})_2$	$\ldots$	$(a_{1p})_2$
A_2	$(a_{21})_2$	$(a_{22})_2$	$\ldots$	$(a_{2p})_2$
.	.	.	$\ldots$	.
.	.	.	$\ldots$	.
.	.	.	$\ldots$	.
A_m	$(a_{m1})_2$	$(a_{m2})_2$	$\ldots$	$(a_{mp})_2$
.	.	.	.	.
.	.	.	.	.
.	.	.	.	.
D_n alternatives	B_1	B_2	$\ldots$	B_p
A_1	$(a_{11})_n$	$(a_{12})_n$	$\ldots$	$(a_{1p})_n$
A_2	$(a_{21})_n$	$(a_{22})_n$	$\ldots$	$(a_{2p})_n$
.	.	.	$\ldots$	.
.	.	.	$\ldots$	.
.	.	.	$\ldots$	.
A_m	$(a_{m1})_n$	$(a_{m2})_n$	$\ldots$	$(a_{mp})_n$

(a_{ij}) are single-valued neutrosophic numbers.

Step 3(b): Neutrosophic Operations For $j = 1, 2, \ldots, m$, find $C_{j1} = \{(a_{j1})_1, (a_{j2})_1, \ldots, (a_{jp})_1\}$ and $D_{j1} = \{(a_{j1})_1 \cup (a_{j2})_1, \ (a_{j1})_1 \cup (a_{j3})_1, \ldots, (a_{j1})_1 \cup (a_{jp})_1, (a_{j2})_1 \cup (a_{j3})_1, \ldots, (a_{j2})_1 \cup (a_{jp})_1, \ldots, (a_{j(p-1)})_1 \cup (a_{jp})_1\} \ldots, (a_{j2})_1 \cup (a_{jp})_1, \ldots, (a_{j(p-1)})_1 \cup (a_{jp})_1\}$ such that $(a_{jk})_1 \cup (a_{jl})_1 \notin C_{j1}$.

Step 3(c): Finding Single-Valued Neutrosophic Score Functions For $j = 1, 2, \ldots, m$, find single-valued neutrosophic score functions of C_{j1} and D_{j1} that are defined as follows: $SVNSF\left(C_{j1}\right) = 1/3p \left[\sum_{i=1}^{p} \left[2 + \mu_{ji} - \sigma_{ji} - \gamma_{ji}\right]\right]$, and $SVNSF\left(D_{j1}\right) = 1/3q \left[\sum_{i=1}^{q} \left[2 + \mu_{ji} - \sigma_{ji} - \gamma_{ji}\right]\right]$, where q is the number of elements of D_{j1}. $SVNSF(A_j) = \{SVNSF(C_{j1})$, if $SVNSF(D_{j1}) = 0$. Otherwise, $1/2[\ SVNSF(C_{j1}) + SVNSF(D_{j1})\}$.

Step 3(d): Arrangement For $j = 1, 2, \ldots, m$, arrange all the single-valued neutrosophic score function's values for the alternatives $A_1, A_2, \ldots, A_m$ in ascending order.

Step 3(e): Repetition Repeat steps 3(a)–3(d) for each subproblem of the corresponding region.

Step 3(f): Direct Rank Tabulate all the direct ranks $DR(A_j)$ of the alternatives $A_1, A_2, \ldots, A_m$ by giving ranks $1, 2, \ldots, m-1, m$ in the ascending order to the alternatives for each region $D_1, D_2, \ldots, D_n$. Find the total direct rank $TDR(A_j)$

of each alternative A_j by using $TDR\left(A_j\right) = \sum_{j=1}^{n} DR\left(A_j\right)$, for each $j = 1, 2,$
$\ldots, m$.

Step 4: Reverse Direction

Step 4(a): Subproblem Selection Select first subproblem for the corresponding region.

Step 4(b): Neutrosophic Operations For $j = 1, 2, \ldots, m$, find $E_{j1} = \{(a_{j1})_1{}^c,$
$(a_{j2})_1{}^c, \ldots,$
$(a_{jp})_1{}^c\}$ and $F_{j1} = \{(a_{j1})_1{}^c \cap (a_{j2})_1{}^c, (a_{j1})_1{}^c \cap (a_{j3})_1{}^c, \ldots, (a_{j1})_1{}^c \cap (a_{jp})_1{}^c, (a_{j2})_1{}^c$
$\cap (a_{j3})_1{}^c, \ldots, (a_{j2})_1{}^c \cap (a_{jp})_1{}^c, \ldots, (a_{j(p-1)})_1{}^c \cap (a_{jp})_1{}^c\}$ such that $(a_{jk})_1{}^c \cap$
$(a_{jl})_1{}^c \notin E_{j1}$.

Step 4(c): Finding Single-Valued Neutrosophic Score Functions For $j = 1,$
$2, \ldots, m$, find single-valued neutrosophic score functions of E_{j1} and F_{j1} that
are defined as follows: $SVNSF\left(E_{j1}\right) = 1/3p\left[\sum_{i=1}^{p}\left[2 + \mu_{ji} - \sigma_{ji} - \gamma_{ji}\right]\right]$,
and $SVNSF\left(F_{j1}\right) = 1/3q\left[\sum_{i=1}^{q}\left[2 + \mu_{ji} - \sigma_{ji} - \gamma_{ji}\right]\right]$, where q is the
number of elements of F_{j1}. $SVNSF(A_j) = \{SVNSF(E_{j1})$, if $SVNSF(F_{j1}) = 0$.
Otherwise, $1/2[SVNSF(E_{j1}) + SVNSF(F_{j1})\}$.

Step 4(d): Arrangement For $j = 1, 2, \ldots, m$, arrange all the single-valued
neutrosophic score function's values for the alternatives $A_1, A_2, \ldots, A_m$ in
ascending order.

Step 4(e): Repetition Repeat steps 4(a) to 4(d) for each subproblem of the
corresponding region.

Step 4(f): Reverse Rank Tabulate all the reverse ranks $RR(A_j)$of the alternatives
$A_1, A_2, \ldots, A_m$ by giving ranks $1, 2, \ldots, m - 1, m$ in the ascending order to the
alternatives for each region $D_1, D_2, \ldots, D_n$. Find the total reverse rank $TRR(A_j)$
of each alternative A_j by using $TRR\left(A_j\right) = \sum_{j=1}^{n} RR\left(A_j\right)$, for each $j = 1, 2,$
$\ldots, m$.

Step 5: Decision of Subproblems From the table of steps 3(f) and step 4(f),
calculate the rank difference from direct direction to reverse direction for
each region. Decide the alternative with the highest positive value is the best
alternative in that region and the alternative with the least value is the worst
alternative in that region; here take 0 as positive.

Step 6: Determination For each $j = 1, 2, \ldots, m$, calculate the values of
$TDR(A_j) - TRR(A_j)$ from the table of steps 3(f) and 4(f). Arrange all these
values in order as highest positive $\geq$ highest negative $\geq$ second-highest positive $\geq$ second-highest negative $\geq \ldots \geq$ least number; here take 0 as positive.

Step 7: Final Decision From the order arrangement of the values, decide the
alternative which has the highest positive value is the best alternative and
the alternative of the highest negative value places in second and so on. The
alternative with the least value is the worst alternative in the entire region.

4   The Summary of Process

The summary of the proposed process is demonstrated in Fig. 1.

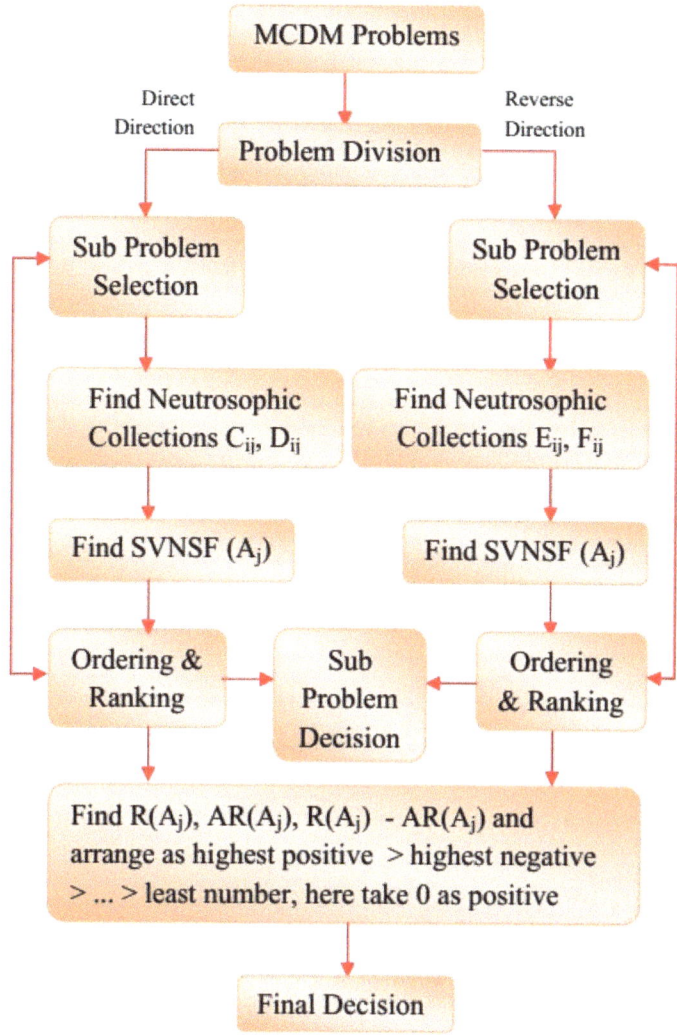

Fig. 1 The summary of the proposed process

5 Numerical Examples in Multi-criteria Decision-Making Problem

Agriculture has increased the volume of information available from modern technologies and comprises uncertainties in the increased yield and grain qualities of the plant hybridization. In the hybridization process, the difficult task is identifying the different hybrid plants with their quantity of yield under the environmental factors. This section demonstrates an agricultural problem for the effectiveness and applicability of the above-proposed approach.

Step 1: Problem Selection Consider the following table giving information about four kinds of hybrid plants (alternatives) such as wheat P_1, grasses rice P_2, cotton P_3, and maize P_4 and three environmental factors (attributes) such as fertilizer B_1, water B_2, and sunlight B_3 when farmers cultivated these plants in three different countries (regions) D_1, D_2, and D_3. We need to enquire which hybrid plant yields the high quantity in three environmental factors soil, water, and sunlight. Table 2 shows the membership, the indeterminacy, and the nonmembership functions of hybrid plants P_1, P_2, P_3, and P_4 under environmental factors fertilizer B_1, water B_2, sunlight B_3. From Table 2, we can observe that the hybrid *wheat* plant P_1 gives high yield ($\mu = 0.8$, $\sigma = 0.3$, $\gamma = 0.1$) under the environmental factor *fertilizer* in the region D_1, but the hybrid *grasses rice* plant gives low yield ($\mu = 0.1$, $\sigma = 0.2$, $\gamma = 0.3$) under the environmental factor *water* in region D_2.

Step 2: Problem Division Here there are three regions D_1, D_2, and D_3; therefore divide the MCDM problem into three subproblems for each region as shown in Tables 3, 4, and 5.

Step 3: Direct Direction

Step 3(a): Subproblem Selection Select the subproblem 1 for the corresponding country D_1.

Table 2 Problem selection

D_1	Attributes		
alternatives	B_1	B_2	B_3
P_1	(0.8,0.3,0.1)	(0.7,0.2,0.1)	(0.5,0.4,0.3)
P_2	(0.9,0.1,0.1)	(0.6,0.1,0.2)	(0.7,0.3,0.1)
P_3	(0.7,0.2,0.3)	(0.1,0.2,0.9)	(0.2,0.6,0.1)
P_4	(0.6,0.1,0.4)	(0.4,0.4,0.4)	(0.2,0.3,0.4)
D_2	Attributes		
alternatives	B_1	B_2	B_3
P_1	(0.7,0.2,0.3)	(0.1,0.3,0.7)	(0.5,0.2,0.3)
P_2	(0.4,0.1,0.1)	(0.1,0.2,0.3)	(0.4,0.1,0.3)
P_3	(0.2,0.3,0.4)	(0.4,0.3,0.2)	(0.2,0.4,0.1)
P_4	(0.7,0.1,0.1)	(0.6,0.2,0.2)	(0.8,0.2,0.3)
D_3	Attributes		
alternatives	B_1	B_2	B_3
P_1	(0.6,0.1,0.2)	(0.7,0.2,0.2)	(0.7,0.3,0.1)
P_2	(0.9,0.1,0.1)	(0.2,0.3,0.1)	(0.8,0.6,0.1)
P_3	(0.8,0.2,0.6)	(0.3,0.4,0.1)	(0.6,0.5,0.4)
P_4	(0.7,0.4,0.3)	(0.4,0.6,0.1)	(0.2,0.4,0.5)

Table 3 Subproblem 1 for the country D_1

D_1	Attributes		
alternatives	B_1	B_2	B_3
P_1	(0.8,0.3,0.1)	(0.7,0.2,0.1)	(0.5,0.4,0.3)
P_2	(0.9,0.1,0.1)	(0.6,0.1,0.2)	(0.7,0.3,0.1)
P_3	(0.7,0.2,0.3)	(0.1,0.2,0.9)	(0.2,0.6,0.1)
P_4	(0.6,0.1,0.4)	(0.4,0.4,0.4)	(0.2,0.3,0.4)

Step 3(b): Neutrosophic Operations

i. $C_{11} = \{(0.8,0.3,0.1), (0.7,0.2,0.1), (0.5,0.4,0.3)\}$ and $D_{11} = \{(0.8,0.4,0.1), (0.7,0.4,0.1)\}$.

ii. $C_{21} = \{(0.9,0.1,0.1), (0.6,0.1,0.2), (0.7,0.3,0.1)\}$ and $D_{21} = \{(0.9,0.3,0.1)\}$.

iii. $C_{31} = \{(0.7,0.2,0.3), (0.1,0.2,0.9), (0.2,0.6,0.1)\}$ and $D_{31} = \{(0.7,0.6,0.1)\}$.

iv. $C_{41} = \{(0.6,0.1,0.4), (0.4,0.4,0.4), (0.2,0.3,0.4)\}$ and $D_{41} = \{(0.6,0.4,0.4), (0.6,0.3,0.4)\}$.

Step 3(c): Finding Single-Valued Neutrosophic Score Functions

The single-valued neutrosophic score functions (shortly SVNSF) of C_{j1} and D_{j1} are defined as follows:

i. SVNSF $(C_{11}) = 0.7333$, SVNSF $(D_{11}) = 0.75$, where $q = 2$. SVNSF $(P_1) = 0.7417$.

ii. SVNSF $(C_{21}) = 0.8111$, SVNSF $(D_{21}) = 0.8333$, where $q = 1$. SVNSF $(P_2) = 0.8222$.

iii. SVNSF $(C_{31}) = 0.5222$, SVNSF $(D_{31}) = 0.6667$, where $q = 1$. SVNSF $(P_3) = 0.5945$.

iv. SVNSF $(C_{41}) = 0.5667$, SVNSF $(D_{41}) = 0.6167$, where $q = 2$. SVNSF $(P_4) = 0.5917$.

Step 3(d): Arrangement $P_2 \geq P_1 \geq P_3 \geq P_4$.

Step 3(e): Repetition

Step 3e1(a): Subproblem Selection Select the subproblem 2 to find the best alternatives for the corresponding country D_2.

Step 3e1(b): Neutrosophic Operations

i. $C_{12} = \{(0.7,0.2,0.3), (0.1,0.3,0.7), (0.5,0.2,0.3)\}$ and $D_{12} = \{(0.7,0.3,0.3), (0.5,0.3,0.3)\}$.

ii. $C_{22} = \{(0.4,0.1,0.1), (0.1,0.2,0.3), (0.4,0.1,0.3)\}$ and $D_{22} = \{(0.4,0.2,0.1), (0.4,0.2,0.3)\}$.

iii. $C_{32} = \{(0.2,0.3,0.4), (0.4,0.3,0.2), (0.2,0.4,0.1)\}$ and $D_{32} = \{(0.4,0.4,0.1)\}$.

iv. $C_{42} = \{(0.7,0.1,0.1), (0.6,0.2,0.2), (0.8,0.2,0.3)\}$ and $D_{42} = [(0.7,0.2,0.1), (0.8,0.2,0.1), (0.8,0.2,0.2)\}$.

Step 3e1(c): Finding Single-Valued Neutrosophic Score Functions

The single-valued neutrosophic score functions of C_{j2} and D_{j2} are defined as follows:

i. SVNSF $(C_{12}) = 0.5889$, SVNSF $(D_{12}) = 0.6667$, where $q = 2$. SVNSF $(P_1) = 0.6278$.

Table 4 Subproblem 2 for the country D_2

D_2 alternatives	Attributes		
	B_1	B_2	B_3
P_1	(0.7,0.2,0.3)	(0.1,0.3,0.7)	(0.5,0.2,0.3)
P_2	(0.4,0.1,0.1)	(0.1,0.2,0.3)	(0.4,0.1,0.3)
P_3	(0.2,0.3,0.4)	(0.4,0.3,0.2)	(0.2,0.4,0.1)
P_4	(0.7,0.1,0.1)	(0.6,0.2,0.2)	(0.8,0.2,0.3)

Table 5 Subproblem 3 for the country D_3

D_3 alternatives	Attributes B_1	B_2	B_3
P_1	(0.6,0.1,0.2)	(0.7,0.2,0.2)	(0.7,0.3,0.1)
P_2	(0.9,0.1,0.1)	(0.2,0.3,0.1)	(0.8,0.6,0.1)
P_3	(0.8,0.2,0.6)	(0.3,0.4,0.1)	(0.6,0.5,0.4)
P_4	(0.7,0.4,0.3)	(0.4,0.6,0.1)	(0.2,0.4,0.5)

    ii. SVNSF (C_{22}) = 0.6444, SVNSF (D_{22}) = 0.6667, where q = 2. SVNSF (P_2) = 0.6555.

    iii. SVNSF (C_{32}) = 0.5667, SVNSF (D_{32}) = 0.6333, where q = 1. SVNSF (P_3) = 0.6.

    iv. SVNSF (C_{42}) = 0.7778, SVNSF (D_{42}) = 0.8111, where q = 3. SVNSF (P_4) = 0.7389.

Step 3e1(d): Arrangement $P_4 \geq P_2 \geq P_1 \geq P_3$.

Step 3e: Repetition

Step 3e2(a): Subproblem Selection Select the subproblem 3 to find the best alternatives for the corresponding country D_3.

Step 3e2(b): Neutrosophic Operations

    i. C_{13} = {(0.6,0.1,0.2), (0.7,0.2,0.2), (0.7,0.3,0.1)} and D_{13} = $\varnothing$.

    ii. C_{23} = {(0.9,0.1,0.1), (0.2,0.3,0.1), (0.8,0.6,0.1)} and D_{23} = {(0.9,0.3,0.1), (0.9,0.6,0.1)}.

    iii. C_{33} = {(0.8,0.4,0.1), (0.3,0.4,0.1), (0.6,0.5,0.4)} and D_{33} = {(0.8,0.4,0.1), (0.8,0.5,0.4)(0.6,0.5,0.1)}.

    iv. C_{43} = {(0.7,0.4,0.3), (0.4,0.6,0.1), (0.2,0.4,0.5)} and D_{43} = {(0.7,0.6,0.1)}.

Step 3e2(c): Finding Single-Valued Neutrosophic Score Functions

The single-valued neutrosophic score functions (shortly SVNSF) of C_{j3} and D_{j3} are defined as follows:

    i. SVNSF (C_{13}) = 0.7667, SVNSF (D_{13}) = 0, where q = 0. SVNSF (P_1) = 0.3834.

    ii. SVNSF (C_{23}) = 0.7333, SVNSF (D_{23}) = 0.7833, where q = 2. SVNSF (P_2) = 0.7583.

    iii. SVNSF (C_{33}) = 0.6222, SVNSF (D_{33}) = 0.6889, where q = 3. SVNSF (P_3) = 0.6556.

    iv. SVNSF (C_{43}) = 0.5556, SVNSF (D_{43}) = 0.6667, where q = 1. SVNSF (P_4) = 0.6112.

Step 3e2(d): Arrangement $P_2 \geq P_3 \geq P_4 \geq P_1$.

Step 3(f): Direct Rank Table 6 tabulates all the rank $DR(P_j)$ of the alternatives P_1, P_2, P_3, P_4 in ascending order for the regions D_1, D_2, D_3.

Step 4: Reverse Direction

Step 4(a): Subproblem Selection Select the subproblem 1 for the corresponding country D_1.

Table 6 Direct rank table

Country	Alternatives			
	P_1	P_2	P_3	P_4
D_1	3	4	2	1
D_2	2	3	1	4
D_3	1	4	3	2
$\text{TDR}(P_j) = \sum \text{DR}(P_j)$	6	11	6	7

Step 4(b): Neutrosophic Operations

i. $E_{11} = \{(0.2,0.7,0.9),(0.3,0.8,0.9),(0.5,0.6,0.7)\}$ and $F_{11} = \{(0.2,0.6,0.9),(0.3,0.6,0.9)\}$.

ii. $E_{21} = \{(0.1,0.9,0.9),(0.4,0.9,0.8),(0.3,0.7,0.9)\}$ and $F_{21} = \{(0.1,0.7,0.9)\}$.

iii. $E_{31} = \{(0.3,0.8,0.7),(0.9,0.8,0.1),(0.8,0.4,0.9)\}$ and $F_{31} = \{(0.3,0.4,0.9)\}$.

iv. $E_{41} = \{(0.4,0.9,0.6),(0.6,0.6,0.6),(0.8,0.7,0.6)\}$ and $F_{41} = \{(0.4,0.6,0.6),(0.4,0.7,0.6)\}$.

Step 4(c): Finding Single-Valued Neutrosophic Score Functions

The single-valued neutrosophic score functions of E_{j1} and F_{j1} are defined as follows:

i. SVNSF $(E_{11}) = 0.2667$, SVNSF $(F_{11}) = 0.25$, where $q = 2$. SVNSF $(P_1) = 0.2584$.

ii. SVNSF $(E_{21}) = 0.1889$, SVNSF $(F_{21}) = 0.1667$, where $q = 1$. SVNSF $(P_2) = 0.1778$.

iii. SVNSF $(E_{31}) = 0.4778$, SVNSF $(F_{31}) = 0.3333$, where $q = 1$. SVNSF $(P_3) = 0.4056$.

iv. SVNSF $(E_{41}) = 0.3556$, SVNSF $(F_{41}) = 0.3833$, where $q = 2$. SVNSF $(P_4) = 0.3695$.

Step 4(d): Arrangement $P_3 \geq P_4 \geq P_1 \geq P_2$.

Step 4(e): Repetition

Step 4e1(a): Subproblem Selection Select the subproblem 2 for the corresponding country D_2.

Step 4e1(b): Neutrosophic Operations

i. $E_{12} = \{(0.3,0.8,0.7),(0.9,0.7,0.3),(0.5,0.8,0.7)\}$ and $F_{12} = \{(0.3,0.7,0.7),(0.5,0.7,0.7)\}$.

ii. $E_{22} = \{(0.6,0.9,0.9),(0.9,0.8,0.7),(0.6,0.9,0.7)\}$ and $F_{22} = \{(0.6,0.8,0.9),(0.6,0.8,0.7)\}$.

iii. $E_{32} = \{(0.8,0.7,0.6),(0.6,0.7,0.8),(0.8,0.6,0.9)\}$ and $F_{32} = \{(0.6,0.6,0.9)\}$.

iv. $E_{42} = \{(0.3,0.9,0.9),(0.4,0.8,0.8),(0.2,0.8,0.7)\}$ and $F_{42} = \{(0.3,0.8,0.9),(0.2,0.8,0.9),(0.2,0.8,0.8)\}$.

Step 4e1(c): Finding Single-Valued Neutrosophic Score Functions

The single-valued neutrosophic score functions of E_{j2} and F_{j2} are defined as follows:

i. SVNSF $(E_{12}) = 0.4111$, SVNSF $(F_{12}) = 0.3333$, where $q = 2$. SVNSF $(P_1) = 0.3722$.

ii. SVNSF $(E_{22}) = 0.3556$, SVNSF $(F_{22}) = 0.3333$, where $q = 2$. SVNSF $(P_2) = 0.3445$.

iii. SVNSF $(E_{32}) = 0.4333$, SVNSF $(F_{32}) = 0.3667$, where $q = 1$. SVNSF $(P_3) = 0.4$.

iv. SVNSF $(E_{42}) = 0.2222$, SVNSF $(F_{42}) = 0.1889$, where $q = 3$. SVNSF $(P_4) = 0.2056$.

Step 4e1(d): Arrangement $P_3 \geq P_1 \geq P_2 \geq P_4$.

Step 4(e): Repetition

Step 4e2(a): Subproblem Selection Select the subproblem 3 for the corresponding country D_3.

Step 4e2(b): Neutrosophic Operations

i. $E_{13} = \{(0.4,0.9,0.8),(0.3,0.8,0.8),(0.3,0.7,0.9)\}$ and $F_{13} = \varnothing$.

ii. $E_{23} = \{(0.1,0.9,0.9),(0.8,0.7,0.9),(0.2,0.4,0.9)\}$ and $F_{23} = \{(0.1,0.7,0.9),$
$(0.1,0.4,0.9)\}$.

iii. $E_{33} = \{(0.2,0.6,0.9),(0.7,0.6,0.9),(0.4,0.5,0.6)\}$ and $F_{33} = \{(0.2,0.6,0.9),$
$(0.2,0.5,0.6),(0.4,0.5,0.9)\}$.

iv. $E_{43} = \{(0.3,0.6,0.7),(0.6,0.4,0.9),(0.8,0.6,0.5)\}$ and $F_{43} = \{(0.3,0.4,0.9)\}$.

Step 4e2(c): Finding Single-Valued Neutrosophic Score Functions

The single-valued neutrosophic score functions of E_{j3} and F_{j3} are defined as follows:

i. SVNSF $(E_{13}) = 0.2333$, SVNSF $(F_{13}) = 0$, where $q = 0$. SVNSF $(P_1) = 0.1167$.

ii. SVNSF $(E_{23}) = 0.2667$, SVNSF $(F_{23}) = 0.05$, where $q = 2$. SVNSF $(P_2) = 0.1584$.

iii. SVNSF $(E_{33}) = 0.3889$, SVNSF $(F_{33}) = 0.3511$, where $q = 3$. SVNSF $(P_3) = 0.37$.

iv. SVNSF $(E_{43}) = 0.4444$, SVNSF $(D_{43}) = 0.3333$, where $q = 1$. SVNSF $(P_4) = 0.3889$.

Step 4e2(d): Arrangement $P_4 \geq P_3 \geq P_2 \geq P_1$.

Step 4(f): Reverse Rank Table 7 tabulates all the rank $RR(P_j)$ of the alternatives P_1, P_2, P_3, P_4 in ascending order for the regions D_1, D_2, D_3.

Table 7 Reverse rank table

	Alternatives			
Country	P_1	P_2	P_3	P_4
D_1	2	1	4	3
D_2	3	2	4	1
D_3	1	2	3	4
$TRR(P_j) = \sum RR(P_j)$	6	5	11	8

Step 5(1): Decision of Subproblem 1 From the table of step 3(f) and step 4(f), the rank differences of P_1, P_2, P_3, and P_4 from direct direction to reverse direction for each region are, respectively, 1, 3, -2, and -2. Therefore for the country D_1, the best yield is given by the hybrid plant grasses rice, and the worst hybrid plant is wheat. The hybrid cotton and maize plants give an equal quantity of yield.

Step 5(2): Decision of Subproblem 2 The rank difference of P_1, P_2, P_3, and P_4 from direct direction to reverse direction for each region is, respectively, -1, 1, -3, and 3. Therefore for the country D_2, the best hybrid plant is maize. The second and third best yields are, respectively, given by the hybrid plants cotton and grasses rice. The hybrid wheat plant is the worst hybrid plant.

Step 5(3): Decision of Subproblem 3 The rank difference of P_1, P_2, P_3, and P_4 from direct direction to reverse direction for each region is, respectively, 0,2, 0, and -2. Therefore for the country D_3, the best yield is given by the hybrid grasses rice plant, and the second-best yield hybrid plant is maize. The lowest yield is given by the hybrid plants wheat and cotton which are giving an equal quantity of yield.

Step 6: Determination From Tables 6 and 7, $TDR(P_1) - TRR(P_1) = 0$, $TDR(P_2) - TRR(P_2) = 6$, $TDR(P_3) - TRR(P_3) = -5$, $TDR(P_4) - TRR(P_4) = -1$, and here $P_2 \geq P_3 \geq P_1 \geq P_4$.

Step 7: Final Decision For the entire region, the best yield hybrid plant is grasses rice, and the worst yield hybrid plant is maize. The second-best yield hybrid plant is cotton and the hybrid wheat plant is the third-best yield plant.

6 Results and Discussion

This section analyzes the result of the above problem for the proposed method:

i. From the decision of subproblems, the high quantity of yield for the countries D_1 and D_3 is given by the hybrid plant grasses rice, but the high quantity of yield for the country D_2 is given by the hybrid plant maize. That is, the best yield hybrid plant for the countries D_1 and D_3 is grasses rice, but the best yield hybrid plant for the country D_2 is maize.

ii. The worst yield hybrid plant for the countries D_1 and D_2 is wheat, but the lowest yield for the country D_3 is given by the hybrid plants wheat and cotton.

iii. The hybrid cotton and maize plants give the second quantity of yield for the country D_1. The second and third quantities of yield for the country D_2 are given by the hybrid plants cotton and grasses rice. For the country D_3, the second yield hybrid plant is maize.

iv. From the observations the result of the country D_1 may be not the same for the countries D_2 and D_3. That is, the result varies from country to country.

v. Therefore for the entire region, the best yield hybrid plant is grasses rice and the worst yield hybrid plant is maize. The second yield hybrid plant is cotton and the hybrid wheat plant is the third quantity plant.

7 Limitations and Advantages

This section states some limitations and advantages of the proposed method:

i. The novelty of the present work is the first method to solve the plant hybridization problem in the neutrosophic environment.

ii. This method is new, because it uses only single-valued neutrosophic score functions (SVNSF) with single-valued neutrosophic numbers in both direct and reverse directions to find the best alternatives in each and the entire regions.

iii. It considers the neutrosophic sets and their unions from multi-criteria decision-making (MCDM) problems in the direct direction. The complements of these neutrosophic sets and their intersections are considered in the reverse direction.

iv. The proposed method divides the MCDM problems into small subproblems for the corresponding region.

v. From each subproblem, this method can find the best alternatives in each corresponding region. This is one of the advantages of this method.

vi. The proposed method can always find the best alternatives if the MCDM problem has only one region. This is another advantage. This new method in the MCDM problem does not use any similarity measures such as cosine similarity measures [23], tangent similarity measures [21], cotangent similarity measures [25], distance functions such as Euclidean distances [19], hamming distance [24], rank matrices [34], etc.

vii. This method needs single-valued neutrosophic sets and single-valued neutrosophic score function [19]. Hence this method will be applicable in many real-life situations.

viii. This new method gives the same result of [36] even though not using neutrosophic distance functions, similarity measures.

ix. My previous method [35] funded the patients (alternatives) with their caused disease (alternatives), but the present method is an extension of the previous work. Here the method deals with plant hybridization problems in the neutrosophic environment to find the best and the worst alternatives not only in each region but also in the entire regions.

x. The decision of the proposed method is equally consistent, dependable, and reliable, and the method may also be suitable for a large amount of data.

8 Future and Summary of Work

This section discusses the future and summary of the proposed method:

i. Neutrosophic set theory is a new structure considering three independent membership functions to deal with the concept of incompleteness, uncertainty, and vagueness. The method of multi-criteria decision-making (MCDM) problem is an important key in the existence of multiple criteria and alternatives in solving sophisticated and complicated decision problems.

ii. This chapter derived a new neutrosophic method in multi-criteria decision-making problems to find the best alternatives in each and the entire regions under the neutrosophic environment.

iii. This method considered neutrosophic sets with their unions and the complements with their intersections.

iv. The single-valued neutrosophic score functions are computed to find the best alternatives not only in each region but also in the entire region.

v. This chapter solved the plant hybridization problem as a real-life application of neutrosophic set theory to demonstrate the effectiveness of the proposed method.

vi. The present methodology may further be applied in content-based image retrieval (CBIR), dimensionality reduction in dimensional space, multimedia databases, manufacturing systems, personal selection in academia, project evaluation, and supply chain management.

vii. The proposed method can alternatively be used for other multi-criteria decision-making methods such as ELECTRE, DEMTEL, PROMOTEE, TOPSIS, and VIKOR methods.

viii. The techniques of this method may also be applied in fuzzy and intuitionistic fuzzy environments.

Acknowledgments The authors are thankful to the referees' suggestions and valuable commands to improve the quality of this chapter work.

References

1. Zadch, L. A. (1968). Probability measures of fuzzy events. *Journal of Mathematical Analysis and Applications, 23*, 421–427.
2. Adlassnig, K. P. (1986). Fuzzy set theory in medical diagnosis. *IEEE Transactions on Systems, Man, and Cybernetics, 16*(2), 260–265.
3. Sugeno, M. (1985). An Introductory survey of fuzzy control. *Information Sciences, 36*, 56–83.
4. Innocent, P. R., & John, R. I. (2004). Computer aided fuzzy medical diagnosis. *Information Sciences, 162*, 81–104.
5. Roos, T. J. (1994). *Fuzzy logic with engineering applications*. New York: McGraw Hill P.C.
6. Hyung, L. K., Song, Y. S., & Lee, K. M. (1994). Similarity measure between fuzzy sets and between elements. *Fuzzy Sets and Systems, 62*, 291–293.
7. Chen, S. M., Yeh, S. M., & Hsiao, P. H. (1995). A comparison of similarity measures of fuzzy values. *Fuzzy Sets and Systems, 72*(1), 79–89.
8. Wang, W. J. (1997). New similarity measures on fuzzy sets and elements. *Fuzzy Sets and Systems, 85*(3), 305–309.
9. Atanassov, K. (1986). Intuitionistic fuzzy sets. *Fuzzy Sets and Systems, 20*, 87–96.
10. De, S. K., Biswas, A., & Roy, R. (2001). An application of intuitionistic fuzzy sets in medical diagnosis. *Fuzzy Sets and Systems, 117*(2), 209–213.

11. Szmidt, E., & Kacprzyk, J. (2001). Intuitionistic fuzzy sets in some medical applications. In E. Szmidt & J. Kacprzyk (Eds.), *International conference on computational intelligence* (pp. 148–151). Berlin, Heidelberg: Springer.
12. Biswas, P., Pramanik, S., & Giri, B. C. (2014). A study on information technology professionals' health problem based on intuitionistic fuzzy cosine similarity measure. *Swiss Journal of Statistical and Applied Mathematics, 2*(1), 44–50.
13. Khatibi, V., & Montazer, G. A. (2009). Intuitionistic fuzzy set vs. fuzzy set application in medical pattern recognition. *Artificial Intelligence in Medicine, 47*(1), 43–52.
14. Hung, K. C., & Tuan, H. W. (2013). Medical diagnosis based on intuitionistic fuzzy sets revisited. *Journal of Interdisciplinary Mathematics, 16*(6), 385–395.
15. Smarandache, F. (1998). *A unifying field of logics. Neutrosophy: Neutrosophic probability, set and logic*. Rehoboth: American Research Press.
16. Smarandache, F. (2005). Neutrosophic set, a generalization of the intuitionistic fuzzy sets. *Journal of Pure and Applied Mathematics, 24*, 287–297.
17. Wang, W. J., Smarandache, F., Zhang, Y., & Sunderraman, R. (2010). Single valued neutrosophic sets. *Multi-Space and Multi-Structure, 4*, 410–413.
18. Majumdar, P., & Samanta, S. K. (2014). On similarity and entropy of neutrosophic sets. *Journal of Intelligent Fuzzy Systems, 26*(3), 1245–1252.
19. Smarandache, F., & Pramanik, S. (2016). *New trends in neutrosophic theory and applications*. Brussels, Belgium, EU: Pons Editions.
20. Ye, J., & Zhang, Q. (2014). Single valued neutrosophic similarity measures for multiple attribute decision-making. *Neutrosophic Sets and Systems, 2*, 48–54.
21. Ye, J. (2015). Neutrosophic tangent similarity measure and its application to multiple attribute decision making. *Neutrosophic Sets and Systems, 9*, 85–92.
22. Ye, J., & Ye, S. (2015). Medical diagnosis using distance-based similarity measures of single valued neutrosophic multisets. *Neutrosophic Sets and Systems, 7*, 47–54.
23. Biswas, P., Pramanik, S., & Giri, B. C. (2015). Cosine similarity measure based multi attribute decision-making with trapezoidal fuzzy neutrosophic numbers. *Neutrosophic Sets and Systems, 8*, 47–57.
24. Broumi, S., & Smarandache, F. (2013). Several similarity measures of neutrosophic sets. *Neutrosophic Sets and Systems, 1*, 54–62.
25. Pramanik, S., & Mondal, K. (2015). Cotangent similarity measure of rough neutrosophic sets and its application to medical diagnosis. *Journal of New Theory, 4*, 90–102.
26. Ulucay, V., Kilic, A., Sahin, M., & Deniz, H. (2019). A new hybrid distance-based similarity measure for refined neutrosophic sets and its application in medical diagnosis. *MATEMATIKA: Malaysian Journal of Industrial and Applied Mathematics, 35*(1), 83–94.
27. Nabeeh, N. A., Abdel-Basset, M., El-Ghareeb, H. A., & Aboelfetouh, A. (2019). Neutrosophic multi-criteria decision making approach for iot-based enterprises. *IEEE Access, 7*, 59559–59574.
28. Nabeeh, N. A., Smarandache, F., Abdel-Basset, M., El-Ghareeb, H. A., & Aboelfetouh, A. (2019). An integrated neutrosophic-TOPSIS approach and its application to personnel selection: A new trend in brain processing and analysis. *IEEE Access, 7*, 29,734–29,744.
29. Abdel-Basset, M., Saleh, M., Gamal, A., & Smarandache, F. (2019). An approach of TOPSIS technique for developing supplier selection with group decision making under type-2 neutrosophic number. *Applied Soft Computing, 77*, 438–452.
30. Abdel-Basset, M., Manogaran, G., Gamal, A., & Smarandache, F. (2019). A group decision making framework based on neutrosophic TOPSIS approach for smart medical device selection. *Journal of Medical Systems, 43*(2), 1–13.
31. Abdel-Basset, M., Chang, V., & Gamal, A. (2019). Evaluation of the green supply chain management practices: A novel neutrosophic approach. *Computers in Industry, 108*, 210–220.
32. Karaaslan, F. (2018). Gaussian single-valued neutrosophic numbers and its application in multi-attribute decision making. *Neutrosophic Sets and Systems, 22*, 101–117.
33. Giri, B. C., Molla, M. U., & Biswas, P. (2018). TOPSIS method for MADM based on interval trapezoidal neutrosophic number. *Neutrosophic Sets and Systems, 22*, 151–167.

34. Aal, S. I. A., Ellatif, M. M. A. A., & Hassan, M. M. (2018). Two ranking methods of single valued triangular neutrosophic numbers to rank and evaluate information systems quality. *Neutrosophic Sets and Systems, 19*, 132–141.
35. Jayaparthasarathy, G., Little Flower, V. F., & Arockia Dasan, M. (2019). Neutrosophic supra topological applications in data mining process. *Neutrosophic Sets and Systems, 27*, 80–97.
36. Tooranloo, H. S., Zanjirchi, S. M., & Tavangar, M. (2020). ELECTRE approach for multi-attribute decision-making in refined neutrosophic environment. *Neutrosophic Sets and Systems, 31*, 101–119.
37. Mendel, G. (1865). Experiments in plant hybridization. Verhandlungen des naturforschenden Ver-eines in Brünn, Bd. IV für das Jahr, Abhand-lungen, pp. 3–47.

Multicriteria Decision-Making Based on Quadripartitioned Neutrosophic Cross-Entropy

M. Mohanasundari and K. Mohana

1 Introduction

Many real-world problems consist of various uncertainties in it and we are in need to solve that uncertainty measure effectively. To overcome this many methods are used in mathematics. The problems related to ambiguous, imprecise judgments are handled by fuzzy set theory which was introduced by Zadeh [1, 2] in 1965. It allows membership function value in the interval [0,1]. Membership function along with nonmembership defined a new set called intuitionistic fuzzy set introduced by Atanassov [3] in 1983. Many real-life problem parameters lie within a range of numbers that is in intervals and handling the intervals along with the sets fuzzy and intuitionistic fuzzy, respectively, defined by Turksen [4] and Atanassov [5]. Neutrosophic set in which indeterminacy is specified explicitly was introduced by Smarandache [6], and it is a generalization of aforementioned sets. Therefore we get truth, indeterminacy, and falsity membership function for each element in the universe of neutrosophic sets. The indeterminacy value in neutrosophic set is independent of both truth and falsity values, but in intuitionistic fuzzy set, it is dependent of both. Since neutrosophic set handles indeterminate and inconsistent information effectively, it is applied in many fields like decision support system, semantic web services, new economy's growth, etc. Later Wang et al. [7] introduced the concept of single-valued neutrosophic set (SVNS) which is used in many real scientific and engineering fields. Many researchers [8–14] immensely show

M. Mohanasundari (✉)
Department of Mathematics, Nirmala College For Women, Coimbatore, Tamil Nadu, India
e-mail: mohanadhianesh@gmail.com

K. Mohana
Department of Mathematics, Nirmala College for Women, Coimbatore, Tamil Nadu, India
e-mail: riyaraju1116@gmail.com

© The Author(s), under exclusive license to Springer Nature Switzerland AG 2021
F. Smarandache, M. Abdel-Basset (eds.), *Neutrosophic Operational Research*,
https://doi.org/10.1007/978-3-030-57197-9_2

19

their research interest in application of multicriteria decision-making problems under SVNS environment. A new concept of four-valued logic was introduced by Belnap [15] in which any information is represented by four parameters T, F, None, and Both which, respectively, denote true, false, neither true nor false, and both true and false. Based on this Belnap's four-valued logic, Smarandache [16] introduced four numerical-valued neutrosophic logic in which indeterminacy is splitted into two terms, namely, Unknown (U) and Contradiction (C). Hence a new set quadripartitioned single-valued neutrosophic set (QSVNS) was introduced by Rajashi Chatterjee et al. [17] in which we have four components T,C,U,F in real unit interval [0,1]. Mohanasundari M and Mohana K [18] studied the improved correlation coefficients of quadripartitioned single-valued neutrosophic sets and interval quadripartitioned neutrosophic sets and also used the proposed concept in solving multicriteria decision-making problems.

The measure "entropy" was used by Zadeh [19] in fuzzy set theory to know about some uncertain information which present in real-life problems. Entropy plays a vital role in measuring uncertain information in fuzzy set theory, intuitionistic fuzzy set theory, neutrosophic set, etc. Many researchers [5, 20–23] focused entropy measure in fuzzy set, intuitionistic fuzzy set, interval fuzzy set, and interval-valued intuitionistic fuzzy set and used it in a multicriteria decision-making problem under its respective environments. Based on Shannon's [24, 25] inequality, a new concept of cross-entropy is used to identify the discrimination information between objects. Cross-entropy between two fuzzy sets, intuitionistic fuzzy sets, interval-valued fuzzy set, and interval-valued intuitionistic fuzzy set was introduced by Shang and Jiang [24], Vlachos and Sergiadis [26], and Zhang [27], respectively. Single-valued neutrosophic cross-entropy was defined by Ye [28] in 2014 which is an extension of cross-entropy of fuzzy sets. Under single-valued neutrosophic environment, many authors [29, 30] used entropy in solving multicriteria decision-making problems. Cross-entropy measure on interval neutrosophic sets and its applications in multicriteria decision-making was proposed by Sahin [31] in which he defined the conversion of interval neutrosophic set into fuzzy set and single-valued neutrosophic set, respectively. Further he also explained the concept of interval neutrosophic cross-entropy based on fuzzy and single-valued neutrosophic cross-entropy, respectively.

In this paper Sect. 2 deals about some basic definitions of QSVNS and IQNS. In Sect. 3 we define the concept of cross-entropy between two QSVNSs as an extension of cross-entropy of SVNSs and also studied weighted cross-entropy of QSVNSs. Section 4 illustrates the application of the above said proposed method in multicriteria decision-making problem. In Sect. 5 we discuss the cross-entropy of IQNS as an extension of the cross-entropy of QSVNSs and also define its weighted cross-entropy. Section 6 deals an illustrative example of the cross-entropy of IQNS in multicriteria decision-making problem. Section 7 concludes the paper.

2 Preliminaries

In this section we recall the basic definitions of QSVNS and IQNS which will be used in proving the rest of the paper.

Definition 2.1 [17] Let X be a non-empty set. A quadripartitioned single-valued neutrosophic set (QSVNS) A over X characterizes each element x in X by a truth membership function T_A, a contradiction membership function C_A, an ignorance membership function U_A, and a falsity membership function F_A such that for each $x \in X, T_A, C_A, U_A, F_A \in [0, 1]$ and $0 \leq T_A(x) + C_A(x) + U_A(x) + F_A(x) \leq 4$ when X is discrete, A is represented as $A = \sum_{i=1}^{n} \langle T_A(x_i), C_A(x_i), U_A(x_i), F_A(x_i) \rangle / x_i$, $x_i \in X$.

However, when the universe of discourse is continuous, A is represented as

$$A = \int_X \langle T_A(x), C_A(x), U_A(x), F_A(x) \rangle / x, x \in X$$

Definition 2.2 [17] Consider two QSVNSs A and B, over X. A is said to be contained in B, denoted by $A \subseteq B$ iff $T_A(x) \leq T_B(x)$, $C_A(x) \leq C_B(x)$, $U_A(x) \geq U_B(x)$ and $F_A(x) \geq F_B(x)$.

Definition 2.3 [17] The complement of a QSVNS A is denoted by A^C and is defined as

$$A^C = \sum_{i=1}^{n} \langle F_A(x_i), U_A(x_i), C_A(x_i), T_A(x_i) \rangle / x_i, x_i \in X$$

i.e., $T_{A^C}(x_i) = F_A(x_i)$, $C_{A^C}(x_i) = U_A(x_i)$, $U_{A^C}(x_i) = C_A(x_i)$ and $F_{A^C}(x_i) = T_A(x_i)$, $x_i \in X$

Definition 2.4 [17] The union of two QSVNSs A and B is denoted by $A \cup B$ and is defined as

$$A \cup B = \sum_{i=1}^{n} \langle T_A(x_i) \vee T_B(x_i), C_A(x_i) \vee C_B(x_i), U_A(x_i) \wedge U_B(x_i),$$

$$F_A(x_i) \wedge F_B(x_i) \rangle / x_i, x_i \in X$$

Definition 2.5 [17] The intersection of two QSVNSs A and B is denoted by $A \cap B$ and is defined as

$$A \cap B = \sum_{i=1}^{n} \langle T_A(x_i) \wedge T_B(x_i), C_A(x_i) \wedge C_B(x_i), U_A(x_i) \vee U_B(x_i),$$

$$F_A(x_i) \vee F_B(x_i) \rangle / x_i, x_i \in X$$

Definition 2.6 [18] An interval-quadripartitioned neutrosophic set (IQNS) A in X is denoted by a truth membership function $T_A(x)$, contradiction membership function $C_A(x)$, an unknown membership function $U_A(x)$, and a falsity membership function $F_A(x)$. For each point x in X, there are

$$T_A(x) = \left[\inf\ T_A(x), \sup\ T_A(x)\right] \subseteq [0, 1],$$

$$C_A(x) = \left[\inf\ C_A(x), \sup\ C_A(x)\right] \subseteq [0, 1],$$

$$U_A(x) = \left[\inf\ U_A(x), \sup\ U_A(x)\right] \subseteq [0, 1],$$

$$F_A(x) = \left[\inf\ F_A(x), \sup\ F_A(x)\right] \subseteq [0, 1]$$

Therefore an IQNS A can be denoted as

$$A = \{\langle x, T_A(x), C_A(x), U_A(x), F_A(x) | x \in X \rangle\}$$
$$= \left\{ \left\langle \begin{matrix} x, \left[\inf\ T_A(x), \sup\ T_A(x)\right], \left[\inf\ C_A(x), \sup\ C_A(x)\right], \\ \left[\inf\ U_A(x), \sup\ U_A(x)\right], \left[\inf\ F_A(x), \sup\ F_A(x)\right] \end{matrix} \right\rangle | x \in X \right\}$$

Then the sum of $T_A(x)$, $C_A(x)$, $U_A(x)$, $F_A(x)$ satisfies the condition $0 \leq\ \sup T_A(x) +\ \sup C_A(x) +\ \sup U_A(x) +\ \sup F_A(x) \leq 4$. If the upper and lower ends of the interval values of $T_A(x)$, $C_A(x)$, $U_A(x)$, and $F_A(x)$ in an IQNS are equal, then IQNS reduces to the QSVNS. Both QSVNSs and IQNSs are all the subclasses of quadripartitioned neutrosophic sets (QNSs).

Definition 2.7 [18] The complement of an IQNS A is denoted by A^C and is defined as

$$\inf T_{A^C}(x) = 1 - \sup\ T_A(x), \quad \sup T_{A^C}(x) = 1 - \inf\ T_A(x),$$
$$\inf C_{A^C}(x) = 1 - \sup\ C_A(x), \quad \sup C_{A^C}(x) = 1 - \inf\ C_A(x),$$
$$\inf U_{A^C}(x) = 1 - \sup\ U_A(x), \quad \sup U_{A^C}(x) = 1 - \inf\ U_A(x),$$
$$\inf F_{A^C}(x) = 1 - \sup\ F_A(x), \quad \sup F_{A^C}(x) = 1 - \inf\ F_A(x)$$

for any x in X.

Definition 2.8 [18] An IQNS A is contained in the other IQNS B, $A \subseteq B$ if and only if

$$\inf T_A(x) \leq \inf T_B(x), \quad \sup T_A(x) \leq \sup T_B(x),$$
$$\inf C_A(x) \leq \inf C_B(x), \quad \sup C_A(x) \leq \sup C_B(x),$$
$$\inf U_A(x) \geq \inf U_B(x), \quad \sup U_A(x) \geq \sup U_B(x),$$
$$\inf F_A(x) \geq \inf F_B(x), \quad \sup F_A(x) \geq \sup F_B(x)$$

for any x in X.

Definition 2.9 [18] Two IQNSs A and B are equal, i.e., $A=B$ if and only if $A \subseteq B$ and $B \subseteq A$.

3   Cross-Entropy of QSVNS

In this section we define a cross-entropy of quadripartitioned single-valued neutrosophic set (QSVNS) as an extension of cross-entropy between single-valued neutrosophic sets. Before that we discuss the cross-entropy and symmetric discrimination information under SVNS environment which was introduced by Ye [28].

Definition 3.1 [28] Assume that A and B are two SVNSs in a universe of discourse $X = \{x_1, x_2, \ldots, x_n\}$ which are denoted by

$A = \{\langle x_i, T_A(x_i), I_A(x_i), F_A(x_i)\rangle \mid x_i \in X\}$ and $B = \{\langle x_i, T_B(x_i), I_B(x_i), F_B(x_i)\rangle \mid x_i \in X\}$

where $T_A(x_i), I_A(x_i), F_A(x_i), T_B(x_i), I_B(x_i), F_B(x_i) \in [0, 1]$ for every $x_i \in X$. Hence a novel single-valued neutrosophic cross-entropy measure between A and B is

$$E(A, B) = \sum_{i=1}^{n} \left[T_A(x_i) \log_2 \frac{T_A(x_i)}{\frac{1}{2}(T_A(x_i) + T_B(x_i))} + (1 - T_A(x_i)) \log_2 \frac{1 - T_A(x_i)}{1 - \frac{1}{2}(T_A(x_i) + T_B(x_i))} \right]$$
$$+ \sum_{i=1}^{n} \left[I_A(x_i) \log_2 \frac{I_A(x_i)}{\frac{1}{2}(I_A(x_i) + I_B(x_i))} + (1 - I_A(x_i)) \log_2 \frac{1 - I_A(x_i)}{1 - \frac{1}{2}(I_A(x_i) + I_B(x_i))} \right]$$
$$+ \sum_{i=1}^{n} \left[F_A(x_i) \log_2 \frac{F_A(x_i)}{\frac{1}{2}(F_A(x_i) + F_B(x_i))} + (1 - F_A(x_i)) \log_2 \frac{1 - F_A(x_i)}{1 - \frac{1}{2}(F_A(x_i) + F_B(x_i))} \right]$$

It also indicates discrimination degree of A from B. According to Shannon's inequality [ ], one can easily prove that $E(A, B) \geq 0$ and $E(A, B) = 0$ if and only if $T_A(x_i) = T_B(x_i), I_A(x_i) = I_B(x_i)$ and $F_A(x_i) = F_B(x_i)$ for any $x_i \in X$. And also we can see that $E(A^C, B^C) = E(A, B)$ where A^C and B^C are the complement of SVNSs A and B, respectively.

Then $E(A, B)$ is not symmetric. So it should be modified to a symmetric discrimination information measure for SVNSs as $D(A, B) = E(A, B) + E(B, A)$. The larger the difference between A and B is, the larger D(A,B) is.

Definition 3.2 Let us consider the two quadripartitioned single-valued neutrosophic sets A and B in the universe of discourse $X = \{x_1, x_2, \ldots, x_n\}$ where $A = \{\langle x_i, T_A(x_i), C_A(x_i), U_A(x_i), F_A(x_i)\rangle \mid x_i \in X\}$ and $B = \{\langle x_i, T_B(x_i), C_B(x_i), U_B(x_i), F_B(x_i)\rangle \mid x_i \in X\}$ and also T, C, U, F, respectively, denote truth, contradiction, unknown, and falsity membership function in the real unit interval of [0,1]. Discrimination information of truth membership function between two QSVNSs A and B is denoted by

$$E^T(A, B; x_i) = T_A(x_i) \log_2 \frac{T_A(x_i)}{T_B(x_i)} + (1 - T_A(x_i)) \log_2 \frac{1 - T_A(x_i)}{1 - \frac{1}{2}(T_A(x_i) + T_B(x_i))}$$

Then the following equation gives the expected discrimination information of truth membership function $T_A(x_i)$ against $T_A(x_i)$:

$$E^T(A, B) = \sum_{i=1}^{n} \left[T_A(x_i) \log_2 \frac{T_A(x_i)}{T_B(x_i)} + (1 - T_A(x_i)) \log_2 \frac{1 - T_A(x_i)}{1 - \frac{1}{2}(T_A(x_i) + T_B(x_i))} \right]$$

The following equations, respectively, denote the expected discrimination information of contradiction, unknown, and falsity membership function which is given by

$$E^C(A, B) = \sum_{i=1}^{n} \left[C_A(x_i) \log_2 \frac{C_A(x_i)}{C_B(x_i)} + (1 - C_A(x_i)) \log_2 \frac{1 - C_A(x_i)}{1 - \frac{1}{2}(C_A(x_i) + C_B(x_i))} \right]$$

$$E^U(A, B) = \sum_{i=1}^{n} \left[U_A(x_i) \log_2 \frac{U_A(x_i)}{U_B(x_i)} + (1 - U_A(x_i)) \log_2 \frac{1 - U_A(x_i)}{1 - \frac{1}{2}(U_A(x_i) + U_B(x_i))} \right]$$

$$E^F(A, B) = \sum_{i=1}^{n} \left[F_A(x_i) \log_2 \frac{F_A(x_i)}{F_B(x_i)} + (1 - F_A(x_i)) \log_2 \frac{1 - F_A(x_i)}{1 - \frac{1}{2}(F_A(x_i) + F_B(x_i))} \right]$$

Sum of the above four amounts defines a quadripartitioned single-valued neutrosophic cross-entropy measure and is also given by

$$
\begin{aligned}
H_{QSVNS}(A, B) = &\sum_{i=1}^{n} \left[T_A(x_i) \log_2 \frac{T_A(x_i)}{\frac{1}{2}(T_A(x_i) + T_B(x_i))} + (1 - T_A(x_i)) \log_2 \frac{1 - T_A(x_i)}{1 - \frac{1}{2}(T_A(x_i) + T_B(x_i))} \right] \\
&+ \sum_{i=1}^{n} \left[C_A(x_i) \log_2 \frac{C_A(x_i)}{\frac{1}{2}(C_A(x_i) + C_B(x_i))} + (1 - C_A(x_i)) \log_2 \frac{1 - C_A(x_i)}{1 - \frac{1}{2}(C_A(x_i) + C_B(x_i))} \right] \\
&+ \sum_{i=1}^{n} \left[U_A(x_i) \log_2 \frac{U_A(x_i)}{\frac{1}{2}(U_A(x_i) + U_B(x_i))} + (1 - U_A(x_i)) \log_2 \frac{1 - U_A(x_i)}{1 - \frac{1}{2}(U_A(x_i) + U_B(x_i))} \right] \\
&+ \sum_{i=1}^{n} \left[F_A(x_i) \log_2 \frac{F_A(x_i)}{\frac{1}{2}(F_A(x_i) + F_B(x_i))} + (1 - F_A(x_i)) \log_2 \frac{1 - F_A(x_i)}{1 - \frac{1}{2}(F_A(x_i) + F_B(x_i))} \right]
\end{aligned}
\tag{1}
$$

Based on Shannon's inequality, it is easy to prove $H_{QSVNS}(A, B) \geq 0$ and $H_{QSVNS}(A, B) = 0$ if and only if $T_A(x_i) = T_B(x_i)$, $C_A(x_i) = C_B(x_i)$, $U_A(x_i) = U_B(x_i)$ $F_A(x_i) = F_B(x_i)$ for any $x_i \in X$. And also we can see that $H_{QSVNS}(A^C, B^C) = H_{QSVNS}(A, B)$ where A^C and B^C, respectively, denote the complement of QSVNSs A and B. Since $H_{QSVNS}(A, B)$ is not symmetric, it is modified to get a symmetric discrimination information measure of two QSVNSs A and B as

$$D_{QSVNS}(A, B) = H_{QSVNS}(A, B) + H_{QSVNS}(B, A) \tag{2}$$

$D_{QSVNS}(A, B)$ is higher when the difference between A and B is higher.

4 Multicriteria Decision-Making Method Based on the Cross-Entropy of QSVNSs

Decision-making plays a vital role in everyone's life nowadays. There are many important factors included in decision-making. It is one of the primary skills which everyone needs to lead a final choice that should be a good one. There are many techniques being used in decision-making. One of the techniques is multicriteria decision-making (MCDM) which includes the essential steps of defining the context, deciding the objectives, and selecting the right criteria that represent the value. The advantage of using MCDM is open and explicit. It is possible to compare many different factors with one another and also the chosen criteria can be adjusted. Many decision-making scenarios include indefinite, uncertainty, insufficient, and inconsistent information which should be handled in an effective manner to make the decisions in a better way. This multicriteria decision-making problem includes the process of identifying the best alternative among suitable alternatives which can be evaluated based on the number of criteria or attributes that are used in the particular problem.

This section deals about the multicriteria decision-making problem under quadripartitioned single-valued neutrosophic environment with the proposed cross-entropy measure of QSVNSs. Let $A = \{A_1, A_2, \ldots, A_n\}$ and $C = \{C_1, C_2, \ldots, C_n\}$ be sets of alternative and criteria, respectively. Let $W_j (j = 1, 2, \ldots, n)$ denote the weight of the criterion C_j and it belongs to the interval of $[0,1]$ provided $\sum_{i=1}^{n} w_j = 1$. Here the characteristic of the alternative $A_i (i = 1, 2, \ldots, m)$ is denoted by QSVNS as $A_i = \{\langle C_j, T_{A_i}(C_j), C_{A_i}(C_j), U_{A_i}(C_j), F_{A_i}(C_j)\rangle | C_j \in C\}$ where $T_{A_i}(C_j), C_{A_i}(C_j), U_{A_i}(C_j), F_{A_i}(C_j) \in [0, 1], i = 1, 2, \ldots, m$ and $j = 1, 2, \ldots, n$.

To write the criterion value $\langle C_j, T_{A_i}(C_j), C_{A_i}(C_j), U_{A_i}(C_j), F_{A_i}(C_j)\rangle$ in a simple way is denoted by $\alpha_{ij} = \langle T_{ij}, C_{ij}, U_{ij}, F_{ij}\rangle$ $(i = 1, 2, \ldots, m$ and $j = 1, 2, \ldots, n)$. The concept of ideal point plays a vital role in multicriteria decision-making environments which helps to identify the best alternative in the decision set. Though it is not possible to get ideal alternative in real world, it needs to construct the theoretical approach against evaluating the alternatives. Hence the ideal alternative A^* is denoted by the criterion value $\alpha_{j^*} = \langle T_{j^*}, C_{j^*}, U_{j^*}, F_{j^*}\rangle = \langle 1, 1, 0, 0\rangle$. Hence by using (1) and (2), we get the weighted cross-entropy between an alternative A_i and the ideal alternative A^*:

$$
\begin{aligned}
D_{QSVNS}\left(A^*, A_i\right) = \sum_{j=1}^{n} w_j & \left[\log_2 \frac{1}{\frac{1}{2}(1+T_{ij})} + \log_2 \frac{1}{\frac{1}{2}(1+C_{ij})} + \log_2 \frac{1}{1-\frac{1}{2}(U_{ij})} + \log_2 \frac{1}{1-\frac{1}{2}(F_{ij})}\right] \\
+ \sum_{j=1}^{n} w_j & \left[T_{ij}\log_2 \frac{T_{ij}}{\frac{1}{2}(1+T_{ij})} + \left(1-T_{ij}\right)\log_2 \frac{1-T_{ij}}{1-\frac{1}{2}(1+T_{ij})}\right] \\
+ \sum_{j=1}^{n} w_j & \left[C_{ij}\log_2 \frac{C_{ij}}{\frac{1}{2}(1+C_{ij})} + \left(1-C_{ij}\right)\log_2 \frac{1-C_{ij}}{1-\frac{1}{2}(1+C_{ij})}\right] \\
+ \sum_{j=1}^{n} w_j & \left[U_{ij} + \left(1-U_{ij}\right)\log_2 \frac{1-U_{ij}}{1-\frac{1}{2}(U_{ij})}\right] + \sum_{j=1}^{n} w_j \left[F_{ij} + \left(1-F_{ij}\right)\log_2 \frac{1-F_{ij}}{1-\frac{1}{2}(F_{ij})}\right]
\end{aligned}
$$

$$(3)$$

Based on this weighted cross-entropy $D_{QSVNS}(A^*, A_i)\,(i = 1, 2, \ldots, m)$, we get the ranking order of all alternatives, and the lesser value of $D_{QSVNS}(A^*, A_i)$ gives us the best alternative A_i.

4.1 Practical Example

In our everyday life, maintaining good health is very important than any other thing in this world. Healthy habits help to prevent certain disease like heart disease, stroke, and high blood pressure. Though we maintain healthy habits, the increasing of population in this world leads to spread the disease in air, unclean water, etc. Now people are in need of selecting a good health center according to some criteria used. Let us consider a set of health centers $A = \{A_1, A_2, A_3\}$ whose primary ability is evaluated by the following four criteria: $C = \{C_1, C_2, C_3, C_4\}$ where C_1, C_2, C_3, C_4 denote (i) medical expertise, (ii) efficiency and caring, (iii) transparency in price, and (iv) hospital reputation, respectively. The weight vector for the above four criteria is given as $w = (0.35, 0.25, 0.25, 0.15)$. By using questionnaire of a domain expert, we evaluate an alternative $A_i(i = 1, 2, 3)$ based upon the above said four criteria $C_j(j = 1, 2, 3, 4)$. When we ask an expert to get an opinion of an alternative A_i with respect to a particular criterion C_j, he or she responds that the degree in which the statement is true is 0.6, not true is 0.1, both true and false is 0.3, and neither true nor false is 0.2. By using neutrosophic notation, it can be written as $\alpha_{11} = \langle 0.6, 0.3, 0.2, 0.1 \rangle$. Proceeding in this manner, we get the following quadripartitioned single-valued neutrosophic decision matrix A, when we ask about all three possible alternatives with respect to above four criteria to the expert:

$$
A = \begin{bmatrix}
(0.6, 0.3, 0.2, 0.1) & (0.6, 0.3, 0.4, 0.2) & (0.5, 0.3, 0.2, 0.1) & (0.4, 0.2, 0.5, 0.6) \\
(0.5, 0.4, 0.3, 0.2) & (0.3, 0.4, 0.2, 0.1) & (0.6, 0.5, 0.2, 0.4) & (0.3, 0.2, 0.5, 0.4) \\
(0.5, 0.4, 0.2, 0.1) & (0.6, 0.4, 0.2, 0.1) & (0.5, 0.3, 0.1, 0.4) & (0.6, 0.5, 0.2, 0.1)
\end{bmatrix}
$$

We get the following cross-entropy values by using Eq. (3) between an alternative A_i and the ideal alternative A^*:

$$D_1\left(A^*, A_1\right) = 2.02997, \qquad D_2\left(A^*, A_2\right) = 2.48202,$$

$$D_3\left(A^*, A_3\right) = 1.78736$$

Based on these values, we rank the order of three health centers as A_3, A_1, and A_2. Hence the best health center is A_3.

5 Cross-Entropy for IQNS

This section deals about cross-entropy of an interval quadripartitioned neutrosophic set which is based on the cross-entropy of QSVNSs discussed in the previous section.

Definition 5.1 Let us consider an interval quadripartitioned neutrosophic set

$$A = \langle [T_A{}^L(x), T_A{}^U(x)], [C_A{}^L(x), C_A{}^U(x)], [U_A{}^L(x), U_A{}^U(x)], [F_A{}^L(x), F_A{}^U(x)] \rangle,$$

and $g_\delta : IQNS(X) \to QSVNS(X)$ is a mapping given by

$$g_\delta(A) = \Big\langle T_A{}^L(x) + \delta\ \Delta T_A(x), C_A{}^L(x) + \delta\ \Delta C_A(x), U_A{}^L(x) + (1-\delta)\ \Delta U_A(x),$$

$$F_A{}^L(x) + (1-\delta)\ \Delta F_A(x) \Big\rangle \qquad (4)$$

where $\Delta T_A(x) = T_A{}^U(x) - T_A{}^L(x)$, $\Delta C_A(x) = C_A{}^U(x) - C_A{}^L(x)$, $\Delta U_A(x) = U_A{}^U(x) - U_A{}^L(x)$, and $\Delta F_A(x) = F_A{}^U(x) - F_A{}^L(x)$ for $x \in X$ and $\delta \in [0, 1]$. Here g_δ is known as reduction operator which helps to assign an interval quadripartitioned neutrosophic set to quadripartitioned single-valued neutrosophic set. Hence we get quadripartitioned single-valued neutrosophic set A_δ in universe X.

Example 5.2 Let $A = \langle (x_1, [0.3, 0.5], [0.1, 0.2], [0.3, 0.6], [0.4, 0.7]) : x_1 \in X \rangle$ be an interval quadripartitioned neutrosophic set in the universe set $X = \{x_1\}$. For $\delta = 0.5$, we get the following QSVNS $g_{0.5}(A)$ as $g_{0.5}(A) = \langle (x_1, 0.4, 0.15, 0.45, 0.55) : x_1 \in X \rangle$.

Proposition 5.3 Consider two IQNSs A and B in the universe of discourse $X = \{x_1, x_2, \ldots, x_n\}$ where $A = \langle [T_A{}^L(x), T_A{}^U(x)], [C_A{}^L(x), C_A{}^U(x)], [U_A{}^L(x), U_A{}^U(x)], [F_A{}^L(x), F_A$

$^U(x)] \rangle$, $B = \langle [T_B{}^L(x), T_B{}^U(x)], [C_B{}^L(x), C_B{}^U(x)], [U_B{}^L(x), U_B{}^U(x)], [F_B{}^L(x), F_B{}^U(x)] \rangle$
Let $g_\delta : IQNS(X) \to QSVNS(X)$ be a reduction operator and $\delta, \rho \in [0, 1]$. Then

i.   *if $0 \le \delta \le \rho$ then $g_\delta(A) \subseteq g_\rho(A)$*
ii. *if $A \subseteq B$ then $g_\delta(A) \subseteq g_\delta(B)$*
iii. $g_\delta(g_\rho(A)) = g_\rho(A)$
iv. $g_\delta(A^C))^C = g_{1-\delta}(A)$

Proof (i)   Let us assume that $0 \le \delta \le \rho$ for $\delta, \rho \in [0, 1]$. Then we get

$$g_\delta(A) = \left\langle T_A{}^L(x) + \delta\ \Delta T_A(x),\ C_A{}^L(x) + \delta\ \Delta C_A(x),\ U_A{}^L(x) \right.$$
$$\left. + (1-\delta)\ \Delta U_A(x),\ F_A{}^L(x) + (1-\delta)\ \Delta F_A(x) \right\rangle$$

$$g_\rho(A) = \left\langle T_A{}^L(x) + \rho\ \Delta T_A(x),\ C_A{}^L(x) + \rho\ \Delta C_A(x),\ U_A{}^L(x) \right.$$
$$\left. + (1-\rho)\ \Delta U_A(x),\ F_A{}^L(x) + (1-\rho)\ \Delta F_A(x) \right\rangle$$

where,

$$T_A{}^L(x) + \delta\ \Delta T_A(x) \le T_A{}^L(x) + \rho\ \Delta T_A(x)$$

$$C_A{}^L(x) + \delta\ \Delta C_A(x) \le C_A{}^L(x) + \rho\ \Delta C_A(x)$$
$$U_A{}^L(x) + (1-\rho)\ \Delta U_A(x) \le U_A{}^L(x) + (1-\delta)\ \Delta U_A(x)$$
$$F_A{}^L(x) + (1-\rho)\ \Delta F_A(x) \le F_A{}^L(x) + (1-\delta)\ \Delta F_A(x)$$

Hence we get $g_\delta(A) \subseteq g_\rho(A)$.

(ii)   If $A \subseteq B$, then we get

$$T_A{}^L(x) \le T_B{}^L(x),\ T_A{}^U(x) \le T_B{}^U(x),\ \ C_A{}^L(x) \le C_B{}^L(x),\ C_A{}^U(x) \le C_B{}^U(x)$$
$$U_A{}^L(x) \ge U_B{}^L(x),\ U_A{}^U(x) \ge U_B{}^U(x),\ F_A{}^L(x) \ge F_B{}^L(x),\ F_A{}^U(x) \ge F_B{}^U(x) \text{ for all } x \in X$$

It shows that

$$T_A{}^L(x) + \delta\ \Delta T_A(x) \le T_B{}^L(x) + \delta\ \Delta T_B(x)$$

$$C_A{}^L(x) + \delta\ \Delta C_A(x) \le C_B{}^L(x) + \delta\ \Delta C_B(x)$$
$$U_B{}^L(x) + (1-\delta)\ \Delta U_B(x) \le U_A{}^L(x) + (1-\delta)\ \Delta U_A(x)$$
$$F_B{}^L(x) + (1-\delta)\ \Delta F_B(x) \le F_A{}^L(x) + (1-\delta)\ \Delta F_A(x)$$

Hence we get $g_\delta(A) \subseteq g_\delta(B)$.

(iii)   It is easy to prove that $g_\delta(g_\rho(A)) = g_\rho(A)$.

(iv)   If $A \in IQNS(X)$, then its complement

$$A^C = \left\langle \left[F_A{}^L(x), F_A{}^U(x) \right], \left[U_A{}^L(x), U_A{}^U(x) \right], \left[C_A{}^L(x), C_A{}^U(x) \right], \left[T_A{}^L(x), T_A{}^U(x) \right] \right\rangle \text{ for each } x \in X$$

Then we get

$$g_\delta\left(A^C \right) = \left\langle F_A{}^L(x) + \delta\ \Delta F_A(x),\ U_A{}^L(x) + \delta\ \Delta U_A(x),\ C_A{}^L(x) \right.$$

$$+ (1 - \delta) \ \Delta C_A(x), T_A{}^L(x) + (1 - \delta) \ \Delta T_A(x) \Big) \quad \text{for each } x \in X.$$

Hence,

$$\left(g_\delta\left(A^C\right)\right)^C = \Big(T_A{}^L(x) + (1-\delta) \ \Delta T_A(x), C_A{}^L(x) + (1-\delta) \ \Delta C_A(x), U_A{}^L(x) + \delta \ \Delta U_A(x), F_A{}^L(x) + \delta \ \Delta F_A(x)\Big)$$
$$= g_{1-\delta}(A)$$

Based on this reduction of IQNS, we define a cross-entropy of IQNS as a generalization of QSVNS cross-entropy. $\square$

Definition 5.4 Let us consider two IQNSs A and B in the universe of discourse $X = \{x_1, x_2, \ldots, x_n\}$. Then interval quadripartitioned neutrosophic cross-entropy based on quadripartitioned single-valued neutrosophic cross-entropy is defined by

$$
\begin{aligned}
H_{IQNS}(A, B) = & \sum_{i=1}^{n} \left[\overline{T}_A(x_i) \log_2 \frac{\overline{T}_A(x_i)}{\frac{1}{2}(\overline{T}_A(x_i) + \overline{T}_B(x_i))} + \left(1 - \overline{T}_A(x_i)\right) \log_2 \frac{1 - \overline{T}_A(x_i)}{1 - \frac{1}{2}(\overline{T}_A(x_i) + \overline{T}_B(x_i))} \right] \\
& + \sum_{i=1}^{n} \left[\overline{C}_A(x_i) \log_2 \frac{\overline{C}_A(x_i)}{\frac{1}{2}(\overline{C}_A(x_i) + \overline{C}_B(x_i))} + \left(1 - \overline{C}_A(x_i)\right) \log_2 \frac{1 - \overline{C}_A(x_i)}{1 - \frac{1}{2}(\overline{C}_A(x_i) + \overline{C}_B(x_i))} \right] \\
& + \sum_{i=1}^{n} \left[\overline{U}_A(x_i) \log_2 \frac{\overline{U}_A(x_i)}{\frac{1}{2}(\overline{U}_A(x_i) + \overline{U}_B(x_i))} + \left(1 - \overline{U}_A(x_i)\right) \log_2 \frac{1 - \overline{U}_A(x_i)}{1 - \frac{1}{2}(\overline{U}_A(x_i) + \overline{U}_B(x_i))} \right] \\
& + \sum_{i=1}^{n} \left[\overline{F}_A(x_i) \log_2 \frac{\overline{F}_A(x_i)}{\frac{1}{2}(\overline{F}_A(x_i) + \overline{F}_B(x_i))} + \left(1 - \overline{F}_A(x_i)\right) \log_2 \frac{1 - \overline{F}_A(x_i)}{1 - \frac{1}{2}(\overline{F}_A(x_i) + \overline{F}_B(x_i))} \right]
\end{aligned}
$$
$$\text{(5)}$$

where

$$
\begin{aligned}
\overline{T}_A(x_i) &= T_A{}^L(x) + \delta \ \Delta T_A(x), \\
\overline{C}_A(x_i) &= C_A{}^L(x) + \delta \ \Delta C_A(x), \\
\overline{U}_A(x_i) &= U_A{}^L(x) + (1 - \delta) \ \Delta U_A(x), \\
\overline{F}_A(x_i) &= F_A{}^L(x) + (1 - \delta) \ \Delta F_A(x)
\end{aligned}
$$

Similarly,

$$
\begin{aligned}
\overline{T}_B(x_i) &= T_B{}^L(x) + \delta \ \Delta T_B(x), \\
\overline{C}_B(x_i) &= C_B{}^L(x) + \delta \ \Delta C_B(x), \\
\overline{U}_B(x_i) &= U_B{}^L(x) + (1 - \delta) \ \Delta U_B(x), \\
\overline{F}_B(x_i) &= F_B{}^L(x) + (1 - \delta) \ \Delta F_B(x)
\end{aligned}
$$

for all $x \in X$ and $\delta \in [0, 1]$ where δ is a threshold value. We always choose $\delta = 0.5$ unless its value is mentioned. We can easily prove the following results based on Shannon's inequality.

i. $H_{IQNS}(A, B) \geq 0$
ii. $H_{IQNS}(A, B) = 0$ if and only if

$$\left[T_A{}^L(x), T_A{}^U(x)\right] = \left[T_B{}^L(x), T_B{}^U(x)\right], \left[C_A{}^L(x), C_A{}^U(x)\right] = \left[C_B{}^L(x), C_B{}^U(x)\right],$$
$$\left[U_A{}^L(x), U_A{}^U(x)\right] = \left[U_B{}^L(x), U_B{}^U(x)\right], \left[F_A{}^L(x), F_A{}^U(x)\right] = \left[F_B{}^L(x), F_B{}^U(x)\right]$$

iii. Since $H_{IQNS}(A, B)$ does not satisfy the symmetry property, it is modified to get a symmetric discrimination information for IQNS as

$$K_{IQNS}(A, B) = H_{IQNS}(A, B) + H_{IQNS}(B, A)$$

$K_{IQNS}(A, B)$ is higher when the difference between A and B is higher.

Hence we get the following weighted cross-entropy measure between an alternative A_i and the ideal alternative A^*:

$$
\begin{aligned}
K_{IQNS}(A^*, A_i) = &\sum_{j=1}^{n} w_j \left[\log_2 \frac{1}{\frac{1}{2}(1+\overline{T}_{ij})} + \log_2 \frac{1}{\frac{1}{2}(1+\overline{C}_{ij})} + \log_2 \frac{1}{1-\frac{1}{2}(\overline{U}_{ij})} + \log_2 \frac{1}{1-\frac{1}{2}(\overline{F}_{ij})}\right] \\
&+ \sum_{j=1}^{n} w_j \left[\overline{T}_{ij}\log_2 \frac{\overline{T}_{ij}}{\frac{1}{2}(1+\overline{T}_{ij})} + (1-\overline{T}_{ij})\log_2 \frac{1-\overline{T}_{ij}}{1-\frac{1}{2}(1+\overline{T}_{ij})}\right] \\
&+ \sum_{j=1}^{n} w_j \left[\overline{C}_{ij}\log_2 \frac{\overline{C}_{ij}}{\frac{1}{2}(1+\overline{C}_{ij})} + (1-\overline{C}_{ij})\log_2 \frac{1-\overline{C}_{ij}}{1-\frac{1}{2}(1+\overline{C}_{ij})}\right] \\
&+ \sum_{j=1}^{n} w_j \left[\overline{U}_{ij} + (1-\overline{U}_{ij})\log_2 \frac{1-\overline{U}_{ij}}{1-\frac{1}{2}(\overline{U}_{ij})}\right] + \sum_{j=1}^{n} w_j \left[\overline{F}_{ij} + (1-\overline{F}_{ij})\log_2 \frac{1-\overline{F}_{ij}}{1-\frac{1}{2}(\overline{F}_{ij})}\right]
\end{aligned}
$$
$$(6)$$

6 Multicriteria Decision-Making Method Based on the Cross-Entropy of IQNSs

Since IQNS deals with set of real numbers in the real unit interval instead of deals a particular number, it plays a vital role in handling uncertain and inconsistent information that exists in real world. In this section we discuss about cross-entropy of IQNS with an illustrative example adopted by [18]. Hence the proposed approach helps to prove its feasibility and benefits while dealing real-world decision-making problem.

Let $A = \{A_1, A_2, \ldots, A_m\}$ and $C = \{C_1, C_2, \ldots, C_n\}$, respectively, denote alternative and criteria sets. Let $w_j (j = 1, 2, \ldots, n)$ denote the weight of the criterion c_j which belongs to the interval of $[0,1]$ such that $\sum_{j=1}^{n} w_j = 1$. Here the characteristic of the alternative A_i $(i = 1, 2, \ldots, m)$ is expressed by interval quadripartitioned neutrosophic set

$$A_i = \{\langle x, \left[T_{A_i}{}^L(C_j), T_{A_i}{}^U(C_j)\right], \left[C_{A_i}{}^L(C_j), C_{A_i}{}^U(C_j)\right], \left[U_{A_i}{}^L(C_j), U_{A_i}{}^U(C_j)\right], \left[F_{A_i}{}^L(C_j), F_{A_i}{}^U(C_j)\right]\rangle : C_j \in C\}$$ where $0 \leq T_{A_i}{}^U(C_j) + C_{A_i}{}^U(C_j) + U_{A_i}{}^U(C_j) + F_{A_i}{}^U(C_j) \leq 4$ for $i = 1, 2, \ldots, m, j = 1, 2, \ldots, n$.

Here $\left[T_{A_i}^{\ L}(C_j), T_{A_i}^{\ U}(C_j)\right], \left[C_{A_i}^{\ L}(C_j), C_{A_i}^{\ U}(C_j)\right]$,
$\left[U_{A_i}^{\ L}(C_j), U_{A_i}^{\ U}(C_j)\right]$ and $\left[F_{A_i}^{\ L}(C_j), F_{A_i}^{\ U}(C_j)\right]$,
respectively, denote interval of the degree in which the alternative A_i is true, both true and false, neither true nor false, and false under the particular criterion C_j given by the decision-maker. Hence we get a decision matrix $A = (\alpha_{ij})_{m \times n}$. Furthermore the concept of ideal point is used here to identify the best alternative in the decision set. Here the ideal alternative A^* is denoted by the criterion value $\alpha_{j^*} = \langle T_{j^*}, C_{j^*}, U_{j^*}, F_{j^*} \rangle = \langle 1, 1, 0, 0 \rangle$. The procedure to calculate the cross-entropy of IQNS based on the cross-entropy of QSVNSs is given below in detail:

i. Compute the QSVNS from given IQNS based on mentioned threshold value.
ii. From obtained QSVNS in previous step, calculate the weighted cross-entropy $K_{IQNS}(A^*, A_i)$ by using Eq. (6).
iii. Select the best alternative based on the rank of the alternatives.

6.1 Practical Example

The example illustrated here is adapted from [18]. This example deals about the best mobile phone among all available alternatives based on various criteria. The alternatives A_1, A_2, A_3, respectively, denote the Mobile 1, Mobile 2, and Mobile 3. The customer must take a decision according to the following four attributes, that is, (1) C_1 is the cost, (2) C_2 is the storage space, (3) C_3 is the camera quality, and (4) C_4 is the looks. According to these attributes, we will derive the ranking order of all alternatives, and based on this ranking order, customer will select the best one. Here the weight vector of the above attributes is given by $w = (0.18, 0.25, 0.37, 0.20)^T$. In general the evaluation of an alternative A_i with respect to an attribute $C_j (i = 1,2,3; j = 1,2,3,4)$ will be done by the questionnaire of a domain expert. Therefore we get the following interval quadripartitioned neutrosophic decision matrix R:

$$R = \begin{bmatrix} \langle [0.3, 0.5] & [0.1, 0.2] & [0.3, 0.6] & [0.4, 0.7] \rangle \\ \langle [0.1, 0.2] & [0.3, 0.4] & [0.7, 0.8] & [0.4, 0.6] \rangle \\ \langle [0.6, 0.8] & [0.6, 0.7] & [0.4, 0.6] & [0.3, 0.6] \rangle \end{bmatrix},$$

$$\begin{array}{cccc} \langle [0.3, 0.5] & [0.1, 0.6] & [0.4, 0.7] & [0.3, 0.7] \rangle \\ \langle [0.1, 0.5] & [0.2, 0.3] & [0.5, 0.6] & [0.4, 0.6] \rangle \\ \langle [0.4, 0.5] & [0.6, 0.7] & [0.4, 0.6] & [0.2, 0.8] \rangle \end{array},$$

$$\begin{array}{cccc} \langle [0.3, 0.6] & [0.2, 0.6] & [0.1, 0.3] & [0.4, 0.7] \rangle \\ \langle [0.4, 0.5] & [0.2, 0.3] & [0.4, 0.7] & [0.2, 0.5] \rangle \\ \langle [0.1, 0.3] & [0.5, 0.7] & [0.2, 0.3] & [0.7, 0.8] \rangle \end{array},$$

$$\begin{bmatrix} \langle [0.3, 0.5] & [0.1, 0.6] & [0.4, 0.7] & [0.3, 0.7] \rangle \\ \langle [0.4, 0.7] & [0.1, 0.5] & [0.2, 0.8] & [0.1, 0.8] \rangle \\ \langle [0.2, 0.7] & [0.1, 0.6] & [0.2, 0.4] & [0.2, 0.8] \rangle \end{bmatrix}$$

Then we follow the method which we have discussed in previous section to find the best alternative.

Step 1: Find the QSVNS from IQNS based on the threshold value as $\delta = 0.5$ by using Eq. (4) as follows:

$A_1 = \{(C_1, 0.4, 0.15, 0.45, 0.55), (C_2, 0.4, 0.35, 0.55, 0.5), (C_3, 0.45, 0.4, 0.2, 0.55), (C_4, 0.4, 0.35, 0.55, 0.5)\}$
$A_2 = \{(C_1, 0.15, 0.35, 0.75, 0.5), (C_2, 0.3, 0.25, 0.55, 0.5), (C_3, 0.45, 0.25, 0.55, 0.35), (C_4, 0.55, 0.3, 0.5, 0.45)\}$
$A_3 = \{(C_1, 0.7, 0.65, 0.5, 0.45), (C_2, 0.45, 0.65, 0.5, 0.5), (C_3, 0.2, 0.6, 0.25, 0.75), (C_4, 0.45, 0.35, 0.3, 0.5)\}$

Writing the above values in matrix form, we get

$$A = \begin{bmatrix} (0.4, 0.15, 0.45, 0.55) & (0.4, 0.35, 0.55, 0.5) & (0.45, 0.4, 0.2, 0.55) & (0.4, 0.35, 0.55, 0.5) \\ (0.15, 0.35, 0.75, 0.5) & (0.3, 0.25, 0.55, 0.5) & (0.45, 0.25, 0.55, 0.35) & (0.55, 0.3, 0.5, 0.45) \\ (0.7, 0.65, 0.5, 0.45) & (0.45, 0.65, 0.5, 0.5) & (0.2, 0.6, 0.25, 0.75) & (0.45, 0.35, 0.3, 0.5) \end{bmatrix}$$

Step 2: Calculate $K_{IQNS}(A^*, A_i)\, (i = 1, 2, \ldots, m)$ by using equation (6) as

$$K_{IQNS}\left(A^*, A_1\right) = 2.85762, \qquad K_{IQNS}\left(A^*, A_2\right) = 3.18306,$$

$$K_{IQNS}\left(A^*, A_3\right) = 2.57690$$

Step 3: Based on these values, we rank the order of three mobiles as A_3, A_1, and A_2. Hence the best mobile is A_3 which gives the same result obtained in [18]. Hence the proposed approach is consistent with the result obtained in [18].

7 Conclusion

Entropy is a very important tool to measure uncertain information in neutrosophic set theory. Based on Shannon's inequality, cross-entropy helps to measure the discrimination information between objects. In this paper we have defined the cross-entropy of QSVNS as an extension of cross-entropy of SVNS and also studied the cross-entropy of IQNS as an extension of cross-entropy of QSVNS. And also we illustrated an example of multicriteria decision-making problem for the proposed method which helps to identify the best alternative according to the rank of the given alternatives.

References

1. Zadeh, L. A. (1965). Fuzzy sets. *Information and Control, 8*, 338–353.
2. Zadeh, L. A. (1965). Fuzzy sets. *Information and Control, 8*, 87–96.
3. Atanassov, K. (1986). Intuitionistic fuzzy sets. *Fuzzy Sets and Systems, 20*, 87–96.
4. Turksen, B. (1986). Interval valued fuzzy sets based on normal forms. *Fuzzy Sets and Systems, 20*, 191–210.
5. Atanassov, K., & Gargov, G. (1989). Interval-valued intuitionistic fuzzy sets. *Fuzzy Sets and Systems, 31*(3), 343–349.
6. Smarandache, F. (1999). A unifying field in logics. In F. Smarandache (Ed.), *Neutrosophy: Neutrosophic probability, set and logic*. Rehoboth: American Research Press.
7. Wang, H., Smarandache, F., Zhang, Y. Q., & Sunderraman, R. (2010). Single valued neutrosophic sets. *Multispace and Multistructure, 4*, 410–413.
8. Ye, J. (2013). Multicriteria decision-making method using the correlation coefficient under single-valued neutrosophic environment. *International Journal of General Systems, 42*(4), 386–394.
9. Ye, J., & Zhang, Q. (2014). Single valued neutrosophic similarity measures for multiple attribute decision-making. *Neutrosophic Sets and Systems, 2*, 48–54.
10. Ye, J. (2014). Improved correlation coefficients of single valued neutrosophic sets and interval neutrosophic sets for multiple attribute decision making. *Journal of Intelligent Fuzzy Systems, 27*, 24532462.
11. Ye, J. (2015). *Another form of correlation coefficient between single valued neutrosophic sets and its multiple attribute decision-making method* (pp. 1). doi: https://doi.org/10.5281/zenodo.22428.
12. Ye, J., & Smarandache, F. (2016). Similarity measure of refined single-valued neutrosophic sets and its multicriteria decision making method. *Neutrosophic Sets and Systems, 12*, 41–44.
13. Mohana, K., Christy, V., & Smarandache, F. (2019). On multi-criteria decision making problem via bipolar single-valued neutrosophic settings. *Neutrosophic Sets and Systems, 25*, 125–135.
14. Mohana, K., & Mohanasundari, M. (2019). On some similarity measures of single valued neutrosophic rough sets. *Neutrosophic Sets and Systems, 24*, 10–22.
15. Belnap, N. D., Jr. (1977). A useful four valued logic. In J. Dunn & G. Epstein (Eds.), *Modern uses of multiple valued logic* (pp. 9–37). Dordrecht: D. Reidel Publishing Company.
16. Smarandache, F. (2014). n-valued refined neutrosophic logic and its applications to physics, *arXivpreprintarXiv*:1407.1041
17. Chatterjee, R., Majumdar, P., & Samanta, S. K. (2016). On some similarity measures and entropy on quadripartitioned single valued neutrosophic sets. *Journal of Intelligent Fuzzy Systems, 30*, 2475–2485.

18. Mohanasundari, M., & Mohana, K. (2019). Chapter 14, Improved correlation coefficients of quadripartitioned single-valued neutrosophic sets and interval-quadripartitioned neutrosophic sets. In M. Abdel-Basset & F. Smarandache (Eds.), *Neutrosophic sets in decision analysis and operations research* (pp. 331–363). Hershey, PA: IGI Global Publisher.
19. Zadeh, L. A. (1968). Probability measures of fuzzy events. *Journal of Mathematical Analysis and Applications, 23*, 421–427.
20. Burillo, P., & Bustince, H. (1996). Entropy on intuitionistic fuzzy sets and on interval-valued fuzzy sets. *Fuzzy Sets and Systems, 78*, 305–316.
21. DeLuca, A., & Termini, S. (1972). A definition of nonprobabilistic entropy in the setting of fuzzy sets theory. *Information and Control, 20*, 301–312.
22. Szmidt, E., & Kacprzyk, J. (2001). Entropy for intuitionistic fuzzy sets. *Fuzzy Sets and Systems, 118*, 467–477.
23. Wei, C. P., Wang, P., & Zhang, Y. Z. (2011). Entropy, similarity measure of interval-valued intuitionistic fuzzy sets and their applications. *Information Sciences, 181*, 4273–4286.
24. Shang, X. G., & Jiang, W. S. (1997). A note on fuzzy information measures. *Pattern Recognition Letters, 18*, 425–432.
25. Shannon, C. E. (1948). A mathematical theory of communication. *Bell System Technical Journal, 27*, 379–423.
26. Vlachos, I. K., & Sergiadis, G. D. (2007). Intuitionistic fuzzy information-applications to pattern recognition. *Pattern Recognition Letters, 28*, 197–206.
27. Zhang, Q. S., Jiang, S., Jia, B., & Luo, S. (2010). Some information measures for interval-valued intuitionistic fuzzy sets. *Information Sciences, 180*, 5130–5145.
28. Ye, J. (2014). Single valued neutrosophic cross-entropy for multicriteria decision making problems. *Applied Mathematical Modelling, 38*, 1170–1175.
29. Biswas, P., Pramanik, S., & Giri, B. C. (2014). Entropy based grey relational analysis method for multi-attribute decision making under single valued neutrosophic assessments. *Neutrosophic Sets and Systems, 2*, 102–110.
30. Nirmal, N. P., & Bhatt, M. G. (2016). Selection of automated guided vehicle using single valued neutrosophic entropy based novel multi attribute decision making technique. In F. Smarandache & S. Pramanik (Eds.), *New trends in neutrosophic theory and applications* (pp. 104–116). Brussells: Pons asbl.
31. Şahin, R. (2017). Cross-entropy measure on interval neutrosophic sets and its applications in multicriteria decision making. *Neural Computing and Applications, 28*, 1177–1187.

New Entropy, Similarity Measures of Interval-Valued Neutrosophic Sets, and Application in Supplier Selection

Truong Thi Thuy Duong, Nguyen Xuan Thao, and Florentin Smarandache

1 Introduction

Neutrosophic set proposed by Smarandache [1] is an extension of fuzzy set [2, 3], intuitionistic fuzzy sets (IFSs) [4], and interval-valued intuitionistic fuzzy sets (IVIFSs) [5]. A neutrosophic set A in a universal set X has three membership functions, a truth-membership function T_A, an indeterminacy-membership function I_A, and a falsity-membership function F_A. For each element x of X, then $T_A(x)$, $I_A(x)$, $F_A(x)$ are real standard or nonstandard subsets of $]^-0, 1^+[$. With practical problems, the above values of membership functions will be difficult to apply. Because of this, many individual cases of neutrosophic sets have been proposed and applied to practical problems. Wang et al. [6, 7] introduced the single-valued neutrosophic sets (SVNSs) and interval-valued neutrosophic sets (IVNSs). In IVNS, for each element of the universal set, then their membership function is the standard subsets of the unit interval [0, 1]. So far, there have been many applications of IVNS in decision-making problem. Broumi and Smarandache [8] gave some distance measures and similarity measures of IVNS. Ye [9] investigated some similarity

T. T. T. Duong (✉)
Banking Academy, Hanoi, Vietnam
e-mail: thuyduongktv@yahoo.com.vn

N. X. Thao
Faculty of Information Technology, Vietnam National University of Agriculture, Hanoi, Vietnam
e-mail: nxthao@vnua.edu.vn

F. Smarandache
Department of Mathematics and Science, University of New Mexico, Gallup, NM, USA
e-mail: fsmarandache@gmail.com

© The Author(s), under exclusive license to Springer Nature Switzerland AG 2021   35
F. Smarandache, M. Abdel-Basset (eds.), *Neutrosophic Operational Research*,
https://doi.org/10.1007/978-3-030-57197-9_3

measures based on the Hamming and Euclidean distance measures between IVNS and their applications in multicriteria decision-making (MCDM). After that, the similarity measures improved based on the cosine and cotangent functions to apply in the real problems [10, 11]. Aiwu et al. [12] studied the generalized weight aggregation operator for IVNS and applied in MCDM.

Entropy measure is an important tool for measuring uncertain information [13]. Over time, with the development of fuzzy sets, then their entropy measures are also studied and applied. Entropy of fuzzy set was first introduced by Zadeh [3]. In 1996, Bustince and Burillo [14] investigated the entropy of intuitionistic fuzzy sets as an extension of the entropy of fuzzy sets. The set of axiomatic requirements for entropy measures of IVIFSs is proposed by Liu et al. [15]. Hung and Yang exploited the concept of probability for defining the fuzzy entropy of intuitionistic fuzzy sets [16]. The sine and cosine entropy of interval-valued intuitionistic fuzzy set is introduced by Ye [17]. Relationship between entropy and similarity measure of interval-valued intuitionistic fuzzy sets is also studied and applied by many researchers [18–20]. As the extension of the concept of entropy measures of the intuitionistic fuzzy sets and interval-valued intuitionistic fuzzy sets, the entropy measures of SVNS and IVNS are also studied by some researchers [21–23]. In which, Aydoğdu's [21] entropy measure is not effective in some cases because of the negative value (see Sect. 3). Ye and Du [23] gave some entropy measures based on the distance functions. This is fine, but it is very difficult to directly test a formula that has to be an entropy measure. Overcoming these limitations, we introduce a new concept for the entropy measure of IVNS. This new concept is developed based on the set of axiomatic requirements for entropy measures of IVIFSs [15] and the concept entropy of IFSs of Hung and Yang [16]. This concept has four conditions that allow us to directly construct the entropy measure formulas of the SVNSs. Following the idea of constructing similarity measures from the entropy measure, we also construct similarity measures for IVNS based on the proposed entropy measures and apply them in the MCDM problem.

Supplier selection has recognized a strategy role due to its influence on the success of firms. Bevilacqua et al. [24] believed that the relationship between suppliers and buyers has changed in terms of development of the long-term cooperation instead of short-term negotiation one. In addition, firms tend to prefer deep cooperation to wide one to take more advantages. This creates a sustainable buyer-vendor relationship which contributes to reduce the operation cost and enhance the quality of production. Thus, it is very important to select the reasonable suppliers. The process of supplier selection is complex by the fact that it requires a variety of criteria including quantitative and qualitative criteria which might be conflict. The criteria are related to the price [25–28], delivery [27, 29, 30], techniques [26, 27, 29], quality [28, 31, 32], service, and performance. Thereby, supplier selection process is an MCDM problem. In an MCDM problem, it raises a vital matter to evaluate for criteria as well as alternatives by exactly real numbers. Furthermore, the human thinking sometimes is subjective and thus imprecise. To overcome this restriction, the neutrosophic set is an appropriate choice to show the judgments in the sense of linguistic variables for criteria and alternatives.

The rest of this chapter is organized as follows: In Sect 2, we recall the definition of IVNS together with some operator of IVNSs, comparison of two closing interval. In Sect. 3, we introduce the new concept of entropy measure of IVNSs and give some formulas to define the entropy measure of IVNSs. In this section, we also compare the proposed entropy measure with some existing entropy measures of IVNSs. In Sect. 4, we investigate some similarity measures of IVNSs which are induced by the proposed entropy measure of IVNSs in Sect. 3. We also practice the comparison of the new similarity measure with some existing similarity measures of IVNSs in this Sect. 4. Finally, we apply the new similarity measures of IVNSs to the MCDM problem. A case study to select the best supplier is employed to illustrate the efficiency of our model.

2 Preliminaries

We denote an universal $X = \{x_1, x_2, \ldots, x_n\}$ and $[UI]$ is the collection of standard subsets of the unit interval $[0, 1]$.

Definition 1 Let $[a, b], [c, d] \in [UI]$; then we define the following relationships:

$$[a, b] \leq [c, d] \ \text{iff} \ a \leq c, b \leq d, \ \ [a, b] \prec [c, d] \ \text{iff} \ a \leq c, b \geq d$$

$$[a, b] = [c, d] \ \text{iff} \ a = c, b = d, \ \ [a, b] \vee [c, d] = [a \vee c, b \vee d]$$

$$[a, b] \wedge [c, d] = [a \wedge c, b \wedge d]$$

Definition 2 An interval-valued neutrosophic set A in X can be expressed by the form

$$A = \left\{ \left(x_i, T_A(x_i), I_A(x_i), F_A(x_i) \right) | x_i \in X \right\}$$

where $T_A(x_i) = \left[T_A^L(x_i), T_A^U(x_i) \right] \in [UI], I_A(x_i) = \left[I_A^L(x_i), I_A^U(x_i) \right] \in [UI]$, and $F_A(x_i) = \left[F_A^L(x_i), F_A^U(x_i) \right] \in [UI]$ are the truth-membership, indeterminacy-membership, and falsity-membership function, respectively, for all $x_i \in X$.

- Complement of A: $A^C = \left\{ \left(x_i, T_{A^C}(x_i), I_{A^C}(x_i), F_{A^C}(x_i) \right) | x_i \in X \right\}$
  where $T_{A^C}(x_i) = F_A(x_i), F_{A^C}(x_i) = T_A(x_i), I_{A^C}(x_i)$
  $= \left[1 - I_A^U(x_i), 1 - I_A^L(x_i) \right]$ for all $x_i \in X$.
- Subset: $A \subset B$ if and only if $T_A(x_i) \leq T_B(x_i), I_A(x_i) \geq I_B(x_i)$, and $F_A(x_i) \geq F_B(x_i)$ for all $x_i \in X$.
- Equality: $A = B$ if and only if $A \subset B$ and $B \subset A$.

– Union: $A \cup B = \{\langle x_i, T_A(x_i) \vee T_B(x_i), I_A(x_i) \wedge I_B(x_i), F_A(x_i) \wedge F_B(x_i)\rangle \mid x_i \in X\}$.
– Intersection: $A \cap B = \{\langle x_i, T_A(x_i) \wedge T_B(x_i), I_A(x_i) \vee I_B(x_i), F_A(x_i) \vee F_B(x_i)\rangle \mid x_i \in X\}$.

For two IVNSs A, B and the positive real number $\lambda \geq 0$, we define the operator law [23]

$$
A^{\lambda} = \left\{ \left(x_i, \left[\left(T_A^L(x_i) \right)^{\lambda}, \left(T_A^L(x_i) \right)^{\lambda} \right], \left[1 - \left(1 - I_A^L(x_i) \right)^{\lambda}, 1 - \left(1 - I_A^L(x_i) \right)^{\lambda} \right], \right. \right.
$$
$$
\left. \left. \left[1 - \left(1 - F_A^L(x_i) \right)^{\lambda}, 1 - \left(1 - F_A^L(x_i) \right)^{\lambda} \right] \right) \mid x_i \in X \right\}
\tag{1}
$$

3   Entropy on the Interval-Valued Neutrosophic Set

In this section we will introduce the new concept of an IVNS. This concept is an extension of the concept of an intuitionistic fuzzy set which was known in the literatures. The entropy of IVNS is a useful tool to measure the uncertainty of information. Unlike the entropy of fuzzy sets, the entropy of the IVNS must show the effect of the indefinite of the indeterminacy in the IVNS.

Definition 3 An entropy on $IVNS(X)$ is a function $EN : IVNS(X) \rightarrow [0,1]$, satisfying all the following conditions:

(E1) $E(A) = 0$ if A is a crisp set, i.e., $A_i = (1,0,0)$ or $A_i = (0,0,1)$ for all $x_i \in X$.

(E2) $E(A) = 0$ if $A = \{(x_i, [0.5, 0.5], [0.5, 0.5], [0.5, 0.5]) \mid x_i \in X\}$.

(E3) $E(A) \leq E(B)$ if $T_A(x_i) \leq T_B(x_i)$, $I_A(x_i) \leq I_B(x_i)$, $F_A(x_i) \leq F_B(x_i)$ if $\max_{\leq} \{T_B(x_i), I_B(x_i), F_B(x_i)\} \leq [0.5, 0.5]$; $T_A(x_i) \geq T_B(x_i)$, $I_A(x_i) \geq I_B(x_i)$, $F_A(x_i) \geq F_B(x_i)$ if $\min_{\leq} \{T_B(x_i), I_B(x_i), F_B(x_i)\} \geq [0.5, 0.5]$;

$T_A(x_i) \prec T_B(x_i)$, $I_A(x_i) \prec I_B(x_i)$, $F_A(x_i) \prec F_B(x_i)$ if $\max_{\prec} \{T_B(x_i), I_B(x_i), F_B(x_i)\} \prec [0.5, 0.5]$; $T_A(x_i) \succ T_B(x_i)$, $I_A(x_i) \succ I_B(x_i)$, $F_A(x_i) \succ F_B(x_i)$ if $\min_{\prec} \{T_B(x_i), I_B(x_i), F_B(x_i)\} \succ [0.5, 0.5]$, where max, min are performed by operator $\leq$ and $\max_{\leq}$, $\min_{\leq}$ are performed by operator $\prec$ in Definition 1.

(E4) $E(A) = E(A^C)$, for all $A \in IVNS(X)$.

It can easily gain the following lemma.

Lemma 1 If the following conditions hold:

$T_A(x_i) \leq T_B(x_i)$, $I_A(x_i) \leq I_B(x_i)$, $F_A(x_i) \leq F_B(x_i)$ if $\min_{\leq} \{T_B(x_i), I_B(x_i), F_B(x_i)\} \geq [0.5, 0.5]$; $T_A(x_i) \geq T_B(x_i)$, $I_A(x_i) \geq I_B(x_i)$, $F_A(x_i) \geq F_B(x_i)$ if $\min\{T_B(x_i), I_B(x_i), F_B(x_i)\} \geq [0.5, 0.5]$; $T_A(x_i) \prec T_B(x_i)$, $I_A(x_i) \prec I_B(x_i)$, $F_A(x_i) \prec F_B(x_i)$ if $\max_{\prec} \{T_B(x_i), I_B(x_i), F_B(x_i)\} \prec [0.5, 0.5]$, and $T_A(x_i) \succ T_B(x_i)$, $I_A(x_i) \succ I_B(x_i)$, $F_A(x_i) \succ F_B(x_i)$ if $\min_{\prec} \{T_B(x_i), I_B(x_i), F_B(x_i)\} \succ [0.5, 0.5]$

then

$$
\left| T_B^L(x_i) - 0.5 \right| \leq \left| T_A^L(x_i) - 0.5 \right|, \left| T_B^U(x_i) - 0.5 \right| \leq \left| T_A^U(x_i) - 0.5 \right|,
$$

$$
\left| I_B^L(x_i) - 0.5 \right| \leq \left| I_A^L(x_i) - 0.5 \right|, \left| I_B^U(x_i) - 0.5 \right| \leq \left| I_A^U(x_i) - 0.5 \right|,
$$

$$
\left| F_B^L(x_i) - 0.5 \right| \leq \left| F_A^L(x_i) - 0.5 \right|, \left| F_B^U(x_i) - 0.5 \right| \leq \left| F_A^U(x_i) - 0.5 \right|
$$

for all $x_i \in X$.

Now, we give some the expression to determine the entropy of an IVNS on X.

Definition 4 Let $A = \{(x_i, T_A(x_i), I_A(x_i), F_A(x_i)) | x_i \in X\}$ be an IVNS set on X. We define

$$
E_M(A) = 1 - \frac{2}{n} \sum_{i=1}^{n} \max \left\{ \begin{array}{l} \left| T_A^L(x_i) - 0.5 \right|, \left| T_A^U(x_i) - 0.5 \right|, \left| I_A^L(x_i) - 0.5 \right|, \left| I_A^U(x_i) - 0.5 \right|, \\ \left| I_{A^C}^L(x_i) - 0.5 \right|, \left| I_{A^C}^U(x_i) - 0.5 \right|, \left| F_A^L(x_i) - 0.5 \right|, \left| F_A^U(x_i) - 0.5 \right| \end{array} \right\}
\tag{2}
$$

and

$$
E_T(A) = 1 - \frac{1}{n} \sum_{i=1}^{n} \frac{\left\{ \begin{array}{l} \left| T_A^L(x_i) - 0.5 \right| + \left| T_A^U(x_i) - 0.5 \right| + \left| I_A^L(x_i) - 0.5 \right| + \left| I_A^U(x_i) - 0.5 \right| \\ + \left| I_{A^C}^L(x_i) - 0.5 \right| + \left| I_{A^C}^U(x_i) - 0.5 \right| + \left| F_A^L(x_i) - 0.5 \right| + \left| F_A^U(x_i) - 0.5 \right| \end{array} \right\}}{4}
\tag{3}
$$

$$
E_{MD}(A) = \frac{1}{n} \sum_{i=1}^{n} \frac{1 - 2 * \max \left\{ \begin{array}{l} \left| T_A^L(x_i) - 0.5 \right|, \left| T_A^U(x_i) - 0.5 \right|, \left| I_A^L(x_i) - 0.5 \right|, \left| I_A^U(x_i) - 0.5 \right|, \\ \left| I_{A^C}^L(x_i) - 0.5 \right|, \left| I_{A^C}^U(x_i) - 0.5 \right|, \left| F_A^L(x_i) - 0.5 \right|, \left| F_A^U(x_i) - 0.5 \right| \end{array} \right\}}{1 + 2 * \max \left\{ \begin{array}{l} \left| T_A^L(x_i) - 0.5 \right|, \left| T_A^U(x_i) - 0.5 \right|, \left| I_A^L(x_i) - 0.5 \right|, \left| I_A^U(x_i) - 0.5 \right|, \\ \left| I_{A^C}^L(x_i) - 0.5 \right|, \left| I_{A^C}^U(x_i) - 0.5 \right|, \left| F_A^L(x_i) - 0.5 \right|, \left| F_A^U(x_i) - 0.5 \right| \end{array} \right\}}
\tag{4}
$$

$$
E_{TD}(A) = \frac{1}{n} \sum_{i=1}^{n} \frac{1 - \frac{1}{4} \left\{ \begin{array}{l} |T_A^L(x_i) - 0.5| + |T_A^U(x_i) - 0.5| + |I_A^L(x_i) - 0.5| + |I_A^U(x_i) - 0.5| \\ + |I_{A^C}^L(x_i) - 0.5| + |I_{A^C}^U(x_i) - 0.5| + |F_A^L(x_i) - 0.5| + |F_A^U(x_i) - 0.5| \end{array} \right\}}{1 - \frac{1}{4} \left\{ \begin{array}{l} |T_A^L(x_i) - 0.5| + |T_A^U(x_i) - 0.5| + |I_A^L(x_i) - 0.5| + |I_A^U(x_i) - 0.5| \\ + |I_{A^C}^L(x_i) - 0.5| + |I_{A^C}^U(x_i) - 0.5| + |F_A^L(x_i) - 0.5| + |F_A^U(x_i) - 0.5| \end{array} \right\}}
\tag{5}
$$

Theorem 1 The function defined by Eq. (2) is an entropy of an IVNS.

Proof (E1) If A is a crisp set, then for all $x_i \in X$, we have

$$E^i_M (A) = \max \left\{ \begin{array}{l} \left|T^L_A (x_i) - 0.5\right|, \left|F^L_A (x_i) - 0.5\right|, \left|I^L_A (x_i) - 0.5\right|, \left|I^L_{A^C} (x_i) - 0.5\right|, \\ \left|T^U_A (x_i) - 0.5\right|, \left|F^U_A (x_i) - 0.5\right|, \left|I^U_A (x_i) - 0.5\right|, \left|I^U_{A^C} (x_i) - 0.5\right| \end{array} \right\} = 0.5 \quad (6)$$

It implies that $E_M(A) = 0$.

(E2) If $A = \{(x_i, [0.5, 0.5], [0.5, 0.5], [0.5, 0.5]) | \, x_i \in X\}$, then $E^i_M (A) = 0$. It implies that $E_M(A) = 1$.

(E3) From Lemma 1, we have $E_M(A) \leq E_M(B)$ if:

$T_A(x_i) \leq T_B(x_i), I_A(x_i) \leq I_B(x_i), F_A(x_i) \leq F_B(x_i)$ if $\max_{\leq} \{T_B (x_i), I_B (x_i), F_B (x_i)\}$
$\leq [0.5, 0.5]$;

$T_A(x_i) \geq T_B(x_i), I_A(x_i) \geq I_B(x_i), F_A(x_i) \geq F_B(x_i)$ if $\min_{\leq} \{T_B (x_i), I_B (x_i), F_B (x_i)\}$
$\geq [0.5, 0.5]$;

$T_A(x_i) \prec T_B(x_i), I_A(x_i) \prec I_B(x_i), F_A(x_i) \prec F_B(x_i)$ if $\max_{\prec} \{T_B (x_i), I_B (x_i), F_B (x_i)\}$
$\prec [0.5, 0.5]$;

$T_A(x_i) \succ T_B(x_i), I_A(x_i) \succ I_B(x_i), F_A(x_i) \succ F_B(x_i)$ if $\min_{\prec} \{T_B (x_i), I_B (x_i), F_B (x_i)\}$
$\succ [0.5, 0.5]$.

(E4) It is easy to verify that $E(A) = E(A^C)$ for all $A \in IVNS(X)$. □

Theorem 2 The function defined by Eq. (3) is an entropy of an IVNS.

Proof (E1) If A is a crisp set, then for all $x_i \in X$, then

$$E^i_T (A) = \left|T^L_A (x_i) - 0.5\right| + \left|T^U_A (x_i) - 0.5\right| + \left|I^L_A (x_i) - 0.5\right| + \left|I^U_A (x_i) - 0.5\right|$$
$$+ \left|I^L_{A^C} (x_i) - 0.5\right| + \left|I^U_{A^C} (x_i) - 0.5\right| + \left|F^L_A (x_i) - 0.5\right| + \left|F^U_A (x_i) - 0.5\right|$$

Hence $E^i_T (A) = 4$. It implies that $E_T(A) = 0$.

(E2) If $A = \{(x_i, [0.5, 0.5], [0.5, 0.5], [0.5, 0.5]) | \, x_i \in X\}$, then $E^i_T (A) = 0$. It implies that $E_T(A) = 1$.

(E3) From Lemma 1, we have $E_T(A) \leq E_T(B)$ if:

$T_A(x_i) \leq T_B(x_i), I_A(x_i) \leq I_B(x_i), F_A(x_i) \leq F_B(x_i)$ if $\max_{\leq} \{T_B (x_i), I_B (x_i), F_B (x_i)\}$
$\leq [0.5, 0.5]$;

$T_A(x_i) \geq T_B(x_i), I_A(x_i) \geq I_B(x_i), F_A(x_i) \geq F_B(x_i)$ if $\min_{\leq} \{T_B (x_i), I_B (x_i), F_B (x_i)\}$
$\geq [0.5, 0.5]$;

$T_A(x_i) \prec T_B(x_i), I_A(x_i) \prec I_B(x_i), F_A(x_i) \prec F_B(x_i)$ if $\max_{\prec} \{T_B (x_i), I_B (x_i), F_B (x_i)\}$
$\prec [0.5, 0.5]$;

$T_A(x_i) \succ T_B(x_i), I_A(x_i) \succ I_B(x_i), F_A(x_i) \succ F_B(x_i)$ if $\min_{\prec} \{T_B (x_i), I_B (x_i), F_B (x_i)\}$
$\succ [0.5, 0.5]$.

(E4) It can easily verify that $E(A) = E(A^C)$ for all $A \in IVNS(X)$. □

By the same way, we also verify Eqs. (4) and (5) to be two entropy measures on IVNSs.

In applications, the weight of elements is to be considered. So here we also determine the entropy over the weights assigned to the components of the universal set.

Let $X = \{x_1, x_2, \ldots, x_n\}$ be a universal set. In which, each element x_i in X will be assigned with a weight $\omega_i \in [0, 1]$ and $\sum_{i=1}^{n} \omega_i = 1$; $\omega = (\omega_1, \omega_2, \ldots, \omega_n)$ is called the vector weight on X. We have the entropy measure as follows.

Definition 5 Let $A = \{(x_i, T_A(x_i), I_A(x_i), F_A(x_i))| x_i \in X\}$ be an IVNS set on X. The following functions define the entropy measures of IVNS:

$$E_M^\omega(A) = \sum_{i=1}^{n} \omega_i \left\{ 1 - 2 \times max \left\{ \begin{array}{l} \left|T_A^L(x_i) - 0.5\right|, \left|T_A^U(x_i) - 0.5\right|, \left|I_A^L(x_i) - 0.5\right|, \left|I_A^U(x_i) - 0.5\right|, \\ \left|I_{A^C}^L(x_i) - 0.5\right|, \left|I_{A^C}^U(x_i) - 0.5\right|, \left|F_A^L(x_i) - 0.5\right|, \left|F_A^U(x_i) - 0.5\right| \end{array} \right\} \right\}$$

$$(7)$$

$$E_T(A) = \sum_{i=1}^{n} \omega_i \times \left\{ 1 - \frac{\begin{array}{l} \left|T_A^L(x_i) - 0.5\right| + \left|T_A^U(x_i) - 0.5\right| + \left|I_A^L(x_i) - 0.5\right| + \left|I_A^U(x_i) - 0.5\right| \\ + \left|I_{A^C}^L(x_i) - 0.5\right| + \left|I_{A^C}^U(x_i) - 0.5\right| + \left|F_A^L(x_i) - 0.5\right| + \left|F_A^U(x_i) - 0.5\right| \end{array}}{4} \right\}$$

$$(8)$$

$$E_{MD}(A) = \sum_{i=1}^{n} \omega_i \frac{1 - 2 * max \left\{ \begin{array}{l} \left|T_A^L(x_i) - 0.5\right|, \left|T_A^U(x_i) - 0.5\right|, \left|I_A^L(x_i) - 0.5\right|, \left|I_A^U(x_i) - 0.5\right|, \\ \left|I_{A^C}^L(x_i) - 0.5\right|, \left|I_{A^C}^U(x_i) - 0.5\right|, \left|F_A^L(x_i) - 0.5\right|, \left|F_A^U(x_i) - 0.5\right| \end{array} \right\}}{1 + 2 * max \left\{ \begin{array}{l} \left|T_A^L(x_i) - 0.5\right|, \left|T_A^U(x_i) - 0.5\right|, \left|I_A^L(x_i) - 0.5\right|, \left|I_A^U(x_i) - 0.5\right|, \\ \left|I_{A^C}^L(x_i) - 0.5\right|, \left|I_{A^C}^U(x_i) - 0.5\right|, \left|F_A^L(x_i) - 0.5\right|, \left|F_A^U(x_i) - 0.5\right| \end{array} \right\}}$$

$$(9)$$

$$E_{TD}(A) = \sum_{i=1}^{n} \omega_i \frac{1 - \frac{1}{4} \left\{ \begin{array}{l} \left|T_A^L(x_i) - 0.5\right| + \left|T_A^U(x_i) - 0.5\right| + \left|I_A^L(x_i) - 0.5\right| + \left|I_A^U(x_i) - 0.5\right| \\ + \left|I_{A^C}^L(x_i) - 0.5\right| + \left|I_{A^C}^U(x_i) - 0.5\right| + \left|F_A^L(x_i) - 0.5\right| + \left|F_A^U(x_i) - 0.5\right| \end{array} \right\}}{1 - \frac{1}{4} \left\{ \begin{array}{l} \left|T_A^L(x_i) - 0.5\right| + \left|T_A^U(x_i) - 0.5\right| + \left|I_A^L(x_i) - 0.5\right| + \left|I_A^U(x_i) - 0.5\right| \\ + \left|I_{A^C}^L(x_i) - 0.5\right| + \left|I_{A^C}^U(x_i) - 0.5\right| + \left|F_A^L(x_i) - 0.5\right| + \left|F_A^U(x_i) - 0.5\right| \end{array} \right\}}$$

$$(10)$$

Compare with Some Existing Entropies of IVNSs

For convenient comparison, we introduce the existing information entropy measures:

+ Entropy of Aydoğdu [21]:

$$E_{Ay}(A) = \frac{1}{n} \sum_{i=1}^{n} \frac{2 - \left|T_A^L(x_i) - F_A^L(x_i)\right| - \left|I_A^L(x_i) - I_A^U(x_i)\right| - \left|T_A^U(x_i) - F_A^U(x_i)\right|}{2 + \left|T_A^L(x_i) - F_A^L(x_i)\right| + \left|I_A^L(x_i) - I_A^U(x_i)\right| + \left|T_A^U(x_i) - F_A^U(x_i)\right|}$$

This measure violates the valued domain of entropy. Indeed, let $X = \{x\}$ be a universal and $A = \{(x, [0.1, 0.2], [0.1, 0.9], [0.9, 1])\}$ be an IVNS; then $E_{Ay}(A) = -\frac{1}{11}$ is not in $[0, 1]$. So $E_{Ay}(A)$ is not suitable in practice.

+ Entropy of Majumder and Samanta is introduced by Ye and Du [23]:

$$E_{MM}(A) = 1 - \frac{1}{2n} \sum_{i=1}^{n} \left\{ \left[T_A^L(x_i) + F_A^L(x_i) \right] \left| I_A^L(x_i) - I_{AC}^L(x_i) \right| + \left[T_A^U(x_i) + F_A^U(x_i) \right] \left| I_A^U(x_i) - I_{AC}^U(x_i) \right| \right\}$$

This entropy measure is not good. Indeed, we choose an IVNS on $X = \{x\}$ as $A = \{(x, [0.9, 1], [0, 1], [0.9, 1])\}$; then $E_{MM}(A) = 1 - \frac{1}{2}[(0.9 + 0.9) |0 - 0.9| + (1 + 1) |0 - 1|] = -0.7$ is not in $[0, 1]$. So $E_{MM}(A)$ is also not suitable in practice.

+ Entropy of Ye and Du [23]:

$$E_{YD1}(A) = 1 - \frac{2}{3n} \sum_{i=1}^{n} \left\{ \begin{array}{l} \max\left[\left| T_A^L(x_i) - 0.5 \right|, \left| T_A^U(x_i) - 0.5 \right| \right] + \max\left[\left| I_A^L(x_i) - 0.5 \right|, \left| I_A^U(x_i) - 0.5 \right| \right] \\ + \max\left[\left| F_A^L(x_i) - 0.5 \right|, \left| F_A^U(x_i) - 0.5 \right| \right] \end{array} \right\}$$

$$E_{YD2}(A) = 1 - \frac{1}{3n} \sum_{i=1}^{n} \left\{ \begin{array}{l} \left| T_A^L(x_i) - 0.5 \right| + \left| T_A^U(x_i) - 0.5 \right| + \left| I_A^L(x_i) - 0.5 \right| + \left| I_A^U(x_i) - 0.5 \right| \\ + \left| F_A^L(x_i) - 0.5 \right| + \left| F_A^U(x_i) - 0.5 \right| \end{array} \right\}$$

In the following example, we will compare the performance of our proposed entropy measures E_M, E_T with entropy measures E_{YD1}, E_{YD2} and E_{Ay}. To illustrate the significance of the new entropy measure, let us consider an example that applies the entropy measures for decision-making. According to Ye [11], in terms of choice of alternative, if one has a lower entropy value, that alternative has a higher rating in decision-making problem. Then, we apply the proposed entropy measures to the decision-making problem in the following example with data in Ye [11].

Example 1 Suppose that an investment company wants to invest a sum of money for the best project. There are four investment projects (alternatives) as A_1, a car company; A_2, a food company; A_3, a computer company; and A_4, an arms company. The investment company must choose the best project based on three following criteria as the risk (C_1), the growth (C_2), and the environment impact (C_3). Each alternative is considered as an IVNS on the set of criteria $C = \{C_1, C_2, C_3\}$ which is shown in Table 1.

We use four new entropy measures E_M, E_T, E_{MD}, and E_{TD} to evaluate all projects. For convenience of comparison, their entropy values along with the values of E_{YD1} and E_{YD2} are shown in Table 2.

Based on Table 2, we find that A_4 is the best alternative (project). This outcome is as the same as with one in Ye [9]. Thus, our measures are effective.

Table 1 Relationship between alternatives and criteria in terms of IVNS

	C_1	C_2	C_3
A_1	([0.4, 0.5], [0.2, 0.3], [0.3, 0.4])	([0.4, 0.6], [0.1, 0.3], [0.2, 0.4])	([0.7, 0.9], [0.2, 0.3], [0.4, 0.5])
A_2	([0.6, 0.7], [0.1, 0.2], [0.2, 0.3])	([0.6, 0.7], [0.1, 0.2], [0.2, 0.3])	([0.3, 0.6], [0.3, 0.5], [0.8, 0.9])
A_3	([0.3, 0.6], [0.2, 0.3], [0.3, 0.4])	([0.5, 0.6], [0.2, 0.3], [0.3, 0.4])	([0.4, 0.5], [0.2, 0.4], [0.7, 0.9])
A_4	([0.7, 0.8], [0.0, 0.1], [0.1, 0.2])	([0.6, 0.7], [0.1, 0.2], [0.2, 0.3])	([0.6, 0.7], [0.3, 0.4], [0.8, 0.9])

Table 2 Comparison of the decision results with entropy measures E_{YD1} and E_{YD2} of IVNSs

	E_M	E_T	E_{MD}	E_{TD}	E_{YD1}	E_{YD2}
A_1	0.267	0.592	0.157	0.421	0.511	0.633
A_2	0.2	0.517	0.111	0.354	0.422	0.533
A_3	0.333	0.625	0.204	0.454	0.533	0.656
A_4	0.133	0.425	0.074	0.280	0.333	0.444
Priority ranking	$A_4 \succ A_2 \succ A_1 \succ A_3$					

4 Similarity Measure of IVNS Set Based on Entropy

In this section, we introduce a new way to construct the similarity measure of IVNS. In this way, we use the entropy of IVNS to determine the similarity measure of IVNS.

For two given IVNSs, A, $B \in IVNS(X)$. Setting a new IVNS $N(A, B)$ as the following:

$$T^L_{N(A,B)}(x_i) = \frac{1 - \left| T^U_A(x_i) - T^U_B(x_i) \right|}{2}, \quad T^U_{N(A,B)}(x_i) = \frac{1 + \left| T^L_A(x_i) - T^L_B(x_i) \right|}{2},$$

$$I^L_{N(A,B)}(x_i) = \frac{1 - \left| I^U_A(x_i) - I^U_B(x_i) \right|}{2}, \quad I^U_{N(A,B)}(x_i) = \frac{1 + \left| I^L_A(x_i) - I^L_B(x_i) \right|}{2},$$

$$F^L_{N(A,B)}(x_i) = \frac{1 - \left| F^U_A(x_i) - F^U_B(x_i) \right|}{2},$$

$$F^U_{N(A,B)}(x_i) = \frac{1 + \left| F^L_A(x_i) - F^L_B(x_i) \right|}{2}, \tag{10}$$

for all $x_i \in X$.

It implies

$$I^L_{N(A,B)^C}(x_i) = 1 - \frac{1 + \left| I^L_A(x_i) - I^L_B(x_i) \right|}{2} = \frac{1 - \left| I^L_A(x_i) - I^L_B(x_i) \right|}{2}$$

$$I^U_{N(A,B)^C}(x_i) = 1 - \frac{1 - \left|I^U_A(x_i) - I^U_B(x_i)\right|}{2} = \frac{1 + \left|I^U_A(x_i) - I^U_B(x_i)\right|}{2}$$

Then $N(A, B)$ is an IVNS set on X. In particular $T_{N(A,B)}(x_i) \prec [0.5, 0.5]$, $I_{N(A,B)}(x_i) \prec [0.5, 0.5]$, and $F_{N(A,B)}(x_i) \prec [0.5, 0.5]$ for all $x_i \in X$.

Theorem 3 If E is an entropy measure on $IVNS(X)$, then $S(A, B) = E(N(A, B))$ will be a similarity measure of two IVNSs A and B on X.

Proof (S1) Since $0 \le E(C) \le 1$ for all $C \in IVNS(X)$, then $0 \le S(A, B) = E(N(A, B)) \le 1$ for all $A, B \in IVNS(X)$.

(S2) It is obvious that $S(A, B) = S(B, A)$ for all $A, B \in IVNS(X)$.

(S3) Since $T_{N(A,A)}(x_i) = I_{N(A,A)}(x_i) = F_{N(A,A)}(x_i) = [0.5, 0.5]$ for all $x_i \in X$, then $S(A, A) = E(N(A, A)) = 1$ for all $A \in IVNS(X)$.

(S4) For all $A, B, C \in NS(X)$ and $A \subset B \subset C$, we have:

$$T^L_A(x_i) \le T^L_B(x_i) \le T^L_C(x_i), \quad T^U_A(x_i) \le T^U_B(x_i) \le T^U_C(x_i),$$

$$I^L_A(x_i) \ge I^L_B(x_i) \ge I^L_C(x_i), \quad I^U_A(x_i) \ge I^U_B(x_i) \ge I^U_C(x_i),$$

$$F^L_A(x_i) \ge F^L_B(x_i) \ge F^L_C(x_i), \quad F^U_A(x_i) \ge F^U_B(x_i) \ge F^U_C(x_i),$$

for all $x_i \in X$.

These imply that

$$T_{N(A,C)}(x_i) \prec T_{N(A,B)}(x_i) \prec [0.5, 0.5], \quad I_{N(A,C)}(x_i) \prec I_{N(A,B)}(x_i) \prec [0.5, 0.5],$$

and

$$F_{N(A,C)}(x_i) \prec F_{N(A,B)}(x_i) \prec [0.5, 0.5]$$

for all $x_i \in X$.

Hence, $S(A, C) = E(N(A, C)) \le S(A, B) = E(N(A, B))$. In the same way, we also have $S(A, C) = E(N(A, C)) \le S(B, C) = E(N(B, C))$. $\square$

From Theorem 3, using the entropy measures defined by Eqs. (2) and (3), we obtain two similarity measures of two IVNSs $A, B \in NS(X)$ as follows:

—   with the entropy

$$E_M(A) = 1 - \frac{2}{n}\sum_{i=1}^{n}\max\left\{\begin{array}{l}\left|T^L_A(x_i) - 0.5\right|, \left|T^U_A(x_i) - 0.5\right|, \left|I^L_A(x_i) - 0.5\right|, \left|I^U_A(x_i) - 0.5\right|, \\ \left|I^L_{AC}(x_i) - 0.5\right|, \left|I^U_{AC}(x_i) - 0.5\right|, \left|F^L_A(x_i) - 0.5\right|, \left|F^U_A(x_i) - 0.5\right|\end{array}\right\} \quad (11)$$

we have a similarity measure

$$S_1(A, B) = 1 - \frac{1}{n} \sum_{i=1}^{n} \max \left\{ \begin{array}{l} \left| T_A^L(x_i) - T_B^L(x_i) \right|, \left| T_A^U(x_i) - T_B^U(x_i) \right|, \left| I_A^L(x_i) - I_B^L(x_i) \right|, \\ \left| I_A^U(x_i) - I_B^U(x_i) \right|, \left| F_A^L(x_i) - F_B^L(x_i) \right|, \left| F_A^U(x_i) - F_B^U(x_i) \right| \end{array} \right\} \tag{12}$$

− with the entropy

$$E_T(A) = 1 - \frac{1}{n} \sum_{i=1}^{n} \frac{\left\{ \begin{array}{l} \left| T_A^L(x_i) - 0.5 \right| + \left| T_A^U(x_i) - 0.5 \right| + \left| I_A^L(x_i) - 0.5 \right| + \left| I_A^U(x_i) - 0.5 \right| \\ + \left| I_{A^C}^L(x_i) - 0.5 \right| + \left| I_{A^C}^U(x_i) - 0.5 \right| + \left| F_A^L(x_i) - 0.5 \right| + \left| F_A^U(x_i) - 0.5 \right| \end{array} \right\}}{4} \tag{13}$$

we have a similarity measure

$$s_T(A, B) = 1 - \frac{1}{n} \sum_{i=1}^{n} \frac{\left\{ \begin{array}{l} \left| T_A^L(x_i) - T_B^L(x_i) \right| + \left| T_A^U(x_i) - T_B^U(x_i) \right| + 2 \left| I_A^L(x_i) - I_B^L(x_i) \right| \\ + 2 \left| I_A^U(x_i) - I_B^U(x_i) \right| + \left| F_A^L(x_i) - F_B^L(x_i) \right| + \left| F_A^U(x_i) - F_B^U(x_i) \right| \end{array} \right\}}{8} \tag{14}$$

+ with entropy

$$E_{MD}(A) = \frac{1}{n} \sum_{i=1}^{n} \frac{1 - 2 * \max \left\{ \begin{array}{l} \left| T_A^L(x_i) - 0.5 \right|, \left| T_A^U(x_i) - 0.5 \right|, \left| I_A^L(x_i) - 0.5 \right|, \left| I_A^U(x_i) - 0.5 \right|, \\ \left| I_{A^C}^L(x_i) - 0.5 \right|, \left| I_{A^C}^U(x_i) - 0.5 \right|, \left| F_A^L(x_i) - 0.5 \right|, \left| F_A^U(x_i) - 0.5 \right| \end{array} \right\}}{1 + 2 * \max \left\{ \begin{array}{l} \left| T_A^L(x_i) - 0.5 \right|, \left| T_A^U(x_i) - 0.5 \right|, \left| I_A^L(x_i) - 0.5 \right|, \left| I_A^U(x_i) - 0.5 \right|, \\ \left| I_{A^C}^L(x_i) - 0.5 \right|, \left| I_{A^C}^U(x_i) - 0.5 \right|, \left| F_A^L(x_i) - 0.5 \right|, \left| F_A^U(x_i) - 0.5 \right| \end{array} \right\}} \tag{15}$$

we have

$$S_{MD}(A, B) = \frac{1}{n} \sum_{i=1}^{n} \frac{1 - \max \left\{ \begin{array}{l} \left| T_A^L(x_i) - T_B^L(x_i) \right|, \left| T_A^U(x_i) - T_B^U(x_i) \right|, \left| I_A^L(x_i) - I_B^L(x_i) \right|, \\ \left| I_A^U(x_i) - I_B^U(x_i) \right|, \left| F_A^L(x_i) - F_B^L(x_i) \right|, \left| F_A^U(x_i) - F_B^U(x_i) \right| \end{array} \right\}}{1 + \max \left\{ \begin{array}{l} \left| T_A^L(x_i) - T_B^L(x_i) \right|, \left| T_A^U(x_i) - T_B^U(x_i) \right|, \left| I_A^L(x_i) - I_B^L(x_i) \right|, \\ \left| I_A^U(x_i) - I_B^U(x_i) \right|, \left| F_A^L(x_i) - F_B^L(x_i) \right|, \left| F_A^U(x_i) - F_B^U(x_i) \right| \end{array} \right\}} \tag{16}$$

+ with entropy

$$E_{TD}(A) = \frac{1}{n} \sum_{i=1}^{n} \frac{1 - \frac{1}{4} \left\{ \begin{array}{l} \left|T_A^L(x_i) - 0.5\right| + \left|T_A^U(x_i) - 0.5\right| + \left|I_A^L(x_i) - 0.5\right| + \left|I_A^U(x_i) - 0.5\right| \\ + \left|I_{A^C}^L(x_i) - 0.5\right| + \left|I_{A^C}^U(x_i) - 0.5\right| + \left|F_A^L(x_i) - 0.5\right| + \left|F_A^U(x_i) - 0.5\right| \end{array} \right\}}{1 + \frac{1}{4} \left\{ \begin{array}{l} \left|T_A^L(x_i) - 0.5\right| + \left|T_A^U(x_i) - 0.5\right| + \left|I_A^L(x_i) - 0.5\right| + \left|I_A^U(x_i) - 0.5\right| \\ + \left|I_{A^C}^L(x_i) - 0.5\right| + \left|I_{A^C}^U(x_i) - 0.5\right| + \left|F_A^L(x_i) - 0.5\right| + \left|F_A^U(x_i) - 0.5\right| \end{array} \right\}} \tag{17}$$

we have

$$S_{TD}(A, B) = \frac{1}{n} \sum_{i=1}^{n} \frac{1 - \frac{1}{8} \left\{ \begin{array}{l} \left|T_A^L(x_i) - T_B^L(x_i)\right| + \left|T_A^U(x_i) - T_B^U(x_i)\right| + 2\left|I_A^L(x_i) - I_B^L(x_i)\right| \\ + 2\left|I_A^U(x_i) - T_B^U(x_i)\right| + \left|F_A^L(x_i) - F_B^L(x_i)\right| + \left|F_A^U(x_i) - F_B^U(x_i)\right| \end{array} \right\}}{1 + \frac{1}{8} \left\{ \begin{array}{l} \left|T_A^L(x_i) - T_B^L(x_i)\right| + \left|T_A^U(x_i) - T_B^U(x_i)\right| + 2\left|I_A^L(x_i) - I_B^L(x_i)\right| \\ + 2\left|I_A^U(x_i) - T_B^U(x_i)\right| + \left|F_A^L(x_i) - F_B^L(x_i)\right| + \left|F_A^U(x_i) - F_B^U(x_i)\right| \end{array} \right\}} \tag{18}$$

In general, with $\omega = (\omega_1, \omega_2, \ldots, \omega_n)$ as the vector weight on X, we have two similarity measures generated from Eqs. (10) and (11) as follows:

$$S_M^{\omega}(A, B) = \sum_{i=1}^{n} \omega_i \times \left\{ 1 - \max \left\{ \begin{array}{l} \left|T_A^L(x_i) - T_B^L(x_i)\right|, \left|T_A^U(x_i) - T_B^U(x_i)\right|, \left|I_A^L(x_i) - I_B^L(x_i)\right|, \\ \left|I_A^U(x_i) - I_B^U(x_i)\right|, \left|F_A^L(x_i) - F_B^L(x_i)\right|, \left|F_A^U(x_i) - F_B^U(x_i)\right| \end{array} \right\} \right\} \tag{19}$$

$$S_D^{\omega}(A, B) = \sum_{i=1}^{n} \omega_i \times \left\{ 1 - \frac{\left\{ \begin{array}{l} \left|T_A^L(x_i) - T_B^L(x_i)\right| + \left|T_A^U(x_i) - T_B^U(x_i)\right| + 2\left|I_A^L(x_i) - I_B^L(x_i)\right| \\ + 2\left|I_A^U(x_i) - I_B^U(x_i)\right| + \left|F_A^L(x_i) - F_B^L(x_i)\right| + \left|F_A^U(x_i) - F_B^U(x_i)\right| \end{array} \right\}}{8} \right\} \tag{20}$$

$$S_{MD}^{\omega}(A, B) = \sum_{i=1}^{n} \omega_i \frac{1 - \max \left\{ \begin{array}{l} \left|T_A^L(x_i) - T_B^L(x_i)\right|, \left|T_A^U(x_i) - T_B^U(x_i)\right|, \left|I_A^L(x_i) - I_B^L(x_i)\right|, \\ \left|I_A^U(x_i) - I_B^U(x_i)\right|, \left|F_A^L(x_i) - F_B^L(x_i)\right|, \left|F_A^U(x_i) - F_B^U(x_i)\right| \end{array} \right\}}{1 + \max \left\{ \begin{array}{l} \left|T_A^L(x_i) - T_B^L(x_i)\right|, \left|T_A^U(x_i) - T_B^U(x_i)\right|, \left|I_A^L(x_i) - I_B^L(x_i)\right|, \\ \left|I_A^U(x_i) - I_B^U(x_i)\right|, \left|F_A^L(x_i) - F_B^L(x_i)\right|, \left|F_A^U(x_i) - F_B^U(x_i)\right| \end{array} \right\}} \tag{21}$$

and

$$S_{TD}^{\omega}(A, B) = \sum_{i=1}^{n} \omega_i \frac{1 - \frac{1}{8} \left\{ \begin{array}{l} |T_A^L(x_i) - T_B^L(x_i)| + |T_A^U(x_i) - T_B^U(x_i)| + 2|I_A^L(x_i) - I_B^L(x_i)| \\ + 2|I_A^U(x_i) - T_B^U(x_i)| + |F_A^L(x_i) - F_B^L(x_i)| + |F_A^U(x_i) - F_B^U(x_i)| \end{array} \right\}}{1 + \frac{1}{8} \left\{ \begin{array}{l} |T_A^L(x_i) - T_B^L(x_i)| + |T_A^U(x_i) - T_B^U(x_i)| + 2|I_A^L(x_i) - I_B^L(x_i)| \\ + 2|I_A^U(x_i) - T_B^U(x_i)| + |F_A^L(x_i) - F_B^L(x_i)| + |F_A^U(x_i) - F_B^U(x_i)| \end{array} \right\}}$$

$$(22)$$

Compare with Some Existing Similarity Measures Let $X = \{x_1, x_2, \ldots, x_n\}$ be a universal set; let A, B be two *IVNS(X)*. We recall some existing similarity measures of IVNS sets.

Aydoğdu [21] investigated a similarity measure off IVNS:

$$S_{AA}(A, B) = 1 - \frac{1}{2} \left[\begin{array}{l} \max_{i=1,\ldots,n} |T_A^L(x_i) - T_B^L(x_i)| + \max_{i=1,\ldots,n} |T_A^U(x_i) - T_B^U(x_i)| + \max_{i=1,\ldots,n} |I_A^L(x_i) - I_B^L(x_i)| \\ + \max_{i=1,\ldots,n} |I_A^U(x_i) - I_B^U(x_i)| + \max_{i=1,\ldots,n} |F_A^L(x_i) - F_B^L(x_i)| + \max_{i=1,\ldots,n} |F_A^U(x_i) - F_B^U(x_i)| \end{array} \right]$$

$$(23)$$

Ye [9] proposed some similarity measures of IVNSs and applied them into the pattern recognition:

$$S_{Y1}(A, B) = 1 - \frac{1}{6n} \sum_{i=1}^{n} \left[\begin{array}{l} |T_A^L(x_i) - T_B^L(x_i)| + |T_A^U(x_i) - T_B^U(x_i)| + |I_A^L(x_i) - I_B^L(x_i)| \\ + |I_A^U(x_i) - I_B^U(x_i)| + |F_A^L(x_i) - F_B^L(x_i)| + |F_A^U(x_i) - F_B^U(x_i)| \end{array} \right]$$

$$(24)$$

$$S_{Y2}(A, B) = 1 - \frac{1}{6n} \sum_{i=1}^{n} \left[\begin{array}{l} |T_A^L(x_i) - T_B^L(x_i)|^2 + |T_A^U(x_i) - T_B^U(x_i)|^2 + |I_A^L(x_i) - I_B^L(x_i)|^2 \\ + |I_A^U(x_i) - I_B^U(x_i)|^2 + |F_A^L(x_i) - F_B^L(x_i)|^2 + |F_A^U(x_i) - F_B^U(x_i)|^2 \end{array} \right]^{1/2}$$

$$(25)$$

Ye and Du [23] gave some similarity measures of IVNS:

$$S_{YD1}(A, B) = 1 - \frac{1}{3n} \sum_{i=1}^{n} \left[\begin{array}{l} \max\left(|T_A^L(x_i) - T_B^L(x_i)|, |T_A^U(x_i) - T_B^U(x_i)|\right) + \max\left(\begin{array}{l} |I_A^L(x_i) - I_B^L(x_i)|, \\ |I_A^U(x_i) - I_B^U(x_i)| \end{array}\right) + \\ \max\left(|F_A^L(x_i) - F_B^L(x_i)|, |F_A^U(x_i) - F_B^U(x_i)|\right) \end{array} \right]$$

$$(26)$$

$$S_{YD1}(A, B) = 1 - \frac{1}{n} \sum_{i=1}^{n} \max \left[\begin{array}{l} \frac{1}{2}\left(|T_A^L(x_i) - T_B^L(x_i)| + |T_A^U(x_i) - T_B^U(x_i)|\right), \\ \frac{1}{2}\left(|I_A^L(x_i) - I_B^L(x_i)| + |I_A^U(x_i) - I_B^U(x_i)|\right), \\ \frac{1}{2}\left(|F_A^L(x_i) - F_B^L(x_i)| + |F_A^U(x_i) - F_B^U(x_i)|\right) \end{array} \right]$$

To compare our proposed similarity measures S_M, S_T, S_{MD}, and S_{TD} with five above similarity measures, we consider some example as follows:

Example 2 Given two pattern IVNSs A, B on the universal set $X = \{x\}$ by $A = \{(x, [0,0], [0,0], [0.5,0.5])\}$, $B = \{(x, [0,0], [0.5,0.5], [0,0])\}$ and a sample $C = \{(x, [0.5,0.5], [0,0], [0,0])\}$.

Question: What pattern does sample C belong to?

Using the similarity measures, we have:

+ with the similarity measure of Ali Aydoğdu [21]: $S_{AA}(A,C) = 0$, $S_{AA}(B,C) = -1 \notin [0,1]$. This is impractical.
+ with some similarity measures of Ye [9]: $S_{Y1}(A,C) = S_{Y1}(B,C) = 0.6667$ and $S_{Y2}(A,C) = S_{Y2}(B,C) = 0.833$. These results show that the similarity measures are not used to classify in this case.
+ with some similarity measures of Ye and Du [23]: $S_{YD1}(A,C) = S_{YD1}(B,C) = 0.6667$ and $S_{YD2}(A,C) = S_{YD2}(B,C) = 0.8333$. These results show that the similarity measures are invalid to classify in this case.
+ with our proposed similarity measure, we have $S_M(A,C) = S_M(B,C) = 0.75$, $S_{MD}(A,C) = S_{MD}(B,C) = 0.3333$. These results show that the similarity measures S_M, S_{MD} are invalid to classify in this case.

However, we have $S_T(A,C) = 0.875 > S_T(B,C) = 0.75$ and $S_{TD}(A,C) = 0.6 > S_{TD}(B,C) = 0.3333$. Therefore, we can put the sample C belongs to the pattern A. This can also be explained because $C = A^C$ so that they can belong to the same class, which has the indeterminacy-membership function equals to zero. Of course, B belongs to the remaining class, which has the indeterminacy-membership function not equal to zero and the truth-membership and falsity-membership functions equal to zero.

Example 3 Given two pattern IVNSs A, B on the universal set $X = \{x\}$ by $A = \{(x, [0,0], [0.25,0.25], [0.25,0.25])\}$, $B = \{(x, [0.5,0.5], [0,0], [0,0])\}$ and a sample $C = \{(x, [0.25,0.25], [0.75,0.75], [0,0])\}$.

Question: What pattern does sample C belong to?

Using the similarity measures, we have:

+ with the similarity measure of Ali Aydoğdu [21]: $S_{AA}(A,C) = 0.5$, $S_{AA}(B,C) = 0.4167$.
+ with some similarity measures of Ye [9]:
$S_{Y1}(A,C) = S_{Y1}(B,C) = 0.6667$, $S_{Y2}(A,C) = 0.8557 > S_{Y2}(B,C) = 0.8137$.
+ with some similarity measures of Ye and Du [23]:
$S_{YD1}(A,C) = S_{YD1}(B,C) = 0.6667$, $S_{YD2}(A,C) = 0.8333 > S_{YD2}(B,C) = 0.75$.
+ with our proposed similarity measure, we have
$S_M(A,C) = 0.75 > S_M(B,C) = 0.625$, $S_{MD}(A,C) = 0.333 > S_{MD}(B,C) = 0.143$.
$S_T(A,C) = 0.813 > S_T(B,C) = 0.781$, $S_{TD}(A,C) = 0.455 > S_{TD}(B,C) = 0.391$.

According to these results, we do not know how to put sample C on pattern A or B if we use the similarity measures S_{Y1} and S_{YD1}, but we can put the sample C belongs to the pattern A if we use the remaining similarity measures. This proves that the new measure is better than some old measures in such cases.

5 Applications in Supplier Selection

In this section, we applied our proposed similarity measures of IVNS in the multiple criteria decision-making to select the best supplier. In MCDM problem, we have to find an optimal alternative from set of all feasible alternatives. Assume that we have a set of m suppliers $A = \{A_1, A_2, \ldots, A_m\}$ and a set of n criteria $C = \{C_1, C_2, \ldots, C_n\}$. In this problem, we use the interval neutrosophic numbers to present the values of alternatives under criteria. We denote the decision matrix for the evaluation of alternatives with respect to criteria by $D_{ij} = (T_{ij}, I_{ij}, F_{ij})$ in which $T_{ij} = \left[T_{ij}^L (C_j), T_{ij}^U (C_j) \right]$, $I_{ij} = \left[I_{ij}^L (C_j), I_{ij}^U (C_j) \right]$, $F_{ij} = \left[F_{ij}^L (C_j), F_{ij}^U (C_j) \right]$ for all $i = 1, 2, \ldots, m$.

Procedure of proposed MCDM concludes five steps as follows:

Step 1. Determine the weight ω_j of each criterion C_j. To do this we express each C_j as a set of IVNS C_j for all $j = 1, 2, \ldots, n$ and $i = 1, 2, \ldots, m$ to show the importance of each criterion. Compute entropy e_j of this IVNS set C_j for all $j = 1, 2, \ldots, n$.

The weight ω_j of each criterion C_j is determined by

$$\omega_j = \frac{e_j}{\sum\limits_{j=1}^{n} e_j} \tag{27}$$

for all $j = 1, 2, \ldots, n$.

Step 2. Determine the IVNS best solution A^* as follows:

$$+ \; A^* = \left\{ \left(C_j, \left[T_j^{*L}, T_j^{*U} \right], \left[I_j^{*L}, I_j^{*U} \right], \left[F_j^{*L}, F_j^{*U} \right] \right) \mid C_j \in C \right\}$$

where

$$T_j^{*L} = \begin{cases} \min\limits_{i=1,\ldots,m} T_{ij}^{L} & \text{if } C_j \text{ is a cost criteria} \\ \max\limits_{i=1,\ldots,m} T_{ij}^{L} & \text{if } C_j \text{ is a benefit criteria} \end{cases}$$

$$T_j^{*U} = \begin{cases} \min\limits_{i=1,\ldots,m} T_{ij}^{U} & \text{if } C_j \text{ is a cost criteria} \\ \max\limits_{i=1,\ldots,m} T_{ij}^{U} & \text{if } C_j \text{ is a benefit criteria} \end{cases}$$

$$I_j^{*L} = \begin{cases} \max\limits_{i=1,\ldots,m} I_{ij}^{L} & \text{if } C_j \text{ is a cost criteria} \\ \min\limits_{i=1,\ldots,m} I_{ij}^{L} & \text{if } C_j \text{ is a benefit criteria} \end{cases}$$

$$I_j^{*U} = \begin{cases} \max\limits_{i=1,\ldots,m} I_{ij}^U & \text{if } C_j \text{ is a cost criteria} \\ \min\limits_{i=1,\ldots,m} I_{ij}^U & \text{if } C_j \text{ is a benefit criteria} \end{cases}$$

$$F_j^{*L} = \begin{cases} \max\limits_{i=1,\ldots,m} F_{ij}^L & \text{if } C_j \text{ is a cost criteria} \\ \min\limits_{i=1,\ldots,m} F_{ij}^L & \text{if } C_j \text{ is a benefit criteria} \end{cases}$$

$$F_j^{*U} = \begin{cases} \max\limits_{i=1,\ldots,m} F_{ij}^U & \text{if } C_j \text{ is a cost criteria} \\ \min\limits_{i=1,\ldots,m} F_{ij}^U & \text{if } C_j \text{ is a benefit criteria} \end{cases} \tag{28}$$

Step 3. Calculate the similarity measures S_i^+ from A_i to A^*.

Step 4. Ranking $A = \{A_1, A_2, \ldots, A_m\}$ with $A_i \succ A_k$ if $S_i^+ \succ S_k^+$ for all $i, k = 1, 2, \ldots, m$.

Example 4 A construction company wants to buy the material for upcoming project. There are six criteria $C = \{C_1, C_2, C_3, C_4, C_5, C_6\}$ (where C_1 is the quality of material, C_2 is the price, C_3 is the services, C_4 is the delivery, C_5 is the technical support required, C_6 is the behavior of material) and five suppliers $A = \{A_1, A_2, A_3, A_4, A_5\}$. In which, the criterion C_2 (price) is the cost criterion, and other ones are benefit criteria. The relationship between the choices and criteria is shown in Table 3.

Table 3 The decision matrix

	C_1	C_2	C_3
A_1	([0.4,0.4],[0.5,0.5],[0.4,0.4])	([0.8,0.8],[0.1,0.1], 0.1,0.1])	([0.7,0.7],[0.3,0.3], [0.3,0.3])
A_2	([0.7,0.7],[0.2,0.2],[0.2,0.2])	([0.5,0.5],[0.3,0.3], 0.3,0.3])	([0.3,0.3],[0.4,0.4],[0.3,0.3])
A_3	([0.6,0.6],[0.1,0.1],[0.1,0.1])	([0.7,0.7],[0.3,0.3],[0.3,0.3])	([0.6,0.6],[0.2,0.2],[0.2,0.2])
A_4	([0.5,0.5],[0.4,0.4],[0.4,0.4])	([0.3,0.3],[0.4,0.4],[0.3,0.3])	([0.8,0.8],[0.1,0.1], [0.1,0.1])
A_5	([0.4,0.4],[0.3,0.3],[0.3,0.3])	([0.7,0.7],[0.1,0.1], 0.1,0.1])	([0.5,0.5],[0.2,0.2],[0.2,0.2])
	C_4	C_5	C_6
A_1	([0.6,0.6],[0.2,0.2],[0.2,0.2])	([0.5,0.5],[0.4,0.4],[0.4,0.4])	([0.3,0.3],[0.4,0.4],[0.3,0.3])
A_2	([0.8,0.8],[0.1,0.1], 0.1,0.1])	([0.2,0.2],[0.6,0.6],[0.2,0.2])	([0.4,0.4],[0.5,0.5],[0.4,0.4])
A_3	([0.4,0.4],[0.1,0.1],[0.1,0.1])	([0.3,0.3],[0.4,0.4],[0.3,0.3])	([0.8,0.8],[0.2,0.2],[0.2,0.2])
A_4	([0.7,0.7],[0.2,0.2],[0.2,0.2])	([0.6,0.6],[0.1,0.1],[0.1,0.1])	([0.7,0.7],[0.1,0.1], [0.1,0.1])
A_5	([0.9,0.9],[0.1,0.1],[0.1,0.1])	([0.8,0.8],[0,0],[0,0])	([0.6,0.6],[0.4,0.4], [0.4,0.4])

Table 4 The best solution A^*

	C_1	C_2	C_3
A^*	([0.7,0.7],[0.1,0.1], [0.1,0.1])	([0.3,0.3],[0.4,0.4],[0.3,0.3])	([0.8,0.8],[0.1,0.1], [0.1,0.1])
	C_4	C_5	C_6
A^*	([0.9,0.9],[0.1,0.1], [0.1,0.1])	([0.8,0.8],[0,0],[0,0])	([0.8,0.8],[0.1,0.1], [0.1,0.1])

Table 5 The similarity measures S_i^+ from A_i to A^*

$S_M(A_i, A_*)$	A_1	A_2	A_3	A_4	A_5
$S(A_i, A^*)$	0.6168	0.6856	0.7016	0.8672	0.7816

- Case 1. Using thes entropy E_M and the similarity measure S_M

  Procedure of proposed MCDM conclusion four steps as follows:

  Step 1. Using Eqs. (2) and (27), the weight vector of criteria is $\omega = (0.168, 0.168, 0.168, 0.168, 0.16, 0.168)$.

  Step 2. Using Eq. (28), we get the best solution A^* which is shown in Table 4.

  Step 3. We calculate the similarity measures S_i^+ from A_i to A^* using Eq. (6). These results are shown in Table 5.

  Step 4. Ranking $A_4 \succ A_5 \succ A_3 \succ A_2 \succ A_1$. Thus A_4 is the best alternative.

- Case 2. Using the entropy E_T and the similarity measure S_T

  Procedure of proposed MCDM conclusion four steps as follows:

  Step 1. Using Eqs. (3) and (27), the weight vector of criteria is $\omega = (0.195, 0.195, 0.158, 0.106, 0.158, 0.188)$.

  Step 2. Using Eq. (28), we get the best solution A^* which is shown in Table 6.

  Step 3. We calculate the similarity measures S_i^+ from A_i to A^* using Eq. (7). These results are shown in Table 7.

  Step 4. Ranking $A_4 \succ A_3 \succ A_5 \succ A_2 \succ A_1$. Thus A_4 is the best alternative.

- Case 3. Using the entropy E_{MD} and the similarity measure S_{MD}

  Procedure of proposed MCDM conclusion four steps as follows:

  Step 1. Using Eqs. (4) and (27), we have the weight vector of criteria as $\omega = (0.223, 0.159, 0.154, 0.088, 0.153, 0.223)$.

  Step 2. Using Eq. (28), we get the best solution A^* which is shown in Table 8.

  Step 3. We calculate the similarity measures S_i^+ from A_i to A^* using Eq. (8). These results are shown in Table 9.

  Step 4. Ranking $A_4 \succ A_5 \succ A_3 \succ A_2 \succ A_1$. Thus A_4 is the best alternative.

- Case 4. Using the entropy E_{TD} and the similarity measure S_{TD}

  Procedure of proposed MCDM conclusion four steps as follows:

  Step 1. Using Eqs. (5) and (27), the weight vector of criteria is $\omega = (0.205, 0.204, 0.147, 0.088, 0.159, 0.197)$.

Table 6 The best solution A^*

	C_1	C_2	C_3
A^*	([0.7,0.7],[0.1,0.1], [0.1,0.1])	([0.3,0.3],[0.4,0.4],[0.3,0.3])	([0.8,0.8],[0.1,0.1],[0 .1,0.1])
	C_4	C_5	C_6
A^*	([0.9,0.9],[0.1,0.1], [0.1,0.1])	([0.8,0.8],[0,0],[0,0])	([0.8,0.8],[0.1,0.1], [0.1,0.1])

Table 7 The similarity measures S_i^+ from A_i to A^*

$S_M(A_i,A_*)$	A_1	A_2	A_3	A_4	A_5
$S(A_i,A^*)$	0.7045	0.7624	0.8556	0.9087	0.8222

Table 8 The best solution A^*

	C_1	C_2	C_3
A^*	([0.7,0.7],[0.1,0.1], [0.1,0.1])	([0.3,0.3],[0.4,0.4],[0.3,0.3])	([0.8,0.8],[0.1,0.1], [0.1,0.1])
	C_4	C_5	C_6
A^*	([0.9,0.9],[0.1,0.1], [0.1,0.1])	([0.8,0.8],[0,0],[0,0])	([0.8,0.8],[0.1,0.1], [0.1,0.1])

Table 9 The similarity measures S_i^+ from A_i to A^*

$S_M(A_i,A_*)$	A_1	A_2	A_3	A_4	A_5
$S(A_i,A^*)$	0.4385	0.5456	0.6161	0.7762	0.6322

Table 10 The best solution A^*

	C_1	C_2	C_3
A^*	([0.7,0.7],[0.1,0.1], [0.1,0.1])	([0.3,0.3],[0.4,0.4],[0.3,0.3])	([0.8,0.8],[0.1,0.1], [0.1,0.1])
	C_4	C_5	C_6
A^*	([0.9,0.9],[0.1,0.1], [0.1,0.1])	([0.8,0.8],[0,0],[0,0])	([0.8,0.8],[0.1,0.1], [0.1,0.1])

Table 11 The similarity measures S_i^+ from A_i to A^*

$S_M(A_i,A_*)$	A_1	A_2	A_3	A_4	A_5
$S(A_i,A^*)$	0.5435	0.6444	0.7662	0.8471	0.7072

Step 2. Using Eq. (28), we get the best solution A^* which is shown in Table 10.

Step 3. We calculate the similarity measures S_i^+ from A_i to A^* using Eq. (9). These results are shown in Table 11.

Step 4. Ranking $A_4 \succ A_3 \succ A_5 \succ A_2 \succ A_1$. Thus A_4 is the best alternative.

Thus, four different measures of IVNS have the same result that A_4 is the best alternative.

6 Conclusion

In this chapter, we introduce the new concept of entropy measure of interval-valued neutrosophic sets, in which the conditions for an entropy measure are proposed. These conditions allow us to easily examine an entropy measure of IVNS. At the same time, we give some formulas to define the entropy measures of IVNS and compare to some existing entropy measures by numerical examples. After that, we introduce the similarity measures based on the new entropy measures and comparison with the existing similarity measures. Finally, we make a MCDM model using the proposed entropy and similarity measures. A case study for choosing the best suppliers is performed to illustrate our measures. In the future, we continue to find a link between the entropy measures and other measures of IVNS. At the same time, we also find more entropy measures of IVNS and apply them to handle the real problems.

Disclosure of potential conflicts of interest: The authors declare that they do not have any conflict of interests.

References

1. Smarandache, F. (1998). *Neutrosophy. IVSN probability, set, and logic, ProQuest information & learning* (p. 105), Ann Arbor, Michigan, USA. Retrieved from http://fs.gallup.unm.edu/eBook-IVSNs6.pdf (last edition online).
2. Zadeh, L. (1965a). Fuzzy sets. *Information and Control, 8,* 338–353.
3. Zadeh, L. (1965b). Fuzzy sets and systems. In *Proceedings of the symposium on systems* (pp. 29–37). New York: Theory Polytechnic Institute of Brooklyn.
4. Atanassov, K. (1986). Intuitionistic fuzzy sets. *Fuzzy Sets and Systems, 20,* 87–96.
5. Atanassov, K., & Gargov, G. (1989). Interval-valued intuitionistic fuzzy sets. *Fuzzy Sets and Systems, 31,* 343–349.
6. Wang, H., Smarandache, F., Zhang, Y. Q., & Sunderraman, R. (2005). *Interval neutrosophic sets and logic: Theory and applications in computing.* Phoenix, USA: Hexis.
7. Wang, H., Smarandache, F., Zhang, Y. Q., & Sunderraman, R. (2010). Single valued neutrosophic sets. *Multispace and Multistructure, 4,* 410–413.
8. Broumi, S., & Smarandache, F. (2014). New distance and similarity measures of interval neutrosophic sets. In *17th International Conference on Information Fusion (FUSION)* (pp. 1–7), Salamanca.
9. Ye, J. (2014). Similarity measures between interval neutrosophic sets and their applications in multicriteria decision-making. *Journal of Intelligent Fuzzy Systems, 26*(1), 165–172.
10. Ye, J. (2015). Improved cosine similarity measures of simplified IVSN sets for medical diagnoses. *Artificial Intelligence in Medicine, 63*(3), 171–179.
11. Ye, J. (2017). Single-valued IVSN similarity measures based on cotangent function and their application in the fault diagnosis of steam turbine. *Soft Computing, 21*(3), 817–825.
12. Aiwu, Z., Jianguo, D., & Hongjun, G. (2015). Interval valued neutrosophic sets and multi-attribute decision-making based on generalized weighted aggregation operator. *Journal of Intelligent Fuzzy Systems, 29*(6), 2697–2706.
13. Shannon, C. E. (1948). A mathematical theory of communications. *Bell System Technical Journal, 27,* 379–423.

14. Bustince, H., & Burillo, P. (1996). Entropy on intuitionistic fuzzy sets and on interval-valued fuzzy sets. *Fuzzy Sets and Systems, 78*, 305–316.
15. Liu, X. D., Zheng, S. H., & Xiong, F. L. (2005). Entropy and subsethood for general interval-valued intuitionistic fuzzy sets. In *International conference on fuzzy systems and knowledge discovery* (pp. 42–52). Berlin, Heidelberg: Springer.
16. Hung, W. L., & Yang, M. S. (2006). Fuzzy entropy on intuitionistic fuzzy sets. *International Journal of Intelligent Systems, 21*(4), 443–451.
17. Ye, J. (2010). Multicriteria fuzzy decision-making method using entropy weights-based correlation coefficients of interval-valued intuitionistic fuzzy sets. *Applied Mathematical Modelling, 34*, 3864–3870.
18. Meng, F., & Chen, X. (2016). Entropy and similarity measure for Atannasov's interval-valued intuitionistic fuzzy sets and their application. *Fuzzy Optimization and Decision Making, 15*(1), 75–101.
19. Wei, C. P., Wang, P., & Zhang, Y. Z. (2011). Entropy, similarity measure of interval-valued intuitionistic fuzzy sets and their applications. *Information Sciences, 181*(19), 4273–4286.
20. Zhang, Q., & Jiang, S. (2010). Relationships between entropy and similarity measure of interval valued intuitionistic fuzzy sets. *International Journal of Intelligent Systems, 25*(11), 1121–1140.
21. Aydoğdu, A. (2015). On entropy and similarity measure of interval valued neutrosophic sets. *Neutrosophic Sets and Systems, 9*, 47–49.
22. Majumder, P., & Samanta, S. K. (2014). On similarity and entropy of neutrosophic sets. *Journal of Intelligent Fuzzy Systems, 26*(3), 1245–1252.
23. Ye, J., & Du, S. (2017). Some distances, similarity and entropy measures for interval-valued neutrosophic sets and their relationship. *International Journal of Machine Learning and Cybernetics, 10*, 1–9.
24. Bevilacqua, M., Ciarapica, E. F., & Giacchetta, G. (2006). A fuzzy-QFD approach to supplier selection. *Journal of Purchasing and Supply Management, 1*, 14–27.
25. Abdollahi, M., Arvan, M., & Razmi, J. (2015). An integrated approach for supplier portfolio selection: Lean or agile. *Expert Systems with Applications, 42*, 679–690.
26. Hashemi, S. H., Karimi, A., & Tavana, M. (2015). An integrated green supplier selection approach with analytic network process and improved Grey relational analysis. *International Journal of Production Economics, 159*, 178–191.
27. Heidarzade, A., Mahdavi, I., & Mahdavi-Amiri, N. (2016). Supplier selection using a clustering method based on a new distance for interval Type-2 fuzzy sets: A case study. *Applied Soft Computing, 38*, 213–231.
28. Junior, F. R. L., & Carpinetti, L. C. R. (2016). A multicriteria approach based on fuzzy QFD for choosing criteria for supplier selection. *Computers and Industrial Engineering, 101*, 269–285.
29. Chen, Y. J. (2011). Structured methodology for supplier selection and evaluation in a supply chain. *Information Sciences, 181*, 1651–1670.
30. Memon, M. S., Lee, Y. H., & Mari, S. I. (2015). Group multi-criteria supplier selection using combined grey systems theory and uncertainty theory. *Expert Systems with Applications, 42*, 7951–7959.
31. Büyüközkan, G., & Çifçi, G. (2012). Evaluation of the green supply chain management practices: A fuzzy ANP approach. *Production Planning and Control, 23*, 405–418.
32. Kuo, R. J., Wang, Y. C., & Tien, F. C. (2010). Integration of artificial neural network and MADA methods for green supplier selection. *Journal of Cleaner Production, 18*, 1161–1170.

Control Chart for Monitoring Variation Using Multiple Dependent State Sampling Under Neutrosophic Statistics

Nasrullah Khan, Liaquat Ahmad, Muhammad Azam, Muhammad Aslam, and Florentin Smarandache

1 Introduction

Variation is a law of nature. The beauty of things lies in its variability. Manufacturing industry also exhibits variation which is attributable to many factors such as variation in raw material, adjustment of machine, expertise of the operators, temperature, etc. Broadly speaking, in statistical process monitoring (SPM) system, two types of causes are associated with the production process, i.e., common causes of variation and special causes of variations [1]. The first type of variation is known as inherent and natural which cannot be controlled and bearable. The latter type of variation, in which the production process is declared as out-of-control process, is challenging to the production processes and required to be detected as early as possible to avoid losses to the firm in the form of monetary losses, and the most important one is the repute of the firm which faces unbearable losses in the case of any delay.

N. Khan
Department of Statistics, College of Veterinary and Animal Sciences, Jhang, University of Veterinary and Animal Sciences, Lahore, Pakistan
e-mail: nasrullah.khan@uvas.edu.pk

L. Ahmad · M. Azam
Department of Statistics and Computer Science, University of Veterinary and Animal Sciences, Lahore, Pakistan
e-mail: liaquatahmad@uvas.edu.pk; mazam@uvas.edu.pk

M. Aslam (✉)
Department of Statistics, Faculty of Science, King Abdulaziz University, Jeddah, Saudi Arabia
e-mail: aslam_ravian@hotmail.com

F. Smarandache
Department of Mathematics and Science, University of New Mexico, Gallup, NM, USA
e-mail: fsmarandache@gmail.com

© The Author(s), under exclusive license to Springer Nature Switzerland AG 2021
F. Smarandache, M. Abdel-Basset (eds.), *Neutrosophic Operational Research*,
https://doi.org/10.1007/978-3-030-57197-9_4

Walter A. Shewhart, an American physicist and statistician also known as the father of industrial statistics, presented a tremendous and fantastic notion of control chart during the 1920s for monitoring the production process, and it was the formal beginning of statistical control chart [1]. The graphical presentation of the interested quality characteristic, with two limits known as the upper control limit (UCL) and the lower control limit (LCL) and having a central line, quickly describes any unusual change in the production process. Two phases are employed: phase I control limits, initially collected data from the trials-basis processes, and after rectification a stable and improved phase II control limits are established, for the efficient monitoring of the production process [2]. Phase II control limits are expected to run for a long time, but the continuous threat of unusual changes in the production process cannot be avoided.

According to the nature of the data, two types of control charts are established: the first one is the variable control chart, when the observations are based on the measurable scale, and the other one is the attribute control chart, when the data are collected on the categorical scale. The commonly used variable control charts are the mean and range chart, mean and variance or standard deviation charts, etc., in which the location and the dispersion of the interested quality characteristic are considered. The variance chart S^2 is used to monitor the variation within a sample of size $n \geq 2$, while the X-bar charts are used to monitor the variation in different samples of size ≥ 2.

Acosta-Mejia et al. [3] claimed that the analysis of dispersion charts is important for any production process because it is assumed that the variance is constant when location charts are applied. Calzada and Scariano [4] studied combined charts of (X-bar, S^2), (EWMA, S^2), and (CUSUM, S^2) for known and unknown process parameters, and run length distribution was employed for comparison between these three combined charts using average, second moment, and standard deviation. Lee and Park [5] proposed maximum likelihood estimator for monitoring process mean and/or variance of normally distributed data. Grabov and Ingman [6] analyzed the joint distribution of the measures of central tendency and the dispersion of the variable control charts. Elam and Case [7] developed and executed a computer program for the combined monitoring of the process mean and variance. Nazir et al. [8] stated that standard deviation of any process is monitored first to develop any monitoring scheme. They focused upon the standard deviation chart using the estimated parameters and developed a stepwise procedure for developing the standard deviation chart. Maravelakis et al. [9] studied the effects of the estimated process parameters of univariate control charts for dispersion. Mahmoud et al. [10] provided the estimation of the standard deviation in the monitoring of the manufacturing process. Psarakis et al. [11] analyzed some modern techniques for the estimation of the standard deviation. Yang et al. [12] developed a mean and variance control chart for the variables appropriate for many SPC applications. Chen et al. [13] developed EWMA control chart for monitoring the process mean and variance. Li et al. [14] presented the simultaneous monitoring of the mean and variance chart using likelihood ratio test for the unknown process parameters. Reynolds Jr. and Stoumbos [15] developed adoptive EWMA control chart for effective monitoring of the process variance and assessing the effect of inertia. Control charts for variance monitoring have been explored by several researchers including Wu et al. [16], Yeh et al. [17], and Zhang and Chang [18].

Multiple dependent state (MDS) sampling scheme also known as the multiple deferred sampling scheme which is the conditional sampling scheme originally developed by Wortham and Baker [19] has comparative advantage over the conventional single sampling scheme as it utilizes the current information of the item as well as considers the previous item information to reach any decision. Hsu et al. [20] studied the MDS sampling scheme for developing variable acceptance sampling based upon process capability index. Aslam et al. [21] developed MDS sampling control chart for the exponentially distributed random variables to detect out-of-control process more speedily. Zhou et al. [22] developed attribute adaptive control chart for using MDS sampling scheme for quick and efficient monitoring of the processes. Aldosari et al. [23] presented an efficient monitoring scheme for attribute control chart using the combined sampling schemes of the MDS and the repetitive sampling scheme. Aslam et al. [24] developed np control chart for quick detection of small process shifts. Aslam et al. [25] proposed variable control chart for efficient monitoring using the MDS sampling. The MDS sampling scheme has been explored by several quality control researchers including Aslam et al. [26–28], Balamurali et al. [29], Balamurali and Jun [30], Govindaraju and Subramani [31], Kuralmani and Govlndaraju [32], Soundararajan and Vijayaraghavan [33, 34], Wu et al. [35], Wu and Wang [36], and Yan et al. [37].

Neutrosophic statistics has attracted the attention of many quality control researchers due its nice and the most appropriate application to the ground realities during the last few years. Neutrosophic statistics is the extension of the classical statistics introduced by Smarandache [38] to handle the observations which are vague, uncertain, unclear, incomplete, indefinite, and imprecise. Neutrosophic Extended Triplets were introduced by F. Smarandache in 2016 [39]: Zhang et al. [40] introduced the notion of singular neutrosophic extended triplet group and proved some useful extensions. Abdel-Basset et al. [41] introduced a decision-making approach using hybrid neutrosophic statistics. Albassam et al. [42] developed W/S test for the neutrosophic statistics for its normality. Khan et al. [43] developed the S-chart for monitoring dispersion of the indeterminate values. Panthong and Pongpullponsak [44] proposed X-bar chart for monitoring variable parameters of the non-crispy data. Aslam and Raza [45] developed neutrosophic sampling plan using the neutrosophic optimization technique. Hsieh et al. [46] presented control charts for monitoring the defectives in the manufacturing industry using the fuzzy theory. Aslam et al. [47] developed a variable control chart for neutrosophic statistics using the gamma distribution. Aslam et al. [48] developed attribute control chart for neutrosophic statistical interval method. Afshari and Sadeghpour Gildeh [49] developed neutrosophic statistical plan for MDS sampling for attribute data. Aslam et al. [50] developed dispersion chart for monitoring the neutrosophic data using neutrosophic interval method. Aslam et al. [51] developed control chart for failure-censored reliable data using the neutrosophic statistics.

A widely used performance indicator is the average run length (ARL) which is defined as the average number of samples before the process indicates an out-of-control process. Ahmad et al. [52] used the average run length for evaluating the coal quality monitoring control chart. Ahmad et al. [53] utilized the ARL for measuring the performance of the X-bar control chart for process capability index. Several researchers used the ARL for performance evaluation of the proposed chart including Ahmad et al. [54–56], and Azam et al. [57]. Therefore, in this chapter a MDS sampling scheme has been developed under the neutrosophic statistics which has not been explored by any researcher. The rest of the chapter is designed as follows: The design of the proposed MDS sampling chart is given in Sect. 2. In Sect. 3 the performance of the proposed chart is discussed. In Sect. 4 a comparison of the proposed chart with the existing chart is given. A real-word example has been discussed in Sect. 5. At the end conclusion and some recommendations have been described.

2 Designing of the Proposed Chart

Let a neutrosophic random sample be defined as a random sample selected from the population or the sample from the neutrosophic statistic. Let $X_{Ni} \epsilon \{X_L, X_U\}$, $i = 1, 2, 3, \ldots, n_N$ be a neutrosophic random variable where X_L is a determinate amount and X_U is an indeterminate amount. Further the mean and variances of these two amounts of X_L, X_U can be defined as $\mu_N = \sum_{i=1}^{N_N} X_N / N_N$, and $\mu_N \epsilon \{\mu_L, \mu_U\}$ is defined as the mean of indeterminate population; and μ_L is the mean of determinate amount and μ_U is the mean of the indeterminate amount. Further we represent the variances of the determinate amount and the indeterminate amount as $\sigma_N^2 = \sum_{i=1}^{nN} (X_N - \mu_N)^2 / N_N - 1$ and $\sigma_N^2 \epsilon \{\sigma_L^2, \sigma_U^2\}$, and σ_L^2 is the variance of the determinate amount and σ_U^2 is the variance of the indeterminate amount. Further let $\overline{X}_N = \sum_{i=1}^{nN} X_N / n_N$ and $\overline{X}_N \epsilon \{\overline{X}_L, \overline{X}_U\}$; then we can define $S_N^2 = \sum_{i=1}^{nN} (X_N - \overline{X}_N)^2 / n_N - 1$ and $S_N^2 \epsilon \{S_L^2, S_U^2\}$.

Here we will develop variance control chart for the neutrosophic statistics using neutrosophic interval method.

The operational steps for developing the variance control charts are:

Step 1: Choose a neutrosophic random sample of size n_N from the manufacturing process and calculate S_N^2.

Step 2: If the interested quality characteristic S_N^2 is such that $LCL_N \leq S_N^2 \leq UCL_N$, then declare the process as in-control, otherwise out-of-control.

Using the neutrosophic interval method, the LCL_N and the UCL_N can be defined as

$$\text{LCL}_N = \sigma_N^2 - K_N \frac{}{\sqrt{\frac{2(\sigma_N^2)^2}{n_N} - 1}}; \sigma_N^2 \in \left\{\sigma_L^2, \sigma_U^2\right\} \text{ and } K_N \in \{K_L, K_U\} \qquad (1)$$

$$\text{UCL}_N = \sigma_N^2 + K_N \frac{}{\sqrt{2(\sigma_N^2)^2/n_N - 1}}; \sigma_N^2 \in \left\{\sigma_L^2, \sigma_U^2\right\} \text{ and } K_N \in \{K_L, K_U\} \qquad (2)$$

The probability that a process is declared as out-of-control under the neutrosophic interval method is computed as

$$P_{\text{in},N}^{(0)} = P\left(S_N^2 \geq \text{UCL}_N\right) + P\left(S_N^2 \leq \text{LCL}_N\right) \text{ and } S_N^2 \in \left\{S_L^2, S_U^2\right\} \qquad (3)$$

$$P\left(S_N^2 \geq \text{UCL}_n\right) = 1 - G_N\left(\frac{(n_N - 1)\,\text{UCL}_N}{\sigma_N^2}\right) = 1 - G_N\left((n_N - 1)\left(1 + k_N\sqrt{2/(n_N - 1)}\right)\right); n_N \in \{n_L, n_U\}, k_N \in \{k_L, k_U\} \qquad (4)$$

Similarly,

$$P\left(S_N^2 \leq \text{LCL}_n\right) = G_N\left(\frac{(n_N - 1)\,\text{UCL}_N}{\sigma_N^2}\right) = G_N\left((n_N - 1)\left(1 - k_N\sqrt{2/(n_N - 1)}\right)\right); n_N \in \{n_L, n_U\}, k_N \in \{k_L, k_U\}$$

$$P_{\text{in},N}^{(0)} = P\left(\text{LCL2}_n \leq S_N^2 \leq \text{UCL2}_n\right) + \left[P\left(\text{LCL1}_n \leq S_N^2 \leq \text{LCL2}_n\right) + P\left(\text{UCL2}_n \leq S_N^2 \leq \text{UCL1}_n\right)\right]$$
$$* \left[P\left(\text{LCL2}_n \leq S_N^2 \leq \text{UCL2}_n\right)\right]^m \qquad (5)$$

where

$$P\left(\text{LCL2}_n \leq S_N^2 \leq \text{UCL2}_n\right) = P\left(S_N^2 \leq \text{UCL2}_n\right) - P\left(S_N^2 \leq \text{LCL2}_n\right)$$

$$= G_N\left(\frac{(n_N - 1)\,\text{UCL2}_N}{\sigma_N^2}\right) - G_N\left(\frac{(n_N - 1)\,\text{LCL1}_N}{\sigma_N^2}\right)$$

$$= G_N\left((n_N - 1)\left(1 + k_{2N}\sqrt{2/(n_N - 1)}\right)\right) - G_N\left((n_N - 1)\left(1 - k_{2N}\sqrt{2/(n_N - 1)}\right)\right)$$

$$P\left(\text{LCL1}_n \leq S_N^2 \leq \text{LCL2}_n\right) = G_N\left((n_N - 1)\left(1 - k_{2N}\sqrt{2/(n_N - 1)}\right)\right) - G_N\left((n_N - 1)\left(1 - k_{1N}\sqrt{2/(n_N - 1)}\right)\right)$$

$$P\left(\text{UCL2}_n \leq S_N^2 \leq \text{UCL1}_n\right) = G_N\left((n_N - 1)\left(1 + k_{1N}\sqrt{2/(n_N - 1)}\right)\right) - G_N\left((n_N - 1)\left(1 + k_{2N}\sqrt{2/(n_N - 1)}\right)\right)$$

The probability of in-control process under the neutrosophic interval method is given by

$$n_N \in \{n_L, n_U\}, k_N \in \{k_L, k_U\} \qquad (6)$$

3   The Performance of the Proposed Chart

As mentioned above the ARL is used as performance measure of the proposed control chart. The ARL indicates that when on the average the process will be out-of-control when it is actually an in-control state. The neutrosophic average run length (NARL) under the neutrosophic interval method is defined by

$$\text{NARL}_{0N} = \frac{1}{P_{\text{outN}}^{(0)}}; \text{ARL}_{0N} \in \{\text{ARL}_{0L}, \text{ARL}_{0U}\} \qquad (7)$$

Using the abovementioned equation, an R-language program was written and run to calculate the ARLs. Tables 1, 2, 3, 4, 5, and 6 have been generated for different process settings. From these tables the following findings are pointed out:

1. As shift levels increase, the neutrosophic NARL is going to decrease.
2. As the value of n increases, the NARL decreases rapidly.

4   The Comparison of the Proposed Chart with the Existing MDS Sampling Chart

In this section the proposed control chart has been compared with the existing MDS sampling chart. Using the same parameters, MDS sampling chart was applied for

Table 1   The neutrosophic average run length values when $m = [3, 5]$; $n = [3, 5]$

	$m = [3,5]$; $n = [3,5]$		
$k1$	[5.65542,6.47574]	[6.83087,6.83742]	[6.71614,6.88626]
$k2$	[2.31749,2.3277]	[2.46326,2.50699]	[2.5882,2.63133]
Shift	ARLs		
1.0	[201.95,206.07]	[311.13,312.56]	[378.78,414.93]
1.1	[112.57,95.02]	[166.6,137.17]	[198.57,176.06]
1.2	[69.52,50.93]	[99.39,70.51]	[116.41,87.96]
1.3	[46.46,30.64]	[64.44,40.92]	[74.39,49.82]
1.4	[33.03,20.16]	[44.61,26.1]	[50.87,31.11]
1.5	[24.67,14.24]	[32.54,17.94]	[36.72,21.00]
1.6	[19.17,10.64]	[24.77,13.09]	[27.7,15.09]
1.7	[15.40,8.32]	[19.53,10.03]	[21.66,11.4]
1.8	[12.71,6.75]	[15.84,8.00]	[17.45,8.98]
1.9	[10.72,5.65]	[13.17,6.58]	[14.42,7.32]
2.0	[9.23,4.85]	[11.18,5.57]	[12.17,6.13]
3.0	[3.78,2.18]	[4.21,2.32]	[4.43,2.43]
4.0	[2.55,1.62]	[2.73,1.69]	[2.83,1.73]

Table 2 The neutrosophic average run length values when $m = [3, 5]$; $n = [5, 7]$

	$m = [3,5]$; $n = [5,7]$		
$k1$	[4.8711,4.91632]	[5.95187,6.17866]	[6.00234,6.00887]
$k2$	[2.21687,2.30595]	[2.3325,2.4171]	[2.40878,2.4977]
Shift	ARLs		
1.0	[200.23,206.14]	[313.37,307.49]	[370.12,370.94]
1.1	[95.82,84.43]	[142.11,119.15]	[164.66,140.18]
1.2	[52.74,41.62]	[74.77,56.05]	[85.26,64.57]
1.3	[32.3,23.63]	[44.07,30.58]	[49.57,34.6]
1.4	[21.5,14.97]	[28.38,18.71]	[31.56,20.85]
1.5	[15.29,10.33]	[19.61,12.53]	[21.58,13.78]
1.6	[11.46,7.62]	[14.33,9.01]	[15.64,9.8]
1.7	[8.96,5.93]	[10.97,6.86]	[11.87,7.39]
1.8	[7.26,4.82]	[8.71,5.47]	[9.37,5.84]
1.9	[6.06,4.05]	[7.14,4.52]	[7.63,4.79]
2.0	[5.17,3.5]	[6.01,3.85]	[6.39,4.06]
3.0	[2.21,1.71]	[2.36,1.77]	[2.43,1.81]
4.0	[1.62,1.34]	[1.67,1.37]	[1.7,1.39]

Table 3 The neutrosophic average run length values when $m = [3, 5]$; $n = [8, 10]$

	$m = [3,5]$; $n = [8,10]$		
$k1$	[6.32059,6.53739]	[4.73303,6.63674]	[6.00835,6.49874]
$k2$	[2.03247,2.18696]	[2.28824,2.32664]	[2.28568,2.40161]
Shift	ARLs		
1.0	[200.07,206.19]	[302.61,302.52]	[377.53,371.58]
1.1	[81.4,73.17]	[116.64,101.4]	[140.73,120.81]
1.2	[39.88,32.74]	[54.55,43.22]	[64.09,50.18]
1.3	[22.49,17.46]	[29.57,22.11]	[33.99,25.11]
1.4	[14.13,10.64]	[17.98,13.02]	[20.27,14.51]
1.5	[9.66,7.2]	[11.95,8.54]	[13.26,9.37]
1.6	[7.07,5.27]	[8.54,6.1]	[9.34,6.6]
1.7	[5.45,4.11]	[6.46,4.66]	[6.97,4.98]
1.8	[4.39,3.37]	[5.11,3.75]	[5.46,3.97]
1.9	[3.66,2.86]	[4.2,3.14]	[4.44,3.3]
2.0	[3.14,2.5]	[3.56,2.71]	[3.73,2.83]
3.0	[1.52,1.37]	[1.61,1.41]	[1.62,1.43]
4.0	[1.23,1.15]	[1.27,1.17]	[1.27,1.17]

the comparison purposes, and the chart coefficients were determined given Tables 7 and 8. The smaller values of ARL show the better detecting ability of the proposed chart. Comparing Tables 7 and 8 with the tables having $m = 2$ and 3 with $n = 3, 8$, and 9 shows smaller ARL which concluding shows that the proposed chart is better in detecting quickly the out-of-control process. The comparison has also extended

Table 4 The neutrosophic average run length values when $m = [2,3]$; $n = [3,5]$

	$m = [2,3]$; $n = [3,5]$		
$k1$	[6.51047,6.5629]	[6.48253,6.66778]	[6.61961,6.95228]
$k2$	[2.04171,2.11625]	[2.29109,2.30004]	[2.38891,2.39305]
Shift	ARLs		
1.0	[202.1,204.09]	[308.66,309.57]	[370.75,386.9]
1.1	[115.87,97.78]	[169.08,141.22]	[199.45,172.24]
1.2	[73.02,53.82]	[102.64,74.61]	[119.26,89.14]
1.3	[49.5,32.93]	[67.44,44.1]	[77.37,51.77]
1.4	[35.55,21.89]	[47.16,28.46]	[53.52,32.89]
1.5	[26.73,15.53]	[34.67,19.68]	[38.97,22.45]
1.6	[20.86,11.62]	[26.54,14.39]	[29.58,16.22]
1.7	[16.8,9.06]	[21,11.01]	[23.24,12.29]
1.8	[13.87,7.33]	[17.09,8.75]	[18.79,9.67]
1.9	[11.71,6.09]	[14.23,7.17]	[15.56,7.86]
2.0	[10.06,5.19]	[12.09,6.03]	[13.15,6.56]
3.0	[4.02,2.19]	[4.5,2.35]	[4.73,2.45]
4.0	[2.64,1.6]	[2.86,1.67]	[2.97,1.7]

Table 5 The neutrosophic average run length values when $m = [2,3]$; $n = [5,7]$

	$m = [2,3]$; $n = [5,7]$		
$k1$	[4.70745,4.71498]	[5.68318,5.71529]	[6.388,6.7641]
$k2$	[2.08866,2.14748]	[2.16587,2.24509]	[2.22752,2.30063]
Shift	ARLs		
1.0	[200.13,207.93]	[300.06,308.99]	[373.32,375.66]
1.1	[98.18,88.16]	[140.56,124.37]	[171.6,148.35]
1.2	[55.01,44.43]	[75.77,60.01]	[90.97,70.39]
1.3	[34.12,25.54]	[45.47,33.24]	[53.78,38.4]
1.4	[22.91,16.25]	[29.67,20.49]	[34.62,23.35]
1.5	[16.37,11.2]	[20.68,13.74]	[23.84,15.46]
1.6	[12.3,8.23]	[15.21,9.85]	[17.34,10.96]
1.7	[9.63,6.36]	[11.68,7.45]	[13.18,8.21]
1.8	[7.8,5.12]	[9.29,5.89]	[10.39,6.42]
1.9	[6.49,4.26]	[7.62,4.83]	[8.45,5.22]
2.0	[5.53,3.65]	[6.4,4.07]	[7.05,4.37]
3.0	[2.28,1.69]	[2.44,1.75]	[2.56,1.8]
4.0	[1.63,1.32]	[1.69,1.34]	[1.73,1.36]

for Shewhart-type charts to prove the better detecting ability of the proposed chart. Table 9 has been generated to calculate the ARL with the same parameters of the proposed chart. By comparing the ARL values of Table 9, we conclude that the proposed chart has better detecting ability as compared to the Shewhart-type charts.

Table 6 The neutrosophic average run length values when $m = [2, 3]$; $n = [8, 10]$

	$m = [2,3]; n = [8,10]$		
$k1$	[3.95429,3.97508]	[5.41679,5.41986]	[5.29237,5.31014]
$k2$	[2.15749,2.19068]	[2.06833,2.16853]	[2.1716,2.25394]
Shift	ARLs		
1.0	[200.68,203.73]	[300.01,300.68]	[375,370.35]
1.1	[86.35,77.7]	[118.98,105.06]	[144.3,125.61]
1.2	[43.79,36.22]	[56.82,45.97]	[67.22,53.61]
1.3	[25.2,19.71]	[31.24,23.84]	[36.19,27.22]
1.4	[16.02,12.09]	[19.14,14.09]	[21.79,15.81]
1.5	[11.01,8.16]	[12.78,9.22]	[14.33,10.19]
1.6	[8.06,5.93]	[9.12,6.54]	[10.1,7.13]
1.7	[6.2,4.57]	[6.88,4.94]	[7.54,5.33]
1.8	[4.97,3.69]	[5.42,3.92]	[5.88,4.19]
1.9	[4.12,3.09]	[4.43,3.25]	[4.77,3.44]
2.0	[3.51,2.67]	[3.73,2.77]	[3.99,2.92]
3.0	[1.59,1.38]	[1.6,1.38]	[1.65,1.4]
4.0	[1.25,1.15]	[1.25,1.15]	[1.26,1.16]

Table 7 Comparison of the proposed chart with the MDS sampling chart when $m = 2$

	$m = 2$		
n	8	8	9
$k1$	3.65299	3.83555	4.75295
$k2$	2.53058	3.32585	2.20076
Shift	ARLs		
1.0	200.0001	300.000	370.0002
1.1	92.27657	139.473	136.4418
1.2	48.96576	74.51973	61.64113
1.3	28.98442	44.26452	32.48786
1.4	18.71636	28.55974	19.28777
1.5	12.96402	19.68258	12.57132
1.6	9.50713	14.30954	8.82355
1.7	7.30614	10.8707	6.57178
1.8	5.8358	8.5656	5.1352
1.9	4.81316	6.95935	4.17249
2.0	4.07725	5.80265	3.50055
3.0	1.7394	2.14252	1.51838
4.0	1.31706	1.49065	1.20279

4.1 Simulation Data

First 20 observations are generated from in-control process, next 20 observations are generated from shifted process for $c = 1.6$, from Table 1 the ARLs are [27.7,15.09], the shift is not detected for upper limit, and similarly shift is not detected for MDS

Table 8 Comparison of the proposed chart with the MDS sampling chart when $m = 3$

	$m = 3$		
n	3	3	3
$k1$	4.30176	4.84778	4.91951
$k2$	4.12439	3.28169	4.51432
Shift	ARLs		
1.0	200.00	300.00	370.00
1.1	123.33	171.24	215.42
1.2	82.40	107.06	137.13
1.3	58.55	71.86	93.51
1.4	43.68	51.04	67.30
1.5	33.87	37.94	50.58
1.6	27.11	29.29	39.38
1.7	22.27	23.32	31.57
1.8	18.70	19.07	25.93
1.9	15.99	15.93	21.74
2.0	13.89	13.57	18.55
3.0	5.70	5.07	6.80
4.0	3.66	3.21	4.14

Table 9 Comparison of the proposed chart with the Shewhart-type control chart

n	3	3	3
$k1$	4.29832	4.7038	4.91359
Shift	ARLs		
1.0	200.00	300.00	370.03
1.1	123.55	178.62	216.15
1.2	82.70	115.95	138.10
1.3	58.89	80.44	94.53
1.4	44.01	58.80	68.31
1.5	34.20	44.81	51.54
1.6	27.42	35.33	40.29
1.7	22.57	28.65	32.41
1.8	18.98	23.78	26.72
1.9	16.26	20.13	22.48
2.0	14.14	17.32	19.24
3.0	5.85	6.69	7.18
4.0	3.76	4.16	4.39

and Shewhart-type charts. Figure 1 shows the plotting of the observation from the simulated data for the parameters of the proposed MDS sampling and the Shewhart-type control charts. It can be observed very easily that the proposed chart has better detecting ability as compared to the MDS sampling chart and the Shewhart-type chart.

Table 10 Simulated data

Sr. #	S_N^2
1	[0.8063436,7.970054]
2	[0.9633025,9.471769]
3	[10.6847528,7.21224]
4	[0.7615773,9.798711]
5	[0.0734741,8.909672]
6	[8.7829873,1.963683]
7	[3.6309931,1.989987]
8	[6.0204344,2.873147]
9	[1.7357131,3.641832]
10	[0.2504555,4.603007]
11	[0.6512268,3.514267]
12	[6.5609188,7.56021]
13	[2.861941,5.419543]
14	[2.3644659,5.234254]
15	[1.2758567,22.856254]
16	[0.4992012,2.267944]
17	[7.6322441,10.04702]
18	[17.2252988,7.406326]
19	[0.2995892,5.465167]
20	[5.2564763,7.509622]
21	[4.1514453,1.729163]
22	[0.6768823,1.476735]
23	[18.3957442,1.465412]
24	[17.4447431,9.871106]
25	[0.4397932,4.086531]
26	[13.7336304,31.719196]
27	[14.7077476,11.114637]
28	[5.3451325,5.113941]
29	[2.098338,17.085518]
30	[10.2480202,9.963991]
31	[4.2298041,8.559895]
32	[1.5763058,2.659397]
33	[1.2928805,16.989177]
34	[5.5657452,7.962122]
35	[1.0437075,41.571827]
36	[8.6505908,6.986083]
37	[9.9608503,0.865519]
38	[5.6856885,1.838416]
39	[0.5786713,14.907056]
40	[13.5634625,29.407623]

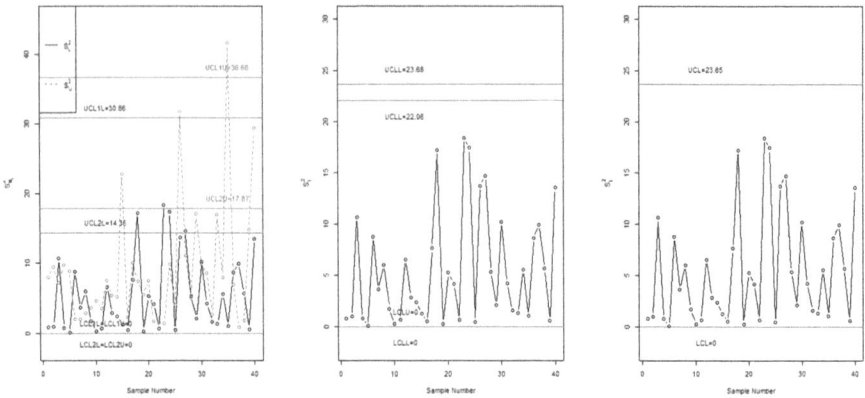

Fig. 1 Simulated example of the proposed MDS sampling and the Shewhart-type chart

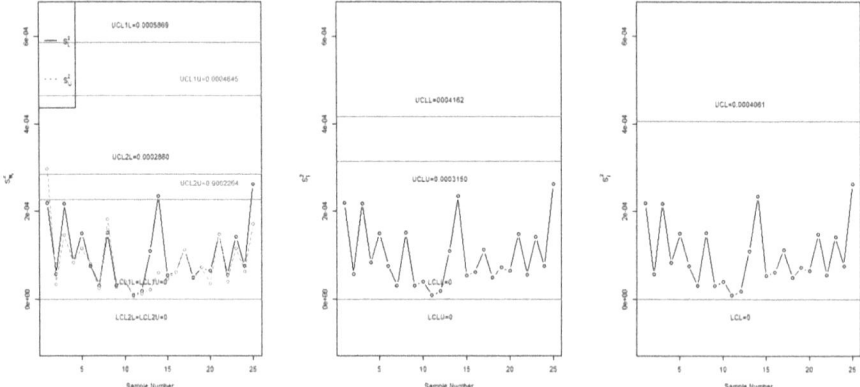

Fig. 2 Plotting of real example data

5 Real-Word Example

For practical application of the proposed chart, a real-word example of the automobile industry has been included for easy and quick understanding of the quality control personnel. Piston ring size is an important variable of the automobile industry as mentioned in Montgomery [1] on page 251. Figure 2 has been generated for the plotting of the inside diameters of piston rings using the limits LCL1 = [0, 0], LCL2 = [0, 0], UCL2 = [0.0002264, 0002860], and UCL1 = [0.0004645, 0.0005868]. From Fig. 2 it can be observed that the proposed chart has better detecting ability as compared to the other two charts of the MDS sampling and the Shewhart-type chart.

6   Concluding Remarks

In this chapter the variance control chart using the multiple dependent state sampling for the neutrosophic statistic has been presented. The control chart coefficients for different process settings have been included. On comparing the proposed chart with existing charts as the multiple dependent state sampling chart and the Shewhart-type chart, it can be concluded that the proposed chart has better detecting ability of the out-of-control process. It is concluded that the proposed chart is a valuable addition in the toolkit of the quality control personnel. The proposed chart can further be extended to the multivariate case on the same lines.

References

1. Montgomery, C. D. (2009). *Introduction to statistical quality control* (6th ed.). New York: John Wiley & Sons.
2. Oakland, J. S. (2008). *Statistical process control* (6th ed.). Oxford, UK: Butterworth-Heinemann.
3. Acosta-Mejia, C. A., Pignatiello, J. J., & Rao, B. V. (1999). A comparison of control charting procedures for monitoring process dispersion. *IIE Transactions, 31*(6), 569–579.
4. Calzada, M. E., & Scariano, S. M. (2007). Joint monitoring of the mean and variance of combined control charts with estimated parameters. *Communications in Statistics: Simulation and Computation, 36*(5), 1115–1134. https://doi.org/10.1080/03610910701540052.
5. Lee, J., & Park, C. (2007). Estimation of the change point in monitoring the process mean and variance. *Communications in Statistics: Simulation and Computation, 36*(6), 1333–1345. https://doi.org/10.1080/03610910701569028.
6. Grabov, P., & Ingman, D. (1996). Adaptive control limits for bivariate process monitoring. *Journal of Quality Technology, 28*(3), 320–330.
7. Elam, M. E., & Case, K. E. (2003). A computer program to calculate two-stage short-run control chart factors for (X‾,v) and (X‾,√v) charts. *Quality Engineering, 15*(4), 609–638. https://doi.org/10.1081/qen-120018393.
8. Nazir, H. Z., Schoonhoven, M., Riaz, M., & Does, R. J. M. M. (2013). Quality quandaries: How to set up a robust Shewhart control chart for dispersion? *Quality Engineering, 26*(1), 130–136. https://doi.org/10.1080/08982112.2013.848367.
9. Maravelakis, P., Panaretos, J., & Psarakis, S. (2002). Effect of estimation of the process parameters on the control limits of the univariate control charts for process dispersion. *Communications in Statistics: Simulation and Computation, 31*(3), 443–461.
10. Mahmoud, M. A., Henderson, G. R., Epprecht, E. K., & Woodall, W. H. (2010). Estimating the standard deviation in quality-control applications. *Journal of Quality Technology, 42*(4), 348.
11. Psarakis, S., Vyniou, A. K., & Castagliola, P. (2014). Some recent developments on the effects of parameter estimation on control charts. *Quality and Reliability Engineering International, 30*(8), 1113–1129.
12. Yang, M., Wu, Z., Lee, K. M., & Khoo, M. B. C. (2012). The X control chart for monitoring process shifts in mean and variance. *International Journal of Production Research, 50*(3), 893–907.
13. Chen, G., Cheng, S. W., & Xie, H. (2001). Monitoring process mean and variability with one EWMA chart. *Journal of Quality Technology, 33*(2), 223.
14. Li, Z., Zhang, J., & Wang, Z. (2010). Self-starting control chart for simultaneously monitoring process mean and variance. *International Journal of Production Research, 48*(15), 4537–4553.

15. Reynolds, M. R., Jr., & Stoumbos, Z. G. (2006). Comparisons of some exponentially weighted moving average control charts for monitoring the process mean and variance. *Technometrics, 48*(4), 550–567.
16. Wu, Z., Yang, M., Khoo, M. B., & Yu, F.-J. (2010). Optimization designs and performance comparison of two CUSUM schemes for monitoring process shifts in mean and variance. *European Journal of Operational Research, 205*(1), 136–150.
17. Yeh, A. B., Huwang, L., & Wu, C.-W. (2005). A multivariate EWMA control chart for monitoring process variability with individual observations. *IIE Transactions, 37*(11), 1023–1035. https://doi.org/10.1080/07408170500232263.
18. Zhang, G., & Chang, S. (2008). Multivariate EWMA control charts using individual observations for process mean and variance monitoring and diagnosis. *International Journal of Production Research, 46*(24), 6855–6881. https://doi.org/10.1080/00207540701197028.
19. Wortham, A. W., & Baker, R. C. (1976). Multiple deferred state sampling inspection. *International Journal of Production Research, 14*(6), 719–731.
20. Hsu, B.-M., Shu, M.-H., & Wang, T.-C. (2020). Variables adjustable multiple dependent state sampling plans with a loss-based capability index. *The International Journal of Advanced Manufacturing Technology, 107*, 2163–2175. https://doi.org/10.1007/s00170-020-05137-9.
21. Aslam, M., Azam, M., Khan, N., & Jun, C.-H. (2015). A control chart for an exponential distribution using multiple dependent state sampling. *Quality & Quantity, 49*(2), 455–462. https://doi.org/10.1007/s11135-014-0002-2.
22. Zhou, W., Wan, Q., Zheng, Y., & Zhou, Y.-w. (2017). A joint-adaptive np control chart with multiple dependent state sampling scheme. *Communications in Statistics-Theory and Methods, 46*(14), 6967–6979.
23. Aldosari, M. S., Aslam, M., & Jun, C.-H. (2017). A new attribute control chart using multiple dependent state repetitive sampling. *IEEE Access, 5*, 6192–6197.
24. Aslam, M., Nazir, A., & Jun, C.-H. (2015). A new attribute control chart using multiple dependent state sampling. *Transactions of the Institute of Measurement and Control, 37*(4), 569–576.
25. Aslam, M., Khan, N., & Jun, C.-H. (2014). A multiple dependent state control chart based on double control limit. *Research Journal of Applied Sciences, Engineering and Technology, 7*(21), 4490–4493.
26. Aslam, M., Azam, M., & Jun, C.-H. (2013). Multiple dependent state sampling plan based on process capability index. *Journal of Testing and Evaluation, 41*(2), 1–7.
27. Aslam, M., Balamurali, S., & Jun, C.-H. (2019). A new multiple dependent state sampling plan based on the process capability index. *Communications in Statistics: Simulation and Computation*, 1–17. https://doi.org/10.1080/03610918.2019.1588307.
28. Aslam, M., Yen, C.-H., Chang, C.-H., & Jun, C.-H. (2014). Multiple dependent state variable sampling plans with process loss consideration. *The International Journal of Advanced Manufacturing Technology, 71*(5-8), 1337–1343.
29. Balamurali, S., Jeyadurga, P., & Usha, M. (2016). Designing of Bayesian multiple deferred state sampling plan based on Gamma–Poisson distribution. *American Journal of Mathematical and Management Sciences, 35*(1), 77–90.
30. Balamurali, S., & Jun, C.-H. (2007). Multiple dependent state sampling plans for lot acceptance based on measurement data. *European Journal of Operational Research, 180*(3), 1221–1230. https://doi.org/10.1016/j.ejor.2006.05.025.
31. Govindaraju, K., & Subramani, K. (1993). Selection of multiple deferred (dependent) state sampling plans for given acceptable quality level and limiting quality level. *Journal of Applied Statistics, 20*(3), 423–428.
32. Kuralmani, V., & Govlndaraju, K. (1992). Selection of multiple deferred (dependent) state sampling plans. *Communications in Statistics—Theory and Methods, 21*(5), 1339–1366. https://doi.org/10.1080/03610929208830851.
33. Soundararajan, V., & Vijayaraghavan, R. (1989). On designing multiple deferred state sampling (MDS-1 (0, 2)) plans involving minimum risks. *Journal of Applied Statistics, 16*(1), 87–94. https://doi.org/10.1080/02664768900000010.

34. Soundararajan, V., & Vijayaraghavan, R. (1990). Construction and selection of multiple dependent (deferred) state sampling plan. *Journal of Applied Statistics, 17*(3), 397–409. https://doi.org/10.1080/02664769000000012.

35. Wu, C.-W., Liu, S.-W., & Lee, A. H. (2015). Design and construction of a variables multiple dependent state sampling plan based on process yield. *European Journal of Industrial Engineering, 9*(6), 819–838.

36. Wu, C.-W., & Wang, Z.-H. (2017). Developing a variables multiple dependent state sampling plan with simultaneous consideration of process yield and quality loss. *International Journal of Production Research, 55*(8), 2351–2364.

37. Yan, A., Liu, S., & Dong, X. (2016). Designing a multiple dependent state sampling plan based on the coefficient of variation. *Springerplus, 5*(1), 1447. https://doi.org/10.1186/s40064-016-3087-3.

38. Smarandache, F. (1998). *Neutrosophy: Neutrosophic probability, set, and logic: Analytic synthesis & synthetic analysis*. Rehoboth, NM: American Research Press.

39. Smarandache, F. (2016). *Neutrosophic extended triplets,*. Special Collections Tempe, AZ: Arizona State University.

40. Zhang, X., Wang, X., Smarandache, F., Jaíyéolá, T. G., & Lian, T. (2019). Singular neutrosophic extended triplet groups and generalized groups. *Cognitive Systems Research, 57*, 32–40.

41. Abdel-Basset, M., Atef, A., & Smarandache, F. (2019). A hybrid neutrosophic multiple criteria group decision making approach for project selection. *Cognitive Systems Research, 57*, 216–227.

42. Albassam, M., Khan, N., & Aslam, M. (2020). The W/S test for data having neutrosophic numbers: An application to USA village population. *Complexity, 2020*, 3690879. https://doi.org/10.1155/2020/3690879.

43. Khan, Z., Gulistan, M., Hashim, R., Yaqoob, N., & Chammam, W. (2020). Design of S-control chart for neutrosophic data: An application to manufacturing industry. *Journal of Intelligent Fuzzy Systems, 38*(4), 4743–4751. https://doi.org/10.3233/JIFS-191439.

44. Panthong, C., & Pongpullponsak, A. (2016). Non-normality and the fuzzy theory for variable parameters control charts. *Thai Journal of Mathematics, 14*(1), 203–213.

45. Aslam, M., & Raza, M. A. (2019). Design of new sampling plans for multiple manufacturing lines under uncertainty. *International Journal of Fuzzy Systems, 21*(3), 978–992.

46. Hsieh, K.-L., Tong, L.-I., & Wang, M.-C. (2007). The application of control chart for defects and defect clustering in IC manufacturing based on fuzzy theory. *Expert Systems with Applications, 32*(3), 765–776.

47. Aslam, M., Bantan, R. A., & Khan, N. (2019a). Design of a control chart for gamma distributed variables under the indeterminate environment. *IEEE Access, 7*, 8858–8864.

48. Aslam, M., Bantan, R. A., & Khan, N. (2019b). Design of a new attribute control chart under neutrosophic statistics. *International Journal of Fuzzy Systems, 21*(2), 433–440.

49. Afshari, R., & Sadeghpour Gildeh, B. (2017). Designing a multiple deferred state attribute sampling plan in a fuzzy environment. *American Journal of Mathematical and Management Sciences, 36*(4), 328–345.

50. Aslam, M., Khan, N., & Khan, M. (2018). Monitoring the variability in the process using neutrosophic statistical interval method. *Symmetry, 10*(11), 562.

51. Aslam, M., Khan, N., & Albassam, M. (2018). Control chart for failure-censored reliability tests under uncertainty environment. *Symmetry, 10*(12), 690.

52. Ahmad, L., Aslam, M., & Jun, C.-H. (2014a). Coal quality monitoring with improved control charts. *European Journal of Scientific Research, 125*(2), 427–434.

53. Ahmad, L., Aslam, M., & Jun, C.-H. (2014b). Designing of X-bar control charts based on process capability index using repetitive sampling. *Transactions of the Institute of Measurement and Control, 36*(3), 367–374.

54. Ahmad, L., Aslam, M., Arif, O., & Jun, C.-H. (2016). Dispersion chart for some popular distributions under repetitive sampling. *Journal of Advanced Mechanical Design, Systems, and Manufacturing, 10*(4), 1–18. https://doi.org/10.1299/jamdsm.2016jamdsm0058.

55. Aslam, M., Ahmad, L., Jun, C.-H., & Arif, O. H. (2016). A control chart for COM–Poisson distribution using multiple dependent state sampling. *Quality and Reliability Engineering International, 32*(8), 2803–2812. https://doi.org/10.1002/qre.1965.
56. Aslam, M., Khan, N., Ahmad, L., Jun, C.-H., & Hussain, J. (2017). A mixed control chart using process capability index. *Sequential Analysis, 36*(2), 278–289. https://doi.org/10.1080/07474946.2017.1319690.
57. Azam, M., Ahmad, L., & Aslam, M. (2016). Design of X-bar chart for burr distribution under the repetitive sampling. *Science International (Lahore), 28*(4), 3265–3271.

Three-Way Decisions with Single-Valued Neutrosophic Uncertain Linguistic Decision-Theoretic Rough Sets Based on Generalized Maclaurin Symmetric Mean Operators

Zeeshan Ali, Tahir Mahmood, and Florentin Smarandache

1 Introduction

In 1965, the theory of fuzzy set (FS) was presented by Zadeh [1]. FS contains the grade of supporting which is not exceeded from unit interval. FS has resolved various types of awkward and complicated information in real-life decisions. But there was a problem, when a decision-maker faced a situation like yes or no. To address effectively such kind of issues, the theory of intuitionistic FS (IFS) was explored by Atanassove [2]. IFS contains two functions called the grade of supporting and the grade of supporting against with a condition that is the sum of both is not exceeded from unit interval. IFS is a proficient technique to resolve unreliable and vague information. IFS has received wide-ranging attention from scholars, and various scholars have utilized it in different fields like similarity measures [3], aggregation operators [4], and hybrid aggregation operators [5]. The grade of a neutral cannot be discussed in IFS. However, a neutral grade is used in many real-life scenarios, such as polling stations; human beings give opinions having more answers of a kind: positive, abstinence, negative, and refusal/neutral. For example, in a democratic voting system, 100 people have appeared in the election and the election commission issued only 100 ballot papers. One person can take only one ballot paper for giving his/her vote and is only one candidate. Basically, the result is divided into four groups like vote for candidate 50, abstinence in vote 20, vote negative candidate 20, and refusal vote is 10. The "abstinence in

Z. Ali · T. Mahmood (✉)
Department of Mathematics and Statistics, International Islamic University, Islamabad, Pakistan
e-mail: zeeshan.phdma102@iiu.edu.pk; tahirbakhat@iiu.edu.pk

F. Smarandache (✉)
Department of Mathematics and Science, University of New Mexico, Gallup, NM, USA
e-mail: smarand@unm.edu

© The Author(s), under exclusive license to Springer Nature Switzerland AG 2021   71
F. Smarandache, M. Abdel-Basset (eds.), *Neutrosophic Operational Research*,
https://doi.org/10.1007/978-3-030-57197-9_5

vote" shows that ballot paper is white which contradicts both "vote for a candidate" and "vote negative candidate," but it considers the vote, and the "refusal of voting" means bypassing the vote. Therefore, Smarandache [6] presented the theory of neutrosophic set (NS), which is the modified version of the IFS characterized by the grade of supporting, the grade of abstinence, and the grade of supporting against. NS is beneficial technique in the environment of fuzzy sets theory to cope with uncertain and awkward realistic decision problems. Later, Wang et al. [7] explored the novel idea of single-valued NS (S-VNS), which is the generalization of the IFS to cope with unreliable and uncertain information. NS and S-VNS have received extensive attention from scholars and many researchers have applied them in different fields. For instance, various scholars explored similarity measures S-VNS and NS [8, 9], aggregation operators [10–15], and hybrid aggregation operators [16–21].

In various realistic decision problems, since there are fuzziness and vulnerability, the assessment estimations of properties are simpler spoken to by linguistic terms (LTs) instead of crisp numbers, particularly for subjective attributes. Zadeh [22] discovered another idea called linguistic variables (LVs) to portray the subjective data for realistic decision problems. So as to stay away from the downside of the data mutilation in the operational procedure, Herrera and Martinez [23] proposed a two-tuple linguistic (2TL) model as for processing with words. Herrera and Martinez [24] proposed a way to deal with setting up the change connection among LTs and numbers dependent on the 2TL model in decision. Martinez and Herrera [25] additionally gave a thorough review of the exploration on the 2TL models in decision. In addition, some collection administrators for 2TL data have been introduced [26].

Decision-theoretic rough set (D-TRS) is one of the proficient techniques to evaluate awkward and complicated information in real-decision problems. Various scholars have modified it to different ways like loss function (LF) [27], attribute reduction [28], and further improved models using D-TRS [29]. Yao [30] presented the three-way decision (T-WD), which is modified from D-TRS to cope with realistic decision problems. T-WD can separate all universal sets into three different parts: the positive region (POS(C)), the negative region (NEG(C)), and the boundary region (BND(C)). As a blend of D-TRSs and the Bayesian decision procedure, the techniques of T-WD have effectively managed bunches of issues of order. This hypothesis has been applied in numerous real fields [31, 32]. A large portion of the previous research works for the most part center around two issues to extend the model of T-WD: one is the restrictive probability and the other is the LF. Further, Liu and Yang [33] presented the loss function in T-WDs based on intuitionistic uncertain linguistic sets. Based on the above analysis, the objective of this manuscript is as follows: to explore three-way decision (T-WD) based on decision-theoretic rough set (D-TRS) for S-VNULV, and their some operational laws are also presented. Further, based on the S-VNULV, the generalized MSM operators are called S-VNULGMSM operator and S-VNULWGMSM operator, and their special cases are also discussed. Based on the explored operator and D-TRS, we explored an algorithm to evaluate the numerical example for finding the reliability and

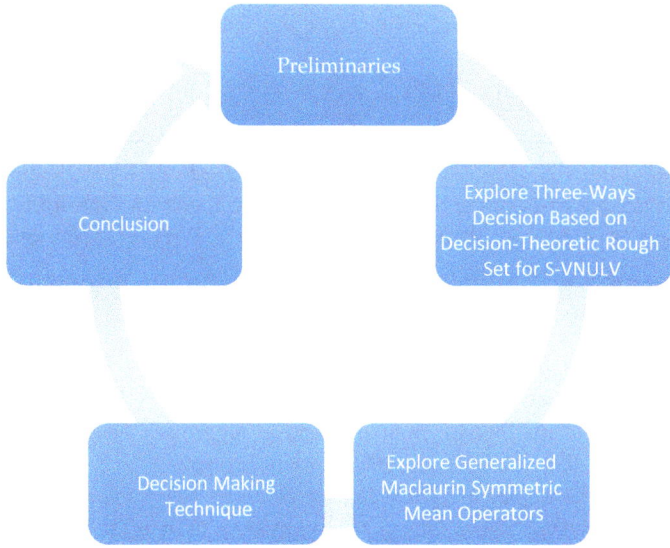

Fig. 1 Geometrical interpretation of the explored work in this manuscript

effectiveness of the presented approaches. The graphical representation of the explored works in this manuscript is discussed with the help of Fig. 1.

The rest of this manuscript is summarized in the following ways: In Sect. 2, we review some basic notions like S-VNS, uncertain linguistic variables (ULVs), single-valued neutrosophic uncertain linguistic variables (S-VNULS), generalized Maclaurin symmetric mean (GMSM) operator, and their basic operational laws. The T-WDs based on D-TRS are also discussed in detail. In Sect. 3, we explore three-way decision (T-WD) based on decision-theoretic rough set (D-TRS) for S-VNULV, and their some operational laws are also presented. In Sect. 4, based on the S-VNULV, the generalized MSM operators are called S-VNULGMSM operator and S-VNULWGMSM operator, and their special cases are also discussed. In Sect. 5, based on the explored operator and D-TRS, we explored an algorithm to evaluate the numerical example for finding the reliability and effectiveness of the presented approaches. The conclusion of this paper is discussed in Sect. 6.

2　Preliminaries

In this study, we review basic notions like S-VNS, ULVs, S-VNULS, GMSM operator, and their basic operational laws. The T-WDs based on D-TRS are also discussed in detail.

Definition 1 [7] The S-VNS $\hat{A}$ is stated by

$$\hat{A} = \left\{ \left(\underline{T}(\mathfrak{r}), \mathsf{I}(\mathfrak{r}), F(\mathfrak{r}) \right) \mid \mathfrak{r} \in \mathfrak{R} \right\} \tag{1}$$

where $\underline{T}(\mathfrak{r})$, $\mathsf{I}(\mathfrak{r})$, and $F(\mathfrak{r})$ represent the degree of membership, degree of abstinence, and degree of nonmembership, respectively, satisfying the following: $0 \leq \underline{T}(\mathfrak{r}) + \mathsf{I}(\mathfrak{r}) + F(\mathfrak{r}) \leq 3$ and $0 \leq \underline{T}(\mathfrak{r}), \mathsf{I}(\mathfrak{r}), F(\mathfrak{r}) \leq 1$. Further, $\hat{A} = \left(\underline{T}(\mathfrak{r}), \mathsf{I}(\mathfrak{r}), F(\mathfrak{r}) \right)$ represents the single-valued neutrosophic number (S-VNN); simply we write $\hat{A} = \left(\underline{T}, \mathsf{I}, F \right)$.

Definition 2 [22] Let set $\mathsf{\S} = \{\S_i \mid i = 0, 1, 2, \ldots, l-1\}$ be a linguistic term set (LTS) with odd number of elements, and $\S_i$ signifies a feasible value of the LTS. It is noted that the inferior and superior values of LTS are $\S_0$ and $\S_{l-1}$, respectively. For example, a LTS $\mathsf{\S}$ can be given as $\mathsf{\S} = \{\S_0 = \text{veryshort}, \S_1 = \text{short}, \S_2 = \text{slightlyshort}, \S_3 = \text{fair}, \S_4 = \text{slightlyhigh}, \S_5 = \text{high}, \S_6 = \text{veryhigh}\}$ satisfying the properties:

1. The set is ordered: $\S_i > \S_j \leftrightarrow i > j$.
2. There is a negation operator: $(\S_i) = \S_{l-1-i}$.
3. Maximum operator: $\max \left(\S_i, \S_j \right)$, if $i \geq j$.
4. Minimum operator: $\min \left(\S_i, \S_j \right)$, if $i \leq j$.

For discrete LTSs, in order to minimize the loss of linguistic decision information, Xu extends $\mathsf{\S} = \{\S_i \mid i = 0, 1, 2, \ldots, l-1\}$ to a continuous set $\acute{\mathsf{\S}} = \{\S_{\mathfrak{z}} \mid \mathfrak{z} \in R\}$..

Definition 3 [34] Let $\mathsf{s} = [\S_\alpha, \S_\beta]$ with $\S_\alpha$, $\S_\beta \in \mathsf{\S}$ and $\alpha \leq \beta$; $\S_\alpha$ and $\S_\beta$ are the inferior and superior limits of s, separately; and then s is an ULV. In addition we can use $\acute{\mathsf{s}}$ to express all ULVs.

Let $\mathsf{s} = [\S_\alpha, \S_\beta]$, $\S_1 = [\S_{\alpha_1}, \S_{\beta_1}]$, and $\S_2 = [\S_{\alpha_2}, \S_{\beta_2}]$ be any three ULVs, and λ, $\lambda_1 > 0$. Then, operational laws of ULVs are as follows:

$$\S_1 \oplus \S_2 = [\S_{\alpha_1}, \S_{\beta_1}] \oplus [\S_{\alpha_2}, \S_{\beta_2}] = [\S_{\alpha_1+\alpha_2}, \S_{\beta_1+\beta_2}] \tag{2}$$

$$\S_1 \otimes \S_2 = [\S_{\alpha_1}, \S_{\beta_1}] \otimes [\S_{\alpha_2}, \S_{\beta_2}] = [\S_{\alpha_1\alpha_2}, \S_{\beta_1\beta_2}] \tag{3}$$

$$\lambda\S = \lambda[\S_\alpha, \S_\beta] = [\S_{\lambda\alpha}, \S_{\lambda\beta}] \tag{4}$$

$$(\S)^\lambda = [\S_\alpha, \S_\beta]^\lambda = [\S_{\alpha^\lambda}, \S_{\beta^\lambda}] \tag{5}$$

Definition 4 [35] The S-VNULV $\hat{A}$ is stated by

$$\hat{A} = \left\{ \left(\left[s_{\phi(\mathfrak{x})}, s_{\phi(\mathfrak{x})} \right], \left(\underline{T}(\mathfrak{x}), \mathfrak{l}(\mathfrak{x}), F(\mathfrak{x}) \right) \right) \mid \mathfrak{x} \in \mathfrak{R} \right\} \tag{6}$$

where $\underline{T}(\mathfrak{x})$, $\mathfrak{l}(\mathfrak{x})$, and $F(\mathfrak{x})$ represent the degree of membership, degree of abstinence, and degree of nonmembership, respectively, satisfying the following: $0 \leq \underline{T}(\mathfrak{x}) + \mathfrak{l}(\mathfrak{x}) + F(\mathfrak{x}) \leq 3$ and $0 \leq \underline{T}(\mathfrak{x}), \mathfrak{l}(\mathfrak{x}), F(\mathfrak{x}) \leq 1$, where $[s_{\phi(\mathfrak{x})}, s_{\phi(\mathfrak{x})}] \in \dot{\mathfrak{S}}$ denoted ULV. Further, $\hat{A} = \left(\left[s_{\phi(\mathfrak{x})}, s_{\phi(\mathfrak{x})} \right], \left(\underline{T}(\mathfrak{x}), \mathfrak{l}(\mathfrak{x}), F(\mathfrak{x}) \right) \right)$ represents the single-valued neutrosophic uncertain linguistic variable (S-VNULV); simply we write $\hat{A} = \left(\left[s_{\phi}, s_{\phi} \right], \left(\underline{T}, \mathfrak{l}, F \right) \right)$. Moreover, $\pi(\mathfrak{x}) = 1 - \underline{T}(\mathfrak{x}) - \mathfrak{l}(\mathfrak{x}) - F(\mathfrak{x})$ is the indeterminacy term of $\mathfrak{x}$ to ULV $[s_{\phi(\mathfrak{x})}, s_{\phi(\mathfrak{x})}]$.

For any two S-VNULVs $\hat{A}_1 = \left(\left[s_{\phi(\mathfrak{x}_1)}, s_{\phi(\mathfrak{x}_1)} \right], \left(\underline{T}(\mathfrak{x}_1), \mathfrak{l}(\mathfrak{x}_1), F(\mathfrak{x}_1) \right) \right)$ and $\hat{A}_2 = \left(\left[s_{\phi(\mathfrak{x}_2)}, s_{\phi(\mathfrak{x}_2)} \right], \left(\underline{T}(\mathfrak{x}_2), \mathfrak{l}(\mathfrak{x}_2), F(\mathfrak{x}_2) \right) \right)$, then

$$\hat{A}_1 + \hat{A}_2 = \left(\left[s_{\phi(\mathfrak{x}_1) + \phi(\mathfrak{x}_2)}, s_{\phi(\mathfrak{x}_1) + \phi(\mathfrak{x}_2)} \right], \right.$$
$$\left. \left(\underline{T}(\mathfrak{x}_1) + \underline{T}(\mathfrak{x}_2) - \underline{T}(\mathfrak{x}_1) \underline{T}(\mathfrak{x}_2), \mathfrak{l}(\mathfrak{x}_1) \mathfrak{l}(\mathfrak{x}_2), F(\mathfrak{x}_1) F(\mathfrak{x}_2) \right) \right) \tag{7}$$

$$\hat{A}_1 \times \hat{A}_2 = \left(\begin{array}{c} \left[s_{\phi(\mathfrak{x}_1) \times \phi(\mathfrak{x}_2)}, s_{\phi(\mathfrak{x}_1) \times \phi(\mathfrak{x}_2)} \right], \\ \left(\underline{T}(\mathfrak{x}_1) \underline{T}(\mathfrak{x}_2), \mathfrak{l}(\mathfrak{x}_1) + \mathfrak{l}(\mathfrak{x}_2) - \mathfrak{l}(\mathfrak{x}_1) \mathfrak{l}(\mathfrak{x}_2), F(\mathfrak{x}_1) + F(\mathfrak{x}_2) - F(\mathfrak{x}_1) F(\mathfrak{x}_2) \right) \end{array} \right) \tag{8}$$

$$\lambda \hat{A}_1 = \left(\left[s_{\lambda \times \phi(\mathfrak{x}_1)}, s_{\lambda \times \phi(\mathfrak{x}_1)} \right], \left(1 - \left(1 - \underline{T}(\mathfrak{x}_1) \right)^{\lambda}, \mathfrak{l}(\mathfrak{x}_1)^{\lambda}, F(\mathfrak{x}_1)^{\lambda} \right) \right) \tag{9}$$

$$\hat{A}_1^{\lambda} = \left(\left[s_{\lambda \times \phi(\mathfrak{x}_1)}, s_{\lambda \times \phi(\mathfrak{x}_1)} \right], \left(\underline{T}(\mathfrak{x}_1)^{\lambda}, 1 - \left(1 - \mathfrak{l}(\mathfrak{x}_1) \right)^{\lambda}, 1 - \left(1 - F(\mathfrak{x}_1) \right)^{\lambda} \right) \right) \tag{10}$$

Definition 5 [35] For any two S-VNULVs $\hat{A}_1 = \left(\left[s_{\phi(\mathfrak{x}_1)}, s_{\phi(\mathfrak{x}_1)} \right], \left(\underline{T}(\mathfrak{x}_1), \mathfrak{l}(\mathfrak{x}_1), F(\mathfrak{x}_1) \right) \right)$, then the expected value $Q\left(\hat{A}_1 \right)$ and accuracy value of $G\left(\hat{A}_1 \right)$ are stated by

$$Q\left(\hat{A}_1\right) = \frac{1}{2} \times \left(\mathbb{T}\left(\varsigma_1\right) + \mathbb{I}\left(\varsigma_1\right) + 1 - F\left(\varsigma_1\right)\right) \times \S_{\frac{\left(\phi\left(\varsigma_1\right)+\phi\left(\varsigma_1\right)\right)}{2}}$$

$$= \S_{\left(\frac{\left(\left(\phi\left(\varsigma_1\right)+\phi\left(\varsigma_1\right)\right)\times\left(\mathbb{T}\left(\varsigma_1\right)+\mathbb{I}\left(\varsigma_1\right)+1-F\left(\varsigma_1\right)\right)\right)}{4}\right)} \tag{17}$$

$$G\left(\hat{A}_1\right) = \left(\mathbb{T}\left(\varsigma_1\right) + \mathbb{I}\left(\varsigma_1\right) + F\left(\varsigma_1\right)\right) \times \S_{\frac{\left(\phi\left(\varsigma_1\right)+\phi\left(\varsigma_1\right)\right)}{2}}$$

$$= \S_{\left(\frac{\left(\left(\phi\left(\varsigma_1\right)+\phi\left(\varsigma_1\right)\right)\times\left(\mathbb{T}\left(\varsigma_1\right)+\mathbb{I}\left(\varsigma_1\right)+F\left(\varsigma_1\right)\right)\right)}{4}\right)} \tag{18}$$

Definition 6 [35] For any two S-VNULVs $\hat{A}_1 = \left(\left[\S_{\phi\left(\varsigma_1\right)}, \S_{\phi\left(\varsigma_1\right)}\right], \left(\mathbb{T}\left(\varsigma_1\right), \mathbb{I}\left(\varsigma_1\right), F\left(\varsigma_1\right)\right)\right)$ and $\hat{A}_2 = \left(\left[\S_{\phi\left(\varsigma_2\right)}, \S_{\phi\left(\varsigma_2\right)}\right], \left(\mathbb{T}\left(\varsigma_2\right), \mathbb{I}\left(\varsigma_2\right), F\left(\varsigma_2\right)\right)\right)$,

1. If $Q\left(\hat{A}_1\right) > Q\left(\hat{A}_2\right)$, then $\hat{A}_1 > \hat{A}_2$.
2. If $Q\left(\hat{A}_1\right) = Q\left(\hat{A}_2\right)$, then:

    (a) If $G\left(\hat{A}_1\right) > G\left(\hat{A}_2\right)$, then $\hat{A}_1 > \hat{A}_2$.
    (b) If $G\left(\hat{A}_1\right) = G\left(\hat{A}_2\right)$, then $\hat{A}_1 = \hat{A}_2$.

Definition 7 [36] For any group of positive numbers $\hat{A}_i$ $(i = 1, 2, \ldots, n)$, then the GMSM operator is stated by

$$GMSM^{(k,p_1,p_2,\ldots,p_k)}\left(\hat{A}_1, \hat{A}_2, \ldots, \hat{A}_n\right) = \left(\frac{\sum_{1 \leq i_1 < \cdots < i_k \leq i_n} \prod_{j=1}^{k} \hat{A}_{i_j}^{p_j}}{C_n^k}\right)^{\frac{1}{p_1+p_2+\cdots+p_k}} \tag{19}$$

where $k = 1, 2, \ldots, n$, $p_1, p_2, \ldots, p_m \geq 0$, $(i_1, i_2, \ldots, i_k)$ and C_n^k is the coefficient binomial, which holds the following conditions:

1. $GMSM^{(k,p_1,p_2,\ldots,p_k)}\left(\hat{A}_1, \hat{A}_2, \ldots, \hat{A}_n\right) = \hat{A}$, $GMSM^{(k,p_1,p_2,\ldots,p_k)}(0, 0, \ldots, 0) = 0$.
2. $GMSM^{(k,p_1,p_2,\ldots,p_k)}\left(\hat{A}_1, \hat{A}_2, \ldots, \hat{A}_n\right) \leq GMSM^{(k,p_1,p_2,\ldots,p_k)}\left(\hat{A}_{*1}, \hat{A}_{*2}, \ldots, \hat{A}_{*n}\right)$, if $\hat{A}_i \leq \hat{A}_{*i}$ for all i.
3. $\min_i \hat{A}_i \leq GMSM^{(k,p_1,p_2,\ldots,p_k)}\left(\hat{A}_1, \hat{A}_2, \ldots, \hat{A}_n\right) \leq \max_i \hat{A}_i$.

Table 1 Loss functions and their representations

Symbols	$\mathcal{F}_B\left(P_{AC}\right)$ = correctly solved	$\sim \mathcal{F}_B\left(N_{AC}\right)$ = wrongly solved
$\chi_{P_{AC}}$	$\hat{\overline{A}}_{SVN-P_{AC}P_{AC}}$	$\hat{\overline{A}}_{SVN-P_{AC}N_{AC}}$
$\chi_{B_{AC}}$	$\hat{\overline{A}}_{SVN-B_{AC}P_{AC}}$	$\hat{\overline{A}}_{SVN-B_{AC}N_{AC}}$
$\chi_{N_{AC}}$	$\hat{\overline{A}}_{SVN-N_{AC}P_{AC}}$	$\hat{\overline{A}}_{SVN-N_{AC}N_{AC}}$

$\hat{\overline{A}}_{SVN-P_{AC}P_{AC}}, \hat{\overline{A}}_{SVN-P_{AC}N_{AC}}$ = Costs of the correct sort and error sort of the object ŗ in accepted decision

$\hat{\overline{A}}_{SVN-B_{AC}P_{AC}}, \hat{\overline{A}}_{SVN-B_{AC}N_{AC}}$ = Costs of the correct sort and error sort of the object ŗ in delay decision

$\hat{\overline{A}}_{SVN-N_{AC}P_{AC}}, \hat{\overline{A}}_{SVN-N_{AC}N_{AC}}$ = Costs of the correct sort and error sort of the object ŗ in rejected decision

2.1 Review of the Three-Way Decisions Based on D-TRSs

Due to D-TRS procedure based on Ref. [37], it contains two status and three actions, whose information is as follows: $\Omega_S = \{\mathcal{F}_B, \sim \mathcal{F}_B\}$ and $A_{AC} = \{\chi_{P_{AC}}, \chi_{B_{AC}}, \chi_{N_{AC}}\}$. The symbols of the set Ω_S are used for belonging and for not belonging, and the symbols of the set A_{AC} are used for positive, boundary, and negative. The symbols which are used for positive, its mean accepted event object, for negative its mean rejected event object, and for boundary its mean does not promise or delay the decision event object.

From Table 1, the inequality is as follows:

$$\hat{\overline{A}}_{SVN-P_{AC}P_{AC}} < \hat{\overline{A}}_{SVN-B_{AC}P_{AC}} < \hat{\overline{A}}_{SVN-N_{AC}P_{AC}} \tag{20}$$

$$\hat{\overline{A}}_{SVN-N_{AC}N_{AC}} < \hat{\overline{A}}_{SVN-B_{AC}N_{AC}} < \hat{\overline{A}}_{SVN-P_{AC}N_{AC}} \tag{21}$$

These all informations are taken from Ref. [37] and it is called Bayesian risk decision theory; we have

$$Pr\left(\mathcal{F}_B\mid[\mathfrak{r}]\right) + Pr\left(\sim \mathcal{F}_B\mid[\mathfrak{r}]\right) = 1 \tag{22}$$

Further, we explained the expected losses $Y_{EL}\left(\chi_{j_{AC}}\mid[\mathfrak{r}]\right)$, $j = P, B, N$, by using Eq. (22); the different actions are expressed below:

$$Y_{EL}\left(\chi_{P_{AC}}\mid[\mathfrak{r}]\right) = \hat{\overline{A}}_{SVN-P_{AC}P_{AC}}Pr\left(\mathcal{F}_B\mid[\mathfrak{r}]\right) + \hat{\overline{A}}_{SVN-P_{AC}N_{AC}}Pr\left(\sim \mathcal{F}_B\mid[\mathfrak{r}]\right)$$
$$\tag{23}$$

$$Y_{EL}\left(\chi_{B_{AC}}|\left[\mathfrak{s}\right]\right) = \overline{\hat{A}}_{SVN-B_{AC}P_{AC}} Pr\left(\mathcal{F}_B|\left[\mathfrak{s}\right]\right) + \overline{\hat{A}}_{SVN-B_{AC}N_{AC}} Pr\left(\sim \mathcal{F}_B|\left[\mathfrak{s}\right]\right) \tag{24}$$

$$Y_{EL}\left(\chi_{N_{AC}}|\left[\mathfrak{s}\right]\right) = \overline{\hat{A}}_{SVN-N_{AC}P_{AC}} Pr\left(\mathcal{F}_B|\left[\mathfrak{s}\right]\right) + \overline{\hat{A}}_{SVN-N_{AC}N_{AC}} Pr\left(\sim \mathcal{F}_B|\left[\mathfrak{s}\right]\right) \tag{25}$$

Based on the above analysis, we expose the three-way decisions which are discussed below:

$$P_{AC} : \text{When } Y_{EL}\left(\chi_{P_{AC}}|\left[\mathfrak{s}\right]\right) \leq Y_{EL}\left(\chi_{B_{AC}}|\left[\mathfrak{s}\right]\right) \text{ and } Y_{EL}\left(\chi_{P_{AC}}|\left[\mathfrak{s}\right]\right)$$
$$\leq Y_{EL}\left(\chi_{N_{AC}}|\left[\mathfrak{s}\right]\right), \text{ then } \mathfrak{s} \in POS\left(\mathcal{F}_P\right) \tag{26}$$

$$B_{AC} : \text{When } Y_{EL}\left(\chi_{B_{AC}}|\left[\mathfrak{s}\right]\right) \leq Y_{EL}\left(\chi_{P_{AC}}|\left[\mathfrak{s}\right]\right) \text{ and } Y_{EL}\left(\chi_{B_{AC}}|\left[\mathfrak{s}\right]\right)$$
$$\leq Y_{EL}\left(\chi_{N_{AC}}|\left[\mathfrak{s}\right]\right), \text{ then } \mathfrak{s} \in BUN\left(\mathcal{F}_B\right) \tag{27}$$

$$N_{AC} : \text{When } Y_{EL}\left(\chi_{N_{AC}}|\left[\mathfrak{s}\right]\right) \leq Y_{EL}\left(\chi_{P_{AC}}|\left[\mathfrak{s}\right]\right) \text{ and } Y_{EL}\left(\chi_{N_{AC}}|\left[\mathfrak{s}\right]\right)$$
$$\leq Y_{EL}\left(\chi_{B_{AC}}|\left[\mathfrak{s}\right]\right), \text{ then } \mathfrak{s} \in NRG\left(\mathcal{F}_N\right) \tag{28}$$

3   A Novel D-TRS Model Based on Single-Valued Neutrosophic Uncertain Linguistic Variables

The purpose of this communication is to explore the novel D-TRS model by using the *SVNULV* based on the three-way decisions. The LF matrix for *SVNULV* is presented in the form of Table 2.

From Table 2, the inequality is as follows:

$$\hat{A}_{SVN-\overline{\hat{A}}_{SVN-P_{AC}P_{AC}}} < \hat{A}_{SVN-\overline{\hat{A}}_{SVN-B_{AC}P_{AC}}} < \hat{A}_{SVN-\overline{\hat{A}}_{SVN-N_{AC}P_{AC}}} \tag{29}$$

$$\hat{A}_{SVN-\overline{\hat{A}}_{SVN-N_{AC}N_{AC}}} < \hat{A}_{SVN-\overline{\hat{A}}_{SVN-B_{AC}N_{AC}}} < \hat{A}_{SVN-\overline{\hat{A}}_{SVN-P_{AC}N_{AC}}} \tag{30}$$

Further, we explained the expected losses $Y_{EL}\left(\chi_{j_{AC}}|\left[\mathfrak{s}\right]\right), j = P, B, N$, by using Eq. (22); the different actions are expressed below:

Table 2 Loss functions and their representations for single-valued neutrosophic uncertain linguistic variables

Symbols	$\mathcal{F}_B\left(P_{AC}\right)$ = correctly solved	$\sim \mathcal{F}_B\left(N_{AC}\right)$ = wrongly solved
XP_{AC}	$\hat{A}_{-\bar{A}_{SVN}-P_{AC}P_{AC}} = \left[{}^{\$}_{\phi LT}\left(\bar{\hat{A}}_{SVN}-P_{AC}P_{AC}\right) \cdot {}^{\$}_{\phi LT}\left(\bar{\hat{A}}_{SVN}-P_{AC}P_{AC}\right), \left(\begin{array}{l} T_{\hat{A}_{SVN}}\left(\bar{\hat{A}}_{SVN}-P_{AC}P_{AC}\right) \cdot \\ I_{\hat{A}_{SVN}}\left(\bar{\hat{A}}_{SVN}-P_{AC}P_{AC}\right) \cdot \\ F_{\hat{A}_{SVN}}\left(\bar{\hat{A}}_{SVN}-P_{AC}P_{AC}\right) \end{array}\right) \right]$	$\hat{A}_{SVN}-\bar{\hat{A}}_{SVN}-P_{AC}N_{AC} = \left[{}^{\$}_{\phi LT}\left(\bar{\hat{A}}_{SVN}-P_{AC}N_{AC}\right) \cdot {}^{\$}_{\phi LT}\left(\bar{\hat{A}}_{SVN}-P_{AC}N_{AC}\right), \left(\begin{array}{l} T_{\hat{A}_{SVN}}\left(\bar{\hat{A}}_{SVN}-P_{AC}N_{AC}\right) \cdot \\ I_{\hat{A}_{SVN}}\left(\bar{\hat{A}}_{SVN}-P_{AC}N_{AC}\right) \cdot \\ F_{\hat{A}_{SVN}}\left(\bar{\hat{A}}_{SVN}-P_{AC}N_{AC}\right) \end{array}\right) \right]$
XB_{AC}	$\hat{A}_{SVN}-\bar{\hat{A}}_{SVN}-B_{AC}P_{AC} = \left[{}^{\$}_{\phi LT}\left(\bar{\hat{A}}_{SVN}-E_{AC}P_{AC}\right) \cdot {}^{\$}_{\phi LT}\left(\bar{\hat{A}}_{SVN}-B_{AC}P_{AC}\right), \left(\begin{array}{l} T_{\hat{A}_{SVN}}\left(\bar{\hat{A}}_{SVN}-B_{AC}P_{AC}\right) \cdot \\ I_{\hat{A}_{SVN}}\left(\bar{\hat{A}}_{SVN}-B_{AC}P_{AC}\right) \cdot \\ F_{\hat{A}_{SVN}}\left(\bar{\hat{A}}_{SVN}-B_{AC}P_{AC}\right) \end{array}\right) \right]$	$\hat{A}_{SVN}-\bar{\hat{A}}_{SVN}-B_{AC}N_{AC} = \left[{}^{\$}_{\phi LT}\left(\bar{\hat{A}}_{SVN}-B_{AC}N_{AC}\right) \cdot {}^{\$}_{\phi LT}\left(\bar{\hat{A}}_{SVN}-B_{AC}N_{AC}\right), \left(\begin{array}{l} T_{\hat{A}_{SVN}}\left(\bar{\hat{A}}_{SVN}-B_{AC}N_{AC}\right) \cdot \\ I_{\hat{A}_{SVN}}\left(\bar{\hat{A}}_{SVN}-B_{AC}N_{AC}\right) \cdot \\ F_{\hat{A}_{SVN}}\left(\bar{\hat{A}}_{SVN}-B_{AC}N_{AC}\right) \end{array}\right) \right]$
XN_{AC}	$\hat{A}_{SVN}-\bar{\hat{A}}_{SVN}-N_{AC}P_{AC} = \left[{}^{\$}_{\phi LT}\left(\bar{\hat{A}}_{SVN}-N_{AC}P_{AC}\right) \cdot {}^{\$}_{\phi LT}\left(\bar{\hat{A}}_{SVN}-N_{AC}P_{AC}\right), \left(\begin{array}{l} T_{\hat{A}_{SVN}}\left(\bar{\hat{A}}_{SVN}-N_{AC}P_{AC}\right) \cdot \\ I_{\hat{A}_{SVN}}\left(\bar{\hat{A}}_{SVN}-N_{AC}P_{AC}\right) \cdot \\ F_{\hat{A}_{SVN}}\left(\bar{\hat{A}}_{SVN}-N_{AC}P_{AC}\right) \end{array}\right) \right]$	$\hat{A}_{SVN}-\bar{\hat{A}}_{SVN}-N_{AC}N_{AC} = \left[{}^{\$}_{\phi LT}\left(\bar{\hat{A}}_{SVN}-N_{AC}N_{AC}\right) \cdot {}^{\$}_{\phi LT}\left(\bar{\hat{A}}_{SVN}-N_{AC}N_{AC}\right), \left(\begin{array}{l} T_{\hat{A}_{SVN}}\left(\bar{\hat{A}}_{SVN}-N_{AC}N_{AC}\right) \cdot \\ I_{\hat{A}_{SVN}}\left(\bar{\hat{A}}_{SVN}-N_{AC}N_{AC}\right) \cdot \\ F_{\hat{A}_{SVN}}\left(\bar{\hat{A}}_{SVN}-N_{AC}N_{AC}\right) \end{array}\right) \right]$

$\hat{A}_{SVN}-\bar{\hat{A}}_{SVN}-P_{AC}P_{AC}$, $\hat{A}_{SVN}-\bar{\hat{A}}_{SVN}-P_{AC}N_{AC}$ = Costs of the correct sort and error sort of the object ç in accepted decision

$\hat{A}_{SVN}-\bar{\hat{A}}_{SVN}-B_{AC}P_{AC}$, $\hat{A}_{SVN}-\bar{\hat{A}}_{SVN}-B_{AC}N_{AC}$ = Costs of the correct sort and error sort of the object ç in delay decision

$\hat{A}_{SVN}-\bar{\hat{A}}_{SVN}-N_{AC}P_{AC}$, $\hat{A}_{SVN}-\bar{\hat{A}}_{SVN}-N_{AC}N_{AC}$ = Costs of the correct sort and error sort of the object ç in rejected decision

$$Y_{EL}\left(\chi_{P_{AC}}|\left[\mathfrak{r}\right]\right) = \hat{A}_{SVN-\overline{\hat{A}}_{SVN-P_{AC}P_{AC}}} Pr\left(\mathcal{F}_B|\left[\mathfrak{r}\right]\right) \oplus_{SVN} \hat{A}_{SVN-\overline{\hat{A}}_{SVN-P_{AC}P_{AC}}} Pr\left(\sim \mathcal{F}_B|\left[\mathfrak{r}\right]\right)$$
$$(31)$$

$$Y_{EL}\left(\chi_{B_{AC}}|\left[\mathfrak{r}\right]\right) = \hat{A}_{SVN-\overline{\hat{A}}_{SVN-B_{AC}P_{AC}}} Pr\left(\mathcal{F}_B|\left[\mathfrak{r}\right]\right) \oplus_{SVN} \hat{A}_{SVN-\overline{\hat{A}}_{SVN-B_{AC}N_{AC}}} Pr\left(\sim \mathcal{F}_B|\left[\mathfrak{r}\right]\right)$$
$$(32)$$

$$Y_{EL}\left(\chi_{N_{AC}}|\left[\mathfrak{r}\right]\right) = \hat{A}_{SVN-\overline{\hat{A}}_{SVN-N_{AC}P_{AC}}} Pr\left(\mathcal{F}_B|\left[\mathfrak{r}\right]\right) \oplus_{SVN} \hat{A}_{SVN-\overline{\hat{A}}_{SVN-N_{AC}N_{AC}}} Pr\left(\sim \mathcal{F}_B|\left[\mathfrak{r}\right]\right)$$
$$(33)$$

For simplicity, we consider the values of $Pr\left(\mathcal{F}_B|\left[\mathfrak{r}\right]\right) = \delta_B$ and $Pr\left(\sim \mathcal{F}_B|\left[\mathfrak{r}\right]\right) = \delta'_B$; then Eq. (22) is as follows: $\delta_B + \delta'_B = 1$. By using the above idea, we explored the following results.

Theorem 1 By using Eqs. (31), (32), and (33), we get

$$Y_{EL}\left(\chi_{P_{AC}}|\left[\mathfrak{r}\right]\right)$$

$$= \left(\begin{array}{c} \left[{}^{\$}_{\delta_B \times \phi_{LT}}\left(\overline{\hat{A}}_{SVN-P_{AC}P_{AC}}\right) + \delta'_B \times \phi_{LT}\left(\overline{\hat{A}}_{SVN-P_{AC}N_{AC}}\right), {}^{\$}_{\delta_B \times \phi_{LT}}\left(\overline{\hat{A}}_{SVN-P_{AC}P_{AC}}\right) + \delta'_B \times \phi_{LT}\left(\overline{\hat{A}}_{SVN-P_{AC}N_{AC}}\right) \right], \\ \left(\begin{array}{c} \left(1-\left(1-\mathrm{T}_{\hat{A}_{SVN}}\left(\overline{\hat{A}}_{SVN-P_{AC}P_{AC}}\right)\right)^{\delta_B}\left(1-\mathrm{T}_{\hat{A}_{SVN}}\left(\overline{\hat{A}}_{SVN-P_{AC}N_{AC}}\right)\right)^{\delta'_B}\right), \\ \left(\mathrm{I}_{\hat{A}_{SVN}}\left(\overline{\hat{A}}_{SVN-P_{AC}P_{AC}}\right)\right)^{\delta_B}\left(\mathrm{I}_{\hat{A}_{SVN}}\left(\overline{\hat{A}}_{SVN-P_{AC}N_{AC}}\right)\right)^{\delta'_B}, \\ \left(\mathrm{F}_{\hat{A}_{SVN}}\left(\overline{\hat{A}}_{SVN-P_{AC}P_{AC}}\right)\right)^{\delta_B}\left(\mathrm{F}_{\hat{A}_{SVN}}\left(\overline{\hat{A}}_{SVN-P_{AC}N_{AC}}\right)\right)^{\delta'_B} \end{array} \right) \end{array} \right)$$
$$(34)$$

$$Y_{EL}\left(\chi_{B_{AC}}|\left[\mathfrak{r}\right]\right)$$

$$= \left(\begin{array}{c} \left[{}^{\$}_{\delta_B \times \phi_{LT}}\left(\overline{\hat{A}}_{SVN-B_{AC}P_{AC}}\right) + \delta'_B \times \phi_{LT}\left(\overline{\hat{A}}_{SVN-B_{AC}N_{AC}}\right), {}^{\$}_{\delta_B \times \phi_{LT}}\left(\overline{\hat{A}}_{SVN-B_{AC}P_{AC}}\right) + \delta'_B \times \phi_{LT}\left(\overline{\hat{A}}_{SVN-B_{AC}N_{AC}}\right) \right], \\ \left(\begin{array}{c} \left(1-\left(1-\mathrm{T}_{\hat{A}_{SVN}}\left(\overline{\hat{A}}_{SVN-B_{AC}P_{AC}}\right)\right)^{\delta_B}\left(1-\mathrm{T}_{\hat{A}_{SVN}}\left(\overline{\hat{A}}_{SVN-B_{AC}N_{AC}}\right)\right)^{\delta'_B}\right), \\ \left(\mathrm{I}_{\hat{A}_{SVN}}\left(\overline{\hat{A}}_{SVN-B_{AC}P_{AC}}\right)\right)^{\delta_B}\left(\mathrm{I}_{\hat{A}_{SVN}}\left(\overline{\hat{A}}_{SVN-B_{AC}N_{AC}}\right)\right)^{\delta'_B}, \\ \left(\mathrm{F}_{\hat{A}_{SVN}}\left(\overline{\hat{A}}_{SVN-B_{AC}P_{AC}}\right)\right)^{\delta_B}\left(\mathrm{F}_{\hat{A}_{SVN}}\left(\overline{\hat{A}}_{SVN-B_{AC}N_{AC}}\right)\right)^{\delta'_B} \end{array} \right) \end{array} \right)$$
$$(35)$$

$$Y_{EL}\left(\chi_{N_{AC}}|\left[\varsigma\right]\right)$$

$$= \left(\begin{array}{l} \left[\S_{\delta_B \times \phi_{LT}\left(\overline{\hat{A}}_{SVN-N_{AC}P_{AC}}\right)+\delta'_B \times \phi_{LT}\left(\overline{\hat{A}}_{SVN-N_{AC}N_{AC}}\right)}, \S_{\delta_B \times \phi_{LT}\left(\overline{\hat{A}}_{SVN-N_{AC}P_{AC}}\right)+\delta'_B \times \phi_{LT}\left(\overline{\hat{A}}_{SVN-N_{AC}N_{AC}}\right)}\right], \\ \left(\begin{array}{l} \left(1-\left(1-\underline{T}_{\hat{A}_{SVN}}\left(\overline{\hat{A}}_{SVN-N_{AC}P_{AC}}\right)\right)^{\delta_B}\left(1-\underline{T}_{\hat{A}_{SVN}}\left(\overline{\hat{A}}_{SVN-N_{AC}N_{AC}}\right)\right)^{\delta'_B}\right), \\ \left(I_{\hat{A}_{SVN}}\left(\overline{\hat{A}}_{SVN-N_{AC}P_{AC}}\right)\right)^{\delta_B}\left(I_{\hat{A}_{SVN}}\left(\overline{\hat{A}}_{SVN-N_{AC}N_{AC}}\right)\right)^{\delta'_B}, \\ \left(F_{\hat{A}_{SVN}}\left(\overline{\hat{A}}_{SVN-N_{AC}P_{AC}}\right)\right)^{\delta_B}\left(F_{\hat{A}_{SVN}}\left(\overline{\hat{A}}_{SVN-N_{AC}N_{AC}}\right)\right)^{\delta'_B} \end{array}\right) \end{array}\right) \tag{36}$$

Proof First, we proved Eq. (34); the proof of Eqs. (35) and (36) are similar to Eq. (35):

$$Y_{EL}\left(\chi_{P_{AC}}|\left[\varsigma\right]\right) = \hat{A}_{SVN-\overline{\hat{A}}_{SVN-P_{AC}P_{AC}}} Pr\left(\mathcal{F}_B|\left[\varsigma\right]\right) \oplus_{SVN} \hat{A}_{SVN-\overline{\hat{A}}_{SVN-P_{AC}P_{AC}}} Pr\left(\sim\mathcal{F}_B|\left[\varsigma\right]\right)$$

$$= \hat{A}_{SVN-\overline{\hat{A}}_{SVN-P_{AC}P_{AC}}} \delta_B + \hat{A}_{SVN-\overline{\hat{A}}_{SVN-P_{AC}P_{AC}}} \delta_B$$

$$= \left(\begin{array}{l} \left(\begin{array}{l} \left[\S_{\delta_B \times \phi_{LT}\left(\overline{\hat{A}}_{SVN-P_{AC}P_{AC}}\right)}, \S_{\delta_B \times \phi_{LT}\left(\overline{\hat{A}}_{SVN-P_{AC}P_{AC}}\right)}\right], \\ \left(\begin{array}{l} \left(1-\left(1-\underline{T}_{\hat{A}_{SVN}}\left(\overline{\hat{A}}_{SVN-P_{AC}P_{AC}}\right)\right)^{\delta_B}\right), \\ \left(I_{\hat{A}_{SVN}}\left(\overline{\hat{A}}_{SVN-P_{AC}P_{AC}}\right)\right)^{\delta_B}, \\ \left(F_{\hat{A}_{SVN}}\left(\overline{\hat{A}}_{SVN-P_{AC}P_{AC}}\right)\right)^{\delta_B} \end{array}\right) \end{array}\right) + \\ \left(\begin{array}{l} \left[\S_{\delta'_B \times \phi_{LT}\left(\overline{\hat{A}}_{SVN-P_{AC}N_{AC}}\right)}, \S_{\delta'_B \times \phi_{LT}\left(\overline{\hat{A}}_{SVN-P_{AC}N_{AC}}\right)}\right], \\ \left(\begin{array}{l} \left(1-\left(1-\underline{T}_{\hat{A}_{SVN}}\left(\overline{\hat{A}}_{SVN-P_{AC}N_{AC}}\right)\right)^{\delta'_B}\right), \\ \left(I_{\hat{A}_{SVN}}\left(\overline{\hat{A}}_{SVN-P_{AC}N_{AC}}\right)\right)^{\delta'_B}, \\ \left(F_{\hat{A}_{SVN}}\left(\overline{\hat{A}}_{SVN-P_{AC}N_{AC}}\right)\right)^{\delta'_B} \end{array}\right) \end{array}\right) \end{array}\right)$$

$$
= \left(\begin{array}{l} \left[\S_{\delta_B \times \phi_{LT}\left(\widehat{\overline{A}}_{SVN-P_{AC}P_{AC}}\right)+\delta'_B \times \phi_{LT}\left(\widehat{\overline{A}}_{SVN-P_{AC}N_{AC}}\right)}, \, \S_{\delta_B \times \phi_{LT}\left(\widehat{\overline{A}}_{SVN-P_{AC}P_{AC}}\right)+\delta'_B \times \phi_{LT}\left(\widehat{\overline{A}}_{SVN-P_{AC}N_{AC}}\right)} \right], \\ \left(\begin{array}{l} \left(1-\left(1-\underset{\widehat{A}_{SVN}}{T}\left(\widehat{\overline{A}}_{SVN-P_{AC}P_{AC}}\right)\right)^{\delta_B}\left(1-\underset{\widehat{A}_{SVN}}{T}\left(\widehat{\overline{A}}_{SVN-P_{AC}N_{AC}}\right)\right)^{\delta'_B}\right), \\ \left(\mathsf{I}_{\widehat{A}_{SVN}}\left(\widehat{\overline{A}}_{SVN-P_{AC}P_{AC}}\right)\right)^{\delta_B}\left(\mathsf{I}_{\widehat{A}_{SVN}}\left(\widehat{\overline{A}}_{SVN-P_{AC}N_{AC}}\right)\right)^{\delta'_B}, \\ \left(\mathsf{F}_{\widehat{A}_{SVN}}\left(\widehat{\overline{A}}_{SVN-P_{AC}P_{AC}}\right)\right)^{\delta_B}\left(\mathsf{F}_{\widehat{A}_{SVN}}\left(\widehat{\overline{A}}_{SVN-P_{AC}N_{AC}}\right)\right)^{\delta'_B} \end{array} \right) \end{array} \right)
$$

The proof is completed. Further, for the relationships between any two expected losses, we explore the notions of expected values, which are stated below:

$$
Q_{EV}\left(Y_{EL}\left(\chi_{P_{AC}}\vert\,[\mathfrak{x}]\right)\right)
$$

$$
= \S \left(\begin{array}{l} \delta_B \times \phi_{LT}\left(\widehat{\overline{A}}_{SVN-P_{AC}P_{AC}}\right)+\delta_B \times \phi_{LT}\left(\widehat{\overline{A}}_{SVN-P_{AC}N_{AC}}\right)+ \\ \delta'_B \times \phi_{LT}\left(\widehat{\overline{A}}_{SVN-P_{AC}P_{AC}}\right)+\delta'_B \times \phi_{LT}\left(\widehat{\overline{A}}_{SVN-P_{AC}N_{AC}}\right) \end{array} \right)
$$

$$
\times \frac{\left(\begin{array}{l} \left(1-\left(1-\underset{\widehat{A}_{SVN}}{T}\left(\widehat{\overline{A}}_{SVN-P_{AC}P_{AC}}\right)\right)^{\delta_B}\left(1-\underset{\widehat{A}_{SVN}}{T}\left(\widehat{\overline{A}}_{SVN-P_{AC}N_{AC}}\right)\right)^{\delta'_B}\right)- \\ \left(\mathsf{I}_{\widehat{A}_{SVN}}\left(\widehat{\overline{A}}_{SVN-P_{AC}P_{AC}}\right)\right)^{\delta_B}\left(\mathsf{I}_{\widehat{A}_{SVN}}\left(\widehat{\overline{A}}_{SVN-P_{AC}N_{AC}}\right)\right)^{\delta'_B}- \\ \left(\mathsf{F}_{\widehat{A}_{SVN}}\left(\widehat{\overline{A}}_{SVN-P_{AC}P_{AC}}\right)\right)^{\delta_B}\left(\mathsf{F}_{\widehat{A}_{SVN}}\left(\widehat{\overline{A}}_{SVN-P_{AC}N_{AC}}\right)\right)^{\delta'_B} \end{array} \right)}{4}
$$

$$(37)$$

$$
Q_{EV}\left(Y_{EL}\left(\chi_{B_{AC}}\vert\,[\mathfrak{x}]\right)\right)
$$

$$
= \S \left(\begin{array}{l} \delta_B \times \phi_{LT}\left(\widehat{\overline{A}}_{SVN-B_{AC}P_{AC}}\right)+\delta_B \times \phi_{LT}\left(\widehat{\overline{A}}_{SVN-B_{AC}N_{AC}}\right)+ \\ \delta'_B \times \phi_{LT}\left(\widehat{\overline{A}}_{SVN-B_{AC}P_{AC}}\right)+\delta'_B \times \phi_{LT}\left(\widehat{\overline{A}}_{SVN-B_{AC}N_{AC}}\right) \end{array} \right)
$$

$$
\times \frac{\left(\begin{array}{l} \left(1-\left(1-\underset{\widehat{A}_{SVN}}{T}\left(\widehat{\overline{A}}_{SVN-B_{AC}P_{AC}}\right)\right)^{\delta_B}\left(1-\underset{\widehat{A}_{SVN}}{T}\left(\widehat{\overline{A}}_{SVN-B_{AC}N_{AC}}\right)\right)^{\delta'_B}\right)- \\ \left(\mathsf{I}_{\widehat{A}_{SVN}}\left(\widehat{\overline{A}}_{SVN-B_{AC}P_{AC}}\right)\right)^{\delta_B}\left(\mathsf{I}_{\widehat{A}_{SVN}}\left(\widehat{\overline{A}}_{SVN-B_{AC}N_{AC}}\right)\right)^{\delta'_B}- \\ \left(\mathsf{F}_{\widehat{A}_{SVN}}\left(\widehat{\overline{A}}_{SVN-B_{AC}P_{AC}}\right)\right)^{\delta_B}\left(\mathsf{F}_{\widehat{A}_{SVN}}\left(\widehat{\overline{A}}_{SVN-B_{AC}N_{AC}}\right)\right)^{\delta'_B} \end{array} \right)}{4}
$$

$$(38)$$

$$Q_{EV}\left(Y_{EL}\left(\chi_{N_{AC}}|\left[\mathfrak{x}\right]\right)\right)$$

$$= \S \begin{pmatrix} \delta_B \times \phi_{LT}\left(\widehat{\overline{A}}_{SVN-N_{AC}P_{AC}}\right) + \delta_B \times \phi_{LT}\left(\widehat{\overline{A}}_{SVN-N_{AC}N_{AC}}\right) + \\ \delta'_B \times \phi_{LT}\left(\widehat{\overline{A}}_{SVN-N_{AC}P_{AC}}\right) + \delta'_B \times \phi_{LT}\left(\widehat{\overline{A}}_{SVN-N_{AC}N_{AC}}\right) \end{pmatrix}$$

$$\times \frac{\begin{pmatrix} \left(\left(1-\left(1-T_{\widehat{A}_{SVN}}\left(\widehat{\overline{A}}_{SVN-N_{AC}P_{AC}}\right)\right)^{\delta_B}\left(1-T_{\widehat{A}_{SVN}}\left(\widehat{\overline{A}}_{SVN-N_{AC}N_{AC}}\right)\right)^{\delta'_B}\right) - \\ \left(I_{\widehat{A}_{SVN}}\left(\widehat{\overline{A}}_{SVN-N_{AC}P_{AC}}\right)\right)^{\delta_B}\left(I_{\widehat{A}_{SVN}}\left(\widehat{\overline{A}}_{SVN-N_{AC}N_{AC}}\right)\right)^{\delta'_B} - \\ \left(F_{\widehat{A}_{SVN}}\left(\widehat{\overline{A}}_{SVN-N_{AC}P_{AC}}\right)\right)^{\delta_B}\left(F_{\widehat{A}_{SVN}}\left(\widehat{\overline{A}}_{SVN-N_{AC}N_{AC}}\right)\right)^{\delta'_B} \end{pmatrix}}{4}$$

$$(39)$$

When the expected values if failed to find the relationships between any two expected losses. For these kinds of issues, we explore the notions of accuracy function, which is stated below:

$$G_{AF}\left(Y_{EL}\left(\chi_{P_{AC}}|\left[\mathfrak{x}\right]\right)\right)$$

$$= \S \begin{pmatrix} \delta_B \times \phi_{LT}\left(\widehat{\overline{A}}_{SVN-P_{AC}P_{AC}}\right) + \delta_B \times \phi_{LT}\left(\widehat{\overline{A}}_{SVN-P_{AC}N_{AC}}\right) + \\ \delta'_B \times \phi_{LT}\left(\widehat{\overline{A}}_{SVN-P_{AC}P_{AC}}\right) + \delta'_B \times \phi_{LT}\left(\widehat{\overline{A}}_{SVN-P_{AC}N_{AC}}\right) \end{pmatrix}$$

$$\times \frac{\begin{pmatrix} \left(\left(1-\left(1-T_{\widehat{A}_{SVN}}\left(\widehat{\overline{A}}_{SVN-P_{AC}P_{AC}}\right)\right)^{\delta_B}\left(1-T_{\widehat{A}_{SVN}}\left(\widehat{\overline{A}}_{SVN-P_{AC}N_{AC}}\right)\right)^{\delta'_B}\right) + \\ \left(I_{\widehat{A}_{SVN}}\left(\widehat{\overline{A}}_{SVN-P_{AC}P_{AC}}\right)\right)^{\delta_B}\left(I_{\widehat{A}_{SVN}}\left(\widehat{\overline{A}}_{SVN-P_{AC}N_{AC}}\right)\right)^{\delta'_B} + \\ \left(F_{\widehat{A}_{SVN}}\left(\widehat{\overline{A}}_{SVN-P_{AC}P_{AC}}\right)\right)^{\delta_B}\left(F_{\widehat{A}_{SVN}}\left(\widehat{\overline{A}}_{SVN-P_{AC}N_{AC}}\right)\right)^{\delta'_B} \end{pmatrix}}{4}$$

$$(40)$$

$$G_{AF}\left(Y_{EL}\left(\chi_{B_{AC}}|\,[\mathfrak{x}]\right)\right)$$

$$= \S \begin{pmatrix} \delta_B \times \phi_{LT}\left(\widehat{\overline{A}}_{SVN-B_{AC}P_{AC}}\right) + \delta_B \times \phi_{LT}\left(\widehat{\overline{A}}_{SVN-B_{AC}N_{AC}}\right) + \\ \delta'_B \times \phi_{LT}\left(\widehat{\overline{A}}_{SVN-B_{AC}P_{AC}}\right) + \delta'_B \times \phi_{LT}\left(\widehat{\overline{A}}_{SVN-B_{AC}N_{AC}}\right) \end{pmatrix}$$

$$\times \frac{\begin{pmatrix} \left(\left(1-\left(1-\underset{\widehat{A}_{SVN}}{T}\left(\widehat{\overline{A}}_{SVN-B_{AC}P_{AC}}\right)\right)^{\delta_B}\left(1-\underset{\widehat{A}_{SVN}}{T}\left(\widehat{\overline{A}}_{SVN-B_{AC}N_{AC}}\right)\right)^{\delta'_B}\right)+ \\ \left(\mathsf{I}_{\widehat{A}_{SVN}}\left(\widehat{\overline{A}}_{SVN-B_{AC}P_{AC}}\right)\right)^{\delta_B}\left(\mathsf{I}_{\widehat{A}_{SVN}}\left(\widehat{\overline{A}}_{SVN-B_{AC}N_{AC}}\right)\right)^{\delta'_B}+ \\ \left(\mathsf{F}_{\widehat{A}_{SVN}}\left(\widehat{\overline{A}}_{SVN-B_{AC}P_{AC}}\right)\right)^{\delta_B}\left(\mathsf{F}_{\widehat{A}_{SVN}}\left(\widehat{\overline{A}}_{SVN-B_{AC}N_{AC}}\right)\right)^{\delta'_B} \end{pmatrix}}{4} \tag{41}$$

$$G_{AF}\left(Y_{EL}\left(\chi_{D_{AC}}|\,[\mathfrak{x}]\right)\right)$$

$$= \S \begin{pmatrix} \delta_B \times \phi_{LT}\left(\widehat{\overline{A}}_{SVN-N_{AC}P_{AC}}\right) + \delta_B \times \phi_{LT}\left(\widehat{\overline{A}}_{SVN-N_{AC}N_{AC}}\right) + \\ \delta'_B \times \phi_{LT}\left(\widehat{\overline{A}}_{SVN-N_{AC}P_{AC}}\right) + \delta'_B \times \phi_{LT}\left(\widehat{\overline{A}}_{SVN-N_{AC}N_{AC}}\right) \end{pmatrix}$$

$$\times \frac{\begin{pmatrix} \left(\left(1-\left(1-\underset{\widehat{A}_{SVN}}{T}\left(\widehat{\overline{A}}_{SVN-N_{AC}P_{AC}}\right)\right)^{\delta_B}\left(1-\underset{\widehat{A}_{SVN}}{T}\left(\widehat{\overline{A}}_{SVN-N_{AC}N_{AC}}\right)\right)^{\delta'_B}\right)+ \\ \left(\mathsf{I}_{\widehat{A}_{SVN}}\left(\widehat{\overline{A}}_{SVN-N_{AC}P_{AC}}\right)\right)^{\delta_B}\left(\mathsf{I}_{\widehat{A}_{SVN}}\left(\widehat{\overline{A}}_{SVN-N_{AC}N_{AC}}\right)\right)^{\delta'_B}+ \\ \left(\mathsf{F}_{\widehat{A}_{SVN}}\left(\widehat{\overline{A}}_{SVN-N_{AC}P_{AC}}\right)\right)^{\delta_B}\left(\mathsf{F}_{\widehat{A}_{SVN}}\left(\widehat{\overline{A}}_{SVN-N_{AC}N_{AC}}\right)\right)^{\delta'_B} \end{pmatrix}}{4} \tag{42}$$

Based on the above analysis, we exposed the three-way decision rules which are discussed below:

$$P_{AC-1}: \text{When } Q_{EV}\left(Y_{EL}\left(\chi_{P_{AC}}|\,[\mathfrak{x}]\right)\right) < Q_{EV}\left(Y_{EL}\left(\chi_{B_{AC}}|\,[\mathfrak{x}]\right)\right) \vee Q_{EV}\left(Y_{EL}\left(\chi_{P_{AC}}|\,[\mathfrak{x}]\right)\right)$$

$$= Q_{EV}\left(Y_{EL}\left(\chi_{B_{AC}}|\,[\mathfrak{x}]\right)\right) \bigwedge G_{AF}\left(Y_{EL}\left(\chi_{P_{AC}}|\,[\mathfrak{x}]\right)\right) \leq G_{AF}\left(Y_{EL}\left(\chi_{B_{AC}}|\,[\mathfrak{x}]\right)\right) \bigwedge Q_{EV}\left(Y_{EL}\left(\chi_{P_{AC}}|\,[\mathfrak{x}]\right)\right)$$

$$< Q_{EV}\left(Y_{EL}\left(\chi_{N_{AC}}|\,[\mathfrak{x}]\right)\right) \vee Q_{EV}\left(Y_{EL}\left(\chi_{P_{AC}}|\,[\mathfrak{x}]\right)\right) = Q_{EV}\left(Y_{EL}\left(\chi_{N_{AC}}|\,[\mathfrak{x}]\right)\right) \bigwedge G_{AF}\left(Y_{EL}\left(\chi_{P_{AC}}|\,[\mathfrak{x}]\right)\right)$$

$$\leq G_{AF}\left(Y_{EL}\left(\chi_{N_{AC}}|\,[\mathfrak{x}]\right)\right), \text{ then } \mathfrak{x} \in POS\left(\mathcal{F}_P\right) \tag{43}$$

$$B_{AC-1} : \text{When } Q_{EV}\left(Y_{EL}\left(\chi_{B_{AC}}|\,[\mathfrak{x}]\right)\right) < Q_{EV}\left(Y_{EL}\left(\chi_{P_{AC}}|\,[\mathfrak{x}]\right)\right) \vee Q_{EV}\left(Y_{EL}\left(\chi_{B_{AC}}|\,[\mathfrak{x}]\right)\right)$$

$$= Q_{EV}\left(Y_{EL}\left(\chi_{P_{AC}}|\,[\mathfrak{x}]\right)\right) \bigwedge G_{AF}\left(Y_{EL}\left(\chi_{B_{AC}}|\,[\mathfrak{x}]\right)\right) \leq G_{AF}\left(Y_{EL}\left(\chi_{P_{AC}}|\,[\mathfrak{x}]\right)\right) \bigwedge Q_{EV}\left(Y_{EL}\left(\chi_{B_{AC}}|\,[\mathfrak{x}]\right)\right)$$

$$< Q_{EV}\left(Y_{EL}\left(\chi_{N_{AC}}|\,[\mathfrak{x}]\right)\right) \vee Q_{EV}\left(Y_{EL}\left(\chi_{B_{AC}}|\,[\mathfrak{x}]\right)\right) = Q_{EV}\left(Y_{EL}\left(\chi_{N_{AC}}|\,[\mathfrak{x}]\right)\right) \bigwedge G_{AF}\left(Y_{EL}\left(\chi_{B_{AC}}|\,[\mathfrak{x}]\right)\right)$$

$$\leq G_{AF}\left(Y_{EL}\left(\chi_{N_{AC}}|\,[\mathfrak{x}]\right)\right), \text{ then } \mathfrak{x} \in BUN\,(\mathcal{F}_P) \tag{44}$$

$$N_{AC-1} : \text{When } Q_{EV}\left(Y_{EL}\left(\chi_{N_{AC}}|\,[\mathfrak{x}]\right)\right) < Q_{EV}\left(Y_{EL}\left(\chi_{P_{AC}}|\,[\mathfrak{x}]\right)\right) \vee Q_{EV}\left(Y_{EL}\left(\chi_{N_{AC}}|\,[\mathfrak{x}]\right)\right)$$

$$= Q_{EV}\left(Y_{EL}\left(\chi_{P_{AC}}|\,[\mathfrak{x}]\right)\right) \bigwedge G_{AF}\left(Y_{EL}\left(\chi_{N_{AC}}|\,[\mathfrak{x}]\right)\right) \leq G_{AF}\left(Y_{EL}\left(\chi_{P_{AC}}|\,[\mathfrak{x}]\right)\right) \bigwedge Q_{EV}\left(Y_{EL}\left(\chi_{N_{AC}}|\,[\mathfrak{x}]\right)\right)$$

$$< Q_{EV}\left(Y_{EL}\left(\chi_{B_{AC}}|\,[\mathfrak{x}]\right)\right) \vee Q_{EV}\left(Y_{EL}\left(\chi_{N_{AC}}|\,[\mathfrak{x}]\right)\right) = Q_{EV}\left(Y_{EL}\left(\chi_{B_{AC}}|\,[\mathfrak{x}]\right)\right) \bigwedge G_{AF}\left(Y_{EL}\left(\chi_{N_{AC}}|\,[\mathfrak{x}]\right)\right)$$

$$\leq G_{AF}\left(Y_{EL}\left(\chi_{B_{AC}}|\,[\mathfrak{x}]\right)\right), \text{ then } \mathfrak{x} \in NEG\,(\mathcal{F}_P) \tag{45}$$

Therefore,

$$G_{AF}\left(\hat{A}_{SVN-\bar{\bar{A}}_{SVN-P_{AC}P_{AC}}}\right) < G_{AF}\left(\hat{A}_{SVN-\bar{\bar{A}}_{SVN-B_{AC}P_{AC}}}\right) < G_{AF}\left(\hat{A}_{SVN-\bar{\bar{A}}_{SVN-N_{AC}P_{AC}}}\right) \tag{46}$$

$$G_{AF}\left(\hat{A}_{SVN-\bar{\bar{A}}_{SVN-N_{AC}N_{AC}}}\right) < G_{AF}\left(\hat{A}_{SVN-\bar{\bar{A}}_{SVN-B_{AC}N_{AC}}}\right) < G_{AF}\left(\hat{A}_{SVN-\bar{\bar{A}}_{SVN-P_{AC}N_{AC}}}\right) \tag{47}$$

$\square$

4 Some Single-Valued Neutrosophic Uncertain Linguistic Generalized Maclaurin Symmetric Mean Operators

The purpose of this communication is to explore the GMSM operators based on S-VNULVs which are called single-valued neutrosophic uncertain linguistic generalized Maclaurin symmetric mean (S-VNULGMSM) operator and single-valued neutrosophic uncertain linguistic weighted generalized Maclaurin symmetric mean (S-VNULWGMSM) operator. The special cases of the explored operators are also presented.

Definition 8 For any group of S-VNULVs $\hat{A}_i$ $(i = 1, 2, \ldots, n)$, then the S-VNULGMSM operator is stated by

$$S - \text{VNULGMSM}^{(k,p_1,p_2,\ldots,p_k)} \left(\hat{A}_1, \hat{A}_2, \ldots, \hat{A}_n\right) = \left(\frac{\sum_{1 \le i_1 < \cdots < i_k \le i n} \prod_{j=1}^{k} \hat{A}_{i_j}^{p_j}}{C_n^k}\right)^{\frac{1}{p_1 + p_2 + \cdots + p_k}}$$

(48)

where $k = 1, 2, \ldots, n$, $p_1, p_2, \ldots, p_m \ge 0$, $(i_1, i_2, \ldots, i_k)$ and C_n^k is the coefficient binomial.

Theorem 1 For any group of S-VNULVs $\hat{A}_i = \left(\left[\$_{\phi(\mathfrak{s}_i)}, \$_{\phi(\mathfrak{s}_i)}\right], \left(T(\mathfrak{s}_i), I(\mathfrak{s}_i), F(\mathfrak{s}_i)\right)\right)$ $(i = 1, 2, \ldots, n)$, then we obtain the S-VNULGMSM operator:

$$S - \text{VNULGMSM}^{(k,p_1,p_2,\ldots,p_k)} \left(\hat{A}_1, \hat{A}_2, \ldots, \hat{A}_n\right)$$

$$= \left(\begin{array}{c} \left[\$\left[\left(\frac{\sum_{1 \le i_1 < \ldots < i_k \le i n} \prod_{j=1}^{k} \left(\phi_{(\mathfrak{s}_{i_j})}\right)^{p_j}}{C_n^k}\right)^{\frac{1}{p_1 + p_2 + \cdots + p_k}}\right], \$\left[\left(\frac{\sum_{1 \le i_1 < \ldots < i_k \le i n} \prod_{j=1}^{k} \left(\phi_{(\mathfrak{s}_{i_j})}\right)^{p_j}}{C_n^k}\right)^{\frac{1}{p_1 + p_2 + \cdots + p_k}}\right]\right], \\[4em] \left(\left(1 - \prod_{1 \le i_1 < \ldots < i_k \le i n} \left(1 - \prod_{j=1}^{k} \left(T_{(\mathfrak{s}_{i_j})}\right)^{p_j}\right)^{\frac{1}{C_n^k}}\right)\right)^{\frac{1}{p_1 + p_2 + \cdots + p_k}}, \\[3em] 1 - \left(\left(1 - \left(\prod_{1 \le i_1 < \ldots < i_k \le i n} \left(1 - \prod_{j=1}^{k} \left(1 - I_{(\mathfrak{s}_{i_j})}\right)^{p_j}\right)^{\frac{1}{C_n^k}}\right)\right)\right)^{\frac{1}{p_1 + \cdots + p_k}}, \\[3em] 1 - \left(\left(1 - \left(\prod_{1 \le i_1 < \ldots < i_k \le i n} \left(1 - \prod_{j=1}^{k} \left(1 - F_{(\mathfrak{s}_{i_j})}\right)^{p_j}\right)^{\frac{1}{C_n^k}}\right)\right)\right)^{\frac{1}{p_1 + \cdots + p_k}} \end{array}\right)$$

(49)

Proof We consider the Definition 4, such that

$$
\prod_{j=1}^{k} \hat{A}_{ij}^{p_j} = \left(\left[\S_{\Pi_{j=1}^{k} \left(\phi_{\left(\mathfrak{r}_{ij} \right)} \right)^{p_j}}, \S_{\Pi_{j=1}^{k} \left(\phi_{\left(\mathfrak{r}_{ij} \right)} \right)^{p_j}} \right], \left(\left(\prod_{j=1}^{k} \left(\underline{T}_{\left(\mathfrak{r}_{ij} \right)} \right) \right)^{p_j}, 1 - \prod_{j=1}^{k} \left(I_{\left(\mathfrak{r}_{ij} \right)} \right)^{p_j}, 1 - \prod_{j=1}^{k} \left(1 - \left(F_{\left(\mathfrak{r}_{ij} \right)} \right)^{p_j} \right) \right) \right)
$$

Then

$$
\sum_{1 \le i_1 < \dots < i_k \le n} \prod_{j=1}^{k} \hat{A}_{ij}^{p_j}
$$

$$
= \left(\left[\S_{\sum_{1 \le i_1 < \dots < i_k \le n} \Pi_{j=1}^{k} \left(\phi_{\left(\mathfrak{r}_{ij} \right)} \right)^{p_j}}, \S_{\sum_{1 \le i_1 < \dots < i_k \le n} \Pi_{j=1}^{k} \left(\phi_{\left(\mathfrak{r}_{ij} \right)} \right)^{p_j}} \right], \left(1 - \prod_{1 \le i_1 < \dots < i_k \le n} \left(1 - \prod_{j=1}^{k} \left(\underline{T}_{\left(\mathfrak{r}_{ij} \right)} \right)^{p_j} \right), \right. \right.
$$
$$
\left. \left. \prod_{1 \le i_1 < \dots < i_k \le n} \left(1 - \prod_{j=1}^{k} \left(1 - I_{\left(\mathfrak{r}_{ij} \right)} \right)^{p_j} \right), \right. \right.
$$
$$
\left. \left. \prod_{1 \le i_1 < \dots < i_k \le n} \left(1 - \prod_{j=1}^{k} \left(1 - F_{\left(\mathfrak{r}_{ij} \right)} \right)^{p_j} \right) \right) \right)
$$

Further,

$$\frac{\sum_{1\leq i_1<\ldots<i_k\leq n}\prod_{j=1}^k \hat{A}_{i_j}^{p_j}}{C_n^k}$$

$$= \left(\left[\frac{\S_{\sum_{1\leq i_1<\ldots<i_k\leq n}\prod_{j=1}^k \left(\phi_{(\mathfrak{r}_{i_j})}\right)^{p_j}}}{C_n^k}, \frac{\S_{\sum_{1\leq i_1<\ldots<i_k\leq n}\prod_{j=1}^k \left(\phi_{(\mathfrak{r}_{i_j})}\right)^{p_j}}}{C_n^k} \right], \left(1-\left(\prod_{1\leq i_1<\ldots<i_k\leq n}\left(1-\prod_{j=1}^k \left(\mathrm{T}_{(\mathfrak{r}_{i_j})}\right)^{p_j}\right)\right)^{\frac{1}{C_n^k}}, \right. \right.$$

$$\left.\left. \left(\prod_{1\leq i_1<\ldots<i_k\leq n}\left(1-\prod_{j=1}^k \left(1-\mathrm{I}_{(\mathfrak{r}_{i_j})}\right)^{p_j}\right)\right)^{\frac{1}{C_n^k}}, \right.\right.$$

$$\left.\left. \left(\prod_{1\leq i_1<\ldots<i_k\leq n}\left(1-\prod_{j=1}^k \left(1-\mathrm{F}_{(\mathfrak{r}_{i_j})}\right)^{p_j}\right)\right)^{\frac{1}{C_n^k}} \right)\right)$$

Therefore,

$$\mathrm{S-VNULGMSM}^{(k,p_1,p_2,\ldots,p_k)}\left(\hat{A}_1,\hat{A}_2,\ldots,\hat{A}_n\right)$$

$$= \left(\left[\S\left[\left(\frac{\sum_{1\leq i_1<\ldots<i_k\leq n}\prod_{j=1}^k \left(\phi_{(\mathfrak{r}_{i_j})}\right)^{p_j}}{C_n^k}\right)^{\frac{1}{p_1+p_2+\ldots+p_k}}\right], \S\left[\left(\frac{\sum_{1\leq i_1<\ldots<i_k\leq n}\prod_{j=1}^k \left(\phi_{(\mathfrak{r}_{i_j})}\right)^{p_j}}{C_n^k}\right)^{\frac{1}{p_1+p_2+\ldots+p_k}}\right] \right], \right.$$

$$\left(\left(1-\prod_{1\leq i_1<\ldots<i_k\leq n}\left(1-\prod_{j=1}^k \left(\mathrm{T}_{(\mathfrak{r}_{i_j})}\right)^{p_j}\right)^{\frac{1}{C_n^k}}\right)\right)^{\frac{1}{p_1+p_2+\ldots+p_k}},$$

$$1-\left(\left(1-\left(\prod_{1\leq i_1<\ldots<i_k\leq n}\left(1-\prod_{j=1}^k \left(1-\mathrm{I}_{(\mathfrak{r}_{i_j})}\right)^{p_j}\right)^{\frac{1}{C_n^k}}\right)\right)\right)^{\frac{1}{p_1+p_2+\ldots+p_k}},$$

$$\left. 1-\left(\left(1-\left(\prod_{1\leq i_1<\ldots<i_k\leq n}\left(1-\prod_{j=1}^k \left(1-\mathrm{F}_{(\mathfrak{r}_{i_j})}\right)^{p_j}\right)^{\frac{1}{C_n^k}}\right)\right)\right)^{\frac{1}{p_1+p_2+\ldots+p_k}} \right)$$

Hence the proof is completed. $\square$

Further, we examine some properties of the explored approaches like idempotency, commutativity, monotonicity, and boundedness which are stated below:

Property 1 For any group of S-VNULVs $\hat{A}_i = \left(\left[s_{\phi(\varsigma_i)}, s_{\phi(\varsigma_i)} \right], \left(T(\varsigma_i), I(\varsigma_i), F(\varsigma_i) \right) \right)$ $(i = 1, 2, \ldots, n)$, if $\hat{A}_i = \hat{A}$, $(i = 1, 2, \ldots, n)$, then

$$S - VNULGMSM^{(k, p_1, p_2, \ldots, p_k)} \left(\hat{A}, \hat{A}, \ldots, \hat{A} \right) = \hat{A}. \tag{50}$$

Proof If $\hat{A}_i = \left(\left[s_{\phi(\varsigma_i)}, s_{\phi(\varsigma_i)} \right], \left(T(\varsigma_i), I(\varsigma_i), F(\varsigma_i) \right) \right) = \hat{A} = \left(\left[s_{\phi(\varsigma)}, s_{\phi(\varsigma)} \right], \left(T(\varsigma), I(\varsigma), F(\varsigma) \right) \right)$, then

$$S - VNULGMSM^{(k, p_1, p_2, \ldots, p_k)} \left(\hat{A}, \hat{A}, \ldots, \hat{A} \right) =$$

$$\left(\left[s_{\left[\left(\frac{\sum_{1 \leq i_1 < \ldots < i_k \leq i_n} \prod_{j=1}^k \left(\phi_{(\varsigma)} \right)^{p_j}}{C_n^k} \right)^{\frac{1}{p_1 + p_2 + \ldots + p_k}} \right]}, s_{\left[\left(\frac{\sum_{1 \leq i_1 < \ldots < i_k \leq i_n} \prod_{j=1}^k \left(\phi_{(\varsigma)} \right)^{p_j}}{C_n^k} \right)^{\frac{1}{p_1 + p_2 + \ldots + p_k}} \right]} \right], \right.$$

$$\left(\left(1 - \prod_{1 \leq i_1 < \ldots < i_k \leq i_n} \left(1 - \prod_{j=1}^k \left(T_{(\varsigma)} \right)^{p_j} \right)^{\frac{1}{C_n^k}} \right)^{\frac{1}{p_1 + p_2 + \ldots + p_k}}, \right.$$

$$1 - \left(\left(1 - \left(\prod_{1 \leq i_1 < \ldots < i_k \leq i_n} \left(1 - \prod_{j=1}^k \left(1 - I_{(\varsigma)} \right)^{p_j} \right) \right)^{\frac{1}{C_n^k}} \right) \right)^{\frac{1}{p_1 + p_2 + \ldots + p_k}},$$

$$\left. 1 - \left(\left(1 - \left(\prod_{1 \leq i_1 < \ldots < i_k \leq i_n} \left(1 - \prod_{j=1}^k \left(1 - F_{(\varsigma)} \right)^{p_j} \right) \right)^{\frac{1}{C_n^k}} \right) \right)^{\frac{1}{p_1 + p_2 + \ldots + p_k}} \right)$$

$$
= \left(\left[\sqrt[\S]{\left[\left(\frac{\sum_{1 \leq i_1 < \ldots < i_k \leq i_n} \left(\phi_{(\mathfrak{r})} \right)^{p_1 + p_2 + \ldots + p_k}}{C_n^k} \right)^{\frac{1}{p_1 + p_2 + \ldots + p_k}} \right]}, \sqrt[\S]{\left[\left(\frac{\sum_{1 \leq i_1 < \ldots < i_k \leq i_n} \left(\phi_{(\mathfrak{r})} \right)^{p_1 + p_2 + \ldots + p_k}}{C_n^k} \right)^{\frac{1}{p_1 + p_2 + \ldots + p_k}} \right]} \right],
\right.
$$

$$
\left.
\left(\begin{array}{c}
\left(\left(1 - \prod_{1 \leq i_1 < \ldots < i_k \leq i_n} \left(1 - \left(\underline{T}_{(\mathfrak{r})} \right)^{p_1 + p_2 + \ldots + p_k} \right)^{\frac{1}{C_n^k}} \right) \right)^{\frac{1}{p_1 + p_2 + \ldots + p_k}}, \\[12pt]
1 - \left(\left(1 - \left(\prod_{1 \leq i_1 < \ldots < i_k \leq i_n} \left(1 - \left(1 - I_{(\mathfrak{r})} \right)^{p_1 + p_2 + \ldots + p_k} \right)^{\frac{1}{C_n^k}} \right) \right) \right)^{\frac{1}{p_1 + p_2 + \ldots + p_k}}, \\[12pt]
1 - \left(\left(1 - \left(\prod_{1 \leq i_1 < \ldots < i_k \leq i_n} \left(1 - \left(1 - F_{(\mathfrak{r})} \right)^{p_1 + p_2 + \ldots + p_k} \right)^{\frac{1}{C_n^k}} \right) \right) \right)^{\frac{1}{p_1 + p_2 + \ldots + p_k}}
\end{array} \right)
\right)
$$

$$
= \left(\left[\sqrt[\S]{\left[\left(\frac{C_n^k \left(\phi_{(\mathfrak{r})} \right)^{p_1 + p_2 + \ldots + p_k}}{C_n^k} \right)^{\frac{1}{p_1 + p_2 + \ldots + p_k}} \right]}, \sqrt[\S]{\left[\left(\frac{C_n^k \left(\phi_{(\mathfrak{r})} \right)^{p_1 + p_2 + \ldots + p_k}}{C_n^k} \right)^{\frac{1}{p_1 + p_2 + \ldots + p_k}} \right]} \right],
\right.
$$

$$
\left.
\left(\begin{array}{c}
\left(\left(1 - \left(1 - \left(\underline{T}_{(\mathfrak{r})} \right)^{p_1 + p_2 + \ldots + p_k} \right)^{\frac{1}{C_n^k}} \right) \right)^{\frac{1}{p_1 + p_2 + \ldots + p_k}}, \\[12pt]
1 - \left(1 - \left(\left(1 - \left(1 - I_{(\mathfrak{r})} \right)^{p_1 + p_2 + \ldots + p_k} \right)^{\frac{1}{C_n^k}} \right) \right)^{\frac{1}{p_1 + p_2 + \ldots + p_k}}, \\[12pt]
1 - \left(1 - \left(\left(1 - \left(1 - F_{(\mathfrak{r})} \right)^{p_1 + p_2 + \ldots + p_k} \right)^{\frac{1}{C_n^k}} \right) \right)^{\frac{1}{p_1 + p_2 + \ldots + p_k}}
\end{array} \right)
\right)
$$

$$
= \left(
\begin{bmatrix}
{}^\$\left[\left(\left(\phi_{(\mathfrak{x})}\right)^{p_1+p_2+\ldots+p_k}\right)^{\frac{1}{p_1+p_2+\ldots+p_k}}\right], {}^\$\left[\left(\left(\phi_{(\mathfrak{x})}\right)^{p_1+p_2+\ldots+p_k}\right)^{\frac{1}{p_1+p_2+\ldots+p_k}}\right]
\end{bmatrix},
\begin{pmatrix}
\left(1-\left(1-\left(\underline{T}_{(\mathfrak{x})}\right)^{p_1+p_2+\ldots+p_k}\right)\right)^{\frac{1}{p_1+p_2+\ldots+p_k}}, \\
1-\left(\left(1-\left(1-\left(1-I_{(\mathfrak{x})}\right)^{p_1+p_2+\ldots+p_k}\right)\right)\right)^{\frac{1}{p_1+p_2+\ldots+p_k}} \\
1-\left(\left(1-\left(1-\left(1-F_{(\mathfrak{x})}\right)^{p_1+p_2+\ldots+p_k}\right)\right)\right)^{\frac{1}{p_1+p_2+\ldots+p_k}}
\end{pmatrix}
\right)
$$

$$
= \left(\left[s_{\phi(\mathfrak{x})}, s_{\phi(\mathfrak{x})}\right], \left(\underline{T}_{(\mathfrak{x})}, I_{(\mathfrak{x})}, F_{(\mathfrak{x})}\right)\right) = \hat{A}
$$

Hence the proof is completed. $\square$

Property 2 For any group of S-VNULVs $\hat{A}_i = \left(\left[s_{\phi(\mathfrak{x}_i)}, s_{\phi(\mathfrak{x}_i)}\right], \left(\underline{T}(\mathfrak{x}_i), I(\mathfrak{x}_i), F(\mathfrak{x}_i)\right)\right)$ $(i = 1, 2, \ldots, n)$, if $\left(\hat{A}_1, \hat{A}_2, \ldots, \hat{A}_n\right)$ is any substitution of $\left(\hat{A}'_1, \hat{A}'_2, \ldots, \hat{A}'_n\right)$, then

$$
\text{S} - \text{VNULGMSM}^{(k, p_1, p_2, \ldots, p_k)}\left(\hat{A}'_1, \hat{A}'_2, \ldots, \hat{A}'_n\right)
$$
$$
= \text{S} - \text{VNULGMSM}^{(k, p_1, p_2, \ldots, p_k)}\left(\hat{A}_1, \hat{A}_2, \ldots, \hat{A}_n\right). \tag{51}
$$

Proof By hypothesis given that $\left(\hat{A}_1, \hat{A}_2, \ldots, \hat{A}_n\right)$ is any substitution of $\left(\hat{A}'_1, \hat{A}'_2, \ldots, \hat{A}'_n\right)$, then

$$S - \text{VNULGMSM}^{(k,p_1,p_2,\dots,p_k)}\left(\hat{A}'_1, \hat{A}'_2, \dots, \hat{A}'_n\right)$$

$$= \left(\frac{\sum_{1 \le i_1 < \cdots < i_k \le n} \prod_{j=1}^k \hat{A}'_{i_j}{}^{p_j}}{C_n^k}\right)^{\frac{1}{p_1+p_2+\cdots+p_k}}$$

$$= \left(\frac{\sum_{1 \le i_1 < \cdots < i_k \le n} \prod_{j=1}^k \hat{A}_{i_j}^{p_j}}{C_n^k}\right)^{\frac{1}{p_1+p_2+\cdots+p_k}}$$

$$= S - \text{VNULGMSM}^{(k,p_1,p_2,\dots,p_k)}\left(\hat{A}_1, \hat{A}_2, \dots, \hat{A}_n\right).$$

Hence the proof is completed. $\square$

Property 3 For any group of S-VNULVs $\hat{A}_i = \left(\left[\$_{\phi(\varsigma_i)}, \$_{\phi(\varsigma_i)}\right], \left(\mathcal{T}(\varsigma_i), \mathsf{I}(\varsigma_i),\right.\right.$ $\left.\left. F(\varsigma_i)\right)\right)$ $(i = 1, 2, \dots, n)$, where $\left(\hat{A}_1, \hat{A}_2, \dots, \hat{A}_n\right)$ and $\left(\hat{A}'_1, \hat{A}'_2, \dots, \hat{A}'_n\right)$ are two collections of S-VNULVs, if $\hat{A}_i \le \hat{A}'_i$, then

$$\begin{aligned} &S - \text{VNULGMSM}^{(k,p_1,p_2,\dots,p_k)}\left(\hat{A}_1, \hat{A}_2, \dots, \hat{A}_n\right) \\ &\le S - \text{VNULGMSM}^{(k,p_1,p_2,\dots,p_k)}\left(\hat{A}'_1, \hat{A}'_2, \dots, \hat{A}'_n\right). \end{aligned} \tag{52}$$

Proof By hypothesis, given that $\hat{A}_i \le \hat{A}'_i$, then

$$\prod_{j=1}^k \hat{A}_{i_j}^{p_j} \le \prod_{j=1}^k \hat{A}'_{i_j}{}^{p_j} \implies \frac{\sum_{1 \le i_1 < \cdots < i_k \le n} \prod_{j=1}^k \hat{A}_{i_j}^{p_j}}{C_n^k} \le \frac{\sum_{1 \le i_1 < \cdots < i_k \le n} \prod_{j=1}^k \hat{A}'_{i_j}{}^{p_j}}{C_n^k}$$

$$\implies \left(\frac{\sum_{1 \le i_1 < \cdots < i_k \le n} \prod_{j=1}^k \hat{A}_{i_j}^{p_j}}{C_n^k}\right)^{\frac{1}{p_1+p_2+\cdots+p_k}} \le \left(\frac{\sum_{1 \le i_1 < \cdots < i_k \le n} \prod_{j=1}^k \hat{A}'_{i_j}{}^{p_j}}{C_n^k}\right)^{\frac{1}{p_1+p_2+\cdots+p_k}}$$

Therefore,

$$\begin{aligned} S - \text{VNULGMSM}^{(k,p_1,p_2,\dots,p_k)}\left(\hat{A}_1, \hat{A}_2, \dots, \hat{A}_n\right) \le S \\ - \text{VNULGMSM}^{(k,p_1,p_2,\dots,p_k)}\left(\hat{A}'_1, \hat{A}'_2, \dots, \hat{A}'_n\right). \end{aligned}$$

Hence the proof is completed. $\square$

Property 4 For any group of S-VNULVs $\hat{A}_i = \left(\left[\$_{\phi(\varsigma_i)}, \$_{\phi(\varsigma_i)}\right], \left(\mathcal{T}(\varsigma_i), \mathsf{I}(\varsigma_i), F(\varsigma_i)\right)\right)$ $(i = 1, 2, \dots, n)$, then

$$\min\left(\hat{A}_1, \hat{A}_2, \ldots, \hat{A}_n\right) \leq \text{PIULGMSM}^{(k, p_1, p_2, \ldots, p_k)}\left(\hat{A}_1, \hat{A}_2, \ldots, \hat{A}_n\right)$$

$$\leq \max\left(\hat{A}_1, \hat{A}_2, \ldots, \hat{A}_n\right). \tag{53}$$

Proof Consider that

$\hat{A}^- = \min\left(\hat{A}_1, \hat{A}_2, \ldots, \hat{A}_n\right)$, $\hat{A}^+ = \max\left(\hat{A}_1, \hat{A}_2, \ldots, \hat{A}_n\right)$, and by properties 1 and 3,

$\text{S} - \text{VNULGMSM}^{(k, p_1, p_2, \ldots, p_k)}\left(\hat{A}_1, \hat{A}_2, \ldots, \hat{A}_n\right)$

$\geq \text{S} - \text{VNULGMSM}^{(k, p_1, p_2, \ldots, p_k)}\left(\hat{A}^-, \hat{A}^-, \ldots, \hat{A}^- = \hat{A}^-\right)$ and

$\text{S}-\text{VNULGMSM}^{(k, p_1, p_2, \ldots, p_k)}\left(\hat{A}_1, \hat{A}_2, \ldots, \hat{A}_n\right) \leq \text{S}-\text{VNULGMSM}^{(k, p_1, p_2, \ldots, p_k)}$

$\left(\hat{A}^+, \hat{A}^+, \ldots, \hat{A}^+ = \hat{A}^+\right)$. Thus the proof is done. Therefore we get

$$\min\left(\hat{A}_1, \hat{A}_2, \ldots, \hat{A}_n\right) \leq \text{PIULGMSM}^{(k, p_1, p_2, \ldots, p_k)}\left(\hat{A}_1, \hat{A}_2, \ldots, \hat{A}_n\right)$$

$$\leq \max\left(\hat{A}_1, \hat{A}_2, \ldots, \hat{A}_n\right).$$

Hence the proof is completed. $\square$

Definition 9 For any group of S-VNULVs $\hat{A}_i$ $(i = 1, 2, \ldots, n)$, then the S-VNULWGMSM operator is stated by

$$\text{S} - \text{VNULWGMSM}^{(k, p_1, p_2, \ldots, p_k)}\left(\hat{A}_1, \hat{A}_2, \ldots, \hat{A}_n\right)$$

$$= \left(\frac{\sum_{1 \leq i_1 < \cdots < i_k \leq n} \prod_{j=1}^{k}\left(\omega_i \hat{A}_{i_j}\right)^{p_j}}{C_n^k}\right)^{\frac{1}{p_1 + p_2 + \cdots + p_k}} \tag{54}$$

where $k = 1, 2, \ldots, n$, $p_1, p_2, \ldots, p_m \geq 0$, $(i_1, i_2, \ldots, i_k)$ and C_n^k is coefficient binomial. Further, $\omega = (\omega_1, \omega_2, \ldots, \omega_n)^T$ is the weight vector of $\hat{\bar{a}}_i (i = 1, 2, \ldots, n)$ such that $\omega_i \in [0, 1]$ is the weight vector of $\hat{\bar{a}}_i$, and $\sum_{i=1}^{n} \omega_i = 1$.

Theorem 2 For any group of S-VNULVs
$\hat{A}_i = \left(\left[\$_{\phi(\mathfrak{x}_i)}, \$_{\phi(\mathfrak{x}_i)}\right], \left(\underline{T}(\mathfrak{x}_i), \mathsf{I}(\mathfrak{x}_i), F(\mathfrak{x}_i)\right)\right)$ $(i = 1, 2, \ldots, n)$, then we obtain the S-VNULWGMSM operator:

$$\text{S}-\text{VNULWGMSM}^{(k,p_1,p_2,\dots,p_k)}\left(\hat{A}_1,\hat{A}_2,\dots,\hat{A}_n\right)$$

$$=\begin{pmatrix}\left[\left[\left(\dfrac{\sum_{1\le i_1<\dots<i_k\le i_n}\prod_{j=1}^{k}\left(m_j\,\phi_{(\zeta_{i_j})}\right)^{p_j}}{c_n^k}\right)^{\frac{1}{p_1+p_2+\dots+p_k}}\right]^{\$}\cdot\left[\left(\dfrac{\sum_{1\le i_1<\dots<i_k\le i_n}\prod_{j=1}^{k}\left(m_j\,\phi_{(\zeta_{i_j})}\right)^{p_j}}{c_n^k}\right)^{\frac{1}{p_1+p_2+\dots+p_k}}\right]^{\$}\right],\\ \left(\left(\left(1-\prod_{1\le i_1<\dots<i_k\le i_n}\left(1-\prod_{j=1}^{k}\left(1-\left(\left(1-\underline{\underline{T}}_{(\zeta_{i_j})}\right)\right)^{m_j}\right)^{p_j}\right)^{\frac{1}{c_n^k}}\right)\right)^{\frac{1}{p_1+p_2+\dots+p_k}}\right),\\ 1-\left(\left(1-\left(\prod_{1\le i_1<\dots<i_k\le i_n}\left(1-\prod_{j=1}^{k}\left(1-\left(I_{(\zeta_{i_j})}\right)^{m_j}\right)^{p_j}\right)^{\frac{1}{c_n^k}}\right)\right)^{\frac{1}{p_1+p_2+\dots+p_k}}\right),\\ 1-\left(\left(1-\left(\prod_{1\le i_1<\dots<i_k\le i_n}\left(1-\prod_{j=1}^{k}\left(1-\left(F_{(\zeta_{i_j})}\right)^{m_j}\right)^{p_j}\right)^{\frac{1}{c_n^k}}\right)\right)^{\frac{1}{p_1+p_2+\dots+p_k}}\right)\end{pmatrix} \tag{55}$$

Proof Straightforward (the proof of this theorem is similar to the proof of the Theorem 1) □

Further, we examine some properties of the explored approaches like idempotency, commutativity, monotonicity, and boundedness which are stated below:

Property 5 For any group of S-VNULVs $\hat{A}_i = \left(\left[\$_{\phi(\zeta_i)},\$_{\phi(\zeta_i)}\right],\left(\underline{\underline{T}}(\zeta_i),I(\zeta_i),F(\zeta_i)\right)\right)$ $(i=1,2,\dots,n)$, if $\hat{A}_i = \hat{A},(i=1,2,\dots,n)$, then

$$\text{S}-\text{VNULGMSM}^{(k,p_1,p_2,\dots,p_k)}\left(\hat{A},\hat{A},\dots,\hat{A}\right)=\hat{A}. \tag{56}$$

Proof Straightforward (the proof of this theorem is similar to the proof of the Property 1) □

Property 6 For any group of S-VNULVs $\hat{A}_i = \left(\left[\$_{\phi(\zeta_i)},\$_{\phi(\zeta_i)}\right],\left(\underline{\underline{T}}(\zeta_i),I(\zeta_i),F(\zeta_i)\right)\right)$ $(i=1,2,\dots,n)$, if $\left(\hat{A}_1,\hat{A}_2,\dots,\hat{A}_n\right)$ is any substitution of $\left(\hat{A}'_1,\hat{A}'_2,\dots,\hat{A}'_n\right)$, then

$$\text{S}-\text{VNULGMSM}^{(k,p_1,p_2,\dots,p_k)}\left(\hat{A}'_1,\hat{A}'_2,\dots,\hat{A}'_n\right)=\text{S}-\text{VNULGMSM}^{(k,p_1,p_2,\dots,p_k)}$$
$$\left(\hat{A}_1,\hat{A}_2,\dots,\hat{A}_n\right). \tag{57}$$

Proof Straightforward (the proof of this theorem is similar to the proof of the Property 2) □

Property 7 For any group of S-VNULVs $\hat{A}_i = \left(\left[\$_{\phi(\zeta_i)},\$_{\phi(\zeta_i)}\right],\right.$ $\left.\left(\underline{\underline{T}}(\zeta_i),I(\zeta_i),F(\zeta_i)\right)\right)$ $(i=1,2,\dots,n)$, where $\left(\hat{A}_1,\hat{A}_2,\dots,\hat{A}_n\right)$ and

$\left(\hat{A}'_1, \hat{A}'_2, \ldots, \hat{A}'_n\right)$ are two collections of S-VNULVs, if $\hat{A}_i \leq \hat{A}'_i$, then

$$\text{S}-\text{VNULGMSM}^{(k, p_1, p_2, \ldots, p_k)}\left(\hat{A}_1, \hat{A}_2, \ldots, \hat{A}_n\right) \leq \text{S}-\text{VNULGMSM}^{(k, p_1, p_2, \ldots, p_k)}$$

$$\left(\hat{A}'_1, \hat{A}'_2, \ldots, \hat{A}'_n\right).$$

(58)

Proof Straightforward (the proof of this theorem is similar to the proof of the Property 3) □

Property 8 For any group of S-VNULVs $\hat{A}_i = \left(\left[\S_{\phi(\mathfrak{r}_i)}, \S_{\phi(\mathfrak{r}_i)}\right],\right.$ $\left.\left(\underline{T}\left(\mathfrak{r}_i\right), \mathsf{I}\left(\mathfrak{r}_i\right), F\left(\mathfrak{r}_i\right)\right)\right) (i = 1, 2, \ldots, n)$, then

$$\min\left(\hat{A}_1, \hat{A}_2, \ldots, \hat{A}_n\right) \leq \text{PIULGMSM}^{(k, p_1, p_2, \ldots, p_k)}\left(\hat{A}_1, \hat{A}_2, \ldots, \hat{A}_n\right)$$

$$\leq \max\left(\hat{A}_1, \hat{A}_2, \ldots, \hat{A}_n\right).$$

(59)

Proof Straightforward (the proof of this theorem is similar to the proof of the Property 4) □

5 Three-Way Decisions Based on the S-VULVs

By using the LF based on S-VULVs, we explore algorithm which contains the following steps for examining the 3WD rules. Basically, these explorations that Combination the LF based on S-VULVs and the 3WD for S-VULVs are also presented in this section. For solving this algorithm, we have revised the information about actions and state and related to probability vectors, which is stated by $A_{AC} = \left\{\chi_{PAC}, \chi_{BAC}, \chi_{NAC}\right\}$, $\Omega_S = \{\mathcal{F}_B, \sim \mathcal{F}_B\}$ and $\mathcal{D} = \left\{Pr\left(\mathcal{F}_B | [\mathfrak{r}]\right), Pr\left(\sim \mathcal{F}_B | [\mathfrak{r}]\right)\right\}$ with a condition that is $Pr\left(\mathcal{F}_B | [\mathfrak{r}]\right) + Pr\left(\sim \mathcal{F}_B | [\mathfrak{r}]\right) = 1$. Then the steps of the algorithm are summarized as follows:

Step 1: By using the equations of Table 2, we examine the LFs.
Step 2: By using Eq. (55), we aggregate the decision matrix which is constructed by decision experts.
Step 3: By using Eqs. (34), (35), and (36), we examine the expected losses $Y_{EL}\left(\chi_{jAC} | [\mathfrak{r}]\right), j = P, B, N$.
Step 4: By using Eqs. (37), (38), and (39), we examine the expected values.
When the expected values if failed to find the relationships between any two expected losses. For these kinds of issues, we explore the notions of accuracy function, which are stated in Eqs. (40), (41), and (42).

Step 5: By using Eq. (43), Eq. (44), and Eq. (45), we examine the three-way decision
   rules.
Step 6: The end.

Example 1 h

We consider the business of a large state-owned company which is to give and sell
nonferrous items, coping to invest in world mining productions to explore its main
business scopes. After, in-large investigation, four alternative and three decision
experts, whose representations are as follows: $\{\Gamma_1, \Gamma_2, \Gamma_3\}$ and $D_{DE-k}(k = 1, 2, 3)$
with their weight vectors $\Omega_{WV} = (0.3, 0.1, 0.6)^T$. To address these problems
effectively, we choose the values of condition probability as $Pr\left(\mathcal{F}_B | [\Gamma_j]\right) =$
$0.3, j = 1, 2, 3, 4$; the decision results are discussed below:

Then the steps of the algorithm are summarized as follows:

Step 1: By using the equations of Table 2, we examine the LFs of Tables 3, 4, and
   5, which are stated below.
Step 2: By using Eq. (55), we aggregate the decision matrix which is
   constructed by decision experts in *Step 1*, for $K_{SC} = 3$, and the values of
   $\alpha_{SC-1} = \alpha_{SC-2} = \alpha_{SC-3} = 1$; the aggregated values are discussed in Table 6.
Step 3: By using Eqs. (34), (35), and (36), we examine the expected losses
   $Y_{EL}\left(\chi_{jAC} | [\Gamma]\right), j = P, B, N$, for $\delta_{B-j} = 0.4, j = 1, 2, 3$, and $q_{SC} = 1$; the
   different actions are expressed and discussed in Table 7.
Step 4: By using Eqs. (37), (38), and (39), we examine the expected values, which
   are stated in Table 8.

When the expected values if failed to find the relationships between any two
   expected losses. For these kinds of issues, we explore the notions of accuracy
   function, which is stated in Table 9.
Step 5: By using Eqs. (43), (44), and (45), we examine the three-way decision rules
   which are discussed in Table 10.
Step 6: The end.

From the above analysis, we obtain the result that these all alternatives are
belonging to positive opinions, which is P_{AC-1}.

Table 3 Decision matrix which is represented by risk and given by $\mathcal{D}_1$

Γ_1			Γ_2	
	$\mathcal{F}_B$	$\sim \mathcal{F}_B$	$\mathcal{F}_B$	$\sim \mathcal{F}_B$
$\chi_{P_{AC}}$	$([s_1, s_2], (0.6, 0.4, 0.4))$	$([s_4, s_5], (0.5, 0.4, 0.4))$	$([s_0, s_1], (0.7, 0.1, 0.1))$	$([s_4, s_5], (0.6, 0.2, 0.2))$
$\chi_{B_{AC}}$	$([s_3, s_4], (0.7, 0.2, 0.2))$	$([s_2, s_3], (0.6, 0.4, 0.4))$	$([s_3, s_5], (0.6, 0.3, 0.3))$	$([s_2, s_3], (0.8, 0.2, 0.2))$
$\chi_{N_{AC}}$	$([s_4, s_5], (0.8, 0.1, 0.1))$	$([s_1, s_2], (0.7, 0.2, 0.2))$	$([s_4, s_5], (0.7, 0.2, 0.2))$	$([s_1, s_2], (0.7, 0.3, 0.3))$
	Γ_3		Γ_4	
$\chi_{P_{AC}}$	$([s_0, s_1], (0.7, 0.3, 0.3))$	$([s_4, s_5], (0.5, 0.4, 0.4))$	$([s_1, s_2], (0.6, 0.4, 0.4))$	$([s_4, s_5], (0.6, 0.2, 0.2))$
$\chi_{B_{AC}}$	$([s_2, s_3], (0.7, 0.2, 0.2))$	$([s_2, s_3], (0.7, 0.3, 0.3))$	$([s_2, s_3], (0.7, 0.3, 0.3))$	$([s_2, s_3], (0.8, 0.1, 0.1))$
$\chi_{N_{AC}}$	$([s_3, s_4], (0.8, 0.1, 0.1))$	$([s_1, s_2], (0.8, 0.2, 0.2))$	$([s_3, s_4], (0.8, 0.1, 0.1))$	$([s_1, s_2], (0.7, 0.2, 0.2))$

Table 4 Decision matrix which is represented by risk and given by $\mathcal{D}_2$

$\mathfrak{r}_1$			$\mathfrak{r}_2$	
	$\mathcal{F}_B$	$\sim \mathcal{F}_{NB}$	$\mathcal{F}_B$	$\sim \mathcal{F}_B$
$\chi_{P_{AC}}$	$([s_0,s_1],$ $(0.8,0.1,0.5))$	$([s_4,s_5],(0.6,0.2,0.5))$	$([s_0,s_1],(0.5,0.4,0.3))$	$([s_4,s_5],(0.7,0.3,0.2))$
$\chi_{B_{AC}}$	$([s_1,s_2],$ $(0.81,0.11,0.6))$	$([s_3,s_4],(0.61,0.21,0.3))$	$([s_1,s_2],(0.1,0.41,0.4))$	$([s_4,s_5],(0.7,0.2,0.1))$
$\chi_{N_{AC}}$	$([s_2,s_3],$ $(0.82,0.12,0.8))$	$([s_2,s_3],(0.62,0.22,0.2))$	$([s_2,s_3],(0.7,0.2,0.5))$	$([s_1,s_2],(0.7,0.3,0.2))$
	$\mathfrak{r}_3$		$\mathfrak{r}_4$	
$\chi_{P_{AC}}$	$([s_0,s_1],$ $(0.7,0.1,0.5))$	$([s_4,s_5],(0.6,0.2,0.4))$	$([s_4,s_5],(0.6,0.2,0.1))$	$([s_0,s_1],(0.5,0.4,0.4))$
$\chi_{B_{AC}}$	$([s_4,s_5],$ $(0.6,0.3,0.7))$	$([s_2,s_3],(0.8,0.2,0.3))$	$([s_3,s_4],(0.61,0.21,0.3))$	$([s_1,s_2],(0.1,0.41,0.3))$
$\chi_{N_{AC}}$	$([s_4,s_5],$ $(0.7,0.2,0.4))$	$([s_1,s_2],(0.7,0.3,0.8))$	$([s_2,s_3],(0.62,0.22,0.4))$	$([s_2,s_3],(0.7,0.2,0.1))$

Table 5 Decision matrix which is represented by risk and given by $\mathcal{D}_3$

$\mathfrak{r}_1$			$\mathfrak{r}_2$	
	$\mathcal{F}_B$	$\sim \mathcal{F}_{NB}$	$\mathcal{F}_B$	$\sim \mathcal{F}_{NB}$
$\chi_{P_{AC}}$	$([s_0,s_1],(0.7,0.3,0.1))$	$([s_4,s_5],(0.5,0.4,0.1))$	$([s_1,s_2],(0.6,0.4,0.1))$	$([s_4,s_5],$ $(0.6,0.2,0.1))$
$\chi_{B_{AC}}$	$([s_2,s_3],(0.7,0.2,0.2))$	$([s_2,s_3],(0.7,0.3,0.3))$	$([s_2,s_3],(0.7,0.3,0.6))$	$([s_2,s_3],$ $(0.8,0.1,0.2))$
$\chi_{N_{AC}}$	$([s_0,s_1],(0.7,0.1,0.3))$	$([s_4,s_5],(0.6,0.2,0.2))$	$([s_4,s_5],(0.6,0.2,0.5))$	$([s_0,s_1],$ $(0.5,0.4,0.3))$
	$\mathfrak{r}_3$		$\mathfrak{r}_4$	
$\chi_{P_{AC}}$	$([s_0,s_1],(0.7,0.3,0.5))$	$([s_4,s_5],(0.5,0.4,0.1))$	$([s_1,s_2],(0.6,0.4,0.5))$	$([s_4,s_5],$ $(0.6,0.2,0.4))$
$\chi_{B_{AC}}$	$([s_0,s_1],(0.8,0.1,0.4))$	$([s_4,s_5],(0.6,0.2,0.2))$	$([s_0,s_1],(0.5,0.4,0.4))$	$([s_4,s_5],$ $(0.7,0.3,0.2))$
$\chi_{N_{AC}}$	$([s_1,s_2],(0.81,0.11,0.6))$	$([s_3,s_4],(0.61,0.21,0.3))$	$([s_1,s_2],(0.1,0.41,0.1))$	$([s_4,s_5],$ $(0.71,0.21,0.1))$

The presented approaches in this manuscript are more beneficial and more reliable than existing drawbacks [37]. The T-WD based on intuitionistic fuzzy uncertain linguistic D-TRS was explored by Liu and Yang [33]. But when a decision-maker is faced more answer of types like yes, abstinence, and no, then the IFS cannot deal it with effectively. For coping with such kinds of issues, the theory of single-valued neutrosophic set was explored by [7]. To improve the reliability and proficiency of the explored work in this manuscript, we presented the T-WDs based on S-VNULD-TRS with GMSM operators and their application in decision-making. The existing work [33, 37] is the special case of the proposed work for choosing the value of abstinence will be zero. The explored work is more reliable and more accurate to resolve various kinds of issues.

Table 6 By using Eq. (55), we aggregate the information of Tables 3, 4, and 5

Γ_1			Γ_2	
	$\mathcal{F}_B$	$\sim \mathcal{F}_{NB}$	$\mathcal{F}_B$	$\sim \mathcal{F}_{NB}$
$\chi_{P_{AC}}$	$([s_0, s_{0.48}],$ $(0.37,0.68,0.74))$	$([s_{1.51}, s_{1.88}], (0.25,0.75,0.75))$	$([s_0, s_{0.47}],$ $(0.30,0.72,0.58))$	$([s_{1.51}, s_{1.89}],$ $(0.32,0.66,0.59))$
$\chi_{B_{AC}}$	$([s_{0.69}, s_{1.09}],$ $(0.40,0.59,0.75))$	$([s_{0.86}, s_{1.24}], (0.32,0.73,0.75))$	$([s_{0.69}, s_{1.17}],$ $(0.16,0.75,0.82))$	$([s_{0.95}, s_{1.34}],$ $(0.43,0.59,0.59))$
$\chi_{N_{AC}}$	$([s_0, s_{0.93}],$ $(0.44,0.50,0.82))$	$([s_{0.75}, s_{1.17}], (0.33,0.64,0.63))$	$([s_{1.2}, s_{1.59}],$ $(0.35,0.63,0.80))$	$([s_0, s_{0.6}],$ $(0.32,0.75,0.7))$
	Γ_3		Γ_4	
$\chi_{P_{AC}}$	$([s_0, s_{0.37}],$ $(0.3,0.71,0.76))$	$([s_{1.51}, s_{1.89}], (0.25,0.75,0.72))$	$([s_{0.6}, s_{1.02}],$ $(0.3,0.75,0.75))$	$([s_0, s_{1.1}],$ $(0.27,0.69,0.75))$
$\chi_{B_{AC}}$	$([s_0, s_{0.93}],$ $(0.37,0.65,0.74))$	$([s_{0.95}, s_{1.34}], (0.37,0.66,0.70))$	$([s_0, s_{0.87}],$ $(0.32,0.73,0.75))$	$([s_{0.76}, s_{1.17}],$ $(0.19,0.69,0.62))$
$\chi_{N_{AC}}$	$([s_{0.87}, s_{1.29}],$ $(0.4,0.62,0.81))$	$([s_{0.55}, s_{0.95}], (0.37,0.67,0.84))$	$([s_{0.69}, s_{1.09}],$ $(0.18,0.67,0.62))$	$([s_{0.76}, s_{1.17}],$ $(0.38,0.63,0.54))$

Table 7 By using again Eqs. (34–36), we aggregate the information of Table 6

Symbols	$Y_{EL}\left(\chi_{P_{AC}} \mid [\Gamma]\right)$	$Y_{EL}\left(\chi_{B_{AC}} \mid [\Gamma]\right)$	$Y_{EL}\left(\chi_{N_{AC}} \mid [\Gamma]\right)$
Γ_1	$([s_{0.91}, s_{1.32}],$ $(0.3,0.72,0.75))$	$([s_{0.79}, s_{1.18}], (0.36,0.67,0.75))$	$([s_{0.45}, s_{1.07}], (0.37,0.58,0.7))$
Γ_2	$([s_{0.90}, s_{1.32}],$ $(0.31,0.68,0.58))$	$([s_{0.85}, s_{1.28}], (0.34,0.65,0.67))$	$([s_{0.48}, s_1], (0.33,0.7,0.73))$
Γ_3	$([s_{0.91}, s_{1.28}],$ $(0.3,0.71,0.76))$	$([s_{0.57}, s_{1.18}], (0.37,0.65,0.74))$	$([s_{0.67}, s_{1.09}], (0.4,0.62,0.81))$
Γ_4	$([s_{0.24}, s_{1.07}],$ $(0.28,0.72,0.75))$	$([s_{0.45}, s_{1.05}], (0.25,0.71,0.67))$	$([s_{0.73}, s_{1.14}], (0.3,0.65,0.57))$

Table 8 Expected values of the information, which are shown in Table 7

Symbols	$Q_{EV}\left(Y_{EL}\left(\chi_{P_{AC}} \mid [\Gamma]\right)\right)$	$Q_{EV}\left(Y_{EL}\left(\chi_{B_{AC}} \mid [\Gamma]\right)\right)$	$Q_{EV}\left(Y_{EL}\left(\chi_{N_{AC}} \mid [\Gamma]\right)\right)$
Γ_1	-0.65	-0.52	-0.35
Γ_2	-0.53	-0.52	-0.40
Γ_3	-0.64	-0.44	-0.45
Γ_4	-0.38	-0.42	-0.43

6 Conclusion

Single-valued neutrosophic uncertain linguistic variable (S-VNULV) is a mixture of two different notions like single-valued neutrosophic set (S-VNS) and uncertain linguistic variable (ULV) to cope with complicated and awkward information in realistic decision problems. The aims of this manuscript are to explore three-way decision (T-WD) based on decision-theoretic rough set (D-TRS) for S-VNULV, and their some operational laws are also presented. Further, based on the S-VNULV, the generalized Maclaurin symmetric mean operators are called single-valued neutrosophic uncertain linguistic generalized Maclaurin symmetric mean

Table 9 Accuracy values of the information, which is shown in Table 6

Symbols	$G_{AF}\left(Y_{EL}\left(\chi_{P_{AC}} \mid [\mathfrak{r}]\right)\right)$	$G_{AF}\left(Y_{EL}\left(\chi_{B_{AC}} \mid [\mathfrak{r}]\right)\right)$	$G_{AF}\left(Y_{EL}\left(\chi_{N_{AC}} \mid [\mathfrak{r}]\right)\right)$
$\mathfrak{r}_1$	0.99	0.87	0.63
$\mathfrak{r}_2$	0.8819	0.8814	0.65
$\mathfrak{r}_3$	0.97	0.77	0.80
$\mathfrak{r}_4$	0.57	0.61	0.71

Table 10 Three-way decision based on Eqs. (43–45)

Enterprises	Decision rule
$\mathfrak{r}_1$	P_{AC-1}
$\mathfrak{r}_2$	P_{AC-1}
$\mathfrak{r}_3$	P_{AC-1}
$\mathfrak{r}_4$	P_{AC-1}

(S-VNULGMSM) operator and single-valued neutrosophic uncertain linguistic weighted generalized Maclaurin symmetric mean (S-VNULWGMSM) operator, and their special cases are also discussed. Based on the explored operator and D-TRS, we explored an algorithm to evaluate the numerical example for finding the reliability and effectiveness of the presented approaches.

Acknowledgments *Conflicts of Interest:* We declare that we do not have any commercial or associative interests that represent conflicts of interest in connection with this manuscript. There are no professional or other personal interests that can inappropriately influence our submitted work.

Research Involving Human Participants and/or Animals: This paper does not include any researches with human participants or animals performed by any of the authors.

References

1. Zadeh, L. A. (1965). Fuzzy sets. *Information and Control, 8*(3), 338–353.
2. Atanassov, K. T. (1999). *Intuitionistic fuzzy sets* (pp. 1 137). Heidelberg: Physica.
3. Liang, Z., & Shi, P. (2003). Similarity measures on intuitionistic fuzzy sets. *Pattern Recognition Letters, 24*(15), 2687–2693.
4. Xu, Z. (2007). Intuitionistic fuzzy aggregation operators. *IEEE Transactions on Fuzzy Systems, 15*(6), 1179–1187.
5. Liu, P., & Chen, S. M. (2016). Group decision making based on Heronian aggregation operators of intuitionistic fuzzy numbers. *IEEE Transactions on Cybernetics, 47*(9), 2514–2530.
6. Smarandache, F. (2005). Neutrosophic set-a generalization of the intuitionistic fuzzy set. *International Journal of Pure and Applied Mathematics, 24*(3), 287.
7. Wang, H., Smarandache, F., Zhang, Y., & Sunderraman, R. (2010). Single valued neutrosophic sets. *Multispace and Multistructure, 4*, 410–413.
8. Peng, X., & Dai, J. (2018). Approaches to single-valued neutrosophic MADM based on MABAC, TOPSIS and new similarity measure with score function. *Neural Computing and Applications, 29*(10), 939–954.
9. Liu, P. (2016). The aggregation operators based on Archimedean t-conorm and t-norm for single-valued neutrosophic numbers and their application to decision making. *International Journal of Fuzzy Systems, 18*(5), 849–863.

10. Sodenkamp, M. A., Tavana, M., & Di Caprio, D. (2018). An aggregation method for solving group multi-criteria decision-making problems with single-valued neutrosophic sets. *Applied Soft Computing, 71*, 715–727.
11. Lu, Z., & Ye, J. (2017). Single-valued neutrosophic hybrid arithmetic and geometric aggregation operators and their decision-making method. *Information, 8*(3), 84.
12. Ashraf, S., Abdullah, S., & Smarandache, F. (2019). Logarithmic hybrid aggregation operators based on single valued neutrosophic sets and their applications in decision support systems. *Symmetry, 11*(3), 364.
13. Deli, I., & Subas, Y. (2014). Single valued neutrosophic numbers and their applications to multicriteria decision making problem. *Neutrosophic Sets and Systems, 2*(1), 1–13.
14. Jana, C., & Pal, M. (2019). A robust single-valued neutrosophic soft aggregation operators in multi-criteria decision making. *Symmetry, 11*(1), 110.
15. Garg, H. (2018). New logarithmic operational laws and their applications to multiattribute decision making for single-valued neutrosophic numbers. *Cognitive Systems Research, 52*, 931–946.
16. Chen, J., & Ye, J. (2017). Some single-valued neutrosophic Dombi weighted aggregation operators for multiple attribute decision-making. *Symmetry, 9*(6), 82.
17. Wei, G., & Zhang, Z. (2019). Some single-valued neutrosophic Bonferroni power aggregation operators in multiple attribute decision making. *Journal of Ambient Intelligence and Humanized Computing, 10*(3), 863–882.
18. Wang, J. Q., Yang, Y., & Li, L. (2018). Multi-criteria decision-making method based on single-valued neutrosophic linguistic Maclaurin symmetric mean operators. *Neural Computing and Applications, 30*(5), 1529–1547.
19. Li, Y., Liu, P., & Chen, Y. (2016). Some single valued neutrosophic number heronian mean operators and their application in multiple attribute group decision making. *Informatica, 27*(1), 85–110.
20. Liu, P., & Wang, Y. (2014). Multiple attribute decision-making method based on single-valued neutrosophic normalized weighted Bonferroni mean. *Neural Computing and Applications, 25*(7–8), 2001–2010.
21. Zhang, H., Wang, F., & Geng, Y. (2019). Multi-criteria decision-making method based on single-valued neutrosophic Schweizer–Sklar Muirhead mean aggregation operators. *Symmetry, 11*(2), 152.
22. La, Z. (1975). The concept of a linguistic variable and its application to approximate reasoning-III. *Information Sciences, 9*(1), 43–80.
23. Herrera, F., & Martínez, L. (2000). A 2-tuple fuzzy linguistic representation model for computing with words. *IEEE Transactions on Fuzzy Systems, 8*(6), 746–752.
24. Herrera, F., & Martinez, L. (2000). An approach for combining linguistic and numerical information based on the 2-tuple fuzzy linguistic representation model in decision-making. *International Journal of Uncertainty, Fuzziness and Knowledge-Based Systems, 8*(05), 539–562.
25. Martı, L., & Herrera, F. (2012). An overview on the 2-tuple linguistic model for computing with words in decision making: Extensions applications and challenges. *Information Sciences, 207*, 1–18.
26. Li, Y., & Liu, P. (2015). Some Heronian mean operators with 2-tuple linguistic information and their application to multiple attribute group decision making. *Technological and Economic Development of Economy, 21*(5), 797–814.
27. Liang, D., Liu, D., Pedrycz, W., & Hu, P. (2013). Triangular fuzzy decision-theoretic rough sets. *International Journal of Approximate Reasoning, 54*(8), 1087–1106.
28. Jia, X., Liao, W., Tang, Z., & Shang, L. (2013). Minimum cost attribute reduction in decision-theoretic rough set models. *Information Sciences, 219*, 151–167.
29. Liu, D., Li, T., & Ruan, D. (2011). Probabilistic model criteria with decision-theoretic rough sets. *Information Sciences, 181*(17), 3709–3722.
30. Yao, Y., & Zhao, Y. (2008). Attribute reduction in decision-theoretic rough set models. *Information Sciences, 178*(17), 3356–3373.

31. Liu, D., Yao, Y., & Li, T. (2011). Three-way investment decisions with decision-theoretic rough sets. *International Journal of Computational Intelligence Systems, 4*(1), 66–74.
32. Liu, D., Li, T., & Liang, D. (2012). Three-way government decision analysis with decision-theoretic rough sets. *International Journal of Uncertainty, Fuzziness and Knowledge-Based Systems, 20*(supp01), 119–132.
33. Liu, P., & Yang, H. (2020). Three-way decisions with intuitionistic uncertain linguistic decision-theoretic rough sets based on generalized Maclaurin symmetric mean operators. *International Journal of Fuzzy Systems, 22*(2), 653–667.
34. Ye, J. (2017). Linguistic neutrosophic cubic numbers and their multiple attribute decision-making method. *Information, 8*(3), 110.
35. Fan, C., Fan, E., & Hu, K. (2018). New form of single valued neutrosophic uncertain linguistic variables aggregation operators for decision-making. *Cognitive Systems Research, 52*, 1045–1055.
36. Liu, P., & Li, Y. (2019). Multi-attribute decision making method based on generalized Maclaurin symmetric mean aggregation operators for probabilistic linguistic information. *Computers & Industrial Engineering, 131*, 282–294.
37. Liang, D., Xu, Z., & Liu, D. (2017). Three-way decisions with intuitionistic fuzzy decision-theoretic rough sets based on point operators. *Information Sciences, 375*, 183–201.

Novel Distance Measures Over SVTN-Numbers and Their Application to Multi-Criteria Decision Making Problems

Şerif Özlü and İrfan Deli

1 Introduction

Fuzzy set structure was proposed by Zadeh in 1965 [51], due to lack of exact uncertainty and vagueness and then new set structures have been proposed over time such as Turksen [28] introduced interval-valued fuzzy set (IVFS) by extending fuzzy sets, Atanasov [1] defined intuitionistic fuzzy set (IFS), and interval valued intuitionistic fuzzy set (IVIFS) [2] to overcome uncertainness in 1986. These cluster structures included a truth-membership function and falsity- membership function with range [0,1] which is the sum of these values in [0,1] and the structures studied in [8, 10, 12, 15, 17, 23, 32, 33, 36–40]. After then, these structures were extended to neutrosophic set (NS) by Smarandache [25], which took much attention by some researchers. This set structure contains a truth-membership function(T), an indeterminacy-membership function(I), and a falsity-membership function(F) with $T, I, F \in [0, 1]$ such that $T(x) + I(x) + F(x) \in [0, 3]$ for all x in the universe set. The NS has taken much attention by authors and revealed relationship with many disciplines like graph theory, prospect theory, image thresholding, and image segmentation [7, 9, 13, 24, 26, 29–31]. Also, Wang et al. [35] developed a single valued neutrosophic set (SVNS) and interval neutrosophic set (INS) [34] which each membership consisted of interval values in [0,1]. The correlation coefficient of INSs was studied in [5, 6, 18]. Ye [44, 45] put forwarded cross entropy and similarity over SVNSs and INSs, respectively. Biswas et al. [3] gave neutrosophic subset

Ş. Özlü (✉)
Nizip Vocational High School, Gaziantep University, Gaziantep, Turkey

I. Deli
Muallim Rıfat Faculty of Education, Kilis 7 Aralık University, Kilis, Turkey
e-mail: irfandeli@kilis.edu.tr

hood based on distance measure obtained using single valued neutrosophic sets. Furthermore, some structures are defined over neutrosophic sets by many authors in [22, 41–43, 47, 48, 52, 53].

Similarity measure is an essential tool for determining the best alternative in decision making problems such as medical diagnosis problems, etc. The similarity measure theory is widely studied by many researchers under NSs and INs [4, 14, 16, 46, 50]. While the values are real numbers in clasical multi- criteria decision making problems in real life problems the values generally are linguistic terms like very unsitable, unsuitable, medium unsuitable, medium suitable, suitable, very suitable, extremely suitable to express uncertainty and vagueness. Also, it has been observed that all the papers given above used methods including distance and similarity measures between NNs on finite universe of discourses only. But construction of methods between NSs for countable and uncountable universe of discourse is also necessary. To do this, the concept of single valued trapezoidal neutrosophic number(SVTN-number) with the universe of discourse as the real line was introduced and studied in [11, 49]. Here, the SVTN-number is a number which has a truth-membership, indeterminacy-membership, and a falsity-membership on the real number set R. Then, Biswas [3] introduced some operational rules of SVTN-number and expected value of SVTN-number.

Recently, Ngan et al. [19] developed a new distance measure called H-max distance measures of intuitionistic fuzzy sets. But this distance measure is not effective for uncountable universe. With this point of view in this paper a new approach is developed to calculate the distance measure between two SVTN-numbers. Therefore, the paper is organized as follows. Section 2 briefly describes the basic definition and notations of Ns and SVTN-numbers. Section 3 introduces the new distance measure for SVTN-numbers. Section 4 gives a multi-criteria decision making method. Section 5 presents a real example to show whether the proposed method works or not. Section 6 concludes the paper.

2 Preliminary

In this section, we recall some basic notions of intuitionistic fuzzy sets [1, 19], neutrosophic sets [25, 35], and single valued trapezoidal neutrosophic number [11, 49].

Definition 1 ([1]) Let E be a universe. An intuitionistic fuzzy set A on E can be defined as follows:

$$A = \{< x, \mu(x), \gamma(x) >: \ x \in E\}$$

where $\mu : E \rightarrow [0, 1]$ and $\gamma : E \rightarrow [0, 1]$ such that $0 \leq \mu(x) + \gamma(x) \leq 1$ for any $x \in E$.

Here, $\mu_K(x)$ and $\gamma_K(x)$ are the degree of membership and degree of non-membership of the element x, respectively.

Definition 2 ([27]) Let $A = \{< x, \mu_1(x), \gamma_1(x) >: x \in E\}$ and $B = \{< x, \mu_2(x), \gamma_2(x) >: x \in E\}$ be two $IFSs$. Then, Hamming distance between A and B is defined as

$$d^*_{Hm}(A, B) = \frac{1}{2m} \sum_{i=1}^{m} (|\mu_1(x_i) - \mu_2(x_i)| + |\nu_1(x_i) - \nu_2(x_i)| + |\pi_1(x_i) - \pi_2(x_i)|)$$

Definition 3 ([27]) Let $A = \{< x, \mu_1(x), \gamma_1(x) >: x \in E\}$ and $B = \{< x, \mu_2(x), \gamma_2(x) >: x \in E\}$ be two $IFSs$. Then, Euclidean distance between A and B is defined as

$$d^*_E(A, B) = \frac{1}{2m} \sum_{i=1}^{m} ((\mu_1(x_i) - \mu_2(x_i))^2 + (\nu_1(x_i) - \nu_2(x_i))^2 + (\pi_1(x_i) - \pi_2(x_i))^2)^{\frac{1}{2}}.$$

Definition 4 ([21]) Let $A = \{< x, \mu_1(x), \gamma_1(x) >: x \in E\}$ and $B = \{< x, \mu_2(x), \gamma_2(x) >: x \in E\}$ be two $IFSs$. Then, Distance measure between A and B is defined as

$$d_j(A, B) = \frac{1}{4m} \sum_{i=1}^{m} |\mu_1(x_i) - \mu_2(x_i)| + |\nu_1(x_i) - \nu_2(x_i)| + |\pi_1(x_i) - \pi_2(x_i)| + 2max\{|\mu_1(x_i) - \mu_2(x_i)|, |\nu_1(x_i) - \nu_2(x_i)|, |\pi_1(x_i) - \pi_2(x_i)|\}$$

Definition 5 ([19]) Let $A = \{< x, \mu_1(x), \gamma_1(x) >: x \in E\}$ and $B = \{< x, \mu_2(x), \gamma_2(x) >: x \in E\}$ be two $IFSs$. Then, Distance measure between A and B is defined as

$$d_{Hm}(A, B) = \frac{1}{3m} \sum_{i=1}^{m} (|\mu_1(x_i) - \mu_2(x_i)| + |\nu_1(x_i) - \nu_2(x_i)| + |max\{\mu_1(x_i), \nu_2(x_i)\} - max\{\mu_2(x_i), \nu_1(x_i)\}|)$$

Definition 6 ([35]) Let E be a universe. A single valued neutrosophic set (SVNS) A, which can be used in real scientific and engineering applications, in E is characterized by a truth-membership function T_A, an indeterminacy-membership function I_A, and a falsity-membership function F_A. $T_A(x)$, $I_A(x)$, and $F_A(x)$ are real standard elements of $[0, 1]$. It can be written as

$$A = \{< x, (T_A(x), I_A(x), F_A(x)) >: x \in E, \ T_A(x), I_A(x), F_A(x) \in [0, 1]\}.$$

There is no restriction on the sum of $T_A(x)$, $I_A(x)$, and $F_A(x)$, so $0 \leq T_A(x) + I_A(x) + F_A(x) \leq 3$.

Definition 7 ([11]) A single valued trapezoidal neutrosophic number $\tilde{a} = \langle(a_1, b_1, c_1, d_1); w_{\tilde{a}}, u_{\tilde{a}}, y_{\tilde{a}}\rangle$ is a special neutrosophic set on the real number set R, whose truth-membership, indeterminacy-membership, and a falsity-membership are given as follows:

$$\mu_{\tilde{a}}(x) = \begin{cases} (x - a_1)w_{\tilde{a}}/(b_1 - a_1) & (a_1 \le x < b_1) \\ w_{\tilde{a}} & (b_1 \le x \le c_1) \\ (d_1 - x)w_{\tilde{a}}/(d_1 - c_1) & (c_1 < x \le d_1) \\ 0 & otherwise, \end{cases}$$

$$v_{\tilde{a}}(x) = \begin{cases} (b_1 - x + u_{\tilde{a}}(x - a_1))/(b_1 - a_1) & (a_1 \le x < b_1) \\ u_{\tilde{a}} & (b_1 \le x \le c_1) \\ (x - c_1 + u_{\tilde{a}}(d_1 - x))/(d_1 - c_1) & (c_1 < x \le d_1) \\ 1 & otherwise \end{cases}$$

and

$$\lambda_{\tilde{a}}(x) = \begin{cases} (b_1 - x + y_{\tilde{a}}(x - a_1))/(b_1 - a_1) & (a_1 \le x < b_1) \\ y_{\tilde{a}} & (b_1 \le x \le c_1) \\ (x - c_1 + y_{\tilde{a}}(d_1 - x))/(d_1 - c_1) & (c_1 < x \le d_1) \\ 1 & otherwise \end{cases}$$

respectively.

Note that if $a, b, c, d \in [0, 1]$ then $\tilde{a} = \langle (a_1, b_1, c_1, d_1); w_{\tilde{a}}, u_{\tilde{a}}, y_{\tilde{a}} \rangle$ is normal SVTN-numbers

From now on we use normal SVTN-numbers.

Definition 8 ([11]) Let $\tilde{a} = \langle (a_1, b_1, c_1, d_1); w_{\tilde{a}}, u_{\tilde{a}}, y_{\tilde{a}} \rangle$ and $\tilde{b} = \langle (a_2, b_2, c_2, d_2); w_{\tilde{b}}, u_{\tilde{b}}, y_{\tilde{b}} \rangle$ be two single valued trapezoidal neutrosophic numbers and $\gamma \neq 0$, then

1. $\tilde{a} + \tilde{b} = \langle (a_1 + a_2, b_1 + b_2, c_1 + c_2, d_1 + d_2); w_{\tilde{a}} \wedge w_{\tilde{b}}, u_{\tilde{a}} \vee u_{\tilde{b}}, y_{\tilde{a}} \vee y_{\tilde{b}} \rangle$
2. $\tilde{a} - \tilde{b} = \langle (a_1 - d_2, b_1 - c_2, c_1 - b_2, d_1 - a_2); w_{\tilde{a}} \wedge w_{\tilde{b}}, u_{\tilde{a}} \vee u_{\tilde{b}}, y_{\tilde{a}} \vee y_{\tilde{b}} \rangle$
3. $\tilde{a}\tilde{b} = \begin{cases} \langle (a_1 a_2, b_1 b_2, c_1 c_2, d_1 d_2); w_{\tilde{a}} \wedge w_{\tilde{b}}, u_{\tilde{a}} \vee u_{\tilde{b}}, y_{\tilde{a}} \vee y_{\tilde{b}} \rangle & (d_1 > 0, d_2 > 0) \\ \langle (a_1 d_2, b_1 c_2, c_1 b_2, d_1 a_2); w_{\tilde{a}} \wedge w_{\tilde{b}}, u_{\tilde{a}} \vee u_{\tilde{b}}, y_{\tilde{a}} \vee y_{\tilde{b}} \rangle & (d_1 < 0, d_2 > 0) \\ \langle (d_1 d_2, c_1 c_2, b_1 b_2, a_1 a_2); w_{\tilde{a}} \wedge w_{\tilde{b}}, u_{\tilde{a}} \vee u_{\tilde{b}}, y_{\tilde{a}} \vee y_{\tilde{b}} \rangle & (d_1 < 0, d_2 < 0) \end{cases}$
4. $\tilde{a}/\tilde{b} = \begin{cases} \langle (a_1/d_2, b_1/c_2, c_1/b_2, d_1/a_2); w_{\tilde{a}} \wedge w_{\tilde{b}}, u_{\tilde{a}} \vee u_{\tilde{b}}, y_{\tilde{a}} \vee y_{\tilde{b}} \rangle & (d_1 > 0, d_2 > 0) \\ \langle (d_1/d_2, c_1/c_2, b_1/b_2, a_1/a_2); w_{\tilde{a}} \wedge w_{\tilde{b}}, u_{\tilde{a}} \vee u_{\tilde{b}}, y_{\tilde{a}} \vee y_{\tilde{b}} \rangle & (d_1 < 0, d_2 > 0) \\ \langle (d_1/a_2, c_1/b_2, b_1/c_2, a_1/d_2); w_{\tilde{a}} \wedge w_{\tilde{b}}, u_{\tilde{a}} \vee u_{\tilde{b}}, y_{\tilde{a}} \vee y_{\tilde{b}} \rangle & (d_1 < 0, d_2 < 0) \end{cases}$
5. $\gamma \tilde{a} = \begin{cases} \langle (\gamma a_1, \gamma b_1, \gamma c_1, \gamma d_1); w_{\tilde{a}}, u_{\tilde{a}}, y_{\tilde{a}} \rangle & (\gamma > 0) \\ \langle (\gamma d_1, \gamma c_1, \gamma b_1, \gamma a_1); w_{\tilde{a}}, u_{\tilde{a}}, y_{\tilde{a}} \rangle & (\gamma < 0) \end{cases}$
6. $\tilde{a}^{\gamma} = \begin{cases} \langle (a_1^{\gamma}, b_1^{\gamma}, c_1^{\gamma}, d_1^{\gamma}); w_{\tilde{a}}, u_{\tilde{a}}, y_{\tilde{a}} \rangle & (\gamma > 0) \\ \langle (d_1^{\gamma}, c_1^{\gamma}, b_1^{\gamma}, a_1^{\gamma}); w_{\tilde{a}}, u_{\tilde{a}}, y_{\tilde{a}} \rangle & (\gamma < 0) \end{cases}$
7. $\tilde{a}^{-1} = \langle (1/d_1, 1/c_1, 1/b_1, 1/a_1); w_{\tilde{a}}, u_{\tilde{a}}, y_{\tilde{a}} \rangle \ (\tilde{a} \neq 0)$.

Note that if $a_1, b_1, c_1, d_1 \in [0, 1]$ then $\tilde{a} = \langle (a_1, b_1, c_1, d_1); w_{\tilde{a}}, u_{\tilde{a}}, y_{\tilde{a}} \rangle$ is normal SVTN-numbers

3   Distance Measures Over the SVTN-Numbers

In this section, various distance measures are defined over neutrosophic sets based on uncountable universe of discourse as the real line.

Definition 9   Let $m = \langle (a_1, b_1, c_1, d_1); w_m, u_m, \gamma_m \rangle, n = \langle (a_2, b_2, c_2, d_2); w_n, u_n, \gamma_n \rangle$ be two normal SVTN-numbers. Then, the Hamming distance between m and n is denoted by $d_H(m, n)$ and is defined as

$$
\begin{aligned}
d_H^*(m, n) = \tfrac{1}{4}[& |[\int_{a_1}^{b_1} (w_m(x - a_1))/(b_1 - a_1)dx + \int_{b_1}^{c_1} w_m dx + \\
& \int_{c_1}^{d_1} (w_m(d_1 - x))/(d_1 - c_1)dx] \\
& -[\int_{a_2}^{b_2} (w_n(x - a_2))/(b_2 - a_2)dx + \\
& \int_{b_2}^{c_2} w_n dx + \int_{c_2}^{d_2} ((w_n(d_2 - x))/(d_2 - c_2)dx]|+ \\
& |[\int_{a_1}^{b_1} ((b_1 - x) + u_m(x - a_1))/(b_1 - a_1)dx + \int_{b_1}^{c_1} u_m dx + \\
& \int_{c_1}^{d_1} ((x - c_1) + u_m(d_1 - x))/(d_1 - c_1)dx] \\
& -[\int_{a_2}^{b_2} ((b_2 - x) + u_n(x - a_2))/(b_2 - a_2)dx + \\
& \int_{b_2}^{c_2} u_n dx + \int_{c_2}^{d_2} ((x - c_2) + u_n(d_2 - x))/(d_2 - c_2)dx]| \\
& +|[\int_{a_1}^{b_1} ((b_1 - x) + \gamma_m(x - a_1))/(b_1 - a_1)dx + \\
& \int_{b_1}^{c_1} \gamma_m dx + \int_{c_1}^{d_1} ((x - c_1) + \gamma_m(d_1 - x))/(d_1 - c_1)dx] \\
& -[\int_{a_2}^{b_2} ((b_2 - x) + \gamma_n(x - a_2))/(b_2 - a_2)dx + \\
& \int_{b_2}^{c_2} \gamma_n dx + \int_{c_2}^{d_2} ((x - c_2) + \gamma_n(d_2 - x))/(d_2 - c_2)dx]|]
\end{aligned}
$$

Lemma 1   *Let* $m = \langle (a_1, b_1, c_1, d_1); w_m, u_m, \gamma_m \rangle$, $n = \langle (a_2, b_2, c_2, d_2); w_n, u_n, \gamma_n \rangle$ *be two normal SVTN-numbers. Then, the Hamming distance between m and n is calculated as*

$$
\begin{aligned}
d_H^*(m, n) = \tfrac{1}{4}[\tfrac{1}{2} & |[(w_m(c_1 + d_1 - b_1 - a_1) - (w_n(c_2 + d_2 - b_2 - a_2)]| \\
& +\tfrac{1}{2}|[(u_m(c_1 + d_1 - b_1 - a_1) + (b_1 - a_1 + d_1 - c_1)) - (u_n(c_2 + \\
& d_2 - b_2 - a_2) + (b_2 - a_2 + d_2 - c_2))]| + \tfrac{1}{2}|[(\gamma_m(c_1 + d_1 - b_1 - a_1) \\
& +(b_1 - a_1 + d_1 - c_1)) - (\gamma_n(c_2 + d_2 - b_2 - a_2) + (b_2 - a_2 + d_2 \\
& -c_2))]|]
\end{aligned}
$$

Proof   If we solve integrals as follows

$$
\begin{aligned}
d_H^*(m, n) = \tfrac{1}{4}[& |[\int_{a_1}^{b_1} (w_m(x - a_1))/(b_1 - a_1)dx + \\
& \int_{b_1}^{c_1} w_m dx + \int_{c_1}^{d_1} (w_m(d_1 - x))/(d_1 - c_1)dx] \\
& -[\int_{a_2}^{b_2} (w_n(x - a_2))/(b_2 - a_2)dx + \\
& \int_{b_2}^{c_2} w_n dx + \int_{c_2}^{d_2} ((w_n(d_2 - x))/(d_2 - c_2)dx]|+ \\
& |[\int_{a_1}^{b_1} ((b_1 - x) + u_m(x - a_1))/(b_1 - a_1)dx +
\end{aligned}
$$

$$\int_{b_1}^{c_1} u_m dx + \int_{c_1}^{d_1} ((x - c_1) + u_m(d_1 - x))/(d_1 - c_1)dx]$$
$$-[\int_{a_2}^{b_2} ((b_2 - x) + u_n(x - a_2))/(b_2 - a_2)dx + \int_{b_2}^{c_2} u_n dx +$$
$$\int_{c_2}^{d_2} ((x - c_2) + u_n(d_2 - x))/(d_2 - c_2)dx]|$$
$$+|[\int_{a_1}^{b_1} ((b_1 - x) + \gamma_m(x - a_1))/(b_1 - a_1)dx +$$
$$\int_{b_1}^{c_1} \gamma_m dx + \int_{c_1}^{d_1} ((x - c_1) + \gamma_m(d_1 - x))/(d_1 - c_1)dx]$$
$$-[\int_{a_2}^{b_2} ((b_2 - x) + \gamma_n(x - a_2))/(b_2 - a_2)dx +$$
$$\int_{b_2}^{c_2} \gamma_n dx + \int_{c_2}^{d_2} ((x - c_2) + \gamma_n(d_2 - x))/(d_2 - c_2)dx]|]$$
$$= \frac{1}{4}[|[1/(b_1 - a_1)(w_m x^2/2 - w_m a_1 x)|_{a_1}^{b_1} + w_m x|_{b_1}^{c_1} +$$
$$1/(d_1 - c_1)(w_m d_1 x - w_m x^2/2)|_{c_1}^{d_1}]$$
$$-[1/(b_2 - a_2)(w_n x^2/2 - w_n a_2)|_{a_2}^{b_2} +$$
$$w_n x|_{b_2}^{c_2} + 1/(d_2 - c_2)(w_n d_2 x - w_n x^2/2)|_{c_2}^{d_2}]|$$
$$+ |[1/(b_1 - a_1)((u_m - 1)x^2/2 +$$
$$(b_1 - u_m a_1)x)|_{a_1}^{b_1} + u_m x|_{b_1}^{c_1} + 1/(d_1 - c_1)((1 - u_m)x^2/2$$
$$+ (u_m d_1 - c_1)x)|_{c_1}^{d_1}] - [1/(b_2 - a_2)((u_n - 1)x^2/2 +$$
$$(b_2 - u_n a_2)x)|_{a_2}^{b_2} + u_n x|_{b_2}^{c_2} + 1/(d_2 - c_2)$$
$$((1 - u_n)x^2/2 + (u_n d_2 - c_2)x)|_{c_2}^{d_2}]|$$
$$+ |[1/(b_1 - a_1)((\gamma_m - 1)x^2/2 + (b_1 - \gamma_m a_1)x)|_{a_1}^{b_1} +$$
$$\gamma_m x|_{b_1}^{c_1} + 1/(d_1 - c_1)((1 - \gamma_m)x^2/2$$
$$+ (\gamma_m d_1 - c_1)x)|_{c_1}^{d_1}] - [1/(b_2 - a_2)((\gamma_n - 1)x^2/2 +$$
$$(b_2 - \gamma_n a_2)x)|_{a_2}^{b_2} + \gamma_n x|_{b_2}^{c_2} + 1/(d_2 - c_2)$$
$$((1 - \gamma_n)x^2/2 + (\gamma_n d_2 - c_2)x)|_{c_2}^{d_2}]|]$$
$$= \frac{1}{4}[\frac{1}{2}|[(w_m(c_1 + d_1 - b_1 - a_1) - (w_n(c_2 + d_2 - b_2 - a_2)]|$$
$$+ \frac{1}{2}|[(u_m(c_1 + d_1 - b_1 - a_1) +$$
$$(b_1 - a_1 + d_1 - c_1)) - (u_n(c_2 + d_2 - b_2 - a_2) + (b_2 - a_2 + d_2 - c_2))]|$$
$$+ \frac{1}{2}|[(\gamma_m(c_1 + d_1 - b_1 - a_1) +$$
$$(b_1 - a_1 + d_1 - c_1)) - (\gamma_n(c_2 + d_2 - b_2 - a_2) + (b_2 - a_2 + d_2 - c_2))]|]$$

then, from here, is obtained as following;

$$d_H^*(m, n) = \frac{1}{4}[\frac{1}{2}|[(w_m(c_1 +$$
$$d_1 - b_1 - a_1) - (w_n(c_2 + d_2 - b_2 - a_2)]|$$
$$+ \frac{1}{2}|[(u_m(c_1 + d_1 - b_1 - a_1) +$$
$$(b_1 - a_1 + d_1 - c_1)) - (u_n(c_2 + d_2 - b_2 - a_2)$$
$$+ (b_2 - a_2 + d_2 - c_2))]| + \frac{1}{2}|[(\gamma_m(c_1 + d_1 - b_1 - a_1) +$$
$$(b_1 - a_1 + d_1 - c_1)) - (\gamma_n(c_2 + d_2 - b_2 - a_2)$$
$$+ (b_2 - a_2 + d_2 - c_2))]|]$$

Example 1 Let $m = \langle(0.1, 0.2, 0.6, 0.7); 0.3, 0.4, 0, 5\rangle$ and $n = \langle(0.3, 0.4, 0.5, 0.7); 0.5, 0.6, 0, 9\rangle$ be two normal SVTN-numbers. Then, Hamming distance is computed as

$$
\begin{aligned}
d_H^*(m, n) = \tfrac{1}{4}[\tfrac{1}{2}|[0.3(0.6 + 0.7 - 0.2 - 0.1) - (0.5(0.5 + 0.7 - 0.4 - 0.3)]|+ \\
\tfrac{1}{2}|[0.4(0.6 + 0.7 - 0.2 - 0.1) + (0.2 - 0.1 + 0.7 - 0.6)) - \\
(0.6(0.5 + 0.7 - 0.4 - 0.3) + (0.4 - 0.3 + 0.7 - 0.5))| \\
+ \tfrac{1}{2}|[(0.5(0.6 + 0.7 - 0.2 - 0.1) + (0.2 - 0.1 + 0.7 - 0.6)) \\
- (0.9(0.5 + 0.7 - 0.4 - 0.3) + (0.4 - 0.3 + 0.7 - 0.5))]|] = 0,0125.
\end{aligned}
$$

Definition 10 Let $m = \langle(a_1, b_1, c_1, d_1); w_m, u_m, \gamma_m\rangle, n = \langle(a_2, b_2, c_2, d_2); w_n, u_n, \gamma_n\rangle$ be two normal SVTN-numbers. If $T_m < T_n, I_m > I_n$ and $F_m > F_n$, then m is smaller than n, denoted by $m < n$
where

$$
T_m = \frac{1}{2}[(w_m(c_1 + d_1 - b_1 - a_1),
$$

$$
T_n = \frac{1}{2}[w_n(c_2 + d_2 - b_2 - a_2))],
$$

$$
I_m = \frac{1}{2}[(c_1 + d_1 - b_1 - a_1) + (b_1 - a_1 + d_1 - c_1)],
$$

$$
I_n = \frac{1}{2}[(u_n(c_2 + d_2 - b_2 - a_2) + (b_2 - a_2 + d_2 - c_2))],
$$

$$
F_m = \frac{1}{2}[(\gamma_m(c_1 + d_1 - b_1 - a_1) + (b_1 - a_1 + d_1 - c_1)]
$$

and

$$
F_n = \frac{1}{2}[(\gamma_n(c_2 + d_2 - b_2 - a_2) + (b_2 - a_2 + d_2 - c_2))].
$$

Theorem 1 *Let m, n, p be three normal SVTN-numbers. Then the Hamming Distance Measure satisfies the following properties:*

1. $0 \leq d_H^*(m, n) \leq 1$,
2. $d_H^*(m, n) = d_H^*(n, m)$,
3. $d_H^*(m, n) = 0 \Leftrightarrow m = n$,
4. $d_H^*(m, n) \leq d_H^*(m, p)$ and $d_H^*(n, p) \leq d_H^*(m, p)$ for $m \geq n \geq p$.

Proof Firstly, we know that

$$
\begin{aligned}
d_H^*(m,n) = \tfrac{1}{4}[|[&\int_{a_1}^{b_1}((b_1-x)+w_m(x-a_1))\setminus(b_1-a_1)dx+ \\
&\int_{b_1}^{c_1} w_m dx + \int_{c_1}^{d_1}((x-c_1)+w_m(d_1-x))\setminus(d_1-c_1)dx] \\
-[&\int_{a_2}^{b_2}((b_2-x)+w_n(x-a_2))\setminus(b_2-a_2)dx+ \\
&\int_{b_2}^{c_2} w_n dx + \int_{c_2}^{d_2}((x-c_2)+w_n(d_2-x))\setminus(d_2-c_2)dx]| \\
+|[&\int_{a_1}^{b_1}((x-a_1)u_m\setminus(b_1-a_1)dx+ \\
&\int_{b_1}^{c_1} u_m dx + \int_{c_1}^{d_1} u_m(d_1-x))\setminus(d_1-c_1)dx] \\
-[&\int_{a_2}^{b_2}((x-a_2)u_n\setminus(b_2-a_2)dx+ \\
&\int_{b_2}^{c_2} u_n dx + \int_{c_2}^{d_2} u_n(d_2-x))\setminus(d_2-c_2)dx]| \\
+|[&\int_{a_1}^{b_1}((x-a_1)\gamma_m\setminus(b_1-a_1)dx+ \\
&\int_{b_1}^{c_1} \gamma_m dx + \int_{c_1}^{d_1} \gamma_m(d_1-x))\setminus(d_1-c_1)dx] \\
-[&\int_{a_2}^{b_2}((x-a_2)\gamma_n\setminus(b_2-a_2)dx+ \\
&\int_{b_2}^{c_2} \gamma_n dx + \int_{c_2}^{d_2} \gamma_n(d_2-x))\setminus(d_2-c_2)dx]|] \\
= \tfrac{1}{4}[\tfrac{1}{2}&|[(w_m(c_1+d_1-b_1-a_1)-(w_n(c_2+ \\
&d_2-b_2-a_2))]|+\tfrac{1}{2} \\
&|[(u_m(c_1+d_1-b_1-a_1)+(b_1-a_1+d_1-c_1))-(u_n(c_2+ \\
&d_2-b_2-a_2)+(b_2-a_2+d_2-c_2))]| \\
+\tfrac{1}{2}&|[(\gamma_m(c_1+d_1-b_1-a_1)+(b_1-a_1+d_1-c_1))-(\gamma_n(c_2+ \\
&d_2-b_2-a_2)+(b_2-a_2+d_2-c_2))]|]
\end{aligned}
$$

1. We can write that

$$
0 \le (c_1+d_1)-(b_1+a_1) \le 2
$$

so

$$
0 \le \frac{1}{2}[(w_m(c_1+d_1-b_1-a_1)] \le 1
$$

and also

$$
-1 \le -\frac{1}{2}[(w_n(c_1+d_1-b_1-a_1)] \le 0
$$

from here it is understood

$$
0 \le |T_m - T_n| \le 1 \dots (1)
$$

where $T_m = \frac{1}{2}[(w_m(c_1+d_1-b_1-a_1)]$ and $T_n = \frac{1}{2}[(w_n(c_1+d_1-b_1-a_1)]$.
And also, it is open for the first equality that $0 \le u_m((c_1+d_1)-(b_1+a_1)) \le 2$.

Let us look at the second equality, it is clear that $0 \le (b_1 - a_1) + (d_1 - c_1) \le 1$ where $a_1, b_1, c_1, d_1 \in [0, 1]$ and $a_1 \le b_1 \le c_1 \le d_1$ also then it clearly follows that

$$0 \le \frac{1}{2}[(u_m(c_1 + d_1 - b_1 - a_1) + (b_1 - a_1 + d_1 - c_1))] \le \frac{3}{2}$$

$$-\frac{3}{2} \le -\frac{1}{2}[u_n(c_2 + d_2 - b_2 - a_2) + (b_2 - a_2 + d_2 - c_2))] \le 0$$

where $a_i \le b_i \le c_i \le d_i$ for $i = 1, 2$ and $a_i, b_i, c_i, d_i \in [0, 1]$ $w, u, \pi \in [0, 1]$.
Thus,

$$-\frac{3}{2} \le I_m - I_n \le \frac{3}{2}$$

and also

$$0 \le |I_m - I_n| \le \frac{3}{2} \dots (2)$$

where $I_m = \frac{1}{2}[(w_m(c_1 + d_1 - b_1 - a_1) + (b_1 - a_1 + d_1 - c_1))]$ and $I_n = \frac{1}{2}[w_n(c_2 + d_2 - b_2 - a_2) + (b_2 - a_2 + d_2 - c_2))]$.
Similarly,

$$0 \le F_m \le \frac{3}{2}$$

$$0 \le F_n \le \frac{3}{2}$$

from here

$$0 \le |F_m - F_n| \le \frac{3}{2} \dots (3)$$

where $F_m = \frac{1}{2}[(\gamma_m(c_1 + d_1 - b_1 - a_1) + (b_1 - a_1 + d_1 - c_1))]$ and $F_n = \frac{1}{2}[\gamma_n(c_2 + d_2 - b_2 - a_2) + (b_2 - a_2 + d_2 - c_2))]$.
Finally, if (1), (2), and (3) equalities are collected side to side, then we have

$$[(1) + (2) + (3)] = 4$$

and

$$\frac{1}{4}4 = 1$$

2. It is clear that

$$d_H^*(m, n) = |T_m - T_n| + |I_m - I_n| + |F_m - F_n| = |T_n - T_m| + |I_n - I_m| + |F_n - F_m| = d_H^*(n, m).$$

where $T_m = \frac{1}{2}[(w_m(c_1 + d_1 - b_1 - a_1)], T_n = \frac{1}{2}[(w_n(c_1 + d_1 - b_1 - a_1)],$
$I_m = \frac{1}{2}[(u_m(c_1 + d_1 - b_1 - a_1) + (b_1 - a_1 + d_1 - c_1))], I_n = \frac{1}{2}[u_n(c_2 + d_2 - b_2 - a_2) + (b_2 - a_2 + d_2 - c_2))], F_m = \frac{1}{2}[(\gamma_m(c_1 + d_1 - b_1 - a_1) + (b_1 - a_1 + d_1 - c_1))]$
and $F_n = \frac{1}{2}[\gamma_n(c_2 + d_2 - b_2 - a_2) + (b_2 - a_2 + d_2 - c_2))]$. thus we have

$$d_H^*(m, n) = d_H^*(n, m)$$

3. It is obvious that if $d_H^*(m, n) = 0$,

$$|T_m - T_n| + |I_m - I_n| + |F_m - F_n| = 0$$

from here $T_m = T_n, I_m = I_n$, and $F_m = F_n$. Thus, $m = n$.

4. If $m \geq n \geq p$, then we have $T_m \geq T_n \geq T_p, I_m \leq I_n \leq I_p, F_m \leq F_n \leq F_p$ from definition 3.5 where
$T_m = \frac{1}{2}[(w_m(c_1 + d_1 - b_1 - a_1)], T_n = \frac{1}{2}[(w_n(c_1 + d_1 - b_1 - a_1)], I_m = \frac{1}{2}[(u_m(c_1 + d_1 - b_1 - a_1) + (b_1 - a_1 + d_1 - c_1))], I_n = \frac{1}{2}[u_n(c_2 + d_2 - b_2 - a_2) + (b_2 - a_2 + d_2 - c_2))], F_m = \frac{1}{2}[(\gamma_m(c_1 + d_1 - b_1 - a_1) + (b_1 - a_1 + d_1 - c_1))],$
and $F_n = \frac{1}{2}[\gamma_n(c_2 + d_2 - b_2 - a_2) + (b_2 - a_2 + d_2 - c_2))]$. Therefore, we have

$$|T_m - T_n| \leq |T_m - T_p|$$

$$|I_m - I_n| \leq |I_m - I_p|$$

$$|F_m - F_n| \leq |F_m - F_p|$$

and then

$$d_H^*(m, n) = |T_m - T_n| + |I_m - I_n| + |F_m - F_n|$$

$$d_H^*(m, p) = |T_m - T_p| + |I_m - I_p| + |F_m - F_p|$$

Finally, $d_H^*(m, n) \leq d_H^*(m, p)$.
We can make similarly that

$$d_H^*(n, p) \leq d_H^*(m, p).$$

Definition 11 Let $m = \langle(a_1, b_1, c_1, d_1); w_m, u_m, \gamma_m\rangle, n = \langle(a_2, b_2, c_2, d_2); w_n, u_n, \gamma_n\rangle$ be two normal SVTN-numbers. Then, the Euclidean distance $d_E^*(m, n)$ between m and n is defined as

$$d_E^*(m, n) = \frac{\sqrt{2}}{\sqrt{11}}[(T_m - T_n)^2 + (I_m - I_n)^2 + (F_m - F_n)^2]^{\frac{1}{2}}$$

where $T_m = \frac{1}{2}[(w_m(c_1 + d_1 - b_1 - a_1)]$, $T_n = \frac{1}{2}[(w_n(c_1 + d_1 - b_1 - a_1)]$, $I_m = \frac{1}{2}[(u_m(c_1 + d_1 - b_1 - a_1) + (b_1 - a_1 + d_1 - c_1))]$, $I_n = \frac{1}{2}[u_n(c_2 + d_2 - b_2 - a_2) + (b_2 - a_2 + d_2 - c_2))]$, $F_m = \frac{1}{2}[(\gamma_m(c_1 + d_1 - b_1 - a_1) + (b_1 - a_1 + d_1 - c_1))]$ and $F_n = \frac{1}{2}[\gamma_n(c_2 + d_2 - b_2 - a_2) + (b_2 - a_2 + d_2 - c_2))]$.

Lemma 2 Let $m = \langle (a_1, b_1, c_1, d_1); w_m, u_m, \gamma_m \rangle$, $n = \langle (a_2, b_2, c_2, d_2); w_n, u_n, \gamma_n \rangle$ be two normal SVTN-numbers. Then, the Euclidean distance between m and n is calculated as

$$d_E^*(m, n) = \frac{\sqrt{2}}{\sqrt{11}}([\frac{1}{2}[(w_m(c_1 + d_1 - b_1 - a_1)]) - \frac{1}{2}[(w_n(c_1 + d_1 - b_1 - a_1)])])^2 +$$
$$([\frac{1}{2}(w_m(c_1 + d_1 - b_1 - a_1) + (b_1 - a_1 + d_1 - c_1))] - \frac{1}{2}[w_n(c_2 + d_2 - b_2 - a_2) + (b_2 - a_2 + d_2 - c_2))])^2 +$$
$$(\frac{1}{2}[(\gamma_m(c_1 + d_1 - b_1 - a_1) + (b_1 - a_1 + d_1 - c_1))] - \frac{1}{2}[\gamma_n(c_2 + d_2 - b_2 - a_2) + (b_2 - a_2 + d_2 - c_2))])^2)^{\frac{1}{2}}$$

Proof It is obvious from proof 3.3.

Example 2 Let $m = \langle (0.1, 0.2, 0.6, 0.7); 0.3, 0.4, 0, 5 \rangle$ and $n = \langle (0.3, 0.4, 0.5, 0.7); 0.5, 0.6, 0, 9 \rangle$ be two normal SVTN-numbers. Then, we computed the Euclidean distance between m and n as

$$d_E^*(m, n) = \frac{\sqrt{2}}{\sqrt{11}}[(\frac{1}{2}[0.3(0.6 + 0.7 - 0.2 - 0.1) - (0.5(0.5 + 0.7 - 0.4 - 0.3))^2]$$
$$+ (\frac{1}{2}[0.4(0.6 + 0.7 - 0.2 - 0.1) + (0.2 - 0.1 + 0.7 - 0.6)) -$$
$$(0.6(0.5 + 0.7 - 0.4 - 0.3) + (0.4 - 0.3 + 0.7 - 0.5))])^2 + (\frac{1}{2}[(0.5$$
$$(0.6 + 0.7 - 0.2 - 0.1) + (0.2 - 0.1 + 0.7 - 0.6)) - (0.9(0.5 + 0.7$$
$$- 0.4 - 0.3) + (0.4 - 0.3 + 0.7 - 0.5))])^2]^{\frac{1}{2}} = 0.027455024.$$

Theorem 2 Let m, n, p be three normal SVTN-numbers. Then the Euclidean Distance Measure satisfies the following properties:

1. $0 \leq d_E^*(m, n) \leq 1$,
2. $d_E^*(m, n) = d_E^*(n, m)$,
3. $d_E^*(m, n) = 0 \Leftrightarrow m = n$,
4. $d_E^*(m, n) \leq d_E^*(m, p)$ and $d_E^*(n, p) \leq d_E^*(m, p)$ for $m \geq n \geq p$.

Proof Assume that the Euclidean distance $d_E^*(m, n)$ between m and n is given as

$$d_E^*(m, n) = \frac{\sqrt{2}}{\sqrt{11}}[(T_m - T_n)^2 + (I_m - I_n)^2 + (F_m - F_n)^2]^{\frac{1}{2}}$$

1. From Proof 3.6., we have

$$0 \le (T_m - T_n)^2 \le 1$$

$$0 \le (I_m - I_n)^2 \le \frac{9}{4}$$

$$0 \le (F_m - F_n)^2 \le \frac{9}{4}$$

from here

$$0 \le \frac{\sqrt{2}}{\sqrt{11}}[(T_m - T_n)^2 + (I_m - I_n)^2 + (F_m - F_n)^2]^{\frac{1}{2}} \le 1$$

2. it is clear that

$$d_E^*(m, n) = \frac{\sqrt{2}}{\sqrt{11}}[(T_m - T_n)^2 + (I_m - I_n)^2 + (F_m - F_n)^2]^{\frac{1}{2}}$$
$$= \frac{\sqrt{2}}{\sqrt{11}}[(T_n - T_m)^2 + (I_n - I_m)^2 + (F_n - F_m)^2]^{\frac{1}{2}} = d_E^*(n, m).$$

3. It is obvious that if $d_E^*(m, n) = 0$,

$$d_E^*(m, n) = \frac{\sqrt{2}}{\sqrt{11}}[(T_m - T_n)^2 + (I_m - I_n)^2 + (F_m - F_n)^2]^{\frac{1}{2}} = 0$$

from here $T_m = T_n$, $I_m = I_n$ and $F_m = F_n$.
4. It is obvious from the previous theorem.

Definition 12 Let $m = \langle (a_1, b_1, c_1, d_1); w_m, u_m, \gamma_m \rangle$, $n = \langle (a_2, b_2, c_2, d_2); w_n, u_n, \gamma_n \rangle$ be two normal SVTN-numbers. Then, the hybrid distance between two normal SVTNs is denoted by $d_h^*(m, n)$ and is defined as

$$d_h^*(m, n) = \frac{2}{11}[|T_m - T_n| + |I_m - I_n| + |F_m - F_n| + max\{|T_m - T_n|, |I_m - I_n|, |F_m - F_n|\}]$$

where $T_m = \frac{1}{2}[(w_m(c_1 + d_1 - b_1 - a_1)]$, $T_n = \frac{1}{2}[(w_n(c_1 + d_1 - b_1 - a_1)]$, $I_m = \frac{1}{2}[(u_m(c_1 + d_1 - b_1 - a_1) + (b_1 - a_1 + d_1 - c_1))]$, $I_n = \frac{1}{2}[u_n(c_2 + d_2 - b_2 - a_2) + (b_2 - a_2 + d_2 - c_2))]$, $F_m = \frac{1}{2}[(\gamma_m(c_1 + d_1 - b_1 - a_1) + (b_1 - a_1 + d_1 - c_1))]$, and $F_n = \frac{1}{2}[\gamma_n(c_2 + d_2 - b_2 - a_2) + (b_2 - a_2 + d_2 - c_2))]$.

Lemma 3 Let $m = \langle (a_1, b_1, c_1, d_1); w_m, u_m, \gamma_m \rangle$, $n = \langle (a_2, b_2, c_2, d_2); w_n, u_n, \gamma_n \rangle$ be two normal SVTN-numbers. Then, the distance measure between m and n is calculated as

$$d_h^*(m, n) = \frac{2}{11}[|\frac{1}{2}[(w_m(c_1 + d_1 - b_1 - a_1)] - \frac{1}{2}[(w_n(c_1 + d_1 - b_1 - a_1)]|+$$
$$|\frac{1}{2}[(u_m(c_1 + d_1 - b_1 - a_1) + (b_1 - a_1 + d_1 - c_1))]-$$
$$\frac{1}{2}[u_n(c_2 + d_2 - b_2 - a_2) + (b_2 - a_2 + d_2 - c_2))]|+$$
$$|\frac{1}{2}[(\gamma_m(c_1 + d_1 - b_1 - a_1)+$$
$$(b_1 - a_1 + d_1 - c_1))] - \frac{1}{2}[\gamma_n(c_2 + d_2 - b_2 - a_2)+$$
$$(b_2 - a_2 + d_2 - c_2))]| + max\{|\frac{1}{2}[(w_m(c_1 + d_1 - b_1 - a_1)]-$$
$$\frac{1}{2}[(w_n(c_1 + d_1 - b_1 - a_1)]|, |\frac{1}{2}[(u_m(c_1 + d_1 - b_1 - a_1)$$
$$+(b_1 - a_1 + d_1 - c_1))]$$
$$-\frac{1}{2}[u_n(c_2 + d_2 - b_2 - a_2) + (b_2 - a_2 + d_2 - c_2))]|,$$
$$|\frac{1}{2}[(\gamma_m(c_1 + d_1 - b_1 - a_1) + (b_1 - a_1 + d_1 - c_1))]$$
$$-\frac{1}{2}[\gamma_n(c_2 + d_2 - b_2 - a_2)$$
$$+(b_2 - a_2 + d_2 - c_2))].$$

Proof It can be made similarly to Lemma 3.2.

Example 3 Let $m = \langle(0.1, 0.2, 0.6, 0.7); 0.3, 0.4, 0, 5\rangle$ and $n = \langle(0.3, 0.4, 0.5, 0.7); 0.5, 0.6, 0, 9\rangle$ be two single valued trapezoidal neutrosophic numbers. Then, we have,

$$d_h^*(m, n) = \frac{2}{11}[|\frac{1}{2}[(0.3(0.6 + 0.7 - 0.2 - 0.1)]] - \frac{1}{2}[(0.5(0.7 + 0.5 - 0.4 - 0.3)]|$$
$$+|\frac{1}{2}[0.4(0.6 + 0.7 - 0.2 - 0.1) + (0.2 - 0.1 + 0.7 - 0.6))]-$$
$$\frac{1}{2}[(0.6(0.5 + 0.7 - 0.4 - 0.3) + (0.4 - 0.3 + 0.7 - 0.5))]|+$$
$$|\frac{1}{2}[(0.5(0.6 + 0.7 - 0.2 - 0.1)+$$
$$(0.2 - 0.1 + 0.7 - 0.6))] - \frac{1}{2}[(0.9(0.5 + 0.7 - 0.4 - 0.3)+$$
$$(0.4 - 0.3 + 0.7 - 0.5))]| + max\{|[\frac{1}{2}[(0.3(0.6 + 0.7 - 0.2 - 0.1)]]-$$
$$\frac{1}{2}[(0.5(0.7 + 0.5 - 0.4 - 0.3)]|, |[\frac{1}{2}[0.4(0.6 + 0.7 - 0.2 - 0.1)+$$
$$(0.2 - 0.1 + 0.7 - 0.6))]] - \frac{1}{2}[(0.6(0.5 + 0.7 - 0.4 - 0.3)+$$
$$(0.4 - 0.3 + 0.7 - 0.5))]|, |\frac{1}{2}[(0.5(0.6 + 0.7 - 0.2 - 0.1) + (0.2$$
$$-0.1 + 0.7 - 0.6))] - \frac{1}{2}[(0.9(0.5 + 0.7 - 0.4 - 0.3)+$$
$$(0.4 - 0.3 + 0.7 - 0.5))]|\}]$$
$$= 0,013636364.$$

Theorem 3 *Let m, n, p be three normal SVTN-numbers. Then the Hybrid Distance Measure satisfies the following properties:*

1. $0 \leq d_h^*(m, n) \leq 1$,
2. $d_h^*(m, n) = d_h^*(n, m)$,
3. $d_h^*(m, n) = 0 \Leftrightarrow m = n$,
4. $d_h^*(m, n) \leq d_h^*(m, p)$ and $d_h^*(n, p) \leq d_h^*(m, p)$ for $m \geq n \geq p$.

Proof Assume that the hybrid distance $d_h^*(m, n)$ between m and n is given as

$$d_h^*(m, n) = \frac{2}{11}[|T_m - T_n| + |I_m - I_n| + |F_m - F_n| + max\{|T_m - T_n|, |I_m - I_n|, |F_m - F_n|\}]$$

1. We can write from proof 3.6 as follows:

$$0 \leq |T_m - T_n| \leq 1$$

$$0 \leq |I_m - I_n| \leq \frac{3}{2}$$

$$0 \leq |F_m - F_n| \leq \frac{3}{2}$$

we have

$$0 \leq \frac{2}{11}[|T_m - T_n| + |I_m - I_p| + (F_m - F_n) + max\{|T_m - T_n|, |I_m - I_n|, |F_m - F_n|\}] \leq 1$$

2. We know that

$$\begin{aligned}
d_h^*(m, n) &= \tfrac{2}{11}[|T_m - T_n| + |I_m - I_n| + |F_m - F_n| \\
&\quad + max\{|T_m - T_n|, |I_m - I_n|, |F_m - F_n|\}] \\
&= \tfrac{2}{11}[|T_n - T_m| + |I_n - I_m| + |F_n - F_m| \\
&\quad + max\{|T_n - T_m|, |I_n - I_m|, |F_n - F_m|\}] \\
&= d_h^*(n, m).
\end{aligned}$$

where $T_m = \frac{1}{2}[(w_m(c_1 + d_1 - b_1 - a_1)]$, $T_n = \frac{1}{2}[(w_n(c_1 + d_1 - b_1 - a_1)]$, $I_m = \frac{1}{2}[(u_m(c_1 + d_1 - b_1 - a_1) + (b_1 - a_1 + d_1 - c_1))]$, $I_n = \frac{1}{2}[u_n(c_2 + d_2 - b_2 - a_2) + (b_2 - a_2 + d_2 - c_2))]$, $F_m = \frac{1}{2}[(\gamma_m(c_1 + d_1 - b_1 - a_1) + (b_1 - a_1 + d_1 - c_1))]$ and $F_n = \frac{1}{2}[\gamma_n(c_2 + d_2 - b_2 - a_2) + (b_2 - a_2 + d_2 - c_2))]$.

3. It is obvious that if $d_h^*(m, n) = 0$,

$$\begin{aligned}
d_h^*(m, n) &= \tfrac{2}{11}[|T_m - T_n| + |I_m - I_n| + |F_m - F_n| + max\{|T_m - T_n|, \\
&\quad |I_m - I_n|, |F_m - F_n|\}] \\
&= 0.
\end{aligned}$$

from here $T_m = T_n$, $I_m = I_n$, and $F_m = F_n$. Thus, $m = n$.

4. It is obvious from the previous theorem.

Definition 13 Let $m = \langle(a_1, b_1, c_1, d_1); w_m, u_m, \gamma_m\rangle$, $n = \langle(a_2, b_2, c_2, d_2); w_n, u_n, \gamma_n\rangle$ be two normal SVTN-numbers. Then, the H-max measure distance between m and n is denoted by $d_{Hm}^*(m, n)$ and is defined as

$$d_{Hm}^*(m, n) = \frac{2}{11}[|T_m - T_n| + |I_m - I_n| + |F_m - F_n| + |max\{T_m, I_n, F_n\} - max\{T_n, I_m, F_m\}|]$$

where $T_m = \frac{1}{2}[(w_m(c_1 + d_1 - b_1 - a_1)]$, $T_n = \frac{1}{2}[(w_n(c_1 + d_1 - b_1 - a_1)]$, $I_m = \frac{1}{2}[(u_m(c_1 + d_1 - b_1 - a_1) + (b_1 - a_1 + d_1 - c_1))]$, $I_n = \frac{1}{2}[u_n(c_2 + d_2 - b_2 - a_2) +$

$(b_2 - a_2 + d_2 - c_2))]$, $F_m = \frac{1}{2}[(\gamma_m(c_1 + d_1 - b_1 - a_1) + (b_1 - a_1 + d_1 - c_1))]$, and $F_n = \frac{1}{2}[\gamma_n(c_2 + d_2 - b_2 - a_2) + (b_2 - a_2 + d_2 - c_2))]$.

Lemma 4 *Let* $m = \langle(a_1, b_1, c_1, d_1); w_m, u_m, \gamma_m\rangle$, $n = \langle(a_2, b_2, c_2, d_2); w_n, u_n, \gamma_n\rangle$ *be two normal SVTN-numbers. Then, the H-max measure between* m *and* n *is calculated as*

$$
\begin{aligned}
d_{Hm}^*(m, n) = \frac{2}{11}[\,|\,&\tfrac{1}{2}[(w_m(c_1 + d_1 - b_1 - a_1)] - \tfrac{1}{2}[(w_n(c_1 + d_1 - b_1 - a_1)]\,|+ \\
&|\tfrac{1}{2}[(u_m(c_1 + d_1 - b_1 - a_1) + (b_1 - a_1 + d_1 - c_1))] - \\
&\tfrac{1}{2}[u_n(c_2 + d_2 - b_2 - a_2) + (b_2 - a_2 + d_2 - c_2))]\,| + |\tfrac{1}{2}[(\gamma_m(c_1 + \\
= d_1 &- b_1 - a_1) + (b_1 - a_1 + d_1 - c_1))] - \\
&\tfrac{1}{2}[\gamma_n(c_2 + d_2 - b_2 - a_2) + (b_2 - a_2 + d_2 - c_2))]\,| + \\
&|max\{\tfrac{1}{2}[(w_m(c_1 + d_1 - b_1 - a_1)], \tfrac{1}{2}[u_n(c_2 + d_2 - b_2 - a_2) \\
&+(b_2 - a_2 + d_2 - c_2))], \tfrac{1}{2}[\gamma_n(c_2 + d_2 - b_2 - a_2) + (b_2 - a_2 + d_2 - c_2))]\} - \\
&max\{\tfrac{1}{2}[(w_n(c_1 + d_1 - b_1 - a_1)], \tfrac{1}{2}[(u_m(c_1 + d_1 - b_1 - a_1) + (b_1 - a_1 + \\
&d_1 - c_1))], \tfrac{1}{2}[(\gamma_m(c_1 + d_1 - b_1 - a_1) + (b_1 - a_1 + d_1 - c_1))]\}.
\end{aligned}
$$

Proof It is obvious from Lemma 3.2.

Example 4 Let $\tilde{a} = \langle(0.1, 0.2, 0.6, 0.7); 0.3, 0.4, 0, 5\rangle$ and $\tilde{b} = \langle(0.3, 0.4, 0.5, 0.7); 0.5, 0.6, 0, 9\rangle$ be two normal SVTN-numbers. Then, the H-max measure between m and n is calculated as

$$
\begin{aligned}
d_{Hm}^*(m, n) = \frac{2}{11}[\,|\,&\tfrac{1}{2}[(0.3(0.6 + 0.7 - 0.2 - 0.1)]] - \tfrac{1}{2}[(0.5(0.7 + 0.5 - 0.4 - 0.3)]\,|+ \\
&|\tfrac{1}{2}[0.4(0.6 + 0.7 - 0.2 - 0.1) + (0.2 - 0.1 + 0.7 - 0.6))] - \\
&\tfrac{1}{2}[(0.6(0.5 + 0.7 - 0.4 - 0.3) + (0.4 - 0.3 + 0.7 - 0.5))]\,|+ \\
&|\tfrac{1}{2}[(0.5(0.6 + 0.7 - 0.2 - 0.1) + (0.2 - 0.1 + 0.7 - 0.6))] - \\
&\tfrac{1}{2}[(0.9(0.5 + 0.7 - 0.4 - 0.3) + (0.4 - 0.3 + 0.7 - 0.5))]\,|+ \\
= |max\{&\tfrac{1}{2}[(0.3(0.6 + 0.7 - 0.2 - 0.1)]], \tfrac{1}{2}[(0.6(0.5 + 0.7 - 0.4 - 0.3) \\
&+(0.4 - 0.3 + 0.7 - 0.5))], \tfrac{1}{2}[(0.9(0.5 + 0.7 - 0.4 - 0.3) + \\
&(0.4 - 0.3 + 0.7 - 0.5))] - max\{\tfrac{1}{2}[(0 \ 5(0.7 + 0.5 \ - 0.4 - 0.3)], \\
= \tfrac{1}{2}[&0.4(0.6 + 0.7 - 0.2 - 0.1) + (0.2 - 0.1 + 0.7 - 0.6))] \\
&, \tfrac{1}{2}[0.5(0.6 + 0.7 - 0.2 - 0.1) + (0.2 - 0.1 + 0.7 - 0.6))]\}\}\,|] \\
= 0, &013636364.
\end{aligned}
$$

Theorem 4 *Let* m, n, p *be three normal SVTN-numbers. Then the H-max Measure Distance Measure satisfies the following properties:*

1. $0 \le d_{Hm}^*(m, n) \le 1$,
2. $d_{Hm}^*(m, n) = d_{Hm}^*(n, m)$,
3. $d_{Hm}^*(m, n) = 0 \Leftrightarrow m = n$,
4. $d_{Hm}^*(m, n) \le d_{Hm}^*(m, p)$ and $d_{Hm}^*(n, p) \le d_{Hm}^*(m, p)$ *for* $m \ge n \ge p$.

Proof Assume that the H-max distance $d^*_{Hm}(m, n)$ between m and n is given as

$$d^*_{Hm}(m, n) = \frac{2}{11}[|T_m - T_n| + |I_m - I_n| + |F_m - F_n| + |max\{T_m, I_n, F_n\} - max\{T_n, I_m, F_m\}|]$$

1. Since

$$0 \le |T_m - T_n| \le 1$$

$$0 \le |I_m - I_n| \le \frac{3}{2}$$

$$0 \le |F_m - F_n| \le \frac{3}{2}$$

and also

$$0 \le T_m \le 1 \quad \text{and} \quad 0 \le T_n \le 1$$

$$0 \le I_m \le \frac{3}{2} \quad \text{and} \quad 0 \le I_n \le \frac{3}{2}$$

$$0 \le F_m \le \frac{3}{2} \quad \text{and} \quad 0 \le F_n \le \frac{3}{2}$$

we have

$$0 \le d^*_{Hm}(m, n) \le 1.$$

2. it is clear that

$$d^*_{Hm}(m, n) = d^*_{Hm}(n, m).$$

3. It is obvious that

$$d^*_{Hm}(m, n) = 0, m = n.$$

4. It is obvious from the previous theorem.

Definition 14 Let $\tilde{a}_1, \tilde{a}_2$ be two normal SVTN-numbers and $\tilde{a^+}$ and $\tilde{a^-}$ be positive and negative ideal solutions, respectively. Then, we can compare the $\tilde{a}_1$ and $\tilde{a}_2$ as

1. If $d^+(\tilde{a}, \tilde{a^+}_1) < d^+(\tilde{a^+}, \tilde{a}_2)$, then $\tilde{a}_1$ is bigger than $\tilde{a}_2$, denoted by $\tilde{a}_1 > \tilde{a}_2$;
2. If $d^+(\tilde{a^+}, \tilde{a}_1) = d^+(\tilde{a^+}, \tilde{a}_2)$;

   (a) If $d^-(\tilde{a^-}, \tilde{a}_1) > d^-(\tilde{a^-}, \tilde{a}_2)$, then $\tilde{a}_1$ is smaller than $\tilde{a}_2$, denoted by $\tilde{a}_1 < \tilde{a}_2$;

(b) If $d^-(\tilde{a^-}, \tilde{a}_1) = d^-(\tilde{a^-}, \tilde{a}_2)$, then $\tilde{a}_1$ and $\tilde{a}_2$ are the same, denoted by $\tilde{a}_1 = \tilde{a}_2$

where $\tilde{a^+} = \langle (0, 0, 1, 1); 1, 0, 0 \rangle$ and $\tilde{a^-} = \langle (0, 0, 1, 1); 0, 1, 1 \rangle$

Example 5 Let $\tilde{a}_1, \tilde{a}_2, \tilde{a}_3, \tilde{a}_4$, and $\tilde{a}_5$ be the normal SVTN-numbers as follows;

$$\tilde{a}_1 = \langle (0.57, 0.69, 0.72, 0.78); 0.80, 0.40, 0.35 \rangle$$
$$\tilde{a}_2 = \langle (0.03, 0.05, 0.08, 0.09); 0.10, 0.90, 0.97 \rangle$$
$$\tilde{a}_3 = \langle (0.40, 0.50, 0.64, 0.65); 0.50, 0.60, 0.70 \rangle$$
$$\tilde{a}_4 = \langle (0.72, 0.82, 0.89, 0.90); 0.70, 0.20, 0.20 \rangle$$
$$\tilde{a}_5 = \langle (0.83, 0.90, 0.93, 0.94); 0.90, 0.11, 0.19 \rangle$$

then we rank the normal SVTN-numbers by using Hamming distance as Since;

$$d^+(\tilde{a^+}, \tilde{a}_1) = 0,258285714$$
$$d^+(\tilde{a^+}, \tilde{a}_2) = 0,304185714$$
$$d^+(\tilde{a^+}, \tilde{a}_3) = 0,314571429$$
$$d^+(\tilde{a^+}, \tilde{a}_4) = 0,259285714$$
$$d^+(\tilde{a^+}, \tilde{a}_5) = 0,262285714$$

we have

$$d^+(\tilde{a^+}, \tilde{a}_1) < d^+(\tilde{a^+}, \tilde{a}_4) < d^+(\tilde{a^+}, \tilde{a}_5) < d^+(\tilde{a^+}, \tilde{a}_2) < d^+(\tilde{a^+}, \tilde{a}_3)$$

Therefore, the ranking order of the alternatives x_j ($j = 1, 2, 3, 4$) is generated as follows:

$$\tilde{a}_3 < \tilde{a}_2 < \tilde{a}_5 < \tilde{a}_4 < \tilde{a}_1.$$

4 Multi-Criteria Decision Making Based on Normal SVTN-Numbers

In this section, we propose multi-criteria decision making based on normal SVTN-numbers.

Let $U = \{u_i : i = 1 \ldots m\}$ be set of alternatives based on a set of criteria $C = \{c_j : j = 1 \ldots n\}$ and $w = (w_1, w_2, \ldots, w_n)^T$ is a weight vector of criterions such that $w_k \in [0, 1]$ and $\sum_{k=1}^{n} w_k = 1$ which is created by distance measure based normal SVTN-numbers. Then, the decision maker can utilize to make an assessment of linguistic terms for alternatives based on criterions and weight vector of criterions as indicated in Tables 1 and 2, respectively (The tables are constructed based on Definition 14).

Table 1 Linguistic values for alternatives based on criterions

Linguistic terms	Linguistic values of normal SVTN-numbers
Very unsitable(vu)	$\langle(0.1, 0.2, 0.4, 0.5); 0.3, 0.83, 0.87\rangle$
Unsuitable(u)	$\langle(0.3, 0.4, 0.56, 0.6); 0.4, 0.7, 0.79\rangle$
Medium unsuitable(mu)	$\langle(0.4, 0.5, 0.64, 0.65); 0.5, 0.6, 0.7\rangle$
Medium suitable(ms)	$\langle(0.57, 0.69, 0.72, 0.78); 0.8, 0.4, 0.35\rangle$
Suitable(s)	$\langle(0.03, 0.05, 0.08, 0.09); 0.1, 0.9, 0.97\rangle$
Very suitable (vs)	$\langle(0.72, 0.82, 0.89, 0.9); 0.7, 0.2, 0.2\rangle$
Extremely suitable (es)	$\langle(0.83, 0.9, 0.93, 0.94); 0.9, 0.11, 0.19\rangle$

Table 2 Linguistic values for weight vector of criterions

Linguistic terms	Linguistic values of normal SVTN-numbers	$d_H^+(\tilde{\alpha}, lt)$
Very noneffective(vn)	$\langle(0.0, 0.3, 0.5, 0.6); 0.1, 0.9, 0.89\rangle$	0.519
Noneffective(n)	$\langle(0.2, 0.35, 0.4, 0.5); 0.3, 0.84, 0.87\rangle$	0.3741875
Medium noneffective(mn)	$\langle(0.41, 0.48, 0.5, 0.7); 0.33, 0.72, 0.79\rangle$	0.363225
Medium(m)	$\langle(0.5, 0.53, 0.69, 0.8); 0.4, 0.76, 0.65\rangle$	0.343075
Medium effective(me)	$\langle(0.6, 0.73, 0.75, 0.85); 0.6, 0.6, 0.6\rangle$	0.32775
Effective (e)	$\langle(0.72, 0.8, 0.83, 0.88); 0.8, 0.3, 0.2\rangle$	0.275375
Very effective (ve)	$\langle(0.89, 0.9, 0.93, 0.99); 0.96, 0.1, 0.1\rangle$	0.25515

Definition 15 ([11]) Let $X = (x_1, x_2, \ldots, x_m)$ be a set of alternatives, $U = (u_1, u_2, \ldots, u_n)$ be the set of attributes. If $\tilde{a}_{ij} = \langle(a_{ij}, b_{ij}, c_{ij}, d_{ij}); w_{ij}, u_{ij}, y_{ij}\rangle$ be normal SVTN-numbers, then

$$[\tilde{a}_{ij}]_{m \times n} = \begin{matrix} & \begin{matrix} u_1 & u_2 & \cdots & u_n \end{matrix} \\ \begin{matrix} x_1 \\ x_2 \\ \vdots \\ x_m \end{matrix} & \begin{pmatrix} \tilde{a}_{11} & \tilde{a}_{12} & \cdots & \tilde{a}_{1n} \\ \tilde{a}_{21} & \tilde{a}_{22} & \cdots & \tilde{a}_{2n} \\ \vdots & \vdots & \vdots & \vdots \\ \tilde{a}_{m1} & \tilde{a}_{m2} & \cdots & \tilde{a}_{mn} \end{pmatrix} \end{matrix}$$

is called an SVTN-multi-criteria decision making matrix of the decision maker.

Now, we can give an algorithm of the normal SVTN-multi-criteria decision making method as follows:

Algorithm

Step 1. Give the SVTN-multi-criteria decision making matrix based on Table 1 as

$$[\tilde{a}_{ij}]_{m \times n} = \langle(a_{ij}, b_{ij}, c_{ij}, d_{ij}); \tilde{w}_{ij}, \tilde{u}_{ij}, \tilde{y}_{ij}\rangle \ (i = 1 \ldots m; \ j = 1 \ldots n)$$

Step 2. Give weight vectors $W_j = \{W_1, W_2, \ldots, W_n\}$ based on Table 2 and normalize as $w_j = \{w_1, w_2, \ldots, w_n\}$ such that $w_k \in [0, 1]$ and $\sum_{k=1}^{n} w_k = 1$ with

$$w_j = \frac{d(W_j, \tilde{a}^+)}{\sum_{j=1}^{n} d(W_j, \tilde{a}^+)}$$

where $\tilde{a}^+ = \langle (0, 0, 1, 1); 1, 0, 0 \rangle$

Step 3.   Obtain $[m_{ij}]_{m \times n}$ matrix where $m_{ij} = w_j \times \tilde{a}_{ij}$ (i=1...m; j=1...n);

Step 4.   Compute $S_i = \sum_{j=1}^{k} m_{ij}$ (i=1...m; j=1...n);

Step 5.   Obtain $d(S_i, \tilde{a}^+)$ and $d(S_i, \tilde{a}^-)$;

Step 6.   Rank all alternatives u_i (i=1...m) based on $d(S_i, \tilde{a}^+)$ and $d(S_i, \tilde{a}^-)$.

5   An Illustrative Example

Example 6 (It is Adapted from [20]) Think that a company wants to invest in a city. This food company evaluates to make investment in five cities which are Ankara x_1, Istanbul x_2, Gaziantep x_3, Trabzon x_4, and Kayseri x_5. This company uses to choose the best alternative three criterions: (u_1) raw material and labor; (u_2) transportation and marketing; (u_3) labor and energy need which are defined based on normal SVTN-numbers in Tables 1 and 2. Thus, the company will select the best alternative among the five cities with the help of the following algorithm.

Step 1.   We gave the SVTN-multi-criteria decision making matrix based on Table 1 as

$$[\tilde{a}_{ij}]_{5 \times 3} = \begin{pmatrix} \langle (0.57, 0.69, 0.72, 0.78); 0.80, 0.40, 0.35 \rangle \\ \langle (0.03, 0.05, 0.08, 0.09); 0.10, 0.90, 0.97 \rangle \\ \langle (0.40, 0.50, 0.64, 0.65); 0.50, 0.60, 0.70 \rangle \\ \langle (0.72, 0.82, 0.89, 0.90); 0.70, 0.20, 0.20 \rangle \\ \langle (0.83, 0.90, 0.93, 0.94); 0.90, 0.11, 0.19 \rangle \end{pmatrix}$$

$$\begin{pmatrix} \langle (0.40, 0.50, 0.64, 0.65); 0.50, 0.60, 0.70 \rangle \\ \langle (0.10, 0.20, 0.40, 0.50); 0.30, 0.83, 0.87 \rangle \\ \langle (0.57, 0.69, 0.72, 0.78); 0.80, 0.40, 0.35 \rangle \\ \langle (0.83, 0.90, 0.93, 0.94); 0.90, 0.11, 0.19 \rangle \\ \langle (0.72, 0.82, 0.89, 0.90); 0.70, 0.20, 0.20 \rangle \end{pmatrix}$$

$$\begin{pmatrix} \langle (0.30, 0.40, 0.56, 0.60); 0.40, 0.70, 0.79 \rangle \\ \langle (0.72, 0.82, 0.89, 0.90); 0.70, 0.20, 0.20 \rangle \\ \langle (0.83, 0.90, 0.93, 0.94); 0.90, 0.11, 0.19 \rangle \\ \langle (0.40, 0.50, 0.64, 0.65); 0.50, 0.60, 0.70 \rangle \\ \langle (0.57, 0.69, 0.72, 0.78); 0.80, 0.40, 0.35 \rangle \end{pmatrix}$$

Step 2. We gave weight vectors $W = (W_1, W_2, W_3) = (mn, n, e)$ based on Table 2 and normalize as $w = (w_1, w_2, w_3) = (0.147786859, 0.152247217, 0.112042966)$ with

$$w_j = \frac{d(W_j, \tilde{a}^+)}{\sum_{j=1}^{3} d(W_j, \tilde{a}^+)}$$

where $\tilde{a}^+ = \langle (0, 0, 1, 1); 1, 0, 0 \rangle$

Step 3. We obtained $[m_{ij}]_{m \times n}$ matrix where $m_{ij} = w_j \times \tilde{a}_{ij}$ ($i = 1, 2, 3, 4, 5;$ $j = 1, 2, 3$) as

$$[m_{ij}]_{5 \times 3} = \begin{pmatrix} \langle (0.0842, 0.1019, 0.1064, 0, 1152); 0.80, 0.40, 0.35 \rangle \\ \langle (0.0044, 0.0073, 0.0118, 0.0133); 0.10, 0.90, 0.97 \rangle \\ \langle (0.0591, 0.0738, 0.0945, 0.0960); 0.50, 0.60, 0.70 \rangle \\ \langle (0.1064, 0.1211, 0.1315, 0.1330); 0.70, 0.20, 0.20 \rangle \\ \langle (0.1226, 0.1330, 0.1374, 0.1389); 0.90, 0.11, 0.19 \rangle \end{pmatrix}$$

$$\langle (0.0608, 0.0761, 0.0974, 0.0989); 0.50, 0.60, 0.70 \rangle \\ \langle (0.0152, 0.0304, 0.0608, 0.0761); 0.30, 0.83, 0.87 \rangle \\ \langle (0.0867, 0.1050, 0.1096, 0.1187); 0.80, 0.40, 0.35 \rangle \\ \langle (0.1263, 0.1370, 0.1415, 0.1431); 0.90, 0.11, 0.19 \rangle \\ \langle (0.1096, 0.1248, 0.13553, 0.1370); 0.70, 0.20, 0.20 \rangle$$

$$\langle (0.0336, 0.0448, 0.0627, 0.0672); 0.40, 0.70, 0.79 \rangle \\ \langle (0.0806, 0.0918, 0.0997, 0.1008); 0.70, 0.20, 0.20 \rangle \\ \langle (0.0929, 0.1008, 0.1041, 0.1053); 0.90, 0.11, 0.19 \rangle \\ \langle (0.0448, 0.0560, 0.0717, 0.0728); 0.50, 0.60, 0.70 \rangle \\ \langle (0.0638, 0.0773, 0.0806, 0.0873); 0.80, 0.40, 0.35 \rangle$$

Step 5. We computed $S_i = \sum_{j=1}^{3} m_{ij}$ (i=1,2,3,4,5; j=1,2,3) as ;

$$S_1 = \langle (0.1787, 0.2229, 0.2665, 0, 2814); 0.40, 0.70, 0.79 \rangle$$

$$S_2 = \langle (0.1003, 0.1297, 0.1724, 0.1902); 0.10, 0.90, 0.97 \rangle$$

$$S_3 = \langle (0.2388, 0.2797, 0.3084, 0.3201); 0.50, 0.60, 0.70 \rangle$$

$$S_4 = \langle (0.2775, 0.3142, 0.3448, 0.3489); 0.50, 0.60, 0.70 \rangle$$

$$S_5 = \langle (0.2961, 0.3351, 0.35368, 0.3633); 0.70, 0.40, 0.35 \rangle$$

Step 6. We obtained $d(S_i, \tilde{a}^+)$ as Table 3;

Table 3 Rank all alternatives x_i (i=1,2,3,4,5)

i	$d_H^*(S_i, \alpha^+)$	$d_E^*(S_i, \alpha^+)$	$d_h^*(S_i, \alpha^+)$	$d_{Hm}^*(S_i, \alpha^+)$
1	0.2847	0.7589	0.3835	0.3839
2	0.2911	0.7771	0.3923	0.3926
3	0.2741	0.7583	0.3762	0.3762
4	0.2703	0.7581	0.373	0.374
5	0.2627	0.7540	0.3674	0.3676
The best alternative	S_5	S_5	S_5	S_5
The worst alternative	S_2	S_2	S_2	S_2

6 Conclusion

In this paper, the basic definitions and properties of single valued trapezoidal neutrosophic numbers (SVTN-numbers) are given. Then, two distance measures based on SVTN-numbers are proposed and their properties are examined. Also, a multi-criteria decision making method based on the proposed distance measures under SVTN-numbers is developed. Finally, a numerical example in order to indicate the validity of obtained distance measures in real world is introduced.

Compliance with Ethical Standards

Conflict of Interest The authors declare that there is no conflict of interest with other organizations or people on this article.

Human and Animal Rights This article does not contain any studies with human participants or animals performed by the authors.

References

1. Atanassov, K. (1986). Intuitionistic fuzzy sets. *Fuzzy Sets and Systems, 20,* 87–96.
2. Atanassov, K. (1994). Operators over interval-valued intuitionistic fuzzy sets. *Fuzzy Sets and Systems, 64*(2), 159–174.
3. Biswas, P., Pramanik, S., & Giri, B. C. (2015). Cosine similarity measure based multi-attribute decision making with trapezoidal fuzzy neutrosophic numbers. *Neutrosophic Sets and System, 8,* 47–57.
4. Broumi, S., & Smarandache F. (2013). Several similarity measures of neutrosophic sets. *Neutrosophic Sets and Systems, 1,* 54–65.
5. Broumi, S., & Smarandache, F. (2013). Correlation coefficient of interval neutrosophic set. *Applied Mechanics and Materials, 436,* 511–517.
6. Broumi, S., & Smarandache, F. (2015). New operations on interval neutrosophic sets. *Journal of New Theory, 1,* 24–37.
7. Broumi, S., Talea, M., & Smarandache, F. (2016). Single Valued Neutrosophic Graphs: Degree, Order and Size. In *2016 IEEE international conference on fuzzy systems (FUZZ-IEEE).* New York: IEEE WCCI.

8. Chen, T. (2015). The inclusion-based TOPSIS method with interval-valued intuitionistic fuzzy sets for multiple criteria group decision making. *Applied Soft Computing, 26*, 57–73.
9. Cheng, H. D. & Guo, Y. (2008). A new neutrosophic approach to image thresholding. *New Mathematical Natural Computer, 4*(3), 291–308.
10. De Miguel, L., Bustince, H., Fernandez, J., Indura, E., Kolesa, A., & Mesiar, R. (2016). Construction of admissible linear orders for interval-valued Atanassov intuitionistic fuzzy sets with an application to decision making. *Information Fusion, 27*, 189–197.
11. Deli, I. & Subas, Y. (2016). A ranking method of single valued neutrosophic numbers and its applications to multi-attribute decision making problems. *International Journal of Machine Learning and Cybernatics, 8*(4), 1309–1322. https://doi.org/10.1007/s13042-016-0505-3.
12. Garg, H., Rani, M., Sharma, A., & Vishwakarma, Y. (2014). Intuitionistic fuzzy optimization technique for solving multi-objective reliability optimization problems in interval environment. *Expert Systems with Applications, 41*, 3157–3167.
13. Guo, Y., & Cheng, H.D. (2009). New neutrosophic approach to image segmentation. *Pattern Recognit, 42*, 587–595.
14. Khatibi, V. & Montazer, G. A. (2009). Intuitionistic fuzzy set vs. fuzzy set application in medical pattern recognition. *Artificial Intelligence in Medicine, 47*(1), 43–52.
15. Krohling, R., Pacheco, A., & Siviero, A. (2013). IF-TODIM: An intuitionistic fuzzy TODIM to multi-criteria decision making. *Knowledge-Based Systems, 53*, 142–146.
16. Li, D. & Cheng, C. (2002). New similarity measures of intuitionistic fuzzy sets and application to pattern recognition. *Pattern Recognition Letters, 23*, 221–225.
17. Liang, C., Zhao, S., & Zhang J. (2014). Aggregation operators on triangular intuitionistic fuzzy numbers and its application to multi-criteria decision making problems. *Foundations of Computing and Decision Sciences, 39*(3), 189–208.
18. Lupianez, F. G. (2009). Interval neutrosophic sets and topology. *Kybernetes, 38*(3/4), 621–624.
19. Ngan, R. T., Son, L. H., Cuong, B. C., & Ali, M. (2018). H-max distance measure of intuitionistic fuzzy sets in decision making. *Applied Soft Computing, 69*, 393–425.
20. Öztürk, E. K. (2018). Some new approaches on single valued trapezoidal neutrosophic numbers and their applications to multiple criteria decision making problems. (In Turkish) (Masters Thesis, Kilis 7 Aralik University, Graduate School of Natural and Applied Science).
21. Park, J. H., Lim, K. M., & Kwun, Y. C. (2009). Distance measure between intuitionistic fuzzy sets and its application to pattern recognition. *Korean Institute of Intelligent Systems, 19*(4), 556–561.
22. Peng, J. J., Wang, J. Q., Wu, X. H., Wang, J., & Chen, X. H. (2014). Multi-valued neutrosophic sets and power aggregation operators with their applications in multi-criteria group decision-making problems. *International Journal of Computational Intelligence Systems, 8*(2), 345–363.
23. Puri, J. & Yadav, S. P. (2015). Intuitionistic fuzzy data envelopment analysis an application to the banking sector in India. *Expert System Application, 42*(11), 4982–4998.
24. Rivieccio, U. (2008). Neutrosophic logics: Prospects and problems. *Fuzzy sets and systems, 159*, 1860–1868.
25. Smarandache, F. (1998). A unifying field in logics. In *Neutrosophy: Neutrosophic probability, set and logic*. Rehoboth: American Research Press.
26. Smarandache, F. (2015). *Types of Neutrosophic Graphs and neutrosophic algebraic structures together with their applications in Technology, seminar*. Brasov, Romania: Universitatea Transilvania din Brasov/Facultatea de Design de Produs si Mediu.
27. Szmidt, E. & Kacprzyk, J. (2000). Distances between intuitionistic fuzzy sets. *Fuzzy Sets and Systems, 114*(3), 505–518.
28. Turksen, B. (1986). Interval valued fuzzy sets based on normal forms. *Fuzzy Sets and Systems, 20*, 191–210.
29. Vasantha Kandasamy, W. B. & Smarandache, F. (2003). Fuzzy Cognitive Maps and Neutrosophic Cognitive Maps. arXiv:math/0311063v1.

30. Vasantha Kandasamy, W. B. & Smarandache, F. (2004). Analysis of social aspects of migrant laborers living with HIV/AIDS using Fuzzy Theory and Neutrosophic Cognitive Maps. Xiquan: Phoenix.

31. Vasantha Kandasamy, W. B., Ilanthenral, K., & Smarandache, F. (2015). *Neutrosophic graphs: A new dimension to graph theory*, Kindle edn.

32. Wan, S. & Dong, J. (2014). A possibility degree method for interval-valued intuitionistic fuzzy multi-attribute group decision making. *Journal of Computer and System Sciences, 80*, 237–256.

33. Wan, S. & Dong, J. (2015). Power geometric operators of trapezoidal intuitionistic fuzzy numbers and application to multi-attribute group decision making. *Applied Soft Computing Journal, 29*, 153–168. http://dx.doi.org/10.1016/j.asoc.2014.12.031.

34. Wang, H., Smarandache, F., Zhang, Y. Q., & Sunderraman, R. (2005). Interval Neutrosophic Sets and Logic, Theory and Applications in Computing, Hexis, Phoenix, AZ.

35. Wang, H., Smarandache, F., Zhang, Q., & Sunderraman, R. (2010). Single valued neutrosophic sets. *Multispace and Multistructure, 4*, 410–413.

36. Wang, W. & Liu, X. (2013). The multi-attribute decision making method based on interval-valued intuitionistic fuzzy Einstein hybrid weighted geometric operator. *Computers and Mathematics with Applications, 66*, 1845–1856.

37. Wang, J. Q., Zhou, P., Li, K. J., Zhang, H. Y., & Chen, X. H. (2014). Multicriteria decision-making method based on normal intuitionistic fuzzy-induced generalized aggregation operator. *TOP, 22*, 1103–1122.

38. Wang, J. Q., Han, Z. Q., & Zhang, H. Y. (2014). Multi-criteria group decision- making method based on intuitionistic interval fuzzy information. *Group Decision and Negotiation, 23*(4), 715–733.

39. Xu, Z. & Cai, X. (2015). Group decision making with incomplete interval-valued intuitionistic preference relations. *Group Decision and Negotiation, 24*(2), 193–215.

40. Xu, J. & Shen, F. (2014). A new outranking choice method for group decision making under Atanassov's interval-valued intuitionistic fuzzy environment. *Knowledge-Based Systems, 70*, 177–188.

41. Ye, J. (2014). A multicriteria decision-making method using aggregation operators for simplified neutrosophic sets. *Journal of Intelligent and Fuzzy Systems, 26*(5), 2459–2466.

42. Ye, J. (2014). Vector similarity measures of simplified neutrosophic sets and their application in multicriteria decision making. *International Journal of Fuzzy System, 16*(2), 2204–2211.

43. Ye, J. (2014). Single-valued neutrosophic minimum spanning tree and its clustering method. *Journal of Intelligent Systems, 23*(3), 311–324.

44. Ye, J. (2014). Single valued neutrosophic cross-entropy for multicriteria decision making problems. *Application of Mathematical Model, 38*(3), 1170–1175.

45. Ye, J. (2014). A multicriteria decision-making method using aggregation operators for simplified neutrosophic sets. *Journal of Intelligent and Fuzzy Systems, 26*, 2459–2466.

46. Ye, J. (2014). Similarity measures between interval neutrosophic sets and their multicriteria decision making method. *Journal of Intelligent and Fuzzy Systems, 26*(1), 165–172. https://doi.org/10.3233/IFS-120724.

47. Ye, J. (2015). Improved cosine similarity measures of simplified neutrosophic sets for medical diagnoses. *Artificial Intelligence in Medicine, 63*(3), 171–179. https://doi.org/10.1016/j.artmed.2014.

48. Ye, J. (2015). Trapezoidal neutrosophic set and its application to multiple attribute decision-making. *Neural Computing and Applications, 26*(5), 1157–1166. https://doi.org/10.1007/s00521-014-1787-6.

49. Ye, J. (2015). Trapezoidal fuzzy neutrosophic set and its application to multiple attribute decision making. *Neural Computing and Applications, 26*(5), 1157–1166. https://doi.org/10.1007/s00521-014-1787-6.

50. Ye, S., Fu, J., & Ye, J. (2015). Medical diagnosis using distance-based similarity measures of single valued neutrosophic multisets. *Neutrosophic Sets and Systems, 7*, 47–52.

51. Zadeh, L. A. (1965). Fuzzy sets. *Information and Control, 8*, 338–353.
52. Zhang, H. Y., Ji, P., Wang, J., & Chen, X. H. (2015). Improved weighted correlation coefficient based on integrated weight for interval neutrosophic sets and its application in multi-criteria decision making problems. *International Journal of Computational Intelligence Systems, 8*(6), 1027–1043.
53. Zhang, H. Y., Wang, J., & Chen, X. H. (2016). An outranking approach for multi-criteria decision-making problems with interval-valued neutrosophic sets. *Neural Computing and Applications, 27*(3), 615–627.

The Determinant and Adjoint of an Interval-Valued Neutrosophic Matrix

Faruk Karaaslan, Khizar Hayat, and Chiranjibe Jana

1 Introduction

The concept of neutrosophic sets was introduced by Smarandache [12] in 1999. Smarandache characterized a neutrosophic set with three independent membership functions called truth-membership function (T), indeterminacy-membership function (I), and falsity-membership function (F). Even though neutrosophic set is a useful tool for dealing with problems including indeterminate and inconsistent information, in some science and engineering applications, it has some difficulties in the modeling of problems. To overcome these difficulties, Wang et al. [15] defined the single-valued neutrosophic set by assigning to membership functions T, I, and F values from real-standard interval [0, 1] instead of non-standard subset $]^-0, 1^+[$.

Matrix theory has an important role in many areas such as science, engineering, and social sciences. However, in the modeling of the problems involving incompleteness of knowledge the classical matrix theory may not be sufficient. Therefore, Thomason [13] defined the concept of the fuzzy matrix in order to represent fuzzy relation in a system under fuzzy environment. He also studied on the convergence of powers of fuzzy matrix. Kim et al. [8] studied on determinants of fuzzy square matrices including some Boolean matrices and obtained some properties of determinants of the fuzzy square matrices. Ragab and Emam [11]

F. Karaaslan (✉)
Department of Mathematics, Faculty of Sciences, Çankiri Karatekin University, Çankiri, Turkey
e-mail: fkaraaslan@karatekin.edu.tr

K. Hayat
Department of Mathematics, University of Kotli, AJK, Pakistan

C. Jana
Department of Applied Mathematics with Oceanology and Computer programming, Vidyasagar University, West Bengal, Midnapore, India

introduced the concept of adjoint matrix of fuzzy square matrix, and investigated some properties of adjoint matrices. In this way, intuitionistic fuzzy sets [2] have possible many applications for modeling of concepts in artificial intelligence and computer science. The idea of intuitionistic fuzzy matrix was defined by Pal et al. [10]. This contribution of intuitionistic fuzzy matrix allows the decision makers to specify some uncertainties in assigning non-membership degrees and leaves compass to hesitation. Khan and Pal [7] studied on operations of intuitionistic fuzzy matrices. In 2001, Pal [9] introduced notion of the determinant of intuitionistic fuzzy matrices. Concept of the neutrosophic matrix and square neutrosophic fuzzy matrices of which elements are belong to a neutrosophic field $K(I)$ was defined by Kandasamy and Smarandache [5], and they investigated some properties of them. They also for the first time introduced notions of super neutrosophic matrices and quasisuper- neutrosophic matrices, and studied on their properties [6]. Dhar et al. [4] defined operations of addition and multiplication between two neutrosophic fuzzy matrices based on definitions given by Kandasamy and Smarandache, and obtained some properties of these operations. In 2014, Arockiarani and Sumathi [1] defined the notion of fuzzy neutrosophic soft matrices. They also developed a decision-making technique by establishing a new score function which allows evaluating the proper of the alternatives. Deli and Broumi [3] introduced concept of neutrosophic soft matrices and operations between two neutrosophic soft matrices. They also proposed a decision-making method called NSM-decision-making based on neutrosophic soft matrices. Uma et al. [14] defined determinant and adjoint of a fuzzy neutrosophic soft matrices and obtained some of their properties.

The use of interval values has an important role in obtaining more reasonable results in mathematical models. Whereas in several real applications, it is difficult to measure the membership (Truth or indeterminacy or False) values as a single point. Therefore, we consider the membership (Truth, indeterminacy, and False) values as Intervals, respectively. Precisely, such interval frameworks can deal incompleteness of data entire domain of fuzzification.

The matrix theory and determinant theory have many applications in some areas of sciences such as economy, engineering, and environmental sciences. Also, a matrix is an important tool to store data in the computer. Coping with uncertainty has been a key point in the solution of decision-making problems since ancient times. The IVN-set is a useful devise to model uncertainty and inconsistency information. In literature there do not exist a study related to determinant of the IVN-matrices. We would like to carry the advantages of X sets in uncertainty modeling to matrix theory and determinant theory. With this motivation, in this study, some new definitions and results related to IVN-matrices are given. Also, the determinant of a square matrix is defined and some properties existing in the classical determinant theory are investigated for the determinant of IVN-matrices, and some new results are obtained. Furthermore, concepts of complement, constant, reflexive, symmetric, transitive, and idempotent IVN-matrices are defined, and some properties of them related to determinant and adjoint are obtained. The remainder of this paper is organized as follows. In Sect. 2, basic definition and operations are presented related to IVN-sets and IVN-matrices. In Sect. 3, some properties of determinant of an

IVN-matrix are obtained. In Sect. 4, the concept of adjoint is introduced for square IVN-matrix, and some results are obtained related to adjoint of square IVN-matrix. Also, some new concepts are defined for IVN-matrices. In Sect. 5, some conclusions and directions for future work are given.

2 Preliminaries

In this section, we recall some definitions and operations to be required in the next sections.

2.1 Neutrosophic Sets

A neutrosophic set (NS) A on the universe of discourse X is expressed as follows:

$$A = \{\langle x, T_A(x), I_A(x), F_A(x) \rangle : x \in X\},$$

where $T_A, I_A, F_A : X \to]^-0, 1^+[$ and $^-0 \leq T_A(x) + I_A(x) + F_A(x) \leq 3^+$ [12].

Note that the image of an element in NS is a standard or non-standard subsets of $]^-0, 1^+[$. In some practical applications, standard or non-standard subsets of $]^-0, 1^+[$ may not be easy modeling of problems. Therefore the concept of single-valued neutrosophic set (SVN-set) was introduced by Wang et al. [15] as follows.

Let X be a non-empty set and generally its element is denoted as x. A single-valued neutrosophic set (SVN-set) A is identified by three functions T_A, I_A, and F_A from X to $[0, 1]$ and they called as truth-membership, indeterminacy-membership, falsity-membership functions, respectively. Formally, X may be continuous or discrete.

- If X is continuous, representation of an SVN-set A is expressed as follows:

$$A = \int_X \langle T_A(x), I_A(x), F_A(x) \rangle / x, \quad \text{for all } x \in X.$$

- If X is a crisp set, an SVN-set A can be written as follows:

$$A = \sum_x \langle T_A(x), I_A(x), F_A(x) \rangle / x, \quad \text{for all } x \in X.$$

Here $0 \leq T_A(x) + I_A(x) + F_A(x) \leq 3$ for all $x \in X$.

2.2   Interval-Valued Neutrosophic Sets

An interval neutrosophic set A on the universe of discourse X can be identified as follows:

$$[A] = \left\{ \langle x, [T_A^L(x), T_A^U(x)], [I_A^L(x), I_A^U(x)], [F_A^L(x), F_A^U(x)] \rangle : x \in X \right\},$$

where $[T_A^L(x), T_A^U(x)], [I_A^L(x), I_A^U(x)], [F_A^L(x), F_A^U(x)] \subseteq [0,1]$, and $0 \leq T_A^U(x) + I_A^U(x) + F_A^U(x) \leq 3$ [16].

Let $[A]$ and $[B]$ be two interval neutrosophic sets over X. Operations and relations between two interval neutrosophic sets are defined by Wang et al.[15] which are listed as in the following:

1. $[A] \subseteq [B]$ if and only if $T_A^L(x) \leq T_B^L(x)$, $I_A^L(x) \geq I_B^L(x)$, $F_A^L(x) \geq F_B^L(x)$ and $T_A^U(x) \leq T_B^U(x)$, $I_A^U(x) \geq I_B^U(x)$, $F_A^U(x) \geq F_B^U(x)$ for all $x \in X$,
2. $[A] = [B]$ if and only if $[A] \subseteq [B]$ and $[B] \subseteq [A]$ for all $x \in X$,
3. $[A]^c = \{ \langle x, [F_A^L(x), F_A^U(x)], [1 - I_A^U(x), 1 - I_A^L(x)][T_A^L(x), T_A^U(x)] \rangle : x \in X \}$,
4. $[A] \cup [B] = \{ \langle x, ([T_A^L(x) \vee T_B^L(x), T_A^U(x) \vee T_B^U(x)], [I_A^L(x) \wedge I_B^L(x), I_A^U(x) \wedge I_B^U(x)], [F_A^L(x) \wedge F_B^L(x), F_A^U(x) \wedge F_B^U(x)]) \rangle : x \in X \}$,
5. $[A] \cap [B] = \{ \langle x, ([T_A^L(x) \wedge T_B^L(x), T_A^U(x) \wedge T_B^U(x)], [I_A^L(x) \vee I_B^L(x), I_A^U(x) \vee I_B^U(x)], [F_A^L(x) \vee F_B^L(x), F_A^U(x) \vee F_B^U(x)]) \rangle : x \in X \}$.

Let I^2 denote the set of closed subinterval of $[0,1]$, $J = I^2 \times I^2 \times I^2$, and $N(J) = \{ \langle [\alpha_1^L, \alpha_1^U], [\alpha_2^L, \alpha_2^U], [\alpha_3^L, \alpha_3^U] \rangle : \alpha_1^L \leq \alpha_1^U, \alpha_2^L \leq \alpha_2^U, \alpha_3^L \leq \alpha_3^U \}$. Then, $N(J)$ is a lattice together with partial ordered relation $\preceq$, where order relation $\preceq$ on $N(J)$ can be seen as follows: $\langle [0,0], [1,1], [1,1] \rangle \preceq \langle [\alpha_1^L, \alpha_1^U], [\alpha_2^L, \alpha_2^U], [\alpha_3^L, \alpha_3^U] \rangle \preceq \langle [\beta_1^L, \beta_1^U], [\beta_2^L, \beta_2^U], [\beta_3^L, \beta_3^U] \rangle \preceq \langle [1,1], [0,0], [0,0] \rangle \Leftrightarrow \alpha_1^L \leq \beta_1^L, \alpha_1^U \leq \beta_1^U, \alpha_2^L \geq \beta_2^L, \alpha_2^U \geq \beta_2^U, \alpha_3^L \geq \beta_3^L, \alpha_3^U \geq \beta_3^U$, where $\langle [\alpha_1^L, \alpha_1^U], [\alpha_2^L, \alpha_2^U], [\alpha_3^L, \alpha_3^U] \rangle, \langle [\beta_1^L, \beta_1^U], [\beta_2^L, \beta_2^U], [\beta_3^L, \beta_3^U] \rangle \in N(J)$.

2.3   Interval-Valued Neutrosophic Matrices

Let $U = \{c_1, c_2, \ldots, c_m\}$ be the universal set, $E = \{e_1, e_2, \ldots, e_n\}$ be a set of parameters, and let $A \subseteq E$. A pair (F, A) be an interval neutrosophic soft set over U [3]. Then the subset of $U \times E$ is defined by $R_A = \{(u, e) : e \in A, u \in f_A(e)\}$ which is called a relation form of (f_A, E). $T_{R_A} : U \times E \to [0,1], I_{R_A} : U \times E \to [0,1]$ and $F_{R_A} : U \times E \to [0,1]$ are truth-membership function, indeterminacy-membership function, and non-membership function, respectively.

It is noted that $T_{R_A}(u, e) \in I^2$, $I_{R_A}(u, e) \in I^2$, and $F_{R_A}(u, e) \in I^2$. If $\left[[T_{ij}^L, T_{ij}^U], [I_{ij}^L, I_{ij}^U], [F_{ij}^L, F_{ij}^U] \right] = \left[[T_{ij}^L(u_i, e_j), T_{ij}^U(u_i, e_j)], [I_{ij}^L(u_i, e_j), I_{ij}^U(u_i, e_j)], [F_{ij}^L(u_i, e_j), F_{ij}^U(u_i, e_j)] \right]$, then $\left[\langle [T_{ij}^L, T_{ij}^U], [I_{ij}^L, I_{ij}^U], [F_{ij}^L, F_{ij}^U] \rangle \right]$

formed a matrix of $m \times n$. This matrix is called an $m \times n$ interval neutrosophic soft matrix of the interval neutrosophic soft set (F_A, E) over U. Set of all $m \times n$ interval neutrosophic soft matrices is denoted by $I\mathcal{N}_{m \times n}$.

The interval neutrosophic matrix is a representation of an interval neutrosophic soft set. To construct an interval valued neutrosophic matrix we need an interval neutrosophic soft set.

The interval neutrosophic matrix can be defined as a general concept as follows.

Definition 1 Let $M_{m \times n}(J) = \left\{ \left[\langle [T_{ij}^L, T_{ij}^U], [I_{ij}^L, I_{ij}^U], [F_{ij}^L, F_{ij}^U] \rangle \right]_{m \times n} : \langle [T_{ij}^L, T_{ij}^U], [I_{ij}^L, I_{ij}^U], [F_{ij}^L, F_{ij}^U] \rangle \in N(J) \right\}$, where $J = I^2 \times I^2 \times I^2$. Any matrix $\hat{A}$ in $M_{m \times n}$ is called an interval neutrosophic matrix (IN-matrix).

From now on we use interval-valued neutrosophic matrix (INV-matrix) instead of interval neutrosophic matrix (IN-matrix).

Some basic concepts and operations of IVN-matrices can be defined with a similar way to interval neutrosophic soft matrices.

Definition 2 Let $\hat{A} = \left[\langle [T_{A_{ij}}^L, T_{A_{ij}}^U], [I_{A_{ij}}^L, I_{A_{ij}}^U], [F_{A_{ij}}^L, F_{A_{ij}}^U] \rangle \right]_{m \times m}$, $\hat{B} = \left[\langle [T_{B_{ij}}^L, T_{B_{ij}}^U], [I_{B_{ij}}^L, I_{B_{ij}}^U], [F_{B_{ij}}^L, F_{B_{ij}}^U] \rangle \right]_{m \times n}$, and $\hat{C} = \left[\langle [T_{C_{ij}}^L, T_{C_{ij}}^U], [I_{C_{ij}}^L, I_{C_{ij}}^U], [F_{C_{ij}}^L, F_{C_{ijX}}^U] \rangle \right]_{m \times n}$. Then, addition between IVN-matrices $\hat{B}$ and $\hat{C}$ and product operations between IVN-matrices $\hat{A}$, and $\hat{B}$, and some concepts related to IVN-matrices are defined as follows:

1. $\hat{B} + \hat{C} = \left[\langle [T_{B_{ij}}^L + T_{C_{ij}}^L, T_{B_{ij}}^U + T_{C_{ij}}^U], [I_{B_{ij}}^L + I_{C_{ij}}^L, I_{B_{ij}}^U + I_{C_{ij}}^U], [F_{B_{ij}}^L + F_{C_{ij}}^L, F_{B_{ij}}^U + F_{C_{ij}}^U] \rangle \right]$, where $\begin{aligned} T_{B_{ij}}^L + T_{C_{ij}}^L &= T_{B_{ij}}^L \vee T_{C_{ij}}^L, & T_{B_{ij}}^U + T_{C_{ij}}^U &= T_{B_{ij}}^U \vee T_{C_{ij}}^U, \\ I_{B_{ij}}^L + I_{C_{ij}}^L &= I_{B_{ij}}^L \wedge I_{C_{ij}}^L, & I_{B_{ij}}^U + I_{C_{ij}}^U &= I_{B_{ij}}^U \wedge I_{C_{ij}}^U, \\ F_{B_{ij}}^L + F_{C_{ij}}^L &= F_{B_{ij}}^L \wedge F_{C_{ij}}^L, & F_{B_{ij}}^U + F_{C_{ij}}^U &= F_{B_{ij}}^U \wedge F_{C_{ij}}^U. \end{aligned}$

2. $\hat{A}\hat{B} = [\langle [T_{AB_{ij}}^L, T_{AB_{ij}}^U], [I_{AB_{ij}}^L, I_{AB_{ij}}^U], [F_{AB_{ij}}^L, F_{AB_{ij}}^U] \rangle]_{m \times n}$, where $\begin{aligned} T_{AB_{ij}}^L &= \bigvee_{k=1}^m (T_{A_{ik}}^L \wedge T_{B_{kj}}^L), & T_{AB_{ij}}^U &= \bigvee_{k=1}^m (T_{A_{ik}}^U \wedge T_{B_{ki}}^U), \\ I_{AB_{ij}}^L &= \bigvee_{k=1}^m (I_{A_{ik}}^L \vee I_{B_{kj}}^L), & I_{AB_{ij}}^U &= \bigvee_{k=1}^m (I_{A_{ik}}^U \vee I_{B_{kj}}^U), \\ F_{AB_{ij}}^L &= \bigvee_{k=1}^m (F_{A_{ik}}^L \vee F_{B_{kj}}^L), & F_{AB_{ij}}^U &= \bigvee_{k=1}^m (F_{A_{ik}}^U \vee F_{B_{kj}}^U). \end{aligned}$

3. Transpose of IVN-matrix $\hat{B}$ is defined as follows:

$$\hat{B}^t = \left[\langle [T_{B_{ji}}^L, T_{B_{ji}}^U], [I_{B_{ji}}^L, I_{B_{ji}}^U], [F_{B_{ji}}^L, F_{B_{ji}}^U] \rangle \right]_{n \times m}.$$

4.

$$\hat{A}^k = \left[\langle [T_{A_{ij}}^{L\ (k)}, T_{A_{ij}}^{U\ (k)}], [I_{A_{ij}}^{L\ (k)}, I_{A_{ij}}^{U\ (k)}], [F_{A_{ij}}^{L\ (k)}, F_{A_{ij}}^{U\ (k)}] \rangle \right], \quad \hat{A}^{k+1}$$
$$= \hat{A}^k \hat{A}, \ (k = 0, 1, 2, \ldots).$$

5.

$$\hat{A}^0 = I_m = \left[\langle [T^L_{I_{mij}}, T^U_{I_{mij}}], [I^L_{I_{mij}}, I^U_{I_{mij}}], [F^L_{I_{mij}}, F^U_{I_{mij}}] \rangle \right],$$

where

$$\langle [T^L_{I_{mij}}, T^U_{I_{mij}}], [I^L_{I_{mij}}, I^U_{I_{mij}}], [F^L_{I_{mij}}, F^U_{I_{mij}}] \rangle = \begin{cases} \langle [1, 1], [0, 0], [0, 0] \rangle, & i = j \\ \langle [0, 0], [1, 1], [1, 1] \rangle, & i \neq j \end{cases}$$

and I_n is called IN-unit matrix.

6. If $\langle [T^L_{B_{ij}}, T^U_{B_{ij}}], [I^L_{B_{ij}}, I^U_{B_{ij}}], [F^L_{B_{ij}}, F^U_{B_{ij}}] \rangle \preceq \langle [T^L_{C_{ij}}, T^U_{C_{ij}}], [I^L_{C_{ij}}, I^U_{C_{ij}}], [F^L_{C_{ij}}, F^U_{C_{ij}}] \rangle$
   for all $1 \leq i \leq m$, and $1 \leq j \leq n$, then IVN-matrix $\hat{B}$ is smaller than IVN-matrix
   $\hat{C}$, and denoted by $\hat{B} \tilde{\preceq} \hat{C}$.

Remark 1 $M_{m \times n}(J)$ M(J) is a semiring on the operations of addition and multiplication (defined above). This semiring is called IVN-matrix semiring.

3   Determinant of the Square IVN-Matrix

In this section, determinant of $m \times m$ an IVN-matrix is defined and some properties of the determinant of the square IVN-matrix are obtained.

Definition 3   Determinant of an $m \times m$ IVN-matrix $\hat{A}$, denoted by $|\hat{A}|$, is defined as follows:

$$|\hat{A}| = \Big\langle \Big[\bigvee_{\sigma \in S_m} (\wedge^m_{i=1} T^L_{A_{i\sigma(i)}}), \bigvee_{\sigma \in S_m} (\wedge^m_{i=1} T^U_{A_{i\sigma(i)}}) \Big],$$

$$\Big[\bigwedge_{\sigma \in S_m} (\vee^m_{i=1} I^L_{A_{i\sigma(i)}}), \bigwedge_{\sigma \in S_m} (\vee^m_{i=1} I^U_{A_{i\sigma(i)}}) \Big],$$

$$\Big[\bigwedge_{\sigma \in S_m} (\vee^m_{i=1} F^L_{A_{i\sigma(i)}}), \bigwedge_{\sigma \in S_m} (\vee^m_{i=1} F^U_{A_{i\sigma(i)}}) \Big] \Big\rangle. \tag{1}$$

Example 1   Let us consider IVN-matrix $\hat{A}$ given as follows:

$$\hat{A} = \begin{bmatrix} \langle [0.7, 0.9], [0.6, 0.7], [0.8, 1.0] \rangle & \langle [0.3, 0.6], [0.5, 0.5], [0.4, 0.7] \rangle & \langle [0.6, 0.8], [0.5, 0.6], [0.2, 0.4] \rangle \\ \langle [0.4, 0.5], [0.4, 0.4], [0.8, 0.9] \rangle & \langle [0.9, 1.0], [0.6, 0.7], [0.8, 0.8] \rangle & \langle [0.1, 0.3], [0.4, 0.7], [0.5, 0.5] \rangle \\ \langle [0.8, 0.9], [0.6, 0.8], [0.1, 0.2] \rangle & \langle [0.5, 0.9], [0.2, 0.4], [0.3, 0.6] \rangle & \langle [0.8, 0.9], [0.7, 0.9], [0.2, 0.6] \rangle \end{bmatrix}.$$

Then, we calculate the determinant of IVN-matrix $\hat{A}$ as follows:

$$|\hat{A}| = \langle [0.7, 0.9], [0.6, 0.7], [0.8, 1.0] \rangle \begin{vmatrix} \langle [0.9, 0.1], [0.6, 0.7], [0.8, 0.8] \rangle & \langle [0.1, 0.3], [0.4, 0.7], [0.5, 0.5] \rangle \\ \langle [0.5, 0.9], [0.2, 0.4], [0.3, 0.6] \rangle & \langle [0.8, 0.9], [0.7, 0.9], [0.2, 0.6] \rangle \end{vmatrix}$$

$+ \langle [0.3, 0.6], [0.5, 0.5], [0.4, 0.7] \rangle \begin{vmatrix} \langle [0.4, 0.5], [0.4, 0.4], [0.8, 0.9] \rangle & \langle [0.1, 0.3], [0.4, 0.7], [0.5, 0.5] \rangle \\ \langle [0.8, 0.9], [0.6, 0.8], [0.1, 0.2] \rangle & \langle [0.8, 0.9], [0.7, 0.9], [0.2, 0.6] \rangle \end{vmatrix}$

$+ \langle [0.6, 0.8], [0.5, 0.6], [0.2, 0.4] \rangle \begin{vmatrix} \langle [0.4, 0.5], [0.4, 0.4], [0.8, 0.9] \rangle & \langle [0.9, 1.0], [0.6, 0.7], [0.8, 0.8] \rangle \\ \langle [0.8, 0.9], [0.6, 0.8], [0.1, 0.2] \rangle & \langle [0.5, 0.9], [0.2, 0.4], [0.3, 0.6] \rangle \end{vmatrix}$

$= \langle [0.7, 0.9], [0.6, 0.7], [0.8, 1.0] \rangle \langle [0.8, 0.9], [0.4, 0.7], [0.5, 0.6] \rangle$

$+ \langle [0.3, 0.6], [0.5, 0.5], [0.4, 0.7] \rangle \langle [0.4, 0.5], [0.6, 0.8], [0.5, 0.5] \rangle$

$+ \langle [0.6, 0.8], [0.5, 0.6], [0.2, 0.4] \rangle \langle [0.8, 0.9], [0.4, 0.4], [0.8, 0.8] \rangle$

$= \langle [0.7, 0.9], [0.6, 0.7], [0.8, 1.0] \rangle + \langle [0.3, 0.5], [0.6, 0.8], [0.5, 0.7] \rangle + \langle [0.6, 0.8], [0.5, 0.6], [0.8, 0.8] \rangle$

$= \langle [0.7, 0.9], [0.5, 0.6], [0.5, 0.7]. \rangle)$

Proposition 1 *Let $\hat{A}$ and $\hat{B}$ be two $m \times m$ IVN-matrices. If $\hat{A} \tilde{\leq} \hat{B}$, then $|\hat{A}| \preceq |\hat{B}|$.*

Proof Let $([T^L_{A_{ij}}, T^U_{A_{ij}}], [I^L_{A_{ij}}, I^U_{A_{ij}}], [F^L_{A_{ij}}, F^U_{A_{ij}}]) \preceq ([T^L_{B_{ij}}, T^U_{B_{ij}}], [I^L_{B_{ij}}, I^U_{B_{ij}}], [F^L_{B_{ij}}, F^U_{B_{ij}}])$ and $\hat{A} \tilde{\leq} \hat{B}$ for all $1 \leq i, j \leq m$,

Then, $T^L_{A_{ij}} \leq T^L_{B_{ij}}$, $T^U_{A_{ij}} \leq T^U_{B_{ij}}$, $I^L_{A_{ij}} \geq I^L_{B_{ij}}$, $I^U_{A_{ij}} \geq I^U_{B_{ij}}$ and $F^L_{A_{ij}} \geq F^L_{B_{ij}}$, $F^U_{A_{ij}} \geq F^U_{B_{ij}}$. It is clear that $\wedge^m_{i=1} T^L_{A_{ij}} \leq \wedge^m_{i=1} T^L_{B_{ij}}$, $\wedge^m_{i=1} T^U_{A_{ij}} \leq \wedge^m_{i=1} T^U_{B_{ij}}$, $\vee^m_{i=1} I^L_{A_{ij}} \geq \vee^m_{i=1} I^L_{B_{ij}}$, $\vee^m_{i=1} I^U_{A_{ij}} \geq \vee^m_{i=1} I^U_{B_{ij}}$ and $\vee^m_{i=1} F^L_{A_{ij}} \geq \vee^m_{i=1} F^L_{B_{ij}}$, $\vee^m_{i=1} F^U_{A_{ij}} \geq \vee^m_{i=1} F^U_{B_{ij}}$. So, $\bigvee_{\sigma \in S_m} (\wedge^m_{i=1} T^L_{A_{i\sigma(i)}}) \leq \bigvee_{\sigma \in S_m} (\wedge^m_{i=1} T^L_{B_{i\sigma(i)}})$, $\bigvee_{\sigma \in S_m} (\wedge^m_{i=1} T^U_{A_{i\sigma(i)}}) \leq \bigvee_{\sigma \in S_m} (\wedge^m_{i=1} T^U_{B_{i\sigma(i)}})$, $\bigwedge_{\sigma \in S_m} (\vee^m_{i=1} I^L_{A_{i\sigma(i)}}) \geq \bigwedge_{\sigma \in S_m} (\vee^m_{i=1} I^L_{B_{i\sigma(i)}})$, $\bigwedge_{\sigma \in S_m} (\vee^m_{i=1} I^U_{A_{i\sigma(i)}}) \geq \bigwedge_{\sigma \in S_m} (\vee^m_{i=1} I^U_{B_{i\sigma(i)}})$ and $\bigwedge_{\sigma \in S_m} (\vee^m_{i=1} F^L_{A_{i\sigma(i)}}) \geq \bigwedge_{\sigma \in S_m} (\vee^m_{i=1} F^L_{B_{i\sigma(i)}})$, $\bigwedge_{\sigma \in S_m} (\vee^m_{i=1} F^U_{A_{i\sigma(i)}}) \geq \bigwedge_{\sigma \in S_m} (\vee^m_{i=1} F^U_{B_{i\sigma(i)}})$.

$$|\hat{A}| = \left([\bigvee_{\sigma \in S_m} (\wedge^m_{i=1} T^L_{A_{i\sigma(i)}}), \bigvee_{\sigma \in S_m} (\wedge^m_{i=1} T^U_{A_{i\sigma(i)}})], [\bigwedge_{\sigma \in S_m} (\vee^m_{i=1} I^L_{A_{i\sigma(i)}}), \bigwedge_{\sigma \in S_m} (\vee^m_{i=1} I^U_{A_{i\sigma(i)}})], \right.$$
$$\left. [\bigwedge_{\sigma \in S_m} (\vee^m_{i=1} F^L_{A_{i\sigma(i)}}), \bigwedge_{\sigma \in S_m} (\vee^m_{i=1} F^U_{A_{i\sigma(i)}})] \right)$$
$$\preceq \left([\bigvee_{\sigma \in S_m} (\wedge^m_{i=1} T^L_{B_{i\sigma(i)}}), \bigvee_{\sigma \in S_m} (\wedge^m_{i=1} T^U_{B_{i\sigma(i)}})], [\bigwedge_{\sigma \in S_m} (\vee^m_{i=1} I^L_{B_{i\sigma(i)}}), \bigwedge_{\sigma \in S_m} (\vee^m_{i=1} I^U_{B_{i\sigma(i)}})], \right.$$
$$\left. [\bigwedge_{\sigma \in S_m} (\vee^m_{i=1} F^L_{B_{i\sigma(i)}}), \bigwedge_{\sigma \in S_m} (\vee^m_{i=1} F^U_{B_{i\sigma(i)}})] \right)$$
$$= |\hat{B}|.$$

□

Proposition 2 *Let $\hat{A}$ and $\hat{B}$ be two $m \times m$ IVN-matrices. If $\hat{A} \tilde{\leq} \hat{B}$ or $\hat{B} \tilde{\leq} \hat{A}$, then $|\hat{A}| + |\hat{B}| \preceq |\hat{A} + \hat{B}|$.*

Proof Suppose that $\hat{A} \tilde{\leq} \hat{B}$. From Definition 3, it is clear that $|\hat{A} + \hat{B}| = |\hat{B}|$ and $|\hat{A}| \preceq |\hat{A} + \hat{B}|$. If it is added $|\hat{B}|$ to both sides of the second inequality, we get

$|\hat{A}| + |\hat{B}| \preceq |\hat{A} + \hat{B}| + |\hat{B}|$. Since $|\hat{A} + \hat{B}| = |\hat{B}|$, $|\hat{A}| + |\hat{B}| \preceq |\hat{A} + \hat{B}| + |\hat{A} + \hat{B}|$. Also it is clear that $|\hat{A} + \hat{B}| + |\hat{A} + \hat{B}| = |\hat{A} + \hat{B}|$. So $|\hat{A}| + |\hat{B}| \preceq |\hat{A} + \hat{B}|$. Thus,   □

Note that, in generally $|\hat{A} + \hat{B}| \neq |\hat{A}| + |\hat{B}|$.

Example 2 Let $\hat{A} = \begin{bmatrix} \langle [0.5, 0.6], [0.4, 0.7], [0.2, 0.6] \rangle & \langle [0.6, 0.9], [0.3, 0.7], [0.1, 0.5] \rangle \\ \langle [0.8, 0.9], [0.7, 1.0], [0.4, 0.8] \rangle & \langle [0.3, 0.5], [0.2, 0.4], [0.5, 0.6] \rangle \end{bmatrix}$ and

$\hat{B} = \begin{pmatrix} \langle [0.3, 0.4], [0.2, 0.5], [0.1, 0.4] \rangle & \langle [0.4, 0.6], [0.7, 0.7], [0.8, 0.9] \rangle \\ \langle [0.6, 0.7], [0.4, 0.4], [0.2, 0.5] \rangle & \langle [0.5, 0.8], [0.8, 0.9], [0.9, 0.9] \rangle \end{pmatrix}$ be

two 2×2 IVN-matrices. Then,

$$|\hat{A}| + |\hat{B}| = \langle [0.6, 0.9], [0.4, 0.7], [0.4, 0.6] \rangle$$

and

$$|\hat{A} + \hat{B}| = \langle [0.6, 0.9], [0.2, 0.5], [0.2, 0.5] \rangle.$$

Thus $|\hat{A}| + |\hat{B}| \neq |\hat{A} + \hat{B}|$.

Proposition 3 Let $\hat{A} = \left[\langle [T_{A_{ij}}^L, T_{A_{ij}}^U], [I_{A_{ij}}^L, I_{A_{ij}}^U], [F_{A_{ij}}^L, F_{A_{ij}}^U] \rangle \right]$ be an $m \times m$ IVN-matrix.

1. If an IVN-matrix $\hat{B}$ is obtained from $\hat{A}$ by multiplying the all of rows (all of columns) of $\hat{A}$ by $\lambda = \langle [T_\lambda^L, T_\lambda^U], [I_\lambda^L, I_\lambda^U], [F_\lambda^L, F_\lambda^U] \rangle \in I^3$, then $\lambda |\hat{A}| = |\hat{B}|$.
2. If $\hat{A}$ contain zero rows (column) (i.e., if all of elements of any row or column are $\bar{0} = \langle [0, 0], [1, 1], [1, 1] \rangle$), then $|\hat{A}| = \bar{0}$.

Proof

1. Let $\hat{A} = \left[\langle [T_{A_{ij}}^L, T_{A_{ij}}^U], [I_{A_{ij}}^L, I_{A_{ij}}^U], [F_{A_{ij}}^L, F_{A_{ij}}^U] \rangle \right]$ be an $m \times m$ IVN-matrix. Then $\lambda \hat{A} = \hat{B}$ is as follows:

$$\hat{B} = [\langle [T_\lambda^L \wedge T_{A_{ij}}^L, T_\lambda^U \wedge T_{A_{ij}}^U], [I_\lambda^L \vee I_{A_{ij}}^L, I_\lambda^U \vee I_{A_{ij}}^U], [F_\lambda^L \vee F_{A_{ij}}^L, F_\lambda^U \vee F_{A_{ij}}^U] \rangle].$$

Then,

$$|\hat{B}| = \Big\langle [\bigvee_{\sigma \in S_m} (\wedge_{i=1}^m (T_\lambda^L \wedge T_{A_{i\sigma(i)}}^L)), \bigvee_{\sigma \in S_m} (\wedge_{i=1}^m (T_\lambda^U \wedge T_{A_{i\sigma(i)}}^U))],$$

$$[\bigwedge_{\sigma \in S_m} (\vee_{i=1}^m (I_\lambda^L \vee I_{A_{i\sigma(i)}}^L)), \bigwedge_{\sigma \in S_m} (\vee_{i=1}^m (I_\lambda^U \vee I_{A_{i\sigma(i)}}^U))],$$

$$[\bigwedge_{\sigma \in S_m} (\vee_{i=1}^m (F_\lambda^L \vee F_{A_{i\sigma(i)}}^L)), \bigwedge_{\sigma \in S_m} (\vee_{i=1}^m (F_\lambda^U \vee F_{A_{i\sigma(i)}}^U))] \Big\rangle$$

$$= \Big\langle [\bigvee_{\sigma \in S_m} T^L_\lambda(\wedge^m_{i=1}(T^L_{A_{i\sigma(i)}})), \bigvee_{\sigma \in S_m} T^U_\lambda(\wedge^m_{i=1}(T^U_{A_{i\sigma(i)}}))],$$

$$[\bigwedge_{\sigma \in S_m} I^L_\lambda(\vee^m_{i=1}(I^L_{A_{i\sigma(i)}})), \bigwedge_{\sigma \in S_m} I^U_\lambda(\vee^m_{i=1}(I^U_{A_{i\sigma(i)}}))],$$

$$[\bigwedge_{\sigma \in S_m} F^L_\lambda(\vee^m_{i=1}(F^L_{A_{i\sigma(i)}})), \bigwedge_{\sigma \in S_m} F^U_\lambda(\vee^m_{i=1}(F^U_{A_{i\sigma(i)}}))]\Big\rangle$$

$$= \Big\langle [T^L_\lambda \bigvee_{\sigma \in S_m} (\wedge^m_{i=1}(T^L_{A_{i\sigma(i)}})), T^U_\lambda \bigvee_{\sigma \in S_m} (\wedge^m_{i=1}(T^U_{A_{i\sigma(i)}}))],$$

$$[I^L_\lambda \bigwedge_{\sigma \in S_m} (\vee^m_{i=1}(I^L_{A_{i\sigma(i)}})), I^U_\lambda \bigwedge_{\sigma \in S_m} (\vee^m_{i=1}(I^U_{A_{i\sigma(i)}}))],$$

$$[F^L_\lambda \bigwedge_{\sigma \in S_m} (\vee^m_{i=1}(F^L_{A_{i\sigma(i)}})), F^U_\lambda \bigwedge_{\sigma \in S_m} (\vee^m_{i=1}(F^U_{A_{i\sigma(i)}}))]\Big\rangle$$

$$= \langle [T^L_\lambda, T^U_\lambda], [I^L_\lambda, I^U_\lambda], [F^L_\lambda, F^U_\lambda] \rangle \Big\langle [\bigvee_{\sigma \in S_m} (\wedge^m_{i=1} T^L_{A_{i\sigma(i)}}), \bigvee_{\sigma \in S_m} (\wedge^m_{i=1} T^U_{A_{i\sigma(i)}})],$$

$$[\bigwedge_{\sigma \in S_m} (\vee^m_{i=1} I^L_{A_{i\sigma(i)}}), \bigwedge_{\sigma \in S_m} (\vee^m_{i=1} I^U_{A_{i\sigma(i)}})],$$

$$[\bigwedge_{\sigma \in S_m} (\vee^m_{i=1} F^L_{A_{i\sigma(i)}}), \bigwedge_{\sigma \in S_m} (\vee^m_{i=1} \vee F^U_{A_{i\sigma(i)}})]\Big\rangle$$

$$= \lambda|\hat{A}|.$$

2. Let $\hat{A} = \begin{vmatrix} \langle[0,0],[1,1],[1,1]\rangle & \langle[0,0],[1,1],[1,1]\rangle \\ \langle[T^L_{A_{21}}, T^U_{A_{21}}],[I^L_{A_{21}}, I^U_{A_{21}}],[F^L_{A_{21}}, F^U_{A_{21}}]\rangle & \langle[T^L_{A_{22}}, T^U_{A_{22}}],[I^L_{A_{22}}, I^U_{A_{22}}],[F^L_{A_{22}}, F^U_{A_{22}}]\rangle \\ \vdots & \vdots \\ \langle[T^L_{A_{m1}}, T^U_{A_{m1}}],[I^L_{A_{m1}}, I^U_{A_{m1}}],[F^L_{A_{m1}}, F^U_{A_{m1}}]\rangle & \langle[T^L_{A_{m2}}, T^U_{A_{m2}}],[I^L_{A_{m2}}, I^U_{A_{m2}}],[F^L_{A_{m2}}, F^U_{A_{m2}}]\rangle \end{vmatrix}$

$\cdots \quad \langle[0,0],[1,1],[1,1]\rangle$

$\cdots \quad \langle[T^L_{A_{2m}}, T^U_{A_{2m}}],[I^L_{A_{2m}}, I^U_{A_{2m}}],[F^L_{A_{2m}}, F^U_{A_{2m}}]\rangle$

$\ddots \qquad \vdots$

$\cdots \quad \langle[T^L_{A_{mm}}, T^U_{A_{mm}}],[I^L_{A_{mm}}, I^U_{A_{mm}}],[F^L_{A_{mm}}, F^U_{A_{mm}}]\rangle.$

By definition of determinant of IVN-matrix,

$$|\hat{A}| = \langle[0,0],[1,1],[1,1]\rangle \begin{vmatrix} \langle[T^L_{A_{22}}, T^U_{A_{22}}],[I^L_{A_{22}}, I^U_{A_{22}}],[F^L_{A_{22}}, F^U_{A_{22}}]\rangle \\ \vdots \\ \langle[T^L_{A_{m2}}, T^U_{A_{m2}}],[I^L_{A_{m2}}, I^U_{A_{m2}}],[F^L_{A_{m2}}, F^U_{A_{m2}}]\rangle \end{vmatrix}$$

$$\cdots \quad \langle [T^L_{A_{2m}}, T^U_{A_{2m}}], [I^L_{A_{2m}}, I^U_{A_{2m}}], [F^L_{A_{2m}}, F^U_{A_{2m}}] \rangle$$
$$\left. \ddots \qquad \vdots \right|$$
$$\cdots \quad \langle [T^L_{A_{mm}}, T^U_{A_{mm}}], [I^L_{A_{mm}}, I^U_{A_{mm}}], [F^L_{A_{mm}}, F^U_{A_{mm}}] \rangle$$

$$+ \langle [0, 0], [1, 1], [1, 1] \rangle \left| \begin{array}{c} \langle [T^L_{A_{21}}, T^U_{A_{21}}], [I^L_{A_{21}}, I^U_{A_{21}}], [F^L_{A_{21}}, F^U_{A_{21}}] \rangle \\ \vdots \\ \langle [T^L_{A_{m1}}, T^U_{A_{m1}}], [I^L_{A_{m1}}, I^U_{A_{m1}}], [F^L_{A_{m1}}, F^U_{A_{m1}}] \rangle \end{array} \right.$$

$$\cdots \quad \langle [T^L_{A_{2m}}, T^U_{A_{2m}}], [I^L_{A_{2m}}, I^U_{A_{2m}}], [F^L_{A_{2m}}, F^U_{A_{2m}}] \rangle$$
$$\left. \ddots \qquad \vdots \right| \cdots$$
$$\cdots \quad \langle [T^L_{A_{mm}}, T^U_{A_{mm}}], [I^L_{A_{mm}}, I^U_{A_{mm}}], [F^L_{A_{mm}}, F^U_{A_{mm}}] \rangle$$

$$+ \langle [0, 0], [1, 1], [1, 1] \rangle \left| \begin{array}{c} \langle [T^L_{A_{21}}, T^U_{A_{21}}], [I^L_{A_{21}}, I^U_{A_{21}}], [F^L_{A_{21}}, F^U_{A_{21}}] \rangle \\ \vdots \\ \langle [T^L_{A_{m1}}, T^U_{A_{m1}}], [I^L_{A_{m1}}, I^U_{A_{m1}}], [F^L_{A_{m1}}, F^U_{A_{m1}}] \rangle \end{array} \right.$$

$$\cdots \quad \langle [T^L_{A_{2(m-1)}}, T^U_{A_{2(m-1)}}], [I^L_{A_{2(m-1)}}, I^U_{A_{2(m-1)}}], [F^L_{A_{2(m-1)}}, F^U_{A_{2(m-1)}}] \rangle$$
$$\left. \ddots \qquad \vdots \right|$$
$$\cdots \quad \langle [T^L_{A_{m(m-1)}}, T^U_{A_{m(m-1)}}], [I^L_{A_{m(m-1)}}, I^U_{A_{m(m-1)}}], [F^L_{A_{m(m-1)}}, F^U_{A_{m(m-1)}}] \rangle$$

$$= \langle [0, 0], [1, 1], [1, 1] \rangle + \langle [0, 0], [1, 1], [1, 1] \rangle + \langle [0, 0], [1, 1], [1, 1] \rangle$$
$$= \langle [0, 0], [1, 1], [1, 1] \rangle. \qquad \qquad \square$$

Corollary 1 *Let $\hat{A}$ be an $m \times m$ IVN-matrix and let $\lambda = \langle [T^L_\lambda, T^U_\lambda], [I^L_\lambda, I^U_\lambda], [F^L_\lambda, F^U_\lambda] \rangle \in I^3$. Then $|\lambda^n \hat{A}| = |\lambda \hat{A}| = \lambda |\hat{A}|$.*

Remark 2 $\hat{A}$ and $\hat{B}$ be two $m \times m$ IVN-matrices. In generally, $|\hat{A}\hat{B}| \neq |\hat{A}||\hat{B}|$.

Example 3 Let us consider IVN-matrices $A = \begin{bmatrix} \langle [0, 0], [1, 1], [1, 1] \rangle & \langle [1, 1], [0, 0], [0, 0] \rangle \\ \langle [0, 0], [1, 1], [1, 1] \rangle & \langle [1, 1], [0, 0], [0, 0] \rangle \end{bmatrix}$

and

$$B = \begin{bmatrix} \langle [0.5, 0.6], [0.4, 0.7], [0.3, 0.5] \rangle & \langle [0.3, 0.5], [0.2, 0.7], [0.4, 0.4] \rangle \\ \langle [0.2, 0.4], [0.4, 0.4], [0.7, 1.0] \rangle & \langle [0.4, 0.8], [0.3, 0.6], [0.5, 0.7] \rangle \end{bmatrix}.$$

Then, $|\hat{A}| = \langle [0, 0], [1, 1], [1, 1] \rangle$ and $|\hat{B}| = \langle [0.4, 0.6], [0.4, 0.7], [0.5, 0.7] \rangle$. So $|\hat{A}||\hat{B}| = \langle [0, 0], [1, 1], [1, 1] \rangle$.

Also $\hat{A}\hat{B} = \begin{bmatrix} \langle [0.2, 0.4], [0, 4, 0.4], [0, 7, 0.1] \rangle & \langle [0.4, 0.8], [0.3, 0.6], [0.5, 0.7] \rangle \\ \langle [0.2, 0.4], [0.4, 0.4], [0.7, 1.0] \rangle & \langle [0.4, 0.8], [0.3, 0.6], [0.5, 0.7] \rangle \end{bmatrix}$ and $|\hat{A}\hat{B}| = \langle [0.2, 0.4], [0.4, 0.6], [0.7, 0.7] \rangle$. This shows that $|\hat{A}\hat{B}| \neq |\hat{A}||\hat{B}|$.

Definition 4 Let $\hat{A}$ be an $m \times m$ IVN-matrix in which all the entries above the main diagonal are IVN-zero. Then, IVN-matrix $\hat{A}$ is called lower triangular IVN-matrix. If all of the entries below the main diagonal of IVN-matrix $\hat{A}$ are IVN-zero, then $\hat{A}$ is called upper triangular IVN-matrix.

Theorem 1 Let $\hat{A} = \left[\langle [T_{A_{ij}}^L, T_{A_{ij}}^U], [I_{A_{ij}}^L, I_{A_{ij}}^U], [F_{A_{ij}}^L, F_{A_{ij}}^U] \rangle \right]$ be an $m \times m$ IVN-matrix. If $\hat{A}$ is triangular, then $|\hat{A}| = \langle [\bigwedge_{i=1}^m T_{A_{ii}}^L, \bigwedge_{i=1}^m T_{A_{ii}}^U], [\bigvee_{i=1}^m I_{A_{ii}}^U, \bigvee_{i=1}^m I_{A_{ii}}^U], [\bigvee_{i=1}^m F_{A_{ii}}^L, \bigvee_{i=1}^m F_{A_{ii}}^U] \rangle$ for all $1 \le i \le m$.

Proof The proof is trivial. □

Definition 5 Let $\hat{A} = \left[\langle [T_{A_{ij}}^L, T_{A_{ij}}^U], [I_{A_{ij}}^L, I_{A_{ij}}^U], [F_{A_{ij}}^L, F_{A_{ij}}^U] \rangle \right]$ be an $m \times n$ IVN-matrix. IVN-matrix $\hat{B} = \left[\langle [F_{A_{ij}}^L, F_{A_{ij}}^U], [1 - I_{A_{ij}}^U, 1 - I_{A_{ij}}^L], [T_{A_{ij}}^L, T_{A_{ij}}^U] \rangle \right]$ is called complement IVN-matrix of IVN-matrix $\hat{A}$, and denoted by $\hat{B} = \hat{A}^c$.

Proposition 4 Let $\hat{A} = \left[\langle [T_{A_{ij}}^L, T_{A_{ij}}^U], [I_{A_{ij}}^L, I_{A_{ij}}^U], [F_{A_{ij}}^L, F_{A_{ij}}^U] \rangle \right]$ be an $m \times m$ IVN-matrix. Then,

$$|\hat{A}|^c \ge |\hat{A}^c|.$$

Proof We have that

$$|\hat{A}| = \left\langle [\bigvee_{\sigma \in S_m} (\bigwedge_{i=1}^m T_{A_{i\sigma(i)}}^L), \bigvee_{\sigma \in S_m} (\bigwedge_{i=1}^m T_{A_{i\sigma(i)}}^U)], [\bigwedge_{\sigma \in S_m} (\bigvee_{i=1}^m I_{A_{i\sigma(i)}}^L), \bigwedge_{\sigma \in S_m} (\bigvee_{i=1}^m I_{A_{i\sigma(i)}}^U)], \right.$$
$$\left. [\bigwedge_{\sigma \in S_m} (\bigvee_{i=1}^m F_{A_{i\sigma(i)}}^L), \bigwedge_{\sigma \in S_m} (\bigvee_{i=1}^m F_{A_{i\sigma(i)}}^U)] \right\rangle.$$

Then

$$|\hat{A}|^c = \left\langle [\bigwedge_{\sigma \in S_m} (\bigvee_{i=1}^m F_{A_{i\sigma(i)}}^L), \bigwedge_{\sigma \in S_m} (\bigvee_{i=1}^m F_{A_{i\sigma(i)}}^U)], [1 - \bigwedge_{\sigma \in S_m} (\bigvee_{i=1}^m I_{A_{i\sigma(i)}}^U), 1 - \bigwedge_{\sigma \in S_m} (\bigvee_{i=1}^m I_{A_{i\sigma(i)}}^L)], \right.$$
$$\left. [\bigvee_{\sigma \in S_m} (\bigwedge_{i=1}^m T_{A_{i\sigma(i)}}^L), \bigvee_{\sigma \in S_m} (\bigwedge_{i=1}^m T_{A_{i\sigma(i)}}^U)] \right\rangle.$$

$$|\hat{A}|^c = \Big\langle [\bigwedge_{\sigma \in S_m} (\vee_{i=1}^m F^L_{A_{i\sigma(i)}}), \bigwedge_{\sigma \in S_m} (\vee_{i=1}^m F^U_{A_{i\sigma(i)}})], [\bigvee_{\sigma \in S_m} (1 - \vee_{i=1}^m I^U_{A_{i\sigma(i)}}), \bigvee_{\sigma \in S_m} (1 - \vee_{i=1}^m I^L_{A_{i\sigma(i)}})],$$

$$[\bigvee_{\sigma \in S_m} (\wedge_{i=1}^m T^L_{A_{i\sigma(i)}}), \bigvee_{\sigma \in S_m} (\wedge_{i=1}^m T^U_{A_{i\sigma(i)}})] \Big\rangle. \tag{2}$$

$$\hat{A}^c = \Big[\langle [F^L_{A_{ij}}, F^U_{A_{ij}}], [1 - I^U_{A_{ij}}, 1 - I^L_{A_{ij}}], [T^L_{A_{ij}}, T^U_{A_{ij}}] \rangle \Big]$$

$$|\hat{A}^c| = \Big\langle [\bigvee_{\sigma \in S_m} (\wedge_{i=1}^m F^L_{A_{i\sigma(i)}}), \bigvee_{\sigma \in S_m} (\wedge_{i=1}^m F^U_{A_{i\sigma(i)}})], [\bigwedge_{\sigma \in S_m} (1 - (\vee_{i=1}^m I^U_{A_{i\sigma(i)}})), \bigwedge_{\sigma \in S_m} (1 - (\vee_{i=1}^m I^L_{A_{i\sigma(i)}}))],$$

$$[\bigwedge_{\sigma \in S_m} (\vee_{i=1}^m T_{A_{i\sigma(i)}} L), \bigwedge_{\sigma \in S_m} (\vee_{i=1}^m T^U_{A_{i\sigma(i)}})] \Big\rangle.$$

Since

$$\bigwedge_{\sigma \in S_m} (\vee_{i=1}^m F^L_{A_{i\sigma(i)}}) \geq \bigvee_{\sigma \in S_m} (\wedge_{i=1}^m F^L_{A_{i\sigma(i)}}) \text{ and } \bigwedge_{\sigma \in S_m} (\vee_{i=1}^m F^U_{A_{i\sigma(i)}}) \geq \bigvee_{\sigma \in S_m} (\wedge_{i=1}^m F^U_{A_{i\sigma(i)}}),$$

$$\bigwedge_{\sigma \in S_m} (1 - (\vee_{i=1}^m I^L_{A_{i\sigma(i)}})) \leq 1 - \bigwedge_{\sigma \in S_m} (\vee_{i=1}^m I^L_{A_{i\sigma(i)}}) \text{ and } \bigwedge_{\sigma \in S_m} (1 - (\vee_{i=1}^m I^U_{A_{i\sigma(i)}})) \leq 1 - \bigwedge_{\sigma \in S_m} (\vee_{i=1}^m I^U_{A_{i\sigma(i)}}),$$

$$\bigwedge_{\sigma \in S_m} (\wedge_{i=1}^m T^L_{A_{i\sigma(i)}}) = \bigwedge_{\sigma \in S_m} (\vee_{i=1}^m T^L_{A_{i\sigma(i)}}) \text{ and } \bigwedge_{\sigma \in S_m} (\wedge_{i=1}^m T^U_{A_{i\sigma(i)}}) = \bigwedge_{\sigma \in S_m} (\vee_{i=1}^m T^U_{A_{i\sigma(i)}}).$$

It is concluded that $|\hat{A}|^c \succeq |\hat{A}^c|$.   □

Note that, in generally $|\hat{A}|^c \neq |\hat{A}^c|$.

Example 4 Let us consider IVN-matrix $\hat{A}$ given in Example 3. Then, $|\hat{A}|^c = \langle [1, 1], [0, 0], [0, 0] \rangle$ and $|\hat{A}^c| = \langle [0, 0], [1, 1], [1, 1] \rangle$. Thus $|\hat{A}|^c \neq |\hat{A}^c|$.

Proposition 5 *Let* $\hat{A} = \Big[\langle [T^L_{A_{ij}}, T^U_{A_{ij}}], [I^L_{A_{ij}}, I^U_{A_{ij}}], [F^L_{A_{ij}}, F^U_{A_{ij}}] \rangle \Big]$ *be an* $m \times m$ *IVN-matrix. Then,*

$$|\hat{A}| = |\hat{A}^t|.$$

Proof The proof is obvious from Definitions 2 and 3.   □

4   Adjoint of a Square IVN-Matrix

Definition 6 Let $\hat{A} = \Big[\langle [T^L_{A_{ij}}, T^U_{A_{ij}}], [I^L_{A_{ij}}, I^U_{A_{ij}}], [F^L_{A_{ij}}, F^U_{A_{ij}}] \rangle \Big]$ be an $m \times m$ IVN-matrix, $\hat{A}_{ij}$ be a $(m - 1) \times (m - 1)$ IVN-matrix obtained from $\hat{A}$ by striking out row i and column j, and $|\hat{A}_{ij}|$ be the determinant of IVN-matrix $\hat{A}_{ij}$. Then adjoint matrix of IVN-matrix $\hat{A}$, denoted by $adj(\hat{A})$, is defined by $adj(\hat{A}) = (|\hat{A}_{ij}|)^t$.

Remark 3 An element $\langle [T_{ij}^L, T_{ij}^U], [I_{ij}^L, I_{ij}^U], [F_{ij}^L, F_{ij}^U] \rangle$ of $adj(\hat{A})$ can be written as follows:

$$
\langle [T_{ij}^L, T_{ij}^U], [I_{ij}^L, I_{ij}^U], [F_{ij}^L, F_{ij}^U] \rangle = \Big\langle [\bigvee_{\sigma \in S_{m_j m_i}} (\wedge_{r \in m_j} T_{|A_{\sigma(r)r}|}^L), \bigvee_{\sigma \in S_{m_j m_i}} (\wedge_{r \in m_j} T_{|A_{\sigma(r)r}|}^U)],
$$

$$
[\bigwedge_{\sigma \in S_{m_j m_i}} (\vee_{r \in m_j} I_{|A_{\sigma(r)r}|}^L), \bigwedge_{\sigma \in S_{m_j m_i}} (\vee_{r \in m_j} I_{|A_{\sigma(r)r}|}^U)],
$$

$$
[\bigwedge_{\sigma \in S_{m_j m_i}} (\vee_{r \in m_j} F_{|A_{\sigma(r)r}|}^L), \bigwedge_{\sigma \in S_{m_j m_i}} (\vee_{r \in m_j} F_{|A_{\sigma(r)r}|}^U)] \Big\rangle.
$$

Here $m_j = \{1, 2, \ldots, m\} \setminus \{j\}$ and $S_{m_j m_i}$ is the set of all permutations of set m_j over the set m_i. Also $[T_{|A_{\sigma(r)r}|}^L, T_{|A_{\sigma(r)r}|}^U]$, $[I_{|A_{\sigma(r)r}|}^L, I_{|A_{\sigma(r)r}|}^U]$, and $[F_{|A_{\sigma(r)r}|}^L, F_{|A_{\sigma(r)r}|}^U]$ are truth-membership, indeterminacy-membership, and falsity-membership intervals of $|\hat{A}_{r\sigma(r)}|$, respectively.

Example 5 Let us consider IVN-matrix $\hat{A}$ in Example 1. Then,

$|\hat{A}_{11}| = \langle [0.8, 0.9], [0.4, 0.7], [0.5, 0.6] \rangle$, $|\hat{A}_{12}| = \langle [0.4, 0.5], [0.6, 0.8], [0.5, 0.5] \rangle$,
$|\hat{A}_{13}| = \langle [0.8, 0.9], [0.4, 0.4], [0.8, 0.8] \rangle$, $|\hat{A}_{21}| = \langle [0.5, 0.8], [0.5, 0.6], [0.3, 0.6] \rangle$,
$|\hat{A}_{22}| = \langle [0.7, 0.9], [0.6, 0.8], [0.2, 0.4] \rangle$, $|\hat{A}_{23}| = \langle [0.5, 0.9], [0.6, 0.7], [0.4, 0.7] \rangle$,
$|\hat{A}_{31}| = \langle [0.6, 0.8], [0.5, 0.7], [0.5, 0.7] \rangle$, $|\hat{A}_{32}| = \langle [0.4, 0.5], [0.5, 0.6], [0.8, 0.9] \rangle$,
$|\hat{A}_{33}| = \langle [0.7, 0.9], [0.5, 0.5], [0.8, 0.9] \rangle$ and

$$
adj(\hat{A}) = (|\hat{A}_{ij}|)^t = \begin{bmatrix} \langle [0.8, 0.9], [0.4, 0.7], [0.5, 0.6] \rangle & \langle [0.5, 0.8], [0.5, 0.6], [0.3, 0.6] \rangle \\ \langle [0.4, 0.5], [0.6, 0.8], [0.5, 0.5] \rangle & \langle [0.7, 0.9], [0.6, 0.8], [0.2, 0.4] \rangle \\ \langle [0.8, 0.9], [0.4, 0.4], [0.8, 0.8] \rangle & \langle [0.5, 0.9], [0.6, 0.7], [0.4, 0.7] \rangle \end{bmatrix}
$$

$$
\begin{bmatrix} \langle [0.6, 0.8], [0.5, 0.7], [0.5, 0.7] \rangle \\ \langle [0.4, 0.5], [0.5, 0.6], [0.8, 0.9] \rangle \\ \langle [0.7, 0.9], [0.5, 0.5], [0.8, 0.9] \rangle. \end{bmatrix}
$$

Lemma 1 *Let* $\hat{A} = \left[\langle [T_{A_{ij}}^L, T_{A_{ij}}^U], [I_{A_{ij}}^L, I_{A_{ij}}^U], [F_{A_{ij}}^L, F_{A_{ij}}^U] \rangle \right]$ *and* $\hat{B} = \left[\langle [T_{B_{ij}}^L, T_{B_{ij}}^U], [I_{B_{ij}}^L, I_{B_{ij}}^U], [F_{B_{ij}}^L, F_{B_{ij}}^U] \rangle \right]$ *be two* $m \times m$ *IVN-matrices. If* $\hat{A} \tilde{\leq} \hat{B}$, *then* $|\hat{A}_{ij}| \preceq |\hat{B}_{ij}|$ *for all* $1 \leq i, j \leq m$.

Proof Suppose that $\hat{A} = \left[\langle [T_{A_{ij}}^L, T_{A_{ij}}^U], [I_{A_{ij}}^L, I_{A_{ij}}^U], [F_{A_{ij}}^L, F_{A_{ij}}^U] \rangle \right]$ and $\hat{B} = \left[\langle [T_{B_{ij}}^L, T_{B_{ij}}^U], [I_{B_{ij}}^L, I_{B_{ij}}^U], [F_{B_{ij}}^L, F_{B_{ij}}^U] \rangle \right]$ be two $m \times m$ IVN-matrices and $\hat{A} \tilde{\leq} \hat{B}$. Then, $T_{A_{ij}}^L \leq T_{B_{ij}}^L$, $T_{A_{ij}}^U \leq T_{B_{ij}}^U$, $I_{A_{ij}}^L \geq I_{B_{ij}}^L$, $I_{A_{ij}}^U \geq I_{B_{ij}}^U$, and $F_{A_{ij}}^L \geq F_{B_{ij}}^L$, $F_{A_{ij}}^U \geq F_{B_{ij}}^U$ for all $1 \leq i, j \leq m$. Therefore, $\hat{A}_{ij} \tilde{\leq} \hat{B}_{ij}$ and $|\hat{A}_{ij}| \preceq |\hat{B}_{ij}|$. □

Lemma 2 *Let* $\hat{A} = \left[\langle[T_{A_{ij}}^L, T_{A_{ij}}^U], [I_{A_{ij}}^L, I_{A_{ij}}^U], [F_{A_{ij}}^L, F_{A_{ij}}^U]\rangle\right]$ *and* $\hat{B} = \left[\langle[T_{B_{ij}}^L, T_{B_{ij}}^U], [I_{B_{ij}}^L, I_{B_{ij}}^U],\right.$
$\left.[F_{B_{ij}}^L, F_{B_{ij}}^U]\rangle\right]$ *be two* $m \times m$ *IVN-matrices. If* $\hat{A} \tilde{\leq} \hat{B}$, *then* $|\hat{A}^t| \tilde{\leq} |\hat{B}^t|$.

Proof The proof is clear from Definition 2 and Proposition 1 $\qquad\square$

Lemma 3 *Let* $\hat{A} = \left[\langle[T_{A_{ij}}^L, T_{A_{ij}}^U], [I_{A_{ij}}^L, I_{A_{ij}}^U], [F_{A_{ij}}^L, F_{A_{ij}}^U]\rangle\right]$, $\hat{B} = \left[\langle[T_{B_{ij}}^L, T_{B_{ij}}^U],\right.$
$\left.[I_{B_{ij}}^L, I_{B_{ij}}^U], [F_{B_{ij}}^L, F_{B_{ij}}^U]\rangle\right]$ *be two* $m \times m$ *IVN-matrices. Then,* $(\hat{A}^c)^t = (\hat{A}^t)^c$

Proof The proof is clear from definitions of complement and transpose of the IVN-matrices. $\qquad\square$

Proposition 6 *Let* $\hat{A} = \left[\langle[T_{A_{ij}}^L, T_{A_{ij}}^U], [I_{A_{ij}}^L, I_{A_{ij}}^U], [F_{A_{ij}}^L, F_{A_{ij}}^U]\rangle\right]$ *and* $\hat{B} = \left[\langle[T_{B_{ij}}^L, T_{B_{ij}}^U], [I_{B_{ij}}^L, I_{B_{ij}}^U], [F_{B_{ij}}^L, F_{B_{ij}}^U]\rangle\right]$ *be two* $m \times m$ *IVN-matrices. Then*

1. $\hat{A} \tilde{\leq} \hat{B}$ *implies* $adj(\hat{A}) \tilde{\leq} adj(\hat{B})$,
2. $adj(\hat{A}) + adj(\hat{B}) \tilde{\leq} adj(\hat{A} + \hat{B})$,
3. $adj(\hat{A}^t) = (adj(\hat{A}))^t$,
4. $adj(\hat{A}^c) = (adj(\hat{A}))^c$.

Proof

1. Let $\hat{A} \tilde{\leq} \hat{B}$. By Lemma 1 $|\hat{A}_{ij}| \preceq |\hat{B}_{ij}|$ for all $1 \leq i, j \leq m$, and from Lemma 2 $(|\hat{A}_{ij}|)^t \tilde{\leq} (|\hat{B}_{ij}|)^t$. Therefore, we have that $adj(\hat{A}) \tilde{\leq} adj(\hat{B})$.
2. Since $\hat{A}, \hat{B} \tilde{\leq} \hat{A} + \hat{B}$, by Lemma 1 $|\hat{A}_{ij}|, |\hat{B}_{ij}| \preceq |\hat{A}_{ij} + \hat{B}_{ij}|$ for all $1 \leq i, j \leq m$. It is clear that $adj(\hat{A}), adj(\hat{B}) \tilde{\leq} adj(\hat{A} + \hat{B})$ and so $adj(\hat{A}) + adj(\hat{B}) \tilde{\leq} adj(\hat{A} + \hat{B})$.
3. We know that $\hat{A}^t = [(T_{A_{ji}}, I_{A_{ji}}, F_{A_{ji}})]$. Then, $adj(\hat{A}^t) = (|\hat{A}_{ji}|)^t = ((|\hat{A}_{ij}|)^t)^t = (adj(\hat{A}_{ij}))^t$.
4. We know that $\hat{A}^c = [(F_{A_{ij}}, 1 - I_{A_{ij}}, T_{A_{ij}})]$. By Definition 6 $adj(\hat{A}^c) = (|\hat{A}_{ij}^c|)^t$. From Lemma 3 $(|\hat{A}_{ij}^c|)^t = (|\hat{A}_{ij}^t|)^c$. Therefore, $adj(\hat{A}^c) = (adj(\hat{A}))^c$.

$\qquad\square$

Proposition 7 *Let* $\hat{A} = \left[\langle[T_{A_{ij}}^L, T_{A_{ij}}^U], [I_{A_{ij}}^L, I_{A_{ij}}^U], [F_{A_{ij}}^L, F_{A_{ij}}^U]\rangle\right]$ *be an* $m \times m$ *IVN-matrix. Then,*

1. $|\hat{A}|I_m \tilde{\leq} \hat{A} adj(\hat{A})$.
2. $|\hat{A}|I_m \tilde{\leq} adj(\hat{A}) \hat{A}$

Proof

1. Let $[T_{|\hat{A}_{kj}|}^L, T_{|\hat{A}_{kj}|}^U] = [\bigvee_{\sigma \in S_{m_j m_k}} (\wedge_{r \in m_j} T_{|A_{\sigma(r)r}|}^L), \bigvee_{\sigma \in S_{m_j m_k}} (\wedge_{r \in m_j} T_{|A_{\sigma(r)r}|}^U)]$,
   $[I_{|\hat{A}_{kj}|}^L, I_{|\hat{A}_{kj}|}^U] = [\bigwedge_{\sigma \in S_{m_j m_k}} (\vee_{r \in m_j} I_{|A_{\sigma(r)r}|}^L), \bigwedge_{\sigma \in S_{m_j m_k}} (\vee_{r \in m_j} I_{|A_{\sigma(r)r}|}^U)]$, and
   $[F_{|\hat{A}_{kj}|}^L, F_{|\hat{A}_{kj}|}^U] = [\bigwedge_{\sigma \in S_{m_j m_k}} (\vee_{r \in m_j} F_{|A_{\sigma(r)r}|}^L), \bigwedge_{\sigma \in S_{m_j m_k}} (\vee_{r \in m_j} F_{|A_{\sigma(r)r}|}^U)]$

denote truth-membership, indeterminacy-membership, and falsity-membership values of elements of $adj(\hat{A})$. Let $\hat{A}adj(\hat{A}) = \hat{C}$ and $\langle [T^L_{C_{ij}}, T^U_{C_{ij}}], [I^L_{C_{ij}}, I^U_{C_{ij}}], [F^L_{C_{ij}}, F^U_{C_{ij}}] \rangle \in \hat{C}$, respectively. Then,

$$\langle [T^L_{C_{ij}}, T^U_{C_{ij}}], [I^L_{C_{ij}}, I^U_{C_{ij}}], [F^L_{C_{ij}}, F^U_{C_{ij}}] \rangle = \Big\langle [\textstyle\bigvee_{k=1}^m (T^L_{A_{ik}} \wedge T^L_{|\hat{A}_{kj}|}), \bigvee_{k=1}^m (T^U_{A_{ik}} \wedge T^U_{|\hat{A}_{kj}|})], [\textstyle\bigwedge_{k=1}^m (I^L_{A_{ik}} \vee I^L_{|\hat{A}_{kj}|}), \bigwedge_{k=1}^m (I^U_{A_{ik}} \vee I^U_{|\hat{A}_{kj}|})], [\textstyle\bigwedge_{k=1}^m (F^L_{A_{ik}} \vee F^L_{|\hat{A}_{kj}|}),$$
$$\textstyle\bigwedge_{k=1}^m (F^U_{A_{ik}} \vee F^U_{|\hat{A}_{kj}|})] \Big\rangle.$$

We know that

$$\Big[\bigvee_{\sigma \in S_m} (\textstyle\bigwedge_{k=1}^m T^L_{A_{k\sigma(k)}}) \; , \; \bigvee_{\sigma \in S_m} (\textstyle\bigwedge_{k=1}^m T^U_{A_{k\sigma(k)}}) \Big] \wedge [T^L_{I_{ij}}, T^U_{I_{ij}}]$$

$$= \begin{cases} [\bigvee_{\sigma \in S_m} (\bigwedge_{k=1}^m T^L_{A_{k\sigma(k)}}), \bigvee_{\sigma \in S_m} (\bigwedge_{k=1}^m T^U_{A_{k\sigma(k)}})], & i = j \\ 0, & i \neq j \end{cases}$$

$$\Big[\bigwedge_{\sigma \in S_m} (\textstyle\bigvee_{k=1}^m I^L_{A_{k\sigma(k)}}) \; , \; \bigwedge_{\sigma \in S_m} (\textstyle\bigvee_{k=1}^m I^U_{A_{k\sigma(k)}}) \Big] \vee [I^L_{I_{ij}}, I^U_{I_{ij}}]$$

$$= \begin{cases} [\bigwedge_{\sigma \in S_m} (\bigvee_{k=1}^m I^L_{A_{k\sigma(k)}}), \bigwedge_{\sigma \in S_m} (\bigvee_{k=1}^m I^U_{A_{k\sigma(k)}})], & i = j \\ 1, & i \neq j \end{cases}$$

$$\Big[\bigwedge_{\sigma \in S_m} (\textstyle\bigvee_{k=1}^m F^L_{A_{k\sigma(k)}}) \; , \; \bigwedge_{\sigma \in S_m} (\textstyle\bigvee_{k=1}^m F^U_{A_{k\sigma(k)}}) \Big] \vee [F^L_{I_{ij}}, F^U_{I_{ij}}]$$

$$= \begin{cases} [\bigwedge_{\sigma \in S_m} (\bigvee_{k=1}^m F^L_{A_{k\sigma(k)}}), \bigwedge_{\sigma \in S_m} (\bigvee_{k=1}^m F^U_{A_{k\sigma(k)}})], & i = j \\ 1, & i \neq j. \end{cases}$$

Also it is clear that for $i = j$,

$$[T^L_{C_{ii}}, T^U_{C_{ii}}] = [\bigvee_{k=1}^m (T^L_{A_{ik}} \wedge T^L_{|\hat{A}_{ki}|}), \bigvee_{k=1}^m (T^U_{A_{ik}} \wedge T^U_{|\hat{A}_{ki}|})] = [\bigvee_{\sigma \in S_m} (\bigwedge_{k=1}^m T^L_{A_{k\sigma(k)}}), \bigvee_{\sigma \in S_m} (\bigwedge_{k=1}^m T^U_{A_{k\sigma(k)}})]$$

$$[I^L_{C_{ii}}, I^U_{C_{ii}}] = [\bigwedge_{k=1}^m (I^L_{A_{ik}} \vee T^L_{|\hat{A}_{ki}|}), \bigwedge_{k=1}^m (I^U_{A_{ik}} \vee T^U_{|\hat{A}_{ki}|})] = [\bigwedge_{\sigma \in S_m} (\bigvee_{k=1}^m I^L_{A_{k\sigma(k)}}), \bigwedge_{\sigma \in S_m} (\bigvee_{k=1}^m I^U_{A_{k\sigma(k)}})],$$

$$[F^L_{C_{ii}}, F^U_{C_{ii}}] = [\bigwedge_{k=1}^m (F^L_{A_{ik}} \vee F^L_{|\hat{A}_{ki}|}), \bigwedge_{k=1}^m (F^U_{A_{ik}} \vee F^U_{|\hat{A}_{ki}|})] = [\bigwedge_{\sigma \in S_m} (\bigvee_{k=1}^m F^L_{A_{k\sigma(k)}}), \bigwedge_{\sigma \in S_m} (\bigvee_{k=1}^m F^U_{A_{k\sigma(k)}})]$$

and for $i \neq j$ $[0, 0] \preceq [T^L_{C_{ij}}, T^U_{C_{ij}}]$, $[1, 1] \succeq [I^L_{C_{ij}}, I^U_{C_{ij}}]$, $[1, 1] \succeq [F^L_{C_{ij}}, F^U_{C_{ij}}]$. Therefore, $|\hat{A}|I_m \tilde{\leq} \hat{A}adj(\hat{A})$.

2. The proof can be made in similar way to proof of 1.

□

Corollary 2 *Let* $\hat{A} = \left[\langle [T^L_{A_{ij}}, T^U_{A_{ij}}], [I^L_{A_{ij}}, I^U_{A_{ij}}], [F^L_{A_{ij}}, F^U_{A_{ij}}] \rangle \right]$ *be an* $m \times m$ *IVN-matrix.* $\hat{A}(adj(\hat{A})) \neq |\hat{A}|I_m$.

Example 6 Consider IVN-matrix $\hat{A}$ in Example 1 and $adj(\hat{A})$ in Example 5. Then,

$$\hat{A}(adj(\hat{A})) = \begin{bmatrix} \langle [0.7, 0.9], [0.5, 0.6], [0.5, 0.7] \rangle & \langle [0.5, 0.8], [0.6, 0.7], [0.4, 0.7] \rangle & \langle [0.6, 0.8], [0.5, 0.6], [0.8, 0.9] \rangle \\ \langle [0.4, 0.5], [0.4, 0.7], [0.8, 0.8] \rangle & \langle [0.7, 0.9], [0.5, 0.6], [0.5, 0.7] \rangle & \langle [0.4, 0.5], [0.5, 0.7], [0.8, 0.9] \rangle \\ \langle [0.8, 0.9], [0.6, 0.8], [0.5, 0.6] \rangle & \langle [0.5, 0.9], [0.6, 0.8], [0.3, 0.6] \rangle & \langle [0.7, 0.9], [0.5, 0.6], [0.5, 0.7] \rangle \end{bmatrix}$$

and

$$|\hat{A}|I_m = \langle [0.7, 0.9], [0.5, 0.6], [0.5, 0.7] \rangle \begin{bmatrix} \langle [1, 1], [0, 0], [0, 0] \rangle & \langle [0, 0], [1, 1], [1, 1] \rangle & \langle [0, 0], [1, 1], [1, 1] \rangle \\ \langle [0, 0], [1, 1], [1, 1] \rangle & \langle [1, 1], [0, 0], [0, 0] \rangle & \langle [0, 0], [1, 1], [1, 1] \rangle \\ \langle [0, 0], [1, 1], [1, 1] \rangle & \langle [0, 0], [1, 1], [1, 1] \rangle & \langle [1, 1], [0, 0], [0, 0] \rangle \end{bmatrix}$$

$$= \begin{bmatrix} \langle [0.7, 0.9], [0.5, 0.6], [0.5, 0.7] \rangle & \langle [0, 0], [1, 1], [1, 1] \rangle & \langle [0, 0], [1, 1], [1, 1] \rangle \\ \langle [0, 0], [1, 1], [1, 1] \rangle & \langle [0.7, 0.9], [0.5, 0.6], [0.5, 0.7] \rangle & \langle [0, 0], [1, 1], [1, 1] \rangle \\ \langle [0, 0], [1, 1], [1, 1] \rangle & \langle [0, 0], [1, 1], [1, 1] \rangle & \langle [0.7, 0.9], [0.5, 0.6], [0.5, 0.7]. \rangle \end{bmatrix}$$

Thus, we show that $\hat{A}(adj(\hat{A})) \neq |\hat{A}|I_m$. Note that, here $\hat{A}(adj(\hat{A})) \tilde{\geq} |\hat{A}|I_m$. Also

$$adj(\hat{A})\hat{A} = \begin{bmatrix} \langle [0.7, 0.9], [0.5, 0.6], [0.5, 0.7] \rangle & \langle [0.5, 0.8], [0.5, 0.7], [0.5, 0.7] \rangle & \langle [0.6, 0.8], [0.5, 0.7], [0.5, 0.6] \rangle \\ \langle [0.4, 0.5], [0.6, 0.8], [0.8, 0.9] \rangle & \langle [0.7, 0.9], [0.5, 0.6], [0.5, 0.7] \rangle & \langle [0.4, 0.5], [0.6, 0.8], [0.5, 0.5] \rangle \\ \langle [0.7, 0.9], [0.6, 0.7], [0.8, 0.9] \rangle & \langle [0.5, 0.9], [0.5, 0.5], [0.8, 0.8] \rangle & \langle [0.7, 0.9], [0.5, 0.6], [0.5, 0.7] \rangle \end{bmatrix}$$

and so $|\hat{A}|I_m \tilde{\leq} adj(\hat{A})\hat{A}$.

Theorem 2 *Let* $\hat{A}$ *be an* $m \times m$ *IVN-matrix. Then,* $|\hat{A}| = |adj(\hat{A})|$.

Proof The proof can be made by using definitions of determinant and adjoint of an IVN-matrix.   □

Definition 7 Let $\hat{A}$ be an $m \times m$ IVN-matrix. Then,

1. $\hat{A}$ is reflexive if and only if $I_m \tilde{\leq} \hat{A}$.
2. $\hat{A}$ is symmetric if and only if $\hat{A} = \hat{A}^t$(with similar way [3]).
3. $\hat{A}$ is transitive if and only if $\hat{A}^2 \tilde{\leq} \hat{A}$.
4. $\hat{A}$ is idempotent if and only if $\hat{A}^2 = \hat{A}$.
5. $\hat{A}$ is similarity if and only if it is reflexive, symmetric, and transitive.

Example 7 Let us consider SVN-matrix $\hat{A}$ given as follows:

$$\hat{A} = \begin{bmatrix} \langle [1, 0], [0, 0], [0, 0] \rangle & \langle [0, 0], [0.4, 0.6], [0.3, 0.5] \rangle & \langle [0, 0], [0.3, 0.7], [0.1, 0.6] \rangle \\ \langle [0, 0], [0.4, 0.6], [0.3, 0.5] \rangle & \langle [1, 0], [0, 0], [0, 0] \rangle & \langle [0, 0], [0.4, 0.6], [0.3, 0.5] \rangle \\ \langle [0, 0], [0.3, 0.7], [0.1, 0.6] \rangle & \langle [0, 0], [0.4, 0.6], [0.3, 0.5] \rangle & \langle [1, 0], [0, 0], [0, 0] \rangle. \end{bmatrix}$$

Then it is clear that $\hat{A}$ is reflexive and symmetric. Since

$$\hat{A}^2 = \begin{bmatrix} \langle [1,0],[0,0],[0,0]\rangle & \langle [0,0],[0.4,0.6],[0.3,0.5]\rangle & \langle [0,0],[0.3,0.7],[0.1,0.6]\rangle \\ \langle [0,0],[0.4,0.6],[0.3,0.5]\rangle & \langle [1,0],[0,0],[0,0]\rangle & \langle [0,0],[0.4,0.6],[0.3,0.5]\rangle \\ \langle [0,0],[0.3,0.7],[0.1,0.6]\rangle & \langle [0,0],[0.4,0.6],[0.3,0.5]\rangle & \langle [1,0],[0,0],[0,0]\rangle \end{bmatrix} = \hat{A},$$

$\hat{A}$ is idempotent and transitive IVN-matrix. Also, by Definition 7, IVN-matrix $\hat{A}$ is similarity.

Definition 8 Let $\hat{A} = \left[\langle [T^L_{A_{ij}}, T^U_{A_{ij}}], [I^L_{A_{ij}}, I^U_{A_{ij}}], [F^L_{A_{ij}}, F^U_{A_{ij}}] \rangle \right]$ be IVN-matrix. If $\langle [T^L_{A_{ik}}, T^U_{A_{ik}}], [I^L_{A_{ik}}, I^U_{A_{ik}}], [F^L_{A_{ik}}, F^U_{A_{ik}}] \rangle \big] = \langle [T^L_{A_{jk}}, T^U_{A_{jk}}], [I^L_{A_{jk}}, I^U_{A_{jk}}], [F^L_{A_{jk}}, F^U_{A_{jk}}] \rangle \big]$, for all i, j, k, then $\hat{A}$ is called constant IVN-matrix.

Proposition 8 *Let $\hat{A}$ be an $m \times m$ constant IVN-matrix. Then,*

1. *$(adj\,\hat{A})^t$ is constant.*
2. *$G = \hat{A}(adj\,\hat{A})$ is constant and $g_{ii} = |\hat{A}|$.*

Proof

1. Let $\hat{B} = adj(\hat{A})$. Then

$$\left\langle [T^L_{B_{ij}}, T^U_{B_{ij}}], [I^L_{B_{ij}}, I^U_{B_{ij}}], [F^L_{B_{ij}}, F^U_{B_{ij}}] \right\rangle = \left\langle \left[\bigvee_{\sigma \in S_{m_j m_i}} (\wedge_{r \in m_j} T^L_{B_{r\sigma(r)}}), \bigvee_{\sigma \in S_{m_j m_i}} (\wedge_{r \in m_j} T^U_{B_{r\sigma(r)}}) \right], \right.$$
$$\left[\bigwedge_{\sigma \in S_{m_j m_i}} (\vee_{r \in m_j} I^L_{B_{r\sigma(r)}}), \bigwedge_{\sigma \in S_{m_j m_i}} (\vee_{r \in m_j} I^U_{B_{r\sigma(r)}}) \right],$$
$$\left. \left[\bigwedge_{\sigma \in S_{m_j m_i}} (\vee_{r \in m_j} F_{B_{r\sigma(r)}}), \bigwedge_{\sigma \in S_{m_j m_i}} (\vee_{r \in m_j} F_{B_{r\sigma(r)}}) \right] \right\rangle$$

and

$$\left\langle [T^L_{B_{ik}}, T^U_{B_{ik}}], [I^L_{B_{ik}}, I^U_{B_{ik}}], [F^L_{B_{ik}}, F^U_{B_{ik}}] \right\rangle = \left\langle \left[\bigvee_{\sigma \in S_{m_k m_i}} (\wedge_{r \in m_j} T^L_{B_{r\sigma(r)}}), \bigvee_{\sigma \in S_{m_k m_i}} (\wedge_{r \in m_k} T^U_{B_{r\sigma(r)}}) \right], \right.$$
$$\left[\bigwedge_{\sigma \in S_{m_k m_i}} (\vee_{r \in m_k} I^L_{B_{r\sigma(r)}}), \bigwedge_{\sigma \in S_{m_k m_i}} (\vee_{r \in m_k} I^U_{B_{r\sigma(r)}}) \right],$$
$$\left. \left[\bigwedge_{\sigma \in S_{m_k m_i}} (\vee_{r \in m_k} F_{B_{r\sigma(r)}}), \bigwedge_{\sigma \in S_{m_k m_i}} (\vee_{r \in m_k} F_{B_{r\sigma(r)}}) \right] \right\rangle.$$

Note that, since numbers $\sigma(r)$ of columns cannot be changed,

$$\left\langle [T^L_{B_{ij}}, T^U_{B_{ij}}], [I^L_{B_{ij}}, I^U_{B_{ij}}], [F^L_{B_{ij}}, F^U_{B_{ij}}] \right\rangle = \left\langle [T^L_{B_{ik}}, T^U_{B_{ik}}], [I^L_{B_{ik}}, I^U_{B_{ik}}], [F^L_{B_{ik}}, F^U_{B_{ik}}] \right\rangle.$$

Therefore, $(adj(\hat{A}))^t$ is constant.

2. Since $\hat{A}$ is constant, $\hat{A}_{jk} = \hat{A}_{ik}$ and so $\langle [T^L_{|\hat{A}_{jk}|}, T^U_{|\hat{A}_{jk}|}], [I^L_{|\hat{A}_{jk}|}, I^U_{|\hat{A}_{jk}|}], [F^L_{|\hat{A}_{jk}|},$
$F^U_{|\hat{A}_{jk}|} \rangle = \langle [T^L_{|\hat{A}_{ik}|}, T^U_{|\hat{A}_{ik}|}], [I^L_{|\hat{A}_{ik}|}, I^U_{|\hat{A}_{ik}|}], [F^L_{|\hat{A}_{ik}|}, F^U_{|\hat{A}_{ik}|}] \rangle$ for all $i, j \in \{1, 2, \ldots, m\}$. Therefore

$$
g_{ij} = \left\langle \left[\bigvee_{k=1}^{m} (T_{A_{ik}}^{L} \wedge T_{|A_{jk}|}^{L}), \bigvee_{k=1}^{m} (T_{A_{ik}}^{U} \wedge T_{|A_{jk}|}^{U}) \right], \left[\bigwedge_{k=1}^{m} (I_{A_{ik}}^{L} \vee I_{|A_{jk}|}^{L}), \bigwedge_{k=1}^{m} (I_{A_{ik}}^{U} \vee I_{|A_{jk}|}^{U}) \right], \right.
$$

$$
\left. \left[\bigwedge_{k=1}^{m} (F_{A_{ik}}^{L} \vee F_{|A_{jk}|}^{L}), \bigwedge_{k=1}^{m} (F_{A_{ik}}^{U} \vee F_{|A_{jk}|}^{U}) \right] \right\rangle
$$

$$
= \left\langle \left[\bigvee_{k=1}^{m} (T_{A_{ik}}^{L} \wedge T_{|A_{ik}|}^{L}), \bigvee_{k=1}^{m} (T_{A_{ik}} \wedge T_{|A_{ik}|}^{U}) \right], \left[\bigwedge_{k=1}^{m} (I_{A_{ik}}^{L} \vee I_{|A_{ik}|}^{L}), \bigwedge_{k=1}^{m} (I_{A_{ik}}^{U} \vee I_{|A_{ik}|}^{U}) \right], \right.
$$

$$
\left. \left[\bigwedge_{k=1}^{m} (F_{A_{ik}}^{L} \vee F_{|A_{ik}|}^{L}), \bigwedge_{k=1}^{m} (F_{A_{ik}}^{U} \vee F_{|A_{ik}|}^{U}) \right] \right\rangle
$$

$$
= |\hat{A}|.
$$

Since $\hat{A}$ is constant, for any $\sigma \in S_m$

$$
|\hat{A}| = \left\langle \left[\bigvee_{\sigma \in S_m} (\wedge_{i=1}^{m} T_{A_{i\sigma(i)}}^{L}), \bigvee_{\sigma \in S_m} (\wedge_{i=1}^{m} T_{A_{i\sigma(i)}}^{U}) \right], \left[\bigwedge_{\sigma \in S_m} (\vee_{i=1}^{m} I_{A_{i\sigma(i)}}^{L}), \bigwedge_{\sigma \in S_m} (\vee_{i=1}^{m} I_{A_{i\sigma(i)}}^{U}) \right], \right.
$$

$$
\left. \left[\bigwedge_{\sigma \in S_m} (\vee_{i=1}^{m} F_{A_{i\sigma(i)}}^{L}), \bigwedge_{\sigma \in S_m} (\vee_{i=1}^{m} F_{A_{i\sigma(i)}}^{U}) \right] \right\rangle
$$

$$
= \left\langle \left[\wedge_{i=1}^{m} T_{A_{i\sigma(i)}}^{L}, \wedge_{i=1}^{m} T_{A_{i\sigma(i)}}^{U} \right], \left[\vee_{i=1}^{m} I_{A_{i\sigma(i)}}^{L}, \vee_{i=1}^{m} I_{A_{i\sigma(i)}}^{U} \right], \left[\vee_{i=1}^{m} F_{A_{i\sigma(i)}}^{L}, \vee_{i=1}^{m} F_{A_{i\sigma(i)}}^{U} \right] \right\rangle.
$$

Taking σ as the identity permutation, we have

$$
|\hat{A}| = \left\langle \left[\wedge_{i=1}^{m} T_{A_{ii}}^{L}, \wedge_{i=1}^{m} T_{A_{ii}}^{U} \right], \left[\vee_{i=1}^{m} I_{A_{ii}}^{L}, \vee_{i=1}^{m} I_{A_{ii}}^{U} \right], \left[\vee_{i=1}^{m} F_{A_{ii}}^{L}, \vee_{i=1}^{m} F_{A_{ii}}^{U} \right] \right\rangle = g_{ii}.
$$

$\square$

Definition 9 Let $\hat{A}$ be $m \times m$ IVN-matrix. If $\hat{A}^n = \hat{A}^n \hat{A}$ for any $n \leq m - 1$, then it is said to be IVN-matrix $\hat{A}$ is converge IVN-matrix $\hat{A}^n$.

Proposition 9 *Let $\hat{A}$ be $m \times m$ IVN-matrix. Then, the power of $\hat{A}$ either converge to idempotent $\hat{A}^n$ for finite c or oscillate with a finite period*

Proof Let $\hat{A}$ converge to $\hat{A}^n$. Then

$$
\hat{A}^n = \hat{A}^n \hat{A} = (\hat{A}^n \hat{A}) \hat{A} = \hat{A}^n (\hat{A} \hat{A}) = \hat{A}^n (\hat{A}^2) = \ldots = \hat{A}^n (\hat{A}^{n-1} \hat{A}) = \hat{A}^n \hat{A}^n.
$$

Thus $\hat{A}^n$ is idempotent.

Let $\hat{A}$ does not converge to $\hat{A}^n$. Since m is finite and properties of max and min operations cannot introduce an IVN-number not in $\hat{A}$ originally, $\hat{A}$ must oscillate with a finite period.

$\square$

Proposition 10 Let $\hat{A} = \left[\langle [T_{A_{ij}}^L, T_{A_{ij}}^U], [I_{A_{ij}}^L, I_{A_{ij}}^U], [F_{A_{ij}}^L, F_{A_{ij}}^U] \rangle \right]_{m \times n}$, $\hat{B} = \left[\langle [T_{B_{ij}}^L, T_{B_{ij}}^U], [I_{B_{ij}}^L, I_{B_{ij}}^U], [F_{B_{ij}}^L, F_{B_{ij}}^U] \rangle \right]_{m \times n}$, and $\hat{C} = \left[\langle [T_{C_{ij}}^L, T_{C_{ij}}^U], [I_{C_{ij}}^L, I_{C_{ij}}^U], [F_{C_{ij}}^L, F_{C_{ij}}^U] \rangle \right]_{n \times p}$ be three IVN-matrices. If $\hat{A} \tilde{\leq} \hat{B}$, then $\hat{A}\hat{C} \tilde{\leq} \hat{B}\hat{C}$.

Proof For $i \leq m$, $j \leq p$, $k \leq n$, $T_{A_{ik}}^L \wedge T_{C_{kj}}^L \leq T_{B_{ik}}^L \wedge T_{C_{kj}}^L$, $T_{A_{ik}}^U \wedge T_{C_{kj}}^U \leq T_{B_{ik}}^U \wedge T_{C_{kj}}^U$, $I_{A_{ik}}^L \vee I_{C_{kj}}^L \geq I_{B_{ik}}^L \vee I_{C_{kj}}^L$, $I_{A_{ik}}^U \vee I_{C_{kj}}^U \geq I_{B_{ik}}^U \vee I_{C_{kj}}^U$, and $F_{A_{ik}}^L \vee F_{C_{kj}}^L \geq F_{B_{ik}}^L \vee F_{C_{kj}}^L$, $F_{A_{ik}}^U \vee F_{C_{kj}}^U \geq F_{B_{ik}}^U \vee F_{C_{kj}}^U$. Thus,

$$(T_{A_{i1}}^L \wedge T_{C_{1j}}^L) \vee \cdots \vee (T_{A_{im}}^L \wedge T_{C_{mj}}^L) \leq (T_{B_{i1}}^L \wedge T_{C_{1j}}^L) \vee \cdots \vee (T_{B_{im}}^L \wedge T_{C_{mj}}^L)$$

$$(T_{A_{i1}}^U \wedge T_{C_{1j}}^U) \vee \cdots \vee (T_{A_{im}}^U \wedge T_{C_{mj}}^U) \leq (T_{B_{i1}}^U \wedge T_{C_{1j}}^U) \vee \cdots \vee (T_{B_{im}}^U \wedge T_{C_{mj}}^U)$$

$$(I_{A_{i1}}^L \vee I_{C_{1j}}^L) \wedge \cdots \wedge (I_{A_{im}}^L \vee I_{C_{mj}}^L) \geq (I_{B_{i1}}^L \vee I_{C_{1j}}^L) \wedge \cdots \wedge (I_{B_{im}}^L \vee I_{C_{mj}}^L)$$

$$(I_{A_{i1}}^U \vee I_{C_{1j}}^U) \wedge \cdots \wedge (I_{A_{im}}^U \vee I_{C_{mj}}^U) \geq (I_{B_{i1}}^U \vee I_{C_{1j}}^U) \wedge \cdots \wedge (I_{B_{im}}^U \vee I_{C_{mj}}^U)$$

$$(F_{A_{i1}}^L \vee F_{C_{1j}}^L) \wedge \cdots \wedge (F_{A_{im}}^L \vee F_{C_{mj}}^L) \geq (F_{B_{i1}}^L \vee F_{C_{1j}}^L) \wedge \cdots \wedge (F_{B_{im}}^L \vee F_{C_{mj}}^L),$$

$$(F_{A_{i1}}^U \vee F_{C_{1j}}^U) \wedge \cdots \wedge (F_{A_{im}}^U \vee F_{C_{mj}}^U) \geq (F_{B_{i1}}^U \vee F_{C_{1j}}^U) \wedge \cdots \wedge (F_{B_{im}}^U \vee F_{C_{mj}}^U),$$

and

$$\hat{A}\hat{C} \tilde{\leq} \hat{B}\hat{C}.$$

$\square$

Proposition 11 Let $\hat{A}$ be $m \times m$ IVN-matrix. If $\hat{A}^p \tilde{\leq} \hat{A}^r$, $p \leq r$, then $\hat{A}$ converges

Proof From Proposition 10, $\hat{A}^p \tilde{\leq} \hat{A}^r \tilde{\leq} \hat{A}^{r+1} \tilde{\leq} \cdots$. However, a finite number of distinct matrices occur in the powers of $\hat{A}$ so that $\hat{A}^p \tilde{\leq} \hat{A}^r \tilde{\leq} \hat{A}^{r+1} \tilde{\leq} \cdots \tilde{\leq} \hat{A}^n = \hat{A}^{n+1} = \cdots$ for some finite n. $\square$

Proposition 12 Let $\hat{A}$ be $m \times m$ reflexive IVN-matrix. Then $adj(\hat{A}) = \hat{A}^n$, where $\hat{A}^n$ is idempotent $n \leq m - 1$.

Proposition 13 Let $\hat{A}$ be an $m \times m$ reflexive IVN-matrix. Then the following assertions are available

1. $adj(\hat{A}^2) = (adj\,\hat{A})^2 = adj(\hat{A})$.
2. If $\hat{A}$ is idempotent, then $adj(\hat{A}) = \hat{A}$.
3. $adj(\hat{A})$ is reflexive.
4. $adj(adj(\hat{A})) = adj(\hat{A})$.
5. $adj(\hat{A}) \tilde{\geq} \hat{A}$.
6. $\hat{A}(adj(\hat{A})) = (adj(\hat{A}))\hat{A} = adj(\hat{A})$.

Proof

1. Since $\hat{A}$ is reflexive, $\hat{A}^2$ is reflexive. By Proposition 12 $adj(\hat{A}^2) = (\hat{A}^2)^n = (\hat{A}^n)^2 = (adj(\hat{A}))^2$. Since $adj(\hat{A}) = \hat{A}^n$ and $\hat{A}^n$ is idempotent, $(adj(\hat{A}))^2 = adj(\hat{A})$.

2. From Proposition 12 $\hat{A}^n = adj(\hat{A})$. Then $(\hat{A}^n)^2 = (adj(\hat{A}))^2$. Since A^n idempotent and $(adj(\hat{A}))^2 = adj(\hat{A})$ from statement 1, $\hat{A} = adj(\hat{A})$.

3. Let $\hat{K} = adj(\hat{A})$. By using the identity permutation $\sigma(r) = r$, we have

$$\langle [T_{K_{ii}}^L, T_{K_{ii}}^U], [I_{K_{ii}}^L, I_{K_{ii}}^U], [F_{K_{ii}}^L, F_{K_{ii}}^U] \rangle$$

$$\geq [T_{A_{11}}^L, T_{A_{11}}^U], [I_{A_{11}}^L, I_{A_{11}}^U], [F_{A_{11}}^L, F_{A_{11}}^U] \rangle \langle [T_{A_{22}}^L, T_{A_{22}}^U], [I_{A_{22}}^L, I_{A_{22}}^U], [F_{A_{22}}^L, F_{A_{22}}^U] \rangle$$

$$\cdots \langle [T_{A_{(i-1)(i-1)}}^L, T_{A_{(i-1)(i-1)}}^U], [I_{A_{(i-1)(i-1)}}^L, I_{A_{(i-1)(i-1)}}^U], [F_{A_{(i-1)(i-1)}}^L, F_{A_{(i-1)(i-1)}}^U] \rangle$$

$$\langle [T_{A_{(i+1)(i+1)}}^L, T_{A_{(i+1)(i+1)}}^U], [I_{A_{(i+1)(i+1)}}^L, I_{A_{(i+1)(i+1)}}^U], [F_{A_{(i+1)(i+1)}}^L, F_{A_{(i+1)(i+1)}}^U] \rangle)$$

$$\cdots \langle [T_{A_{mn}}^L, T_{A_{mn}}^U], [I_{A_{mn}}^L, I_{A_{mn}}^U], [F_{A_{mn}}^L, F_{A_{mn}}^U] \rangle = \langle [1, 1], [0, 0], [0, 0] \rangle.$$

Here $\langle [T_{K_{ii}}^L, T_{K_{ii}}^U], [I_{K_{ii}}^L, I_{K_{ii}}^U], [F_{K_{ii}}^L, F_{K_{ii}}^U] \rangle = \langle [1, 1], [0, 0], [0, 0] \rangle$ and so $adj(\hat{A})$ is reflexive.

4. Since $\hat{A}$ is reflexive, by Proposition 12 $adj(\hat{A})$ is idempotent. By statement 3 It also is reflexive. By using statement 2 it is obtained that $adj(adj(\hat{A})) = adj(\hat{A})$.

5. Let $\hat{K} = adj(\hat{A})$. By using the identity permutation $\sigma(r) = r, \sigma(i) = j, r \neq i$, i.e., the permutation

$$\begin{pmatrix} 1\ 2\ 3 \cdots i \cdots j-1\ j+1 \cdots m \\ 1\ 2\ 3 \cdots j \cdots j-1\ j+1 \cdots m \end{pmatrix},$$

then

$$\langle [T_{K_{ij}}^L, T_{K_{ij}}^U], [I_{K_{ij}}^L, I_{K_{ij}}^U], [F_{K_{ij}}^L, F_{K_{ij}}^U] \rangle$$

$$\geq \langle [T_{A_{11}}^L, T_{A_{11}}^U], [I_{A_{11}}^L, I_{A_{11}}^U], [F_{A_{11}}^L, F_{A_{11}}^U] \rangle \langle [T_{A_{22}}^L, T_{A_{22}}^U], [I_{A_{22}}^L, I_{A_{22}}^U], [F_{A_{22}}^L, F_{A_{22}}^U] \rangle$$

$$\cdots \langle [T_{A_{ij}}^L, T_{A_{ij}}^U], [I_{A_{ij}}^L, I_{A_{ij}}^U], [F_{A_{ij}}^L, F_{A_{ij}}^U] \rangle$$

$$\langle [T_{A_{(j-1)(j-1)}}^L, T_{A_{(j-1)(j-1)}}^U], [I_{A_{(j-1)(j-1)}}^L, I_{A_{(j-1)(j-1)}}^U], [F_{A_{(j-1)(j-1)}}^L, F_{A_{(j-1)(j-1)}}^U] \rangle$$

$$\langle [T_{A_{(j+1)(j+1)}}^L, T_{A_{(j+1)(j+1)}}^U], [I_{A_{(j+1)(j+1)}}^L, I_{A_{(j+1)(j+1)}}^U], [F_{A_{(j+1)(j+1)}}^L, F_{A_{(j+1)(j+1)}}^U] \rangle$$

$$\cdots \langle [T_{A_{mn}}^L, T_{A_{mn}}^U], [I_{A_{mn}}^L, I_{A_{mn}}^U], [F_{A_{mn}}^L, F_{A_{mn}}^U] \rangle$$

$$= \langle [T_{A_{ii}}^L, T_{A_{ii}}^U], [I_{A_{ii}}^L, I_{A_{ii}}^U], [F_{A_{ii}}^L, F_{A_{ii}}^U] \rangle.$$

Thus, $K = adj(\hat{A}) \hat{\geq} \hat{A}$.

6. Let $\hat{B} = \hat{A}(adj(\hat{A}))$ and $\hat{C} = (adj(\hat{A}))\hat{A}$. Then

$$\langle [T_{B_{ij}}^L, T_{B_{ij}}^U], [I_{B_{ij}}^L, I_{B_{ij}}^U], [F_{B_{ij}}^L, F_{B_{ij}}^U] \rangle$$

$$= \left\langle \left[\bigvee_{k=1}^{m} (T_{A_{ik}}^L \wedge T_{|\hat{A}_{jk}|}^L), \bigvee_{k=1}^{m} (T_{A_{ik}}^U \wedge T_{|\hat{A}_{jk}|}^U) \right], \left[\bigwedge_{k=1}^{m} (I_{A_{ik}}^L \vee I_{|\hat{A}_{jk}|}^L), \bigwedge_{k=1}^{m} (I_{A_{ik}}^U \vee I_{|\hat{A}_{jk}|}^U) \right], \right.$$

$$\left. \left[\bigwedge_{k=1}^{m} (F_{A_{ik}}^L \vee F_{|\hat{A}_{jk}|}^L), \bigwedge_{k=1}^{m} (F_{A_{ik}}^U \vee F_{|\hat{A}_{jk}|}^U) \right] \right\rangle$$

$$\geq \langle [T_{A_{ii}}^L, T_{A_{ii}}^U], [I_{A_{ii}}^L, I_{A_{ii}}^U], [F_{A_{ii}}^L, F_{A_{ii}}^U] \rangle \langle [T_{|\hat{A}_{ji}|}^L, T_{|\hat{A}_{ji}|}^U], [I_{|\hat{A}_{ji}|}^L, I_{|\hat{A}_{ji}|}^U], [F_{|\hat{A}_{ji}|}^L, F_{|\hat{A}_{ji}|}^U] \rangle$$

$$= \langle [T_{|\hat{A}_{ji}|}^L, T_{|\hat{A}_{ji}|}^U], [I_{|\hat{A}_{ji}|}^L, I_{|\hat{A}_{ji}|}^U], [F_{|\hat{A}_{ji}|}^L, F_{|\hat{A}_{ji}|}^U] \rangle$$

$$= \langle [T_{|\hat{E}_{ij}|}^L, T_{|\hat{E}_{ij}|}^U], [I_{|\hat{E}_{ij}|}^L, I_{|\hat{E}_{ij}|}^U], [F_{|\hat{E}_{ij}|}^L, F_{|\hat{E}_{ij}|}^U] \rangle$$

and
$$\langle [T_{C_{ij}}^L, T_{C_{ij}}^U], [I_{C_{ij}}^L, I_{C_{ij}}^U], [F_{C_{ij}}^L, F_{C_{ij}}^U] \rangle$$

$$= \left\langle \left[\bigvee_{k=1}^{m} (T_{|\hat{A}_{ki}|}^L \wedge T_{A_{jk}}^L), \bigvee_{k=1}^{m} (T_{|\hat{A}_{ki}|}^U \wedge T_{A_{kj}}^U) \right], \left[\bigwedge_{k=1}^{m} (I_{|\hat{A}_{ki}|}^L \vee I_{A_{kj}}^L), \bigwedge_{k=1}^{m} (I_{|\hat{A}_{ki}|}^U \vee I_{A_{kj}}^U) \right], \right.$$

$$\left. \left[\bigwedge_{k=1}^{m} (F_{|\hat{A}_{ki}|}^L \vee F_{A_{kj}}^L), \bigwedge_{k=1}^{m} (F_{|\hat{A}_{ki}|}^U \vee F_{A_{kj}}^U) \right] \right\rangle$$

$$\geq \langle [T_{|A_{ji}|}^L, T_{|A_{ji}|}^U], [I_{|A_{ji}|}^L, I_{|A_{ji}|}^U], [F_{|A_{ji}|}^L, F_{|A_{ji}|}^U] \rangle \langle [T_{A_{jj}}^L, T_{A_{jj}}^U], [I_{A_{jj}}^L, I_{A_{jj}}^U], [F_{A_{jj}}^L, F_{A_{jj}}^U] \rangle$$

$$= \langle [T_{|A_{ji}|}^L, T_{|A_{ji}|}^U], [I_{|A_{ji}|}^L, I_{|A_{ji}|}^U], [F_{|A_{ji}|}^L, F_{|A_{ji}|}^U] \rangle$$

$$= \langle [T_{|E_{ij}|}^L, T_{|E_{ij}|}^U], [I_{|E_{ij}|}^L, I_{|E_{ij}|}^U], [F_{|E_{ij}|}^L, F_{|E_{ij}|}^U] \rangle.$$

Therefore $\hat{A}(adj(\hat{A})) \overset{\sim}{\geq} adj(\hat{A})$ and $(adj(\hat{A}))\hat{A} \overset{\sim}{\geq} adj(\hat{A})$. By statements 1,5 and Proposition 10, $adj(\hat{A}) = (adj(\hat{A}))(adj(\hat{A})) \overset{\sim}{\geq} \hat{A}adj(\hat{A})$ and thereby $\hat{A}(adj(\hat{A})) = adj(\hat{A})$. Furthermore $adj(\hat{A}) = (adj(\hat{A}))(adj(\hat{A})) \overset{\sim}{\geq} (adj(\hat{A}))\hat{A}$, so that $(adj(\hat{A}))\hat{A} = adj(\hat{A})$. Hence $\hat{A}(adj(\hat{A})) = (adj(\hat{A}))\hat{A} = adj(\hat{A})$.

$\square$

Proposition 14 *Let $\hat{A}$ be an $m \times m$ IVN-matrix.*

1. *If $\hat{A}$ is symmetric, then $\hat{A}$ is symmetric.*
2. *IVN-matrix $\hat{A}(adj(\hat{A}))$ is transitive.*

Proof

1. Let $\hat{B} = adj(\hat{A})$. Then,
$$\langle [T_{B_{ij}}^L, T_{B_{ij}}^U], [I_{B_{ij}}^L, I_{B_{ij}}^U], [F_{B_{ij}}^L, F_{B_{ij}}^U] \rangle$$

$$= \left\langle \left[\bigvee_{\sigma \in S_{m_j m_i}} (\wedge_{r \in m_j} T_{B_{r\sigma(r)}}^L), \bigvee_{\sigma \in S_{m_j m_i}} (\wedge_{r \in m_j} T_{B_{r\sigma(r)}}^U) \right], \right.$$

$$\Big[\bigwedge_{\sigma\in S_{m_j m_i}}(\vee_{r\in m_j}I^L_{B_{r\sigma(r)}}),\ \bigwedge_{\sigma\in S_{m_j m_i}}(\vee_{r\in m_j}I^U_{B_{r\sigma(r)}})\Big],$$

$$\Big[\bigwedge_{\sigma\in S_{m_j m_i}}(\vee_{r\in m_j}F^L_{B_{r\sigma(r)}}),\ \bigwedge_{\sigma\in S_{m_j m_i}}(\vee_{r\in m_j}F^U_{B_{r\sigma(r)}})\Big]\Big\rangle.$$

Since $\hat{A}$ is symmetric

$$\langle[T^L_{B_{ij}},T^U_{B_{ij}}],[I^L_{B_{ij}},I^U_{B_{ij}}],[F^L_{B_{ij}},F^U_{B_{ij}}]\rangle$$

$$=\Big\langle\Big[\bigvee_{\sigma\in S_{m_i m_j}}(\wedge_{r\in m_i}T^L_{B_{\sigma(r)r}}),\ \bigvee_{\sigma\in S_{m_i m_j}}(\wedge_{r\in m_i}T^U_{B_{\sigma(r)r}})\Big],$$

$$\Big[\bigwedge_{\sigma\in S_{m_i m_j}}(\vee_{r\in m_i}I^L_{B_{\sigma(r)r}}),\ \bigwedge_{\sigma\in S_{m_i m_j}}(\vee_{r\in m_i}I^U_{B_{\sigma(r)r}})\Big],$$

$$\Big[\bigwedge_{\sigma\in S_{m_i m_j}}(\vee_{r\in m_i}F^L_{B_{\sigma(r)r}}),\ \bigwedge_{\sigma\in S_{m_i m_j}}(\vee_{r\in m_i}F^U_{B_{\sigma(r)r}})\Big]\Big\rangle$$

$$=\langle[T^L_{B_{ji}},T^U_{B_{ji}}],[I^L_{B_{ji}},I^U_{B_{ji}}],[F^L_{B_{ji}},F^U_{B_{ji}}]\rangle.$$

2. Let $\hat{B}=\hat{A}adj(\hat{A})$. Namely,

$$\langle[T^L_{B_{ij}},T^U_{B_{ij}}],[I^L_{B_{ij}},I^U_{B_{ij}}],[F^L_{B_{ij}},F^U_{B_{ij}}]\rangle$$

$$=\Big\langle\Big[\bigvee_{k=1}^{m}(T^L_{A_{ki}}\wedge T^L_{|A_{kj}|}),\ \bigvee_{k=1}^{m}(T^U_{A_{ki}}\wedge T^U_{|A_{kj}|})\Big],\Big[\bigwedge_{k=1}^{m}(I^L_{A_{ki}}\vee I^L_{|A_{kj}|}),\ \bigwedge_{k=1}^{m}(I^U_{A_{ki}}\vee I^U_{|A_{kj}|})\Big],$$

$$\Big[\bigwedge_{k=1}^{m}(F^L_{A_{ki}}\vee F^L_{|A_{kj}|}),\ \bigwedge_{k=1}^{m}(F^U_{A_{ki}}\vee F^U_{|A_{kj}|})\Big]\Big\rangle$$

$$=\Big\langle\Big[T^L_{A_{ki}}\wedge T^L_{|A_{ju}|},\ T^U_{A_{ki}}\wedge T^U_{|A_{ju}|}\Big],\Big[I^L_{A_{ki}}\vee I^L_{|A_{ju}|},\ I^U_{A_{ki}}\vee I^U_{|A_{ju}|}\Big],\Big[F^L_{A_{ki}}\vee F^L_{|A_{ju}|},\ F^U_{A_{ki}}\vee F^U_{|A_{ju}|}\Big]\Big\rangle$$

for some $u\in\{1,2,\dots,m\}$, and

$$\langle[T^L_{B_{ij}},T^U_{B_{ij}}],[I^L_{B_{ij}},I^U_{B_{ij}}],[F^L_{B_{ij}},F^U_{B_{ij}}]\rangle^2$$

$$=\Big\langle\Big[\bigvee_{p=1}^{m}(T^L_{B_{ip}}\wedge T^L_{B_{pj}}),\ \bigvee_{p=1}^{m}(T^U_{B_{ip}}\wedge T^U_{B_{pj}})\Big],\Big[\bigwedge_{r=1}^{m}(I^L_{B_{ip}}\vee I^L_{B_{pj}}),\ \bigwedge_{r=1}^{m}(I^U_{B_{ip}}\vee I^U_{B_{pj}})\Big],$$

$$\Big[\bigwedge_{r=1}^{m}(F^L_{B_{ip}}\vee F^L_{B_{jp}}),\ \bigwedge_{r=1}^{m}(F^U_{B_{ip}}\vee F^U_{B_{jp}})\Big]\Big\rangle$$

$$\Big[\bigvee_{p=1}^{m}(T_{B_{ip}}\wedge T_{B_{pj}}),\ \bigvee_{p=1}^{m}(T_{B_{ip}}\wedge T_{B_{pj}})\Big]$$

$$= \Big[\bigvee_{p=1}^{m} \Big[\Big(\bigvee_{q=1}^{m} (T_{A_{iq}}^{L} \wedge T_{|A_{pq}|}^{L}) \Big) \wedge \Big(\bigvee_{r=1}^{m} (T_{A_{pr}}^{L} \wedge T_{|A_{jr}|}^{L}) \Big) \Big],$$

$$\bigvee_{p=1}^{m} \Big[\Big(\bigvee_{q=1}^{m} (T_{A_{iq}}^{U} \wedge T_{|A_{pq}|}^{U}) \Big) \wedge \Big(\bigvee_{r=1}^{m} (T_{A_{pr}}^{U} \wedge T_{|A_{jr}|}^{U}) \Big) \Big] \Big]$$

$$= \Big[\bigvee_{p=1}^{m} \big[(T_{A_{is}}^{L} \wedge T_{|A_{ps}|}^{L}) \wedge (T_{A_{pt}}^{L} \wedge T_{|A_{jt}|}^{L}) \big], \bigvee_{p=1}^{m} \big[(T_{A_{is}}^{U} \wedge T_{|A_{ps}|}^{U}) \wedge (T_{A_{pt}}^{U} \wedge T_{|A_{jt}|}^{U}) \big] \Big]$$

$$\leq \Big[T_{A_{is}}^{L} \wedge T_{|A_{jt}|}^{L}, T_{A_{is}}^{U} \wedge T_{|A_{jt}|}^{U} \Big]$$

$$\leq \Big[T_{A_{iu}}^{L} \wedge T_{|A_{ju}|}^{L}, T_{A_{iu}}^{U} \wedge T_{|A_{ju}|}^{U} \Big],$$

$$\Big[\bigwedge_{p=1}^{m} (I_{B_{ip}} \vee I_{B_{pj}}), \bigwedge_{p=1}^{m} (I_{B_{ip}} \vee I_{B_{pj}}) \Big]$$

$$= \Big[\bigwedge_{p=1}^{m} \Big[\Big(\bigwedge_{q=1}^{m} (I_{A_{iq}}^{L} \vee I_{|A_{pq}|}^{L}) \Big) \vee \Big(\bigwedge_{r=1}^{m} (I_{A_{pr}}^{L} \vee I_{|A_{jr}|}^{L}) \Big) \Big],$$

$$\bigwedge_{p=1}^{m} \Big[\Big(\bigwedge_{q=1}^{m} (I_{A_{iq}}^{U} \vee I_{|A_{pq}|}^{U}) \Big) \vee \Big(\bigwedge_{r=1}^{m} (I_{A_{pr}}^{U} \vee I_{|A_{jr}|}^{U}) \Big) \Big] \Big]$$

$$= \Big[\bigwedge_{p=1}^{m} \big[(I_{A_{is}}^{L} \vee I_{|A_{ps}|}^{L}) \vee (I_{A_{pt}}^{L} \vee I_{|A_{jt}|}^{L}) \big], \bigwedge_{p=1}^{m} \big[(I_{A_{is}}^{U} \vee I_{|A_{ps}|}^{U}) \vee (T_{A_{pt}}^{U} \vee T_{|A_{jt}|}^{U}) \big] \Big]$$

$$\geq \Big[I_{A_{is}}^{L} \vee I_{|A_{jt}|}^{L}, I_{A_{is}}^{U} \vee I_{|A_{jt}|}^{U} \Big]$$

$$\geq \Big[I_{A_{iu}}^{L} \vee I_{|A_{ju}|}^{L}, I_{A_{iu}}^{U} \vee I_{|A_{ju}|}^{U} \Big],$$

and $\Big[\bigwedge_{p=1}^{m} (F_{B_{ip}} \vee F_{B_{pj}}), \bigwedge_{p=1}^{m} (F_{B_{ip}} \vee F_{B_{pj}}) \Big]$

$$= \Big[\bigwedge_{p=1}^{m} \Big[\Big(\bigwedge_{q=1}^{m} (F_{A_{iq}}^{L} \vee F_{|A_{pq}|}^{L}) \Big) \vee \Big(\bigwedge_{r=1}^{m} (F_{A_{pr}}^{L} \vee F_{|A_{jr}|}^{L}) \Big) \Big],$$

$$\bigwedge_{p=1}^{m} \Big[\Big(\bigwedge_{q=1}^{m} (F_{A_{iq}}^{U} \vee F_{|A_{pq}|}^{U}) \Big) \vee \Big(\bigwedge_{r=1}^{m} (F_{A_{pr}}^{U} \vee F_{|A_{jr}|}^{U}) \Big) \Big] \Big]$$

$$= \Big[\bigwedge_{p=1}^{m} \big[(F_{A_{is}}^{L} \vee F_{|A_{ps}|}^{L}) \vee (F_{A_{pt}}^{L} \vee F_{|A_{jt}|}^{L}) \big], \bigwedge_{p=1}^{m} \big[(F_{A_{is}}^{U} \vee F_{|A_{ps}|}^{U}) \vee (T_{A_{pt}}^{U} \vee T_{|A_{jt}|}^{U}) \big] \Big]$$

$$\geq \left[F^L_{A_{is}} \vee F^L_{|A_{jt}|}, F^U_{A_{is}} \vee F^U_{|A_{jt}|} \right]$$

$$\geq \left[F^L_{A_{iu}} \vee F^L_{|A_{ju}|}, F^U_{A_{iu}} \vee F^U_{|A_{ju}|} \right],$$

for some $s, u \in \{1, 2, \ldots, m\}$. Hence $(\hat{A}adj(\hat{A}))^2 \leq \hat{A}adj\hat{A}$. $\qquad\square$

Example 8 Let us consider IVN-matrix $\hat{A}$ given as follows:

$$\hat{A} = \begin{bmatrix} \langle[0.5, 0.6], [0.4, 0.8], [0.7, 0.9]\rangle & \langle[0.3, 0.4], [0.2, 0.6], [0.8, 0.9]\rangle & \langle[0.4, 0.5], [0.7, 0.8], [0.1, 0.3]\rangle \\ \langle[0.3, 0.5], [0.5, 0.7], [0.2, 0.4]\rangle & \langle[0.4, 0.6], [0.9, 1.0], [0.6, 0.6]\rangle & \langle[0.6, 0.7], [0.5, 0.8], [0.1, 0.1]\rangle \\ \langle[0.4, 0.9], [0.8, 0.8], [0.3, 0.5]\rangle & \langle[0.5, 0.6], [0.7, 0.8], [0.6, 0.7]\rangle & \langle[0.8, 0.8], [0.3, 0.7], [0.4, 0.4]\rangle \end{bmatrix}.$$

Then,

$$adj(\hat{A}) = \begin{bmatrix} \langle[0.5, 0.6], [0.7, 0.8], [0.6, 0.6]\rangle & \langle[0.4, 0.5], [0.3, 0.7], [0.6, 0.7]\rangle & \langle[0.4, 0.5], [0.5, 0.8], [0.6, 0.6]\rangle \\ \langle[0.4, 0.7], [0.5, 0.7], [0.3, 0.4]\rangle & \langle[0.5, 0.6], [0.4, 0.8], [0.3, 0.5]\rangle & \langle[0.5, 0.6], [0.5, 0.8], [0.2, 0.4]\rangle \\ \langle[0.4, 0.6], [0.7, 0.8], [0.6, 0.6]\rangle & \langle[0.5, 0.6], [0.7, 0.8], [0.7, 0.9]\rangle & \langle[0.4, 0.6], [0.5, 0.7], [0.7, 0.9]\rangle \end{bmatrix},$$

$$\hat{A}adj(\hat{A}) = \begin{bmatrix} \langle[0.5, 0.6], [0.5, 0.7], [0.6, 0.6]\rangle & \langle[0.4, 0.5], [0.4, 0.8], [0.7, 0.9]\rangle & \langle[0.4, 0.5], [0.5, 0.8], [0.7, 0.9]\rangle \\ \langle[0.4, 0.6], [0.7, 0.8], [0.6, 0.6]\rangle & \langle[0.5, 0.6], [0.5, 0.7], [0.6, 0.6]\rangle & \langle[0.4, 0.6], [0.5, 0.8], [0.6, 0.6]\rangle \\ \langle[0.4, 0.6], [0.7, 0.8], [0.6, 0.6]\rangle & \langle[0.5, 0.6], [0.7, 0.8], [0.6, 0.7]\rangle & \langle[0.5, 0.6], [0.5, 0.7], [0.6, 0.6]\rangle \end{bmatrix}$$

and

$$(\hat{A}adj(\hat{A}))^2 = \begin{bmatrix} \langle[0.5, 0.6], [0.5, 0.7], [0.6, 0.6]\rangle & \langle[0.4, 0.5], [0.5, 0.7], [0.7, 0.9]\rangle & \langle[0.4, 0.5], [0.5, 0.8], [0.7, 0.9]\rangle \\ \langle[0.4, 0.6], [0.7, 0.8], [0.6, 0.6]\rangle & \langle[0.5, 0.6], [0.5, 0.7], [0.6, 0.6]\rangle & \langle[0.4, 0.6], [0.5, 0.8], [0.6, 0.6]\rangle \\ \langle[0.4, 0.6], [0.7, 0.8], [0.6, 0.6]\rangle & \langle[0.5, 0.6], [0.7, 0.8], [0.6, 0.7]\rangle & \langle[0.5, 0.6], [0.5, 0.7], [0.6, 0.6]\rangle \end{bmatrix}.$$

It is shown that $(\hat{A}adj(\hat{A}))^2 \tilde{\leq} \hat{A}adj(\hat{A})$. Thus IVN-matrix $\hat{A}adj(\hat{A})$ is transitive.

5 Conclusions

The main contributions of this study can be illustrated and reviewed as follows:

1. Concept of the determinant for the IVN-matrices is defined.
2. Many results are obtained related to the determinant of IVN-matrices.
3. The adjoint of an IVN-matrix is defined and many properties related to the adjoint which exist in classical determinant theory are investigated for IVN-matrices.
4. The concepts of complement, constant, reflexive, symmetric, transitive, and idempotent IVN-matrices are defined, and some properties of them related to determinant and adjoint are obtained.

Importance of this study can be summarized as follows:

1. The determinant of a matrix is an important role in many fields such as physic, engineering, and graph theory, etc. In this paper, we fill a gap in the literature by defining determinant of IVN-matrices.

2. We hope that this paper will be the main reference point for researchers to define and study some new concepts such as eigen-value and eigen vector.

References

1. Arockiarani, I. & Sumathi, I.R. (2014). A fuzzy neutrosophic soft matrix approach in decision making. *JGRMA, 2*(2), 14–23.
2. Atanassov, K. T. (1986). Intuitionistic fuzzy sets. *Fuzzy Set and Systems, 20*, 87–96.
3. Deli, I. & Broumi, S. (2015). Neutrosophic soft matrices and NSM-decision making. *Journal of Intelligent and Fuzzy Systems, 28*(5), 2233–2241.
4. Dhar, M., Broumi, S., & Smarandache, F. (2014). A note on square neutrosophic fuzzy matrices. *Neutrosophic Sets and Systems, 3*, 37–41.
5. Kandasamy, W. B. V. & Smarandache, F. (2004). *Fuzzy relational maps and neutrosophic relational maps*. HEXIS Church Rock.
6. Kandasamy, W. B. V., & Smarandache, F. (2012). Neutrosophic super matrices and quasi super matrices. Grandview Heights: ZIP Publishing.
7. Khan, S. K. & Pal, M. (2002). Some operations on intuitionistic fuzzy matrices. In *Presented in international conference on analysis and discrete structures, 22–24 Dec.* Kharagpur, India: Indian Institute of Technology.
8. Kim, J. B., Baartmans, A., & Sahadin, N. S. (1989). Determinant theory for fuzzy matrices. *Fuzzy Sets and Systems, 29*, 349–356.
9. Pal, M. (2001). Intuitionistic fuzzy determinant. *V.U.J. Physical Sciences, 7*, 87–93
10. Pal, M., Khan, S. K., & Shyamal, A. K. (2002). Intuitionistic fuzzy matrices. *Notes on Intuitionistic Fuzzy Sets, 8*(2), 51–62.
11. Ragab, M. Z. Emam, E. G. (1995). The determinant and Adjoint of a Square Fuzzy Matrix. *Information Sciences, 84*, 209–220.
12. Smarandache, F. (1999). A unifying field in logics. In *Neutrosophy: Neutrosophic probability, set and logic*. Rehoboth: American Research Press.
13. Thomson, M. G. (1977). Convergence of powers of a fuzzy matrix. *Journal of Mathematical Analysis and Applications, 57*, 476–480.
14. Uma, R., Murugadas, P., & Sriram, S. (2016). Determinant theory of fuzzy neutrosophic soft matrices. *Progress in Nonlinear Dynamics and Chaos 4*(2), 85–102.
15. Wang, H., Smarandache, F., Zhang, Y. Q., & Sunderraman, R. (2010). Single valued neutrosophic sets. *Multispace and Multistructure, 4*, 410–413.
16. Wang, H., Smarandache, F., Zhang, Y.Q., & Sunderraman, R. (2005). Interval Neutrosophic Sets and Logic: Theory and Applications in Computing. New York: Hexis Arizona.

Neutrosophic Normal Probability Distribution—A Spine of Parametric Neutrosophic Statistical Tests: Properties and Applications

Rehan Ahmad Khan Sherwani, Muhammad Aslam, Muhammad Ali Raza, Muhammad Farooq, Muhammad Abid, and Muhammad Tahir

1 Introduction

The theory of classical statistics is based on two types of statistical methods called descriptive and inferential statistics. The inferential part of statistical methods depends a lot on normal probability distribution for reliable, valid, and powerful inferences of the experiment. Many parametric statistical tests (z, t, F, etc.) require the assumption of normality to be filled first before applying them. The assumption of normality means the sampling distribution of test statistic under a parametric test should follow a normal probability distribution or the data under consideration must come from a normal distribution. Failing the normality assumption leads to apply nonparametric tests which are less powerful compared to parametric counterpart. A normal probability distribution is a bell-shaped distribution also known as Gaussian distribution is symmetrical about the mean portraying the numerous observations closer to the mean. The distribution has two parameters, namely, mean μ and

R. A. K. Sherwani
College of Statistical and Actuarial Sciences, University of the Punjab, New Campus Lahore, Lahore, Pakistan
e-mail: rehan.stat@pu.edu.pk

M. Aslam (✉)
Department of Statistics, Faculty of Science, King Abdulaziz University, Jeddah, Saudi Arabia
e-mail: aslam_ravian@hotmail.com

M. A. Raza · M. Abid
Department of Statistics, Government College University Faisalabad, Faisalabad, Pakistan
e-mail: ali.raza@gcuf.edu.pk; mabid@gcuf.edu.pk

M. Farooq · M. Tahir
Department of Statistics, Faculty of Science, King Abdulaziz University, Jeddah, Saudi Arabia
e-mail: farooq.stat@gcu.edu.pk

© The Author(s), under exclusive license to Springer Nature Switzerland AG 2021
F. Smarandache, M. Abdel-Basset (eds.), *Neutrosophic Operational Research*,
https://doi.org/10.1007/978-3-030-57197-9_8

153

variance σ^2 with probability density function:

$$f(x) = \frac{1}{\sigma\sqrt{2\pi}}e^{-\frac{1}{2}\left(\frac{x-\mu}{\sigma}\right)^2} \tag{1}$$

Although, in classical statistics, the parameters μ and σ^2 are fixed and exact for a random variable X, in many situations, these parameters may vary due to their probable state. This means in situations we are not sure about exact values of one or both of the parameters and their values may vary from one to another value, i.e., in interval form. In such situations, there will be some indeterminacy in the parameter(s) of the distribution, and therefore, we cannot apply the classical normal distribution (1) in solving the relevant problems. One of the applications of normal probability distribution is commonly seen in asset prices data where asset prices are not in exact form and range in an interval form, e.g., the average price of a plot in ABC society ranges from 10,000\$ to 12000\$ showing that the average price of the plot is not fixed or has single value but varies from 10,000\$ to 12000\$. In this case, indeterminacy exists in the average price of the plot where we are not sure about the exact worth of the plot. The component of this indeterminacy is 2000\$. The other common application of normal probability distribution is seen in the technical stock market analysis, but again the prices of the shares in the stock market vary from starting to closing of the market in a particular day or duration of the time the market is operational, e.g., the average price of the share of XYZ company at the beginning of the share market is 2\$ and closed at 2.5\$ showing again indeterminacy of 0.5\$ in the average price of the share. There are so many examples that exist in the literature where the normal distribution is common to apply, but we are not sure about the exact values of the data, and it is in the form of some indeterminacy, e.g., income distribution in economy, height, IQ, shoe size, average academic performance of students, etc. In all such examples or many others, it will be unfair to apply classical normal probability distribution without considering the component of indeterminacy. Florentane [18] introduced the idea of neutrosophic normal probability distribution in which the component of indeterminacy of the data is incorporated in the classical normal probability distribution (discussed in detail in later sections).

In recent times, several developments have been made to deal effectively with the uncertain and complex phenomena. One of the important developments is the introduction of neutrosophic statistics in which the estimates of statistical methods are derived for imprecise data. Neutrosophic statistics is the extension of the fuzzy theory which deals with the false, indeterminate, vague, unclear data set. It is now evident that neutrosophic logic introduced by Smarandache [1] is more efficient than the fuzzy logic [2, 3]. Alhabib et al. [4] presented some introduction to neutrosophic probability distributions. Patro and Smarandache [5] introduced the concept of a neutrosophic binomial distribution. Alhassan and Smarandache [6] developed the concept of neutrosophic Weibull distribution.

Smarandache [7] introduced the concept of neutrosophic sociology by incorporating the indeterminacy in the sociological vague or incomplete data. Aslam [8, 9] discussed the one-way analysis of variance procedure for uncertain and fuzzy data. [10, 11]) modifies Dixon's test for outliers detection in the presence of some imprecise observations in the data. Aslam et al. [12] designed the Hartley and Bartlett tests for homogeneous variances under an indeterminate environment. Neutrosophic Kolmogorov-Smirnov test for testing the normality assumption under incomplete or ambiguous observations was introduced by Aslam [8, 9]. Several studies have been reported dealing with the neutrosophic statistics for quality control problems [13–17]. Further details on neutrosophic logic and neutrosophic statistics can be seen at Smarandache [18–20], Salama and Rafif [21], Abdel-Basset et al. [22], Alhabib et al. [4], and Aslam et al. [12].

2 Neutrosophic Normal Probability Distribution

In classical normal probability distribution, the parameters are fixed and in determined form as in (1), but if one or both of the parameters are indeterminate, then the random variable X follows a neutrosophic normal probability distribution with some indeterminacy in either or both of the parameters. We shall discuss them briefly as follows:

2.1 Neutrosophic Normal Probability Distribution with Indeterminate Mean

Let the neutrosophic random variable X_N follow a normal probability distribution with some indeterminacy in the first parameter μ and denoted as μ_N, i.e., the mean parameter will be of in interval form $[\mu_N, \mu_{I_N}]$ where μ_N shows the determinate part and μ_{I_N} the indeterminate part in the neutrosophic mean μ_N. Let the other parameter of normal distribution remain to be determined and fixed; then, the neutrosophic normal probability distribution with indeterminacy in the mean only is defined as $X_N \sim N_N(\mu_N, \sigma^2)$:

$$f\left(x_N\right) = \frac{1}{\sigma\sqrt{2\pi}}e^{-1/2\left(\frac{x_N - \mu_N}{\sigma}\right)^2} \qquad -\infty \leq x_N \leq \infty \qquad (2)$$

with neutrosophic cumulative distribution function as

$$F\left(x_N\right) = \Phi\left(\frac{x_N - \mu_N}{\sigma}\right) = \frac{1}{2}\left[1 + \exp\left(\frac{x_N - \mu_N}{\sigma\sqrt{2}}\right)\right] \qquad (3)$$

where N_N means that there is some indeterminacy in the normal distribution and is so-called neutrosophic normal probability distribution [2]. $X_N \in [X_L, X_U]$ and $\sigma > 0$. A graphical display of neutrosophic normal probability distribution

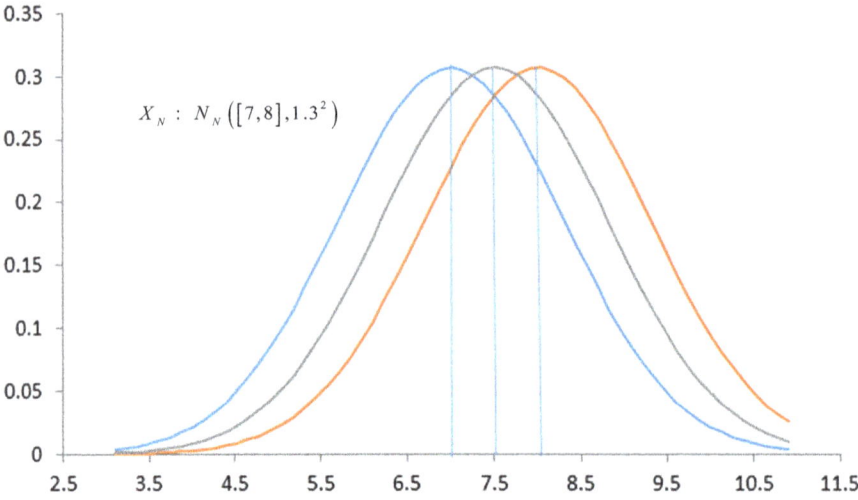

Fig. 1 Neutrosophic normal probability distribution for $X_N \sim N_N([7,8], 1.3^2)$

$X_N \sim N_N([7, 8], 1.3^2)$ is given in Fig. 1 where the parameter μ_N is indeterminate and the level of indeterminacy is $I_{\mu_N} \in (0, 1)$, i.e., the mean value is not exact and varies from 7 to 8.

Example 2.1 Last week ABC corporation has an average share value between 10\$ and 15\$ with a standard deviation of 2\$. If the average share value follows a neutrosophic normal probability distribution, then what will be the probability that next week, the average share value is less than 12\$?

Solution: Let X be an average share value of ABC corporation; then $X_N \sim N_N([10, 15], 2^2)$. Here the parameter μ_N is indeterminate and the level of indeterminacy is $I_{\mu_N} \in (0, 5)$. A neutrosophic normal probability distribution under (2) will become now

$$f(x_N) = \frac{1}{2\sqrt{2\pi}}e^{-1/2\left(\frac{x_N - [10,15]}{\sigma}\right)^2} \qquad -\infty \le x_N \le \infty \qquad (4)$$

and $Z = \frac{x_N - [10,15]}{2} = \frac{12 - [10,15]}{2}$

Now, $P(X_N < 12) = P\left(Z_N < \frac{12 - [10,15]}{2}\right) = [0.07, 0.84]$

This means the chance of average share value less than 12\$ in the next week will be 84% if the average share value is a minimum of 10\$ and reduced to 7% for the maximum average share value of 15\$.

2.2   *Neutrosophic Normal Probability Distribution with Indeterminate Variance*

Let the neutrosophic random variable X_N follow a normal probability distribution with some indeterminacy in the second parameter σ^2 and denoted as σ_N^2, i.e., the variance parameter will be of in interval form $\left[\sigma_N^2, \sigma_{I_N}^2\right]$ where σ_N^2 shows the determinate part and $\sigma_{I_N}^2$ the indeterminate part in the neutrosophic variance σ_N^2. Let the first parameter of normal distribution remain to be determined and fixed; then, the neutrosophic normal probability distribution with indeterminacy in the variance only is defined as $X_N \sim N_N\left(\mu, \sigma_N^2\right)$:

$$f\left(x_N\right) = \frac{1}{\sigma_N \sqrt{2\pi}} e^{-1/2\left(\frac{x_N - \mu}{\sigma_N}\right)^2} \qquad -\infty \le x_N \le \infty \qquad (5)$$

with neutrosophic cumulative distribution function as

$$F\left(x_N\right) = \Phi\left(\frac{x_N - \mu}{\sigma_N}\right) = \frac{1}{2}\left[1 + \exp\left(\frac{x_N - \mu}{\sigma_N \sqrt{2}}\right)\right] \qquad (6)$$

where N_N means that there is some indeterminacy in the normal distribution and is so-called neutrosophic normal probability distribution [2]. $X_N \in [X_L, X_U]$ and $\sigma_N > 0$. A graphical display of neutrosophic normal probability distribution $X_N \sim N_N(7, [4, 9])$ is given in Fig. 2 where the parameter σ_N^2 is indeterminate and the level of indeterminacy is $I_{\sigma_N^2} \in (0, 5)$, i.e., the variance is not exact and varies from 4 to 9.

Example 2.2 Last week ABC corporation has an average share value 10\$ with an indeterminate standard deviation [2\$, 3\$]. If the average share value follows a neutrosophic normal probability distribution, then what will be the probability that next week, the average share value is less than 12\$?

Solution: Let X be an average share value of ABC corporation; then $X_N \sim N_N(10, [4, 9])$. Here the parameter σ_N^2 is indeterminate and the level of indeterminacy is $I_{\sigma_N^2} \in (0, 5)$. A neutrosophic normal probability distribution under (5) will become now

$$f\left(x_N\right) = \frac{1}{[2, 3]\sqrt{2\pi}} e^{-1/2\left(\frac{x_N - 10}{[2,3]}\right)^2} \qquad -\infty \le x_N \le \infty \qquad (7)$$

and $Z = \frac{x_N - 10}{[2,3]} = \frac{12 - 10}{[2,3]}$

Now, $P\left(X_N < 12\right) = P\left(Z_N < \frac{12 - 10}{[2,3]}\right) = [0.74, 0.84]$

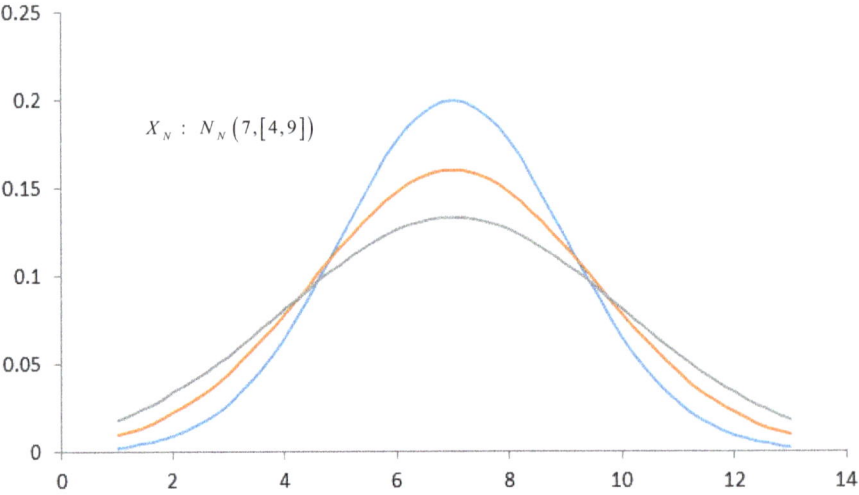

Fig. 2 Neutrosophic normal probability distribution for $X_N \sim N_N(7,[4,9])$

This means the chances of the average share value of less than 12\$ in the next week will be 84% if the variable share value is a minimum of 4\$ and reduced to 74% for the maximum variable share value 9\$.

2.3 Neutrosophic Normal Probability Distribution with Indeterminate Mean and Variance

Let the neutrosophic random variable X_N follow a normal probability distribution with some indeterminacy in both parameters μ and σ^2 denoted as μ_N and σ_N^2, i.e., mean and variance will be of in interval form $[\mu_N, \mu_{I_N}]$ and $[\sigma_N^2, \sigma_{I_N}^2]$. Now, the neutrosophic normal probability distribution with indeterminacy in mean and variance is defined as $X_N \sim N_N\left([\mu_N, \mu_{I_N}], [\sigma_N^2, \sigma_{I_N}^2]\right)$:

$$f(x_N) = \frac{1}{\sigma_N\sqrt{2\pi}}e^{-1/2\left(\frac{x_N - [\mu_N, \mu_{I_N}]}{[\sigma_N, \sigma_{I_N}]}\right)^2} \qquad -\infty \leq x_N \leq \infty \qquad (8)$$

with neutrosophic cumulative distribution function as

$$F(x_N) = \Phi\left(\frac{x_N - [\mu_N, \mu_{I_N}]}{[\sigma_N, \sigma_{I_N}]}\right) = \frac{1}{2}\left[1 + \exp\left(\frac{x_N - [\mu_N, \mu_{I_N}]}{[\sigma_N, \sigma_{I_N}]\sqrt{2}}\right)\right] \qquad (9)$$

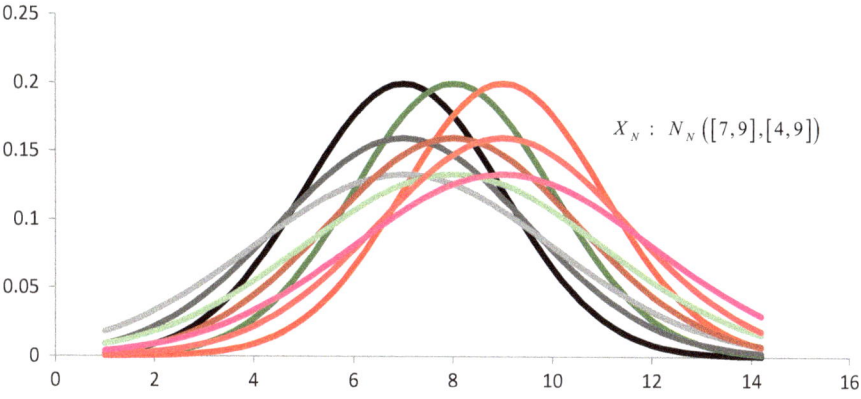

Fig. 3 Neutrosophic normal probability distribution for $X_N \sim N_N([7,9],[4,9])$

where N_N means that there is some indeterminacy in the normal distribution and is so-called neutrosophic normal probability distribution [2]. $X_N \in [X_L, X_U]$ and $\sigma_N > 0$. A graphical display of neutrosophic normal probability distribution $X_N \sim N_N([7,9],[4,9])$ is given in Fig. 3 where the parameters μ_N and σ_N^2 are indeterminate and the level of indeterminacy is $I_{[\mu_N, \sigma_N^2]} \in (0, 5)$.

Example 2.3 Last week ABC corporation has vague average share values [10\$, 15\$] with indeterminate standard deviations [2\$, 3\$]. If the average share value follows a neutrosophic normal probability distribution, then what will be the probability that next week, the average share value is less than 12\$?

Solution: Let X be an average share value of ABC corporation; then $X_N \sim N_N([10, 15], [4, 9])$. Here both the parameters μ_N and σ_N^2 are indeterminate, and the level of indeterminacy is $I_{[\mu_N, \sigma_N^2]} \in (0, 5)$. A neutrosophic normal probability distribution under (8) will become now

$$f(x_N) = \frac{1}{[2,3]\sqrt{2\pi}} e^{-1/2\left(\frac{x_N - [10,15]}{[2,3]}\right)^2} \quad -\infty \leq x_N \leq \infty \quad (10)$$

and $Z = \frac{x_N - [10,15]}{[2,3]} = \frac{12 - [10,15]}{[2,3]}$

Now, $P(X_N < 12) = P\left(Z_N < \frac{12 - [10,15]}{[2,3]}\right) = [0.07, 0.84]$

The minimum and maximum chances for average share value to remain under 12\$ in the next week are 7% and 84% depending upon the different levels of the indeterminacy in the parameters.

Table 1 Quantiles for standard normal distribution and $X_N \sim N_N([10, 15], 2^2)$

p	Quantiles for standard normal distribution	Quantiles for $X_N \sim N_N([10, 15], 2^2)$
0.6	0.84	[11.68, 16.68]
0.7	1.04	[12.08, 17.08]
0.8	1.281	[12.56, 17.56]
0.9	1.645	[13.29, 18.29]
0.95	1.96	[13.92, 18.92]
0.99	2.576	[15.15, 20.15]

3 Quantiles of Neutrosophic Normal Distribution

The quantile function is useful in the construction of confidence interval, drawing Q–Q plot, and hypothesis testing. It is the inverse of the neutrosophic cumulative distribution function or neutrosophic distribution function. For a neutrosophic random variable with mean and variance, the quantile function for a neutrosophic normal probability distribution is defined as

$$F^{-1}(p) = \mu_N + \sigma_N \Phi^{-1}(p) = \mu_N + \sigma_N Z_p, \quad p \in (0, 1) \tag{11}$$

where $Z_p = \Phi^{-1}(p)$ is the quantile of the standard normal distribution. A neutrosophic random variable X_N will exceed $\mu_N + \sigma_N Z_p$ with probability $(1 - p)$ and will fall outside the interval $\mu_N \pm \sigma_N Z_p$ with probability $2(1 - p)$. For larger values of p, we use $Z_p = \Phi^{-1}\left(\frac{p+1}{2}\right)$ in place of $Z_p = \Phi^{-1}(p)$. Let the neutrosophic random variable $X_N \sim N_N([10, 15], 2^2)$, and assume that the variable X_N will fall within the range $\mu_N \pm \sigma_N Z_p$ with a specified probability p; then the values of the neutrosophic quantiles are provided in Table 1.

In Fig. 4, we presented the neutrosophic Q–Q plot for the neutrosophic normal probability distribution. It is a scatter plot of theoretical and observed quantiles. The gap between the two dotted lines shows the level of indeterminacy in the Q–Q plot due to the indeterminate values in the mean of the neutrosophic normal distribution. From the Q–Q plot, it is evident that the supposed neutrosophic random variable X_N with indeterminate mean $\mu_N = [10, 15]$ and variance 4 follows a neutrosophic normal probability distribution.

4 Properties of Neutrosophic Normal Probability Distribution

Property 1 The probability function of the neutrosophic normal distribution is a proper pdf.

For a probability function to be a proper pdf, we need two conditions:

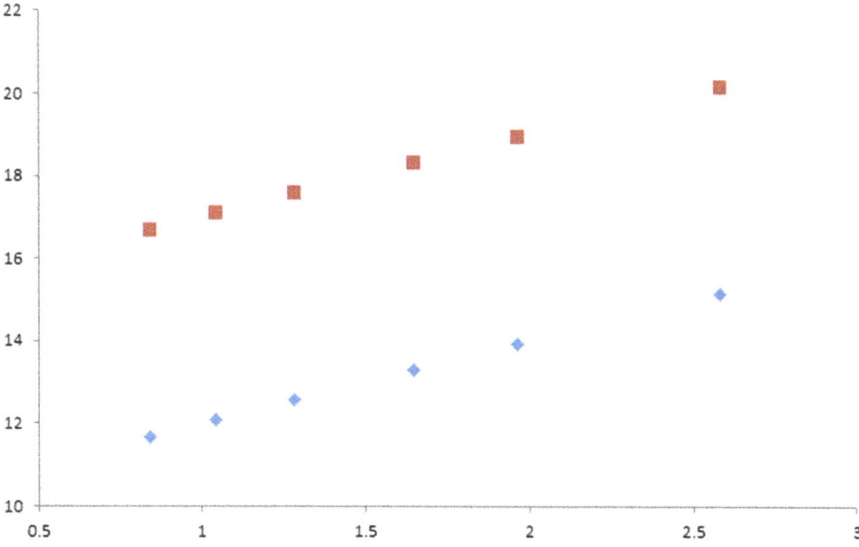

Fig. 4 Neutrosophic Q–Q plot for $X_N \sim N_N([10, 15], 2^2)$

1. $f(x_N) \geq 0$ [Function is nonnegative.]
2. $\int_{-\infty}^{+\infty} f(x_N) \, dx_N = 1$ [Total area is unity.]

For Condition (1), as

$$f(x_N) = \frac{1}{\sigma_N \sqrt{2\pi}} e^{-1/2 \left(\frac{x_N - \mu_N}{\sigma_N} \right)^2} \tag{12}$$

Here, (1) $\sigma > 0$, (2) $\sqrt{2\pi} > 0$, and (3) exponent $(-\text{ve}) > 0$.

As all the elements of pdf are nonnegative which indicates $f(Xn) \geq 0$, so the first condition is satisfied. For Condition (2), we have to prove

$$\int_{-\infty}^{+\infty} f(X_n) \, dX_n = 1 \tag{13}$$

Consider

$$\text{L.H.S.} = \int_{-\infty}^{+\infty} f(X_n) \, dX_n = \frac{1}{\sigma_n \sqrt{2\pi}} e^{-1/2 \left(\frac{X_n - \mu_n}{\sigma_n} \right)^2} dX_n \tag{14}$$

So Eq. (14) becomes

$$= \frac{1}{\sigma_n \sqrt{2\pi}} \int_{-\infty}^{+\infty} e^{-\frac{z_n^2}{2}} \sigma_n dz_n = \frac{1}{\sqrt{2\pi}} \int_{-\infty}^{+\infty} e^{-\frac{z_n^2}{2}} dz_n \qquad (15)$$

Since the integral is an even function, so Eq. (15) can be written as

$$\int_{-\infty}^{+\infty} f(x_n) dx_n = 2 \int_{0}^{+\infty} f(x_n) dx_n = \frac{1}{\sqrt{2\pi}} 2 \int_{0}^{+\infty} e^{-\frac{z_n^2}{2}} dz_n \qquad (16)$$

$$\text{Let } \frac{z_n^2}{2} = w_n \Rightarrow z_n = \sqrt{2w_n}, dz_n = \frac{1}{\sqrt{2w_n}} dw_n$$

so (16) becomes

$$= \frac{2}{\sqrt{2\pi}} \int_{0}^{+\infty} e^{-w_n} \frac{1}{\sqrt{2w_n}} dw_n = \frac{\sqrt{2}}{\sqrt{\pi}} \int_{0}^{+\infty} e^{-w_n} \frac{w_n^{-1/2}}{\sqrt{2}} dw_n = \frac{1}{\sqrt{\pi}} \int_{0}^{+\infty} e^{-w_n} w_n^{-1/2} dw_n$$

Using gamma function,

$$\int_{0}^{+\infty} e^{-x} x^n dx_n = \overline{|n+1|} = \overline{|-1/2+1|} = \left| \frac{1}{2} \right| = \sqrt{\pi}$$

$$\text{so, } = \frac{1}{\sqrt{\pi}} \left| \frac{1}{2} \right| = \frac{1}{\sqrt{\pi}} \sqrt{\pi} = 1 = R.H.S$$

Both conditions are satisfied so neutrosophic normal probability function (12) is a proper neutrosophic pdf.

Property 2 The mean of the neutrosophic normal distribution is μ_n.
By definition,

$$\mu_n = E(X_n) = \int_{-\infty}^{+\infty} X_n f(X_n) dX_n = \frac{1}{\sigma_n \sqrt{2\pi}} \int_{-\infty}^{+\infty} X_n e^{-1/2 \left(\frac{X_n - \mu_n}{\sigma_n} \right)^2} dX_n$$

$$(17)$$

$$\text{Let } z_n = \frac{X_n - \mu_n}{\sigma_n} \Rightarrow X_n = \mu_n + z_n \sigma_n, dX_n = \sigma_n dz_n$$

Now Eq. (17) becomes

$$
E\left(X_n\right) = \frac{1}{\sigma_n\sqrt{2\pi}} \int\limits_{-\infty}^{+\infty} \left(\mu_n + z_n\sigma_n\right) e^{-1/2}\sigma_n dz_n
$$

$$
= \frac{\mu_n}{\sqrt{2\pi}} \int\limits_{-\infty}^{+\infty} e^{-z_n^2/2} dz_n + \frac{\sigma_n}{\sqrt{2\pi}} \int\limits_{-\infty}^{+\infty} z_n e^{-z_n^2} dz_n
$$

(18)

As $= \frac{1}{\sqrt{2\pi}} \int\limits_{-\infty}^{+\infty} e^{-z_n^2/2} dz = 1$ and $\int\limits_{-\infty}^{+\infty} z_n e^{-z_n^2} dz_n = 0$ (Odd Function), so

$$
E\left(X_n\right) = \mu_n(1) + \frac{\sigma_n}{\sqrt{2\pi}}(0) = \mu_n
$$

Property 3 The variance of the neutrosophic normal distribution is σ_N^2.

By definition, $\operatorname{Var}\left(X_N\right) = E(X_N - \mu_N)^2 = \int\limits_{-\infty}^{+\infty} (X_N - \mu_N)^2 f\left(X_N\right) dX_N$

$$
= \frac{1}{\sigma_N\sqrt{2\pi}} \int\limits_{-\infty}^{+\infty} (X_N - \mu_N)^2 e^{-1/2\left(\frac{X_N - \mu_N}{\sigma_N}\right)^2} dX_N
$$

(19)

Let $z_N = \frac{X_N - \mu_N}{\sigma_N} \Rightarrow X_N = \mu_N + z_N\sigma_N, dX_N = \sigma_N dz_N$

so, Eq. (19) becomes

$$
= \frac{\sigma_N^2}{\sqrt{2\pi}} \int\limits_{-\infty}^{+\infty} z_N^2 e^{-z_N^2/2} dz_N
$$

(20)

Let $\frac{z_N^2}{2} = w_N \Rightarrow z_N = \sqrt{2w_N}, dz_N = \frac{1}{\sqrt{2w_N}} dw_N$

and Eq. (20) is also even function so,

$$
= \frac{2\sigma_N^2}{\sqrt{\pi}} \cdot \int\limits_{0}^{+\infty} e^{-w_N} w_N^{1/2} dw_N
$$

(21)

By using gamma function, Eq. (21) becomes

$$= \frac{2\sigma_N^2}{\sqrt{\pi}} * \frac{1}{2}\sqrt{\pi} = \sigma_N^2 = Var\,(X_N)$$

Property 4 The median of the neutrosophic normal distribution is equal to μ_N.
    By definition,

$$M = \text{median} = \int_{-\infty}^{M} f\,(X_N)\,dX_N = \frac{1}{2} \Rightarrow \frac{1}{\sigma_N\sqrt{2\pi}} \int_{-\infty}^{M} e^{-1/2\left(\frac{X_N-\mu_N}{\sigma_N}\right)^2} dX_N = \frac{1}{2}$$

$$(22)$$

$$\text{Let } \frac{X_N - \mu_N}{\sigma_N} = z_N \Rightarrow X_N = \mu_N + \sigma_N dz_N, dX_N = \sigma_N dz_N$$

$$\text{limits, when } -\infty < x < M, \text{ then } -\infty < z < \frac{M - \mu_N}{\sigma_N}$$

$$\frac{1}{\sigma_N\sqrt{2\pi}} \int_{-\infty}^{\frac{M-\mu_N}{\sigma_N}} e^{-z_N^2/2}\sigma_N dz_N = \frac{1}{2} \Rightarrow \frac{1}{\sqrt{2\pi}} \int_{-\infty}^{\frac{M-\mu_N}{\sigma_N}} e^{-z_N^2/2} dz_N = \frac{1}{2} \qquad (23)$$

as

$$\frac{1}{\sqrt{2\pi}} \int_{-\infty}^{0} e^{-z_N^2/2} dz_N = \frac{1}{2} \qquad (24)$$

By comparing Eqs. (23) and (24),

$$\frac{M - \mu_N}{\sigma_N} = 0 \Rightarrow M = \mu_N$$

Property 5 The mean deviation of the neutrosophic normal distribution is approximately 4/5 of its standard deviation.
    By definition,

$$\text{M.D.}=\overset{+\infty}{\underset{-\infty}{\int}}|X_N-\mu_N|\, f\,(X_N)\, dX_N=\frac{1}{\sigma_N\sqrt{2\pi}}\overset{+\infty}{\underset{-\infty}{\int}}|X_N-\mu_N|\, e^{-1/2\left(\frac{X_N-\mu_N}{\sigma_N}\right)^2}\, dX_N$$

$$(25)$$

$$\text{Let } \frac{X_N-\mu_N}{\sigma_N}=z_N \Rightarrow X_N=\mu_N+\sigma_N z_N,\, dX_N=\sigma_N dz_N$$

Now Eq. (25) becomes

$$\text{M.D.}=\frac{1}{\sigma_N\sqrt{2\pi}}\overset{+\infty}{\underset{-\infty}{\int}}|\sigma_N z_N|\, e^{-z_N^2}\sigma_N dz_N$$

$$(26)$$

Now by using even function,

$$\text{M.D.}=\frac{\sigma_N}{\sqrt{2\pi}}2\overset{+\infty}{\underset{0}{\int}}z_N e^{-z_N^2/2}dz_N$$

$$(27)$$

$$\text{Let } \frac{z_N^2}{2}=w_N \Rightarrow z_N=\sqrt{2w_N},\, dz_N=\frac{1}{\sqrt{2w_N}}dw_N$$

$$\text{M.D.}=\frac{\sigma_N}{\sqrt{2\pi}}2\overset{+\infty}{\underset{0}{\int}}\sqrt{2w_N}e^{-w_N}\frac{1}{\sqrt{2w_N}}dw_N$$

$$=\frac{2\sigma_N}{\sqrt{2\pi}}\overset{+\infty}{\underset{0}{\int}}e^{-w_N}dw_N=\frac{2\sigma_N}{\sqrt{2\pi}}.-\left[e^{+\infty}-e^0\right]=\frac{2\sigma_N}{\sqrt{2\pi}}[0+1]$$

$$\text{M.D.}=\sigma_N\sqrt{\frac{2}{\pi}}=0.7979\sigma_N=\frac{4}{5}\sigma_N$$

Property 6 Show that the moment generating function about origin for neutrosophic normal distribution is $M_o(t)=e^{\mu_N t+t^2\sigma_N^2/2}$.

By definition,

$$M_0(t) = E\left(e^{tx}\right) = \int_{-\infty}^{+\infty} e^{tx} f(X_N) dX_N = \frac{1}{\sigma_N \sqrt{2\pi}} \int_{-\infty}^{+\infty} e^{tx}.e^{-1/2\left(\frac{X_N - \mu_N}{\sigma_N}\right)^2} dX_N$$

$$(28)$$

$$\text{Let } z_N = \frac{X_N - \mu_N}{\sigma_N} \Rightarrow X_N = \mu_N + \sigma_N z_N, dX_N = \sigma_N dz_N$$

So,

$$M_0(t) = \frac{1}{\sigma_N \sqrt{2\pi}} \int_{-\infty}^{+\infty} e^{t\left(\mu_N + \sigma_N z_N\right)}.e^{-z_N^2/2} \sigma_N dz_N = \frac{e^{\mu_N t}}{\sqrt{2\pi}} \int_{-\infty}^{+\infty} e^{-1/2\left(z_N^2 - 2t\sigma_N z_N\right)} dz_N$$

$$(29)$$

Adding and subtracting $t^2\sigma_N^2$ in the exponent, we have

$$M_0(t) = \frac{e^{\mu_N t}}{\sqrt{2\pi}} \int_{-\infty}^{+\infty} e^{-1/2\left(z_N^2 - 2t\sigma_N z_N - \sigma_N^2 t^2 + \sigma_N^2 t^2\right)}$$

$$(30)$$

$$dz_N = \frac{e^{\mu_N t + t^2 \sigma_N^2/2}}{\sqrt{2\pi}} \int_{-\infty}^{+\infty} e^{-1/2(z_N - t\sigma_N)^2} dz_N$$

$$\text{Let, } w_N = z_N - t\sigma_N \Rightarrow z_N = w_N + t\sigma_N, dz_N = dw_N$$

Limits remain the same and Eq. (30) becomes

$$\frac{1}{\sqrt{2\pi}} \int_{-\infty}^{+\infty} e^{-w_N^2} dw_N = 1 \Rightarrow M_0(t) = e^{\mu_N t + \frac{1}{2}t^2\sigma_N^2}(1) = e^{\mu_N t + \frac{1}{2}t^2\sigma_N^2} \qquad (31)$$

Note

i. The cumulant generating function is simply the natural log of the moment generating function (31), i.e.,

$$K(t) = \ln M_0(t) = \ln e^{\mu_N t + \frac{1}{2}t^2\sigma_N^2} = \mu_N t + \frac{1}{2}t^2\sigma_N^2 \qquad (32)$$

By comparing coefficient of $\frac{t^r}{r!}$, we have

$$K_1 = \mu_N \, (\text{Mean}) \, , K_2 = \sigma_N^2 \, (\text{Variance}) \, , K_3 = K_4 = \cdots = 0$$

ii. Gamma-ratios of the neutrosophic normal distribution are

$$\gamma_1 = \frac{K_3}{(K_2)^{3/2}} = \frac{0}{\left(\sigma_N^2\right)^{3/2}} = 0 \, (\text{Symmetrical}) \tag{33}$$

$$\gamma_2 = \frac{K_4}{(K_2)^2} = \frac{0}{\left(\sigma_N^2\right)^2} = 0 \, (\text{Mesokurtic}) \tag{34}$$

iii. Beta-ratios of the neutrosophic normal distribution are

$$\beta_1 = \frac{\mu_3^2}{(\mu_2)^3} = \frac{(0)^2}{\left(\sigma_N^2\right)^3} = 0 \, (\text{Symmetrical}) \tag{35}$$

$$\beta_2 = \frac{\mu_4}{(\mu_2)^2} = \frac{3\sigma_N^4}{\left(\sigma_N^2\right)^2} = \frac{3\sigma_N^4}{\sigma_N^4} = 3 \, (\text{Mesokurtic}) \tag{36}$$

iv. The characteristic function of the neutrosophic normal distribution is

$$\text{C.f.} = \phi_o(t) = E\left[e^{itx_N}\right] = \int_{-\infty}^{+\infty} e^{itx_N} f(x_N) \, dX_N = e^{it\mu_N + \frac{1}{2}it^2\sigma_N^2} \tag{37}$$

5   The Entropy of Neutrosophic Normal Distribution

Let the neutrosophic random variable follow a neutrosophic normal distribution with mean μ_N and variance σ_N^2 having the following pdf:

$$f(x_N) = \frac{1}{\sigma_N \sqrt{2\pi}} e^{-\frac{1}{2}\left(\frac{x_N - \mu_N}{\sigma_N}\right)^2}$$

then, the neutrosophic entropy for the neutrosophic normal distribution is

$$H(X_N) = \int_{-\infty}^{\infty} f(x_N) \log f(x_N) \, dx_N = \frac{1}{2}\left[1 + \log\left(2\sigma_N^2\pi\right)\right] \tag{38}$$

6 Maximum Likelihood Estimates of Neutrosophic Normal Distribution

The maximum likelihood estimates of parameters of the neutrosophic normal probability distribution are obtained similar to the MLEs of normal distribution. For this, the log-likelihood function of the neutrosophic normal distribution is

$$
\ln L\left(\mu_N, \sigma_N^2\right) = \sum_{i=1}^{n} \ln f\left(x_i \,\middle|\, \mu_N, \sigma_N^2\right)
$$
$$
= -\frac{n}{2} \ln(2\pi) - \frac{n}{2} \ln \sigma_N^2 - \frac{1}{2\sigma_N^2} \sum_{i=1}^{n} (x_i - \mu_N)^2
$$
(39)

By differentiating (39) with respect to the parameters μ_N and σ_N^2, we obtain the MLEs of μ_N and σ_N^2, i.e.,

$$
\hat{\mu}_N = \overline{x}_N = \frac{1}{n} \sum_{i=1}^{n} x_i \text{ and } \hat{\sigma}_N^2 = \frac{1}{n} \sum_{i=1}^{n} (x_i - \overline{x}_N)^2
$$
(40)

7 Fisher Information Matrix of Neutrosophic Normal Distribution

A Fisher information matrix of neutrosophic normal distribution takes the diagonal form and can be written as

$$
FIM\left[X_N \sim N_N\left(\mu_N, \sigma_N^2\right)\right] = \begin{bmatrix} \frac{1}{\sigma_N^2} & 0 \\ 0 & \frac{1}{2\sigma_N^4} \end{bmatrix}
$$
(41)

References

1. Smarandache F. (1995). *Neutrosophic logic and set*, mss. Retrieved from http://fs.gallup.unm.edu/neutrosophy.htm.
2. Smarandache, F. (2014). *Introduction to neutrosophic statistics*. Conshohocken: Infinite Study.
3. Smarandache, F. (2010). Neutrosophic logic—A generalization of the intuitionistic fuzzy logic. In F. Smarandache (Ed.), *Multispace & multistructure. Neutrosophic transdisciplinarity (100 collected papers of science)* (Vol. 4, p. 396).
4. Alhabib, R., Ranna, M. M., Farah, H., & Salama, A. A. (2018). Some neutrosophic probability distributions. *Neutrosophic Sets and System, 22*, 30–38.

5. Patro, S. K., & Smarandache, F. (2016). The neutrosophic statistics distribution, more problems, more solutions. *Neutrosophic Sets and Systemss, 12*, 73–79.
6. Alhasan, K. F. H., & Smarandache, F. (2019). Neutrosophic Weibull distribution and neutrosophic family Weibull distribution. *Neutrosophic Sets and Systems, 28*, 191–199.
7. Smarandache, F. (2019). Introduction to neutrosophic sociology (neutrosociology) (January 17, 2019). *Infinite Study*. ISBN: 978-1-59973-605-1. SSRN: https://ssrn.com/abstract=3405436.
8. Aslam, M. (2019a). Neutrosophic analysis of variance: application to university students. *Complex & Intelligent Systems, 5*, 403–407. https://doi.org/10.1007/s40747-019-0107-2.
9. Aslam, M. (2019b). Introducing Kolmogorov-Smirnov tests under uncertainty: An application to radioactive data. *ACS Omega, 5*(1), 914–917. https://doi.org/10.1021/acsomega.9b03940.
10. Aslam, M. (2020a). On detecting outliers in complex data using Dixon's test under neutrosophic statistics. *Journal of King Saud University: Science, 32*, 2005–2008. https://doi.org/10.1016/j.jksus.2020.02.003.
11. Aslam, M. (2020b). Design of the Bartlett and Hartley tests for homogeneity of variances under indeterminacy environment. *Journal of Taibah University for Science, 14*(1), 6–10. https://doi.org/10.1080/16583655.2019.1700675.
12. Aslam, M., Arif, O., & Khan, R. A. (2020). New diagnosis test under the neutrosophic statistics: An application to diabetic patients. *BioMed Research International, 2020*, 2086185. https://doi.org/10.1155/2020/2086185.
13. Aslam, M., Bantan, R. A. R., & Khan, N. (2018). Design of a new attribute control chart under neutrosophic statistics. *International Journal of Fuzzy Systems, 21*, 433–440.
14. Aslam, M. (2018). A new sampling plan using neutrosophic process loss consideration. *Symmetry, 10*, 132.
15. Aslam, M., & Arif, O. (2018). Testing of grouped product for the Weibull distribution using neutrosophic statistics. *Symmetry, 10*, 403.
16. Khan, M. Z., Khan, M. F., Aslam, M., & Mughal, A. R. (2019). Design of fuzzy sampling plan using the Birnbaum-Saunders distribution. *Mathematics, 7*, 9.
17. Turanoglu, E., Kaya, I., & Kahraman, C. (2012). Fuzzy acceptance sampling and characteristic curves. *International Journal of Computational Intelligence Systems, 5*, 13–29.
18. Smarandache, F. (2015). *Introduction to neutrosophic measure, integral, probability*. Columbus, OH: Sitech Education Publisher.
19. Smarandache, F. (2002). *Neutrosophy and neutrosophic logic*. First international conference on neutrosophy, neutrosophic logic, set, probability, and statistics. University of New Mexico, Gallup, NM 87301, USA.
20. Smarandache, F. (2005). Neutrosophic set a generalization of the intuitionistic fuzzy sets. *International Journal of Pure and Applied Mathematics, 24*, 287–297.
21. Salama, A., & Rafif, E (2018). *Neutrosophic decision making & neutrosophic decision tree* (Vol. 4), University of Albaath.
22. Abdel-Basset, M., Mohamed, R., Zaied, A. E. N. H., & Smarandache, F. (2019). A hybrid plithogenic decision-making approach with quality function deployment for selecting supply chain sustainability metrics. *Symmetry, 11*(7), 903.

New Algorithms for Bipolar Single-Valued Neutrosophic Hamiltonian Cycle

M. Lathamaheswari, Said Broumi, and Florentin Smarandache

1 Introduction

Zadeh [1] introduced a fuzzy set as a class of objects along with the grades of membership to represent uncertainty. It is characterized by a characteristic function called membership function which allows each object a grade of membership, and these values lie between zero and one. As it deals only with membership function, it could not provide a complete solution, and hence intuitionistic fuzzy sets were introduced as a generalization of fuzzy set by adding a new component called nonmembership of the object [2]. The fuzzy system has been utilized fruitfully in the last few decades in problems which involve approximate reasoning. It has developed into a vast research area in various fields, including social sciences, engineering, graph theory, network analysis, medical sciences, artificial intelligence, pattern recognition, and decision-making process. As the world consists of indeterminacy in nature, the idea of neutrosophic logic to deal with truth, indeterminacy, and falsity of the object was introduced [3]. Operational laws of all the types of sets have been defined by their properties and applied in a real-world problems by many researchers once the new concept was introduced. After introducing neutrosophic set, its special types, namely, single-valued neutrosophic set, interval-valued set, complex neutrosophic set, and bipolar neutrosophic set, have been introduced to their

M. Lathamaheswari (✉)
Department of Mathematics, Hindustan Institute of Technology and Science, Chennai, India

S. Broumi
Laboratory of Information Processing, Faculty of Science Ben M'Sik, University Hassan II, Casablanca, Morocco

F. Smarandache
Department of Mathematics and Science, University of New Mexico, Gallup, NM, USA

© The Author(s), under exclusive license to Springer Nature Switzerland AG 2021
F. Smarandache, M. Abdel-Basset (eds.), *Neutrosophic Operational Research*,
https://doi.org/10.1007/978-3-030-57197-9_9

171

operations and properties. Bipolar fuzzy sets and its relations are the computational framework and for modelling and decision-making analysis. Bipolar fuzzy sets are the extension of fuzzy sets where the range of membership degrees is $[-1,1]$. Here if the membership degree is zero, then the element is not related to the corresponding property. If the membership degree lies in $(0,1]$, then the element partially satisfies the property. If the membership degree lies in $[-1,0)$, then the element partially satisfies the implicit counter property. For example, if we give positive membership values for the sweetness foods, then membership values for bitterness foods are negative, and the other tastes such as sour, salty, and chili are irrelevant to the corresponding property; therefore membership value is zero [4]. The notion of graph theory is an intensively useful tool for dealing with combinatorial problems in various areas such as number theory, operations research, algebra, and topology and computer science [5]. Fuzzy graphs are the graphs with fuzzy edge weights and applied in many real-time problems where uncertainty exists in the path. Fuzzy graph theory has growing number of applications in modelling real-world systems where the level of data fixed in the system differs with various levels of accuracy [6]. Fuzzy models are very useful in reducing the differences between the conventional methods and models used in expert systems and hence applied widely in engineering and the sciences. The notion of complement of fuzzy graph examined some operations on fuzzy graphs [7]. Many concepts such as bipolar fuzzy graphs, regular bipolar fuzzy graphs, complete bipolar fuzzy graphs, and irregular bipolar fuzzy graphs also had been proposed in later decades. If the relations between the vertices are indeterminate in problems, then the notions of fuzzy graphs and its extensions, namely, intuitionistic fuzzy graphs, N-graphs, bipolar fuzzy graphs, and bipolar intuitionistic graphs, are not suitable. To overcome the problem of handling with indeterminacy and bipolarity of the real-world problems, the hybrid concept such as bipolar single-valued graphs, bipolar interval-valued graphs, and bipolar neutrosophic has been introduced.

2 Review of Literature

Samantha and Pal [8] proposed irregular bipolar fuzzy graphs. Akram and Davvaz [6] defined strong intuitionistic fuzzy graphs. Alshehri and Akram [9] introduced Cayley bipolar fuzzy graphs. Karunambigai et al. [10] determined domination in bipolar fuzzy graphs. Rosline and Pathinathan [11] dealt with scheduling job using fuzzy graphs concepts. Akram et al. [7] introduced the concept of balance bipolar fuzzy graphs. Rashmanlou et al. [12, 13] defined new operations on bipolar fuzzy graphs with their properties. Rashmanlou et al. [12, 13] discussed some of the properties of bipolar fuzzy graphs. Akram et al. [14] introduced some of the operations and described dominating and independent sets of bipolar neutrosophic graph and discussed outranking approach for risk analysis using the proposed concepts. Delia et al. [15] introduced the concept of bipolar neutrosophic set and its operations and applied in a decision-making problem. Broumi et al. (2016a) defined the concept of bipolar single-valued neutrosophic graph. Gani and Latha [16] proposed new algorithms to find the Hamiltonian cycle for a fuzzy network. Akram and Akmal [17] introduced some of the operations on intuitionistic fuzzy graph theory. Deli et al. [18] proposed the concept of interval-valued bipolar neutrosophic set and applied in pattern recognition. Sahin et al. [19] presented Jaccard vector similarity measure of bipolar neutrosophic set and solved a decision-making problem using the proposed similarity measure. Pramanik and Mondal [20] developed a hybrid structure called rough bipolar neutrosophic set. Ulucay et al. [21] introduced some of the similarity measure with their properties and applied in a decision-making problem. Broumi et al. [22, 23] examined the properties of different types of degrees, order, and size of a single-valued neutrosophic graph. Mathew et al. [24] introduced the concept of bipolar fuzzy graphs and strong bipolar fuzzy graphs. Gani and Rahman [25] defined lower and upper truncations of intuitionistic fuzzy graphs. Broumi et al. [26] introduced complex neutrosophic graphs of type 1 with its properties and its matrix representation. Gani et al. [27] determined domination on intuitionistic fuzzy graphs. Prabhu et al. [28] introduced finite state machine using the concept of bipolar neutrosophic set. Akram et al. [29] presented some concepts and properties of bipolar single-valued neutrosophic graph structure. Girao et al. [30] determined long cycles in Hamiltonian graphs. Mullai et al. [31] solved a shortest path problem using minimum spanning tree algorithm under bipolar neutrosophic environment. Broumi et al. [32] introduced a matrix algorithm for finding bipolar neutrosophic minimum spanning tree. Akram et al. [33, 34] discussed certain properties of bipolar neutrosophic graphs. Pramanik et al. [35] proposed cross entropy measures of bipolar single- and interval-valued neutrosophic sets and applied in a decision-making problem. Aziz et al. [36] examined k-step Hamiltonian graphs. Topal et al. [37] presented a python tool for the implementation of bipolar neutrosophic matrices. Akram et al. [38, 39] proposed antipodal bipolar fuzzy graphs. Akram et al. [38, 39] applied bipolar neutrosophic sets to incidence graphs. Nagarajan et al. [40, 41] proposed blockchain single- and interval-valued neutrosophic graphs. Nagarajan et al. [40, 41] introduced Dombi interval-valued neutrosophic graphs. Akram et al. [38, 39] introduced single-valued bipolar neutrosophic competition graph with its properties and applied the proposed concept for designations and brands competition. Akram and Shum [42] introduced bipolar neutrosophic planar graphs. Jan et al. [43] proposed cubic bipolar fuzzy graphs with properties and elaborated in a social group. Zeps [44] examined Hamiltonian vertices and edges in uniquely Hamiltonian graphs. Hussain et al. [45] introduced neutrosophic bipolar vague graphs with their properties. Goedebeur et al. [46] described an algorithm to find the Hamiltonian cycles in a graph. Poulik and Ghorai [47] examined the applications of fuzzy graphs. Nagarajan et al. [48] introduced new algorithms to find Hamiltonian cycle for an interval-valued neutrosophic graph. Asratian et al. [49] introduced a new method called localization method in Hamiltonian graph theory. Alnaser et al. [50] introduced and studied bipolar intuitionistic fuzzy graphs and its matrices. Mohamed and Ali [51] defined energy of spherical fuzzy graphs.

It is observed that algorithms have not yet been proposed for finding Hamiltonian cycle for a bipolar single-valued neutrosophic graph till date. Hence, in this paper, we proposed new algorithms based on adjacency matrix and least vertex degree have been proposed and obtained Hamiltonian cycle with the cycle length for a bipolar single-valued neutrosophic graph.

3 Preliminaries

In this section, the definitions of the basic concepts are given which give the better understanding of the present work.

3.1 Fuzzy Set [1]

Let F be a fuzzy set on the universal set X defined by a mapping $m : X \rightarrow [0, 1]$ called membership function, and the fuzzy set is denoted by $F = (X, m)$.

3.2 Fuzzy Graph [5]

Let $G = (X, \phi, \rho)$ be a fuzzy graph which is a nonempty set X together with a pair of functions $\phi : V \rightarrow [0, 1]$ and $\rho : V \times V \rightarrow [0, 1]$ such that for all $x, y \in V$, $\rho(x, y) \leq \min \{\phi(x), \phi(x)\}$

3.3 Bipolar Fuzzy Set [4]

A bipolar fuzzy set B on X is an object which has the form $B = \{(x, m^+(x), m^-(x))/ x \in X\}$, where $m^+(x) : X \rightarrow [0, 1]$ and $m^-(x) : X \rightarrow [0, 1]$ with the following conditions:

If $m^+(x) \neq 0$ and $m^-(x) = 0$, then x has only positive satisfaction for B.

If $m^+(x) = 0$ and $m^-(x) \neq 0$, then x does not satisfy the property of B but somewhat satisfy the counter property of B.

If $m^+(x) \neq 0$ and $m^-(x) \neq 0$, then the membership function of the property overlaps that of its counter property over some portion of X.

3.4 Bipolar Fuzzy Graph [8]

A bipolar fuzzy graph with an elemental set V is defined as the pair $G = (P, Q)$ where $P = \left(m_P^+, m_P^-\right)$ is a bipolar fuzzy set on V and $Q = \left(m_Q^+, m_Q^-\right)$ is a bipolar fuzzy set on $E \subseteq V \times V$ such that $m_Q^+(x, y) \leq \min\left\{m_P^+(x), m_P^+(y)\right\}$ and $m_Q^-(x, y) \geq \max\left\{m_P^-(x), m_P^-(y)\right\}$.

3.5 Intuitionistic Fuzzy Set [17]

An intuitionistic fuzzy set is an object of the form $A = \{(x, \mu_A(x), \nu_A(x))/x \in X\}$ where the mappings $\mu_A : X \rightarrow [0, 1]$ and $\nu_A : X \rightarrow [0, 1]$ denote the degrees of membership and nonmembership, respectively, of all the elements $x \in X$ such that $\mu_A(x) + \nu_A(x) = 1, \forall\, x \in X$.

3.6 Intuitionistic Fuzzy Graph [17]

An intuitionistic fuzzy graph with an elemental set V is defined as the pair $G = (A, B)$, where A is an intuitionistic fuzzy set on V and B is also an intuitionistic fuzzy set on $E \subseteq V \times V$ such that $\mu_B(x, y) \leq \min\{\mu_A(x), \mu_A(y)\}$ and $\nu_B(x, y) \leq \max\{\nu_A(x), \nu_A(y)\}$.

3.7 Bipolar Intuitionistic Fuzzy Graph [17]

A bipolar intuitionistic fuzzy graph $G = (V, E)$ is a pair (P, Q) where $P = \left(\mu_P^+, \mu_P^-, \nu_P^+, \nu_P^-\right)$ is an intuitionistic bipolar fuzzy set on V and $Q = \left(\mu_Q^+, \mu_Q^-, \nu_Q^+, \nu_Q^-\right)$ is an intuitionistic bipolar fuzzy set on $E \subseteq V \times V$ such that

$$\mu_Q^+(x, y) \leq \min\left\{\mu_P^+(x), \mu_P^+(y)\right\}, \forall\, (x, y) \in V \times V$$

$$\mu_Q^-(x, y) \geq \max\left\{\mu_P^-(x), \mu_P^-(y)\right\}, \forall\, (x, y) \in V \times V$$

$$\nu_Q^+(x, y) \geq \max\left\{\mu_P^+(x), \mu_P^+(y)\right\}, \forall\, (x, y) \in V \times V$$

$$\nu_Q^-(x, y) \leq \min\left\{\mu_P^-(x), \mu_P^-(y)\right\}, \forall\, (x, y) \in V \times V$$

$$\mu_Q^+(x, y) = \mu_Q^-(x, y) = 0, \forall (x, y) \in V \times V - E$$

$$\nu_Q^+(x, y) = \nu_Q^-(x, y) = 0, \forall (x, y) \in V \times V - E$$

3.8 Single-Valued Neutrosophic Set [3]

A single-valued neutrosophic set P on X is characterized by truth, indeterminacy, and false membership functions and is defined by $P = \langle x : (T_P(x), I_P(x), F_P(x)) : x \in X \rangle$ where $T_P(x), I_P(x), F_P(x) \in [0, 1]$ for all the values of $x \in X$.

3.9 Bipolar Neutrosophic Set [15]

A bipolar neutrosophic set on X is an object of the form

$$P = \left\{ \left(x, t_P^+(x), i_P^+(x), f_P^+(x) , t_P^-(x), i_P^-(x), f_P^-(x) \right) : x \in X \right\}$$

where $t_P^+, i_P^+, f_P^+ : X \to [0, 1]$ are the truth, indeterminacy, and false membership degrees of the element $x \in X$ and $t_P^-, i_P^-, f_P^- : X \to [0, 1]$ are the implicit counter property of truth, indeterminacy, and false membership degrees of the element $x \in X$ corresponding to the bipolar neutrosophic set P.

3.10 Bipolar Neutrosophic Graph [14]

A bipolar neutrosophic relation on a nonempty set X is a bipolar neutrosophic subset of $X \times X$ of the form

$$Q = \left\{ \left((x, y), t_Q^+\left(x, y\right), i_Q^+\left(x, y\right), f_Q^+\left(x, y\right) , \right. \right.$$
$$\left. \left. t_Q^-(x, y), i_Q^-\left(x, y\right), f_Q^-\left(x, y\right) \right) : (x, y) \in E \subseteq X \times X \right\}$$

where $t_Q^+, i_Q^+, f_Q^+, t_Q^-, i_Q^-, f_Q^-$ are defined by the mappings $t_Q^+, i_Q^+, f_Q^+ : X \times X \to [0, 1]$ and $t_Q^-, i_Q^-, f_Q^- : X \times X \to [-1, 0]$. And therefore $G = (P, Q)$ is a bipolar neutrosophic graph where P is bipolar neutrosophic set on X and Q is bipolar neutrosophic relation on X such that

$$t_Q^+(x, y) \leq \min \left\{ t_P^+(x), t_P^+(y) \right\}, \quad i_Q^+(x, y) \leq \max \left\{ i_P^+(x), i_P^+(y) \right\},$$

$$f_Q^+(x, y) \leq \max\left\{f_P^+(x), f_P^+(y)\right\}, \quad t_Q^-(x, y) \geq \max\left\{t_P^-(x), t_P^-(y)\right\},$$

$$i_Q^-(x, y) \geq \min\left\{i_P^-(x), i_P^-(y)\right\}, \quad f_Q^-(x, y) \geq \min\left\{f_P^-(x), f_P^-(y)\right\}.$$

3.11 Hamiltonian Path and Hamiltonian Cycle [16]

Hamiltonian path is the path passing through all the vertices without repetition, and the corresponding cycle is the Hamiltonian cycle of a graph. A graph which has Hamiltonian cycle is called a Hamiltonian graph. This fact is true for fuzzy graph, intuitionistic fuzzy graph, and neutrosophic graph.

4 Proposed Algorithms to Find Bipolar Single-Valued Neutrosophic Hamiltonian Cycle (BSVNHC)

In this section, we proposed modified form of the score function of bipolar single-valued neutrosophic numbers and new algorithms to find BSVNHC using adjacency matrix and lowest vertex degree in a bipolar single-valued neutrosophic graph (network).

4.1 Modified Form of the Score Function of BSVNNs

The modified form of score function (SF) of BSVNNs is defined by

$$SF(BSVNN) = \frac{1}{6}\left[\left(t^+ + t^-\right) - \left(i^| + i^-\right) - \left(f^+ + f^-\right) + 3\right]$$

4.2 Algorithm to Find BSVNHC Using Adjacency Matrix

Here, a network is adapted from Gani and Latha [16] where the bipolar single-valued neutrosophic graph (BSVNG) (Fig. 1) of the shipyard routes between the cities Cochin (C), Mumbai (M), Vizag (V), Kolkata (K), Goa (G), Chennai (Ch), and New Mangalore (NM) is considered. Here vertices and edges are assigned by bipolar single-valued neutrosophic numbers (BSVNNs).

The proposed algorithm is described by the following steps:

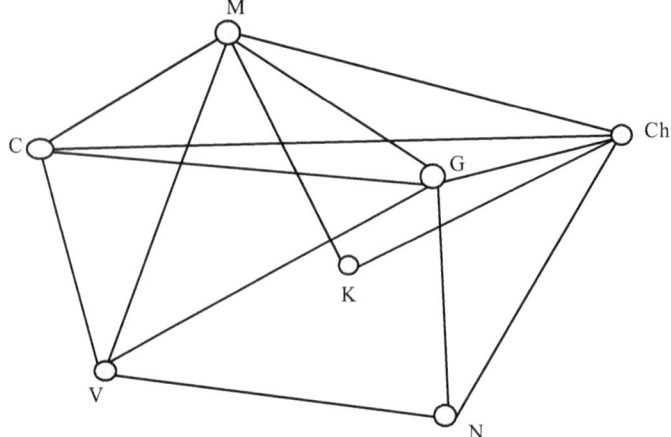

Fig. 1 Bipolar single-valued neutrosophic graph

Table 1 Vertices with BSVNNs

Cities	Bipolar single-valued neutrosophic weights
C	$(0.5, 0.7, 0.2, -0.6, -0.3, -0.7)$
M	$(0.9, 0.7, 0.3, -0.8, -0.6, -0.1)$
V	$(0.3, 0.4, 0.2, -0.6, -0.3, -0.7)$
K	$(0.5, 0.3, 0.2, -0.5, -0.2, -0.2)$
G	$(0.7, 0.5, 0.3, -0.4, -0.2, -0.2)$
Ch	$(0.6, 0.3, 0.2, -0.5, -0.3, -0.3)$
N	$(0.8, 0.2, 0.1, -0.1, -0.9, -0.2)$

Step 1: Using the modified score function of BSVNNs (Sect. 4.1), find the score values of each route.

Step 2: Form the adjacency matrix (AM), using the score values of the edges. Here route from city to the same city is zero.

Step 3: Find the degree and score values of all the bipolar single-valued neutrosophic vertices.

Step 4: Choose the vertex with the lowest degree (LVD) and start the process of finding BSVNHC.

Step 5: Repeat step 4, until finding BSVNHC.

Consider the following bipolar single-valued neutrosophic graph for the process of finding BSVNHC.

In BSVNG, the vertices (cities) are assigned bipolar single-valued neutrosophic numbers (Table 1).

Step 1: Here using a modified score function of BSVNNs, score values of all the routes from all the cities are obtained (Tables 2, 3, 4, 5, 6, 7, 8).

Step 2: Here adjacency matrix has been formed using the score values of each route (Table 9).

Table 2 Cochin to other cities

Cities	Bipolar neutrosophic weights	Score value
C-C	0	0
C-M	$(0.5, 0.7, 0.3, -0.6, -0.6, -0.7)$	0.533
C-V	$(0.3, 0.7, 0.2, -0.6, -0.3, -0.6)$	0.450
C-K	0	0.000
C-G	$(0.4, 0.7, 0.3, -0.4, -0.3, -0.7)$	0.500
C-Ch	$(0.5, 0.7, 0.2, -0.5, -0.2, -0.7)$	0.500
C-N	$(0.4, 0.7, 0.2, -0.1, -0.3, -0.7)$	0.567

Table 3 Mumbai to other cities

Cities	Bipolar neutrosophic weights	Score value
M-C	$(0.5, 0.7, 0.3, -0.6, -0.5, -0.6)$	0.500
M-M	0	0.000
M-V	$(0.3, 0.7, 0.3, -0.5, -0.6, -0.7)$	0.517
M-K	$(0.5, 0.7, 0.3, -0.4, -0.6, -0.2)$	0.483
M-G	$(0.7, 0.6, 0.3, -0.4, -0.6, -0.2)$	0.533
M-Ch	$(0.6, 0.7, 0.3, -0.5, -0.6, -0.3)$	0.500
M-N	0	0.000

Table 4 Vizag to other cities

Cities	Bipolar neutrosophic weights	Score value
V-C	$(0.3, 0.7, 0.2, -0.5, -0.2, -0.7)$	0.467
V-M	$(0.2, 0.7, 0.3, -0.6, -0.5, -0.7)$	0.467
V-V	0	0.000
V-K	0	0.000
V-G	$(0.3, 0.5, 0.3, -0.4, -0.3, -0.7)$	0.517
V-Ch	0	0.000
V-N	$(0.2, 0.4, 0.2, -0.6, -0.9, -0.7)$	0.600

Table 5 Kolkata to other cities

Cities	Bipolar neutrosophic weights	Score value
K-C	0	0.000
K-M	$(0.5, 0.7, 0.3, -0.5, -0.6, -0.2)$	0.467
K-V	0	0.000
K-K	0	0.000
K-G	0	0.000
K-Ch	$(0.4, 0.3, 0.2, -0.5, -0.3, -0.3)$	0.500
K-N	0	0.000

Step 3: In Table 10, the degree and score value of all the vertices have been obtained using Sects. 3.10 and 4.1.

Step 4: Select the lowest nonzero entry (LNE) from the adjacency matrix.

Iteration 1 In AM, $a_{13} = 0.450 \Rightarrow$ the row C reaches V $\Rightarrow$ C-V.

Therefore from V we need to find LNE.

From V, LNE is 0.467 which exists in M, i.e., from V it goes to M [C-V-M].

From M, LNE is 0.483 which exists in K, i.e., from M it goes to K [C-V-M-K].

Table 6 Goa to other cities

Cities	Bipolar neutrosophic weights	Score value
G-C	$(0.4, 0.6, 0.3, -0.4, -0.3, -0.7)$	0.517
G-M	$(0.7, 0.7, 0.3, -0.4, -0.6, -0.2)$	0.517
G-V	$(0.3, 0.5, 0.3, -0.4, -0.3, -0.6)$	0.500
G-K	0	0.000
G-G	0	0.000
G-Ch	$(0.6, 0.5, 0.3, -0.4, -0.3, -0.3)$	0.500
G-N	$(0.7, 0.5, 0.3, -0.1, -0.9, -0.2)$	0.650

Table 7 Chennai to other cities

Cities	Bipolar neutrosophic weights	Score value
Ch-C	$(0.5, 0.7, 0.2, -0.5, -0.3 - 0.7)$	0.517
Ch-M	$(0.6, 0.7, 0.3, -0.5, -0.6, -0.3)$	0.500
Ch-V	0	0.000
Ch-K	$(0.5, 0.3, 0.2, -0.5, -0.3, -0.3)$	0.517
Ch-G	$(0.6, 0.5, 0.3, -0.4, -0.3, -0.3)$	0.500
Ch-Ch	0	0.000
Ch-N	$(0.5, 0.3, 0.2, -0.1, -0.9, -0.3)$	0.683

Table 8 New Mangalore to other cities

Cities	Bipolar neutrosophic weights	Score value
N-C	0	0.000
N-M	0	0.000
N-V	$(0.3, 0.4, 0.2, -0.1, -0.9, -0.7)$	0.700
N-K	0	0.000
N-G	$(0.7, 0.5, 0.3, -0.1, -0.9, -0.2)$	0.650
N-Ch	$(0.6, 0.3, 0.2, -0.1, -0.9, -0.3)$	0.700
N-N	0	0.000

Table 9 Adjacency matrix

	C	M	V	K	G	Ch	N
C	0.000	0.533	0.450	0.000	0.500	0.500	0.567
M	0.500	0.000	0.517	0.483	0.533	0.500	0.000
V	0.467	0.467	0.000	0.000	0.517	0.000	0.600
K	0.000	0.467	0.000	0.000	0.000	0.500	0.000
G	0.517	0.517	0.500	0.000	0.000	0.500	0.650
Ch	0.517	0.500	0.000	0.517	0.500	0.000	0.683
N	0.000	0.000	0.700	0.000	0.650	0.700	0.000

From K, LNE is 0.500 which exists in Ch, i.e., from K it goes to Ch [C-V-M-K-Ch].

From Ch, LNE is 0.517 which exists in C, i.e., from Ch it goes to C.

Hence we get C-V-M-K-Ch-C, rejected as it is a bipolar single-valued neutrosophic cycle.

Iteration 2 Now choose the other LNE from AM.

It is $a_{31} = 0.467 \Rightarrow$ the row V reaches C $\Rightarrow$ V-C.

Table 10 Degree and score value of each vertex (city)

Vertices (cities)	Degree	Score value
C	$(2.1, 3.5, 1.2, -2.2, -1.7, -3.4)$	0.550
M	$(2.6, 3.4, 1.5, -2.4, -2.9, -2.0)$	0.553
V	$(1.0, 2.3, 1.0, -2.1, -1.9, -2.8)$	0.550
K	$(0.9, 1.0, 0.5, -1.0, -0.9, -0.5)$	0.467
G	$(2.7, 2.8, 1.5, -1.7, -2.4, -2.0)$	0.683
Ch	$(2.7, 2.5, 1.2, -2.0, -2.4, -1.9)$	0.717
N	$(1.6, 1.2, 0.7, -0.3, -2.7, -1.2)$	1.050

Therefore from C we need to find LNE.

From C, LNE is 0.500 which exists in G, i.e., from C it goes to G [V-C-G].

From G, LNE is 0.500 which exists in Ch, i.e., from G it goes to Ch [V-C-G-Ch].

From Ch, LNE is 0.500 which exists in M, i.e., from Ch it goes to M [C-V-M-K-Ch-M].

From M, LNE is 0.483 which exists in K, i.e., from M it goes to K.

Hence we get C-V-M-K-Ch-M-K.

It is rejected as it is a bipolar single-valued neutrosophic Hamiltonian path.

Iteration 3 Choose another LNE in AM.

In adjacency matrix, $a_{32} = 0.467 \Rightarrow$ the row V reaches M.

Therefore from M we need to find LNE.

From M, LNE is 0.483 which exists in K, i.e., from V it goes to K [V-M-K].

From K, LNE is 0.500 which exists in Ch, i.e., from K it goes to Ch [V-M-K-Ch].

From Ch, LNE is 0.517 which exists in C, i.e., from Ch it goes to C [V-M-K-Ch-C].

From C, LNE is 0.500 which exists in G, i.e., from C it goes to G [V-M-K-Ch-C-G].

From G, LNE is 0.650 which exists in N, i.e., from G it goes to N [V-M-K-Ch-C-G-N].

From N, LNE is 0.700 which exists in V, i.e., from N it goes to C [V-M-K-Ch-C-G-N-C].

It is a bipolar single-valued neutrosophic Hamiltonian cycle with length 3.817.

4.3 Algorithm to Find BSVNHC Using the Lowest Vertex Degree (LVD)

Here we proposed an algorithm to find BSVNHC using LVD and it is described by the following steps:

Step 1: Find the degrees of each vertex (city).

Step 2: Choose LVD (if repetitions exist, then select anyone).

Step 3: Choose the vertex with next higher degree of the LVD.

Step 4: Identify the adjacency vertex of the LVD.

Step 5: Visit the unvisited adjacent vertices of lowest degree (if repetitions exist, then select anyone).

Step 6: Repeat steps 4 and 5, until getting BSVNHC.

For the bipolar single-valued neutrosophic graph and adjacency matrix, we apply this algorithm:

Step 1: The degrees of the vertices are $d(C) = 0.550$, $d(M) = 0.553$, $d(V) = 0.550$, $d(K) = 0.467$, $d(G) = 0.683$, $d(Ch) = 0.717$, $d(N) = 1.050$

Step 2: Choose the vertex with the lowest degree. Here, $d(K) = 0.467$.

Step 3: Identify the adjacency vertices (AVs) of K.

AVs of K: M and Ch with $d(M) = 0.553$, $d(Ch) = 0.717$
Since the vertex M has the lowest degree, identify the AVs of M. [K-M]

Iteration 1 AVs of M: C, V, K, G, Ch.
Do not choose K; it would form a Hamiltonian cycle.
Hence, choose the vertex with next higher LVD, i.e., C   [K-M-C]

Iteration 2 AVs of C: M, V, K, G, Ch.
The next unvisited vertex with LVD is V.   [K-M-C-V]

Iteration 3 AVs of V: C, M, G, N.
The next unvisited vertex with LVD is G.   [K-M-C-V-G]

Iteration 4 AVs of G: C, M, V, Ch, N.
The next unvisited vertex with LVD is Ch.   [K-M-C-V-G-Ch]

Iteration 5 AVs of Ch: C, M, K, G, N.
The next unvisited vertex with LVD is N.   [K-M-C-V-G-Ch-N]
It is a bipolar single-valued neutrosophic Hamiltonian path.

Iteration 6 The vertices N and K are not connected and hence BSVNHC does not exist in this path.
So repeat the process with the other LVD of K, i.e., the vertex Ch (using step 3).
AVs of Ch: C, M, K, G, N.
Since the vertex K has LVD, find the AVs of it.   [Ch-K]

Iteration 1 AVs of K: M, Ch.
The next unvisited vertex with LVD is M.   [Ch-K-M]

Iteration 2 AVs of M: C, V, K, G, Ch.
The next unvisited vertex with LVD is V.   [Ch-K-M-V]

Iteration 3 AVs of V: C, M, G, N.
The next unvisited vertex with LVD is C.   [Ch-K-M-V-C]

Iteration 4 AVs of C: M, V, G, Ch.
The next unvisited vertex with LVD is G.   [Ch-K-M-V-C-G]

Table 11 Comparative analysis

Existing methods	Graph environment	Cycle length
[18]	Fuzzy set	0.385
[48]	Interval-valued neutrosophic set	4.149

Iteration 5 AVs of G: C, M, V, Ch, N.

The next unvisited vertex with LVD is N.   [Ch-K-M-V-C-G-N]

It is a bipolar single-valued neutrosophic Hamiltonian path.

Since the vertices N and Ch are connected, we get Ch-K-M-V-C-G-N-Ch.

It is a required BSVNHC with length 3.817.

5 Comparative Analysis and Discussion

In this section, comparative analysis is shown in Table 11 to prove the effectiveness of the proposed methods with the existing methods.

In [18], Hamiltonian cycle had been obtained with fuzzy edge weights using the new algorithms based on adjacency matrix and minimum vertex degree with the cycle length 0.385, and the issue of uncertainty can be done for the fuzzy graph where as in [48], Hamiltonian cycle was obtained with interval-valued neutrosophic edge weights, and the issue of dealing with indeterminacy for interval data is possible. But using this present work, Hamiltonian cycle can be obtained using the algorithms based on adjacency matrix, least vertex degree, and the modified score function of bipolar single-valued neutrosophic numbers. Using this proposed method, indeterminacy and issue of disorder can be dealt simultaneously which shows the effectiveness of the present work.

6 Conclusion

Hamiltonian graph plays a vital role in real-time problem such as operations research, computer graphics, mapping genomes, and electronic circuit design. As the real-time problems have indeterminacy and disorder, bipolar single-valued neutrosophic graph has been considered and obtained Hamiltonian cycle using the proposed algorithms based on adjacency matrix and the lowest vertex degree in this present paper. A modified form of the score function also has been proposed and utilized in the proposed algorithms. Also, comparative analysis has been done to manifest the effectiveness of the proposed algorithms. This proposed concept may be applied in the real-world problems which contains the issues of both indeterminacy and bipolarity. In the future, this work shall be extended to complex neutrosophic graph and Plithogenic graph.

References

1. Zadeh, A. (1965). Fuzzy sets. *Information and Control, 8*(**3**), 338–353.
2. Atanassov, K. (1986). Intuitionistic fuzzy sets. *Fuzzy Sets and Systems, 20*(**1**), 87–96.
3. Smarandache, F. (1998). Neutrosophy. Neutrosophic probability, set, and logic. In *ProQuest information & learning*, Ann Arbor, Michigan, USA (p. 105).
4. Zhang, W. R. (1994). Bipolar fuzzy sets and relations: a computational frame work for cognitive modelling and multiagent decision analysis, Proceedings of the First International Joint Conference of The North American Fuzzy Information Processing Society Biannual Conference. *The Industrial Fuzzy Control and Intelligence*, 305–309.
5. Rosenfeld, A. (1975). Fuzzy graphs. In L. A. Zadeh, K. S. Fu, & M. Shimura (Eds.), *Fuzzy sets and their applications* (pp. 77–95). New York: Academic Press.
6. Akram, M., & Davvaz, B. (2012). Strong intuitionistic fuzzy graphs. *Univerzitet u Nišu, 26*(**1**), 177–196.
7. Akram, M., Karunambigai, M. G., Palanivel, K., & Sivasankar, S. (2014). Balanced bipolar fuzzy graphs. *Journal of Advanced Research in Pure Mathematics, 6*(4), 58–71.
8. Samantha, S., & Pal, M. (**2012**). Irregular bipolar fuzzy graphs. *International Journal of Applications of Fuzzy Sets, 2*, 91–102.
9. Alshehri, N. O., & Akram, M. (2013). Cayley bipolar fuzzy graphs. *The Scientific World Journal, 2013*, 1–8.
10. Karunambigai, M. G., Akram, M., & Palanivel, K. (2013). *Domination in bipolar fuzzy graphs*. In 2013 FUZZ-IEEE International Conference on Fuzzy Systems, July 7–10, 2013, Hyderabad, India (pp. 1–6).
11. Rosline, J. J., & Pathinathan, T. (2014). Application of fuzzy graphs in scheduling jobs. *Integrated Intelligent Research, 3*(1), 34–36.
12. Rashmanlou, H., Samanta, S., Borzooei, R. A., & Pal, M. (2015a). A study on bipolar fuzzy graphs. *Journal of Intelligent Fuzzy Systems*, 1–31.
13. Rashmanlou, H., Samanta, S., Borzooei, R. A., & Pal, M. (2015b). Bipolar fuzzy graphs with categorical properties. *International Journal of Computational Intelligent Systems, 8*(5), 808–818.
14. Akram, M., Sarwar, M., & Smarandache, F. (2015). Bipolar neutrosophic graphs with applications. *The Scientific World Journal, 2014*, 1–23.
15. Delia, I., Ali, M., & Smarandache, F. (2015). Bipolar neutrosophic sets and their application based on multi-criteria decision making problems. In *Proceedings of the 2015 international conference on advanced mechatronics systems, Beijing, China*, August 22–24, 2015 (pp. 249–254).
16. Gani, A. N., & Latha, S. R. (2016). A new algorithm to find fuzzy Hamilton cycle in a fuzzy network using adjacency matrix and minimum vertex degree. *Springerplus, 5*, 1–10.
17. Akram, M., & Akmal, R. (**2016**). Operations on intuitionistic fuzzy graph structures. *Fuzzy information and Engineering, 8*, 389–410.
18. Deli, I., Subas, Y., Smarandache, F., & Ali, M. (2016). Interval valued bipolar neutrosophic sets and their application in pattern recognition. *The Scientific World Journal*, 1–8.
19. Sahin, M., Deli, I., & Ulucay, V. (2016). Jaccard vector similarity measure of bipolar neutrosophic set based on multi-criteria decision making. *International Conference on Natural Science Engineering (ICNASE'16)* March 19–20, 2016, Kilis. 1972–1979.
20. Pramanik, S., & Mondal, K. (2016). Rough bipolar neutrosophic set. *Global Journal of Engineering Science and Research Management, 3*(**6**), 71–81.
21. Ulucay, V., Deli, I., & Sahin, M. (2016). Similarity measures of bipolar neutrosophic sets and their application to multiple criteria decision making. *Neutral Computing and Applications*, 1–10. https://doi.org/10.1007/s00521-016-2479-1.
22. Broumi, S., Smarandache, F., Talea, M., & Bakali, A. (2016a). An introduction to bipolar single valued neutrosophic graph theory. *Applied Mechanics and Materials, 841*, 184–191.

23. Broumi, S., Smarandache, F., Talea, M., & Bakali, A. (2016b). Single valued neutrosophic graphs: Degree order and size. *Journal of Algorithms, 3*(**3**), 2444–2451.
24. Mathew, S., Mordeson, J. N., & Malik, D. S. (2017). Bipolar fuzzy graphs. *Studies in Fuzziness and Soft Computing, 363*, 271–306.
25. Gani, A. N., & Rahman, H. S. M. (2017). Truncations on special intuitionistic fuzzy graphs. *International Journal of Pure and Applied Mathematics, 117*(12), 257–263.
26. Broumi, S., Bakali, A., Talea, M.,& Smarandache, F. (2017). Complex neutrosophic graphs of type-I. In *IEEE international conference on Innovations in Intelligent SysTems and Applications (INISTA)* (pp. 432–437), Gdynia Maritime University, Gdynia, Poland, 3–5 July 2017.
27. Gani, A. N., Kavikumar, J., & Anupriya, S. (2017). Edge domination on intuitionistic fuzzy graphs. *International Journal of Applied Engineering Research, 12*(**17**), 6452–6461.
28. Prabhu, S., Devi, R. N., & Vidhya, D. (2017). Finite state machine via bipolar neutrosophic set theory. *The Journal of Fuzzy Mathematics, 25*(**4**), 865–884.
29. Akram, M., Sitara, M., & Smarandache, F. (2017). Graph structures in bipolar neutrosophic environment. *Mathematics, 5*(60), 1–20.
30. Girao, A., Kittipassorn, T., & Narayanan, B. (2017). Long cycles in hamiltonian graphs. *Israel Journal of Mathematics*, 1–5.
31. Mullai, M., Broumi, S., & Stephen, A. (2017). Shortest path problem by minimal spanning tree algorithm using bipolar neutrosophic numbers. *International Journal of Mathematics Trends and Technology, 46*(**2**), 79–87.
32. Broumi, S., Bakali, A., Talea, M., & Smarandache, F. (2018). Bipolar neutrosophic minimum spanning tree. In *The second international conference on smart applications and data analysis for smart cities* (pp. 1–6). 27–28 February 2018.
33. Akram, M., Siddique, S. M., & Shum, K. P. (2018a). Certain properties of bipolar neutrosophic graphs. *Southeast Asian Bulletin of Mathematics, 42*, 463–490.
34. Akram, M., Nasir, M., & Shum, K. P. (2018b). Novel applications of bipolar single-valued neutrosophic competition graphs. *Applied Mathematics-A Journal of Chinese Universities., 33*, 436–467.
35. Pramanik, S., Dey, P. P., Smarandache, F., & Ye, J. (2018). Cross entropy measures of bipolar and interval bipolar neutrosophic sets and their application for multi-attribute decision making. *Axioms, 7*(21), 1–25.
36. Aziz, A. N. A., Rad, N. J., Kamarulhaili, H., & Hasni, R. (2019). A note on k-step Hamiltonian graphs. *Malaysian Journal of Mathematical Sciences, 13*(1), 87–93.
37. Topal, S., Broumi, S., Bakali, A., Talea, M., & Smarandache, F. (2019). A python tool for implementations on bipolar neutrosophic matrices. *Neutrosophic Sets and Systems, 28*, 138–161.
38. Akram, M., Li, S. G., & Shum, K. P. (2019a). Antipodal bipolar fuzzy graphs. *Italian Journal of Pure and Applied Mathematics, 31*(2013), 97–110.
39. Akram, M., Ishfaq, N., Smarandache, F., & Broumi, S. (2019b). Application of bipolar neutrosophic sets to incidence graphs. *Neutrosophic Sets and Systems, 27*, 180–200.
40. Nagarajan, D., Lathamaheswari, M., Broumi, S., & Kavikumar, J. (2019a). Blockchain single and interval valued neutrosophic graphs. *Neutrosophic Sets And Systems, 24*, 23–35.
41. Nagarajan, D., Lathamaheswari, M., Broumi, S., & Kavikumar, J. (2019b). Dombi interval valued neutrosophic graph and its role in traffic control management. *Neutrosophic Sets and Systems, 24*, 114–133.
42. Akram, M., & Shum, K. P. (2017). Bipolar neutrosophic planar graphs. *Journal of Mathematical Research with Applications, 37*(6), 631–648.
43. Jan, N., Zedam, L., Tahir, M., & Kifayat, U. (2019). Cubic bipolar fuzzy graphs with applications. *Journal of Intelligent Fuzzy Systems, 37*(2), 2289–2307.
44. Zeps, D. (2019). Hamiltonian vertices and Hamiltonian edges in uniquely Hamiltonian graphs. *Combinatorial Mathematics*, 1–24. https://doi.org/10.13140/RG.2.2.12333.33765.

45. Hussain, S. S., Hussain, R. J., Jun, Y. B., & Smarandache, F. (2019). Neutrosophic bipolar vague set and its application to neutrosophic bipolar vague graphs. *Neutrosophic Sets and Systems, 28*, 69–86.
46. Goedebeur, J., Meersman, B., & Zamfirescu, C. T. (2020). Graphs with few Hamiltonian cycles. *Mathematics of Computation, 89*, 965–991.
47. Poulik, S., & Ghorai, G. (2020). Note on "Bipolar fuzzy graphs with applications". *Knowledge-Based Systems, 192*, 1–12. https://doi.org/10.1016/j.knosys.2019.105315.
48. Nagarajan, D., Lathamaheswari, M., Broumi, S., Smarandache, F., & Kavikumar, J. (2020). New algorithms for Hamiltonian cycle under interval neutrosophic environment. *Neutrosophic Graph Theory and Algorithms*, 107–130. https://doi.org/10.4018/978-1-7998-1313-2.ch004.
49. Asratian, A. S., Granholm, J. B., & Khachatryan, N. (2020). A localization method in Hamiltonian graph theory. *Journal of Combinatorial Theory Series*. https://doi.org/10.1016/j.jctb.2020.04.005.
50. Alnaser, A. M. A., AlZoubi, W. A., & Massadeh, M. O. (2020). Bipolar intuitionistic fuzzy graphs and its matrices. *Applied Mathematics and Information Sciences, 14*(2), 205–214.
51. Mohamed, S. Y., & Ali, A. M. (2020). Energy of spherical fuzzy graphs. *Advances in Mathematics: Scientific Journal, 9*(1), 321–332.

Improved Cosine Similarity Measures of Simplified Neutrosophic Sets for Medical Diagnoses: Suggested Modifications

Mohamed Abdel-Basset, Mai Mohamed, and Jun Ye

1 Introduction

Although there is amplified volume of information available in medical diagnosis, it contains a lot of incomplete, uncertainty, and inconsistent information. So, if physicians work efficiently with uncertainties and inconsistencies of information, they will be able to make precise decisions in medical diagnosis. The physicians usually make decisions in medical diagnosis problems based on the similarity among unknown samples and the basic diagnosis patterns.

Since similarity measures are significant tools in medical diagnosis, several research papers are presented under the neutrosophic environment for handling uncertainty and inconsistent information [1–4].

Ye [5] illustrated in his research several drawbacks existing in cosine similarity measures, which are defined in vector space [6]. These drawbacks include unmeaningful phenomena and unreasonable results. So, he said that these measures failed to apply in medical diagnosis problems and pattern recognition. For overcoming existing drawbacks, he suggested an improved cosine similarity measure for single and interval neutrosophic sets.

Yc [5] used his improved cosine similarity measures in solving medical diagnosis problems and has shown that the results obtained by his suggested measures of similarity are better than other existing measures and also overcame all existing drawbacks of other existing measures which are defined in vector space.

M. Abdel-Basset (✉) · M. Mohamed
Faculty of Computers and Informatics, Zagazig University, Zagazig, Sharqiyah, Egypt
e-mail: analyst_mohamed@zu.edu.eg; mmgfar@zu.edu.eg

J. Ye
School of Civil and Environmental Engineering, Ningbo University, Ningbo, P. R. China
e-mail: yejun1@nbu.edu.cn

After a deep study of Ye's suggested measures of similarity, it is observed that the suggested measures also have logical errors in some situations.

This research aims to prove the existing drawbacks of improved cosine similarity measures presented by Ye and also suggest new improved cosine similarity measures for single- and interval-valued neutrosophic sets, which are able to handle all existing drawbacks.

2 Drawbacks of Ye's Measures of Similarity

In this section, the drawbacks of Ye's measures of similarity are discussed.

2.1 Drawbacks of Improved Cosine Similarity Measures for Single-Valued Neutrosophic Sets

As we know it, a single-valued neutrosophic set A over X is an object having the form $A = \{\langle x, T_A(x), I_A(x), F_A(x)\rangle : x \in X\}$, where $T_A(x):X \to [0,1]$, $I_A(x):X \to [0,1]$, and $F_A(x):X \to [0,1]$ with $0 \leq TA(x) + IA(x) + FA(x) \leq 3$ for all $x \in X$. (x), $I(x)$, and $F_A(x)$ denote the truth membership degree, the indeterminacy membership degree, and the falsity membership degree of x to A, respectively [7].

Based on the cosine function, Ye proposed two improved cosine similarity measures among single-valued neutrosophic sets (SVNS) and presented their properties.

Let $B = \{\langle x_j, T_B(x_j), I_B(x_j), F_B(x_j)\rangle \mid x_j \in X\}$ and $C = \{\langle x_j, T_C(x_j), I_C(x_j), F_C(x_j)\rangle \mid x_j \in X\}$ be two single-valued neutrosophic sets in $X = \{x_1, x_2, \ldots, x_n\}$, where $T_B(x_j), I_B(x_j), F_B(x_j) \in [0, 1]$ for any $x_j \in X$ in B and $T_C(x_j), I_C(x_j), F_C(x_j) \in [0, 1]$ for any $x_j \in X$ in C. Then, based on cosine function, Ye proposed two improved cosine similarity measures among B and C, respectively, as follows:

$$SC_1(B, C)$$
$$= \frac{1}{n} \sum_{j=1}^{n} \cos \left[\frac{\pi \left(|T_B(x_j) - T_c(x_j)| \vee |I_B(x_j) - I_c(x_j)| \vee |F_B(x_j) - F_c(x_j)| \right)}{2} \right],$$
$$(1)$$

$$SC_2(B, C)$$
$$= \frac{1}{n} \sum_{j=1}^{n} \cos \left[\frac{\pi \left(|T_B(x_j) - T_c(x_j)| + |I_B(x_j) - I_c(x_j)| + |F_B(x_j) - F_c(x_j)| \right)}{6} \right],$$
$$(2)$$

where $\vee$ is the maximum operation.

Ye defined four propositions that the two improved cosine similarity measures must satisfy as follows:

1. $0 \le SC_k(B, C) \le 1$, where $k = 1, 2$.
2. $SC_k(B, C) = 1$ if and only if $B = C$.
3. $SC_k(B, C) = SC_k(C, B)$.
4. If D is a SVNS in X and $B \subseteq C \subseteq D$, then $SC_k(B, D) \le SC_k(B, C)$ and $SC_k(B, D) \le SC_k(C, D)$.

This example illustrates the drawbacks of proposed measures by Ye:

Let $B = \{<x, 1, 0, 0>| x \in X\}$ and $C = \{<x, 0, 0, 0>| x \in X\}$; by applying Eqs. (1) and (2), then $Sc_1(B, C) = 0$. However at the same case, the $SC_2(B, C) = 0.8660$. The low and high similarity value of $SC_1(B, C)$ and $SC_2(B, C)$ produces an unreasonable phenomenon for the similarity measures among B and C. So, they are not appropriate to handle medical diagnosis problems.

2.2 Drawbacks of Improved Cosine Similarity Measures for Interval-Valued Neutrosophic Sets

Similarly, Ye proposed two improved cosine similarity measures among interval-valued neutrosophic sets and presented their properties.

As we know it, the interval-valued neutrosophic set A in X is described by truth $T_A(x)$, indeterminacy $I_A(x)$, and falsity $F_A(x)$ membership degrees, where $T_A(x) = \left[T_A^L(x), T_A^U(x) \subseteq [0, 1]\right]$, $I_A(x) = \left[I_A^L(x), I_A^U(x) \subseteq [0, 1]\right]$, and $F_A(x) = \left[F_A^L(x), F_A^U(x) \subseteq [0, 1]\right]$. Then, we can write interval-valued neutrosophic set as $A = \left\{< x, \left[T_A^L(x), T_A^U(x)\right], \left[I_A^L(x), I_A^U(x)\right], \left[F_A^L(x), F_A^U(x)\right] > |x \in X\right\}$.

Let $B = \{\langle x_j, T_B(x_j), I_B(x_j), F_B(x_j)\rangle| x_j \in X\}$ and $C = \{\langle x_j, T_C(x_j), I_C(x_j), F_C(x_j)\rangle | x_j \in X\}$ be two interval-valued neutrosophic sets in $X = \{x_1, x_2, \ldots, x_n\}$, where $T_B(x_j) = \left[T_B^L(x_j), T_B^U(x_j)\right]$, $I_B(x_j) = \left[I_B^L(x_j), I_B^U(x_j)\right]$, $F_B(x_j) = \left[F_B^L(x_j), F_B^U(x_j)\right] \subseteq [0, 1]$ for any $x_j \in X$ in B and $T_C(x_j) = \left[T_C^L(x_j), T_C^U(x_j)\right]$, $I_C(x_j) = \left[I_C^L(x_j), I_C^U(x_j)\right]$, $F_C(x_j) = \left[F_C^L(x_j), F_C^U(x_j)\right] \subseteq [0, 1]$ for any $x_j \in X$ in C. Then, based on cosine function, Ye proposed two improved cosine similarity measures among B and C, respectively, as follows:

$$SC_3(B, C)$$
$$= \frac{1}{n} \sum_{j=1}^{n} \cos \left[\begin{array}{c} \frac{\pi}{4} \left(|T_B^L(x_j) - T_C^L(x_j)| \vee |I_B^L(x_j) - I_C^L(x_j)| \vee |F_B^L(x_j) - F_C^L(x_j)| \\ + |T_B^U(x_j) - T_C^U(x_j)| \vee |I_B^U(x_j) - I_C^U(x_j)| \vee |F_B^U(x_j) - F_C^U(x_j)| \right) \end{array} \right]$$
$$(3)$$

$SC_4(B, C)$

$$= \frac{1}{n} \sum_{j=1}^{n} \cos \left[\begin{array}{c} \frac{\pi}{12} \left(\mid T_B^L(x_j) - T_C^L(x_j) \mid + \mid I_B^L(x_j) - I_C^L(x_j) \mid + \mid F_B^L(x_j) \\ - F_C^L(x_j) \mid + \mid T_B^U(x_j) - T_C^U(x_j) \mid + \mid I_B^U(x_j) - I_C^U(x_j) \mid + \mid F_B^U(x_j) - F_C^U(x_j) \mid \right) \end{array} \right]$$

(4)

where $\vee$ is the maximum operation.

By putting $B = \{<x, [1, 1], [0, 0], [0, 0]> \mid x \in X\}$ and $C = \{<x, [0, 0], [0, 0], [0, 0]> \mid x \in X\}$, by applying Eqs. (3) and (4), then $SC_3(B, C) = 0$, and $SC_4(B, C) = 0.8660$. The low and high similarity value of $SC_3(B, C)$ and $SC_4(B, C)$ produces an unreasonable phenomenon for similarity measures among B and C. So, they are not appropriate to handle medical diagnosis problems.

3   Suggested Modifications

Ye [5] claimed that cosine similarity measures have been defined in vector space [6] and have some drawbacks in some situations, and his suggested measures have handled these drawbacks and are superior to existing measures. But, the improved cosine similarity measures proposed by Ye also produced unreasonable results in some situations as we illustrated in the previous section. So it fails to apply in medical domain and pattern recognition.

To handle all existing drawbacks of existing similarity measures [6] plus the unreasonable results obtained by Ye [5] in some situations, we proposed enhanced cosine similarity measures for single- and interval-valued neutrosophic sets and proved their properties.

3.1   Suggested Cosine Similarity Measures for Single-Valued Neutrosophic Sets

Improved cosine similarity measures presented by Ye have unreasonable results in some situations, as we illustrated previously. To improve presented measures and make them produce reasonable and harmonious results, we will improve the first similarity measure (i.e., Eq. (1)) and keep the second as it was (i.e., Eq. (2)). The first similarity measure in case $B = \{<x, 1, 0, 0> \mid x \in X\}$ and $C = \{<x, 0, 1, 1> \mid x \in X\}$ will produce zero. Also in case $B = \{<x, 1, 0, 0> \mid x \in X\}$ and $C = \{<x, 0, 0, 0> \mid x \in X\}$, the first similarity measure will produce zero, although two sets are similar in indeterminacy and falsity degrees. There is the same result in case $B = \{<x, 0, 1, 1> \mid x \in X\}$ and $C = \{<x, 0, 1, 0> \mid x \in X\}$. So, in case $B = \{<x, 1, 0, 0> \mid x \in X\}$ and $C = \{<x, 0, 0, 0> \mid x \in X\}$, there exists a large gap between Ye's first and second improved cosine similarity measures. The same thing also appears in case $B = \{<x, 0, 1, 1> \mid x \in X\}$ and $C = \{<x, 0, 1, 0> \mid x \in X\}$. The main reason for these results is that in the first similarity measure of Ye, he depended only on one maximum value of $|T_B(x_j) - T_c(x_j)|$, $|I_B(x_j) - I_c(x_j)|$, $|F_B(x_j) - F_c(x_j)|$. We can note these problems clearly in binary values (i.e., zero and one) of sets.

So, we enhanced only the first similarity measure (i.e., Eq. (1)) and named it $NewSC_1(B, C)$ to handle all situations precisely and remove the large gap between first and second similarity measures of Ye in some cases as follows:

$$NewSC_1(B, C) = \text{Cos}\left(\frac{\pi}{2} \times \left(\frac{M_1 + M_2}{2}\right)\right) \tag{5}$$

Since

$$M_1 = \text{Max}\left(\left|T_B(x_j) - T_c(x_j)\right|, \left|I_B(x_j) - I_c(x_j)\right|, \left| F_B(x_j) - F_c(x_j) \right|\right),$$

$$M_2 = \text{Min}\left(\left|T_B(x_j) - T_c(x_j)\right|, \left|I_B(x_j) - I_c(x_j)\right|, \left|F_B(x_j) - F_c(x_j)\right|\right).$$

Our enhanced similarity measure also satisfies the following properties:

(a) $0 \le NewSC_1(B, C) \le 1$.

Proof

For any single-valued neutrosophic set, $T(x)$, $I(x)$, and $F(x)$ lies within [0,1]. So, the values of the suggested cosine function are within [0,1].

(b) $NewSC_1(B, C) = NewSC_1(C, B)$.

Proof

Since $|T_B(x_j) - T_C(x_j)| = |T_C(x_j) - T_B(x_j)|, |I_B(x_j) - I_C(x_j)| = |I_C(x_j) - I_B(x_j)|, |F_B(x_j) - F_C(x_j)| = |F_C(x_j) - F_B(x_j)|$.

Thus, $NewSC_1(B, C) = NewSC_1(C, B)$.

(c) $NewSC_1(B, C) = 1$, if and only if $B = C$.

Proof

For any single-valued neutrosophic sets B and C, when $B - C$, then $T_B(x_j) = T_C(x_j), I_B(x_j) = I_C(x_j), F_B(x_j) = F_C(x_j)$.

Also, $|T_B(x_j) - T_C(x_j)| = 0, |I_B(x_j) - I_C(x_j)| = 0, |F_B(x_j) - F_C(x_j)| = 0$, and $\cos(0) = 1$.

Thus, $NewSC_1(B, C) = 1$.

Conversely

$NewSC_1(B, C) = 1$; then, $|T_B(x_j) - T_C(x_j)| = 0, |I_B(x_j) - I_C(x_j)| = 0, |F_B(x_j) - F_C(x_j)| = 0$.

Also,

$T_B(x_j) = T_C(x_j), I_B(x_j) = I_C(x_j), F_B(x_j) = F_C(x_j)$.

Then, $B = C$.

Our proposed cosine similarity measure (Eq. (5)) will be more consistent and logical than Ye's first similarity measure (i.e., Eq. (1)). There doesn't exist a large gap between our proposed measure (Eq. (5)) and the second measure of Ye (i.e., Eq. (2)) as appears in Table 1.

From Table 1 the low and high similarity value of $SC_1(A, B)$ and $SC_2(A, B)$ in case 5 and case 6 produces an unreasonable phenomenon for the similarity measures between A and B. This will get the decision-maker into trouble in practical applications, especially in the medical domain. After handling logical drawback of $SC_1(A, B)$ by suggesting a $NewSC_1(A, B)$, we noted that $NewSC_1(A, B)$ and $SC_2(A, B)$ have strong harmony among them and produced reasonable results in all cases.

3.2 Suggested Cosine Similarity Measures for Interval-Valued Neutrosophic Sets

The previous problems which appeared in Ye similarity measures of single-valued neutrosophic sets also appeared in the interval-valued neutrosophic set. By putting $B = \{<x, [1, 1], [0, 0], [0, 0]> | x \in X\}$ and $C = \{<x, [0, 0], [0, 0], [0, 0]> | x \in X\}$ and applying Eqs. (3) and (4), then $SC_3(B, C) = 0$, and $Sc_4(B, C) = 0.866$. Also, by putting $B = \{<x, [0, 0], [1, 1], [1, 1]> | x \in X\}$ and $C = \{<x, [0, 0], [1, 1], [0, 0]> | x \in X\}$ and applying Eqs. (3) and (4), then $SC_3(B, C) = 0$, and $SC_4(B, C) = 0.8660$. The low and high similarity value of $SC_3(B, C)$ and $SC_4(B, C)$ produces unreasonable phenomenon for the similarity measures among B and C. So, they are not appropriate to handle medical diagnosis problems. The main reason for these unreasonable results is that in $SC_3(B, C)$ (i.e., Eq. (3)) of Ye, he depended only on one maximum value of $| T_B^L(x_j) - T_C^L(x_j) |, | I_B^L(x_j) - I_C^L(x_j) |, | F_B^L(x_j) - F_C^L(x_j) |,|$ $T_B^U(x_j) - T_C^U(x_j) |, | I_B^U(x_j) - I_C^U(x_j) |, | F_B^U(x_j) - F_C^U(x_j) |.$

For making precise harmony between two measures of similarity (i.e., $SC_3(B, C)$ and $SC_4(B, C)$), we suggested a modified version of $SC_3(B, C)$ and named it $NewSC_3(B, C)$:

$$NewSC_3(B, C) = \mathrm{Cos}\left(\frac{\pi}{2} \times \left(\frac{M_1 + M_2}{2}\right)\right) \tag{6}$$

Since

$$M_1 = \mathrm{Max}\left(\begin{array}{c} | T_B^L(x_j) - T_C^L(x_j) |, | I_B^L(x_j) - I_C^L(x_j) |, | F_B^L(x_j) - F_C^L(x_j) |, | T_B^U(x_j) \\ - T_C^U(x_j) |, | I_B^U(x_j) - I_C^U(x_j) |, | F_B^U(x_j) - F_C^U(x_j) | \end{array}\right),$$

$$M_2 = \mathrm{Min}\left(\begin{array}{c} | T_B^L(x_j) - T_C^L(x_j) |, | I_B^L(x_j) - I_C^L(x_j) |, | F_B^L(x_j) - F_C^L(x_j) |, | T_B^U(x_j) \\ - T_C^U(x_j) |, | I_B^U(x_j) - I_C^U(x_j) |, | F_B^U(x_j) - F_C^U(x_j) | \end{array}\right).$$

Our enhanced similarity measure also satisfies the following properties:

(a) $0 \le NewSC_3(B, C) \le 1$.

Table 1 Similarity measure values of Eqs. (1), (2), (5)

	Case 1	Case 2	Case 3	Case 4	Case 5	Case 6
A	<x, 0.2,0.3,0.4>	<x, 0.3,0.2,0.4>	<x, 0.4,0.2,0.6>	<x, 1, 0, 0>	<x, 1, 0, 0>	<x, 0, 1, 1>
B	<x, 0.2,0.3,0.4>	<x, 0.4,0.2,0.3>	<x, 0.2,0.1,0.3>	<x, 0, 1, 1>	<x, 0, 0, 0>	<x, 0, 1, 0>
$SC_1(A,B)$	1	0.9877	0.8910	0	0	0
$SC_2(A,B)$	1	0.9945	0.9511	0	0.8660	0.8660
$NewSC_1(A,B)$	1	0.9969	0.9510	0	0.7071	0.7071

Proof

For any interval-valued neutrosophic set, $T^L(x_j)$, $T^U(x_j)$, $I^L(x_j)$, $I^U(x_j)$, and $F^L(x_j), F^U(x_j)$ lie within $[0,1]$. So, the values of the suggested cosine function are within $[0,1]$.

(b) $NewSC_3(B, C) = NewSC_3(C, B)$.

Proof

Since $| T_B^L(x_j) - T_C^L(x_j) | = |T_C^L(x_j) - T_B^L(x_j)|$, $| I_B^L(x_j) - I_C^L(x_j) | = |I_C^L(x_j) - I_B^L(x_j)|$, $| F_B^L(x_j) - F_C^L(x_j) | = |F_C^L(x_j) - F_B^L(x_j)|$, $| T_B^U(x_j) - T_C^U(x_j) | = | T_C^U(x_j) - T_B^U(x_j) | = | I_B^U(x_j) - I_C^U(x_j) | = |I_C^U(x_j) - I_B^U(x_j)|$, $| F_B^U(x_j) - F_C^U(x_j) | = |F_C^U(x_j) - F_B^U(x_j)|$.

Thus, $NewSC_3(B, C) = NewSC_3(C, B)$.

(c) $NewSC_3(B, C) = 1$, if and only if $B = C$.

Proof

For any interval-valued neutrosophic sets *B* and *C*, when $B = C$, then
$T_B^L(x_j) = T_C^L(x_j)$, $I_B^L(x_j) = I_C^L(x_j)$, $F_B^L(x_j) = F_C^L(x_j)$, $T_B^U(x_j) = T_C^U(x_j)$, $I_B^U(x_j) = I_C^U(x_j)$, $F_B^U(x_j) = F_C^U(x_j)$.

Also, $| T_B^L(x_j) - T_C^L(x_j) | = 0, |T_C^L(x_j) - T_B^L(x_j)| = 0, | I_B^L(x_j) - I_C^L(x_j) | = 0, |I_C^L(x_j) - I_B^L(x_j)| = 0, | F_B^L(x_j) - F_C^L(x_j) | = 0, |F_C^L(x_j) - F_B^L(x_j)| = 0, | T_B^U(x_j) - T_C^U(x_j) | = 0, | T_C^U(x_j) - T_B^U(x_j) | = 0, | I_B^U(x_j) - I_C^U(x_j) | = 0, |I_C^U(x_j) - I_B^U(x_j)| = 0, | F_B^U(x_j) - F_C^U(x_j) | = 0, |F_C^U(x_j) - F_B^U(x_j)| = 0$.

and $\cos(0) = 1$.

Thus, $NewSC_3(B, C) = 1$.

Conversely

$NewSC_3(B, C) = 1$; then, $| T_B^L(x_j) - T_C^L(x_j) | = 0, |T_C^L(x_j) - T_B^L(x_j)| = 0, | I_B^L(x_j) - I_C^L(x_j) | = 0, |I_C^L(x_j) - I_B^L(x_j)| = 0, | F_B^L(x_j) - F_C^L(x_j) | = 0, |F_C^L(x_j) - F_B^L(x_j)| = 0, | T_B^U(x_j) - T_C^U(x_j) | = 0, | T_C^U(x_j) - T_B^U(x_j) | = 0, | I_B^U(x_j) - I_C^U(x_j) | = 0, |I_C^U(x_j) - I_B^U(x_j)| = 0, | F_B^U(x_j) - F_C^U(x_j) | = 0, |F_C^U(x_j) - F_B^U(x_j)| = 0$. Also, $T_B^L(x_j) = T_C^L(x_j)$, $I_B^L(x_j) = I_C^L(x_j)$, $F_B^L(x_j) = F_C^L(x_j)$, $T_B^U(x_j) = T_C^U(x_j)$, $I_B^U(x_j) = I_C^U(x_j)$, $F_B^U(x_j) = F_C^U(x_j)$.

Our modified similarity measure for interval-valued neutrosophic sets (i.e., Eq. (6)) produces consistent and harmony results, as appears in Table 2.

From Table 2 the low and high similarity value of $SC_3(A, B)$ and $SC_4(A, B)$ in case 4 and case 5 produces an unreasonable phenomenon for the similarity measures between *A* and *B*. This will get the decision-maker into trouble in practical applications, especially in the medical domain. After handling logical drawback of $SC_3(A, B)$ through proposing $NewSC_3(A, B)$, we noted that $NewSC_3(A, B)$ and $SC_2(A, B)$ have strong harmony among them and produced reasonable results in all cases.

Table 2 Similarity measure values of Eqs. (3), (4), (6)

	Case 1	Case 2	Case 3	Case 4	Case 5
A	<x, [0.3,0.5], [0.2,0.4],[0,0.1]>	<x, [0.3,0.5], [0.2,0.4],[0.4,0.5]>	<x, [1, 1], [0, 0],[0, 0]>	<x, [1, 1], [0, 0],[0, 0]>	<x, [0, 0], [1, 1],[1, 1]>
B	<x, [0.3,0.5], [0.2,0.4],[0,0.1]>	<x, [0.4,0.5], [0.2,0.4],[0.3,0.5]>	<x, [0, 0], [1, 1],[1, 1]>	<x, [0, 0], [0, 0],[0, 0]>	<x, [0, 0], [1, 1],[0, 0]>
$SC_3(A,B)$	1	0.9969	0	0	0
$SC_4(A,B)$	1	0.9986	0	0.8660	0.8660
$NewSC_3(A,B)$	1	0.9569	0	0.7071	0.7071

4   Conclusions

The logical error and unreasonable results, used in Ye's similarity measures, are pointed out. Also, the required changes of Ye's similarity measures for both single- and interval-valued neutrosophic sets are presented. Our suggested measures produced reasonable results in all cases and deleted the large gap between Ye's measures of similarity. So, using the modified version of proposed measures for single- and interval-valued neutrosophic sets will produce precise decisions and will not get the decision-maker into trouble in practical applications, especially in the medical domain.

Acknowledgments

   Conflict of Interest: Authors declare that there is no conflict of interest about the research.
   Funding: This research has no funding source.
   Ethical Approval: This article does not contain any studies with human participants or animals performed by any of the authors.

References

1. Broumi, S., & Smarandache, F. (2013). Several similarity measures of neutrosophic sets. *Neutrosophic Sets and Systems, 1*(1), 54–62.
2. Majumdar, P., & Samanta, S. K. (2014). On similarity and entropy of neutrosophic sets. *Journal of Intelligent Fuzzy Systems, 26*(3), 1245–1252.
3. Ye, J. (2014). Similarity measures between interval neutrosophic sets and their applications in multicriteria decision-making. *Journal of Intelligent Fuzzy Systems, 26*, 165–172.
4. Ye, J. (2014). Multiple attribute group decision-making method with completely unknown weights based on similarity measures under single valued neutrosophic environment. *Journal of Intelligent Fuzzy Systems, 27*, 2927–2935.
5. Ye, J. (2015). Improved cosine similarity measures of simplified neutrosophic sets for medical diagnoses. *Artificial Intelligence in Medicine, 63*(3), 171–179.
6. Ye, J. (2014). Vector similarity measures of simplified neutrosophic sets and their application in multicriteria decision making. *International Journal of Fuzzy Systems, 16*(2), 204–211.
7. Smarandache, F. (1999). *A unifying field in logics. Neutrosophy: Neutrosophic probability, set and logic*. Rehoboth: American Research Press.

Fuzzy Multi-Objective Inventory Model for Deteriorating Items, with Shortages Under Space Constraint: Neutrosophic Hesitant Fuzzy Programming Approach

Sahidul Islam

1 Introduction

In an inventory model, deterioration plays a significant role. Deterioration items are characterized as those items which are damaged, decayed, or expired and not in a condition to convince customers or market demands. Foods, drugs, fruits, etc. are deteriorating items. Shortage in inventory model is another important factor. There are several types of customers in real-life inventory system. At shortage period, some of the customers can wait for genuine product but others do not. For this circumstance, we consider partially backlogging. For different types of items, demand is also different. It also varies for different stage of time. So we cannot always take demand as constant. In inventory research the relation between demand and time is considered by several authors.

Economic order quantity model was first introduced in February 1913 by Harris [1]; afterward many researchers developed EOQ model in inventory systems like Singh et al. [2], Jong and Lee [3] etc. Deterioration of an item is the most important factor in the inventory systems. Ghare and Schrader [4] developed the inventory model by considering the constant demand rate and constant deterioration rate. Jong et al. [5] developed an EOQ inventory model with time-varying demand and Weibull deterioration with shortages. Mishra [6] presented a paper on an inventory model for Weibull deterioration with stock- and price-dependent demand. Jong and Lee [3] discussed an EOQ inventory model for items with Weibull deterioration, shortages, and time-varying demand. Singh et al. [2] studied an EOQ inventory model for deteriorating items with time-dependent deterioration rate, ramp-type demand rate, and shortages. Roy et al. [7] discussed inventory model of deteriorating items with a

S. Islam (✉)

Department of Mathematics, University of Kalyani, Nadia, West Bengal, India

e-mail: sahidul.math@gmail.com; sahidulmath18@klyuniv.ac.in

© The Author(s), under exclusive license to Springer Nature Switzerland AG 2021

F. Smarandache, M. Abdel-Basset (eds.), *Neutrosophic Operational Research*,

https://doi.org/10.1007/978-3-030-57197-9_11

constraint: a geometric programming approach. The concept of fuzzy set theory was first introduced by Zadeh [8]. Afterward Zimmermann [9, 10] applied the fuzzy set theory concept with some useful membership functions to solve the linear programming problem with some objective functions. Then the various ordinary inventory models transformed to fuzzy versions model by various authors such as Roy and Maity [11] presented on a fuzzy inventory model with constraints. Also Smarandache introduced the neutrosophic set. Smarandache [12] presented neutrosophic set, a generalization of the intuitionistic fuzzy set, and also discussed a geometric interpretation of the NS set, a generalization of the intuitionistic fuzzy set. Multi-item and limitations of spaces are important in the business world. Ye [13] studied on multiple-attribute decision-making method under a single-valued neutrosophic hesitant fuzzy environment. Ahmad et al. [14] established single-valued neutrosophic hesitant fuzzy computational algorithm for multi-objective nonlinear optimization problem. Islam and Mandal [15] considered a fuzzy inventory model with unit production cost and time-dependent holding cost, without shortages under a space constraint: a parametric geometric programming approach. Mondal et al. [16] and Nabeeh et al. [17] studied on optimization of EOQ model with limited storage capacity by neutrosophic geometric programming. Mullai and Surya [18] developed neutrosophic EOQ model with price break. Mohana et al. [19] discussed on multi-criteria decision-making problem via bipolar single-valued neutrosophic. Nabeeh et al. discussed neutrosophic multi-criteria decision-making approach for IOT-based enterprises. Biswas et al. [20] presented multi-attribute group decisionmaking based on expected value of neutrosophic trapezoidal numbers. Pramanik et al. [21] discussed contributions of selected Indian researchers to multi-attribute decision-making in neutrosophic environment: an overview. Abdel-Basset et al. [22] discussed the evaluation framework for smart disaster response systems in uncertainty environment. Abdel-Basset et al. [23] used neutrosophic logic to assess uncertainty of linear time-cost tradeoffs. Abdel-Basset et al. [24] analyzed resource levelling problem in construction projects under neutrosophic environment. AbdelBasset et al. [25] considered a bipolar neutrosophic multi-criteria decision-making framework for professional selection. Abdel-Basset et al. [26] solved the supply chain problem using the best-worst method based on a novel Plithogenic model. Ahmad et al. [14] introduced single-valued neutrosophic hesitant fuzzy computational algorithm for multi-objective nonlinear optimization problem. S. Pramanik presented neutrosophic multi-objective linear programming in 2016 [27].

2 Mathematical Model

2.1 Notations

c_{1i}: Ordering cost per order for ith item

$H_i e^{\alpha_i t}$: Holding cost per unit per unit time for ith item, where $0 < \alpha < 1$ and H_i is constant

c_{2i}: Storage cost for backlogged items per unit per unit time for ith item

c_{3i}: Cost of lost sales per unit for ith item

c_{4i}: Deteriorating cost per unit per unit time for ith item

T_i: The length of cycle time for ith item, $T_i > 0$

t_{1i}: The time when the inventory level starts to reduce due to both demand and deterioration *for* ith item, $t_{1i} > 0$

t_{2i}: The time when the inventory level attains zero for ith item, $t_{2i} > 0$

MI: The maximum inventory level at $t = 0$

MB: The maximum backordered units during stock-out period for ith item

$Q_i(MI + MB)$: The order quantity for the duration of a cycle of length T_i for ith item

$TAC_i(t_{1i}, t_{2i}, T_i)$: Total average cost per unit for ith item

w_i: Storage space per unit time for ith item

W: Total area of space

$\tilde{c}_{1i}$: Fuzzy ordering cost per order for ith item

$\tilde{H}_i e^{\alpha_i t}$: Fuzzy holding cost per unit per unit time for ith item

$\tilde{c}_{2i}$: Fuzzy storage cost for backlogged items per unit per unit time for ith item

$\tilde{c}_{3i}$: Fuzzy cost of lost sales per unit for ith item

$\tilde{c}_{4i}$: Fuzzy deteriorating cost per unit per unit time for ith item

$TAC_i(t_{1i}, t_{2i}, T_i)$: Fuzzy total average cost per unit for ith item

2.2   Assumptions

1. The inventory system involves multi-item.
2. The replenishment occurs instantaneously at infinite rate.
3. The lead time is negligible.
4. Deterioration takes place after lifetime of items.
5. Demand rate is time dependent. It is taken as $D(t) = a_i + b_i t$, where $a_i, b_i > 0$.
6. Deterioration rate per unit time per cycle is θ_i for ith item, $0 \leq \theta_i \leq 1$; θ_i is constant.
7. Backlogging rate is $\frac{1}{1+\delta_i(T_i-t)}$ where $t_{2i} \leq t \leq T_i$ and δ_i is the positive backlogging parameter.

2.3   Model Formation

The inventory model for ith item is illustrated in Fig. 1. During the period $[0, t_{1i}]$, the inventory level reduces only because of demand rate for ith item. In this time period, the inventory level is described by the differential equation:

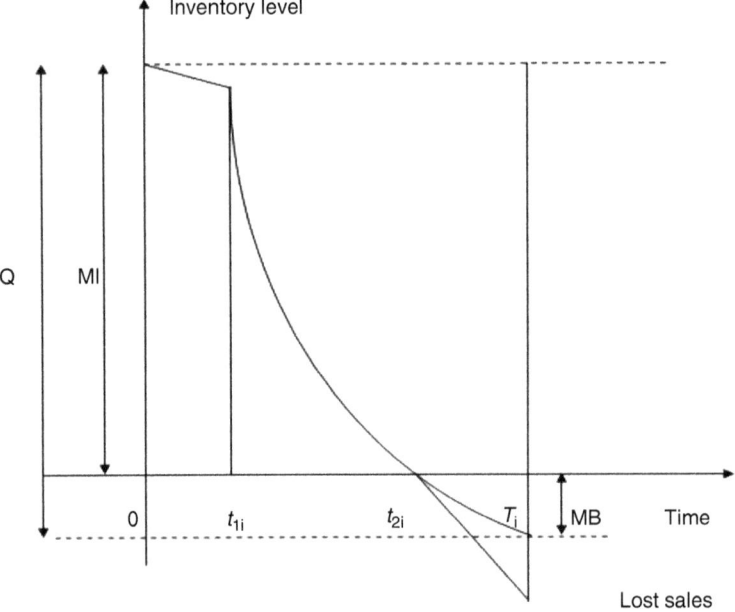

Fig. 1 Inventory model for ith item

$$\frac{dI_1(t)}{dt} = -D(t) = -(a_i + b_i t), 0 \leq t \leq t_{1i} \tag{1}$$

With boundary condition, $I_1(0) = MI$.
Solving (1) we have

$$I_1(t) = -a_i t - \frac{b_i t^2}{2} + MI \tag{2}$$

Now during the period $[t_{1i}, t_{2i}]$, the inventory level decreases due to both demand rate and deterioration for ith item. In this time period, the inventory level is described by the differential equation:

$$\frac{dI_2(t)}{dt} + \theta_i I_2(t) = -(a_i + b_i t), t_{1i} \leq t \leq t_{2i} \tag{3}$$

With boundary condition, $I_2(t_{2i}) = 0$.
Solving (3) we have

$$I_2(t) = a_i (t_{2i} - t) - \frac{a_i \theta_i}{2} \left(t_{2i}^2 - t^2 \right) + \frac{b_i}{2} \left(t_{2i}^2 - t^2 \right) \tag{4}$$

Now at $t = t_{1i}$, we have $I_1(t_{1i}) = I_2(t_{1i})$.
Hence we have

$$MI = \frac{b_i t_{1i}^2}{2} + a_i t_{2i} - \frac{a_i \theta_i}{2} \left(t_{2i}^2 - t_{1i}^2 \right) + \frac{b_i}{2} \left(t_{2i}^2 - t_{1i}^2 \right). \tag{5}$$

Now during the period $[t_{2i}, T_i]$, shortages occur and the demand is partially backlogged. In this time period, the inventory level is described by the differential equation-

$$\frac{dI_3(t)}{dt} = -\frac{(a_i + b_i t)}{1 + \delta_i (T_i - t)}, t_{2i} \le t \le T_i \tag{6}$$

With boundary condition, $I_3(t_{2i}) = 0$.
Solving (6) and neglecting higher power of δ_i we have

$$I_3(t) = (t_{2i} - t) \left\{ a_i + 2b_i T_i + a_i \delta_i T_i + b_i \delta_i T_i^2 \right\}$$
$$+ \left(t^2 - t_{2i}^2 \right) \cdot \left\{ \frac{b_i}{2} + \frac{a_i \delta_i}{2} + \frac{b_i \delta_i T_i}{2} \right\} \tag{7}$$

Now, $-I_3(T_i) = MB$.
So, neglecting higher power of T_i, we have

$$MB = -(t_{2i} - T_i) \{ a_i + 2b_i T_i + a_i \delta_i T_i \} - \left(T_i^2 - t_{2i}^2 \right) \left\{ \frac{b_i}{2} + \frac{a_i \delta_i}{2} \right\} \tag{8}$$

Hence, $Q_i = MI + MB$

$$= \frac{b_i t_{1i}^2}{2} + a_l t_{2i} - \frac{a_i \theta_i}{2} \left(t_{2i}^2 - t_{1i}^2 \right) + \frac{b_i}{2} \left(t_{2i}^2 - t_{1i}^2 \right) - (t_{2i} - T_i)$$
$$\times \{ a_i + 2b_i T_i + a_i \delta_i T_i \} - \left(T_i^2 - t_{2i}^2 \right) \left\{ \frac{b_i}{2} + \frac{a_i \delta_i}{2} \right\} \tag{9}$$

1. Ordering cost per cycle for ith item

$$OC_i = c_{1i} \tag{10}$$

2. Holding cost per cycle for ith item $HC_i = \int_0^{t_{1i}} H_i e^{\alpha_i t} . I_1(t) dt + \int_{t_{1i}}^{t_{2i}} H_i e^{\alpha_i t} . I_2(t) dt$

$$= H_i \left\{ \frac{a_i t_{2i}^2}{2} + \frac{a_i t_{1i} t_{2i}^2}{2} (\theta_i - \alpha_i) + \frac{a_i t_{2i} t_{1i}^2}{2} (\alpha_i - \theta_i) \right\} \tag{11}$$

3. Back-order cost per cycle for ith item $BC_i = c_{2i} \int\limits_{t_{2i}}^{T_i} - I_3(t) dt$

$$= c_{2i} \left\{ \frac{a_i}{2} \left(T_i^2 - t_{2i}^2 \right) + t_{2i}^2 T_i \left(3\frac{b_i}{2} + a_i \delta_i + \frac{b_i \delta_i T_i}{2} \right) - a_i t_{2i} T_i - t_{2i} T_i^2 (2b_i + a_i \delta_i) \right\} \tag{12}$$

4. Cost due to lost sales per cycle for ith item $LS_i = c_{3i} \int\limits_{t_{2i}}^{T_i} \left(1 - \frac{1}{1 + \delta_i (T_i - t)} \right) \cdot$

$(a_i + b_i t) dt$

$$= c_{3i} \left\{ \frac{a_i \delta_i}{2} (T_i - t_{2i})^2 - \frac{b_i \delta_i}{2} T_i t_{2i}^2 \right\} \tag{13}$$

5. Deteriorating cost per cycle for ith item $DC_i = c_{4i} \int\limits_{t_{1i}}^{t_{2i}} \theta_i I_2(t) dt$

$$= c_{4i} \frac{\theta_i}{2} \left\{ a_i (t_{2i} - t_{1i})^2 + \left(\frac{a_i \theta_i}{2} - \frac{b_i}{2} \right) t_{1i} t_{2i}^2 \right\} \tag{14}$$

So, total average cost per cycle for ith ($i = 1, 2, \ldots, n$) item is

$$TAC_i (t_{1i}, t_{2i}, T_i) = \frac{1}{T_i} [OC_i + HC_i + BC_i + LS_i + DC_i]$$

$$= \frac{1}{T_i} \left[c_{1i} + H_i \left\{ \frac{a_i t_{2i}^2}{2} + \frac{a_i t_{1i} t_{2i}^2}{2} (\theta_i - \alpha_i) + \frac{a_i t_{2i} t_{1i}^2}{2} (\alpha_i - \theta_i) \right\} \right.$$

$$+ c_{2i} \left\{ \frac{a_i}{2} \left(T_i^2 - t_{2i}^2 \right) + t_{2i}^2 T_i \left(3\frac{b_i}{2} + a_i \delta_i + \frac{b_i \delta_i T_i}{2} \right) - a_i t_{2i} T_i - t_{2i} T_i^2 (2b_i + a_i \delta_i) \right\}$$

$$\left. + c_{3i} \left\{ \frac{a_i \delta_i}{2} (T_i - t_{2i})^2 - \frac{b_i \delta_i}{2} T_i t_{2i}^2 \right\} + c_{4i} \frac{\theta_i}{2} \left\{ a_i (t_{2i} - t_{1i})^2 + \left(\frac{a_i \theta_i}{2} - \frac{b_i}{2} \right) t_{1i} t_{2i}^2 \right\} \right] \tag{15}$$

Here total average cost of each item can be considered as one objective function. So our proposed multi-objective inventory model (MOIM) can be written as

$$\text{Min} \{ TAC_1 (t_{11}, t_{21}, T_1), TAC_2 (t_{12}, t_{22}, T_2), \ldots\ldots\ldots, TAC_n (t_{1n}, t_{2n}, T_n) \} \tag{16}$$

Subject to, $\sum_{i=1}^{n} w_i Q_i \leq W$
where

$$TAC_i(t_{1i}, t_{2i}, T_i) = \frac{1}{T_i}\left[c_{1i} + H_i \left\{ \frac{a_i t_{2i}^2}{2} + \frac{a_i t_{1i} t_{2i}^2}{2}(\theta_i - \alpha_i) + \frac{a_i t_{2i} t_{1i}^2}{2}(\alpha_i - \theta_i)\right\} \right.$$

$$+c_{2i}\left\{ \frac{a_i}{2}\left(T_i^2 - t_{2i}^2\right) + t_{2i}^2 T_i\left(3\frac{b_i}{2} + a_i\delta_i + \frac{b_i\delta_i T_i}{2}\right) - a_i t_{2i} T_i - t_{2i} T_i^2 (2b_i + a_i\delta_i)\right\}$$

$$\left. + c_{3i}\left\{ \frac{a_i\delta_i}{2}(T_i - t_{2i})^2 - \frac{b_i\delta_i}{2}T_i t_{2i}^2\right\} + c_{4i}\frac{\theta_i}{2}\left\{ a_i(t_{2i} - t_{1i})^2 + \left(\frac{a_i\theta_i}{2} - \frac{b_i}{2}\right)t_{1i} t_{2i}^2\right\}\right]$$

and

$$Q_i = \frac{b_i t_{1i}^2}{2} + a_i t_{2i} - \frac{a_i\theta_i}{2}\left(t_{2i}^2 - t_{1i}^2\right) + \frac{b_i}{2}\left(t_{2i}^2 - t_{1i}^2\right)$$

$$- (t_{2i} - T_i)\{a_i + 2b_i T_i + a_i\delta_i T_i\} - \left(T_i^2 - t_{2i}^2\right)\left\{\frac{b_i}{2} + \frac{a_i\delta_i}{2}\right\}, \ for \ i = 1, 2, \ldots\ldots, n.$$

3 Prerequisite Mathematics

Fuzzy Set: (Zadeh [8]) A *fuzzy set* $\tilde{A}$ in U is a set of ordered pairs

$$\tilde{A} = \left\{\left(x, \mu_{\tilde{A}}(x)\right) \mid x \in U\right\},$$

where U is a collection of objects denoted generically by x and $\mu_{\tilde{A}}(x) : U \to [0, 1]$
is called the membership function or grade of membership of x in $\tilde{A}$.

Fuzzy Number: A fuzzy number is a quantity whose esteem is imprecise, as opposed to correct similar to the case with single-valued number. A fuzzy number measurement does not allude to one single value; it is an associated set of conceivable qualities, where every conceivable quality has a weight somewhere in the range of 0 and 1. This weight is called membership function, which has the form

$$\mu_{\tilde{B}}(x) : R \to [0, 1]$$

where $\mu_{\tilde{B}}(x)$ is a membership function of the fuzzy set $\tilde{B}$.

Generalized Fuzzy Number (GFN): Chen [28, 29] represents a generalized trapezoidal fuzzy number (GTrFN) $\tilde{A}$ as $\tilde{A} = (a_1, a_2, a_3, a_4; w)$, where $0 < w \leq 1$ and a_1, a_2, a_3, and a_4 are real numbers. The generalized fuzzy number (GFN) $\tilde{A}$

is a fuzzy subset of real line R, whose membership function $\mu_{\tilde{A}}(x)$ satisfies the following properties:

1. $\mu_{\tilde{A}}(x)$ is a continuous mapping from R to the closed interval $[0, 1]$.
2. $\mu_{\tilde{A}}(x) = 0$, for all $x \in (-\infty, a_1]$.
3. $\mu_{\tilde{A}}(x)$ is strictly increasing with constant rate on $[a_1, a_2]$.
4. $\mu_{\tilde{A}}(x) = w$ for all $x \in [a_2, a_3]$.
5. $\mu_{\tilde{A}}(x)$ is strictly decreasing with constant rate on $[a_3, a_4]$.
6. $\mu_{\tilde{A}}(x) = 0$ where $x \in [a_4, \infty)$.

Note: $\tilde{A}$ is a normalized fuzzy number when $w = 1$, and it is non-normalized for $w \neq 1$.

Generalized Trapezoidal Fuzzy Number (GTrFN): A generalized trapezoidal fuzzy number (GTrFN) $(a_1, a_2, a_3, a_4; w)$ is a fuzzy set of the real line R whose membership function $\mu_{\tilde{A}}(x) : R \rightarrow [0, w]$ is defined as (Fig. 2)

$$\mu_{\tilde{A}}^{w}(x) = \begin{cases} \mu_{L\tilde{A}}^{w}(x) = w\,\frac{x-a_1}{a_2-a_1} & , \ a_1 \leq x \leq a_2; \\ w & , \ a_2 \leq x \leq a_3; \\ \mu_{R\tilde{A}}^{w}(x) = w\,\frac{x-a_4}{a_3-a_4} & , \ a_3 \leq x \leq a_4 \\ 0 & , \ \text{otherwise} \end{cases} \tag{17}$$

where $\mu_{L\tilde{A}}^{w}$ and $\mu_{R\tilde{A}}^{w}$ are the left and right membership functions of $\tilde{A}$, respectively. And the inverse functions $h_{L\tilde{A}}^{w} : [0, w] \rightarrow [a, b]$ and $h_{R\tilde{A}}^{w} : [0, w] \rightarrow [c, d]$ are defined as

$$h_{L\tilde{A}}^{w}(y) = +\frac{(b-a)}{w}y; \tag{18}$$

$$h_{R\tilde{A}}^{w}(y) = d + \frac{(c-d)}{w}y; \quad y \in [0, w]$$

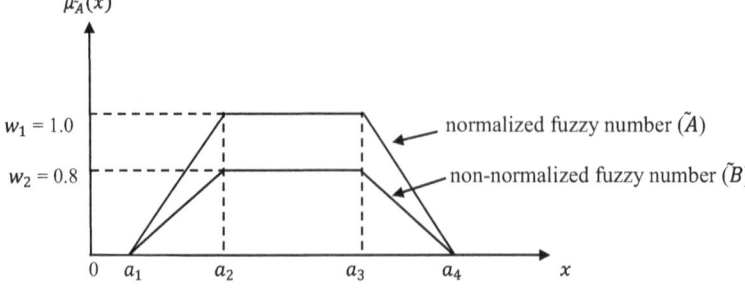

Fig. 2 Two Generalized trapezoidal fuzzy numbers $\tilde{A}$ and $\tilde{B}$

Now according to Liou and Wang [30], integral value method we have for a non-normal fuzzy number $\tilde{A}$, the corresponding membership function $\mu_{\tilde{A}}(x)$ can be normalized by dividing the maximal value of $\mu_{\tilde{A}}(x)$, i.e., w. Let $\overline{\tilde{A}}$ and $\mu_{\overline{\tilde{A}}}$ be the normalized fuzzy number and the corresponding membership function.

Let $\lambda \in [0, 1]$ be the index of optimism which represents the degree of optimism of a decision maker (DM). Then the total λ-integral value is defined as

$$I_\lambda^w\left(\tilde{A}\right) = \left[\lambda I_R^w\left(\tilde{A}\right) + (1 - \lambda) I_L^w\left(\tilde{A}\right)\right]$$

where $I_{wR}\left(\tilde{A}\right)$ and $I_{wL}\left(\tilde{A}\right)$ represents the right and left integral values of $\tilde{A}$ respectively.

Now, when $\tilde{A}$ is being ranked, we have

$$I_L^w\left(\tilde{A}\right) = \int_0^1 h_{L\tilde{A}}^w(y)\, dy = \frac{1}{2}(a+b) \text{ and } I_R^w\left(\tilde{A}\right) = \int_0^1 h_{R\tilde{A}}^w(y)\, dy = \frac{1}{2}(c+d)$$

$$(19)$$

Thus, $I_k^w\left(\tilde{A}\right) = \frac{1}{2}[\lambda\,(c+d) + (1-\lambda)\,(a+b)]$ which does not depend on value of w, i.e, whether $\tilde{A}$ is normal or not. A larger value of k indicates the higher degree of optimism.

Now for $\lambda = 0$, the total λ-integral value is $I_0^w\left(\tilde{A}\right) = \frac{1}{2}(a+b) = I_w^L\left(\tilde{A}\right)$ and represents a pessimistic viewpoint of a DM and optimistic DM's viewpoint, i.e., for $\lambda = 1$, $I_1^w\left(\tilde{A}\right) = \frac{1}{2}(c+d) = I_w^R\left(\tilde{A}\right)$. When $\lambda = 0.5$, the total λ-integral value $I_{0.5}^w\left(\tilde{A}\right) = \frac{1}{2}\left[I_w^R\left(\tilde{A}\right) + I_w^L\left(\tilde{A}\right)\right]$ reflects a moderately optimistic DM's viewpoint.

Property 1.1:

(a) If $\tilde{B} = (b_1, b_2, b_3, b_4; w)$ and $y = qb$, $q > 0$, then $y = q\tilde{b}$ is a fuzzy number $(qb_1, qb_2, qb_3, qb_4; w)$.

(b) If $y = qb$, $q < 0$, then $y = q\tilde{b}$ is a fuzzy number $(qb_4, qb_3, qb_2, qb_1; w)$.

Proof Islam and Roy [7] □

Property 1.2: If $\widetilde{A}_1 = (a_1, b_1, c_1, d_1; w_1)$ and $\widetilde{A}_2 = (a_2, b_2, c_2, d_2; w_2)$, then $\widetilde{A}_1 \oplus \widetilde{A}_2$ is a *fuzzy number* $(a_1 + a_2, b_1 + b_2, c_1 + c_2, d_1 + d_2; \min(w_1, w_2))$.

Proof Islam and Roy [7]   □

4 Fuzzy Model

Due to uncertainty, we assume the costs as generalized trapezoidal fuzzy number (GTrFN). Let us assume $\widetilde{c}_{1i} = \left(c_{1i}{}^1, c_{1i}{}^2, c_{1i}{}^3, c_{1i}{}^4; w_{c_{1i}}\right), 0 < w_{c_{1i}} \leq 1;$

$$\widetilde{H}_i = \left(H_i{}^1, H_i{}^2, H_i{}^3, H_i{}^4; w_{H_i}\right), 0 < w_{H_i} \leq 1;$$

$$\widetilde{c}_{2i} = \left(c_{2i}{}^1, c_{2i}{}^2, c_{2i}{}^3, c_{2i}{}^4; w_{c_{2i}}\right), 0 < w_{c_{2i}} \leq 1;$$

$$\widetilde{c}_{3i} = \left(c_{3i}{}^1, c_{3i}{}^2, c_{3i}{}^3, c_{3i}{}^4; w_{c_{3i}}\right), 0 < w_{c_{3i}} \leq 1;$$

$$\widetilde{c}_{4i} = \left(c_{4i}{}^1, c_{4i}{}^2, c_{4i}{}^3, c_{4i}{}^4; w_{c_{4i}}\right), 0 < w_{c_{4i}} \leq 1.$$

Then fuzzy total average cost is given by

$$T\,AC_i\,(t_{1i}, t_{2i}, T_i) = \frac{1}{T_i}\left[\widetilde{c}_{1i} + \widetilde{H}_i \left\{\frac{a_i t_{2i}{}^2}{2} + \frac{a_i t_{1i} t_{2i}{}^2}{2}(\theta_i - \alpha_i) + \frac{a_i t_{2i} t_{1i}{}^2}{2}(\alpha_i - \theta_i)\right\}\right.$$

$$+ \widetilde{c}_{2i}\left\{\frac{a_i}{2}\left(T_i{}^2 - t_{2i}{}^2\right) + t_{2i}{}^2 T_i\left(3\frac{b_i}{2} + a_i\delta_i + \frac{b_i\delta_i T_i}{2}\right) - a_i t_{2i} T_i - t_{2i} T_i{}^2(2b_i + a_i\delta_i)\right\}$$

$$+ \widetilde{c}_{3i}\left\{\frac{a_i\delta_i}{2}(T_i - t_{2i})^2 - \frac{b_i\delta_i}{2}T_i t_{2i}{}^2\right\}$$

$$+ \widetilde{c}_{4i}\frac{\theta_i}{2}\left\{a_i(t_{2i} - t_{1i})^2 + \left(\frac{a_i\theta_i}{2} - \frac{b_i}{2}\right)t_{1i} t_{2i}{}^2\right]$$

And our MOIM problem (16) becomes fuzzy model as

$$\text{Min}\left\{T\widetilde{A}C_1, T\widetilde{A}C_2, T\widetilde{A}C_3, \ldots\ldots\ldots\ldots, T\widetilde{A}C_n\right\} \tag{20}$$

Subject to, $\sum_{i=1}^n w_i\,Q_i \leq W$ where

$$T\tilde{A}C_i\,(t_{1i}, t_{2i}, T_i) = \frac{1}{T_i}\left[\tilde{c}_{1i} + \tilde{H}_i\left\{\frac{a_i t_{2i}^2}{2} + \frac{a_i t_{1i} t_{2i}^2}{2}(\theta_i - \alpha_i) + \frac{a_i t_{2i} t_{1i}^2}{2}(\alpha_i - \theta_i)\right\}\right.$$

$$+\tilde{c}_{2i}\left\{\frac{a_i}{2}\left(T_i^2 - t_{2i}^2\right) + t_{2i}^2 T_i\left(3\frac{b_i}{2} + a_i\delta_i + \frac{b_i\delta_i T_i}{2}\right) - a_i t_{2i} T_i - t_{2i} T_i^2(2b_i + a_i\delta_i)\right\}$$

$$\left. + \tilde{c}_{3i}\left\{\frac{a_i\delta_i}{2}(T_i - t_{2i})^2 - \frac{b_i\delta_i}{2}T_i t_{2i}^2\right\} + \tilde{c}_{4i}\frac{\theta_i}{2}\left\{a_i(t_{2i} - t_{1i})^2 + \left(\frac{a_i\theta_i}{2} - \frac{b_i}{2}\right)t_{1i} t_{2i}^2\right\}\right]$$

and

$$Q_i = \frac{b_i t_{1i}^2}{2} + a_i t_{2i} - \frac{a_i\theta_i}{2}\left(t_{2i}^2 - t_{1i}^2\right) + \frac{b_i}{2}\left(t_{2i}^2 - t_{1i}^2\right) - (t_{2i} - T_i)\{a_i + 2b_i T_i + a_i\delta_i T_i\}$$

$$- \left(T_i^2 - t_{2i}^2\right)\left\{\frac{b_i}{2} + \frac{a_i\delta_i}{2}\right\}, for\ i = 1, 2, \ldots\ldots, n.$$

Now by λ-integral value method, the total λ-integral value of a GTrFN $\tilde{A} = (a, b, c, d; w)$ is given by $I_\lambda^w\left(\tilde{A}\right) = \lambda w\left(\frac{c+d}{2}\right) + (1 - \lambda)w\left(\frac{a+b}{2}\right)$.

By using this method, we can convert GTrFN costs $\left(\tilde{c}_{1i}, \tilde{H}_i, \tilde{c}_{2i}, \tilde{c}_{3i}, \tilde{c}_{4i}\right)$ into crisp values $(\widehat{c}_{1i}, \widehat{H}_i, \widehat{c}_{2i}, \widehat{c}_{3i}, \widehat{c}_{4i})$. So the above problem (20) becomes crisp model as

$$\text{Min } \left\{\widehat{TAC_1}, \widehat{TAC_2}, \widehat{TAC_3}, \ldots\ldots\ldots, \widehat{TAC_n}\right\} \tag{21}$$

Subject to, $\sum_{i=1}^n w_i Q_i \leq W$ where

$$T\hat{A}C_i\,(t_{1i}, t_{2i}, T_i) = \frac{1}{T_i}\left[\hat{c}_{1i} + \hat{H}_i\left\{\frac{a_i t_{2i}^2}{2} + \frac{a_i t_{1i} t_{2i}^2}{2}(\theta_i - \alpha_i) + \frac{a_i t_{2i} t_{1i}^2}{2}(\alpha_i - \theta_i)\right\}\right.$$

$$+\hat{c}_{2i}\left\{\frac{a_i}{2}\left(T_i^2 - t_{2i}^2\right) + t_{2i}^2 T_i\left(3\frac{b_i}{2} + a_i\delta_i + \frac{b_i\delta_i T_i}{2}\right) - a_i t_{2i} T_i - t_{2i} T_i^2(2b_i + a_i\delta_i)\right\}$$

$$\left. + \hat{c}_{3i}\left\{\frac{a_i\delta_i}{2}(T_i - t_{2i})^2 - \frac{b_i\delta_i}{2}T_i t_{2i}^2\right\} + \tilde{c}_{4i}\frac{\theta_i}{2}\left\{a_i(t_{2i} - t_{1i})^2 + \left(\frac{a_i\theta_i}{2} - \frac{b_i}{2}\right)t_{1i} t_{2i}^2\right\}\right]$$

and

$$Q_i = \frac{b_i t_{1i}^2}{2} + a_i t_{2i} - \frac{a_i\theta_i}{2}\left(t_{2i}^2 - t_{1i}^2\right) + \frac{b_i}{2}\left(t_{2i}^2 - t_{1i}^2\right) - (t_{2i} - T_i)\{a_i + 2b_i T_i + a_i\delta_i T_i\}$$

$$- \left(T_i^2 - t_{2i}^2\right)\left\{\frac{b_i}{2} + \frac{a_i\delta_i}{2}\right\}, for\ i = 1, 2, \ldots\ldots, n.$$

5 Mathematical Analysis

5.1 Neutrosophic Hesitant Fuzzy Non-Linear Programming (NHFNLP) Technique

Solve the MOIM (21) as a single-objective NLP using only one objective at a time and we overlook the others.

From the above results, we find out the corresponding values of every objective function at each solution obtained. With these values the payoff matrix can be prepared as follows:

$$
\begin{array}{cccc}
 & TAC_1\,(t_{11},t_{21},T_1) & TAC_2\,(t_{12},t_{22},T_2) & \ldots & TAC_n\,(t_{1n},t_{2n},T_n) \\
\left(t_{11}{}^1,t_{21}{}^1,T_1{}^1\right) & \left(TAC_1{}^*\left(t_{11}{}^1,t_{21}{}^1,T_1{}^1\right) \right. & TAC_2\left(t_{11}{}^1,t_{21}{}^1,T_1{}^1\right) & \ldots & TAC_n\left(t_{11}{}^1,t_{21}{}^1,T_1{}^1\right) \\
\left(t_{12}{}^2,t_{22}{}^2,T_2{}^2\right) & TAC_1\left(t_{12}{}^2,t_{22}{}^2,T_2{}^2\right) & TAC_2{}^*\left(t_{12}{}^2,t_{22}{}^2,T_2{}^2\right) & \ldots & TAC_n\left(t_{12}{}^2,t_{22}{}^2,T_2{}^2\right) \\
\ldots\ldots & \ldots\ldots & \ldots\ldots\ldots & \ldots\ldots\ldots & \ldots\ldots\ldots \\
\left(t_{1n}{}^n,t_{2n}{}^n,T_n{}^n\right) & TAC_1\left(t_{1n}{}^n,t_{2n}{}^n,T_n{}^n\right) & TAC_2\left(t_{1n}{}^n,t_{2n}{}^n,T_n{}^n\right) & \ldots & \left. TAC_n{}^*\left(t_{1n}{}^n,t_{2n}{}^n,T_n{}^n\right) \right)
\end{array}
$$

Let $U^k = \max\{TAC_k(t_{1i}{}^i,t_{2i}{}^i,T_i{}^i), i = 1,2,\ldots,n\}$ for $k = 1,2,\ldots,n$ and $L^k = TAC_{k^*}(t_{1k}{}^k,t_{2k}{}^k,T_k{}^k)$ for $k = 1,2,\ldots,n$.

Hence U^k, L^k are identified, $L^k \le TAC_k(t_{1i}{}^i,t_{2i}{}^i,T_i{}^i) \le U^k$, for $i = 1,2,\ldots,n$; $k = 1,2,\ldots,n$.

The truth hesitant membership function:

$$
T_{h^-}^{E_1}\left(TAC_k\,(t_{1k},t_{2k},T_k)\right) =
\begin{cases}
1 & if\ TAC_k\,(t_{1k},t_{2k},T_k) < L^k \\[2mm]
\sigma_1 \dfrac{\left(U^k\right)^t - \left(TAC_k(t_{1k},t_{2k},T_k)\right)^t}{\left(U^k\right)^t - \left(L^k\right)^t} & if\ L^k \le TAC_k\,(t_{1k},t_{2k},T_k) \le U^k \\[2mm]
0 & if\ U^k < TAC_k\,(t_{1k},t_{2k},T_k)
\end{cases}
$$

$$
T_{h^-}^{E_2}\left(TAC_k\,(t_{1k},t_{2k},T_k)\right) =
\begin{cases}
1 & if\ TAC_k\,(t_{1k},t_{2k},T_k) < L^k \\[2mm]
\sigma_2 \dfrac{\left(U^k\right)^t - \left(TAC_k(t_{1k},t_{2k},T_k)\right)^t}{\left(U^k\right)^t - \left(L^k\right)^t} & if\ L^k \le TAC_k\,(t_{1k},t_{2k},T_k) \le U^k \\[2mm]
0 & if\ U^k < TAC_k\,(t_{1k},t_{2k},T_k)
\end{cases}
$$

--------------- ------------------   ----------------------------

--------------- ------------------ ------- ----------------------------

$$
T_{h^-}^{E_n}\left(TAC_k\,(t_{1k},t_{2k},T_k)\right) =
\begin{cases}
1 & if\ TAC_k\,(t_{1k},t_{2k},T_k) < L^k \\[2mm]
\sigma_n \dfrac{\left(U^k\right)^t - \left(TAC_k(t_{1k},t_{2k},T_k)\right)^t}{\left(U^k\right)^t - \left(L^k\right)^t} & if\ L^k \le TAC_k\,(t_{1k},t_{2k},T_k) \le U^k \\[2mm]
0 & if\ U^k < TAC_k\,(t_{1k},t_{2k},T_k)
\end{cases}
$$

The indeterminacy hesitant membership function:

$$I_{h^-}^{E_1}\left(TAC_k\left(t_{1k}, t_{2k}, T_k\right)\right) = \begin{cases} 1 & if\ TAC_k\left(t_{1k}, t_{2k}, T_k\right) < L^k - s^k \\ \rho_1 \dfrac{\left(U^k\right)^t - \left(TAC_k\left(t_{1k}, t_{2k}, T_k\right)\right)^t}{\left(s^k\right)^t} & if\ U^k - s^k \leq TAC_k\left(t_{1k}, t_{2k}, T_k\right) \leq U^k \\ 0 & if\ U^k < TAC_k\left(t_{1k}, t_{2k}, T_k\right) \end{cases}$$

$$I_{h^-}^{E_2}\left(TAC_k\left(t_{1k}, t_{2k}, T_k\right)\right) = \begin{cases} 1 & if\ TAC_k\left(t_{1k}, t_{2k}, T_k\right) < L^k - s^k \\ \rho_2 \dfrac{\left(U^k\right)^t - \left(TAC_k\left(t_{1k}, t_{2k}, T_k\right)\right)^t}{\left(s^k\right)^t} & if\ U^k - s^k \leq TAC_k\left(t_{1k}, t_{2k}, T_k\right) \leq U^k \\ 0 & if\ U^k < TAC_k\left(t_{1k}, t_{2k}, T_k\right) \end{cases}$$

---------------   ------------------   ----------------------------

---------------   ------------------   ----------------------------

$$I_{h^-}^{E_n}\left(TAC_k\left(t_{1k}, t_{2k}, T_k\right)\right) = \begin{cases} 1 & if\ TAC_k\left(t_{1k}, t_{2k}, T_k\right) < L^k - s^k \\ \rho_n \dfrac{\left(U^k\right)^t - \left(TAC_k\left(t_{1k}, t_{2k}, T_k\right)\right)^t}{\left(s^k\right)^t} & if\ U^k - s^k \leq TAC_k\left(t_{1k}, t_{2k}, T_k\right) \leq U^k \\ 0 & if\ U^k < TAC_k\left(t_{1k}, t_{2k}, T_k\right) \end{cases}$$

The falsity hesitant membership function:

$$F_{h^-}^{E_1}\left(TAC_k\left(t_{1k}, t_{2k}, T_k\right)\right) = \begin{cases} 0 & if\ TAC_k\left(t_{1k}, t_{2k}, T_k\right) < L^k + v^k \\ \tau_1 \dfrac{\left(TAC_k\left(t_{1k}, t_{2k}, T_k\right)\right)^t - \left(L^k\right)^t - \left(v^k\right)^t}{\left(U^k\right)^t - \left(L^k\right)^t - \left(v^k\right)^t} & if\ L^k + v^k \leq TAC_k\left(t_{1k}, t_{2k}, T_k\right) \leq U^k \\ 1 & if\ U^k < TAC_k\left(t_{1k}, t_{2k}, T_k\right) \end{cases}$$

$$F_{h^-}^{E_2}\left(TAC_k\left(t_{1k}, t_{2k}, T_k\right)\right) = \begin{cases} 0 & if\ TAC_k\left(t_{1k}, t_{2k}, T_k\right) < L^k + v^k \\ \tau_2 \dfrac{\left(TAC_k\left(t_{1k}, t_{2k}, T_k\right)\right)^t - \left(L^k\right)^t - \left(v^k\right)^t}{\left(U^k\right)^t - \left(L^k\right)^t - \left(v^k\right)^t} & if\ L^k + v^k \leq TAC_k\left(t_{1k}, t_{2k}, T_k\right) \leq U^k \\ 1 & if\ U^k < TAC_k\left(t_{1k}, t_{2k}, T_k\right) \end{cases}$$

---------------   ------------------   ----------------------------

---------------   ------------------   ----------------------------

$$F_{h^-}^{E_n} (TAC_k (t_{1k}, t_{2k}, T_k))$$

$$= \begin{cases} 0 & if \ TAC_k (t_{1k}, t_{2k}, T_k) < L^k + v^k \\ \tau_1 \frac{(TAC_k(t_{1k},t_{2k},T_k))^t - (L^k)^t - (v^k)^t}{(U^k)^t - (L^k)^t - (v^k)^t} & if \ L^k + v^k \leq TAC_k (t_{1k}, t_{2k}, T_k) \leq U^k \\ 1 & if \ U^k < TAC_k (t_{1k}, t_{2k}, T_k) \end{cases}$$

where parameters $t > 0$ and s^k, $v^k \epsilon (0, 1) \ \forall \ k = 1, 2, 3, \ldots\ldots, n$ are indeterminacy and falsitytolerance values, which are assigned by decision-making and h^- represents the minimization-type hesitant objective function.

$T_{h^-}^{E_1} (TAC_k (t_{1k}, t_{2k}, T_k))$, $I_{h^-}^{E_1} (TAC_k (t_{1k}, t_{2k}, T_k))$, $F_{h^-}^{E_1} (TAC_k (t_{1k}, t_{2k}, T_k))$ are truth, indeterminacy, and falsity hesitant membership degrees assigned by the first expert.

$T_{h^-}^{E_2} (TAC_k (t_{1k}, t_{2k}, T_k))$, $I_{h^-}^{E_2} (TAC_k (t_{1k}, t_{2k}, T_k))$, $F_{h^-}^{E_2} (TAC_k (t_{1k}, t_{2k}, T_k))$ are truth, indeterminacy, and falsity hesitant membership degrees assigned by the second expert.

$T_{h^-}^{E_n} (TAC_k (t_{1k}, t_{2k}, T_k))$, $I_{h^-}^{E_n} (TAC_k (t_{1k}, t_{2k}, T_k))$, $F_{h^-}^{E_n} (TAC_k (t_{1k}, t_{2k}, T_k))$ are truth, indeterminacy and falsity hesitant membership degrees assigned by nth expert.

Using auxiliary parameters σ_i, ρ_i, τ_i and the above membership function, the multi-objective
nonlinear inventory problem is formulated as follows:

$$\text{Max} \ \frac{\sum_1^n \sigma_i}{n}$$

$$\text{Max} \ \frac{\sum_1^n \rho_i}{n}$$

$$\text{Min} \ \frac{\sum_1^n \tau_i}{n}$$

Subject to

$$T_{h^-}^{E_i} (TAC_k (t_{1k}, t_{2k}, T_k)) \geq \sigma_i, \ I_{h^-}^{E_i} (TAC_k (t_{1k}, t_{2k}, T_k)) \geq \rho_i,$$

$$F_{h^-}^{E_i}\left(TAC_k\left(t_{1k}, t_{2k}, T_k\right)\right) \le \tau_i \sum_{i=1}^{n} w_i Q_i \le W, \sigma_i + \rho_i + \tau_i \le 3,$$

$$\sigma_i \ge \rho_i, \sigma_i \ge \tau_i, Q_i = \frac{b_i t_{1i}{}^2}{2} + a_i t_{2i} - \frac{a_i \theta_i}{2}\left(t_{2i}{}^2 - t_{1i}{}^2\right)$$

$$+ \frac{b_i}{2}\left(t_{2i}{}^2 - t_{1i}{}^2\right) - \left(t_{2i} - T_i\right)\{a_i + 2b_i T_i + a_i \delta_i T_i\}$$

$$- \left(T_i{}^2 - t_{2i}{}^2\right)\left\{\frac{b_i}{2} + \frac{a_i \delta_i}{2}\right\}, \text{ for } i = 1, 2, \ldots\ldots, n \qquad (22)$$

Using above linear membership function, it can be written as

$$\text{Max } \frac{\sigma_1 + \sigma_2 + \ldots\cdots + \sigma_n}{n} + \frac{\rho_1 + \rho_2 + \ldots\ldots\ldots + \rho_n}{n} - \frac{\tau_1 + \tau_2 + \ldots\ldots\cdots + \tau_n}{n}$$

Subject to

$$T_{h^-}^{E_i}\left(TAC_k\left(t_{1k}, t_{2k}, T_k\right)\right) \ge \sigma_i, I_{h^-}^{E_i}\left(TAC_k\left(t_{1k}, t_{2k}, T_k\right)\right) \ge \rho_i,$$

$$F_{h^-}^{E_i}\left(TAC_k\left(t_{1k}, t_{2k}, T_k\right)\right) \le \tau_i \sum_{i=1}^{n} w_i Q_i \le W, \sigma_i + \rho_i + \tau_i \le 3, \sigma_i \ge \rho_i, \sigma_i \ge \tau_i,$$

$$Q_i = \frac{b_i t_{1i}{}^2}{2} + a_i t_{2i} - \frac{a_i \theta_i}{2}\left(t_{2i}{}^2 - t_{1i}{}^2\right) + \frac{b_i}{2}\left(t_{2i}{}^2 - t_{1i}{}^2\right)$$

$$- \left(t_{2i} - T_i\right)\{a_i + 2b_i T_i + a_i \delta_i T_i\}$$

$$- \left(T_i{}^2 - t_{2i}{}^2\right)\left\{\frac{b_i}{2} + \frac{a_i \delta_i}{2}\right\} \qquad t_{ik} > 0, t_{ik} > 0, T_k > 0, \text{ for } i = 1, 2, \ldots\ldots, n,$$

$$k = 1, 2, \ldots, n, 0 \le \sigma_1, \sigma_2, \ldots\ldots, \sigma_n \le 1, 0 \le \rho_1, \rho_2, \ldots\ldots\ldots, \rho_n \le 1, 0 \le \tau_1,$$

$$\tau_2, \ldots\ldots\ldots, t_n \le 1 \qquad (23)$$

Solving the above equations, we get $t_{ik}^*, t_{ik}^*, T_k^* D_i^*, T_i^*$, and then TAC_i^*, for $i = 1, 2, 3, \ldots, n, k = 1, 2, \ldots, n$.

5.2 Fuzzy Programming Technique to Solve MOIM

The fuzzy linear membership functions $\mu_{TAC_k}\left(TAC_k\left(t_{1k}, t_{2k}, T_k\right)\right)$ for the kth objective functions $TAC_k(t_{1k}, t_{2k}, T_k)$, respectively, for $k = 1, 2, \ldots, n$ are defined as follows:

$$\mu_{TAC_k}(TAC_k(t_{1k}, t_{2k}, T_k)) = \begin{cases} 1 & \text{for } TAC_k(t_{1k}, t_{2k}, T_k) < L^k \\ \frac{U^k - TAC_k(t_{1k}, t_{2k}, T_k)}{U^k - L^k} & \text{for } L^k \leq TAC_k(t_{1k}, t_{2k}, T_k) \leq U^k \\ 0 & \text{for } TAC_k(t_{1k}, t_{2k}, T_k) > U^k \end{cases} \quad \text{for } k=1, 2, \ldots, n.$$

(24)

Using the above membership function (24) the fuzzy non-linear programming (FNLP) (based on max-min) is

$$\text{Max } \alpha$$

subject to

$$\alpha\left(U^k - L^k\right) + TAC_k(t_{1k}, t_{2k}, T_k) \leq U^k, \, T_k \geq 0, \, D_k \geq 0, \quad k = 1, 2, \ldots, n. \ 0 \leq \alpha \leq 1$$

(25)

and same as other conditions (21)

6 Numerical Example

Let us consider an inventory model which consists two items with the following parameter values:

$$\text{Min } \{TAC_1(t_{11}, t_{21}, T_1), TAC_2(t_{12}, t_{22}, T_2)\}$$

(26)

Subject to, $\sum_{i=1}^{2} w_i Q_i \leq W$ where

$$TAC_i(t_{1i}, t_{2i}, T_i) = \frac{1}{T_i}\left[\tilde{c}_{1i} + \tilde{H}_i\left\{\frac{a_i t_{2i}^2}{2} + \frac{a_i t_{1i} t_{2i}^2}{2}(\theta_i - \alpha_i) + \frac{a_i t_{2i} t_{1i}^2}{2}(\alpha_i - \theta_i)\right\}\right.$$

$$+ \tilde{c}_{2i}\left\{\frac{a_i}{2}\left(T_i^2 - t_{2i}^2\right) + t_{2i}^2 T_i\left(3\frac{b_i}{2} + a_i \delta_i + \frac{b_i \delta_i T_i}{2}\right) - a_i t_{2i} T_i - t_{2i} T_i^2(2b_i + a_i \delta_i)\right\}$$

$$+ \tilde{c}_{3i}\left\{\frac{a_i \delta_i}{2}(T_i - t_{2i})^2 - \frac{b_i \delta_i}{2}T_i t_{2i}^2\right\}$$

$$\left. + \tilde{c}_{4i}\frac{\theta_i}{2}\left\{a_i(t_{2i} - t_{1i})^2 + \left(\frac{a_i \theta_i}{2} - \frac{b_i}{2}\right)t_{1i} t_{2i}^2\right\}\right]$$

and $Q_i = \frac{b_i t_{1i}^2}{2} + a_i t_{2i} - \frac{a_i \theta_i}{2}\left(t_{2i}^2 - t_{1i}^2\right) + \frac{b_i}{2}\left(t_{2i}^2 - t_{1i}^2\right) - (t_{2i} - T_i)\{a_i + 2b_i T_i + a_i \delta_i T_i\} - \left(T_i^2 - t_{2i}^2\right)\left\{\frac{b_i}{2} + \frac{a_i \delta_i}{2}\right\}, for \ i = 1, 2.$

We consider total storage area W = 400 sq m and

$$\tilde{C}_{11} = (175, 185, 190, 200; 0.8),$$

Table 1 Optimal solutions of MOIM (26)

Methods	$TAC_1\left(t_{11}^*, t_{21}^*, T_1^*\right)$	$TAC_2\left(t_{12}^*, t_{22}^*, T_2^*\right)$
FNLP	162.47	148.54
NHFNLP	162.56	148.5

Table 2 Optimal solutions of MOIM (26) with different weights by WFNLP

Weights	$TAC_1\left(t_{11}^*, t_{21}^*, T_1^*\right)$	$TAC_2\left(t_{12}^*, t_{22}^*, T_2^*\right)$	Type
$\omega_1 = 0.8, \omega_2 = 0.2$	163.46	148.15	I
$\omega_1 = 0.6, \omega_2 = 0.4$	162.81	148.38	II
$\omega_1 = 0.5, \omega_2 = 0.5$	162.47	148.54	III
$\omega_1 = 0.4, \omega_2 = 0.6$	162.17	148.73	IV
$\omega_1 = 0.2, \omega_2 = 0.8$	161.75	149.1	V

$$\widetilde{H}_1 = (3, 4, 6, 7; 0.6),$$

$$\widetilde{C}_{21} = (1, 1.1, 1.2, 1.5; 0.9),$$

$$\widetilde{C}_{31} = (0.5, 0.65, 0.85, 1; 0.8),$$

$$\widetilde{C}_{41} = (0.6, 0.9, 1.1, 1.4; 0.7), a_1 = 40, b_1 = 3, \theta_1 = 0.01, \alpha_1 = 0.3, \delta_1 = 0.5, w_1 = 5.$$

$$\widetilde{C}_{12} = (170, 190, 210, 230; 0.8),$$

$$\widetilde{H}_2 = (1.9, 2.1, 2.3, 2.5; 0.9),$$

$$\widetilde{C}_{22} = (1.4, 1.6, 1.7, 1.8; 0.8),$$

$$\widetilde{C}_{32} = (0.3, 0.45, 0.65, 0.8; 0.9),$$

$$\widetilde{C}_{42} = (0.7, 0.9, 1.1, 1.3; 0.8), a_2 = 40, b_2 = 3.5, \theta_2 = 0.02, \alpha_2 = 0.5, \delta_2 = 0.6, w_2 = 5.$$

Now using ranking fuzzy number with respect to their integral value λ and taking $\lambda = 0.5$, the optimal solutions of the MOIM (21) by FNLP and FAGP are specified in Table 1.

Table 1 shows that both FNLP and NHFNLP techniques give more or less the same results.

The optimal solutions of the MOIM (26) by WFNLP method with different weights when $\lambda = 0.5$ are shown in Table 2.

We see that as ω_1 decreases the value of $TAC_1\left(t_{11}^*, t_{21}^*, T_1^*\right)$ also decreases and as ω_2 decreases, the value of $TAC_2\left(t_{12}^*, t_{22}^*, T_2^*\right)$ decreases.

Table 3 Optimal solutions of MOIM (26) by FNLP and NHFNLP for different values of λ

Methods	λ	$TAC_1\left(t_{11}^*, t_{21}^*, T_1^*\right)$	$TAC_2\left(t_{12}^*, t_{22}^*, T_2^*\right)$
FNLP	0.0	148.56	132.22
	0.3	157.24	141.83
	0.5	162.47	148.54
	0.7	168.87	154.59
	1.0	176.79	164.27
NHFNLP	0.0	148.54	132.23
	0.3	157.28	141.81
	0.5	162.56	148.5
	0.7	169.06	154.5
	1.0	177.13	164.12

Fig. 3 Sensitivity analysis of TAC_1 w.r.t. λ by FNLP and NHFNLP

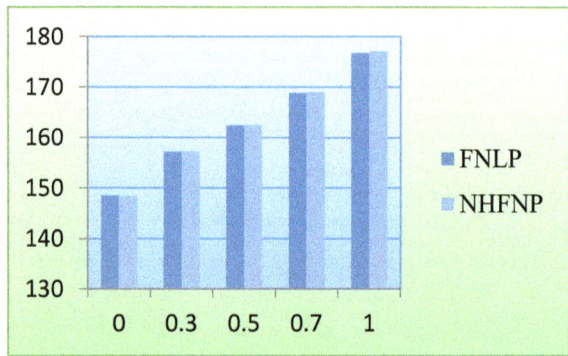

Fig. 4 Sensitivity analysis of TAC_2 w.r.t. λ by FNLP and NHFNLP

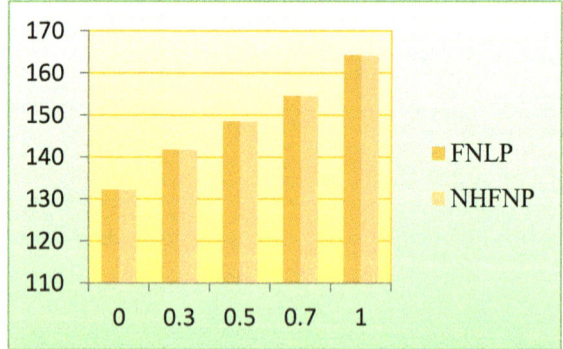

7 Sensitivity Analysis

The optimal solutions of the MOIM (6.1) by FNLP and NHFNLP techniques for different values of λ are given in Table 3.

We see that as λ increases, both $TAC_1\left(t_{11}^*, t_{21}^*, T_1^*\right)$ and $TAC_2\left(t_{12}^*, t_{22}^*, T_2^*\right)$ increase in both methods.

Table 4 Optimal solutions of MOIM (26) by FNLP and NHFNLP for different values of θ

Methods	θ	$TAC_1{}^*$	$TAC_2{}^*$
FNLP	0.05	162.47	148.54
	0.10	167.87	156.75
	0.15	178.83	167.98
	0.20	189.45	175.35
NHFNLP	0.05	162.56	148.5
	0.10	168.22	156.98
	0.15	176.68	164.88
	0.20	188.98	176.35

Table 5 Optimal solutions of MOIM (26) by FNLP and NHFNLP for different values of h_1 and h_2

Methods	h_1	h_2	$TAC_1{}^*$	$TAC_2{}^*$
FNLP	0.63	0.46	162.47	148.54
	1.13	0.96	163.87	149.89
	1.63	1.46	165.12	151.15
	2.13	1.96	167.52	153.54
NHFNLP	0.63	0.46	162.56	148.5
	1.13	0.96	164.78	149.65
	1.63	1.46	166.98	150.85
	2.13	1.96	169.08	152.04

Fig. 5 Sensitivity analysis of $TAC_1{}^*$ w.r.t. θ by both FNLP and NHFNLP methods

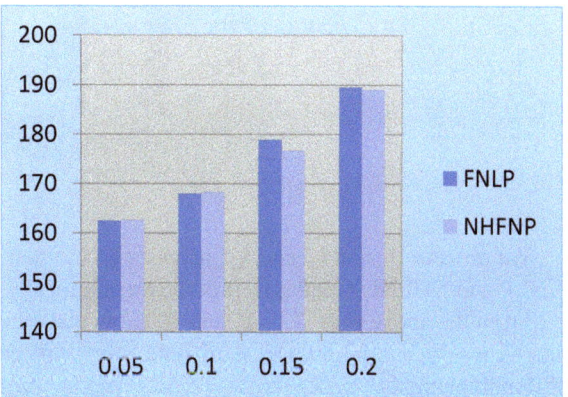

The sensitivity analyses of $TAC_1\left(t_{11}^*, t_{21}^*, T_1^*\right)$ and $TAC_2\left(t_{12}^*, t_{22}^*, T_2^*\right)$ with respect to λ by both FNLP and NHFNLP methods are given graphically in Figs. 3 and 4.

The optimal solutions of the MOIM (26) by FNLP and NHFNLP techniques for different values of θ, h are shown in Tables 4 and 5 respectively.

We observe that as θ increases, $TAC_1{}^*$ and $TAC_2{}^*$ both increase by both FNLP and NHFNLP methods. It is shown graphically below.

From the above Figs. 5 and 6 shows that when θ increases then minimum cost of both items increases in all different methods.

Fig. 6 Sensitivity analysis of $TAC_2{}^*$ w.r.t. θ by both FNLP and NHFNLP methods

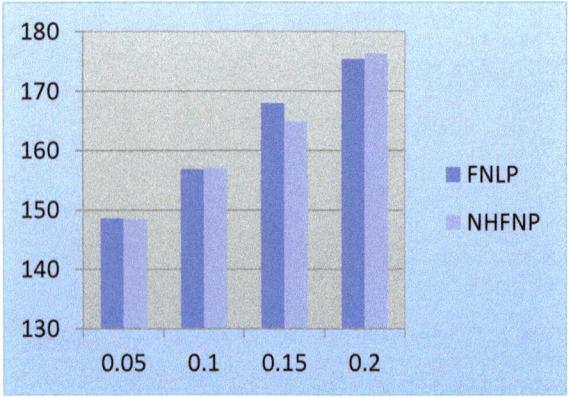

Fig. 7 Sensitivity analysis of $TAC_1{}^*$ w.r.t. h_1 by both FNLP and NHFNLP methods

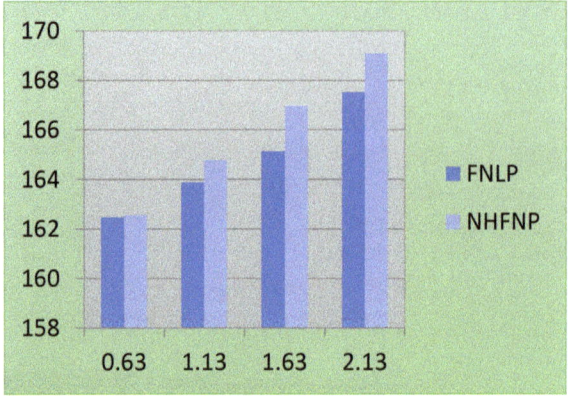

We observe that as h_1 and h_2 increase, $TAC_1{}^*$ and $TAC_2{}^*$ both increase by both FNLP and NHFNLP methods. It is shown graphically below.

From the above Figs. 7, 8, 9, and 10 show that when h_1 and h_2 are continuously increasing then minimum cost of both items are continuously increasing in all different methods.

Fig. 8 Sensitivity analysis of $TAC_2{}^*$ w.r.t. h_1 by both FNLP and NHFNLP methods

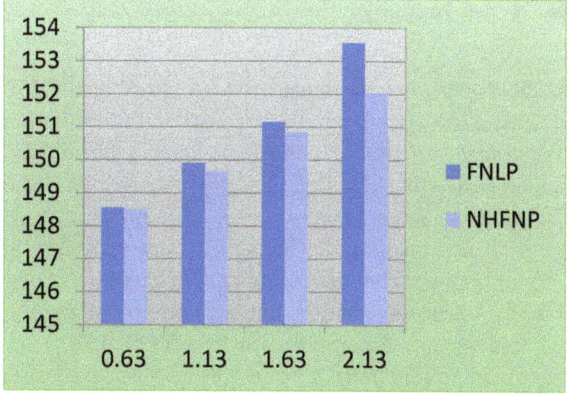

Fig. 9 Sensitivity analysis of $TAC_1{}^*$ w.r.t. h_2 by both FNLP and NHFNLP methods

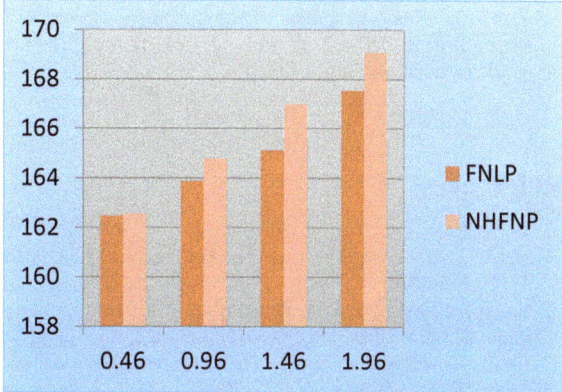

Fig. 10 Sensitivity analysis of $TAC_2{}^*$ w.r.t. h_2 by both FNLP and NHFNLP methods

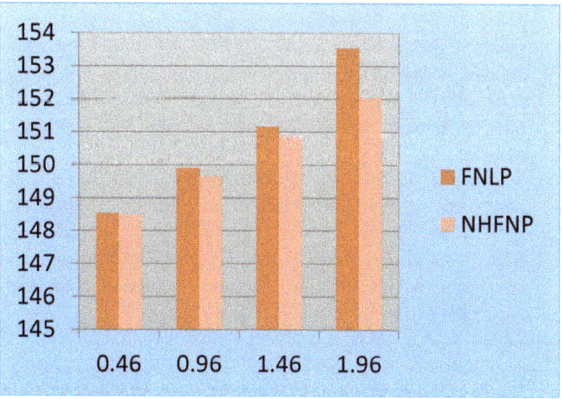

8   Conclusions

In this paper, we have considered a multi-objective inventory problem with constant deterioration after a certain time under the limitation of storage space constraint. We considered time-dependent demand and holding cost. Shortages are allowed under partial backlogging. Due to uncertainty, we consider some parameters of the model as generalized trapezoidal fuzzy number. Our proposed model is solved by both neutrosophic hesitant fuzzy programming approach and fuzzy nonlinear programming technique. Numerical example has been given to illustrate the model. Finally, sensitivity analysis has been presented graphically. In the future, other types of membership functions such as GTFN, PFN, etc. can be considered to build membership functions.

Acknowledgments The author is thankful to the University of Kalyani for providing financial assistance through DST-URSE (Phase-II) Programme. The authors are grateful to the reviewers for their comments and suggestions.

References

1. Harris, F. W. (1913). How many parts to make at once factory. *The Magazine of Management, 10*(2), 135–136.
2. Singh, T., Mishra, P. J., & Pattanayak, H. (2018). An EOQ inventory model for deteriorating items with time-dependent deterioration rate, ramp-type demand rate and shortages. *International Journal of Mathematics in Operational Research, 12*(4), 423–437.
3. Jong, W. W., & Lee, W. C. (2003). An EOQ inventory model for items with Weibull deterioration, shortages and time varying demand. *Journal of Information and Optimization Sciences, 24*(1), 103–122.
4. Ghare, P. N., & Schrader, G. F. (1963). A model for exponentially decaying inventories. *The Journal of Industrial Engineering, 15*(5), 238–243.
5. Jong, W. W., Lin, C., Tan, B., & Lee, W. C. (2000). An EOQ inventory model with time-varying demand and Weibull deterioration with shortages. *International Journal of Systems Science, 31*(6), 677–683.
6. Mishra, U. (2017). An inventory model for weibull deterioration with stock and price dependent demand. *International Journal of Applied and Computational Mathematics, 3*, 1951–1967.
7. Roy, T. K., Maity, M., & Mondal, N. K. (2006). Inventory model of deteriorating items with a constraint: A geometric programming approach. *European Journal of Operational Research, 173*, 199–210.
8. Zadeh, L. A. (1965). Fuzzy sets. *Information and Control, 8*, 338–353.
9. Zimmermann, H. J. (1992). Methods and applications of fuzzy mathematical programming, in: Lotfi A. Zadeh, Ronald R. Yager (eds) An introduction to fuzzy logic application in intelligent systems. Kluwer publishers, Boston, 97- 120.
10. Zimmermann, H. J. (1985). Application of fuzzy set theory to mathematical programming. *Information Sciences, 36*, 29–58.
11. Roy, T. K., & Maity, M. (1995). A fuzzy inventory model with constraints. *Operations Research, 32*(4), 287–298.

12. Smarandache, F. (2005). Neutrosophic set, a generalization of the intuitionistic fuzzy set. *International Journal of Pure and Applied Mathematics, 24*(3), 287–297.
13. Ye, J. (2014). Multiple-attribute decision-making method under a single-valued neutrosophic hesitant fuzzy environment. *Journal of Intelligent Systems, 24*, 23–36.
14. Ahmad, F., Adhami, A. Y., & Smarandache, F. (2018). Single valued neutrosophic hesitant fuzzy computational algorithm for multi objective nonlinear optimization problem. *Neutrosophic Sets and Systems, 22*, 76–86.
15. Islam, S., & Mandal, W. A. (2017). A fuzzy inventory model with unit production cost, time depended holding cost, with-out shortages under a space constraint: A parametric geometric programming approach. *The Journal of Fuzzy Mathematics, 25*(3), 517–532.
16. Mondal, B., Kar, C., Garai, A., & Roy, T. K. (2018). Optimization of EOQ model with limited storage capacity by neutrosophic geometric programming. *Neutrosophic Sets and Systems, 22*, 5–29.
17. Nabeeh, N. A., Abdel-Basset, M., El-Ghareeb, H. A., & Aboelfetouh, A. (2019). Neutrosophic multi-criteria decision making approach for IOT-based enterprises. *IEEE Access, 7*, 59,559–59,574.
18. Mullai, M., & Surya, R. (2018). Neutrosophic EOQ model with price break. *Neutrosophic Sets and Systems, 19*, 24–29.
19. Mohana, K., Christy, V., & Smarandache, F. (2019). On multi-criteria decision making problem via bipolar single-valued neutrosophic settings. *Neutrosophic Sets and Systems, 25*, 125–135.
20. Biswas, P., Pramanik, S., & Giri, B. C. (2018). *Multi-attribute group decision making based on expected value of neutrosophic trapezoidal numbers. New trends in neutrosophic theory and application* (pp. 103–124). Brussells, II: Pons Editions.
21. Pramanik, S., Mallick, R., & Dasgupta, A. (2018). Contributions of selected indian researchers to multi attribute decision making in neutrosophic environment: an over view. *Neutrosophic Sets and Systems, 20*, 108–131.
22. Abdel-Basset, M., Mohamed, R., Elhoseny, M., & Chang, V. (2020a). Evaluation framework for smart disaster response systems in uncertainty environment. *Mechanical Systems and Signal Processing, 145*, 106,941.
23. Abdel-Basset, M., Ali, M., & Atef, A. (2020b). Uncertainty assessments of linear time-cost tradeoffs using neutrosophic set. *Computers & Industrial Engineering, 141*, 106,286.
24. Abdel-Basset, M., Ali, M., & Atef, A. (2020c). Resource levelling problem in construction projects under neutrosophic environment. *The Journal of Supercomputing, 76*(2), 964–988.
25. Abdel-Basset, M., Gamal, A., Son, L. H., & Smarandache, F. (2020d). A bipolar neutrosophic multi criteria decision making framework for professional selection. *Applied Sciences, 10*(4), 1202.
26. Abdel-Basset, M., Mohamed, R., Zaied, A. E. N. H., Gamal, A., & Smarandache, F. (2020e). Solving the supply chain problem using the best-worst method based on a novel Plithogenic model. In F. Smarandache & M. Abdel-Basset (Eds.), *Optimization theory based on neutrosophic and plithogenic sets* (pp. 1–19). London: Academic Press.
27. Pramanik, S. (2016). Neutrosophic multi-objective linear programming. *Global Journal of Engineering Science and Research Management, 3*(8), 36–46.
28. Chen, S. H.(1985). Operations on fuzzy numbers with function principal. *Tamkang Journal of Management Science, 6*(1), 13–25.
29. Chen, S. H. (1999). Ranking generalized fuzzy number with graded mean integration. *Proceedings of 8th International Fuzzy System Association World Congress, Taipei, Taiwan, Republic of China, 2*, 899–902.
30. Liou, T. S., & Wang, M. J. J. (1992), Ranking fuzzy numbers with integral value. *Fuzzy Sets and Systems, 50*, 247–255.

Bulbul Disaster Assessment Using Single-Valued Spherical Hesitant Neutrosophic Dombi Weighted Aggregation Operators

Abhijit Saha, Debjit Dutta, and Said Broumi

1 Introduction

In our daily life, we frequently come across problems that occur mostly in various fields in engineering, social, medical and natural sciences, computer software development, public policy and business analysis, etc., and hence we proceed in quest of having right solution with the help of traditional mathematical methods but because of various uncertainties presented sometimes in these applied problems, it becomes highly essential to look for some new proper mathematical tools in order to execute the inherent inexact information lying in those problems and to achieve of it the best feasible solutions. Fuzzy set theory is one such extremely useful tool that helps us to get so. In 1965, Lotfi A. Zadeh first published the famous research paper on fuzzy sets that originated due to mainly the inclusion of vague human assessments in computing problems. In other words, the fuzzy set theory can deal with the fact evolving from computational perception and cognition that is literally meaning the uncertainty, vagueness, partial trueness, impreciseness, Sharpless boundaries, etc. Basically, the theory of fuzzy set is founded on the concept of relative graded membership which deals with the partial belongings of

A. Saha (✉)
Faculty of Mathematics, Techno College of Engg, Agartala, Tripura, India
e-mail: abhijit84.math@gmail.com

D. Dutta
Faculty of Basic and Applied Science, NIT AP, Papum Pare, Arunachal Pradesh, India
e-mail: debjitdutta.math@gmail.com

S. Broumi
Laboratory of Information Processing, Faculty of Science Ben M'Sik, University Hassan II, Casablanca, Morocco
e-mail: broumisaid78@gmail.com

© The Author(s), under exclusive license to Springer Nature Switzerland AG 2021
F. Smarandache, M. Abdel-Basset (eds.), *Neutrosophic Operational Research*,
https://doi.org/10.1007/978-3-030-57197-9_12

an element in a set in order to process inexact information. Later on, fuzzy sets have been generalized to intuitionistic fuzzy sets through adopting a nonmembership function by Atanassov in 1986 in order to get over problems that possess incomplete information. As advancement, rough set theory, vague set theory, and bipolar fuzzy set theory have been produced therein. In the context of fuzzy sets or intuitionistic fuzzy sets, it is known that the membership (or nonmembership) value of an element in a set admits a unique value in the closed interval [0,1].

But since 2009 researchers thought of getting a discrete finite subset of [0,1] for the membership (nonmembership) value of an element in a set instead of unique value. In view of this fact, as another extension of the fuzzy set, Torra [1] proposed the concept of a hesitant fuzzy set underlying the situation of our hesitation in consideration of few different values lying between the numbers 0 and 1. In this way, the hesitant fuzzy set revealed itself contemplating more accurately of people's hesitancy in embroiling their preferences over objectives when compared to both the fuzzy set and its classical extensions. Subsequently, the work of uniting the concept of intuitionistic fuzzy sets and hesitant fuzzy was carried out by Beg and Rashid [2]. Moreover, considering the high essence of the right decision-making required to take in case various problems occur in our daily life, the researchers working in this domain were able to explore successfully the decision-making problems under proper and powerful mathematical tools, namely, fuzzy, hesitant fuzzy, intuitionistic fuzzy, and intuitionistic hesitant fuzzy environment in Ye [3, 4], Nehi [5], Li [6], Xia and Xu [7], Xu and Xia [8–10], Wei et al. [11], Xu and Zhang [12], Chen et al. [13], Qian et al. [14], Yu [15, 16], Shi et al. [17], Pathinathan and Johnson [18], Joshi and Kumar [19], Zhang [20], Chen and Huang [21], Yang et al. [22], Lan et al. [23], and Zhang et al. [24].

It is known to us that intuitionistic fuzzy sets are vibrant only for hesitancy degree to lead to precarious information, but it fails to give duly administrative indeterminate information because it is entirely subordinated by the membership and nonmembership degrees. In this scenario, to manage the situation, a new type of set called neutrosophic set, a powerful general formal framework, is proposed by Smarandache [25] which is materially a generalization of the notion of the classical set, fuzzy set, and intuitionistic fuzzy set. In particular, the characterization of this neutrosophic set is explicitly done by truth-membership function, indeterminacy-membership function, and falsity-membership function. Its possible application on image segmentation has been studied in Gou and Cheng [26] and Guo and Sensur [27]. Also, we find its probable infliction on clustering analysis in Karaaslan [28] and on medical diagnosis problems in Ansari et al. [29], respectively. Furthermore, the subject of the neutrosophic set theory has been practiced in Wang et al. [30], Gou et al. [31], Wang et al. [32], Ye [33], Sun et al. [34], Ye [35], and Abdel-Basset et al. [36]. In the study of any decision-making problem, the neutrosophic set becomes hardly active to impart uncertain, imprecise, incomplete, and inconsistent information with a few different values imposed by truth-membership degree, indeterminacy-membership degree, and falsity-membership degree on account of doubts that arise from decision-maker. It means that the neutrosophic set-based decision-making algorithms cannot be implemented to all kinds of decision-making problems consisting of hesitancy information. In such a scenario, Ye [37] put forward the concept of hesitant neutrosophic sets which is characterized by three membership degrees, namely, truth-membership degrees, indeterminacy-membership degrees, and falsity-membership degrees, and this set admits few different values lying within the numbers 0 and 1.

On the other hand, it may be recalled that with the usage of aggregation operators, it has become quite easier to compute multi-criteria group decision-making problems that occur during the study of decision-making, supply chain, personnel evaluation, and financial investment. In the papers of Xia et al. [8], Wang et al. [38], Zhao et al. [39], and Peng [40], it has been exercised how to apply various aggregation operators that deal with fuzzy and hesitant fuzzy information in solving decision-making problems. Moreover, in the papers of Xu and Yager [41], Xu [42], Wan and Dong [43], and Wan et al. [44], it has been illustrated both the averaging and geometric aggregation operators for the purpose of aggregating the information underlying the intuitionistic fuzzy sets. Later on, Wang and Liu [45] proposed some Einstein weighted geometric operators for intuitionistic fuzzy sets. Liu et al. [46] also proposed some generalized neutrosophic number Hamacher aggregation operators, and next few neutrosophic normalized, weighted Bonferroni mean operators were developed by Liu and Wang [38]. In the work of Zhao et al. [47], Liu and Shi [48], and Liu and Tang [49], few more aggregation operators on neutrosophic environment have been discussed. Some recent studies in this area can be found in the work of Liu [50], Ye [51], Liang et al. [52], Wei and Wei [53], Ajay et al. [54], Khan et al. [55], Mohansundari and Mohana [56], Tooranloo et al. [57], and Vandhana and Anuradha [58].

Now it is important to mention here that some new operations called Dombi t-norm and Dombi t-conorm, developed by Dombi [59], bear the nature of well precessionness of inconstancy with the operations of parameters. With this aid, Liu et al. [60] could have made it possible to execute Dombi operations into intuitionistic fuzzy sets and at the same time implemented it toward flourishing multi-attribute decision-making problem utilizing Dombi Bonferroni mean operator in intuitionistic fuzzy environment. We refer here that He [61] successfully carried out the operations of Dombi operators to typhoon disaster assessment under hesitant fuzzy environment. Jana et al. [62] successfully examined fuzzy Dombi aggregation operators upon multi-attribute decision-making process. Soon after, Jana et al. [63] formulated Pythagorean fuzzy Dombi aggregation operators to solve multi-attribute decision-making problem. A key role was also played by Chen and Ye [64] to tackle some new challenging multi-criteria decision-making problems using single-valued neutrosophic Dombi weighted aggregation operators. Further, Shi and Ye [65] have led a significant contribution to solve various MADM problems with Dombi operations to neutrosophic cubic sets and developed neutrosophic cubic Dombi weighted aggregation operators.

Very recently Gundogdu and Kahraman [66] suggested a very useful set known as spherical fuzzy set (in short *SFS*) in which the hesitancy of a decision-maker is defined on one's own from membership and nonmembership degrees gratifying the condition $0 \leq \mu_\varsigma^2(x) + \vartheta_\varsigma^2(x) + \pi_\varsigma^2(x) \leq 1 \ \forall \ x \in U$, where $\varsigma = \{<x, \mu_\varsigma(x), \vartheta_\varsigma(x) > \ : x \in U\}$ is an intuitionistic fuzzy set and $\pi_\varsigma(x) = 1 - \mu_\varsigma(x) - \vartheta_\varsigma(x)$ indicates the hesitancy degree of $x \in U$. To this work they also practiced the basic operations such as addition, subtraction, multiplication, and aggregation endowed by *SFS* alongside the multi-criteria decision-making methods too (Gundogdu and Kahraman [66, 67]). The SFS is a notable and an advanced concept which basically provides an extensive domain of priority for decision-makers where the decision-makers are allowed to prescribe their hesitancy information about an alternative on the basis of a fixed measure or criterion. This extra special concept of SFS was further generalized to single-valued spherical neutrosophic set by Smarandache [68].

Motivated by the concepts of Dombi operations, single-valued hesitant neutrosophic set, and single-valued spherical neutrosophic set, we aim at the following points in this article:

1. To present the idea of single-valued spherical hesitant neutrosophic numbers (*SVSHNNs* for short)
2. To define few operations between *SVSHNNs* and study their basic properties
3. To develop the weighted aggregation operators such as single-valued spherical hesitant neutrosophic Dombi weighted arithmetic aggregation operators (*SVSH-NDWAA* for short) and single-valued spherical hesitant neutrosophic Dombi weighted geometric aggregation operators (*SVSHNDWGA* for short) and study their properties
4. To propose a decision-making method based on the single-valued spherical hesitant neutrosophic Dombi weighted aggregation operators to handle multi-criteria decision-making problems with single-valued spherical hesitant neutrosophic information

To do so, the rest of the article is arranged as follows:

In Sect. 2, we review some basic concepts. In Sect. 3, we first define *SVSHNN* and then propose few operations between the *SVSHNNs*. In Sect. 4, we propose various types of single-valued spherical hesitant neutrosophic Dombi weighted aggregation operators to aggregate the single-valued spherical hesitant neutrosophic information. Furthermore, we introduce the score of a *SVSHNN* to ranking the *SVSHNNs*. In Sect. 5, based on the single-valued spherical hesitant neutrosophic Dombi weighted aggregation operators and score of *SVSHNNs*, we develop a multi-attribute decision-making approach, in which the evaluation values of alternatives on the attribute are represented in terms of *SVSHNNs* and the alternatives are ranked according to the values of the score of *SVSHNNs* to select the best (most desirable) one. Also, we present a practical example for Bulbul disaster assessment to demonstrate the application and effectiveness of the proposed method. Sect. 6 is devoted for comparative study. In the final section, we present the conclusion of the study.

2 Preliminaries

In this section, first we recall some basic notions that are relevant to our study.

Definition 2.1 [32] A single-valued neutrosophic set ς on universe set U is given by

$$\varsigma = \left\{ \langle x, T_\varsigma(x), I_\varsigma(x), F_\varsigma(x) \rangle : x \in U \right\}$$

where $T_\varsigma : U \to [0, 1]$, $I_\varsigma : U \to [0, 1]$, and $F_\varsigma : U \to [0, 1]$ satisfy the condition $0 \le T_\varsigma(x) + I_\varsigma(x) + F_\varsigma(x) \le 3$, for every $x \in U$. The functions T_ς, I_ς, and F_ς define the degree of truth-membership function, indeterminacy-membership function, and falsity-membership function, respectively.

Ye [37] thought of merging the single-valued neutrosophic sets and hesitant fuzzy set mathematically that he finally succeeded in providing the notion of the single-valued neutrosophic hesitant fuzzy set, and this might be considered in another sense as the generalized version of so-called old and existing tools like fuzzy set, intuitionistic fuzzy set, single-valued neutrosophic set, etc. Getting inspired with this effective and unprecedented set, Ye [37] also developed both the single-valued neutrosophic hesitant fuzzy weighted averaging operator and single-valued neutrosophic hesitant fuzzy weighted geometric operator, and finally he was able to apply successfully those operators to solve some challenging problems under MADM.

Definition 2.2 [37] A single-valued hesitant neutrosophic set on universe set U is given by

$$\varsigma = \left\{ \left\langle x, \widetilde{T}_\varsigma(x), \widetilde{I}_\varsigma(x), \widetilde{F}_\varsigma(x) \right\rangle : x \in U \right\}$$

in which $\widetilde{T}_\varsigma(x)$, $\widetilde{I}_\varsigma(x)$, and $\widetilde{F}_\varsigma(x)$ are three sets of some values in [0,1], denoting the possible truth-membership hesitant degrees, indeterminacy-membership hesitant degrees, and falsity-membership hesitant degrees of the element $x \in U$, respectively, with the conditions $0 \le \delta, \gamma, \eta \le 1$ and $0 \le \delta^+ + \gamma^+ + \eta^+ \le 3$, where $\delta \in \widetilde{T}_N(x)$, $\gamma \in \widetilde{I}_N(x)$, $\eta \in \widetilde{F}_N(x)$ for $x \in U$.

Definition 2.3 (Smarandache [68]) A single-valued spherical neutrosophic set on the universe set U is defined as follows:

$$\psi = \left\{ < x, \mu_\psi(x), \vartheta_\psi(x), \delta_\psi(x) > : x \in U \right\}$$

where μ_ψ, ϑ_ψ, $\delta_\psi : U \to [0, 1]$ represent truth-membership, indeterminacy-membership, and falsity-membership value, respectively, of $x \in U$ satisfying the condition $0 \le \mu_\psi^2(x) + \vartheta_\psi^2(x) + \delta_\psi^2(x) \le 3$.

3 Single-Valued Spherical Hesitant Neutrosophic Numbers

We shall begin with the definition of single-valued spherical hesitant neutrosophic set and single-valued spherical hesitant neutrosophic number (*SVSHNN*). Next, in this section, we develop few operations between *SVSHNNs* using Dombi operations and understand the underlying properties of proposed one.

The operations of *t*-norm and *t*-conorm, developed by Dombi [59], were generally known as Dombi operations described below.

Definition 3.1 [59] For any two real numbers x and y lying in [0, 1], the Dombi *t*-norm and Dombi *t*-conorm can be defined as the following:

$$\text{Dom}(x, y) = \frac{1}{1 + \left\{ \left(\frac{1-x}{x} \right)^k + \left(\frac{1-y}{y} \right)^k \right\}^{\frac{1}{k}}},$$

$$\text{Dom}^c(x, y) = 1 - \frac{1}{1 + \left\{ \left(\frac{x}{1-x} \right)^k + \left(\frac{y}{1-y} \right)^k \right\}^{\frac{1}{k}}} \quad (k \geq 1)$$

Definition 3.2 A single-valued spherical hesitant neutrosophic set (SVSHNS) on the universe set U is defined as follows:

$$\tilde{a} = \left\{ < x, \left\{ w_{\tilde{a}}^i(x) : i \in I_m \right\}, \left\{ u_{\tilde{a}}^j(x) : j \in I_n \right\}, \left\{ y_{\tilde{a}}^l : l \in I_k \right\} >: x \in U \right\}$$

where $w_{\tilde{a}}^i, u_{\tilde{a}}^j, y_{\tilde{a}}^l : U \rightarrow [0, 1]$ are functions denoting the set of possible hesitant truth-membership degrees, hesitant indeterminacy-membership degrees, and hesitant falsity-membership degrees, respectively, of the element $x \in U$ satisfying condition $0 \leq \alpha^2 + \beta^2 + \lambda^2 \leq 3$ where $\alpha \in \left\{ w_{\tilde{a}}^i(x) : i \in I_m \right\}, \beta \in \left\{ u_{\tilde{a}}^j(x) : j \in I_n \right\}, \lambda \in \left\{ y_{\tilde{a}}^l : l \in I_k \right\}$.

The triplet $< \left\{ w_{\tilde{a}}^i(x) : i \in I_m \right\}, \left\{ u_{\tilde{a}}^j(x) : j \in I_n \right\}, \left\{ y_{\tilde{a}}^l : l \in I_k \right\} >$ is termed as single-valued spherical hesitant neutrosophic number (*SVSHNN*). For the sake of simplicity, throughout the paper we shall use the notation $\tilde{a} = < \left\{ w_{\tilde{a}}^i : i \in I_m \right\}, \left\{ u_{\tilde{a}}^j : j \in I_n \right\}, \left\{ y_{\tilde{a}}^l : l \in I_k \right\} >$ to denote a *SVSHNN*.

Definition 3.3 Let $\tilde{a} = < \left\{ w_{\tilde{a}}^i : i \in I_{m_1} \right\}, \left\{ u_{\tilde{a}}^j : j \in I_{n_1} \right\}, \left\{ y_{\tilde{a}}^l : l \in I_{k_1} \right\} >$ and $\tilde{b} = < \left\{ w_{\tilde{b}}^i : i \in I_{m_2} \right\}, \left\{ u_{\tilde{b}}^j : j \in I_{n_2} \right\}, \left\{ y_{\tilde{b}}^l : l \in I_{k_2} \right\} >$ be two *SVSHNNs* and $\sigma > 0$. Then using the concept of Dombi operations, we define

(i) $\tilde{a} \oplus \tilde{b} = <$

$$\left\{\sqrt{1 - \frac{1}{1 + \left(\left(\frac{\alpha_1^2}{1-\alpha_1^2}\right)^k + \left(\frac{\alpha_2^2}{1-\alpha_2^2}\right)^k\right)^{\frac{1}{k}}}} : \alpha_1 \in \left\{w_{\tilde{a}}^i : i \in I_{m_1}\right\}, \alpha_2 \in \left\{w_{\tilde{b}}^i : i \in I_{m_2}\right\}\right\},$$

$$\left\{\sqrt{\frac{1}{1 + \left(\left(\frac{1-\beta_1^2}{\beta_1^2}\right)^k + \left(\frac{1-\beta_2^2}{\beta_2^2}\right)^k\right)^{\frac{1}{k}}}} : \beta_1 \in \left\{u_{\tilde{a}}^j : j \in I_{n_1}\right\}, \beta_2 \in \left\{u_{\tilde{b}}^j : j \in I_{n_2}\right\}\right\},$$

$$\left\{\sqrt{\frac{1}{1 + \left(\left(\frac{1-\delta_1^2}{\delta_1^2}\right)^k + \left(\frac{1-\delta_2^2}{\delta_2^2}\right)^k\right)^{\frac{1}{k}}}} : \delta_1 \in \left\{y_{\tilde{a}}^l : l \in I_{k_1}\right\}, \delta_2 \in \left\{y_{\tilde{b}}^l : l \in I_{k_2}\right\}\right\} >$$

(ii) $\tilde{a} \otimes \tilde{b} = <$

$$\left\{\sqrt{\frac{1}{1 + \left(\left(\frac{1-\alpha_1^2}{\alpha_1^2}\right)^k + \left(\frac{1-\alpha_2^2}{\alpha_2^2}\right)^k\right)^{\frac{1}{k}}}} : \alpha_1 \in \left\{w_{\tilde{a}}^i : i \in I_{m_1}\right\}, \alpha_2 \in \left\{w_{\tilde{b}}^i : i \in I_{m_2}\right\}\right\},$$

$$\left\{\sqrt{1 - \frac{1}{1 + \left(\left(\frac{\beta_1^2}{1-\beta_1^2}\right)^k + \left(\frac{\beta_2^2}{1-\beta_2^2}\right)^k\right)^{\frac{1}{k}}}} : \beta_1 \in \left\{u_{\tilde{a}}^j : j \in I_{n_1}\right\}, \beta_2 \in \left\{u_{\tilde{b}}^j : j \in I_{n_2}\right\}\right\},$$

$$\left\{\sqrt{1 - \frac{1}{1 + \left(\left(\frac{\delta_1^2}{1-\delta_1^2}\right)^k + \left(\frac{\delta_2^2}{1-\delta_2^2}\right)^k\right)^{\frac{1}{k}}}} : \delta_1 \in \left\{y_{\tilde{a}}^l : l \in I_{k_1}\right\}, \delta_2 \in \left\{y_{\tilde{b}}^l : l \in I_{k_2}\right\}\right\} >$$

(iii) $\lambda \odot \tilde{a} = <$

$$\left\{\sqrt{1 - \frac{1}{1 + \left\{\lambda\left(\frac{\alpha_1^2}{1-\alpha_1^2}\right)^k\right\}^{\frac{1}{k}}}} : \alpha_1 \in \left\{w_{\tilde{a}}^i : i \in I_{m_1}\right\}\right\},$$

$$\left\{\sqrt{\frac{1}{1 + \left\{\lambda\left(\frac{1-\beta_1^2}{\beta_1^2}\right)^k\right\}^{\frac{1}{k}}}} : \beta_1 \in \left\{u_{\tilde{a}}^j : j \in I_{n_1}\right\}\right\},$$

$$\left\{\sqrt{\frac{1}{1 + \left\{\lambda\left(\frac{1-\delta_1^2}{\delta_1^2}\right)^k\right\}^{\frac{1}{k}}}} : \delta_1 \in \left\{y_{\tilde{a}}^l : l \in I_{k_1}\right\}\right\} >$$

$$\text{(iv)}\quad \lambda * \tilde{a} =< \left\{ \sqrt{\cfrac{1}{1+\left\{\lambda\left(\frac{1-\alpha_1^2}{\alpha_1^2}\right)^k\right\}^{\frac{1}{k}}}} : \alpha_1 \in \left\{w_{\tilde{a}}^i : i \in I_{m_1}\right\} \right\},$$

$$\left\{ \sqrt{1-\cfrac{1}{1+\left\{\lambda\left(\frac{\beta_1^2}{1-\beta_1^2}\right)^k\right\}^{\frac{1}{k}}}} : \beta_1 \in \left\{u_{\tilde{a}}^j : j \in I_{n_1}\right\} \right\},$$

$$\left\{ \sqrt{\cfrac{1}{1+\left\{\lambda\left(\frac{\delta_1^2}{1-\delta_1^2}\right)^k\right\}^{\frac{1}{k}}}} : \delta_1 \in \left\{y_{\tilde{a}}^l : l \in I_{k_1}\right\} \right\} >$$

For the sake of simplicity, we use the following notations:

$$\Delta\left(\varsigma_i\right) = \left(\frac{\varsigma_i^2}{1-\varsigma_i^2}\right)^k \quad \left(\varsigma_i \in [0,1], i \in N\right),$$

$$\nabla\left(\varsigma_i\right) = \left(\frac{1-\varsigma_i^2}{\varsigma_i^2}\right)^k \quad \left(\varsigma_i \in [0,1], i \in N\right)$$

Using these notions, the above operations can be expressed as

$$\text{(i)}\quad \tilde{a} \oplus \tilde{b} =< \left\{ \sqrt{1-\frac{1}{1+(\Delta(\alpha_1)+\Delta(\alpha_2))^{\frac{1}{k}}}} : \alpha_1 \in \left\{w_{\tilde{a}}^i : i \in I_{m_1}\right\}, \alpha_2 \in \left\{w_{\tilde{b}}^i : i \in I_{m_2}\right\} \right\},$$

$$\left\{ \sqrt{\frac{1}{1+(\nabla(\beta_1)+\nabla(\beta_2))^{\frac{1}{k}}}} : \beta_1 \in \left\{u_{\tilde{a}}^j : j \in I_{n_1}\right\}, \beta_2 \in \left\{u_{\tilde{b}}^j : j \in I_{n_2}\right\} \right\},$$

$$\left\{ \sqrt{\frac{1}{1+(\nabla(\delta_1)+\nabla(\delta_1))^{\frac{1}{k}}}} : \delta_1 \in \left\{y_{\tilde{a}}^l : l \in I_{k_1}\right\}, \delta_2 \in \left\{y_{\tilde{b}}^l : l \in I_{k_2}\right\} \right\} >$$

$$\text{(ii)}\quad \tilde{a} \otimes \tilde{b} =< \left\{ \sqrt{\frac{1}{1+(\nabla(\alpha_1)+\nabla(\alpha_2))^{\frac{1}{k}}}} : \alpha_1 \in \left\{w_{\tilde{a}}^i : i \in I_{m_1}\right\}, \alpha_2 \in \left\{w_{\tilde{b}}^i : i \in I_{m_2}\right\} \right\},$$

$$\left\{ \sqrt{1-\frac{1}{1+(\Delta(\beta_1)+\Delta(\beta_2))^{\frac{1}{k}}}} : \beta_1 \in \left\{u_{\tilde{a}}^j : j \in I_{n_1}\right\}, \beta_2 \in \left\{u_{\tilde{b}}^j : j \in I_{n_2}\right\} \right\},$$

$$\left\{ \sqrt{1-\frac{1}{1+(\Delta(\delta_1)+\Delta(\delta_2))^{\frac{1}{k}}}} : \delta_1 \in \left\{y_{\tilde{a}}^l : l \in I_{k_1}\right\}, \delta_2 \in \left\{y_{\tilde{b}}^l : l \in I_{k_2}\right\} \right\} >$$

$$\text{(iii)}\quad \lambda \odot \tilde{a} =< \left\{ \sqrt{1-\frac{1}{1+(\lambda\Delta(\alpha_1))^{\frac{1}{k}}}} : \alpha_1 \in \left\{w_{\tilde{a}}^i : i \in I_{m_1}\right\} \right\},$$

$$\left\{ \sqrt{\frac{1}{1+(\lambda\nabla(\beta_1))^{\frac{1}{k}}}} : \beta_1 \in \left\{u_{\tilde{a}}^j : j \in I_{n_1}\right\} \right\},$$

$$\left\{ \sqrt{\frac{1}{1+(\lambda\nabla(\delta_1))^{\frac{1}{k}}}} : \delta_1 \in \left\{y_{\tilde{a}}^l : l \in I_{k_1}\right\} \right\} >$$

$$(iv) \quad \lambda * \tilde{a} = < \left\{ \sqrt{\frac{1}{1+(\lambda\nabla(\alpha_1))^{\frac{1}{k}}}} : \alpha_1 \in \left\{ w_{\tilde{a}}^i : i \in I_{m_1} \right\} \right\},$$

$$\left\{ \sqrt{1 - \frac{1}{1+(\lambda\Delta(\beta_1))^{\frac{1}{k}}}} : \beta_1 \in \left\{ u_{\tilde{a}}^j : j \in I_{n_1} \right\} \right\},$$

$$\left\{ \sqrt{1 - \frac{1}{1+(\lambda\Delta(\delta_1))^{\frac{1}{k}}}} : \delta_1 \in \left\{ y_{\tilde{a}}^l : l \in I_{k_1} \right\} \right\} >$$

Theorem 3.4 Let $\tilde{a} = < \left\{ w_{\tilde{a}}^i : i \in I_{m_1} \right\}, \left\{ u_{\tilde{a}}^j : j \in I_{n_1} \right\}, \left\{ y_{\tilde{a}}^l : l \in I_{k_1} \right\} >, \tilde{b} = <$
$\left\{ w_{\tilde{b}}^i : i \in I_{m_2} \right\}, \left\{ u_{\tilde{b}}^j : j \in I_{n_2} \right\}, \left\{ y_{\tilde{b}}^l : l \in I_{k_2} \right\} >,$ and $\tilde{c} = < \left\{ w_{\tilde{c}}^i : i \in I_{m_3} \right\},$
$\left\{ u_{\tilde{c}}^j : j \in I_{n_3} \right\}, \left\{ y_{\tilde{c}}^l : l \in I_{k_3} \right\} >$ be
three *SVSHNNs* and $\lambda, \lambda_1, \lambda_2 > 0$; then

(i) $\tilde{a} \oplus \tilde{b} = \tilde{b} \oplus \tilde{a}$

(ii) $\tilde{a} \otimes \tilde{b} = \tilde{b} \otimes \tilde{a}$

(iii) $\tilde{a} \oplus \left(\tilde{b} \oplus \tilde{c} \right) = \left(\tilde{a} \oplus \tilde{b} \right) \oplus \tilde{c}$

(iv) $\tilde{a} \otimes \left(\tilde{b} \otimes \tilde{c} \right) = \left(\tilde{a} \otimes \tilde{b} \right) \otimes \tilde{c}$

(v) $\lambda \odot \left(\tilde{a} \oplus \tilde{b} \right) = (\lambda \odot \tilde{a}) \oplus \left(\lambda \tilde{\odot} \tilde{b} \right)$

(vi) $\lambda * \left(\tilde{a} \otimes \tilde{b} \right) = (\lambda * \tilde{a}) \otimes \left(\lambda * \tilde{b} \right)$

(vii) $(\lambda_1 + \lambda_2) \odot \tilde{a} = (\lambda_1 \odot \tilde{a}) \oplus (\lambda_2 \odot \tilde{a})$

(viii) $(\lambda_1 + \lambda_2) * \tilde{a} = (\lambda_1 * \tilde{a}) \tilde{\otimes} (\lambda_2 * \tilde{a})$

Proof Straightforward □

4 Single-Valued Spherical Hesitant Neutrosophic Dombi Weighted Aggregation Operators

This section deals with various types of single-valued spherical hesitant neutrosophic Dombi weighted aggregation operators along with their basic properties.

Definition 4.1 Let $\tilde{a}_j = < \left\{ w_{\tilde{a}_j}^i : i \in I_{m_j} \right\}, \left\{ u_{\tilde{a}_j}^r : r \in I_{n_j} \right\}, \left\{ y_{\tilde{a}_j}^l : l \in I_{k_j} \right\} >$
$(j = 1, 2, 3, \ldots, n)$ be a collection of *SVSHNNs*. Then the single-valued spherical hesitant neutrosophic Dombi weighted arithmetic aggregation operator is denoted by *SVSHNDWAA* and is defined by

$$SVSHNDWAA(\tilde{a}_1, \tilde{a}_2, \tilde{a}_3, \ldots, \tilde{a}_n) = (w_1 \odot \tilde{a}_1) \oplus (w_2 \odot \tilde{a}_2) \oplus (w_3 \odot \tilde{a}_3)$$

$$\oplus \ldots\ldots\cdots \oplus (w_n \odot \tilde{a}_n)$$

where w_j is the weight of $\tilde{a}_j$ ($j = 1, 2, 3, \ldots, n$) such that $w_j \geq 0$ and $\sum\limits_{j=1}^{n} w_j = 1$.

Theorem 4.2 Let $\tilde{a}_j = < \left\{ w_{\tilde{a}_j}^i : i \in I_{m_j} \right\}, \left\{ u_{\tilde{a}_j}^r : r \in I_{n_j} \right\}, \left\{ y_{\tilde{a}_j}^l : l \in I_{k_j} \right\} >$
($j = 1, 2, 3, \ldots, n$) be a collection of *SVSHNNs*. Then *SVSHNDWAA*
($\tilde{a}_1, \tilde{a}_2, \tilde{a}_3, \ldots, \tilde{a}_n$) is a *SVSHNN* and

$$SVSHNDWAA(\tilde{a}_1, \tilde{a}_2, \tilde{a}_3, \ldots, \tilde{a}_n)$$

$$=< \left\{ \sqrt{1 - \frac{1}{1 + \left\{ \sum\limits_{j=1}^{n} w_j \Delta(\alpha_j) \right\}^{\frac{1}{k}}}} : \alpha_j \in \left\{ w_{\tilde{a}_j}^i : i \in I_{m_j} \right\} \right\},$$

$$\left\{ \sqrt{\frac{1}{1 + \left\{ \sum\limits_{j=1}^{n} w_j \nabla(\beta_j) \right\}^{\frac{1}{k}}}} : \beta_j \in \left\{ u_{\tilde{a}_j}^r : r \in I_{n_j} \right\} \right\},$$

$$\left\{ \sqrt{\frac{1}{1 + \left\{ \sum\limits_{j=1}^{n} w_j \nabla(\delta_j) \right\}^{\frac{1}{k}}}} : \delta_j \in \left\{ y_{\tilde{a}_j}^l : l \in I_{k_j} \right\} \right\} >$$

where w_j is the weight of $\tilde{a}_j$ ($j = 1, 2, 3, \ldots, n$) such that $w_j \geq 0$ and $\sum\limits_{j=1}^{n} w_j = 1$.

Proof Straightforward $\square$

Theorem 4.3 Let $\tilde{a}_j = < \left\{ w_{\tilde{a}_j}^i : i \in I_{m_j} \right\}, \left\{ u_{\tilde{a}_j}^r : r \in I_{n_j} \right\}, \left\{ y_{\tilde{a}_j}^l : l \in I_{k_j} \right\} >$
($j = 1, 2, 3, \ldots, n$) be a collection of *SVSHNNs*. Then for any *SVSHNN* $\tilde{a}_\theta$, we
have

(i) $SVSHNDWAA(\tilde{a}_\theta \oplus \tilde{a}_1, \tilde{a}_\theta \oplus \tilde{a}_2, \ldots, \tilde{a}_\theta \oplus \tilde{a}_n) = \tilde{a}_\theta \oplus$
$SVSHNDWAA(\tilde{a}_1, \tilde{a}_2, \tilde{a}_3, \ldots, \tilde{a}_n)$
(ii) $SVSHNDWAA(\tilde{a}_1, \tilde{a}_2, \tilde{a}_3, \ldots, \tilde{a}_n) = \tilde{a}_\theta$ if $\tilde{a}_j = \tilde{a}_\theta$ for each j

Proof Straightforward $\square$

Definition 4.4 Let $\tilde{a} = < \left\{ w_{\tilde{a}}^i : i \in I_m \right\}, \left\{ u_{\tilde{a}}^j : j \in I_n \right\}, \left\{ y_{\tilde{a}}^l : l \in I_k \right\} >$ be a
SVSHNN. Then we define the score of $\tilde{a}$ by

$$S(\tilde{a}) = \frac{1}{3} \left[2 + \frac{1}{m} \sum\limits_{j=1}^{m} \alpha_j - \frac{1}{n} \sum\limits_{j=1}^{n} \beta_j - \frac{1}{k} \sum\limits_{j=1}^{k} \delta_j \right]$$

where $\alpha_j \in \left\{ w_{\tilde{a}_j}^i : i \in I_{m_j} \right\}, \beta_j \in \left\{ u_{\tilde{a}_j}^r : r \in I_{n_j} \right\}, \delta_j \in \left\{ y_{\tilde{a}_j}^l : l \in I_{k_j} \right\}$

If $\tilde{a}_j \ =< \ \left\{ w^i_{\tilde{a}_j} : i \in I_{m_j} \right\}, \left\{ u^r_{\tilde{a}_j} : r \in I_{n_j} \right\}, \left\{ y^l_{\tilde{a}_j} : l \in I_{k_j} \right\} > \ (j = 1, 2)$ are two *SVSHNNs*, then the comparison method is given as

I. If $S(\tilde{a}_1) > S(\tilde{a}_2)$ then $\tilde{a}_1 \succ \tilde{a}_2$

II. If $S(\tilde{a}_1) = S(\tilde{a}_2)$ then $\tilde{a}_1 = \tilde{a}_2$

Theorem 4.5 Let $\tilde{a}_j \ =< \ \left\{ w^i_{\tilde{a}_j} : i \in I_{m_j} \right\}, \left\{ u^r_{\tilde{a}_j} : r \in I_{n_j} \right\}, \left\{ y^l_{\tilde{a}_j} : l \in I_{k_j} \right\} >$ $(j = 1, 2, 3, \ldots, n)$ be a collection of *SVSHNNs*. Then, we have

$$\tilde{a}^- \leq SVSHNDWAA(\tilde{a}_1, \tilde{a}_2, \tilde{a}_3, \ldots, \tilde{a}_n) \leq \tilde{a}^+$$

where

$$\tilde{a}^- =< \left\{ \min_{1 \leq j \leq n} \alpha_j : \alpha_j \in \left\{ w^i_{\tilde{a}_j} : i \in I_{m_j} \right\} \right\}, \left\{ \max_{1 \leq j \leq n} \beta_j : \beta_j \in \left\{ u^r_{\tilde{a}_j} : r \in I_{n_j} \right\} \right\}, \left\{ \max_{1 \leq j \leq n} \delta_j : \delta_j \in \left\{ y^l_{\tilde{a}_j} : l \in I_{k_j} \right\} \right\} >,$$

$$\tilde{a}^+ =< \left\{ \max_{1 \leq j \leq n} \alpha_j : \alpha_j \in \left\{ w^i_{\tilde{a}_j} : i \in I_{m_j} \right\} \right\}, \left\{ \min_{1 \leq j \leq n} \beta_j : \beta_j \in \left\{ u^r_{\tilde{a}_j} : r \in I_{n_j} \right\} \right\}, \left\{ \min_{1 \leq j \leq n} \delta_j : \delta_j \in \left\{ y^l_{\tilde{a}_j} : l \in I_{k_j} \right\} \right\} >.$$

Proof Straightforward $\square$

Theorem 4.6 Let $\tilde{a}_j \ =< \ \left\{ w^i_{\tilde{a}_j} : i \in I_{m_j} \right\}, \left\{ u^r_{\tilde{a}_j} : r \in I_{n_j} \right\}, \left\{ y^l_{\tilde{a}_j} : l \in I_{k_j} \right\} >$ $(j = 1, 2, 3, \ldots, n)$ and $\tilde{\xi}_j =< \left\{ \hat{w}^i_{\tilde{\xi}_j} : i \in I_{m_j} \right\}, \left\{ \hat{u}^r_{\tilde{\xi}_j} : r \in I_{n_j} \right\}, \left\{ \hat{y}^l_{\tilde{\xi}_j} : l \in I_{k_j} \right\} >$ $(j = 1, 2, 3, \ldots, n)$ be two collections of *SVSHNNs* such that $\alpha_j \geq \hat{\alpha}_j, \beta_j \leq \hat{\beta}_j$ and $\delta_j \leq \hat{\delta}_j$ for all j where

$$\alpha_j \in \left\{ w^i_{\tilde{a}_j} : i \in I_{m_j} \right\}, \hat{\alpha}_j \in \left\{ \hat{w}^i_{\tilde{\xi}_j} : i \in I_{m_j} \right\}, \beta_j \in \left\{ u^r_{\tilde{a}_j} : r \in I_{n_j} \right\}, \hat{\beta}_j \in$$

$$\left\{ \hat{u}^r_{\tilde{\xi}_j} : r \in I_{n_j} \right\}, \delta_j \in \left\{ y^l_{\tilde{a}_j} : l \in I_{k_j} \right\}, \hat{\delta}_j \in \left\{ \hat{y}^l_{\tilde{\xi}_j} : l \in I_{k_j} \right\}.$$

$$SVSHNDWAA(\tilde{a}_1, \tilde{a}_2, \tilde{a}_3, \ldots, \tilde{a}_n)$$

Then we have

$$\geq SVSHNDWAA\left(\tilde{\xi}_1, \tilde{\xi}_2, \tilde{\xi}_3, \ldots, \tilde{\xi}_n\right)$$

Proof Straightforward $\square$

Definition 4.7 Let $\tilde{a}_j \ =< \ \left\{ w^i_{\tilde{a}_j} : i \in I_{m_j} \right\}, \left\{ u^r_{\tilde{a}_j} : r \in I_{n_j} \right\}, \left\{ y^l_{\tilde{a}_j} : l \in I_{k_j} \right\} >$ $(j = 1, 2, 3, \ldots, n)$ be a collection of *SVSHNNs*. Then the single-valued spherical hesitant neutrosophic Dombi weighted geometric aggregation operator is denoted by *SVSHNDWGA* and is defined by

$$SVSHNDWGA(\tilde{a}_1, \tilde{a}_2, \tilde{a}_3, \ldots, \tilde{a}_n) = (w_1 * \tilde{a}_1) \otimes (w_2 * \tilde{a}_2) \otimes (w_3 * \tilde{a}_3) \otimes ..$$

$$\cdots\cdots \otimes (w_n * \tilde{a}_n)$$

where w_j is the weight of $\tilde{a}_j$ $(j = 1, 2, 3, \ldots, n)$ such that $w_j \geq 0$ and $\sum_{j=1}^{n} w_j = 1$.

Theorem 4.8 Let $\tilde{a}_j = < \left\{ w^i_{\tilde{a}_j} : i \in I_{m_j} \right\}, \left\{ u^r_{\tilde{a}_j} : r \in I_{n_j} \right\}, \left\{ y^l_{\tilde{a}_j} : l \in I_{k_j} \right\} >$ $(j = 1, 2, 3, \ldots, n)$ be a collection of *SVSHNNs*. Then $SVSHNDWGA$ $(\tilde{a}_1, \tilde{a}_2, \tilde{a}_3, \ldots, \tilde{a}_n)$ is a *SVSHNN* and

$$SVSHNDWGA\,(\tilde{a}_1, \tilde{a}_2, \tilde{a}_3, \ldots, \tilde{a}_n)$$

$$=< \left\{ \sqrt{\frac{1}{1 + \left\{ \sum_{j=1}^{n} w_j \nabla(\alpha_j) \right\}^{\frac{1}{k}}}} : \alpha_j \in \left\{ w^i_{\tilde{a}_j} : i \in I_{m_j} \right\} \right\},$$

$$\left\{ \sqrt{1 - \frac{1}{1 + \left\{ \sum_{j=1}^{n} w_j \Delta(\beta_j) \right\}^{\frac{1}{k}}}} : \beta_j \in \left\{ u^r_{\tilde{a}_j} : r \in I_{n_j} \right\} \right\},$$

$$\left\{ \sqrt{1 - \frac{1}{1 + \left\{ \sum_{j=1}^{n} w_j \Delta\left(\delta_j\right) \right\}^{\frac{1}{k}}}} : \delta_j \in \left\{ y^l_{\tilde{a}_j} : l \in I_{k_j} \right\} \right\} >$$

where w_j is the weight of $\tilde{a}_j$ $(j = 1, 2, 3, \ldots, n)$ such that $w_j \geq 0$ and $\sum_{j=1}^{n} w_j = 1$.

Proof Similar to Theorem 4.2   □

Theorem 4.9 Let $\tilde{a}_j = < \left\{ w^i_{\tilde{a}_j} : i \in I_{m_j} \right\}, \left\{ u^r_{\tilde{a}_j} : r \in I_{n_j} \right\}, \left\{ y^l_{\tilde{a}_j} : l \in I_{k_j} \right\} >$ $(j = 1, 2, 3, \ldots, n)$ be a collection of *SVSHNNs*. Then for any *SVSHNN* $\tilde{a}_\theta$, we have

(i) $SVSHNDWGA\,(\tilde{a}_\theta \otimes \tilde{a}_1, \tilde{a}_\theta \otimes \tilde{a}_2, \tilde{a}_\theta \otimes \tilde{a}_3, \ldots\ldots, \tilde{a}_\theta \otimes \tilde{a}_n) \quad = \quad \tilde{a}_\theta \otimes$
     $SVSHNDWGA\,(\tilde{a}_1, \tilde{a}_2, \tilde{a}_3, \ldots\ldots, \tilde{a}_n)$
(ii) $SVSHNDWGA\,(\tilde{a}_1, \tilde{a}_2, \tilde{a}_3, \ldots\ldots, \tilde{a}_n) = \tilde{a}_\theta$ if $\tilde{a}_j = \tilde{a}_\theta$ for each j

Proof Similar to Theorem 4.3   □

Theorem 4.10 Let $\tilde{a}_j = < \left\{ w^i_{\tilde{a}_j} : i \in I_{m_j} \right\}, \left\{ u^r_{\tilde{a}_j} : r \in I_{n_j} \right\}, \left\{ y^l_{\tilde{a}_j} : l \in I_{k_j} \right\} >$ $(j = 1, 2, 3, \ldots, n)$ be a collection of *SVSHNNs*. Then, we have

$$\tilde{a}^- \leq SVSHNDWGA\,(\tilde{a}_1, \tilde{a}_2, \tilde{a}_3, \ldots\ldots, \tilde{a}_n) \leq \tilde{a}^+$$

where

$$\tilde{a}^- =< \left\{ \min_{1 \leq j \leq n} \alpha_j : \alpha_j \in \left\{ w^i_{\tilde{a}_j} : i \in I_{m_j} \right\} \right\}, \left\{ \max_{1 \leq j \leq n} \beta_j : \beta_j \in \left\{ u^r_{\tilde{a}_j} : r \in I_{n_j} \right\} \right\},$$
$$\left\{ \max_{1 \leq j \leq n} \delta_j : \delta_j \in \left\{ y^l_{\tilde{a}_j} : l \in I_{k_j} \right\} >, \right.$$
$$\tilde{a}^+ =< \left\{ \max_{1 \leq j \leq n} \alpha_j : \alpha_j \in \left\{ w^i_{\tilde{a}_j} : i \in I_{m_j} \right\} \right\}, \left\{ \min_{1 \leq j \leq n} \beta_j : \beta_j \in \left\{ u^r_{\tilde{a}_j} : r \in I_{n_j} \right\} \right\},$$
$$\left\{ \min_{1 \leq j \leq n} \delta_j : \delta_j \in \left\{ y^l_{\tilde{a}_j} : l \in I_{k_j} \right\} > .$$

Proof Similar to theorem 4.5 □

Theorem 4.11 Let $\tilde{a}_j =< \left\{ w^i_{\tilde{a}_j} : i \in I_{m_j} \right\}, \left\{ u^r_{\tilde{a}_j} : r \in I_{n_j} \right\}, \left\{ y^l_{\tilde{a}_j} : l \in I_{k_j} \right\} >$
$(j = 1, 2, 3, \ldots, n)$ and $\tilde{\xi}_j =< \left\{ \hat{w}^i_{\tilde{\xi}_j} : i \in I_{m_j} \right\}, \left\{ \hat{u}^r_{\tilde{\xi}_j} : r \in I_{n_j} \right\}, \left\{ \hat{y}^l_{\tilde{\xi}_j} : l \in I_{k_j} \right\} >$
$(j=1, 2, 3, \ldots, n)$ be two collections of *SVSHNNs* such that $\alpha_j \geq \hat{\alpha}_j, \beta_j \leq$
$\hat{\beta}_j$ and $\delta_j \leq \hat{\delta}_j$ for all j where

$$\alpha_j \in \left\{ w^i_{\tilde{a}_j} : i \in I_{m_j} \right\}, \hat{\alpha}_j \in \left\{ \hat{w}^i_{\tilde{\xi}_j} : i \in I_{m_j} \right\}, \beta_j \in \left\{ u^r_{\tilde{a}_j} : r \in I_{n_j} \right\},$$
$$\hat{\beta}_j \in \left\{ \hat{u}^r_{\tilde{\xi}_j} : r \in I_{n_j} \right\}, \delta_j \in \left\{ y^l_{\tilde{a}_j} : l \in I_{k_j} \right\}, \hat{\delta}_j \in \left\{ \hat{y}^l_{\tilde{\xi}_j} : l \in I_{k_j} \right\}.$$

Then we have $SVSHNDWGA(\tilde{a}_1, \tilde{a}_2, \tilde{a}_3, \ldots, \tilde{a}_n) \geq SVSHNDWGA$
$\left(\tilde{\xi}_1, \tilde{\xi}_2, \tilde{\xi}_3, \ldots, \tilde{\xi}_n \right).$

Proof Similar to theorem 4.6 □

5 Multi-attribute Decision-Making Using SVSHNNs

In this section, we apply the weighted aggregation operators and the score function
of *SVSHNNs* to the multi-attribute decision-making problem.

Let $X = \{A_1, A_2, A_3, \ldots, A_m\}$ be a set of alternatives, $A = \{c_1, c_2, c_3, \ldots, c_n\}$ be a
set of attributes, and $w = \{w_1, w_2, w_3, \ldots, w_n\}$ be a set of weights (w_j is the weight
of attribute c_j ($j = 1, 2, 3, \ldots, n$) such that $w_j \geq 0$ and $\sum_{j=1}^{n} w_j = 1$). In this case, the
characteristic of the alternative $A_i (i = 1, 2, \ldots, m)$ on attribute c_j ($j = 1, 2, \ldots, n$) is
represented by the following form of a *SVSHNN*:

$$A_{ij} =< \left\{ w^p_{\tilde{a}_{ij}} : p \in I_{m_{ij}} \right\}, \left\{ u^r_{\tilde{a}_{ij}} : r \in I_{n_{ij}} \right\}, \left\{ y^l_{\tilde{a}_{ij}} : l \in I_{k_{ij}} \right\} >$$

Now, we construct a multi-attribute decision-making method by the following algorithm:

Algorithm

Step 1: Express the evaluation results of the expert(s) based on the alternative $A_i(i = 1, 2, \ldots, m)$ on attribute c_j $(1, 2, \ldots, n)$ in terms of *SVSHNNs* A_{ij} as a $m \times n$ table.

Step 2: Compute the aggregation values $g_i(i = 1, 2, \ldots, m)$ of A_i $(i = 1, 2, \ldots, m)$ as

$$g_i = SVSHNDWAA\,(A_{i1}, A_{i2}, \ldots, A_{in}) \ (i = 1, 2, \ldots, m) \quad \text{or}$$

$$g_i = SVSHNDWGA\,(A_{i1}, A_{i2}, \ldots, A_{in}) \ (i = 1, 2, \ldots, m).$$

Step 3: Calculate the score values of $g_i(i = 1, 2, \ldots, m)$ which are nothing but the score values of the alternatives $A_i(i = 1, 2, \ldots, m)$ based on Definition 4.4.

Step 4: Rank the alternatives by using Definition 4.4.

5.1 An Illustrative Example

The term "Bulbul," pronounced as Bul bul, is considered as one of the very severe cyclonic storms in the world history that basically in beginning of November 2019 stroked the Indian states of West Bengal, Odisha, as well as Bangladesh surrounded by west central and adjoining east central Bay of Bengal in the form of catastrophic tropical cyclone effectively sparking off heavy rainstorm with surge and flash floods across the areas. The study says, on November 2, 2019, Severe Tropical Storm Matmo's leavings turned in northern Andaman Sea just after going beyond the Indochinese Peninsula and very soon, i.e., on the 5th of November, it converted into a sharp depression upon Bay of Bengal and gradually strengthened into a cyclonic storm due to which the India Meteorological Department (IMD) started to impute the name *Bulbul*. Soon after, it moved its direction toward North with intensifying and increasing in speed for the next 2–3 days, and next it recurved toward northeast and transited West Bengal-Bangladesh Coasts between Sagar Islands of West Bengal and Khepupara (Bangladesh), nearby Sunderban delta by midnight of November 9, 2019, in the form of a severe cyclonic storm (Bulbul) with maximum wind speed recorded approximately 110–120 Kmph and then finally gusted to 135 Kmph and get thrusted into the land with high storm surge. Thereafter, in the next day after hard land interaction, Bulbul quickly weakened into a deep depression as it started to move over Bangladesh with heavy rainfall. During this journey of Bulbul, in spite of IMD's high alert for the people engaged in fishering, beach activity and boating, etc. and locality in surroundings of coastal place (across Bulbul's origin) of Bay of Bengal for Bangladesh and West Bengal including Odisha in India, more than 35 lakhs of people of both these countries got affected directly, and over 7 lakhs homes were damaged by Bulbul cyclone, a big natural disaster.

The above instance made us realize to find the probable way to assess the loss from this kind of disaster that invokes to put an effort to develop some mathematical models which will guide the Bulbul phenomenon toward assessment. As a matter of fact, as on date no mathematical research has been carried out based on severe cyclonic storm like Bulbul disaster. In this regard, we may employ the proposed approach by considering several attributes at the same time in order to characterize the Bulbul disaster in a great deal. Economic loss (C_1), environmental damage (C_2), and social impact (C_3) are three of the most important attributes, and let the weight vector of these three attributes be $w = (0.23, 0.45, 0.32)^T$. Two effected districts (namely, Khulana (A_1) and Bagerhat (A_2)) of Bangladesh and three effected districts (namely, East Midnapore (A_3), North 24 Parganas (A_4), and South 24 Parganas (A_5)) of West Bengal, India, can be evaluated from the point of view of such scenario which is articulated as follows:

Step 1: We express the initial evaluation results of the expert for five possible alternatives based on three attributes by the form of *SVSHNN*s, as shown in Table 1.

Step 2: We compute the aggregation values $g_i(i=1, 2, \ldots, 5)$ of $A_i(i=1, 2, \ldots, 5)$ as

$$g_1 = SVSHNDWAA \ (A_{11}, A_{12}, A_{13})$$

=<{0.719644445, 0.72142341, 0.726099504, 0.771448703, 0.77230834, 0.77460995, 0.825941357, 0.82628133, 0.82720276, 0.838575973, 0.838840644, 0.83955945}, {0.257347606, 0.41428934, 0.260780708, 0.470940213}, {0.142898978, 0.263544682}>,

$$g_2 = SVSHNDWAA \ (A_{21}, A_{22}, A_{23})$$

=<{0.510820607, 0.577623854, 0.582664796, 0.620082732, 0.684682268, 0.698215303}, {0.143570642, 0.143576016, 0.281101199, 0.281265866, 0.492928923, 0.496310067}, {0.239494922, 0.240752921, 0.241299869}>,

Table 1 Evaluation result by the Expert

	C_1	C_2	C_3
A_1	<{0.8,0.9},{0.5}, {0.1, 0.2}}>	<{0.7, 0.8}, {0.4, 0.6}, {0.3}>	<{0.4, 0.5, 0.6}, {0.2, 0.4}, {0.6}>
A_2	<{0.6, 0.7, 0.8}, {0.1, 0.2, 0.4}, {0.5}>	<{0.4, 0.6}, {0.7, 0.8}, {0.2}>	<{0.5}, {0.5}, {0.4, 0.5, 0.7}>
A_3	<{0.5, 0.7}, {0.3, 0.4}, {0.6, 0.8}>	<{0.2, 0.3}, {0.5, 0.7}, {0.4, 0.5}>	<{0.4, 0.5, 0.6}, {0.5}, {0.6, 0.7}>
A_4	<{0.4, 0.5, 0.7}, {0.4}, {0.7, 0.8}>	<{0.6, 0.8}, {0.4}, {0.7, 0.9}>	<{0.5, 0.6}, {0.2, 0.4, 0.5}, {0.6}>
A_5	<{0.8, 0.9}, {0.2, 0.3}, {0.5}>	<{0.3, 0.4, 0.6}, {0.1, 0.2}, {0.5, 0.7}>	<{0.2, 0.3}, {0.5, 0.6}, {0.8, 0.9}>

$$g_3 = SVSHNDWAA \ (A_{31}, A_{32}, A_{33})$$

=<{0.403781348, 0.446340186, 0.513015684, 0.411524764, 0.451434238, 0.515661773, 0.567087628, 0.576829749, 0.599372742, 0.568624999, 0.578220694, 0.600472095}, {0.391702213, 0.402105112, 0.464102569, 0.492928923}, {0.458324945, 0.46269222, 0.541857584, 0.553549139, 0.462621001, 0.46722548, 0.553351476, 0.566674349}>,

$$g_4 = SVSHNDWAA \ (A_{41}, A_{42}, A_{43})$$

=<{0.549075034, 0.578108408, 0.740055606, 0.743710805, 0.557556316, 0.584465614, 0.74103135, 0.74463831, 0.620082732, 0.634878054, 0.75084008, 0.753988832}, {0.25607376, 0.4, 0.420703162}, {0.656232413, 0.684594299, 0.666127908, 0.697913159}>,

$$g_5 = SVSHNDWAA \ (A_{51}, A_{52}, A_{53})$$

=<{0.819557023, 0.819598389, 0.81974836, 0.819789578, 0.821730144, 0.82176986, 0.679014065, 0.67933527, 0.680491563, 0.680807134, 0.69467836, 0.694943883}, {0.120892718, 0.120904964, 0.218930394, 0.219177896, 0.121617394, 0.121630013, 0.236362865, 0.236729345}, {0.534962605, 0.536255132, 0.616560474, 0.619612802}>.

Step 3: We calculate the score values of $g_i(i = 1, 2, 3, 4, 5)$ of $A_i(i = 1, 2, \ldots, 5)$ as

$$S(A_1)$$
$$= \tfrac{1}{3} \times \Big[2 + \tfrac{1}{12} \Big(0.719644445 + 0.72142341 + 0.726099504 + 0.771448703 +$$
$$0.77230834 + 0.77460995 + 0.825941357 + 0.82628133 + 0.82720276$$
$$+ 0.838575973 + 0.838840644 + 0.83955945 \Big) + \tfrac{1}{4} (0.257347606 + 0.41428934$$
$$+ 0.260780708 + 0.470940213) + \tfrac{1}{2} (0.142898978 + 0.263544682) \Big] = 0.7453.$$

Similarly, we have $S(A_2) = 0.6884$, $S(A_3) = 0.5258$, $S(A_4) = 0.5437$, $S(A_5) = 0.6670$.

Step 4: Since $S(A_1) > S(A_2) > S(A_5) > S(A_4) > S(A_3)$, $A_1 \succ A_2 \succ A_5 \succ A_4 \succ A_3$. Thus we conclude that A_1, i.e., Khulana, suffered most due to the Bulbul disaster.

In another aspect, if we apply the other proposed weighted aggregation operator, namely, *SVSHNDWGA*, then the above problem can be solved similarly as above. If we utilize the operator *SVSHNDWAA* or *SVSHNDWGA* for different values of k, then the final score values and the ranking order of the given alternatives are summarized in Table 2. We can conclude from Table 2 that although the ranking orders of the alternatives are slightly different, the most desirable alternative is still A_1 in all cases. In other words, Khulna district is the biggest sufferer of Bulbul disaster in all cases.

Table 2 Ranking order of alternatives for different values of the parameter k

Value of k	Operators used	Overall score values	Ranking order
2	SVSHNDWAA	$S(A_1)=0.7453, S(A_2)=0.6884, S(A_3)=0.5258, S(A_4)=0.5437, S(A_5)=0.6670$	$A_1 \succ A_2 \succ A_5 \succ A_4 \succ A_3$
	SVSHNDWGA	$S(A_1)=0.5378, S(A_2)=0.4450, S(A_3)=0.3821, S(A_4)=0.4568, S(A_5)=0.3573$	$A_1 \succ A_4 \succ A_2 \succ A_5 \succ A_3$
3	SVSHNDWAA	$S(A_1)=0.7609, S(A_2)=0.7075, S(A_3)=0.5455, S(A_4)=0.5554, S(A_5)=0.6880$	$A_1 \succ A_2 \succ A_5 \succ A_4 \succ A_3$
	SVSHNDWGA	$S(A_1)=0.5143, S(A_2)=0.4275, S(A_3)=0.3693, S(A_4)=0.4463, S(A_5)=0.3345$	$A_1 \succ A_4 \succ A_2 \succ A_3 \succ A_5$
5	SVSHNDWAA	$S(A_1)=0.7757, S(A_2)=0.7254, S(A_3)=0.5664, S(A_4)=0.5682, S(A_5)=0.7062$	$A_1 \succ A_2 \succ A_5 \succ A_4 \succ A_3$
	SVSHNDWGA	$S(A_1)=0.4913, S(A_2)=0.4100, S(A_3)=0.3563, S(A_4)=0.4340, S(A_5)=0.3144$	$A_1 \succ A_4 \succ A_2 \succ A_3 \succ A_5$
10	SVSHNDWAA	$S(A_1)=0.7879, S(A_2)=0.7403, S(A_3)=0.5854, S(A_4)=0.5806, S(A_5)=0.7200$	$A_1 \succ A_2 \succ A_5 \succ A_4 \succ A_3$
	SVSHNDWGA	$S(A_1)=0.4713, S(A_2)=0.3944, S(A_3)=0.3450, S(A_4)=0.4218, S(A_5)=0.2988$	$A_1 \succ A_4 \succ A_2 \succ A_3 \succ A_5$
20	SVSHNDWAA	$S(A_1)=0.7940, S(A_2)=0.7480, S(A_3)=0.5955, S(A_4)=0.5875, S(A_5)=0.7267$	$A_1 \succ A_2 \succ A_5 \succ A_3 \succ A_4$
	SVSHNDWGA	$S(A_1)=0.4606, S(A_2)=0.3861, S(A_3)=0.3391, S(A_4)=0.4150, S(A_5)=0.2910$	$A_1 \succ A_4 \succ A_2 \succ A_3 \succ A_5$
50	SVSHNDWAA	$S(A_1)=0.7976, S(A_2)=0.7525, S(A_3)=0.6015, S(A_4)=0.5916, S(A_5)=0.7307$	$A_1 \succ A_2 \succ A_5 \succ A_3 \succ A_4$
	SVSHNDWGA	$S(A_1)=0.4542, S(A_2)=0.3811, S(A_3)=0.3356, S(A_4)=0.4110, S(A_5)=0.2864$	$A_1 \succ A_4 \succ A_2 \succ A_3 \succ A_5$
100	SVSHNDWAA	$S(A_1)=0.7988, S(A_2)=0.7540, S(A_3)=0.6035, S(A_4)=0.5930, S(A_5)=0.7320$	$A_1 \succ A_2 \succ A_5 \succ A_3 \succ A_4$
	SVSHNDWGA	$S(A_1)=0.4521, S(A_2)=0.3794, S(A_3)=0.3344, S(A_4)=0.4096, S(A_5)=0.2848$	$A_1 \succ A_4 \succ A_2 \succ A_3 \succ A_5$

6 Comparative Study

In pursuance of performance comparison of the eloquent method developed by us discussed here with some existing methods [37, 64], a comparative study alongside their corresponding final ranking is summarized in tabular form, numbered by 3. It is very much translucent from Table 3 that in spite of appearance of slight difference that occurs to the respective ranking order of the alternatives, the best, i.e., most desirable, alternative is absolutely the same as found in the existing approaches [37, 64]. Thus, our method presented in this paper enriched with more fuzziness and uncertainties can be regarded as one of the best suitable tools developed so far to solve the multi-attribute decision-making problems.

7 Conclusion

To sum up, in this paper we discussed about the single-valued spherical hesitant neutrosophic numbers (*SVSHNNs*) and their basic properties. Also, various types of Dombi operations between the *SVSHNNs* are studied. Afterward, we proposed two types of single-valued spherical hesitant neutrosophic weighted aggregation operators to aggregate, i.e., amalgamate the single-valued spherical hesitant neutrosophic information. Furthermore, in order to rank the *SVSHNNs*, we suggested the score of *SVSHNNs*. Making use of the single-valued spherical hesitant neutrosophic weighted aggregation operators and score of *SVSHNNs*, we became enable to further develop a multi-attribute decision-making method in which the evaluation values of alternatives on the attribute are expressed in terms of *SVSHNNs* and the alternatives are ranked as per the attained score values of *SVSHNNs* to arrive at the goal, i.e., to select the wished one. Finally, we demonstrated a practical example concerning to the assessment of Bulbul disaster to measure acceptability of the proposed method. Our extensive study in this paper will definitely reflect in improving the decision-making level of both the disaster reduction and disaster prevention. The proposed method presented in this study can be regarded as one of the best suited approaches while solving multi-attribute decision-making problems due to the fact that *SVSHNNs* can manage successfully the indeterminate and inconsistent information and, on the other hand, these are the extended version of hesitant fuzzy numbers, hesitant intuitionistic fuzzy numbers, as well as single-valued neutrosophic numbers, respectively.

Table 3 Comparative study

Value of k	Operators used	Overall score values	Ranking order
4	SVSHNDWAA	$S(A_1) = 0.7806, S(A_2) = 0.7327, S(A_3) = 0.5344, S(A_4) = 0.5627, S(A_5) = 0.7368$	$A_1 \succ A_5 \succ A_2 \succ A_4 \succ A_3$
	SVSHNDWGA	$S(A_1) = 0.4875, S(A_2) = 0.4156, S(A_3) = 0.3730, S(A_4) = 0.4244, S(A_5) = 0.3214$	$A_1 \succ A_4 \succ A_2 \succ A_3 \succ A_5$
	SVNHFWA[37]	$S(A_1) = 0.6648, S(A_2) = 0.5782, S(A_3) = 0.4660, S(A_4) = 0.5099, S(A_5) = 0.5532$	$A_1 \succ A_2 \succ A_5 \succ A_4 \succ A_3$
	SVNHFWG[37]	$S(A_1) = 0.6133, S(A_2) = 0.5122, S(A_3) = 0.4316, S(A_4) = 0.4882, S(A_5) = 0.4639$	$A_1 \succ A_2 \succ A_5 \succ A_4 \succ A_3$
	SVNWDAA [64]	$S(A_1) = 0.7666, S(A_2) = 0.7188, S(A_3) = 0.5184, S(A_4) = 0.5520, S(A_5) = 0.7291$	$A_1 \succ A_5 \succ A_2 \succ A_4 \succ A_3$
	SVNWDGA [64]	$S(A_1) = 0.5030, S(A_2) = 0.4243, S(A_3) = 0.3805, S(A_4) = 0.4321, S(A_5) = 0.3334$	$A_1 \succ A_4 \succ A_2 \succ A_3 \succ A_5$
7	SVSHNDWAA	$S(A_1) = 0.7955, S(A_2) = 0.7461, S(A_3) = 0.5540, S(A_4) = 0.5773, S(A_5) = 0.7499$	$A_1 \succ A_5 \succ A_2 \succ A_4 \succ A_3$
	SVSHNDWGA	$S(A_1) = 0.4656, S(A_2) = 0.4002, S(A_3) = 0.3636, S(A_4) = 0.4116, S(A_5) = 0.3052$	$A_1 \succ A_4 \succ A_2 \succ A_3 \succ A_5$
	SVNHFWA[37]	$S(A_1) = 0.6648, S(A_2) = 0.5782, S(A_3) = 0.4660, S(A_4) = 0.5099, S(A_5) = 0.5532$	$A_1 \succ A_2 \succ A_5 \succ A_4 \succ A_3$
	SVNHFWG[37]	$S(A_1) = 0.6133, S(A_2) = 0.5122, S(A_3) = 0.4316, S(A_4) = 0.4882, S(A_5) = 0.4639$	$A_1 \succ A_2 \succ A_5 \succ A_4 \succ A_3$
	SVNWDAA [64]	$S(A_1) = 0.7868, S(A_2) = 0.7374, S(A_3) = 0.5425, S(A_4) = 0.5699, S(A_5) = 0.7452$	$A_1 \succ A_5 \succ A_2 \succ A_4 \succ A_3$
	SVNWDGA [64]	$S(A_1) = 0.4758, S(A_2) = 0.4064, S(A_3) = 0.3683, S(A_4) = 0.4177, S(A_5) = 0.3127$	$A_1 \succ A_4 \succ A_2 \succ A_3 \succ A_5$
15	SVSHNDWAA	$S(A_1) = 0.8069, S(A_2) = 0.7570, S(A_3) = 0.5695, S(A_4) = 0.5892, S(A_5) = 0.7590$	$A_1 \succ A_2 \succ A_5 \succ A_4 \succ A_3$
	SVSHNDWGA	$S(A_1) = 0.4485, S(A_2) = 0.3883, S(A_3) = 0.3563, S(A_4) = 0.4010, S(A_5) = 0.2934$	$A_1 \succ A_4 \succ A_2 \succ A_3 \succ A_5$
	SVNHFWA [37]	$S(A_1) = 0.6648, S(A_2) = 0.5782, S(A_3) = 0.4660, S(A_4) = 0.5099, S(A_5) = 0.5532$	$A_1 \succ A_2 \succ A_5 \succ A_4 \succ A_3$
	SVNHFWG [37]	$S(A_1) = 0.6133, S(A_2) = 0.5122, S(A_3) = 0.4316, S(A_4) = 0.4882, S(A_5) = 0.4639$	$A_1 \succ A_2 \succ A_5 \succ A_4 \succ A_3$
	SVNWDAA [64]	$S(A_1) = 0.8029, S(A_2) = 0.7528, S(A_3) = 0.5638, S(A_4) = 0.5855, S(A_5) = 0.7568$	$A_1 \succ A_2 \succ A_5 \succ A_4 \succ A_3$
	SVNWDGA [64]	$S(A_1) = 0.4537, S(A_2) = 0.3914, S(A_3) = 0.3586, S(A_4) = 0.4042, S(A_5) = 0.2970$	$A_1 \succ A_4 \succ A_2 \succ A_3 \succ A_5$
40	SVSHNDWAA	$S(A_1) = 0.8130, S(A_2) = 0.7636, S(A_3) = 0.5781, S(A_4) = 0.5959, S(A_5) = 0.7638$	$A_1 \succ A_5 \succ A_2 \succ A_4 \succ A_3$
	SVSHNDWGA	$S(A_1) = 0.4390, S(A_2) = 0.3817, S(A_3) = 0.3523, S(A_4) = 0.3951, S(A_5) = 0.2871$	$A_1 \succ A_4 \succ A_2 \succ A_3 \succ A_5$
	SVNHFWA [37]	$S(A_1) = 0.6648, S(A_2) = 0.5782, S(A_3) = 0.4660, S(A_4) = 0.5099, S(A_5) = 0.5532$	$A_1 \succ A_2 \succ A_5 \succ A_4 \succ A_3$
	SVNHFWG [37]	$S(A_1) = 0.6133, S(A_2) = 0.5122, S(A_3) = 0.4316, S(A_4) = 0.4882, S(A_5) = 0.4639$	$A_1 \succ A_2 \succ A_5 \succ A_4 \succ A_3$
	SVNWDAA [64]	$S(A_1) = 0.8116, S(A_2) = 0.7615, S(A_3) = 0.5760, S(A_4) = 0.5945, S(A_5) = 0.7630$	$A_1 \succ A_5 \succ A_2 \succ A_4 \succ A_3$
	SVNWDGA [64]	$S(A_1) = 0.4409, S(A_2) = 0.3828, S(A_3) = 0.3532, S(A_4) = 0.3963, S(A_5) = 0.2884$	$A_1 \succ A_4 \succ A_2 \succ A_3 \succ A_5$

References

1. Torra, V. (2009). Hesitant fuzzy sets. *International Journal of Intelligent Systems, 25*, 529–539.
2. Beg, I., & Rashid, T. (2014). Group decision making using intuitionistic hesitant fuzzy sets. *International Journal of Fuzzy Logic and Intelligent Systems, 14*(3), 181–187.
3. Ye, J. (2009). Multi-criteria fuzzy decision-making method based on a novel accuracy function under interval-valued intuitionistic fuzzy environment. *Expert Systems with Applications, 36*, 6899–6902.
4. Ye, J. (2014). Correlation coefficient of dual hesitant fuzzy sets and its application to multi attribute decision making. *Applied Mathematical Modelling, 38*, 659–666.
5. Nehi, H. M. (2010). A new ranking method for intuitionistic fuzzy numbers. *International Journal of Fuzzy Systems, 12*(1), 80–86.
6. Li, D. F. (2010). A ratio ranking method of triangular intuitionistic fuzzy numbers and its application to MADM problems. *Computers and Mathematics with Applications, 60*, 1557–1570.
7. Xia, M. M., & Xu, Z. S. (2011). Hesitant fuzzy information aggregation in decision making. *International Journal of Approximate Reasoning, 52*, 395–407.
8. Xia, M. M., Xu, Z. S., & Chen, N. (2013). Some hesitant fuzzy aggregation operators with their application in group decision making. *Group Decision and Negotiation, 22*, 259–279.
9. Xu, Z. S., & Xia, M. M. (2011). Distance and similarity measures for hesitant fuzzy sets. *Information Sciences, 181*, 2128–2138.
10. Xu, Z. S., & Xia, M. M. (2012). Hesitant fuzzy entropy and cross-entropy and their use in multi attribute decision-making. *International Journal of Intelligent Systems, 27*, 799–822.
11. Wei, G., Alsaadi, F. E., Hayat, T., & Alsaedi, A. (2018). Bipolar fuzzy Hamacher aggregation operators in multiple attribute decision making. *International Journal of Fuzzy Systems, 20*(1), 1–12.
12. Xu, Z. S., & Zhang, X. L. (2013). Hesitant fuzzy multi-attribute decision making based on TOPSIS with incomplete weight information. *Knowledge-Based Systems, 52*, 53–64.
13. Chen, N., Xu, Z. S., & Xia, M. M. (2013). Correlation coefficients of hesitant fuzzy sets and their applications to clustering analysis. *Applied Mathematical Modeling, 37*, 2197–2211.
14. Qian, G., Wang, H., & Feng, X. (2013). Generalized of hesitant fuzzy sets and their application in decision support system. *Knowledge-Based Systems, 37*, 357–365.
15. Yu, D. (2013a). Triangular hesitant fuzzy set and its application to teaching quality evaluation. *Journal of Information and Computational Science, 10*(7), 1925–1934.
16. Yu, D. (2013b). Intuitionistic trapezoidal fuzzy information aggregation methods and their applications to teaching quality evaluation. *Journal of Information and Computational Science, 10*(6), 1861–1869.
17. Shi, J., Meng, C., & Liu, Y. (2014). Approach to multiple attribute decision making based on the intelligence computing with hesitant triangular fuzzy information and their application. *Journal of Intelligent Fuzzy Systems, 27*, 701–707.
18. Pathinathan, T., & Johnson, S. S. (2015). Trapezoidal hesitant fuzzy multi-attribute decision making based on TOPSIS. *International Archive of Applied Sciences and Technology, 3*(6), 39–49.
19. Joshi, D., & Kumar, S. (2016). Interval valued intuitionistic hesitant fuzzy Choquet integral based TOPSIS method for multi-criteria group decision making. *European Journal of Operational Research, 248*, 183–191.
20. Zhang, Z. (2017). Hesitant triangular multiplicative aggregation operators and their application to multiple attribute group decision making. *Neural Computing and Applications, 28*, 195–217.
21. Chen, J. J., & Huang, X. J. (2017). Hesitant triangular intuitionistic fuzzy information and its application to multi-attribute decision making problem. *Journal of Non-Linear Science and Applications, 10*, 1012–1029.

22. Yang, Y., Hu, J., An, Q., & Chen, X. (2017). Group decision making with multiplicative triangular hesitant fuzzy preference relations and cooperative games method. *International Journal for Uncertainty Quantification, 3*(7), 271–284.

23. Lan, J., Yang, M., Hu, M., & Liu, F. (2018). Multi attribute group decision making based on hesitant fuzzy sets, Topsis method and fuzzy preference relations. *Technological and Economic Development of Economy, 24*(6), 2295–2317.

24. Zhang, X., Yang, T., Liang, W., & Xiong, M. (2018). Closeness degree based hesitant trapezoidal fuzzy multi-criteria decision making method for evaluating green suppliers with qualitative information. *Hindawi Discrete Dynamics in Nature and Society, 2018*, 3178039. https://doi.org/10.1155/2018/3178039.

25. Smarandache, F. (1999). *A unifying field in logics. Neutrosophy: Neutrosophic probability, set and logic*. Rehoboth: American Research Press.

26. Gou, Y., & Cheng, H. D. (2009). New neutrosophic approach to image segmentation. *Pattern Recognition, 42*, 587–595.

27. Guo, Y. H., & Sensur, A. (2014). A novel image segmentation algorithm based on neutrosophic similarity clustering. *Applied Soft Computing, 25*, 391–398.

28. Karaaslan, F. (2017). Correlation coefficients of single valued neutrosophic refined soft sets and their applications in clustering analysis. *Neural Computing and Applications, 28*, 2781–2793.

29. Ansari, A. Q., Biswas, R., & Aggarwal, S. (2011). Proposal for applicability of neutrosophic set theory in medical AI. *International Journal of Computer Applications, 27*(5), 5–11.

30. Wang, H., Smarandache, F., & Zhang, Y. Q. (2005). *Interval neutrosophic sets and logic: Theory and applications in computing*. Phoenix: Hexis.

31. Gou, Y., Cheng, H. D., Zhang, Y., & Zhao, W. (2009). A new neutrosophic approach to image de-noising. *New Mathematics and Natural Computation, 5*, 653–662.

32. Wang, H., Smarandache, F., Zhang, Y. Q., & Sunderraman, R. (2010). Single valued neutrosophic sets. *Multi-space and Multi-structure, 4*, 410–413.

33. Ye, J. (2013). Multi-criteria decision making method using the correlation coefficient under single valued neutrosophic environment. *International Journal of General Systems, 42*, 386–394.

34. Sun, H. X., Yang, H. X., Wu, J. Z., & Yao, O. Y. (2015). Interval neutrosophic numbers Choquet integral operator for multi-criteria decision making. *Journal of Intelligent Fuzzy Systems, 28*, 2443–2455.

35. Ye, J. (2015b). Trapezoidal neutrosophic set and its application to multiple attribute decision-making. *Neural Computing and Applications, 26*, 1157–1166.

36. Abdel-Basset, M., Chang, V., & Gamal, A. (2019). Evaluation of the green supply chain practices: A novel neutrosophic approach. *Computers in Industry, 108*, 210–220.

37. Ye, J. (2015a). Multiple-attribute decision-making method under a single-valued neutrosophic hesitant fuzzy environment. *Journal of Intelligent Systems, 24*(1), 23–36.

38. Wang, C. Y., Li, Q., Zhou, X. Q., & Yang, T. (2014). Hesitant triangular fuzzy information aggregation operators based on Bonferroni means and their application to multiple attribute decision making. *Scientific World Journal, 2014*, 648516. https://doi.org/10.1155/2014/648516.

39. Zhao, X. F., Lin, R., & Wei, G. (2014). Hesitant triangular fuzzy information aggregation based on Einstein operations and their application to multiple attribute decision making. *Expert Systems with Applications, 41*, 1086–1094.

40. Peng, X. (2017). Hesitant trapezoidal fuzzy aggregation operators based on Archimedean t-norm and t-conorm and their application in MADM with completely unknown weight information. *International Journal for Uncertainty Quantification, 6*(7), 475–510.

41. Xu, Z. S., & Yager, R. R. (2006). Some geometric aggregation operators based on intuitionistic fuzzy sets. *International Journal of General Systems, 35*, 417–433.

42. Xu, Z. S. (2007). Intuitionistic fuzzy aggregation operators. *IEEE Transactions on Fuzzy Systems, 15*, 1179–1187.

43. Wan, S. P., & Dong, J. Y. (2015). Power geometric operators of trapezoidal intuitionistic fuzzy numbers and application to multi attribute group decision making. *Applied Soft Computing, 29*, 153–168.

44. Wan, S. P., Wang, F., Li, L., & Dong, J. Y. (2016). Some generalized aggregation operators for triangular intuitionistic fuzzy numbers and application to multi attribute group decision making. *Computers and Industrial Engineering, 93*, 286–301.
45. Wang, W. Z., & Liu, X. W. (2011). Intuitionistic fuzzy geometric aggregation operators based on Einstein operations. *Internal Journal of Intelligent Systems, 26*, 1049–1075.
46. Liu, P. D., Chu, Y. C., Li, Y. W., & Chen, Y. B. (2014). Some generalized neutrosophic number Hamacher aggregation operators and their applications to group decision making. *International Journal of Fuzzy Systems, 16*(2), 242–255.
47. Zhao, A. W., Du, J. G., & Guan, H. J. (2015). Interval valued neutrosophic sets and multi attribute decision making based on generalized weighted aggregation operators. *Journal of Intelligent Fuzzy Systems, 29*, 2697–2706.
48. Liu, P. D., & Shi, L. L. (2015). The generalized hybrid weighted average operator based on interval neutrosophic hesitant set and its application to multi attribute decision making. *Neural Computing and Applications, 26*, 457–471.
49. Liu, P. D., & Tang, G. L. (2016). Some power generalized aggregation operators based on the interval neutrosophic sets and their applications to decision making. *Journal of Intelligent Fuzzy Systems, 30*, 2517–2528.
50. Liu, P. (2016). The aggregation operators based on Archimedean t-Conorm and t-Norm for single-valued neutrosophic numbers and their application to decision making. *International Journal of Fuzzy Systems, 18*(5), 849–863.
51. Ye, J. (2017). Some weighted aggregation operators of trapezoidal neutrosophic numbers and their multiple attribute decision making method. *Informatica, 28*(2), 387–402.
52. Liang, W., Zhao, G., & Luo, S. (2018). Linguistic neutrosophic Hamacher aggregation operators and the application in evaluating land reclamation schemes for mines. *PLoS One, 13*(11), 1–29. https://doi.org/10.1371/journal.pone.0206178.
53. Wei, G., & Wei, Y. (2018). Some single valued neutrosophic Dombi prioritized weighted aggregation operators in multi attribute decision making. *Journal of Intelligent Fuzzy Systems, 35*(2), 2001–2003.
54. Ajay, D., Broumi, S., & Aldring, J. (2020). An MCDM method under neutrosophic cubic fuzzy sets with geometric Bonferroni mean operator. *Neutrosophic Sets and Systems, 32*, 187–202.
55. Khan, M., Gulistan, M., Hassan, N., & Nasruddin, A. M. (2020). Air pollution model using neutrosophic cubic Einstein averaging operators. *Neutrosophic Sets and Systems, 32*, 372–389.
56. Mohansundari, M., & Mohana, K. (2020). Quadripartitioned single valued neutrosophic Dombi weighted aggregation operators for multi attribute decision making. *Neutrosophic Sets and Systems, 32*, 107–122.
57. Tooranloo, H. S., Zanjirchi, S. M., & Tavangar, M. (2020). Electre approach for multi attribute decision making in refined neutrosophic environment. *Neutrosophic Sets and Systems, 3*, 101–119.
58. Vandhana, S., & Anuradha, J. (2020). Neutrosophic fuzzy Hierarchical clustering for dengue analysis in Sri Lanka. *Neutrosophic Sets and Systems, 31*, 179–199.
59. Dombi, J. (1982). A general class of fuzzy operators, the De Morgan class of fuzzy operators and fuzziness measures induced by fuzzy operators. *Fuzzy Sets and Systems, 8*, 149–163.
60. Liu, P., Liu, J., & Chen, S. M. (2018). Some intuitionistic fuzzy Dombi Bonferroni mean operators and their applications to multi-attribute decision making. *Journal of Operations Research Society, 69*(1), 1–24.
61. He, J. (2017). Typhoon disaster assessment based on Dombi hesitant fuzzy information aggregation operators. *Natural Hazards, 90*, 1153–1175. https://doi.org/10.1007/s11069-017-3091-0.
62. Jana, C., Senapati, T., Pal, M., & Yager, R. R. (2018). Picture fuzzy Dombi aggregation operators: Applications to MADM process. *Applied Soft Computing, 74*, 99–109.
63. Jana, C., Senapati, T., & Pal, M. (2019). Pythagorean fuzzy Dombi aggregation operators and its applications to MADM multi attribute decision making. *International Journal of Intelligent Systems, 34*, 2019–2038. https://doi.org/10.1002/int.22125.

64. Chen, J. Q., & Ye, J. (2017). Some single-valued neutrosophic Dombi weighted aggregation operators for multiple attribute decision making. *Symmetry, 9*, 82. https://doi.org/10.3390/sym9060082.
65. Shi, L., & Ye, J. (2018). Dombi aggregation operators of neutrosophic cubic sets for multi attribute decision making. *Algorithms, 11*(3), 1–15.
66. Gundogdu, F.K., & Kahraman C. (2018). Spherical fuzzy sets and spherical fuzzy TOPSIS method. *Journal of Intelligent and Fuzzy Systems*. https://doi.org/10.3233/JIFS-181401.
67. Gundogdu, F. K., & Kahraman, C. (2019). *Spherical fuzzy sets and decision making*. INFUS 2019, AISC vol. 1029 (pp. 979–987).
68. Smarandache, F. (2019). Neutrosophic set is a generalization of intuitionistic fuzzy set, inconsistent intuitionistic fuzzy set, pythagorean fuzzy set, q-rung orthopair fuzzy set, spherical fuzzy set and n-hyperbolic fuzzy set while neutrosophication is a generalization of regret theory, grey system theory and three ways decision. *Journal of New Theory 29*, 1–35.

Application of Generalized Aggregate Operators on Neutrosophic Hypersoft Set in Decision-Making

Rana Muhammad Zulqarnain, Xiao Long Xin, Muhammad Saqlain, and Florentin Smarandache

1 Introduction

Zadeh developed the notion of fuzzy sets [1] to solve those problems which contain uncertainty and vagueness. It is observed that in some cases circumstances cannot be handled by fuzzy sets; to overcome such types of situations Turksen [2] gave the idea of interval-valued fuzzy set. In some cases, we must deliberate membership unbiased as the non-membership values for the suitable representation of an object in uncertain and indeterminate conditions that could not be handled by fuzzy sets nor interval-valued fuzzy sets. To overcome these difficulties, Atanassov presented the notion of intuitionistic fuzzy sets [3]. The theory which was presented by Atanassov only deals the insufficient data considering both the membership and non-membership values, but the intuitionistic fuzzy set theory cannot handle the incompatible and imprecise information. To deal with such incompatible and imprecise data, the idea of the neutrosophic set (NS) was developed by Smarandache [4].

R. M. Zulqarnain (✉) · X. L. Xin
School of Mathematics, Northwest University Xi'an, Xi'an, China
e-mail: xlxin@nwu.edu.cn

M. Saqlain
Lahore Garrison University, DHA Phase-VI, Sector C, Lahore, Pakistan
e-mail: msaqlain@lgu.edu.pk

F. Smarandache
Department of Mathematics and Science, University of New Mexico, Gallup, NM, USA
e-mail: smarand@unm.edu

© The Author(s), under exclusive license to Springer Nature Switzerland AG 2021
F. Smarandache, M. Abdel-Basset (eds.), *Neutrosophic Operational Research*,
https://doi.org/10.1007/978-3-030-57197-9_13

A general mathematical tool was proposed by Molodtsov [5] to deal indeterminate, fuzzy, and not clearly defined substances known as a soft set (SS). Maji et
al. [6] extended the work on SS and defined some operations and their properties.
In Maji et al. [7], they also used the SS theory for decision making. Ali et al. [8]
revised the Maji approach to SS and developed some new operations with their
properties. De Morgan's law on SS theory was proved in Sezgin & Atagun [9] by
using different operators. Çağman & Enginoğlu [10, 11] developed the concept of
soft matrices with operations and discussed their properties; they also introduced
a decision-making method to resolve those problems which contain uncertainty. In
Çağman & Enginoğlu [10, 11], they revised the operations proposed by Molodtsov's
SS. In Atag & Ayg [12], the authors proposed some new operations on soft matrices
such as soft difference product, soft restricted difference product, soft extended
difference product and soft weak-extended difference product with their properties.

Maji [13] offered the idea of a neutrosophic soft set (NSS) with necessary
operations and properties. The idea of possibility NSS was developed by Karaaslan
[14] and introduced a possibility of neutrosophic soft decision-making method to
solve those problems which contain uncertainty based on And-product. Broumi
[15] developed the generalized NSS with some operations and properties and used
the proposed concept for decision making. To solve MCDM problems with single-
valued neutrosophic numbers presented by Deli and Subas in Deli & Şubaş [16],
they constructed the concept of cut sets of single-valued neutrosophic numbers.
On the base of the correlation of intuitionistic fuzzy sets, the term "correlation
coefficient of SVNSs" Wang et al. [17] is introduced. In Ye [18], the idea of
simplified NSs introduced with some operational laws and aggregation operators
such as real-life neutrosophic weighted arithmetic average operator and weighted
geometric average operator. They constructed an MCDM method on the base of
proposed aggregation operators.

Smarandache [19] generalized the SS to hypersoft set (HSS) by converting the
function to a multi-attribute function to deal with uncertainty. Saqlain et al. [20]
developed the generalization of TOPSIS for the NHSS; by using accuracy function,
they transformed the fuzzy neutrosophic numbers to crisp form. In Rana et al. [21],
the authors proposed the fuzzy plithogenic hypersoft set in matrix form with some
basic operations and properties. In Saqlain et al. [22], the authors proposed the
aggregate operators on NHSS.

This research is organized as follows: In Sect. 2, we recall some basic definitions
used in the following research such as SS, NS, NSS, HSS, and NHSS. We develop
the generalized aggregate operators on NHSS such as extended union, extended
intersection, And-operation, etc., in Sect. 3 with properties. In Sect. 4, the necessity
and possibility of operations are presented with examples and properties.

2 Preliminaries

In this section, we recall some basic definitions such as SS, NSS, and NHSS used in the following sequel.

Definition 2.1: Soft Set The soft set is a pair (F, Λ) over $\acute{U}$ if and only if F: $\Lambda \rightarrow (\acute{U})$ is a mapping. That is the parameterized family of subsets of $\acute{U}$ known as an SS.

Definition 2.2: Neutrosophic Set Let $\acute{U}$ be a universe of discourse and Λ be a NS on $\acute{U}$, which is defined as $\Lambda = \{<u, T_\Lambda(u), I_\Lambda(u), F_\Lambda(u) > \ : u \in \acute{U}\}$, where T, I, $F : \acute{U} \rightarrow]0^-, 1^+[$ and $0^- \leq T_\Lambda(u) + I_\Lambda(u) + F_\Lambda(u) \leq 3^+$.

Definition 2.3: Neutrosophic Soft Set Let $\acute{U}$ and $\breve{E}$ are universal set and set of attributes respectively. Let $P(\acute{U})$ be the set of Neutrosophic values of $\acute{U}$ and $\Lambda \subseteq \breve{E}$. A pair (F, Λ) is called an NSS over $\acute{U}$ and its mapping is given as

$$F : \Lambda \rightarrow \left(\acute{\acute{U}} \right)$$

Definition 2.4: Hypersoft Set Let $\acute{U}$ be a universal set and $P(\acute{U})$ be a power set of $\acute{U}$ and for $n \geq 1$; there are n distinct attributes such as $k_1, k_2, k_3, \ldots, k_n$ and $K_1, K_2, K_3, \ldots, K_n$ are sets for corresponding values attributes, respectively, with following conditions such as $K_i \cap K_j = \varnothing \ (i \neq j)$ and $i, j \in \{1, 2, 3, \ldots, n\}$. Then the pair (F, $K_1 \times K_2 \times K_3 \times \ldots \times K_n$) is said to be hypersoft set over $\acute{U}$ where F is a mapping from $K_1 \times K_2 \times K_3 \times \ldots \times K_n$ to $P(\acute{U})$.

Definition 2.5: Neutrosophic Hypersoft Set (NHSS) Let $\acute{U}$ be a universal set and $P(\acute{U})$ be a power set of $\acute{U}$ and for $n \geq 1$; there are n distinct attributes such as $k_1, k_2, k_3, \ldots, k_n$ and $K_1, K_2, K_3, \ldots, K_n$ are sets for corresponding values attributes, respectively, with following conditions such as $K_i \cap K_j = \varnothing \ (i \neq j)$ and $i, j \in \{1, 2, 3, \ldots, n\}$. Then the pair (F, Λ) is said to be NHSS over $\acute{U}$ if there exists a relation $K_1 \times K_2 \times K_3 \times \ldots \times K_n = \Lambda$. F is a mapping from $K_1 \times K_2 \times K_3 \times \ldots \times K_n$ to $P(\acute{U})$ and $F(K_1 \times K_2 \times K_3 \times \ldots \times K_n) = \{<u, T_\Lambda(u), I_\Lambda(u), F_\Lambda(u) > \ : u \in \acute{U}\}$ where T, I, F are membership values for truthness, indeterminacy, and falsity, respectively, such that $T, I, F : \acute{U} \rightarrow]0^-, 1^+[$ and $0^- \leq T_\Lambda(u) + I_\Lambda(u) + F_\Lambda(u) \leq 3^+$.

Example 1 Assume that a person examines the attractiveness of a house for living. Let $\acute{U}$ be a universe which consists of three choices $\acute{U} = \{u_1, u_2\}$ and $E = \{\acute{\epsilon}_1, \acute{\epsilon}_2, \acute{\epsilon}_3\}$ be a set of decision parameters. Then, the NHSS is given as

$$F_\Lambda = \big\{ < u_1, \big(\acute{\epsilon}_1 \{0.4, 0.7, 0.5\}, \acute{\epsilon}_2 \{0.8, 0.5, 0.3\}, \acute{\epsilon}_3 \{0.6, 0.5, 0.9\}\big) >$$
$$< u_2, \big(\acute{\epsilon}_1 \{0.1, 0.5, 0.7\}, \acute{\epsilon}_2 \{0.5, 0.6, 0.2\}, \acute{\epsilon}_3 \{0.7, 0.4, 0.6\}\big) > \big\}$$

3 Generalized Aggregate Operators on Neutrosophic Hypersoft Set and Properties

In this section, we present the generalized aggregate operations on NHSS with examples. We prove commutative and associative laws by using proposed aggregate operators in the following section.

Definition 3.1 Let $F_\Lambda \in$ NHSS, then its complement is written as $(F_\Lambda)^c = F^c(\Lambda)$ and defined as

$$F^c(\Lambda) = \big\{ < u, T\big(F^c(\Lambda)\big), I\big(F^c(\Lambda)\big), F\big(F^c(\Lambda)\big) >: u \in U \big\} \text{ such that}$$

$$T\big(F^c(\Lambda)\big) = 1 - T_\Lambda(u),$$

$$I\big(F^c(\Lambda)\big) = 1 - I_\Lambda(u),$$

$$F\big(F^c(\Lambda)\big) = 1 - F_\Lambda(u).$$

Example 2: Reconsider Example 1:

$$F^c(\Lambda) = \big\{ < u_1, \big(\acute{\epsilon}_1 \{0.6, 0.3, 0.5\}, \acute{\epsilon}_2 \big\{0.2, 0.5, 0.7\big\}, \acute{\epsilon}_3 \big\{0.4, 0.5, 0.1\big\}\big) >$$
$$< u_2, \big(\acute{\epsilon}_1 \{0.9, 0.5, 0.3\}, \acute{\epsilon}_2 \{0.5, 0.4, 0.8\}, \acute{\epsilon}_3 \{0.3, 0.6, 0.4\}\big) > \big\}$$

Proposition 3.2 If $F_\Lambda \in$ NHSS, then $(F^c(\Lambda))^c = F_\Lambda$.

Proof By using Definition 3.1, we have

$$F^c(\Lambda) = \big\{ < u, T\big(F^c(\Lambda)\big), I\big(F^c(\Lambda)\big), F\big(F^c(\Lambda)\big) >: u \in U \big\}$$

$$= \{ < u, 1 - T(F_\Lambda), 1 - I(F_\Lambda), 1 - F(F_\Lambda) >: u \in U \},$$

Thus

$$\big(F^c(\Lambda)\big)^c = \{ < u, 1 - (1 - T(F_\Lambda)), 1 - (1 - I(F_\Lambda)), 1 - (1 - F(F_\Lambda)) >: u \in U \},$$

$$\big(F^c(\Lambda)\big)^c = \{ < u, T(F_\Lambda), I(F_\Lambda), F(F_\Lambda) >: u \in U \} = F_\Lambda.$$

Which completes the proof. □

Definition 3.3: Extended Union of Two Neutrosophic Hypersoft Set Let $F_{\Lambda 1}, F_{\Lambda 2} \in$ NHSS, then their extended union is

$$
T\ (F_{\Lambda 1} \cup F_{\Lambda 2}) = \begin{cases} T\ (F_{\Lambda 1}) & \text{if } u \in \Lambda_1 - \Lambda_2 \\ T\ (F_{\Lambda 2}) & \text{if } u \in \Lambda_2 - \Lambda_1 \\ \text{Max}\ (T\ (F_{\Lambda 1}), T\ (F_{\Lambda 2})) & \text{if } u \in \Lambda_1 \cap \Lambda_2 \end{cases}
$$

$$
I\ (F_{\Lambda 1} \cup F_{\Lambda 2}) = \begin{cases} I\ (F_{\Lambda 1}) & \text{if } u \in \Lambda_1 - \Lambda_2 \\ I\ (F_{\Lambda 2}) & \text{if } u \in \Lambda_2 - \Lambda_1 \\ \text{Min}\ (I\ (F_{\Lambda 1}), I\ (F_{\Lambda 2})) & \text{if } u \in \Lambda_1 \cap \Lambda_2 \end{cases}
$$

$$
F\ (F_{\Lambda 1} \cup F_{\Lambda 2}) = \begin{cases} F\ (F_{\Lambda 1}) & \text{if } u \in \Lambda_1 - \Lambda_2 \\ F\ (F_{\Lambda 2}) & \text{if } u \in \Lambda_2 - \Lambda_1 \\ \text{Min}\ (F\ (F_{\Lambda 1}), F\ (F_{\Lambda 2})) & \text{if } u \in \Lambda_1 \cap \Lambda_2 \end{cases}
$$

Example 3 Let U $= \{u_1, u_2, u_3, u_4\}$ be a universal set and E $= \{\acute{\epsilon}_1, \acute{\epsilon}_2, \acute{\epsilon}_3, \acute{\epsilon}_4\}$ be a set of decision parameters and $F_{\Lambda 1} = \{u_1, u_4\}$ and $F_{\Lambda 2} = \{u_2, u_4\}$

$$
\begin{aligned}
F_{\Lambda 1} = \{&< u_1, \left(\acute{\epsilon}_1 \{0.4, 0.7, 0.5\}, \acute{\epsilon}_2 \{0.8, 0.5, 0.3\}, \acute{\epsilon}_3 \{0.6, 0.5, 0.9\}, \acute{\epsilon}_4 \{0.3, 0.7, 0.2\}\right) > \\
&< u_4, \left(\acute{\epsilon}_1 \{0.4, 0.7, 0.2\}, \acute{\epsilon}_2 \{0.6, 0.5, 0.3\}, \acute{\epsilon}_3 \{0.8, 0.4, 0.7\}, \acute{\epsilon}_4 \{0.6, 0.4, 0.3\}\right) >\}
\end{aligned}
$$

$$
\begin{aligned}
F_{\Lambda 2} = \{&< u_2, \left(\acute{\epsilon}_1 \{0.7, 0.4, 0.6\}, \acute{\epsilon}_2 \{0.4, 0.6, 0.9\}, \acute{\epsilon}_3 \{0.7, 0.4, 0.6\}, \acute{\epsilon}_4 \{0.7, 0.6, 0.3\}\right) > \\
&< u_4, \left(\acute{\epsilon}_1 \{0.6, 0.2, 0.7\}, \acute{\epsilon}_2 \{0.5, 0.7, 0.3\}, \acute{\epsilon}_3 \{0.4, 0.8, 0.5\}, \acute{\epsilon}_4 \{0.5, 0.6, 0.4\}\right) >\}
\end{aligned}
$$

$$
\begin{aligned}
\{F_{\Lambda 1} \cup F_{\Lambda 2} = \\
\{&< u_1, \left(\acute{\epsilon}_1 \{0.4, 0.7, 0.5\}, \acute{\epsilon}_2 \{0.8, 0.5, 0.3\}, \acute{\epsilon}_3 \{0.6, 0.5, 0.9\}, \acute{\epsilon}_4 \{0.3, 0.7, 0.2\}\right) > \\
&< u_2, \left(\acute{\epsilon}_1 \{0.7, 0.4, 0.6\}, \acute{\epsilon}_2 \{0.4, 0.6, 0.9\}, \acute{\epsilon}_3 \{0.7, 0.4, 0.6\}, \acute{\epsilon}_4 \{0.7, 0.6, 0.3\}\right) > \\
&< u_4, \left(\acute{\epsilon}_1 \{0.6, 0.7, 0.7\}, \acute{\epsilon}_2 \{0.6, 0.7, 0.3\}, \acute{\epsilon}_3 \{0.8, 0.8, 0.7\}, \acute{\epsilon}_4 \{0.6, 0.6, 0.4\}\right) >\}
\end{aligned}
$$

Proposition 3.4: Let $F_{\Lambda 1}, F_{\Lambda 2},$ and $F_{\Lambda 3}$ be NHSSs, then

1. $(F_{\Lambda 1} \cup F_{\Lambda 2}) = (F_{\Lambda 2} \cup F_{\Lambda 1})$ (commutative law)
2. $(F_{\Lambda 1} \cup F_{\Lambda 2}) \cup F_{\Lambda 3} = F_{\Lambda 1} \cup (F_{\Lambda 2} \cup F_{\Lambda 3})$ (associative law)

Proof 1 In the following proof first two cases are trivial; we consider only the third case in this proposition:

$$
\begin{aligned}
(F_{\Lambda 1} \cup F_{\Lambda 2}) = \{&< u, (\max \{T\ (F_{\Lambda 1}), T\ (F_{\Lambda 2})\}, \min \{I\ (F_{\Lambda 1}), I\ (F_{\Lambda 2})\}, \\
&\min \{F\ (F_{\Lambda 1}), F\ (F_{\Lambda 2})\}) >\}
\end{aligned}
$$

$$= \{< u, (\max \{T (F_{\Lambda 2}), T (F_{\Lambda 1})\}, \min \{I (F_{\Lambda 2}), I (F_{\Lambda 1})\},$$
$$\min \{F (F_{\Lambda 2}), F (F_{\Lambda 1})\}) >\}$$
$$= (F_{\Lambda 2} \cup F_{\Lambda 1})$$

$\square$

Proof 2 Let $F_{\Lambda 1}, F_{\Lambda 2,}$ and $F_{\Lambda 3}$ be NHSSs, then

$$F_{\Lambda 1} \cup F_{\Lambda 2} = \{< u, (\text{Max} \{T (F_{\Lambda 1}), T (F_{\Lambda 2})\}, \text{Min} \{I (F_{\Lambda 1}), I (F_{\Lambda 2})\},$$
$$\text{Min} \{F (F_{\Lambda 1}), F (F_{\Lambda 2})\}) >\}$$

$(F_{\Lambda 1} \cup F_{\Lambda 2}) \cup F_{\Lambda 3}$

$$= \left\{ \begin{array}{c} < u, \max \{\max \{T (F_{\Lambda 1}), T (F_{\Lambda 2})\}, T (F_{\Lambda 3})\}, \min \{\min \{I (F_{\Lambda 1}), I (F_{\Lambda 2})\}, I (F_{\Lambda 3})\}, \\ \min \{\min \{F (F_{\Lambda 1}), F (F_{\Lambda 2})\}, F (F_{\Lambda 3})\} > \end{array} \right\}$$

$$= \{u, \max \{T (F_{\Lambda 1}), T (F_{\Lambda 2}), T (F_{\Lambda 3})\}, \min \{\{I (F_{\Lambda 1}), I (F_{\Lambda 2})\}, I (F_{\Lambda 3})\},$$
$$\min \{\{F (F_{\Lambda 1}), F (F_{\Lambda 2})\}, F (F_{\Lambda 3})\} >\}$$

$$= \{< u, \max \{T (F_{\Lambda 1}), \max \{T (F_{\Lambda 2}), T (F_{\Lambda 3})\}\},$$
$$\min \{I (F_{\Lambda 1}), \min \{I (F_{\Lambda 2}), I (F_{\Lambda 3})\}\}, \min \{F (F_{\Lambda 1}), \min \{F (F_{\Lambda 2}), F (F_{\Lambda 3})\}\} >\}$$

$$= F_{\Lambda 1} \cup (F_{\Lambda 2} \cup F_{\Lambda 3})$$

$\square$

Definition 3.5: Extended Intersection of Two Neutrosophic Hypersoft Set Let $F_{\Lambda 1}, F_{\Lambda 2} \in$ NHSS, then their extended intersection is

$$T (F_{\Lambda 1} \cap F_{\Lambda 2}) = \begin{cases} T (F_{\Lambda 1}) & \text{if } u \in \Lambda_1 - \Lambda_2 \\ T (F_{\Lambda 2}) & \text{if } u \in \Lambda_2 - \Lambda_1 \\ \text{Min} (T (F_{\Lambda 1}), T (F_{\Lambda 2})) & \text{if } u \in \Lambda_1 \cap \Lambda_2 \end{cases}$$

$$I (F_{\Lambda 1} \cap F_{\Lambda 2}) = \begin{cases} I (F_{\Lambda 1}) & \text{if } u \in \Lambda_1 - \Lambda_2 \\ I (F_{\Lambda 2}) & \text{if } u \in \Lambda_2 - \Lambda_1 \\ \text{Max} (I (F_{\Lambda 1}), I (F_{\Lambda 2})) & \text{if } u \in \Lambda_1 \cap \Lambda_2 \end{cases}$$

$$F\ (F_{\Lambda 1} \cap F_{\Lambda 2}) = \begin{cases} F\ (F_{\Lambda 1}) & \text{if } u \in \Lambda_1 - \Lambda_2 \\ F\ (F_{\Lambda 2}) & \text{if } u \in \Lambda_2 - \Lambda_1 \\ \text{Max}\ (F\ (F_{\Lambda 1}), F\ (F_{\Lambda 2})) & \text{if } u \in \Lambda_1 \cap \Lambda_2 \end{cases}$$

Proposition 3.6 Let $F_{\Lambda 1}$, $F_{\Lambda 2}$, and $F_{\Lambda 3}$ be NHSSs, then

1. $F_{\Lambda 1} \cap F_{\Lambda 2} = F_{\Lambda 2} \cap F_{\Lambda 1}$ (commutative law)
2. $(F_{\Lambda 1} \cap F_{\Lambda 2}) \cap F_{\Lambda 3} = F_{\Lambda 1} \cap (F_{\Lambda 2} \cap F_{\Lambda 3})$ (associative law)

Proof 1 Similar to Proposition 3.6. □

Proposition 3.7 Let $F_{\Lambda 1}$, $F_{\Lambda 2}$ be NHSSs, then

1. $(F_{\Lambda 1} \cup F_{\Lambda 2})^c = F^c\ (\Lambda_1) \cap F^c\ (\Lambda_2)$
2. $(F_{\Lambda 1} \cap F_{\Lambda 1})^c = F^c\ (\Lambda_1) \cup F^c\ (\Lambda_2)$

Proof 1 Let $F_{\Lambda 1}$ and $F_{\Lambda 1} \in$ NHSS, as follows:
$F_{\Lambda 1} = \{< u, \{T\ (F_{\Lambda 1}), I\ (F_{\Lambda 1}), F\ (F_{\Lambda 1})\} >\}$ and
$F_{\Lambda 2} = \{< u, \{T\ (F_{\Lambda 2}), I\ (F_{\Lambda 2}), F\ (F_{\Lambda 2})\} >\}$

$$(F_{\Lambda 1} \cup F_{\Lambda 2})^c = \Big\{< u, \big(\max\{T\ (F_{\Lambda 1}), T\ (F_{\Lambda 2})\}, \min\{I\ (F_{\Lambda 1}), I\ (F_{\Lambda 2})\}, \\ n\big\{F\ (F_{\Lambda 1}), F\ (F_{\Lambda 2})\big) >\Big]^c$$

$$= \Big\{< u, \big(\min\{1 - T\ (F_{\Lambda 1}), 1 - T\ (F_{\Lambda 2})\}, \max\{1 - I\ (F_{\Lambda 1}), 1 - I\ (F_{\Lambda 2})\}, \\ \max\big\{1 - F\ (F_{\Lambda 1}), 1 - F\ (F_{\Lambda 2})\big) >\Big\}$$

$$= \Big\{< u, \big(\min\{T\ (F^c\ (\Lambda_1)), T\ (F^c\ (\Lambda_2))\}, \max\{I\ (F^c\ (\Lambda_1)), I\ (F^c\ (\Lambda_2))\}, \\ \max\big\{F\ (F^c\ (\Lambda_1)), F\big(F^c\ (\Lambda_2)\big)\big\}) >\Big\}$$

$$= F^c\ (\Lambda_1) \cap F^c\ (\Lambda_2)$$

□

Proof 2 Similarly, we can prove 2. □

Definition 3.8: OR-Operation of Two Neutrosophic Hypersoft Set Let $F_{\Lambda 1}, F_{\Lambda 2} \in$ NHSS. Consider $k_1, k_2, k_3, \ldots, k_n$ for $n \geq 1$, be n well-defined attributes, whose corresponding attributive values are, respectively, the set $K_1, K_2, K_3, \ldots, K_n$ with $K_i \cap K_j = \varnothing$, for $i \neq j$ and $i, j \in \{1, 2, 3, \ldots, n\}$ and their relation $K_1 \times K_2 \times K_3 \times \ldots \times K_n = \Lambda$, then $F_{\Lambda 1} \vee F_{\Lambda 2} = F_{\Lambda 1 \times \Lambda 2}$, then

$$T\ \big(F_{\Lambda 1 \times \Lambda 2}\big) = \text{Max}\ (T\ (F_{\Lambda 1}), T\ (F_{\Lambda 2})),$$

$$I\left(F_{\Lambda_1 \times \Lambda_2}\right) = \text{Min}\left(I\left(F_{\Lambda 1}\right), I\left(F_{\Lambda 2}\right)\right),$$

$$F\left(F_{\Lambda_1 \times \Lambda_2}\right) = \text{Min}\left(F\left(F_{\Lambda 1}\right), F\left(F_{\Lambda 2}\right)\right).$$

Example 4: Reconsider Example 3:

$\{F_{\Lambda 1} \vee F_{\Lambda 2} = F_{\Lambda_1 \times \Lambda_2}$

$=< (u_1, u_2), \left(\acute{\epsilon}_1 \{0.7, 0.4, 0.5\}, \acute{\epsilon}_2 \{0.8, 0.5, 0.3\}, \acute{\epsilon}_3 \{0.7, 0.4, 0.6\}, \acute{\epsilon}_4 \{0.7, 0.6, 0.2\}\right) >$

$< (u_1, u_4), \left(\acute{\epsilon}_1 \{0.6, 0.2, 0.5\}, \acute{\epsilon}_2 \{0.8, 0.5, 0.3\}, \acute{\epsilon}_3 \{0.6, 0.5, 0.5\}, \acute{\epsilon}_4 \{0.5, 0.6, 0.2\}\right) >$

$< (u_4, u_2), \left(\acute{\epsilon}_1 \{0.7, 0.4, 0.2\}, \acute{\epsilon}_2 \{0.6, 0.5, 0.3\}, \acute{\epsilon}_3 \{0.8, 0.4, 0.6\}, \acute{\epsilon}_4 \{0.7, 0.4, 0.3\}\right) >$

$< (u_4, u_4), \left(\acute{\epsilon}_1 \{0.6, 0.2, 0.2\}, \acute{\epsilon}_2 \{0.6, 0.5, 0.3\}, \acute{\epsilon}_3 \{0.8, 0.4, 0.5\}, \acute{\epsilon}_4 \{0.6, 0.4, 0.3\}\right) >\}$

Definition 3.9: AND-Operation of Two Neutrosophic Hypersoft Set Let $F_{\Lambda 1}, F_{\Lambda 2} \in$ NHSS. Consider $k_1, k_2, k_3, \ldots, k_n$ for $n \geq 1$, be n well-defined attributes, whose corresponding attributive values are, respectively, the set $K_1, K_2, K_3, \ldots, K_n$ with $K_i \cap K_j = \varnothing$, for $i \neq j$ and $i, j \in \{1, 2, 3, \ldots n\}$ and their relation $K_1 \times K_2 \times K_3 \times \ldots \times K_n = \Lambda$ then $F_{\Lambda 1} \wedge F_{\Lambda 2} = F_{\Lambda_1 \times \Lambda_2}$, then

$$T\left(F_{\Lambda_1 \times \Lambda_2}\right) = \text{Min}\left(T\left(F_{\Lambda 1}\right), T\left(F_{\Lambda 2}\right)\right),$$

$$I\left(F_{\Lambda_1 \times \Lambda_2}\right) = \text{Max}\left(I\left(F_{\Lambda 1}\right), I\left(F_{\Lambda 2}\right)\right),$$

$$F\left(F_{\Lambda_1 \times \Lambda_2}\right) = \text{Max}\left(F\left(F_{\Lambda 1}\right), F\left(F_{\Lambda 2}\right)\right).$$

Proposition 3.10 Let $F_{\Lambda 1}, F_{\Lambda 2}$ be NHSSs, then

1. $(F_{\Lambda 1} \vee F_{\Lambda 2})^c = F^c(\Lambda_1) \wedge F^c(\Lambda_2)$
2. $(F_{\Lambda 1} \wedge F_{\Lambda 2})^c = F^c(\Lambda_1) \vee F^c(\Lambda_2)$

Proof 1 Let $F_{\Lambda 1}$ and $F_{\Lambda 1} \in$ NHSS, as follows:

$F_{\Lambda 1} = \{< u_i, \{T(F_{\Lambda 1}), I(F_{\Lambda 1}), F(F_{\Lambda 1})\} >: u_i \in U\}$ and
$F_{\Lambda 2} = \{< u_j, \{T(F_{\Lambda 2}), I(F_{\Lambda 2}), F(F_{\Lambda 2})\} >: u_j \in U\}$
By using Definition 3.8, we get

$$F_{\Lambda 1} \vee F_{\Lambda 2} = \left\{< (u_i, u_j), \left[e, \max\left\{T(F_{\Lambda 1}), T(F_{\Lambda 2})\right\}, \min\left\{I(F_{\Lambda 1}), I(F_{\Lambda 2})\right\},\right.\right.$$
$$\left.\left.\min\left\{F(F_{\Lambda 1}), F(F_{\Lambda 2})\right\}\right] >\right\}$$

$$(F_{\Lambda 1} \vee F_{\Lambda 2})^c = \Big\{ < (u_i, u_j), \Big[e, 1 - \max \Big\{T\,(F_{\Lambda 1}), T\,(F_{\Lambda 2})\Big\},$$

$$1 - \min \Big\{I\,(F_{\Lambda 1}), I\,(F_{\Lambda 2})\Big\}, 1 - \min \Big\{F\,(F_{\Lambda 1}), 1 - F\,(F_{\Lambda 2})\Big\}\Big] > \Big\}$$

$$(F_{\Lambda 1} \vee F_{\Lambda 2})^c = \Big\{ < (u_i, u_j), \Big[e, \min \Big\{1 - T\,(F_{\Lambda 1}), 1 - T\,(F_{\Lambda 2})\Big\},$$

$$\max \Big\{1 - I\,(F_{\Lambda 1}), 1 - I\,(F_{\Lambda 2})\Big\}, \max \Big\{1 - F\,(F_{\Lambda 1}), 1 - F\,(F_{\Lambda 2})\Big\}\Big] > \Big\}$$

$$(F_{\Lambda 1} \vee F_{\Lambda 2})^c = \Big\{ < (u_i, u_j), \Big[e, \min \Big\{T\,(F^c\,(\Lambda_1)), T\,(F^c\,(\Lambda_2))\Big\},$$

$$\max \Big\{I\,(F^c\,(\Lambda_1)), I\,(F^c\,(\Lambda_2))\Big\}, \max \Big\{F\,(F^c\,(\Lambda_1)), F\,(F^c\,(\Lambda_2))\Big\}\Big] > \Big\}$$

Since
$F^c(\Lambda_1) = \{<u_i, \{T(F^c(\Lambda_1)), I(F^c(\Lambda_1)), F(F^c(\Lambda_1))\} > : u_i \in U\}$ and
$F^c(\Lambda_2) = \{<u_j, \{T(F^c(\Lambda_2)), I(F^c(\Lambda_2)), F(F^c(\Lambda_2))\} > : u_j \in U\}$
By using Definition 3.9, we get

$$F^c\,(\Lambda_1) \wedge F^c\,(\Lambda_2) = \Big\{ < (u_i, u_j), \Big[e, \min \Big\{T\,(F^c\,(\Lambda_1)), T\,(F^c\,(\Lambda_2))\Big\},$$

$$\max \Big\{I\,(F^c\,(\Lambda_1)), I\,(F^c\,(\Lambda_2))\Big\}, \max \Big\{F\,(F^c\,(\Lambda_1)), F\,(F^c\,(\Lambda_2))\Big\}\Big] > \Big\}$$

So

$$(F_{\Lambda 1} \vee F_{\Lambda 2})^c = F^c\,(\Lambda_1) \wedge F^c\,(\Lambda_2).$$

Similarly, we can prove 2. □

4 Necessity and Possibility Operations

The necessity and possibility operations on NHSS with some properties are presented in the following section.

Definition 4.1: Necessity Operation Let $F_\Lambda \in$ NHSS, then necessity operation on NHSS is represented by $\oplus F_\Lambda$ and defined as follows:
$\oplus F_\Lambda = \{<u, \{T(F_\Lambda), I(F_\Lambda), 1 - T(F_\Lambda)\}>\}$ for all $u \in U$.

Example 5: Reconsider Example 1:

$$\oplus F_\Lambda = \left\{ < u_1, \left(\acute{\epsilon}_1 \{0.4, 0.7, 0.6\}, \acute{\epsilon}_2 \{0.8, 0.5, 0.2\}, \acute{\epsilon}_3 \{0.6, 0.5, 0.4\} \right) > \right.$$
$$\left. < u_2, \left(\acute{\epsilon}_1 \{0.1, 0.5, 0.9\}, \acute{\epsilon}_2 \{0.5, 0.6, 0.5\}, \acute{\epsilon}_3 \{0.7, 0.4, 0.3\} \right) > \right\}$$

Proposition 4.2:

1. $\oplus (F_{\Lambda 1} \cup F_{\Lambda 2}) = \oplus F_{\Lambda 2} \cup \oplus F_{\Lambda 1}$
2. $\oplus (F_{\Lambda 1} \cap F_{\Lambda 2}) = \oplus F_{\Lambda 2} \cap \oplus F_{\Lambda 1}$

Proof 1 Let $F_{\Lambda 1} \cup F_{\Lambda 2} = F_{\Lambda 3}$, then

$$T \ (F_{\Lambda 3}) = \begin{cases} T \ (F_{\Lambda 1}) & \text{if } u \in \Lambda_1 - \Lambda_2 \\ T \ (F_{\Lambda 2}) & \text{if } u \in \Lambda_2 - \Lambda_1 \\ \text{Max} \ \{T \ (F_{\Lambda 1}), T \ (F_{\Lambda 2})\} & \text{if } u \ \in \Lambda_1 \cap \Lambda_2 \end{cases}$$

$$I \ (F_{\Lambda 3}) = \begin{cases} I \ (F_{\Lambda 1}) & \text{if } u \in \Lambda_1 - \Lambda_2 \\ I \ (F_{\Lambda 2}) & \text{if } u \in \Lambda_2 - \Lambda_1 \\ \text{Min} \ \{I \ (F_{\Lambda 1}), I \ (F_{\Lambda 2})\} & \text{if } u \ \in \Lambda_1 \cap \Lambda_2 \end{cases}$$

$$F \ (F_{\Lambda 3}) = \begin{cases} F \ (F_{\Lambda 1}) & \text{if } u \in \Lambda_1 - \Lambda_2 \\ F \ (F_{\Lambda 2}) & \text{if } u \in \Lambda_2 - \Lambda_1 \\ \text{Min} \ \{F \ (F_{\Lambda 1}), F \ (F_{\Lambda 2})\} & \text{if } u \ \in \Lambda_1 \cap \Lambda_2 \end{cases}$$

By using the definition of necessity operation,
$\oplus F_{\Lambda 3} = \{< u, \{\oplus T \ (F_{\Lambda 3}), \oplus I \ (F_{\Lambda 3}), \oplus F \ (F_{\Lambda 3})\} >: u \in U\}$, where

$$\oplus T \ (F_{\Lambda 3}) = \begin{cases} T \ (F_{\Lambda 1}) & \text{if } u \in \Lambda_1 - \Lambda_2 \\ T \ (F_{\Lambda 2}) & \text{if } u \in \Lambda_2 - \Lambda_1 \\ \text{Max} \ \{T \ (F_{\Lambda 1}), T \ (F_{\Lambda 2})\} & \text{if } u \ \in \Lambda_1 \cap \Lambda_2 \end{cases}$$

$$\oplus I \ (F_{\Lambda 3}) = \begin{cases} I \ (F_{\Lambda 1}) & \text{if } u \in \Lambda_1 - \Lambda_2 \\ I \ (F_{\Lambda 2}) & \text{if } u \in \Lambda_2 - \Lambda_1 \\ \text{Min} \ \{I \ (F_{\Lambda 1}), I \ (F_{\Lambda 2})\} & \text{if } u \ \in \Lambda_1 \cap \Lambda_2 \end{cases}$$

$$\oplus F \ (F_{\Lambda 3}) = \begin{cases} 1 - T \ (F_{\Lambda 1}) & \text{if } u \in \Lambda_1 - \Lambda_2 \\ 1 - T \ (F_{\Lambda 2}) & \text{if } u \in \Lambda_2 - \Lambda_1 \\ 1 - \text{Max} \ \{T \ (F_{\Lambda 1}), T \ (F_{\Lambda 2})\} & \text{if } u \ \in \Lambda_1 \cap \Lambda_2 \end{cases}$$

Assume

$\oplus F_{A1} = \{< u, \{T(F_{A1}), I(F_{A1}), 1 - T(F_{A1})\} >: u \in U\}$

$\oplus F_{A2} = \{< u, \{T(F_{A2}), I(F_{A2}), 1 - T(F_{A2})\} >: u \in U\}$

$\oplus F_{A1} \cup \oplus F_{A2} = F_{\delta}$, where

$F_{\delta} = \{<u, \{T(F_{\delta}), I(F_{\delta}), F(F_{\delta})\} > : u \in U\}$, such that

$$T(F_{\delta}) = \begin{cases} T(F_{A1}) & \text{if } u \in \Lambda_1 - \Lambda_2 \\ T(F_{A2}) & \text{if } u \in \Lambda_2 - \Lambda_1 \\ \text{Max}\{T(F_{A1}), T(F_{A2})\} & \text{if } u \in \Lambda_1 \cap \Lambda_2 \end{cases}$$

$$I(F_{\delta}) = \begin{cases} I(F_{A1}) & \text{if } u \in \Lambda_1 - \Lambda_2 \\ I(F_{A2}) & \text{if } u \in \Lambda_2 - \Lambda_1 \\ \text{Min}\{I(F_{A1}), I(F_{A2})\} & \text{if } u \in \Lambda_1 \cap \Lambda_2 \end{cases}$$

$$F(F_{\delta}) = \begin{cases} 1 - T(F_{A1}) & \text{if } u \in \Lambda_1 - \Lambda_2 \\ 1 - T(F_{A2}) & \text{if } u \in \Lambda_2 - \Lambda_1 \\ \text{Min}\{1 - T(F_{A1}), 1 - T(F_{A2})\} & \text{if } u \in \Lambda_1 \cap \Lambda_2 \end{cases}$$

OR

$$F(F_{\delta}) = \begin{cases} 1 - T(F_{A1}) & \text{if } u \in \Lambda_1 - \Lambda_2 \\ 1 - T(F_{A2}) & \text{if } u \in \Lambda_2 - \Lambda_1 \\ 1 - \text{Max}\{T(F_{A1}), T(F_{A2})\} & \text{if } u \in \Lambda_1 \cap \Lambda_2 \end{cases}$$

Consequently $\oplus F_{A3}$ and F_{δ} are same. So

$$\oplus(F_{A1} \cup F_{A2}) = \oplus F_{A2} \cup \oplus F_{A1}.$$

Similarly, we can prove 2. □

Definition 4.3: Possibility Operation Let $F_{\Lambda} \in NHSS$, then possibility operation on NHSS is represented by $\otimes F_{\Lambda}$ and defined as follows:

$\otimes F_{\Lambda} = \{<u, \{1 - F(F_{\Lambda}), I(F_{\Lambda}), F(F_{\Lambda})\}>\}$ for all $u \in U$.

Example 6: Reconsider the Example 1:

$$\otimes F_{\Lambda} = \left\{ < u_1, \left(\acute{\varepsilon}_1 \{0.5, 0.7, 0.5\}, \acute{\varepsilon}_2 \{0.7, 0.5, 0.3\}, \acute{\varepsilon}_3 \{0.1, 0.5, 0.9\} \right) > \right.$$

$$\left. < u_2, \left(\acute{\varepsilon}_1 \{0.3, 0.5, 0.7\}, \acute{\varepsilon}_2 \{0.8, 0.6, 0.2\}, \acute{\varepsilon}_3 \{0.4, 0.4, 0.6\} \right) > \right\}$$

Proposition 4.4:

1. $\otimes (F_{\Lambda 1} \cup F_{\Lambda 2}) = \otimes F_{\Lambda 2} \cup \otimes F_{\Lambda 1}$
2. $\otimes (F_{\Lambda 1} \cap F_{\Lambda 2}) = \otimes F_{\Lambda 2} \cap \otimes F_{\Lambda 1}$

Proof 1 Let $F_{\Lambda 1} \cup F_{\Lambda 2} = F_{\Lambda 3}$, then

$$T\ (F_{\Lambda 3}) = \begin{cases} T\ (F_{\Lambda 1}) & \text{if } u \in \Lambda_1 - \Lambda_2 \\ T\ (F_{\Lambda 2}) & \text{if } u \in \Lambda_2 - \Lambda_1 \\ \text{Max}\ \{T\ (F_{\Lambda 1})\,,\, T\ (F_{\Lambda 2})\} & \text{if } u\ \in \Lambda_1 \cap \Lambda_2 \end{cases}$$

$$I\ (F_{\Lambda 3}) = \begin{cases} I\ (F_{\Lambda 1}) & \text{if } u \in \Lambda_1 - \Lambda_2 \\ I\ (F_{\Lambda 2}) & \text{if } u \in \Lambda_2 - \Lambda_1 \\ \text{Min}\ \{I\ (F_{\Lambda 1})\,,\, I\ (F_{\Lambda 2})\} & \text{if } u\ \in \Lambda_1 \cap \Lambda_2 \end{cases}$$

$$F\ (F_{\Lambda 3}) = \begin{cases} F\ (F_{\Lambda 1}) & \text{if } u \in \Lambda_1 - \Lambda_2 \\ F\ (F_{\Lambda 2}) & \text{if } u \in \Lambda_2 - \Lambda_1 \\ \text{Min}\ \{F\ (F_{\Lambda 1})\,,\, F\ (F_{\Lambda 2})\} & \text{if } u\ \in \Lambda_1 \cap \Lambda_2 \end{cases}$$

By using the definition of necessity operation,
$\otimes F_{\Lambda 3} = \{< u, \{\otimes T\ (F_{\Lambda 3})\,,\, \otimes I\ (F_{\Lambda 3})\,,\, \otimes F\ (F_{\Lambda 3})\} >: u \in U\}$, where

$$\otimes T\ (F_{\Lambda 3}) = \begin{cases} 1 - F\ (F_{\Lambda 1}) & \text{if } u \in \Lambda_1 - \Lambda_2 \\ 1 - F\ (F_{\Lambda 2}) & \text{if } u \in \Lambda_2 - \Lambda_1 \\ 1 - \text{Max}\ \{F\ (F_{\Lambda 1})\,,\, F\ (F_{\Lambda 2})\} & \text{if } u\ \in \Lambda_1 \cap \Lambda_2 \end{cases}$$

$$= \begin{cases} 1 - F\ (F_{\Lambda 1}) & \text{if } u \in \Lambda_1 - \Lambda_2 \\ 1 - F\ (F_{\Lambda 2}) & \text{if } u \in \Lambda_2 - \Lambda_1 \\ \text{Min}\ \{1 - F\ (F_{\Lambda 1})\,,\, 1 - F\ (F_{\Lambda 2})\} & \text{if } u\ \in \Lambda_1 \cap \Lambda_2 \end{cases}$$

$$\otimes I\ (F_{\Lambda 3}) = \begin{cases} I\ (F_{\Lambda 1}) & \text{if } u \in \Lambda_1 - \Lambda_2 \\ I\ (F_{\Lambda 2}) & \text{if } u \in \Lambda_2 - \Lambda_1 \\ \text{Min}\ \{I\ (F_{\Lambda 1})\,,\, I\ (F_{\Lambda 2})\} & \text{if } u\ \in \Lambda_1 \cap \Lambda_2 \end{cases}$$

$$\otimes F\ (F_{\Lambda 3}) = \begin{cases} F\ (F_{\Lambda 1}) & \text{if } u \in \Lambda_1 - \Lambda_2 \\ F\ (F_{\Lambda 2}) & \text{if } u \in \Lambda_2 - \Lambda_1 \\ \text{Min}\ \{F\ (F_{\Lambda 1})\,,\, F\ (F_{\Lambda 2})\} & \text{if } u\ \in \Lambda_1 \cap \Lambda_2 \end{cases}$$

Assume

$$\otimes F_{\Lambda 1} = \{< u, \{1 - F(F_{\Lambda 1}), I(F_{\Lambda 1}), F(F_{\Lambda 1})\} >: u \in U\}$$
$$\otimes F_{\Lambda 2} = \{< u, \{1 - F(F_{\Lambda 2}), I(F_{\Lambda 2}), F(F_{\Lambda 2})\} >: u \in U\}$$
$$\otimes F_{\Lambda 1} \cup \oplus F_{\Lambda 2} = F_{\delta}, \text{ where}$$
$$F_{\delta} = \{<u, \{T(F_{\delta}), I(F_{\delta}), F(F_{\delta})\} > : u \in U\}, \text{ such that}$$

$$\otimes T(F_{\delta}) = \begin{cases} 1 - F(F_{\Lambda 1}) & \text{if } u \in \Lambda_1 - \Lambda_2 \\ 1 - F(F_{\Lambda 2}) & \text{if } u \in \Lambda_2 - \Lambda_1 \\ 1 - \text{Max}\{F(F_{\Lambda 1}), F(F_{\Lambda 2})\} & \text{if } u \in \Lambda_1 \cap \Lambda_2 \end{cases}$$

$$\otimes I(F_{\delta}) = \begin{cases} I(F_{\Lambda 1}) & \text{if } u \in \Lambda_1 - \Lambda_2 \\ I(F_{\Lambda 2}) & \text{if } u \in \Lambda_2 - \Lambda_1 \\ \text{Min}\{I(F_{\Lambda 1}), I(F_{\Lambda 2})\} & \text{if } u \in \Lambda_1 \cap \Lambda_2 \end{cases}$$

$$\otimes F(F_{\delta}) = \begin{cases} F(F_{\Lambda 1}) & \text{if } u \in \Lambda_1 - \Lambda_2 \\ F(F_{\Lambda 2}) & \text{if } u \in \Lambda_2 - \Lambda_1 \\ \text{Min}\{F(F_{\Lambda 1}), F(F_{\Lambda 2})\} & \text{if } u \in \Lambda_1 \cap \Lambda_2 \end{cases}$$

Consequently $\otimes F_{\Lambda 3}$ and F_{δ} are same. So

$$\otimes(F_{\Lambda 1} \cup F_{\Lambda 2}) = \otimes F_{\Lambda 2} \cup \otimes F_{\Lambda 1}$$

Similarly, we can prove 2. $\square$

Proposition 4.5 Let $F_{\Lambda 1}$ and $F_{\Lambda 2} \in$ NHSS, then we have the following:

1. $\oplus(F_{\Lambda 1} \wedge F_{\Lambda 2}) = \oplus F_{\Lambda 1} \wedge \oplus F_{\Lambda 2}$
2. $\oplus(F_{\Lambda 1} \vee F_{\Lambda 2}) = \oplus F_{\Lambda 1} \vee \oplus F_{\Lambda 2}$
3. $\otimes(F_{\Lambda 1} \wedge F_{\Lambda 2}) = \otimes F_{\Lambda 1} \wedge \otimes F_{\Lambda 2}$
4. $\otimes(F_{\Lambda 1} \vee F_{\Lambda 2}) = \otimes F_{\Lambda 1} \vee \otimes F_{\Lambda 2}$

Proof 1 Assume $F_{\Lambda 1} \wedge F_{\Lambda 2} = F_{\Lambda 1 \times \Lambda 2}$, where $(u_i, u_j) \in \Lambda_1 \times \Lambda_2$:

$$F_{\Lambda_1 \times \Lambda_2} = \Big\{< (u_i, u_j), \Big[e, \min\Big\{T(F_{\Lambda 1}), T(F_{\Lambda 2})\Big\}, \max\Big\{I(F_{\Lambda 1}), I(F_{\Lambda 2})\Big\},$$
$$\max\Big\{F(F_{\Lambda 1}), F(F_{\Lambda 2})\Big\}\Big] >\Big\}$$

By using Definition 4.1, we have

$$\oplus(F_{\Lambda 1} \wedge F_{\Lambda 2}) = \Big\{< (u_i, u_j), \Big[e, \min\Big\{T(F_{\Lambda 1}), T(F_{\Lambda 2})\Big\},$$
$$\max\Big\{I(F_{\Lambda 1}), I(F_{\Lambda 2})\Big\}, 1 - \min\Big\{T(F_{\Lambda 1}), T(F_{\Lambda 2})\Big\}\Big] >\Big\}$$

Since
$\oplus F_{\Lambda 1} = \{< u, \{T(F_{\Lambda 1}), I(F_{\Lambda 1}), 1 - T(F_{\Lambda 1})\} >\}$, and
$\oplus F_{\Lambda 2} = \{< u, \{T(F_{\Lambda 2}), I(F_{\Lambda 2}), 1 - T(F_{\Lambda 2})\} >\}$, then by using AND-operation, we get

$$\oplus F_{\Lambda 1} \wedge \oplus F_{\Lambda 2} = \Big\{ < (u_i, u_j), \Big[e, \min \Big\{ T(F_{\Lambda 1}), T(F_{\Lambda 2}) \Big\}, \max \Big\{ I(F_{\Lambda 1}), I(F_{\Lambda 2}) \Big\},$$
$$\max \Big\{ 1 - T(F_{\Lambda 1}), 1 - T(F_{\Lambda 2}) \Big\} \Big] > \Big\}$$

$$= \Big\{ < (u_i, u_j), \Big[e, \min \Big\{ T(F_{\Lambda 1}), T(F_{\Lambda 2}) \Big\}, \max \Big\{ I(F_{\Lambda 1}), I(F_{\Lambda 2}) \Big\},$$
$$1 - \min \Big\{ T(F_{\Lambda 1}), T(F_{\Lambda 2}) \Big\} \Big] > \Big\}$$

$$= \oplus (F_{\Lambda 1} \wedge F_{\Lambda 2})$$

□

Proof 2 Similar to Assertion 1. □

Proof 3 Assume $F_{\Lambda 1} \wedge F_{\Lambda 2} = F_{\Lambda 1 \times \Lambda 2}$, where $(u_i, u_j) \in \Lambda_1 \times \Lambda_2$:

$$F_{\Lambda 1 \times \Lambda 2} = \Big\{ < (u_i, u_j), \Big[e, \min \Big\{ T(F_{\Lambda 1}), T(F_{\Lambda 2}) \Big\}, \max \Big\{ I(F_{\Lambda 1}), I(F_{\Lambda 2}) \Big\},$$
$$\max \Big\{ F(F_{\Lambda 1}), F(F_{\Lambda 2}) \Big\} \Big] > \Big\}$$

By using Definition 4.4, we have

$$\otimes (F_{\Lambda 1} \wedge F_{\Lambda 2}) = \Big\{ < (u_i, u_j), \Big[e, 1 - \max \Big\{ F(F_{\Lambda 1}), F(F_{\Lambda 2}) \Big\},$$
$$\max \Big\{ I(F_{\Lambda 1}), I(F_{\Lambda 2}) \Big\}, \max \Big\{ F(F_{\Lambda 1}), F(F_{\Lambda 2}) \Big\} \Big] > \Big\}$$

Since
$\otimes F_{\Lambda 1} = \{< u, \{1 - F(F_{\Lambda 1}), I(F_{\Lambda 1}), F(F_{\Lambda 1})\} >\}$, and
$\otimes F_{\Lambda 2} = \{< u, \{1 - F(F_{\Lambda 2}), I(F_{\Lambda 2}), F(F_{\Lambda 2})\} >\}$, then by using AND-operation, we get

$$\otimes F_{\Lambda 1} \wedge \oplus F_{\Lambda 2} = \Big\{ < (u_i, u_j), \Big[e, \min \Big\{ 1 - F(F_{\Lambda 1}), 1 - F(F_{\Lambda 2}) \Big\}, \max \Big\{ I(F_{\Lambda 1}), I(F_{\Lambda 2}) \Big\},$$
$$\max \Big\{ F(F_{\Lambda 1}), F(F_{\Lambda 2}) \Big\} \Big] > \Big\}$$

$$= \Big\{ < (u_i, u_j), \Big[e, 1 - \max \Big\{ F(F_{\Lambda 1}), F(F_{\Lambda 2}) \Big\}, \max \Big\{ I(F_{\Lambda 1}), I(F_{\Lambda 2}) \Big\},$$
$$\max \Big\{ F(F_{\Lambda 1}), F(F_{\Lambda 2}) \Big\} \Big] > \Big\}$$

$$\otimes (F_{\Lambda 1} \wedge F_{\Lambda 2})$$

□

Proof 4 Assume $F_{\Lambda 1} \vee F_{\Lambda 2} = F_{\Lambda 1 \times \Lambda 2}$, where $(u_i, u_j) \in \Lambda_1 \times \Lambda_2$:

$$F_{\Lambda 1 \times \Lambda 2} = \Big\{ < (u_i, u_j), \Big[e, \max \Big\{ T(F_{\Lambda 1}), T(F_{\Lambda 2}) \Big\}, \min \Big\{ I(F_{\Lambda 1}), I(F_{\Lambda 2}) \Big\},$$
$$\min \Big\{ F(F_{\Lambda 1}), F(F_{\Lambda 2}) \Big\} \Big] > \Big\}$$

By using Definition 4.4, we have

$$\otimes (F_{\Lambda 1} \vee F_{\Lambda 2}) = \Big\{ < (u_i, u_j), \Big[e, 1 - \min \Big\{ F(F_{\Lambda 1}), F(F_{\Lambda 2}) \Big\}, \min \Big\{ I(F_{\Lambda 1}), I(F_{\Lambda 2}) \Big\},$$
$$\min \Big\{ F(F_{\Lambda 1}), F(F_{\Lambda 2}) \Big\} \Big] > \Big\}$$

Since
$\otimes F_{\Lambda 1} = \{< u, \{1 - F(F_{\Lambda 1}), I(F_{\Lambda 1}), F(F_{\Lambda 1})\} >\}$, and
$\otimes F_{\Lambda 2} = \{< u, \{1 - F(F_{\Lambda 2}), I(F_{\Lambda 2}), F(F_{\Lambda 2})\} >\}$, then by using OR-operation, we get

$$\otimes F_{\Lambda 1} \vee \oplus F_{\Lambda 2} = \Big\{ < (u_i, u_j), \Big[e, \max \Big\{ 1 - F(F_{\Lambda 1}), 1 - F(F_{\Lambda 2}) \Big\},$$
$$\min \Big\{ I(F_{\Lambda 1}), I(F_{\Lambda 2}) \Big\}, \min \Big\{ F(F_{\Lambda 1}), F(F_{\Lambda 2}) \Big\} \Big] > \Big\}$$

$$= \Big\{ < (u_i, u_j), \Big[e, 1 - \min \Big\{ F(F_{\Lambda 1}), F(F_{\Lambda 2}) \Big\}, \min \Big\{ I(F_{\Lambda 1}), I(F_{\Lambda 2}) \Big\},$$
$$\min \Big\{ F(F_{\Lambda 1}), F(F_{\Lambda 2}) \Big\} \Big] > \Big\} = \otimes (F_{\Lambda 1} \wedge F_{\Lambda 2})$$

□

5   Conclusion

In this paper, we studied neutrosophic hypersoft set with some basic definition and propose the generalized aggregate operators on neutrosophic hypersoft sets such as complement, extended union, extended intersection, And-operation, and Or-operation with their properties and proved the commutative and associative laws on NHSS by using extended union and extended intersection. Finally, the concept of necessity and possibility operations on NHSS with suitable numerical examples and properties are presented. For future trends, we can develop the neutrosophic hypersoft matrices by using proposed operations and use for decision making.

Acknowledgments This research is partially supported by a grant of National Natural Science Foundation of China (11971384).

References

1. Zadeh, L. A. (1965). Fuzzy sets. *Information and Control, 8,* 338–353.
2. Turksen, I. B. (1986). Interval valued fuzzy sets based on normal forms. *Fuzzy Sets and Systems, 20,* 191–210.
3. Atanassov, K. (1986). Intultionistic fuzzy sets. *Fuzzy Sets and Systems, 20,* 87–96.
4. Smarandache, F. (2005). Neutrosophic set—A generalization of intuitionistic fuzzy sets. *International Journal of Pure and Applied Mathematics, 24*(3), 287–297.
5. Molodtsov, D. (1999). Soft set theory first results. *Computers & Mathematics with Applications, 37,* 19–31.
6. Maji, P. K., Biswas, R., & Roy, A. R. (2003). Soft set theory. *Computers and Mathematics with Applications, 45*(4–5), 555–562.
7. Maji, P. K., Biswas, R., & Roy, A. R. (2002). An application of soft sets in a decision making problem. *Computers and Mathematics with Applications, 44,* 1077–1083.
8. Ali, M. I., Feng, F., Liu, X., Keun, W., & Shabir, M. (2009). On some new operations in soft set theory. *Computers and Mathematics with Applications, 57*(9), 1547–1553.
9. Sezgin, A., & Atagun, A. O. (2011). On operations of soft sets. *Computers and Mathematics with Applications, 61*(5), 1457–1467.
10. Çağman, N., & Enginoğlu, S. (2010a). Soft matrix theory and its decision making. *Computers and Mathematics with Applications, 59*(10), 3308–3314.
11. Çağman, N., & Enginoğlu, S. (2010b). Soft set theory and uni – int decision making. *European Journal of Operational Research, 207,* 848–855.
12. Atag, O., & Ayg, E. (2019). Difference operations of soft matrices with applications in decision making. *Punjab University Journal of Mathematics, 51*(3), 1–21.
13. Maji, P. K. (2013). Neutrosophic soft set. *Ann. Fuzzy Math. Inform., 5*(1), 157–168.
14. Karaaslan, F. (2016). Possibility neutrosophic soft sets and PNS-decision making method. *Applied Soft Computing Journal, 54,* 403–414.
15. Broumi, S. (2013). Generalized neutrosophic soft set. *International Journal of Computer Science Engineering and Information Technology, 3*(2), 17–30.
16. Deli, I., & Şubaş, Y. (2017). A ranking method of single valued neutrosophic numbers and its applications to multi-attribute decision making problems. *International Journal of Machine Learning and Cybernetics, 8,* 1309–1322.
17. Wang, H., Smarandache, F., & Zhang, Y. (2013). Single valued neutrosophic sets. *International Journal of General Systems, 42,* 386–394.

18. Ye, J. (2014). A multicriteria decision-making method using aggregation operators for simplified neutrosophic sets. *Journal of Intelligent Fuzzy Systems, 26,* 2459–2466.
19. Smarandache, F. (2018). Extension of soft set to hypersoft set, and then to plithogenic hypersoft set. *Neutrosophic Sets and Systems, 22,* 168–170.
20. Saqlain, M., Saeed, M., Ahmad, M. R., & Smarandache, F. (2019). Generalization of TOPSIS for neutrosophic hypersoft set using accuracy function and its application. *Neutrosophic Sets and Systems, 27,* 131–137.
21. Rana, S., Qayyum, M., Saeed, M., & Smarandache, F. (2019). Plithogenic fuzzy whole hypersoft set, construction of operators and their application in frequency matrix multi attribute decision making technique. *Neutrosophic Sets and Systems, 28,* 34–50.
22. Saqlain, M., Moin, S., Jafar, M. N., & Saeed, M. (2020). Aggregate operators of neutrosophic hypersoft set aggregate operators of neutrosophic hypersoft set. *Neutrosophic Sets and Systems, 32,* 294–306.

Neutrosophic Statistics for Grouped Data: Theory and Applications

Rehan Ahmad Khan Sherwani, Muhammad Aslam, Huma Shakeel, Kamran Abbas, and Farrukh Jamal

1 Introduction

For any data obtained under the uncertainty environment, the data analysts are interested to determine the central values of the neutrosophic data, the shape of the neutrosophic data, and the dispersion of the neutrosophic data. As discussed earlier, neutrosophic statistics is the generalization of classical statistics which deals with determined numbers [1]. However, classical statistical tools cannot be applied if there is indeterminacy or uncertainty in the observations under consideration. The neutrosophic center of tendency is one of the important aspects to analyze the data under the NS, see ([1–4], and [5]. There are numerous studies available indicating the significance of the practical usage and application of neutrosophic statistical measures; see for example, [3–7]. Under the uncertainty, the neutrosophic statistics are considered to be more effective and adequate, see for example, [7, 8].

All the techniques that help us in summarizing and describing the nature of the neutrosophic data come under the category of neutrosophic descriptive statistics, see [9]. Practically, graphical representation and tabulation are not ample,

R. A. K. Sherwani · H. Shakeel
College of Statistical and Actuarial Sciences, University of the Punjab New Campus Lahore, Lahore, Pakistan
e-mail: rehan.stat@pu.edu.pk; huma.stat@pu.edu.pk

M. Aslam (✉)
Department of Statistics, Faculty of Science, King Abdulaziz University, Jeddah, Saudi Arabia
e-mail: aslam_ravian@hotmail.com

K. Abbas
Department of Statistics, Faculty of Science, King Abdulaziz University, Jeddah, Saudi Arabia

F. Jamal
Department of Statistics, Govt. S.A P/G College D.N.S BWP, Bahawalpur, Pakistan

F. Smarandache, M. Abdel-Basset (eds.), *Neutrosophic Operational Research*,
https://doi.org/10.1007/978-3-030-57197-9_14

263

especially, when different data sets have to be compared. Neutrosophic measure of central tendency makes it easier to understand and locate the central value of the neutrosophic data set. However, a single value can describe the whole neutrosophic data set which may be located somewhere in the center. Therefore, in this chapter, we will focus on some important neutrosophic measures of central tendency.

The measure of central tendency is a single neutrosophic value, which is used to present the whole neutrosophic data [9]. Therefore, a neutrosophic number expressed in the indeterminacy interval, which represents the whole neutrosophic data under the study, is called the neutrosophic measure of central tendency. Note here that it is not always necessary that the neutrosophic value be always at the center of the data. It may be near the center or may be at the exact center of the neutrosophic data. The neutrosophic measure of central tendency indicates the location or the general positioning of the neutrosophic observations; therefore, such a neutrosophic value is also called neutrosophic measure of location. A method, which is used to find the center of the neutrosophic data, is known as the neutrosophic measures of central tendency. The important neutrosophic measures of central tendency for grouped data are given as follows:

(a) Neutrosophic arithmetic mean
(b) Neutrosophic geometric mean
(c) Neutrosophic harmonic mean
(d) Neutrosophic mode
(e) Neutrosophic median
(f) Neutrosophic quantiles
(g) Neutrosophic deciles
(h) Neutrosophic percentiles
(i) Neutrosophic range
(j) Neutrosophic coefficient of range
(k) Neutrosophic quartile deviation
(l) Neutrosophic variance
(m) Neutrosophic standard deviation
(n) Neutrosophic coefficient of variation
(o) Neutrosophic mean deviation
(p) Neutrosophic empirical relation between mean, median, and mode
(q) Neutrosophic moments about origin
(r) Neutrosophic skewness
(s) Neutrosophic kurtosis
(t) Neutrosophic moment ratios

We will now discuss these neutrosophic measures of central tendency and dispersion in detail with the help of examples.

2   Neutrosophic Arithmetic Mean (NAM)

Suppose that $X_{iN} \epsilon \{X_L, X_U\} = i = 1, 2, 3, \ldots, n_N$ is a neutrosophic random variable (nrv) of sample size n_N, where X_L and X_U denote the lower value and upper value of indeterminacy interval. The sum of all neutrosophic observations divided by the neutrosophic sample size is known as neutrosophic arithmetic mean (NAM). The NAM is defined as follows:

$$\overline{X}_N \epsilon \left[\frac{\sum_{i=1}^{n_L} f_L X_{iL}}{\sum_{i=1}^{n_L} f_L}, \frac{\sum_{i=1}^{n_U} f_U X_{iU}}{\sum_{i=1}^{n_U} f_U} \right]; \overline{X}_N \epsilon \left[\overline{X}_L, \overline{X}_U \right]; n_N \epsilon [n_L, n_U] \qquad (1)$$

Note here that $\overline{X}_L$ and $\overline{X}_U$ represent the arithmetic mean (A.M) of lower values and upper values in the indeterminacy interval. In addition, when $n_L = n_U$ the NAM is given in Eq. (1) reduces to A.M under classical statistics. The NAM can be computed from the grouped and ungrouped data. Here are some formulas in Table 1 to compute NAM for grouped.

We now discuss the calculations of NAM with the help of some examples.

Example 2.1 Calculate the mean for the distribution of Neutrosophic data given below:

Variable (X)	Frequency
0	4
1	[6,5]
2	[7,6]
3	8
4	[13,15]
5	2

Table 1 Methods to compute NAN

Methods under NS	Grouped neutrosophic data
Direct under NS	$\overline{X}_N \epsilon \left[\frac{\sum_{i=1}^{n_L} f_L X_{iL}}{\sum_{i=1}^{n_L} f_L}, \frac{\sum_{i=1}^{n_U} f_U X_{iU}}{\sum_{i=1}^{n_U} f_U} \right]$
Indirect under NS	$\overline{X}_N = A_N + \frac{\sum_{i=1}^{n_N} f_N D_N}{\sum_{i=1}^{n_N} f_N}; \overline{X}_N \epsilon \left[\overline{X}_L, \overline{X}_U \right]$ Here $D_N = X_N - A_N$ and A_N is any assumed neutrosophic mean. $A_N \neq \overline{X}_N$
Step-division under NS	$\overline{X}_N = A_N + \frac{\sum_{i=1}^{n_N} f_N U_N}{\sum_{i=1}^{n_N} f_N} \times c_N; \overline{X}_N \epsilon \left[\overline{X}_L, \overline{X}_U \right]$ Note here $U_N = \frac{X_N - A_N}{c_N \text{ or } h_N}$; where c_N is the common divisor and h_N class interval.

Solution Using the direct method for grouped neutrosophic data, we can calculate the neutrosophic arithmetic mean for given data:

$$\overline{X}_N \in \left[\frac{\sum_{i=1}^{n_L} f_L X_{iL}}{\sum_{i=1}^{n_L} f_L}, \frac{\sum_{i=1}^{n_U} f_U X_{iU}}{\sum_{i=1}^{n_U} f_U} \right]$$

X	f_L	f_U	Xf_L	Xf_U
0	4	4	0	0
1	6	5	6	5
2	7	6	14	12
3	8	10	24	30
4	13	15	52	60
5	2	2	10	10
Total	40	42	106	117

$$\overline{X}_N = \left[\frac{0 + 6 + 14 + 24 + 52 + 10}{40}, \frac{0 + 5 + 12 + 30 + 60 + 10}{42} \right]$$

$$\overline{X}_N \in [3.53, 2.78]$$

It shows that the neutrosophic mean of the given frequency distribution lies between 2.78 and 3.53.

3 Neutrosophic Geometric Mean (NGM)

The neutrosophic geomantic mean (NGM) is another measure of central tendency which is applied when the neutrosophic data is expressed in rate, ratio, and percentage. Mathematically, the NGM is defined as follows.

The neutrosophic geomantic mean is the nth positive root of the product of n_N neutrosophic observations. Mathematically, it is defined for grouped neutrosophic data as follows when the frequency is given for the corresponding of the neutrosophic data:

$$GM_N = \sqrt[n_N]{X_{1N}{}^{f1n} \times X_{2N}{}^{f2n} \cdots \times X_{n_N}{}^{fn_N}}; \quad GM_N \in [GM_L, GM_U] \qquad (2)$$

Note here that $f1n + f2n + \ldots + fn_N = n_N$

Example 3.1 The following data relates to the frequency distribution of heights; find the neutrosophic geometric mean.

Height (inches)	58–60	61–63	64–66	67–69
Frequency	2	[4,5]	[6,7]	[3,2]

Solution Using the formula given in Eq. (2),

$$NGM = \left[\sqrt[15]{59^2 \times 62^4 \times 65^6 \times 68^3}, \ \sqrt[16]{59^2 \times 62^5 \times 65^7 \times 68^2} \right]$$

$$NGM \epsilon \ [63.94, 63.63]$$

The neutrosophic geometric mean shows that there is not a single average due to the indeterminacy part involved in the observations. Therefore, the average height falls between 63.63 and 63.94.

4   Neutrosophic Harmonic Mean (NHM)

The neutrosophic harmonic mean (NHM) is the extension of the harmonic mean (HM) under classical statistics. The NHM is an important average, which is used to find the center of the neutrosophic data when expressed in rate, ratio, and percentage. The NHM is defined as: the reciprocal of NAM and the reciprocal of neutrosophic observations and mathematically, for neutrosophic-grouped data, the NHM is defined as

$$H.M_N = \frac{\sum f_N}{\sum_{i=1}^{n_N} \left(\frac{f_N}{X_N} \right)}; \ H.M_N \epsilon \ [H.M_L, H.M_U] \tag{3}$$

Example 4.1 Calculate HM_N using the grouped data given in Example 3.1.

Solution Using the formula given in Eq. (3) for the grouped neutrosophic data, we can find harmonic mean for the following data set.

Height (inches)	58–60	61–63	64–66	67–69
Frequency	2	[4,5]	[6,7]	[3,2]

$$H.M_N = \frac{\sum f_N}{\sum_{i=1}^{n_N} \left(\frac{f_N}{X_N} \right)}$$

$$H.M_N = \left[\frac{15}{\left(\frac{2}{59} + \frac{4}{62} + \frac{6}{65} + \frac{3}{68}\right)}, \frac{16}{\left(\frac{2}{59} + \frac{5}{62} + \frac{7}{65} + \frac{2}{68}\right)} \right]$$

$$H.M_N \in [63.87, 63.58]$$

The neutrosophic harmonic mean shows that the average height lies between 63.87 and 63.58.

5 Neutrosophic Median (N.Md)

As in classical statistics, N.Md is the center value of the neutrosophic data that is arranged in ascending order partially. In neutrosophic statistics, data cannot be arranged fully because of its two parts that are sure and unsure part.

For grouped data, N.Md can be calculated as follows:

$$\tilde{X}_N = l_N + \frac{h_N}{f_N}\left(\frac{n_N}{2} - c_N\right); \tilde{X}_N \in \left[\tilde{X}_L, \tilde{X}_u\right] \tag{4}$$

where n_N = total neutrosophic frequency, l_n = lower class boundary of the neutrosophic median class, f_N = neutrosophic frequency of median class, h_N = height of neutrosophic median class, c_N = neutrosophic cumulative frequency of neutrosophic median class.

Example 5.1 Compute the median for the grouped data given below.

Time to travel to work	Frequency
1–10	[6,6]
11–20	[14,13]
21–30	[10,8]
31–40	[6,9]

Solution As we have,

$$\tilde{X}_N = l_N + \frac{h_N}{f_N}\left(\frac{n_N}{2} - c_N\right); \tilde{X}_N \in \left[\tilde{X}_L, \tilde{X}_u\right]$$

Here, $n_N = 36, h_N = 10, l_N = 10.5, f_N = [12, 13], c_N = 6$

$$\tilde{X}_N = \left[10.5 + \frac{10}{14}\left(\frac{36}{2} - 6\right), 10.5 + \frac{10}{13}\left(\frac{36}{2} - 6\right)\right]$$

$$\tilde{X}_N \in [19.07, 19.73]$$

The neutrosophic median falls within 19.07 and 19.73 indicating that there is not a single observation dividing the data into two halves. Therefore, the numbers of hours PU students study per week lies between 6 and 9. The result indicates that the neutrosophic median lies between 19.07 and 19.73.

6   Neutrosophic Mode

As in classical statistics, N.Mo is the most frequent value of the neutrosophic data; it attains all the properties of the classical mode. It can be calculated by taking average of most frequent values.

For grouped data it can be calculated as follows:

$$\hat{X}_N = l_N + \left(\frac{f_{1N} - f_{0N}}{2f_{1N} - f_{0N} - f_{2N}} \right) * h_N; \hat{X}_N \in \left[\hat{X}_L, \ \hat{X}_U \right] \tag{5}$$

where l_N = neutrosophic lower C.B of the model class, f_{1N} = neutrosophic frequency of the model class, f_{0N} = neutrosophic frequency of the preceding model class, f_{2N} = neutrosophic frequency of the succeeding model class, h_N = neutrosophic width of the interval of a model class.

Example 6.1 Considering the data given in Example 4.1, calculate neutrosophic mode.

Solution From (5), we have

$$\hat{X}_N = l_N + \left(\frac{f_{1N} - f_{0N}}{2f_{1N} - f_{0N} - f_{2N}} \right) * h_N$$

Here $l_N = 63.5, f_{1N} = [6, 7], f_{0N} = [4, 5], f_{2N} = [3, 2], h_N = 3$

$$\hat{X}_N = \left[63.5 + \left(\frac{6 - 4}{2 \times 6 - 4 - 3} \right) \times 3, 63.5 + \left(\frac{7 - 5}{2 \times 7 - 5 - 2} \right) \times 3 \right]$$

$$\hat{X}_N = [64.7, 64.35]$$

Therefore, the neutrosophic mode of the frequency distribution of heights falls between 64.35 and 64.7.

7 Neutrosophic Quantiles (NQ)

Suppose that for a variable, a set of neutrosophic observation are ordered partially in ascending order. As we are dealing with sets of observations instead of crisp numbers so there is not complete order. As in classical statistics, NQ divides data into 4 parts. First NQ shows 25% data below it. The second NQ represents 50 percent data below it. Third NQ represents 75% data under it and forth NQ displays 100% data below it and usually does not need to be calculated.

For grouped neutrosophic observations, NQ can be calculated from the following formula:

$$Q_{iN} = l_N + \frac{h_N}{f_N} \left(\frac{i * n_N}{4} - c_N \right) \tag{6}$$

where $i = 1, 2, 3$; l_N = neutrosophic lower C.B of the median class, f_N = neutrosophic frequency of the median class, h_N = neutrosophic width of the interval of a median class, c_N = neutrosophic cumulative frequency above-median class.

Example 7.1 Calculate Q_{1N} using the data set given in 3.1.

Solution From (6), we have

$$Q_{iN} = l_N + \frac{h_N}{f_N} \left(\frac{i * n_N}{4} - c_N \right)$$

Here $l_N = 60.5, f_N = [4, 5],\ h_N = 3\ , c_N = 2$

$$Q_{1N} = \left[60.5 + \frac{3}{4} \left(\frac{1 * 15}{4} - 2 \right), 60.5 + \frac{3}{5} \left(\frac{1 * 16}{4} - 2 \right) \right]$$

$$Q_{1N} \in [61.81, 61.7]$$

The first neutrosophic quantile indicates that 25% of the neutrosophic data lies under 61.7 and 61.81.

8 Neutrosophic Deciles (ND)

Same as in the above section of NQ, the neutrosophic set of data is arranged in ascending order partially. As in classical data, ND divides data into 10 parts. First neutrosophic decile represents 10% data below it, second ND shows 20% data below it and up to so on 10th ND represents 100% data below it. Fifth ND is equal to 2nd NQ.

For neutrosophic data in grouped forms, the following formula can be used for calculating deciles:

$$D_{iN} = l_N + \frac{h_N}{f_N}\left(\frac{i * n_N}{10} - c_N\right) \qquad (7)$$

where $i = 1, 2, 3, \ldots, 10$; l_N = neutrosophic lower C.B of the median class, f_N = neutrosophic frequency of the median class, h_N = neutrosophic width of the interval of a median class, c_N = neutrosophic cumulative frequency above-median class.

Example 8.1 For the given neutrosophic data find D_{5N}. Also verify that it is equal to the neutrosophic median.

Time to travel to work	Frequency
1–10	[6,6]
11–20	[14,13]
21–30	[10,8]
31–40	[6,9]

Solution Using the formula in (7), we have

$$D_{5N} = l_N + \frac{h_N}{f_N}\left(\frac{5 * n_N}{10} - c_N\right)$$

Here $l_N = 10.5$, $f_N = [14, 13]$, $h_N = 10$, $c_N = [6, 6]$

$$D_{5N} = \left[10.5 + \frac{10}{14}\left(\frac{5 * 36}{10} - 6\right), 10.5 + \frac{10}{13}\left(\frac{5 * 36}{10} - 6\right)\right]$$

$$D_{5N} \in [19.07, 19.73]$$

From Example 6.1, we can clearly compare and state that 5th neutrosophic decile is equal to the neutrosophic median of the data.

9 Neutrosophic Percentiles (NP)

The neutrosophic set of data for a variable is arranged in ascending order partially. NP divides neutrosophic data into 100 parts the same as in classical statistics. First NP indicates 1% data below it, and NP shows 2% data below it and up to so on 100 NP represents 100% data below it. 25th NP is equal to 1st NQ. 50th NP, 2nd

NQ and 5ND are equal and hence is the median of the neutrosophic data set. 3rd NQ and 75th NP are equal.

For neutrosophic data in grouped forms, the following formula can be used for calculating percentiles:

$$P_{iN} = l_N + \frac{h_N}{f_N} \left(\frac{i * n_N}{100} - c_N \right) \tag{8}$$

where $i = 1, 2, 3, \ldots, 100$; l_N = neutrosophic lower C.B of the median class, f_N = neutrosophic frequency of the median class, h_N = neutrosophic width of the interval of a median class, c_N = neutrosophic cumulative frequency above the median.

Example 9.1 For the given neutrosophic data find P_{40} *and* P_{50}.

Time to travel to work	Frequency
1–10	[6,6]
11–20	[14,13]
21–30	[10,8]
31–40	[6,9]

Solution We have

$$P_{40N} = l_N + \frac{h_N}{f_N} \left(\frac{40 * n_N}{100} - c_N \right)$$

Here $l_N = 10.5, f_N = [14, 13], h_N = 10, c_N = [6, 6]$

$$P_{40N} = \left[10.5 + \frac{10}{14} \left(\frac{40 * 36}{100} - 6 \right), 10.5 + \frac{10}{13} \left(\frac{40 * 36}{100} - 6 \right) \right]$$

$$P_{40N} \in [16.5, 16.96]$$

$$P_{50N} = l_N + \frac{h_N}{f_N} \left(\frac{50 * n_N}{100} - c_N \right)$$

Here $l_N = 10.5, f_N = [14, 13], h_N = 10, c_N = [6, 6]$

$$P_{50N} = \left[10.5 + \frac{10}{14} \left(\frac{50 * 36}{100} - 6 \right), 10.5 + \frac{10}{13} \left(\frac{50 * 36}{100} - 6 \right) \right]$$

$$P_{50N} \in [19.07, 19.73]$$

The 40th and 50th percentile falls between the ranges 16.5–16.96 and 19.07–9.73, respectively. The results indicate that due to the indeterminacy, there is not a single answer to the question. In fact the result falls between the upper and lower range.

10 Neutrosophic Range (NR)

The neutrosophic range is the difference between the highest and lowest observation of neutrosophic data. It is the measure of dispersion as in classical statistics. The applications and limitations of the neutrosophic range are the same as that of classical range. For grouped data, NR can be calculated as the difference between the upper C.B of the highest class and lowest C.B of the lowest class or simply the difference between the highest and lowest midpoints of the interval class.

11 Neutrosophic Coefficient of Range (NCR)

The range can be used in the calculation of the neutrosophic coefficient of range (NCR). It is a relative measure of dispersion to study the spread of the neutrosophic data. It can be calculated as follows:

$$NCR = \frac{X_{MN} - X_{0N}}{X_{MN} + X_{0N}}; C.R_N \in [C.R_L, C.R_U] \tag{9}$$

Example 11.1 Calculate neutrosophic range and coefficient of range for the data given below:

Height (inches)	58–60	61–63	64–66	67–69
Frequency	2	[4,5]	[6,7]	[3,2]

Solution From data, we have

$$X_{0N} = 57.5$$

$$X_{MN} = 69.5$$

We have

$$N.R = 69.5 - 57.5$$

$$N.R = 12$$

$$R_N \in [12, 12]$$

This shows that the neutrosophic range of this data is 12.

For neutrosophic coefficient of range, using formula in (9), we have

$$N.C.R = \frac{69.5 - 57.5}{69.5 + 57.5}$$

$$N.C.R = 0.094$$

$$C.R_N \in [0.094, 0.094]$$

12   Neutrosophic Quartile Deviation (NQD)

It is the relative measure of dispersion based on neutrosophic quartiles. It can be calculated from the following formula:

$$NQD = \frac{Q_{3N} - Q_{1N}}{2}; Q.D_N \in [Q.D_L, Q.D_U]$$

Example 12.1 Using Example 7.1, calculate NQD

Solution We have from (6), we have

$$Q_{iN} = l_N + \frac{h_N}{f_N} \left(\frac{i * n_N}{4} - c_N \right) \tag{10}$$

Here $l_N = 60.5, f_N = [4, 5], \ h_N = 3, c_N = 2$

$$Q_{1N} = \left[60.5 + \frac{3}{4} \left(\frac{1 * 15}{4} - 2 \right), 60.5 + \frac{3}{5} \left(\frac{1 * 16}{4} - 2 \right) \right.$$

$$Q_{1N} \in [61.81, 61.7]$$

For Q_{3N}, we have

$l_N = 63.5, f_N = [6, 7], h_N = 3, c_N = [6, 7]$

$$Q_{3N} = \left[63.5 + \frac{3}{6} \left(\frac{3 * 15}{4} - 6 \right), 63.5 + \frac{3}{7} \left(\frac{3 * 16}{4} - 7 \right) \right.$$

$$Q_{1N} \in [66.125, 65.64]$$

$$Q_{3N} \in [66.125, 65.64] , Q_{1N} \in [61.81, 61.7]$$

Using the formula (10), we have

$$\text{NQD} = \frac{[4.315, 3.94]}{2} = [2.16, 1.97]$$

$$\text{Q.D}_N \in [2.16, 1.97]$$

Hence the neutrosophic quartile deviation lies within the range of 1.97–2.16.

13   Neutrosophic Variance (NV)

As in classical statistics, neutrosophic variance also shows that how much neutrosophic data vary from the neutrosophic mean. If values vary largely, then neutrosophic variance will be large and vice versa. Neutrosophic variance also has all the properties as the classical variance. We can also compute the neutrosophic variance for neutrosophic data. For population data, NV is denoted as $\sigma^2{}_N$.

For sample data, it will be denoted by $S^2{}_N$. For grouped data, NV can be found as follows:

$$S^2{}_N = \sum_{i=1}^{n_N} \frac{f_{iN}\left(X_{iN} - \overline{X}_N\right)^2}{f_{iN}} ; \sigma^2{}_N \in \left[\sigma^2{}_L, \sigma^2{}_u\right] \tag{11}$$

Pooled neutrosophic variance can be calculated when we have more than one neutrosophic data and are assumed to have the same variances but different means. Its formulas are as follows:

$$S^2{}_{PN} = \frac{n_{1N}S^2{}_{1N} + n_{2N}S^2{}_{2N} + \ldots + n_{mN}S^2{}_{mN}}{n_{1N} + n_{2N} + \ldots n_{mN}} ; \sigma^2{}_{PN} \in \left[\sigma^2{}_L, \sigma^2{}_u\right] \tag{12}$$

Example 13.1 The following data relates to the frequency distribution of heights, find the neutrosophic geometric mean.

Height (inches)	58–60	61–63	64–66	67–69
Frequency	2	[4,5]	[6,7]	[3,2]

Find the neutrosophic variance of the distribution.

Solution We have

$$S^2{}_N = \sum_{i=1}^{n_N} \frac{f_{iN}\left(X_{iN} - \overline{X}_N\right)^2}{f_{iN}}$$

Classes	Frequency	X_{iN}	fX_{LN}	fX_{UN}
58–60	2	59	118	118
61–63	[4,5]	62	248	310
64–66	[6,7]	65	390	455
67–69	[3,2]	68	204	136

$$S^2{}_N = \left[\frac{960}{15}, \frac{1019}{16} \right]$$

$$S^2{}_N \in [64, 63.687]$$

The result indicates high variation in the frequency distribution of heights. Due to the presence of indeterminate part, the variance falls within the range of 63.687–64.

14   Neutrosophic Standard deviation (N.S.D)

Neutrosophic standard deviation is the positive square root of neutrosophic variance. It is a measure of dispersion. It tells us how much our data is scattered around its mean. The smaller the value the better the measure is. The smaller N.S.D will produce a better measure of σ_N used for the representation of population neutrosophic standard deviation. S_N is used for the sample N.S.D. For grouped data, following formulas will be used:

$$S_N = \sqrt{\sum_{i=1}^{n_N} \frac{f_{iN}\left(X_{iN} - \overline{X}_N\right)^2}{f_{iN}}} ; \sigma_N \in [\sigma_L, \sigma_u] \tag{13}$$

Hence, pooled neutrosophic S.D can be calculated as follows for more than one different neutrosophic data sets having different means but same variances:

$$S_{PN} = \sqrt{\frac{n_{1N} S^2{}_{1N} + n_{2N} S^2{}_{2N} + \ldots + n_{mN} S^2{}_{mN}}{n_{1N} + n_{2N} + \ldots n_{mN}}} ; \sigma_{PN} \epsilon \left[\sigma_L, \sigma_u\right] \qquad (14)$$

Example 14.1 Considering the Example 13.1, find the neutrosophic standard deviation.

Solution We have

X_{iN}	f	fX_{iN}	$f\left(x_{iN} - \overline{X}_N\right)^2$
59	2	[118,118]	[50, 47.53]
62	[4,5]	[248,310]	[16, 17.57]
65	[6,7]	[390,455]	[6, 8.85]
68	[3,2]	[204,136]	[48, 34.03]

$$S_N = \sqrt{\sum_{i=1}^{n_N} \frac{f_{iN}\left(X_{iN} - \overline{X}_N\right)^2}{f_{iN}}}$$

$$S_N = \sqrt{8, 6.72}$$

$$S_N \in [2.82, 2.59]$$

15 Neutrosophic Coefficient of Variation (N.C.V)

As in classical statistics, the neutrosophic coefficient of variation can be used for comparing positive values on a ratio scale. It can also be used for comparing variability of different measures on different scales. The greater value of N.C.V indicates that the greater the values scatter around its mean. It is also known as a relative neutrosophic standardized measure. It can be calculated from the following formula:

$$N.C.V = \frac{N.S.D}{N.M} \times 100 \qquad (15)$$

We can also write it as

$$C.V_N = \frac{S_N}{\overline{X}_N} * 100; C.V_N \in [C.V_L, C.V_U] \tag{16}$$

Example 15.1 Using N.S.D in Example 14.1 with reference to the data in 13.1, obtain neutrosophic coefficient of variation.

Solution In correspondence to the Examples 13.1 and 14.1, we have following information:

$$N.S.D \in [2.82, 2.59], N.M \in [64, 63.68]$$

Putting values in formula 15, we have

$$N.C.V = \frac{[2.82, 2.59]}{[64, 63.68]} * 100$$

$$N.C.V = [4.41, 4.07]$$

Hence, C. $V_N \in [4.41, 4.07]$

As the data is collected under the uncertain environment, there is not a single answer for the coefficient of variation. In fact, the neutrosophic C.V lies between 4.41 and 4.07.

16 Neutrosophic Mean Deviation (N.M.D)

For a set of neutrosophic data of a variable, the neutrosophic mean deviation is defined to be the sum of the absolute differences from their mean. For grouped data, it can be calculated from the following formula:

$$N.M.D = \sum_{i=1}^{n_N} \frac{f_{iN} | x_{iN} - \overline{X}_N |}{f_{iN}}; M.D_N \in [M.D_L, M.D_U] \tag{17}$$

Example 16.1 For the following distribution, calculate the neutrosophic mean deviation.

Variable (X)	Frequency
0	4
1	[6,5]
2	[7,6]
3	8
4	[13,15]
5	2

Solution

X_{iN}	f	fX_{iN}	$f \mid x_{iN} - \overline{X}_N \mid$
0	4	[0,0]	[10.6, 11.08]
1	[6,5]	[6,5]	[9.9, 8.85]
2	[7,6]	[14,12]	[4.55, 4.62]
3	8	[24,24]	[13.2, 14.16]
4	[13,15]	[52,60]	[34.45, 41.55]
5	2	[10,10]	[7.3, 7.54]

N.Mean $= \left[\frac{106}{40}, \frac{111}{40} \right]$

N. Mean $= [2.65, 2.77]$

$$M.D_N = \frac{[80, \ 87.8]}{40}$$

$$M.D_N = [2, 2.195]$$

$$M.D_N \in [2, 2.195]$$

Hence, the neutrosophic mean deviation for the given data set can be any number between 2 and 2.195.

17 Neutrosophic Empirical Relation Between Mean, Median, and Mode

The empirical relation of median, mean, and mode will be the same neutrosophically as in classical statistics for skewed data.

$$N.Mean - N.Mode = 3 \, (N.Mean - N.Median)$$

We can also write it as follows:

$$\overline{X}_N - \hat{X}_N = 3\left(\overline{X}_N - \tilde{X}_N \right)$$

18 Neutrosophic Moments About Mean

Suppose $X_{1n}, X_{2n}, \ldots, X_{in}$ is the observations of a random variable of sample size n_N, then j-th neutrosophic moment about mean for grouped data, it can be calculated as follows:

$$m_{jN} = \frac{\sum_{i=1}^{n_N} f_{iN}(x_{iN} - \overline{x}_N)^j}{\sum_{i=1}^{n_N} f_{iN}}, \text{ where } i, j = 1, 2, \ldots n \tag{18}$$

$$m_{jN} \in \left[m_{Lj}, m_{Uj} \right]$$

19 Neutrosophic Moments About Origin

Neutrosophic moments about the arbitrary origin is calculated in modifications in neutrosophic moments about mean. In this case, neutrosophic mean will be replaced by any neutrosophic arbitrary number.

For grouped data, it can be calculated as follows:

$$m_{jN}' = \frac{\sum_{i=1}^{n_N} f_{iN}(x_{iN} - A_N)^j}{\sum_{i=1}^{n_N} f_{iN}}, \text{ where } i, j = 1, 2, \ldots n \tag{19}$$

$$m_{jN}' \in \left[m_{Lj}, m_{Uj} \right]$$

When this A is 0, then moments about arbitrary origin became moment about origin.

For grouped data, it can be calculated as follows:

$$m_{jN}' = \frac{\sum_{i=1}^{n_N} f_{iN}(x_{iN})^j}{\sum_{i=1}^{n_N} f_{iN}}, \text{ where } i, j = 1, 2, \ldots n \tag{20}$$

$$m_{jN}' \in \left[m_{Lj}, m_{Uj} \right]$$

Example 19.1 Using the data given in 16.1, estimate the first four moments.

Solution Using the formula in (20), we have

$$m_{jN}' = \frac{\sum_{i=1}^{n_N} f_{iN}(x_{iN})^j}{\sum_{i=1}^{n_N} f_{iN}}$$

X_{iN}	f_{iN}	Xf_{iN}	X^2f_{iN}	X^3f_{iN}	X^4f_{iN}
0	4	[0,0]	[0,0]	[0,0]	[0,0]
1	[6,5]	[6,5]	[6,5]	[6,5]	[6,5]
2	[7,6]	[14,12]	[28,24]	[56,48]	[112,96]
3	8	[24,24]	[72,72]	[216,216]	[648,648]
4	[13,15]	[52,60]	[208,240]	[832,960]	[3328,3840]
5	2	[10,10]	[50,50]	[250,250]	[1250,1250]

$$m_{1N}' = \frac{[106, 111]}{40} = [2.65, 2.77]$$

$$m_{1N}' \in [2.65, 2.77]$$

$$m_{2N}' = \frac{[364, 391]}{40} = [9.1, 9.77]$$

$$m_{2N}' \in [9.1, 9.77]$$

$$m_{3N}' = \frac{[1360, 1479]}{40} = [34, 36.97]$$

$$m_{3N}' \in [34, 36.97]$$

$$m_{4N}' = \frac{[5344, 5839]}{40} = [133.6, 145.975]$$

$$m_{4N}' \in [133.6, 145.975]$$

20 Neutrosophic Skewness

Neutrosophic skewness refers to the term when there is no symmetry in the neutrosophic data or there is non-normality. There are two types of neutrosophic skewness. When there is positive skewness or tail is on the right side, then it is termed as neutrosophic positive skewness. When skewness is negative, or the tail of the graph is on the left side, then it is termed as negative neutrosophic skewness. It can be calculated from the following formula:

$$\text{N.SK} = \frac{\overline{x}_N - \hat{x}_N}{S_N}; \; SK_N \in [SK_L, SK_U] \tag{21}$$

$$\text{N.SK} = \frac{3\overline{x}_N - \widetilde{x}}{S_N}; \; SK_N \in [SK_L, SK_U] \tag{22}$$

21 Neutrosophic Kurtosis

Neutrosophic kurtosis represents the peakedness of the distribution. There are three types of weaknesses as in classical statistics. Mesokurtic, leptokurtic, and platykurtic for peak equal to 3, greater than 3, and less than 3, respectively. It can also be stated as degree of flatness.

22 Neutrosophic Moment Ratios

Neutrosophic coefficient of skewness and kurtosis can be calculated from neutrosophic moment ratios. Its formulas are as follows:

$$b_{1N} = \frac{m_{3N}^2}{m_{2N}^3} \tag{23}$$

The above formula is for the neutrosophic coefficient of skewness. If it is 0, then its symmetrix, if it is greater than 0, then it is positively skewed and vice versa:

$$b_{2N} = \frac{m_{4N}}{m_{2N}^2} \tag{24}$$

The above is for the neutrosophic coefficient of kurtosis. If it is 3, then it is mesokurtic; if it is greater than 3, then it is leptokurtic and platykurtic for less than 3.

Example 22.1 Using Example 19.1, calculate moment ratios.

Solution Using the formula in (23), we have

$$b_{1N} = \frac{[34, 36.97]^2}{[9.1, 9.77]^3} = [1.534, 1.465]$$

$$b_{1N} \in [1.534, 1.465]$$

Hence, the neutrosophic data follows positive distribution.
Using the formula in (24), we have

$$b_{2N} = \frac{[133.6, 145.975]}{[9.1, 9.77]^2} = [1.61, 1.52]$$

$$b_{2N} \in [1.61, 1.52]$$

Hence, it is platykurtic.

Exercise Q1. Multiple Choice Questions

i. A method which is used to find the center of the neutrosophic distribution is called:

    (a) The measure of central tendency
    (b) The measure of dispersion
    (c) Neutrosophic mean
    (d) None of the above

ii. The value that occurs most frequently in a neutrosophic data set is called:

    (a) Neutrosophic mean
    (b) Neutrosophic median
    (c) Neutrosophic range
    (d) Neutrosophic harmonic mean

iii. Which of the following statements is true about the neutrosophic median?

    (a) It is the central value of the neutrosophic observations.
    (b) Fifty percent of the neutrosophic observations are larger than the median.
    (c) It is the middle value of the neutrosophic data.
    (d) The sum of the observations from the neutrosophic median is zero.

iv. The neutrosophic arithmetic mean reduces to A.M under classical statistics if:

$n_L = n_U$
$n_L > n_U$
$n_L < n_U$
$n_L \neq n_U$

v. _____ is the nth positive root of the product of n_N neutrosophic observations.

(a) Neutrosophic arithmetic mean
(b) Neutrosophic geometric mean
(c) Neutrosophic median
(d) Neutrosophic coefficient of variation

vi. ____ is an important average, which is used to find the center of the neutrosophic data when expressed in rate, ratio, and percentage.

(a) Neutrosophic arithmetic mean
(b) Neutrosophic harmonic mean
(c) Neutrosophic median
(d) Neutrosophic coefficient of variation

vii. The neutrosophic quantiles divided the data into four parts and the second NQ represents ____ of data below it.

(a) 25%
(b) 50%
(c) 75%
(d) 100%

viii. The fifth neutrosophic decile is equal to:

(a) Neutrosophic harmonic mean
(b) Second neutrosophic quantile
(c) Zero
(d) None of the above

ix. For skewed data, the neutrosophic empirical relation between mean, median, and mode is:

(a) N.Mean − N.Median = 3(N.Mean − N.Mode)
(b) N.Mean − N.Mode = 3(N.Mean − N.Median)
(c) N.Mean − N.Mode = (N.Mean − N.Median)
(d) N.Mean + N.Mode = 3(N.Mean + N.Median)

x. The lack of symmetry in the neutrosophic data is called:

(a) Neutrosophic moments
(b) Neutrosophic skewness
(c) Neutrosophic kurtosis
(d) Neutrosophic quartile deviation

Q2.

i. What is a neutrosophic measure of central tendency? What is the practical significance of it? Name some different types of neutrosophic measure of central tendency.
ii. What is the difference between classical average and the neutrosophic average?
iii. In what circumstances would you prefer neutrosophic arithmetic mean?

iv. What is the neutrosophic weighted mean? How can you differentiate it with the neutrosophic arithmetic mean?
v. What is the difference between neutrosophic arithmetic and harmonic mean?
vi. Define the neutrosophic median? Describe the manner of computation of it.
vii. Define and explain how to compute the quartile deviation and mean deviation of the neutrosophic data set.
viii. Compare the neutrosophic mean, median, and mode. Give an empirical relation between these neutrosophic measures of central tendency.
ix. Define neutrosophic variance and how it is calculated. Discuss the practical usage of this neutrosophic measure of dispersion.
x. Explain the term "neutrosophic kurtosis" and its different types.

Q3. Following is the monthly income of seven families in dollars ($100) in a certain locality:

Income	Frequency
60–69	[4, 5]
70–79	[9, 12]
80–89	[14, 10]
90–99	[6, 6]

Calculate the neutrosophic mean and also interpret the result.
Q4. Find the average weight of oranges from the given neutrosophic data.

Weight (grams)	Frequency
60–79	[8, 9]
80–99	[11, 13]
100–119	[16, 16]
120–139	[12, 12]
140–159	[4, 3]
160–179	[3, 3]

Q5. For the following frequency distribution of marks, find the neutrosophic mean using direct method. Also, calculate the neutrosophic geometric mean.

Marks	No. of students
30–39	9
40–49	[20, 25]
50–59	49
60–69	[70, 65]
70–79	[38, 35]
80–89	27
90–99	7

Number of goals	Number of matches	
	Team A	Team B
0	24	[14, 12]
1	[6, 7]	6
2	[5, 3]	[3, 5]
3	[3, 1]	2
4	2	2

Q6. In a football season, the goals scored by two teams were as follows:

Find which team can be considered more consistent by calculating the neutrosophic coefficient of variation for both teams.

Q7. Calculate the neutrosophic weighted mean from the following data:

Item	Expenditure	Weights
Food & beverages	[295, 300]	6.5
Housing	[59, 65]	2.5
Clothing	[103, 100]	2.0
Fuel	[80, 80]	1.5
Miscellaneous	[80, 80]	0.5

Q8. The marks obtained by 50 students in IQ test are given below:

Marks	Frequency
3, 5	4
5	6
[6, 5]	12
7	14
8	8
[9, 8]	6

Calculate the neutrosophic median and neutrosophic quartiles. Also find the 7th decile of this neutrosophic data.

Q9. Using the data given in Q8, find the 5th neutrosophic decile and 20th percentile. Also find the neutrosophic quartile deviation.

Q10. From the following distribution of annual death rates, find the neutrosophic arithmetic mean, neutrosophic geometric mean, and neutrosophic harmonic mean. Also calculate the neutrosophic range for this distribution.

Death rate	Frequency
3.5–5.4	5
5.5–7.4	18
7.5–9.4	[30, 32]
9.5–11.4	[12, 10]
11.5–13.4	6

Q11. The following frequency distribution shows the ages (to nearest birthday) of part-time workers in the courier companies of California.

Age	No. of workers
18–19	[8, 6]
20–24	[12, 14]
25–29	[15, 17]
30–34	[10, 8]
35–44	5

Find the neutrosophic median, quartiles, and 12th percentile. Also calculate the quartile deviation for this neutrosophic data set.

Q12. The number of absentees per day in a chocolate factory was recorded over a month. The employer is interested to find out the average absences per month. Find the neutrosophic mean of the data. Also calculate neutrosophic median and neutrosophic mode of the data.

Absentees	No. of days
0	4
1	[8, 6]
2	[10, 12]
3	7
4	[3, 5]
5	2
6	1

Q13. Find the neutrosophic median and mode of the following distribution of weights of 50 football players.

Weight	No. of players
190–209	[4, 5]
210–219	[10, 11]
220–229	[14, 13]
230–239	[12, 11]
240–249	7
250–259	3

Q14. A population of size 10 has the observations 6, [7, 9], [9, 11], 12, [13, 10], [18, 21], 19, 24, 25 and 27. Find its neutrosophic variance and the neutrosophic standard deviation. Also calculate neutrosophic coefficient of variation.

Q15. Calculate the neutrosophic variance for the distribution given below:

Variable	5	[6, 6.5]	7	8	[9, 9.5]	[10, 10.5]
Frequency	2	5	7	13	8	3

Q16. The following distribution shows electricity consumption by residential consumers in a certain area of Saudi Arabia.

Consumption (kw)	No. of consumers
5–19	[4, 5]
21–34	[6, 9]
35–49	[14, 10]
50–64	[22, 19]
65–79	9
80–94	3

Find the neutrosophic standard deviation of this data set. Also estimate the neutrosophic mean deviation.

Q17. The monthly pocket money of 40 students is given below:

Pocket money per week (SR.)	No. of students
10–19	[6, 7]
20–29	[9, 8]
30–39	[14, 12]
40–49	[7, 7]
50–59	[4, 6]

Find the first four neutrosophic moments about the origin. Also calculate the neutrosophic skewness and moment ratios.

Q18. Using the data given in Q15, find the neutrosophic moments about mean. Also calculate the neutrosophic kurtosis and comment on the result.

Q19. The following table gives the frequency distribution of weights of 100 sacks filled with cotton.

x	[25, 25.5]	26	[27, 27.5]	28	[29, 29.5]	30	31
f	4	12	[33, 39]	[36, 30]	[10, 8]	[4, 6]	1

Find the neutrosophic arithmetic mean and the standard deviation. Also calculate the neutrosophic moment ratios.

Q20. Calculate the first four moments about mean and also find the neutrosophic coefficient of skewness for the following data.

x	[1, 1.5]	2	[3, 3.5]	4	[5, 5.5]	6	[7, 7.5]
f	6	19	13	18	20	9	5

References

1. Smarandache, F. (1998). *Neutrosophy. Neutrosophic probability, set, and logic, proquest information & learning*. Michigan: Ann Arbor.
2. Abdel-Basset, M., Manogaran, G., Gamal, A., & Smarandache, F. (2018). A hybrid approach of neutrosophic sets and DEMATEL method for developing supplier selection criteria. *Design Automation for Embedded Systems, 22*, 257–278.
3. Aslam, M. (2018a). A new sampling plan using neutrosophic process loss consideration. *Symmetry, 10*(5), 132.
4. Aslam, M. (2018b). Design of sampling plan for exponential distribution under neutrosophic statistical interval method. *IEEE Access, 6*, 64,153–64,158.
5. Aslam, M., & Alhassam, M. (2019). Application of neutrosophic logic to evaluate correlation between prostate cancer mortality and dietary fat assumption. *Symmetry, 11*(3), 330–336.
6. Broumi, S., & Smarandache, F. (2015). New operations on interval neutrosophic sets. *Journal of New Theory*, 24–37. https://doi.org/10.6084/M9.FIGSHARE.1502590.
7. Chen, J., Ye, J., Du, S., & Yong, R. (2017b). Expressions of rock joint roughness coefficient using neutrosophic interval statistical numbers. *Symmetry, 9*(7), 123.
8. Chen, J., Ye, J., & Du, S. (2017a). Scale effect and anisotropy analyzed for neutrosophic numbers of rock joint roughness. *Symmetry, 9*(10), 208.
9. Smarandache, F. (2014). *Introduction to neutrosophic statistics*. Columbus, OH: Sitech & Edcation Publishing.

A New Approach to Neutrosophic Soft Mappings and Application in Decision Making

Adem Yolcu, Elif Karatas, and Taha Yasin Ozturk

1 Introduction

The fuzzy set theory was introduced by Zadeh in 1965 [1]. After that a lot of new theories have been introduced that treat imprecision and uncertainty. One of these theories is neutrosophic set theory [2]. Samarandache introduced the neutrosophic set theory as a generalization of many theories such as fuzzy set, intuitionistic fuzzy set [3], etc. The neutrosophic logic includes considering the truth-membership, indeterminancy-membership, and falsity-membership values in real world problems. Recent researches on neutrosophic set theory and its applications in various fields are advancing rapidly. A lot of study can be found in this regard in [4–11].

The soft set theory was introduced by Molodtsov [12] as a mathematical tool for dealing with uncertainties. Soft set theory is free from the fuzzy set theory, rough set theory, and probability theory parameterization inadequacy syndrome. The soft set theory has attracted the attention of many researchers because of its rich potential. Many studies have been made using different combinations of soft sets and other set theories like fuzzy soft sets, generalized fuzzy soft sets, intuitionistic fuzzy soft sets, rough soft sets, possibility fuzzy soft sets, generalized intuitionistic fuzzy softs, possibility vague soft sets, and so on [13–16]. All these researches aim to solve most of our real life problems in medical sciences, engineering, management, environment, and social sciences which involve data that are not crisp and precise.

The neutrosophic soft set (NSS) is a combination form of neutrosophic set and soft set. This theory was introduced by Maji [17]. However, there were some structural deficiencies in the study [17]. The concepts of intersection, union, AND, OR, and different operations defined in [17] were insufficient when processing

A. Yolcu (✉) · E. Karatas · T. Y. Ozturk
Department of Mathematic, Kafkas University, Kars, Turkey
e-mail: yolcu.adem@gmail.com

© The Author(s), under exclusive license to Springer Nature Switzerland AG 2021
F. Smarandache, M. Abdel-Basset (eds.), *Neutrosophic Operational Research*,
https://doi.org/10.1007/978-3-030-57197-9_15

neutrosophic soft sets. Therefore, in 2019, Ozturk et al. [18] re-defined the intersection, union, AND, OR, and different operations on the neutrosophic soft sets in contrast to the studies [17, 19]. Using the neutrosophic soft concept, several mathematicians produced their research work in different mathematical structures [17, 18, 20–23].

The concept of mapping in soft classes was proposed by Kharal and Ahmad [24]. They also presented a mapping concept on the classes of fuzzy soft sets [14] and examined the properties of fuzzy soft sets and the properties of fuzzy soft inverse images of fuzzy soft sets and supported them with inverse inconsistency in the examples and examples. The mapping concepts on neutrosophic soft classes have been defined by Alkazaleh et.al. [25]. There are also structural problems in some notions such as null neutrosophic soft set, neutrosophic soft union, neutrosophic soft intersection, "and" operation, "or" operation. The study [25] does not even provide some basic mapping properties such as $(F, A) \subseteq f^{-1}(f((F, A)))$. Therefore, Alkazaleh's study is inadequate in terms of mapping on neutrosophic soft classes. There is a need to have well-defined mapping on neutrosophic soft sets.

In this paper, we re-define the notion of neutrosophic soft mapping and study the properties of neutrosophic soft images and neutrosophic soft inverse images. We give some examples of neutrosophic soft mappings. Finally, we also present an application in decision making with neutrosophic soft mappings.

2 Preliminaries

Definition 2.1 [2] A neutrosophic set A on the universe of discourse X is defined as follows:

$$A = \{\langle x, T_A(x), I_A(x), F_A(x) \rangle : x \in X\},$$

where $T, I, F : X \rightarrow \ ]^-0, 1^+[$ and $^-0 \leq T_A(x) + I_A(x) + F_A(x) \leq 3^+$.

Definition 2.2 [16] Let X be an initial universe, E be a set of all parameters, and $P(X)$ denotes the power set of X. A pair (F, E) is called a soft set over X, where F is a mapping given by $F : E \rightarrow P(X)$.

In other words, the soft set is a parameterized family of subsets of the set X. For $e \in E$, $F(e)$ may be considered as the set of $e-$ elements of the soft set (F, E), or as the set of $e-$ approximate elements of the soft set, i.e.,

$$(F, E) = \left\{ \left(e, F(e) \right) : e \in E, F : E \rightarrow P(X) \right\}.$$

Firstly, neutrosophic soft set defined by Maji [17] and later this concept has been modified by Deli and Bromi [21] as given below.

Definition 2.3 Let X be an initial universe set and E be a set of parameters. Let $P(X)$ denote the set of all neutrosophic sets of X. Then, a neutrosophic soft set $\left(\widetilde{F}, E\right)$ over X is a set defined by a set-valued function $\widetilde{F}$ representing a mapping $\widetilde{F} : E \rightarrow P(X)$ where $\widetilde{F}$ is called approximate function of the neutrosophic soft set $\left(\widetilde{F}, E\right)$. In other words, the neutrosophic soft set is a parameterized family of some elements of the set $P(X)$ and therefore it can be written as a set of ordered pairs:

$$\left(\widetilde{F}, E\right) = \left\{ \left(e, \left\langle x, T_{\widetilde{F}(e)}(x), I_{\widetilde{F}(e)}(x), F_{\widetilde{F}(e)}(x) \right\rangle : x \in X \right) : e \in E \right\}$$

where $T_{\widetilde{F}(e)}(x)$, $I_{\widetilde{F}(e)}(x)$, $F_{\widetilde{F}(e)}(x) \in \left[0, 1 \right]$, respectively, called the truth-membership, indeterminacy-membership, falsity-membership function of $\widetilde{F}(e)$. Since supremum of each T, I, F is 1 so the inequality $0 \leq T_{\widetilde{F}(e)}(x) + I_{\widetilde{F}(e)}(x) + F_{\widetilde{F}(e)}(x) \leq 3$ is obvious.

Definition 2.4 [19] Let $\left(\widetilde{F}, E\right)$ be neutrosophic soft set over the universe set X. The complement of $\left(\widetilde{F}, E\right)$ is denoted by $\left(\widetilde{F}, E\right)^c$ and is defined by:

$$\left(\widetilde{F}, E\right)^c = \left\{ \left(e, \left\langle x, F_{\widetilde{F}(e)}(x), 1 - I_{\widetilde{F}(e)}(x), T_{\widetilde{F}(e)}(x) \right\rangle : x \in X \right) : e \in E \right\}.$$

Obvious that $\left(\left(\widetilde{F}, E\right)^c \right)^c = \left(\widetilde{F}, E\right)$.

Definition 2.5 [17] Let $\left(\widetilde{F}, E\right)$ and $\left(\widetilde{G}, E\right)$ be two neutrosophic soft sets over the universe set X. $\left(\widetilde{F}, E\right)$ is said to be neutrosophic soft subset of $\left(\widetilde{G}, E\right)$ if $T_{\widetilde{F}(e)}(x) \leq T_{\widetilde{G}(e)}(x), I_{\widetilde{F}(e)}(x) \leq I_{\widetilde{G}(e)}(x), F_{\widetilde{F}(e)}(x) \geq F_{\widetilde{G}(e)}(x), \forall e \in E, \forall x \in X$. It is denoted by $\left(\widetilde{F}, E\right) \subseteq \left(\widetilde{G}, E\right)$.

$\left(\widetilde{F}, E\right)$ is said to be neutrosophic soft equal to $\left(\widetilde{G}, E\right)$ if $\left(\widetilde{F}, E\right)$ is neutrosophic soft subset of $\left(\widetilde{G}, E\right)$ and $\left(\widetilde{G}, E\right)$ is neutrosophic soft subset of $\left(\widetilde{F}, E\right)$. It is denoted by $\left(\widetilde{F}, E\right) = \left(\widetilde{G}, E\right)$.

Definition 2.6 [18] Let $\left(\widetilde{F}_1, E\right)$ and $\left(\widetilde{F}_2, E\right)$ be two neutrosophic soft sets over the universe set X. Then their union is denoted by $\left(\widetilde{F}_1, E\right) \widetilde{\cup} \left(\widetilde{F}_2, E\right) = \left(\widetilde{F}_3, E\right)$ and is defined by:

$$\left(\widetilde{F}_3, E\right) = \left\{ \left(e, \left\langle x, T_{\widetilde{F}_3(e)}(x), I_{\widetilde{F}_3(e)}(x), F_{\widetilde{F}_3(e)}(x) \right\rangle : x \in X \right) : e \in E \right\}$$

where

$$T_{\widetilde{F}_3(e)}(x) = \max \left\{ T_{\widetilde{F}_1(e)}(x), T_{\widetilde{F}_2(e)}(x) \right\},$$

$$I_{\widetilde{F}_3(e)}(x) = \max \left\{ I_{\widetilde{F}_1(e)}(x), I_{\widetilde{F}_2(e)}(x) \right\},$$

$$F_{\widetilde{F}_3(e)}(x) = \min \left\{ F_{\widetilde{F}_1(e)}(x), F_{\widetilde{F}_2(e)}(x) \right\}.$$

Definition 2.7 [18] Let $\left(\widetilde{F}_1, E\right)$ and $\left(\widetilde{F}_2, E\right)$ be two neutrosophic soft sets over the universe set X. Then their intersection is denoted by $\left(\widetilde{F}_1, E\right) \widetilde{\cap} \left(\widetilde{F}_2, E\right) = \left(\widetilde{F}_3, E\right)$ and is defined by:

$$\left(\widetilde{F}_3, E\right) = \left\{ \left(e, \left\langle x, T_{\widetilde{F}_3(e)}(x), I_{\widetilde{F}_3(e)}(x), F_{\widetilde{F}_3(e)}(x) \right\rangle : x \in X \right) : e \in E \right\}$$

where

$$T_{\widetilde{F}_3(e)}(x) = \min \left\{ T_{\widetilde{F}_1(e)}(x), T_{\widetilde{F}_2(e)}(x) \right\},$$

$$I_{\widetilde{F}_3(e)}(x) = \min \left\{ I_{\widetilde{F}_1(e)}(x), I_{\widetilde{F}_2(e)}(x) \right\},$$

$$F_{\widetilde{F}_3(e)}(x) = \max \left\{ F_{\widetilde{F}_1(e)}(x), F_{\widetilde{F}_2(e)}(x) \right\}.$$

Definition 2.8 [18] Let $\left(\tilde{F}_1, E\right)$ and $\left(\tilde{F}_2, E\right)$ be two neutrosophic soft sets over the universe set X. Then "$\left(\tilde{F}_1, E\right)$ difference $\left(\tilde{F}_2, E\right)$" operation on them is denoted by $\left(\tilde{F}_1, E\right) \setminus \left(\tilde{F}_2, E\right) = \left(\tilde{F}_3, E\right)$ and is defined by $\left(\tilde{F}_3, E\right) = \left(\tilde{F}_1, E\right) \cap \left(\tilde{F}_2, E\right)^c$ as follows:

$$\left(\tilde{F}_3, E\right) = \left\{\left(e, \left\langle x, T_{\tilde{F}_3(e)}(x), I_{\tilde{F}_3(e)}(x), F_{\tilde{F}_3(e)}(x)\right\rangle : x \in X\right) : e \in E\right\}$$

where

$$T_{\tilde{F}_3(e)}(x) = \min\left\{T_{\tilde{F}_1(e)}(x), F_{\tilde{F}_2(e)}(x)\right\},$$

$$I_{\tilde{F}_3(e)}(x) = \min\left\{I_{\tilde{F}_1(e)}(x), 1 - I_{\tilde{F}_2(e)}(x)\right\},$$

$$F_{\tilde{F}_3(e)}(x) = \max\left\{F_{\tilde{F}_1(e)}(x), T_{\tilde{F}_2(e)}(x)\right\}.$$

Definition 2.9 [18] Let $\left\{\left(\tilde{F}_i, E\right) \middle| i \in I\right\}$ be a family of neutrosophic soft sets over the universe set X. Then

$$\bigcup_{i \in I}\left(\tilde{F}_i, E\right) = \left\{\left(e, \left\langle x, \sup\left[T_{\tilde{F}_i(e)}(x)\right]_{i \in I}, \sup\left[I_{\tilde{F}_i(e)}(x)\right]_{i \in I},\right.\right.\right.$$
$$\left.\left.\left.\inf\left[F_{\tilde{F}_i(e)}(x)\right]_{i \in I}\right\rangle : x \in X\right) : e \in E\right\},$$

$$\bigcap_{i \in I}\left(\tilde{F}_i, E\right) = \left\{\left(e, \left\langle x, \inf\left[T_{\tilde{F}_i(e)}(x)\right]_{i \in I}, \inf\left[I_{\tilde{F}_i(e)}(x)\right]_{i \in I},\right.\right.\right.$$
$$\left.\left.\left.\times \sup\left[F_{\tilde{F}_i(e)}(x)\right]_{i \in I}\right\rangle : x \in X\right) : e \in E\right\}.$$

Definition 2.10 [18]

1. A neutrosophic soft set $\left(\tilde{F}, E\right)$ over the universe set X is said to be null neutrosophic soft set if $T_{\tilde{F}(e)}(x) = 0$, $I_{\tilde{F}(e)}(x) = 0$, $F_{\tilde{F}(e)}(x) = 1$; $\forall e \in E$, $\forall x \in X$. It is denoted by $0_{(X,E)}$.

2. A neutrosophic soft set $\left(\tilde{F}, E\right)$ over the universe set X is said to be absolute neutrosophic soft set if $T_{\tilde{F}(e)}(x) = 1$, $I_{\tilde{F}(e)}(x) = 1$, $F_{\tilde{F}(e)}(x) = 0$; $\forall e \in E$, $\forall x \in X$. It is denoted by $1_{(X,E)}$.

    Clearly, $0^c_{(X,E)} = 1_{(X,E)}$ and $1^c_{(X,E)} = 0_{(X,E)}$.

Definition 2.11 [22] Let $NSS(X, E)$ be the family of all neutrosophic soft sets over the universe set X. Then neutrosophic soft set $x^e_{(\alpha,\beta,\gamma)}$ is called a neutrosophic soft point, for every $x \in X$, $0 < \alpha, \beta, \gamma \leq 1$, $e \in E$, and defined as follows:

$$x^e_{(\alpha,\beta,\gamma)}\left(e'\right)(y) = \begin{cases} (\alpha, \beta, \gamma) \text{ if } e' = e \text{ and } y = x, \\ (0, 0, 1) \text{ if } e' \neq e \text{ or } y \neq x. \end{cases}$$

Definition 2.12 [22] Let $\left(\tilde{F}, E\right)$ be a neutrosophic soft set over the universe set X. We say that $x^e_{(\alpha,\beta,\gamma)} \in \left(\tilde{F}, E\right)$ read as belongs to the neutrosophic soft set $\left(\tilde{F}, E\right)$, whenever $\alpha \leq T_{\tilde{F}(e)}(x)$, $\beta \leq I_{\tilde{F}(e)}(x)$ and $\gamma \geq F_{\tilde{F}(e)}(x)$.

Definition 2.13 [22] Let $x^e_{(\alpha,\beta,\gamma)}$ and $y^{e'}_{(\alpha',\beta',\gamma')}$ be two neutrosophic soft points. For the neutrosophic soft points $x^e_{(\alpha,\beta,\gamma)}$ and $y^{e'}_{(\alpha',\beta',\gamma')}$ over a common universe X, we say that the neutrosophic soft points are distinct points if $x^e_{(\alpha,\beta,\gamma)} \cap y^{e'}_{(\alpha',\beta',\gamma')} = 0_{(X,E)}$. It is clear that $x^e_{(\alpha,\beta,\gamma)}$ and $y^{e'}_{(\alpha_1,\beta_1,\gamma_1)}$ are distinct neutrosophic soft points if and only if $x \neq y$ or $e' \neq e$.

3 Neutrosophic Soft Mappings

In [18] study, the union and intersection operations on neutrosophic soft sets are defined for a single parameter. We have further generalized this definition and redefined it for different parameters.

Throughout this paper X denotes initial universe, E denotes the set of all parameters, and $A, B \subset E$.

Definition 3.1 Let $\left(\widetilde{F}, A\right)$ and $\left(\widetilde{G}, B\right)$ be two neutrosophic soft sets over the universe set X. Then their extended union of $\left(\widetilde{F}, A\right)$ and $\left(\widetilde{G}, B\right)$ is denoted by $\left(\widetilde{F}, A\right) \widetilde{\cup} \left(\widetilde{G}, B\right) = \left(\widetilde{H}, C\right)$ where $C = A \cup B$ and the truth-membership, indeterminancy-membership, falsity-membership of $\left(\widetilde{H}, C\right)$ are as follows:

$$\left(\widetilde{H}, C\right) = \left\{ \left(e, \langle x, T_{H(e)}(x), I_{H(e)}(x), F_{H(e)}(x)\rangle \right) : x \in X \right) : e \in C \right\}$$

where

$$T_{H(e)}(x) = \begin{cases} T_{F(e)}(x), & \text{if } e \in A - B \\ T_{G(e)}(x), & \text{if } e \in B - A \\ \max\left(T_{F(e)}(x), T_{G(e)}(x)\right), & \text{if } e \in A \cap B \end{cases}$$

$$I_{H(e)}(x) = \begin{cases} I_{F(e)}(x), & \text{if } e \in A - B \\ I_{G(e)}(x), & \text{if } e \in B - A \\ \max\left(I_{F(e)}(x), I_{G(e)}(x)\right), & \text{if } e \in A \cap B \end{cases}$$

$$F_{H(e)}(x) = \begin{cases} F_{F(e)}(x), & \text{if } e \in A - B \\ F_{G(e)}(x), & \text{if } e \in B - A \\ \min\left(T_{F(e)}(x), T_{G(e)}(x)\right), & \text{if } e \in A \cap B \end{cases}$$

Definition 3.2 Let $\left(\widetilde{F}, A\right)$ and $\left(\widetilde{G}, B\right)$ be two neutrosophic soft sets over the universe set X. Then their extended intersection of $\left(\widetilde{F}, A\right)$ and $\left(\widetilde{G}, B\right)$ is denoted by $\left(\widetilde{F}, A\right) \widetilde{\cap} \left(\widetilde{G}, B\right) = \left(\widetilde{H}, C\right)$ where $C = A \cup B$ and the truth-membership, indeterminancy-membership, falsity-membership of $\left(\widetilde{H}, C\right)$ are as follows:

$$\left(\widetilde{H}, C\right) = \left\{ \left(e, \langle x, T_{H(e)}(x), I_{H(e)}(x), F_{H(e)}(x)\rangle \right) : x \in X \right) : e \in C \right\}$$

where

$$T_{H(e)}(x) = \begin{cases} T_{F(e)}(x), & \text{if} e \in A - B \\ T_{G(e)}(x), & \text{if} e \in B - A \\ \min\left(T_{F(e)}(x), T_{G(e)}(x)\right), & \text{if} e \in A \cap B \end{cases}$$

$$I_{H(e)}(x) = \begin{cases} I_{F(e)}(x), & \text{if} e \in A - B \\ I_{G(e)}(x), & \text{if} e \in B - A \\ \min\left(I_{F(e)}(x), I_{G(e)}(x)\right), & \text{if} e \in A \cap B \end{cases}$$

$$F_{H(e)}(x) = \begin{cases} F_{F(e)}(x), & \text{if} e \in A - B \\ F_{G(e)}(x), & \text{if} e \in B - A \\ \max\left(T_{F(e)}(x), T_{G(e)}(x)\right), & \text{if} e \in A \cap B \end{cases}$$

Example 3.1 Let $\left(\widetilde{F}, A\right)$ and $\left(\widetilde{G}, B\right)$ be neutrosophic soft sets over the universe set X. Let us consider neutrosophic soft sets $\left(\widetilde{F}, A\right)$ and $\left(\widetilde{G}, B\right)$ as follows:

$$\left(\widetilde{F}, A\right)$$
$$= \begin{cases} (e_1, \{< x_1, 0.6, 0.3, 0.5 >, < x_2, 0.7, 0.3, 0.4 >, < x_3, 0.2, 0.1, 0.7 >\}), \\ (e_2, \{< x_1, 0.3, 0.2, 0.1 >, < x_2, 0.1, 0.5, 0.6 >, < x_3, 0.6, 0.7, 0.3 >\}), \\ (e_3, \{< x_1, 0.5, 0.8, 0.2 >, < x_2, 0.1, 0.4, 0.3 >, < x_3, 0.5, 0.1, 0.2 >\}) \end{cases}$$

$$\left(\widetilde{G}, B\right)$$
$$= \begin{cases} (e_2, \{< x_1, 0.7, 0.4, 0.2 >, < x_2, 0.3, 0.6, 0.2 >, < x_3, 0.1, 0.6, 0.7 >\}), \\ (e_4, \{< x_1, 0.3, 0.5, 0.6 >, < x_2, 0.8, 0.1, 0.9 >, < x_3, 0.6, 0.5, 0.6 >\}), \end{cases}$$

Then the extended union and extended intersection of $\left(\widetilde{F}, A\right)$ and $\left(\widetilde{G}, B\right)$ as follows:

$$\left(\widetilde{F}, A\right) \widetilde{\cup} \left(\widetilde{G}, B\right) = \left(\widetilde{H}, C\right)$$
$$= \begin{cases} (e_1, \{< x_1, 0.6, 0.3, 0.5 >, < x_2, 0.7, 0.3, 0.4 >, < x_3, 0.2, 0.1, 0.7 >\}), \\ (e_2, \{< x_1, 0.7, 0.4, 0.1 >, < x_2, 0.3, 0.6, 0.2 >, < x_3, 0.6, 0.7, 0.3 >\}), \\ (e_3, \{< x_1, 0.5, 0.8, 0.2 >, < x_2, 0.1, 0.4, 0.3 >, < x_3, 0.5, 0.1, 0.2 >\}), \\ (e_4, \{< x_1, 0.3, 0.5, 0.6 >, < x_2, 0.8, 0.1, 0.9 >, < x_3, 0.6, 0.5, 0.6 >\}) \end{cases}$$

$$\left(\tilde{F}, A\right) \tilde{\cap} \left(\tilde{G}, B\right) = \left(\tilde{K}, C\right)$$

$$= \begin{cases} (e_1, \{< x_1, 0.6, 0.3, 0.5 >, < x_2, 0.7, 0.3, 0.4 >, < x_3, 0.2, 0.1, 0.7 >\}), \\ (e_2, \{< x_1, 0.3, 0.2, 0.2 >, < x_2, 0.1, 0.5, 0.6 >, < x_3, 0.1, 0.6, 0.7 >\}), \\ (e_3, \{< x_1, 0.5, 0.8, 0.2 >, < x_2, 0.1, 0.4, 0.3 >, < x_3, 0.5, 0.1, 0.2 >\}), \\ (e_4, \{< x_1, 0.3, 0.5, 0.6 >, < x_2, 0.8, 0.1, 0.9 >, < x_3, 0.6, 0.5, 0.6 >\}) \end{cases}.$$

Remark 3.1 According to [25] study,

1. When defining f and f^{-1}, null neutrosophic soft set taken as $(0,0,0)$. But, null neutrosophic soft set cannot be $(0,0,0)$. Therefore, different definitions previously defined for the null neutrosophic soft set should be used.
2. If $I_o(\alpha, \beta)(m) = \frac{I_H(\alpha)(m)+I_G(\beta)(m)}{2}$ is defined in union, intersection, "or", "and" operations on neutrosophic soft sets and the value of null neutrosophic soft set is $(0,0,0)$, then $\left(\tilde{F}, A\right) \subseteq f^{-1}\left(f\left(\left(\tilde{F}, A\right)\right)\right)$ may not be provided. This condition cannot be provided in the Example 3.1 (page 5 in [25] study). If the necessary calculations are made, it will be as follows:

$$\left(\tilde{F}, A\right)$$

$$= \begin{cases} (e_1, \{< x_1, 0.4, 0.2, 0.3 >, < x_2, 0.7, 0.3, 0.4 >, < x_3, 0.5, 0.2, 0.2 >\}), \\ (e_2, \{< x_1, 0.2, 0.2, 0.7 >, < x_2, 0.3, 0.1, 0.8 >, < x_3, 0.2, 0.3, 0.6 >\}), \\ (e_3, \{< x_1, 0.8, 0.2, 0.1 >, < x_2, 0.9, 0.1, 0.1 >, < x_3, 0.1, 0.4, 0.5 >\}) \end{cases},$$

$$f\left(\left(\tilde{F}, A\right)\right)$$

$$= \begin{cases} (e'_1, \{< y_1, 0.4, 0.2, 0.3 >, < y_2, 0, 0, 0 >, < y_3, 0.7, 0.25, 0.2 >\}), \\ (e'_2, \{< y_1, 0.8, 0.2, 0.1 >, < y_2, 0, 0, 0 >, < y_3, 0.9, 0.25, 0.1 >\}), \\ (e'_3, \{< y_1, 0.2, 0.2, 0.7 >, < y_2, 0, 0, 0 >, < y_3, 0.3, 0.2, 0.6 >\}) \end{cases},$$

$$f^{-1}\left(f\left(\left(\tilde{F}, A\right)\right)\right) =$$

$$\begin{cases} (e_1, \{< x_1, 0.4, 0.2, 0.3 >, < x_2, 0.7, 0.25, 0.2 >, < x_3, 0.7, 0.25, 0.2 >\}), \\ (e_2, \{< x_1, 0.2, 0.2, 0.7 >, < x_2, 0.3, 0.2, 0.6 >, < x_3, 0.3, 0.2, 0.6 >\}), \\ (e_3, \{< x_1, 0.8, 0.2, 0.1 >, < x_2, 0.9, 0.25, 0.1 >, < x_3, 0.9, 0.25, 0.1 >\}) \end{cases}$$

It is clear that $\left(\tilde{F}, A\right) \not\subseteq f^{-1}\left(f\left(\left(\tilde{F}, A\right)\right)\right)$.

Due to such structural deficiencies, a new definition of neutrosophic soft mappings is needed.

Definition 3.3 Let $NSS(X, E)$ and $NSS(Y, E')$ be two neutrosophic soft classes and $u : X \to Y$ and $v : E \to E'$ be mappings. Then a mapping $f = (u, v) : (X, E) \to (Y, E')$ is defined as follows:

For a neutrosophic soft set $\left(\tilde{F}, A \right) \in NSS\,(X, E)$, $f\left(\left(\tilde{F}, A \right) \right)$ is a neutrosophic soft set in $NSS(Y, E)$ obtained as follows:

$$
T_{u(F)(e')}(y) = \begin{cases} \displaystyle\sup_{e \in v^{-1}(e') \cap A, x \in u^{-1}(y)} T_{F(e)}(x), & \text{if } \ u^{-1}(y) \neq \varnothing \\ 0 & , \quad \text{otherwise} \end{cases}
$$

$$
I_{u(F)(e')}(y) = \begin{cases} \displaystyle\sup_{e \in v^{-1}(e') \cap A, x \in u^{-1}(y)} I_{F(e)}(x), & \text{if } \ u^{-1}(y) \neq \varnothing \\ 0 & , \quad \text{otherwise} \end{cases}
$$

$$
F_{u(F)(e')}(y) = \begin{cases} \displaystyle\inf_{e \in v^{-1}(e') \cap A, x \in u^{-1}(y)} F_{F(e)}(x), & \text{if } \ u^{-1}(y) \neq \varnothing \\ 1 & , \quad \text{otherwise} \end{cases}
$$

for $e' \in v(A) \subseteq E', y \in Y$.

Definition 3.4 Let $NSS(X, E)$ and $NSS(Y, E')$ be neutrosophic soft classes and $u : X \to Y$ and $v : E \to E'$ be mappings. Then a mapping $f^{-1} : NSS(Y, E') \to NSS(X, E)$ is defined as follows:

For a neutrosophic soft set $\left(\tilde{G}, B \right)$ in $NSS(Y, E')$, $f^{-1}\left(\left(\tilde{G}, B \right) \right)$ is a neutrosophic soft set in $NSS(X, E)$ obtained as follows:

$$
T_{u^{-1}(G)(e)}(y) = \begin{cases} T_{G(v(e))}\left(u(x) \right), & \text{if } \ v^{-1}(e) \in B \\ 0 & , \quad \text{otherwise} \end{cases}
$$

$$
I_{u^{-1}(G)(e)}(y) = \begin{cases} I_{G(v(e))}\left(u(x) \right), & \text{if } \ v^{-1}(e) \in B \\ 0 & , \quad \text{otherwise} \end{cases}
$$

$$
F_{u^{-1}(G)(e)}(y) = \begin{cases} F_{G(v(e))}\left(u(x) \right), & \text{if } \ v^{-1}(e) \in B \\ 1 & , \quad \text{otherwise} \end{cases}
$$

For $e \in v^{-1}(B) \subseteq E$ and $x \in X$, $f^{-1}\left(\left(\tilde{G}, B\right)\right)$ is called a neutrosophic soft inverse image of the neutrosophic soft set $\left(\tilde{G}, B\right)$.

Example 3.2 Let $X = \{a, b, c\}$, $Y = \{p, q, r\}$, $E = \{e_1, e_2, e_3\}$, $E' = \{e_1', e_2', e_3'\}$ and $NSS(X, E)$, $NSS(Y, E')$ soft classes. Define $u : X \to Y$ and $v : E \to E'$ as follows:

$$u(a) = q, u(b) = r, u(c) = q$$

$$v(e_1) = e_3', v(e_2) = e_1', v(e_3) = e_3'$$

Choose two neutrosophic soft sets over X and Y, respectively, as follows:

$$\left(\tilde{F}, E\right) = \left\{ \begin{array}{l} (e_1, \{< a, 0.2, 0.5, 0.7 >, < b, 0.1, 0.1, 0.3 >, < c, 0.1, 0.2, 0.1 >\}), \\ (e_2, \{< a, 0.1, 0.5, 0.2 >, < b, 0.3, 0.3, 0.3 >, < c, 0.6, 0.9, 0.2 >\}), \\ (e_3, \{< a, 0.1, 0.5, 0.1 >, < b, 0.6, 0.8, 0.1 >, < c, 0.3, 0.2, 0.2 >\}) \end{array} \right\}$$

$$\left(\tilde{G}, E'\right) = \left\{ \begin{array}{l} (e_1', \{< p, 0.5, 0.1, 0.1 >, < q, 0.3, 0.7, 0.3 >, < r, 0.1, 0.5, 0.2 >\}), \\ (e_2', \{< p, 0.3, 0.2, 0.2 >, < q, 0.1, 0.1, 0.1 >, < r, 0.2, 0.3, 0.3 >\}), \\ (e_3', \{< p, 0.1, 0.1, 0.2 >, < q, 0.3, 0.2, 0.3 >, < r, 0.5, 0.4, 0.4 >\}) \end{array} \right\}$$

Then the mapping $f : NSS(X, E) \to NSS(Y, E')$ is given as follows: for a neutrosophic soft set $\left(\tilde{F}, E\right)$ in $NSS(X, E)$, $f\left(\left(\tilde{F}, E\right)\right)$ is a neutrosophic soft set in $NSS(Y, E')$ obtained as follows:

$$T_{u(F)(e_1')}(p) = \sup_{e \in v^{-1}(e_1') \cap A, x \in u^{-1}(p)} T_{F(e)}(x) = 0 \text{ as } u^{-1}(p) = \varnothing$$

$$I_{u(F)(e_1')}(p) = \sup_{e \in v^{-1}(e_1') \cap A, x \in u^{-1}(p)} T_{F(e)}(x) = 0 \text{ as } u^{-1}(p) = \varnothing$$

$$F_{u(F)(e_1')}(p) = \inf_{e \in v^{-1}(e_1') \cap A, x \in u^{-1}(p)} T_{F(e)}(x) = 1 \text{ as } u^{-1}(p) = \varnothing$$

$$T_{u(F)(e_1')}(q) = \sup_{e \in v^{-1}(e_1') \cap A, x \in u^{-1}(q)} T_{F(e)}(x) = \sup \left[T_{F(e_2)}(a), T_{F(e_2)}(c) \right]$$

$$= \max \{0.1, 0.6\} = 0.6$$

$$I_{u(F)(e_1')}(q) = \sup_{e \in v^{-1}(e_1') \cap A, x \in u^{-1}(q)} I_{F(e)}(x)$$

$$= \sup \left[I_{F(e_2)}(a), I_{F(e_2)}(c) \right] = \max \{0.5, 0.9\} = 0.9$$

$$F_{u(F)(e_1')}(q) = \inf_{e \in v^{-1}(e_1') \cap A, x \in u^{-1}(q)} F_{F(e)}(x)$$

$$= \inf \left[F_{F(e_2)}(a), F_{F(e_2)}(c) \right] = \min \{0.2, 0.2\} = 0.2$$

$$T_{u(F)(e_1')}(r) = \sup_{e \in v^{-1}(e_1') \cap A, x \in u^{-1}(r)} T_{F(e)}(x) = \sup \left[T_{F(e_2)}(b) \right] = 0.3$$

$$I_{u(F)(e_1')}(r) = \sup_{e \in v^{-1}(e_1') \cap A, x \in u^{-1}(r)} I_{F(e)}(x) = \sup \left[I_{F(e_2)}(b) \right] = 0.3$$

$$F_{u(F)(e_1')}(r) = \inf_{e \in v^{-1}(e_1') \cap A, x \in u^{-1}(r)} F_{F(e)}(x) = \inf \left[F_{F(e_2)}(b) \right] = 0.3$$

By similar calculations, we get

$$f\left(\left(\tilde{F}, E \right) \right) = \left\{ \begin{array}{l} (e_1', \{< p, 0, 0, 1 >, < q, 0.6, 0.9, 0.2 >, < r, 0.3, 0.3, 0.3 >\}), \\ (e_2', \{< p, 0, 0, 1 >, < q, 0, 0, 1 >, < r, 0, 0, 1 >\}), \\ (e_3', \{< p, 0, 0, 1 >, < q, 0.3, 0.5, 0.1 >, < r, 0.6, 0.8, 0.1 >\}) \end{array} \right\}$$

Next, for the mapping $f^{-1} : NSS(Y, E') \to NSS(X, E)$ and neutrosophic soft set $\left(\tilde{G}, E' \right)$ in $NSS(Y, E')$, we calculate $f^{-1}\left(\left(\tilde{G}, E' \right) \right)$ as follows:

$$T_{u^{-1}(G)(e_1)}(a) = T_{G(v(e_1))}u(a) = 0.3$$

$$I_{u^{-1}(G)(e_1)}(a) = I_{G(v(e_1))}u(a) = 0.2$$

$$F_{u^{-1}(G)(e_1)}(a) = F_{G(v(e_1))}u(a) = 0.3$$

$$T_{u^{-1}(G)(e_1)}(b) = T_{G(v(e_1))}u(b) = 0.5$$

$$I_{u^{-1}(G)(e_1)}(b) = I_{G(v(e_1))}u(b) = 0.4$$

$$F_{u^{-1}(G)(e_1)}(b) = F_{G(v(e_1))}u(b) = 0.4$$

$$T_{u^{-1}(G)(e_1)}(c) = T_{G(v(e_1))}u(c) = 0.3$$

$$I_{u^{-1}(G)(e_1)}(c) = I_{G(v(e_1))}u(c) = 0.2$$

$$F_{u^{-1}(G)(e_1)}(c) = F_{G(v(e_1))}u(c) = 0.3$$

So, it is obtained as follows:

$$f^{-1}\Big((G, E') (e_1) (a) = (0.3, 0.2, 0.3)$$

$$f^{-1}\Big((G, E') (e_1) (b) = (0.5, 0.4, 0.4)$$

$$f^{-1}\Big((G, E') (e_1) (c) = (0.3, 0.2, 0.3)$$

By similar calculations, consequently, we get

$$f^{-1}\Big(\Big(\tilde{G}, E'\Big) (e_2) (a) = (0.3, 0.7, 0.3)$$

$$f^{-1}\Big(\Big(\tilde{G}, E'\Big) (e_2) (b) = (0.1, 0.5, 0.2)$$

$$f^{-1}\Big(\Big(\tilde{G}, E'\Big) (e_2) (c) = (0.3, 0.7, 0.3)$$

$$f^{-1}\Big(\Big(\tilde{G}, E'\Big) (e_3) (a) = (0.3, 0.2, 0.3)$$

$$f^{-1}\Big(\Big(\tilde{G}, E'\Big) (e_3) (b) - (0.5, 0.4, 0.4)$$

$$f^{-1}\Big(\Big(\tilde{G}, E'\Big) (e_3) (c) = (0.3, 0.2, 0.3)$$

Hence, we have neutrosophic soft inverse image as follows:

$$
f^{-1}\Big(\Big(\tilde{G}, E'\Big) \Big)
$$
$$
= \left\{ \begin{array}{l} (e_1, \{< a, 0.3, 0.2, 0.3 >, \, < b, 0.5, 0.4, 0.4 >, \, < c, 0.3, 0.2, 0.3 >\}), \\ (e_2, \{< a, 0.3, 0.7, 0.3 >, \, < b, 0.1, 0.5, 0.2 >, \, < c, 0.3, 0.7, 0.3 >\}), \\ (e_3, \{< a, 0.3, 0.2, 0.3 >, \, < b, 0.5, 0.4, 0.4 >, \, < c, 0.3, 0.2, 0.3 >\}) \end{array} \right\}
$$

Remark 3.2 In Example 3.2, it is clear that the $\left(\widetilde{F}, E\right) \subseteq \left(f^{-1}\left(f\left(\left(\widetilde{F}, E\right)\right)\right)\right)$
properties is provided for $\left(\widetilde{F}, E\right) \in NSS\,(X, E)$.

Theorem 3.1 Let $f : NSS(X, E) \to NSS(Y, E)$ be a neutrosophic soft mapping. Then
for neutrosophic soft sets $\left(\widetilde{F}, A\right)$ and $\left(\widetilde{G}, B\right)$ in the neutrosophic class $NSS(X, E)$,
we have

1. $f(0_{(X, E)}) = 0_{(Y, E)}$
2. $f(1_{(X, E)}) \subseteq 1_{(Y, E)}$
3. $f\left(\left(\widetilde{F}, A\right) \widetilde{\cup} \left(\widetilde{G}, B\right)\right) = f\left(\left(\widetilde{F}, A\right)\right) \widetilde{\cup} f\left(\left(\widetilde{G}, B\right)\right)$
4. $f\left(\left(\widetilde{F}, A\right) \widetilde{\cap} \left(\widetilde{G}, B\right)\right) \subseteq f\left(\left(\widetilde{F}, A\right)\right) \widetilde{\cap} f\left(\left(\widetilde{G}, B\right)\right)$
5. $f\left(\left(\widetilde{F}, A\right) \uplus \left(\widetilde{G}, B\right)\right) = f\left(\left(\widetilde{F}, A\right)\right) \uplus f\left(\left(\widetilde{G}, B\right)\right)$
6. $f\left(\left(\widetilde{F}, A\right) \Cap \left(\widetilde{G}, B\right)\right) \subseteq f\left(\left(\widetilde{F}, A\right)\right) \Cap f\left(\left(\widetilde{G}, B\right)\right)$
7. If $\left(\widetilde{F}, A\right) \subseteq \left(\widetilde{G}, B\right)$, then $f\left(\left(\widetilde{F}, A\right)\right) \subseteq f\left(\left(\widetilde{G}, B\right)\right)$

Proof We only prove (3), (4), and (7). The others can be smilarly proved.
    (3) Suppose that

$$\left(\widetilde{F}, A\right) \cup \left(\widetilde{G}, B\right) = (H, A \cup B)$$

and

$$f\left(\left(\widetilde{F}, A\right)\right) \cup f\left(\left(\widetilde{G}, B\right)\right) = (u(F), v(A)) \cup (u(G), v(B)) = (S, v(A) \cup v(B)).$$

Then $f\left(\left(\widetilde{F}, A\right) \cup \left(\widetilde{G}, B\right)\right) = (u(H), v\,(A \cup B)) = (u(H), v(A) \cup v(B))$.
For any $y \in Y$ and $e' \in v(A) \cup v(B)$, if $u^{-1}(y) = \varnothing$, then
    $T_{S(e')}(y) = T_{u(H)(e')}(y) = 0$, $I_{S(e')}(y) = I_{u(H)(e')}(y) = 0$ and $F_{S(e')}(y) = F_{u(H)(e')}(y) = 1$.
    Otherwise, we consider the following cases:
    Case 1: $e' \in v(A) - v(B)$. Then $S(e') = u(F)(e')$. On the other hand,
$e' \in v(A) - v(B)$ implies that there does not exist $e \in B$ such that $v(e) = e'$,
that is, for any $e = v^{-1}(e') \cap (A \cup B)$, we have $e \in v^{-1}(e') \cap (A - B)$. Hence by
Definition 3.4, we have

$$T_{u(H)(e')}(y) = \sup_{e \in v^{-1}(e') \cap (A \cup B), x \in u^{-1}(y)} T_{H(e)}(x)$$

$$= \sup_{e \in v^{-1}(e') \cap (A-B), x \in u^{-1}(y)} T_{H(e)}(x)$$

$$= \sup_{e \in v^{-1}(e') \cap (A-B), x \in u^{-1}(y)} T_{F(e)}(x) = T_{S(e')}(y)$$

Similarly, we have $I_{u(H)(e')}(y) = I_{S(e')}(y)$ and $F_{u(H)(e')}(y) = F_{S(e')}(y)$.

Case 2: $e' \in v(B) - v(A)$. Analogous to case 1, we have $T_{u(H)(e')}(y) = T_{S(e')}(y)$, $I_{u(H)(e')}(y) = I_{S(e')}(y)$ and $F_{u(H)(e')}(y) = F_{S(e')}(y)$.

Case 3: $e' \in v(A) \cap v(B)$. Then

$$T_{u(H)(e')}(y) = \sup_{e \in v^{-1}(e') \cap (A \cup B), x \in u^{-1}(y)} T_{H(e)}(x)$$

$$= \sup_{e \in (v^{-1}(e') \cap A) \cup (v^{-1}(e') \cap B), x \in u^{-1}(y)} T_{H(e)}(x)$$

$$= \left(\sup_{e \in (v^{-1}(e') \cap A) - (v^{-1}(e') \cap B), x \in u^{-1}(y)} T_{F(e)}(x) \right)$$

$$\vee \left(\sup_{e \in (v^{-1}(e') \cap A) \cap (v^{-1}(e') \cap B), x \in u^{-1}(y)} \max \left\{ T_{F(e)}(x), T_{G(e)}(x) \right\} \right)$$

$$\vee \left(\sup_{e \in (v^{-1}(e') \cap B) - (v^{-1}(e') \cap A), x \in u^{-1}(y)} T_{G(e)}(x) \right)$$

$$= \max \left\{ \sup_{e \in v^{-1}(e') \cap A, x \in u^{-1}(y)} T_{F(e)}(x), \sup_{e \in v^{-1}(e') \cap B, x \in u^{-1}(y)} T_{G(e)}(x) \right\}$$

$$= \max \left\{ T_{u(F)(e')}(y), T_{u(G)(e')}(y) \right\} = T_{S(e')}(y)$$

Similarly, we obtain $I_{u(H)(e')}(y) = I_{S(e')}(y)$ and $F_{u(H)(e')}(y) = F_{S(e')}(y)$.

Thus, in any case, $T_{u(H)(e')}(y) = T_{S(e')}(y)$, $I_{u(H)(e')}(y) = I_{S(e')}(y)$ and $F_{u(H)(e')}(y) = F_{S(e')}(y)$. Therefore, $f\left(\left(\widetilde{F}, A \right) \widetilde{\cup} \left(\widetilde{G}, B \right) \right) = f\left(\left(\widetilde{F}, A \right) \right) \widetilde{\cup} f\left(\left(\widetilde{G}, B \right) \right)$.

(4) Suppose that $\left(\tilde{F}, A\right) \cap \left(\tilde{G}, B\right) = (H, A \cup B)$ and

$$f\left(\left(\tilde{F}, A\right)\right) \cap f\left(\left(\tilde{G}, B\right)\right) = (u(F), v(A)) \cap (u(G), v(B)) = (S, v(A) \cup v(B)).$$

Then $f\left(\left(\tilde{F}, A\right) \cap \left(\tilde{G}, B\right)\right) = (u(H), v\,(A \cup B)) = (u(H), v(A) \cup v(B)).$

For any $y \in Y$ and $e' \in v(A \cup B)$, if $u^{-1}(y) = \varnothing$, then
$T_{S(e')}(y) = T_{u(H)(e')}(y) = 0$, $I_{S(e')}(y) = I_{u(H)(e')}(y) = 0$ and $F_{S(e')}(y) = F_{u(H)(e')}(y) = 1$.

Otherwise, we consider the following cases:

Case 1: $e' \in v(A) - v(B)$. Then $S(e') = u(F)(e')$. On the other hand, $e' \in v(A) - v(B)$ implies that there does not exist $e \in B$ such that $v(e) = e'$, that is, for any $e = v^{-1}(e') \cap (A \cup B)$, we have $e \in v^{-1}(e') \cap (A - B)$. We have

$$T_{u(H)(e')}(y) = \sup_{e \in v^{-1}(e') \cap (A \cap B), x \in u^{-1}(y)} T_{H(e)}(x) = \sup_{e \in v^{-1}(e') \cap (A - B), x \in u^{-1}(y)} T_{H(e)}(x)$$

$$= \sup_{e \in v^{-1}(e') \cap (A - B), x \in u^{-1}(y)} T_{F(e)}(x)$$

$$\leq \min \left\{ \sup_{e \in v^{-1}(e') \cap A, x \in u^{-1}(y)} T_{F(e)}(x), \sup_{e \in v^{-1}(e') \cap B, x \in u^{-1}(y)} T_{G(e)}(x) \right\}$$

$$= \min \left\{ T_{u(F)(e')}(y), T_{u(G)(e')}(y) \right\} = T_{S(e')}(y)$$

Similarly, we have $I_{u(H)(e')}(y) \leq I_{S(e')}(y)$ and $F_{u(H)(e')}(y) \geq F_{S(e')}(y)$.

Case 2: $e' \in v(B) - v(A)$. Analogous to case 1, we have $T_{u(H)(e')}(y) \leq T_{S(e')}(y)$, $I_{u(H)(e')}(y) \leq I_{S(e')}(y)$ and $F_{u(H)(e')}(y) \geq F_{S(e')}(y)$.

Case 3: $e' \in v(A) \cap v(B)$. Then

$$T_{u(H)(e')}(y) = \sup_{e \in v^{-1}(e') \cap (A \cap B), x \in u^{-1}(y)} T_{H(e)}(x)$$

$$= \sup_{e \in v^{-1}(e') \cap (A \cap B), x \in u^{-1}(y)} \max \left\{ T_{F(e)}(x), T_{G(e)}(x) \right\}$$

$$\leq \max \left\{ \sup_{e \in v^{-1}(e') \cap A, x \in u^{-1}(y)} T_{F(e)}(x), \sup_{e \in v^{-1}(e') \cap B, x \in u^{-1}(y)} T_{G(e)}(x) \right\}$$

$$= \max \left\{ T_{u(F)(e')}(y), T_{u(G)(e')}(y) \right\} = T_{S(e')}(y)$$

Similarly, we obtain $I_{u(H)(e')}(y) \leq I_{S(e')}(y)$ and $F_{u(H)(e')}(y) \geq F_{S(e')}(y)$.

Therefore $f\left(\left(\widetilde{F}, A\right) \widetilde{\cap} \left(\widetilde{G}, B\right)\right) \subseteq f\left(\left(\widetilde{F}, A\right)\right) \widetilde{\cap} f\left(\left(\widetilde{G}, B\right)\right)$.

(7) Let $\left(\widetilde{F}, A\right) \subseteq \left(\widetilde{G}, B\right)$. Then $A \subseteq B$ and for any $e \in A$ and $x \in X$, we have $u(A) \subseteq u(B)$, we have

$$T_{u(F)(e')}(y) = \begin{cases} \sup\limits_{e \in v^{-1}(e') \cap A, x \in u^{-1}(y)} T_{F(e)}(x), & \text{if } u^{-1}(y) \neq \varnothing \\ 0 & , \quad \text{otherwise} \end{cases}$$

$$\leq \begin{cases} \sup\limits_{e \in v^{-1}(e') \cap B, x \in u^{-1}(y)} T_{G(e)}(x), & \text{if } u^{-1}(y) \neq \varnothing \\ 0 & , \quad \text{otherwise} \end{cases}$$

$$= T_{u(G)(e')}(y)$$

Similarly $I_{u(F)(e')}(y) \leq I_{u(G)(e')}(y)$ and $F_{u(F)(e')}(y) \geq F_{u(G)(e')}(y)$. Therefore, $f\left(\left(\widetilde{F}, A\right)\right) \subseteq f\left(\left(\widetilde{G}, B\right)\right)$. $\square$

Theorem 3.2 Let $f : NSS(X, E) \to NSS\left(\widetilde{Y}, E\right)$ be a neutrosophic soft mapping. Then for neutrosophic soft sets $\left(\widetilde{F}, A\right)$ and $\left(\widetilde{G}, B\right)$ in the neutrosophic soft class $NSS(X, E)$, we have

1. $f^{-1}(0_{(Y, E)}) = 0_{(X, E)}$
2. $f^{-1}(1_{(Y, E)}) \subseteq 1_{(X, E)}$
3. $f^{-1}\left(\left(\widetilde{F}, A\right) \widetilde{\cup} \left(\widetilde{G}, B\right)\right) = f^{-1}\left(\left(\widetilde{F}, A\right)\right) \widetilde{\cup} f^{-1}\left(\left(\widetilde{G}, B\right)\right)$
4. $f^{-1}\left(\left(\widetilde{F}, A\right) \widetilde{\cap} \left(\widetilde{G}, B\right)\right) = f^{-1}\left(\left(\widetilde{F}, A\right)\right) \widetilde{\cap} f^{-1}\left(\left(\widetilde{G}, B\right)\right)$
5. If $\left(\widetilde{F}, A\right) \subseteq (G, B)$, then $f^{-1}\left(\left(\widetilde{F}, A\right)\right) \subseteq f^{-1}\left(\left(\widetilde{G}, B\right)\right)$

6. $\left(\widetilde{F}, A\right) \subseteq f^{-1}\left(f\left(\left(\widetilde{F}, A\right)\right)\right), \ f\left(f^{-1}\left(\left(\widetilde{G}, B\right)\right)\right) = \left(\widetilde{G}, B\right) \cap$
$f\left(1_{(X,E)}\right)$

Proof It is easily seen similar to the previous theorem. $\square$

Definition 3.5 Let $NSS(X, E)$, $NSS(Y, E)$ be two neutrosophic soft classes, $\left(\widetilde{F}, A\right) \in NSS\,(X, E)$, $\left(\widetilde{G}, B\right) \in NSS\,(Y, E')$. Then $f = (u, v) : NSS(X, E) \rightarrow NSS(Y, E')$ is a neutrosophic soft mapping such that $u : X \rightarrow Y$, $v : E \rightarrow E'$.

1. The neutrosophic soft mapping $f = (u, v)$ is called a neutrosophic soft injective mapping if for every $(x_1)^{e_1}_{(\alpha_1, \beta_1, \gamma_1)}$, $(x_2)^{e_2}_{(\alpha_2, \beta_2, \gamma_2)} \in \left(\widetilde{F}, A\right)$, $(x_1)^{e_1}_{(\alpha_1, \beta_1, \gamma_1)} \neq (x_2)^{e_2}_{(\alpha_2, \beta_2, \gamma_2)}$ implies $f\left((x_1)^{e_1}_{(\alpha_1, \beta_1, \gamma_1)}\right) = (u(x_1), v(e_1)) \neq f\left((x_2)^{e_2}_{(\alpha_2, \beta_2, \gamma_2)}\right) = (u(x_2), v(e_2))$.

2. The neutrosophic soft mapping $f = (u, v)$ is called a neutrosophic soft surjective mapping if there exists a neutrosophic soft point $x^e_{(\alpha, \beta, \gamma)} \in \left(\widetilde{F}, A\right)$, such that $f\left(x^e_{(\alpha, \beta, \gamma)}\right) = y^{e'}_{(\alpha', \beta', \gamma')}$ for every $y^{e'}_{(\alpha', \beta', \gamma')} \in \left(\widetilde{G}, B\right)$.

3. The neutrosophic soft mapping $f = (u, v)$ is called a neutrosophic soft bijective mapping if $f = (u, v)$ is both injective and surjective.

4. The neutrosophic soft mapping $f = (u, v)$ is called a neutrosophic soft constant mapping if $f\left(x^e_{(\alpha, \beta, \gamma)}\right) = y^{e'}_{(\alpha', \beta', \gamma')}$ is provided for $\forall x^e_{(\alpha, \beta, \gamma)} \in \left(\widetilde{F}, A\right)$, $\exists y^{e'}_{(\alpha', \beta', \gamma')} \in \left(\widetilde{G}, B\right)$.

Example 3.3 Let $X = \{x_1, x_2\}$, $Y = \{y_1, y_2\}$, $E = \{e_1, e_2\}$, $E' = \{e'_1, e'_2\}$ and $NSS(X, E)$, $NSS(Y, E')$ be neutrosophic soft classes. Define $f = (u, v)$ neutrosophic soft mapping which $u : X \rightarrow Y$ and $v : E \rightarrow E'$ as follows:

$$u(x_1) = y_1, \quad u(x_2) = y_2$$
$$v(e_1) = e'_2, \quad v(e_2) = e'_1$$

Let us consider two neutrosophic soft sets over X and Y, respectively, following as:

$$\left(\widetilde{F}, A\right) = \left\{ \begin{array}{l} (e_1, \{< x_1, 0.2, 0.3, 0.5 >, \ < x_2, 0.5, 0.3, 0.1 >\}), \\ (e_2, \{< x_1, 0.1, 0.5, 0.3 >, \ < x_2, 0.7, 0.4, 0.6 >\}) \end{array} \right\}$$

$$\left(\widetilde{G}, B\right) = \left\{ \begin{array}{l} (e'_1, \{< y_1, 0.1, 0.5, 0.3 >, \ < y_2, 0.7, 0.4, 0.6 >\}), \\ (e'_2, \{< y_1, 0.2, 0.3, 0.5 >, \ < y_2, 0.5, 0.3, 0.1 >\}) \end{array} \right\}$$

It is clear that $\left(\widetilde{F}, A\right)$ is the union of its neutrosophic soft points $(x_1)^{e_1}_{(0.2,0.3,0.5)}$, $(x_1)^{e_2}_{(0.1,0.5,0.3)}$, $(x_2)^{e_1}_{(0.5,0.3,0.1)}$ and $(x_2)^{e_2}_{(0.7,0.4,0.6)}$. Then the images of this points under f mapping are as follows:

$$f\left((x_1)^{e_1}_{(0.2,0.3,0.5)}\right) = \left\{ \begin{array}{l} (e_1', \{< y_1, 0, 0, 1 >, < y_2, 0, 0, 1 >\}), \\ (e_2', \{< y_1, 0.2, 0.3, 0.5 >, < y_2, 0, 0, 1 >\}) \end{array} \right\} = (y_1)^{e_2'}_{(0.2,0.3,0.5)}$$

$$f\left((x_1)^{e_2}_{(0.1,0.5,0.3)}\right) = \left\{ \begin{array}{l} (e_1', \{< y_1, 0.1, 0.5, 0.3 >, < y_2, 0, 0, 1 >\}), \\ (e_2', \{< y_1, 0, 0, 1 >, < y_2, 0, 0, 1 >\}) \end{array} \right\} = (y_1)^{e_1'}_{(0.1,0.5,0.3)}$$

$$f\left((x_2)^{e_1}_{(0.5,0.3,0.1)}\right) = \left\{ \begin{array}{l} (e_1', \{< y_1, 0, 0, 1 >, < y_2, 0, 0, 1 >\}), \\ (e_2', \{< y_1, 0, 0, 1 >, < y_2, 0.5, 0.3, 0.1 >\}) \end{array} \right\} = (y_2)^{e_2'}_{(0.5,0.3,0.1)}$$

$$f\left((x_2)^{e_2}_{(0.7,0.4,0.6)}\right) = \left\{ \begin{array}{l} (e_1', \{< y_1, 0, 0, 1 >, < y_2, 0.7, 0.4, 0.6 >\}), \\ (e_2', \{< y_1, 0, 0, 1 >, < y_2, 0, 0, 1 >\}) \end{array} \right\} = (y_2)^{e_1'}_{(0.7,0.4,0.6)}$$

It is clear that, for any two different points selected from the neutrosophic soft set $\left(\widetilde{F}, A\right)$, the images of these points under $f = (u, v)$ mapping are different from each other. Therefore this mapping is neutrosophic soft injective mapping.

For any soft neutrosophic soft point from the selected the neutrosophic soft set $\left(\widetilde{G}, B\right)$, there exists a neutrosophic soft point $x^e_{(\alpha,\beta,\gamma)} \in \left(\widetilde{F}, A\right)$ such that $f\left(x^e_{(\alpha,\beta,\gamma)}\right) = y^{e'}_{(\alpha',\beta',\gamma')}$. Therefore this mapping is also neutrosophic soft surjective mapping.

Since the neutrosophic soft mapping $f = (u, v)$ is both injective and surjective, it is bijective.

Definition 3.6 Let $f : NSS(X, E) \rightarrow NSS(Y, E')$, $g : NSS(X, E) \rightarrow NSS(Y, E')$ be neutrosophic soft mappings. Neutrosophic soft mappings f and g are called two neutrosophic soft equal mappings if for every $x^e_{(\alpha,\beta,\gamma)} \in (X, E)$ implies $f\left(x^e_{(\alpha,\beta,\gamma)}\right) = g\left(x^e_{(\alpha,\beta,\gamma)}\right)$. It is denoted by $f = g$.

Definition 3.7 Let $f : NSS(X, E) \rightarrow NSS(Y, E')$, $g : NSS(Y, E') \rightarrow NSS(Z, E'')$ be neutrosophic soft mappings. Then the composition of f and g, denoted by $gof : NSS(X, E) \rightarrow NSS(Z, E'')$, is a neutrosophic soft mapping and defined by $(gof)\left(x^e_{(\alpha,\beta,\gamma)}\right) = g\left(f\left(x^e_{(\alpha,\beta,\gamma)}\right)\right)$ for $x^e_{(\alpha,\beta,\gamma)} \in NSS(X, E)$.

Proposition 3.1 Let $f : NSS(X, E) \rightarrow NSS(Y, E')$, $g : NSS(Y, E') \rightarrow NSS(Z, E'')$ be two neutrosophic soft mappings. We have

1. If the neutrosophic soft mappings f and g are two neutrosophic soft injective mappings, then the composition of f and g, $gof : NSS(X, E) \rightarrow NSS(Z, E'')$ is also neutrosophic soft injective mappings.
2. If the neutrosophic soft mappings f and g are two neutrosophic soft suurjective mappings. then the composition of f and g, $gof : NSS(X, E) \rightarrow NSS(Z, E'')$ is also neutrosophic soft surjective mappings.
3. If the neutrosophic soft mappings f and g are two neutrosophic soft bijective mappings, then the composition of f and g, $gof : NSS(X, E) \rightarrow NSS(Z, E'')$ is also neutrosophic soft bijective mappings.

Proof Straightforward. □

4 An Application in Decision Making

Suppose that a company wants to hire staff. A worker's case may easily be into a neutrosophic soft set. Suppose following is the narration by the company owner:

> "I have three main expectations such that experience, foreign language, and good education. In case the candidates with the necessary qualifications do not apply, computer information, communication ability, and age status will be checked."

The company knowledge may be encoded in the form of a look-up tables. Look-up tables are computer representation of the notion of mapping in mathematics. Suppose following knowledge:

We show candidates as candidate-1, candidate-2, candidate-3, and candidate-4.

$$u \text{ (experience)} = \text{candidate} - 1$$

$$u \text{ (foreignlanguage)} = \text{candidate} - 3$$

$$u \text{ (goodeducation)} = \text{candidate} - 3$$

$$u \text{ (computerinformation)} = \text{candidate} - 2$$

$$u \text{ (communicationability)} = \text{candidate} - 4$$

$$u \text{ (agestatus)} = \text{candidate} - 4$$

and

$$v\,(\text{highimportance}) = \text{Highselectionpotential}$$

$$v\,(\text{mediumimportance}) = \text{Lowselectionpotential}$$

$$v\,(\text{lowimportance}) = \text{Lowselectionpotential}.$$

For the sake of ease in mathematical manipulation, we denote the characteristics of candidates and gradations by symbols as follows:

e = experience
f = foreignlanguage
g = goodeducation
i = computerinformation
c = communicationability
a = agestatus

e_1 = highimportance
e_2 = mediumimportance
e_3 = lowimportance

and

c_1 = candidate $-\,1$
c_2 = candidate $-\,2$
c_3 = candidate $-\,3$
c_4 = candidate $-\,4$

e_1' = highselectionpotential
e_2' = lowselectionpotential

Thus we have two neutrosophic soft classes $NSS(X, E)$ and $NSS(Y, E')$ with $X = \{e, f, g, i, c, a\}$, $E = \{e_1, e_2, e_3\}$ and $Y = \{c_1, c_2, c_3, c_4\}$, $E' = \{e_1', e_2'\}$. $NSS(X, E)$ is the neutrosophic soft class of the characteristics of candidates and $NSS(Y, E')$ represents candidates. The neutrosophic soft set of company owner's narration may be given as follows:

$$\left(\widetilde{F}, A\right)$$
$$= \left(\begin{array}{l} (e_1, \{< e, 0.5, 0.7, 0.2 >, < f, 0.6, 0.8, 0.4 >, < g, 0.8, 0.3, 0.1 >, < i, 0, 0, 1 >, < c, 0, 0, 1 >, < a, 0, 0, 1 >\}) \\ (e_2, \{< e, 0, 0, 1 >, < f, 0, 0, 1 >, < g, 0, 0, 1 >, < i, 0.6, 0.7, 0.4 >, < c, 0.9, 0.1, 0.3 >, < a, 0, 0, 1 >\}) \\ (e_3, \{< e, 0, 0, 1 >, < f, 0, 0, 1 >, < g, 0, 0, 1 >, < i, 0, 0, 1 >, < c, 0, 0, 1 >, < a, 0.6, 0.5, 0.2 >\}) \end{array} \right)$$

As a first task, stored company knowledge is to be applied on the given case. This knowledge, in the language of computer programming, is given as look-up tables. Neutrosophic sof mappings $u : X \to Y$ and $v : E \to E'$ are defined as follows:

$$u(e) = c_1,\, u(f) = c_3,\, u(g) = c_3,\, u(i) = c_2,\, u(c) = c_4,\, u(a) = c_4$$

and

$$v(e_1) = e_1', v(e_2) = e_2', v(e_3) = e_2'$$

Calculations give

$$f\left(\left(\widetilde{F}, A\right)\right)$$

$$= \left\{ \begin{array}{l} (e_1', \{< c_1, 0.5, 0.7, 0.2 >, < c_2, 0, 0, 1 >, < c_3, 0.8, 0.8, 0.1 >, < c_4, 0, 0, 1 >\}) \\ (e_2', \{< c_1, 0, 0, 1 >, < c_2, 0.6, 0.7, 0.4 >, < c_3, 0, 0, 1 >, < c_4, 0.9, 0.1, 0.3 >\}) \end{array} \right\}$$

As seen above, candidates c_2 and c_4 can be eliminated because they do not meet the main expectations. Among candidates c_1 and c_3 should be considered.

5 Conclusion

In the present paper the theoretical point of neutrosophic soft mapping has been discussed. We re-define the notion of neutrosophic soft mapping. Several properties of neutrosophic soft image and neutrosophic soft inverse image are established. We further study some essential features of the initiated neutrosophic soft mapping and support by examples. Finally, these notions have been applied to application in decision making. We hope that these notions will be useful for the researchers to further promote and advanced in neutrosophic soft continuity and decision making problems.

References

1. Zadeh, L. A. (1965). Fuzzy sets. *Information and Control, 8*(3), 338–353.
2. Smarandache, F. (2005). Neutrosophic set, a generalisation of the intuitionistic fuzzy sets. *International Journal of Pure and Applied Mathematics, 24*, 287–297.
3. Atanassov, K. T. (1999). *Intuitionistic fuzzy sets* (pp. 1–137). Heidelberg: Physica.
4. Abobala, M. (2020). On Some Special Substructures of Refined Neutrosophic Rings. *International Journal of Neutrosophic Science, 5*(1), 59–66.
5. Ansari, A. Q., Biswas, R., & Aggarwal, S. (2011). Proposal for applicability of neutrosophic set theory in medical AI. *International Journal of Computer Applications, 27*(5), 5–11.
6. Gallego Lupiáñez, F. (2008). On neutrosophic topology. *Kybernetes, 37*(6), 797–800.
7. Hatip, A. (2020). The special neutrosophic functions. *International Journal of Neutrosophic Science, 4*(2), 104–116.
8. Kharal, A. (2014). A neutrosophic multi-criteria decision making method. *New Mathematics and Natural Computation, 10*(02), 143–162.

9. Nandhini, T., & Vigneshwaran, M. (2020). On neutrosophic nano αg#ψ-closed sets in neutrosophic nano topological spaces. *International Journal of Neutrosophic Science, 5*(2), 67–71.

10. Nordo, G., Mehmood, A., & Broumi, S. (2020). Single valued neutrosophic filters. *International Journal of Neutrosophic Science, 6*(1), 08–21.

11. Ozturk, T. Y., & Ozkan, A. (2019). Neutrosophic bitopological spaces. *Neutrosophic Sets and Systems, 88*, 97–106.

12. Molodtsov, D. (1999). Soft set theory-first results. *Computers & Mathematcs with Applications, 37*, 19–31.

13. Feng, F., Li, C., Davvaz, B., & Ali, M. I. (2010). Soft sets combined with fuzzy sets and rough sets: a tentative approach. *Soft Computing, 14*(9), 899–911.

14. Kharal, A., & Ahmad, B. (2009). Mappings on fuzzy soft classes. *Adv. Fuzzy Syst., 2009*, 407890, 6 pages.

15. Ozturk, T. Y. (2018). On bipolar soft topological spaces. *Journal of New Theory, 20*, 64–75.

16. Yin, Y., Li, H., & Jun, Y. B. (2012). On algebraic structure of intuitionistic fuzzy soft sets. *Computers and Mathematics with Applications, 64*(9), 2896–2911.

17. Maji, P. K. (2013). Neutrosophic soft set, Ann. *Fuzzy Mathematics and Informatics, 5*(1), 157–168.

18. Ozturk, T., Gunduz Aras, C., & Bayramov, S. (2019). A new approach to operations on neutrosophic soft sets and to neutrosophic soft topological spaces. *Communications in Mathematics and Applications, 10*(3), 481–493. https://doi.org/10.26713/cma.v10i3.1068.

19. Bera, T., & Mahapatra, N. K. (2017). Introduction to neutrosophic soft topological space. *Opsearch, 54*(4), 841–867.

20. Bera, T., & Mahapatra, N. K. (2016). On neutrosophic soft function. *Ann. Fuzzy Math. Inform., 12*(1), 101–119.

21. Deli, I., & Broumi, S. (2015). Neutrosophic soft relations and some properties, Ann. *Fuzzy Mathematics and Informatics, 9*(1), 169–182.

22. Gunduz, C., Ozturk, T. Y., & Bayramov, S. (2019). Separation axioms on neutrosophic soft topological spaces. *Turkish Journal of Mathematics, 43*(1), 498–510.

23. Ozturk, T. Y., & Simsekler (Dizman), T. (2019). A new approach to operations on bipolar neutrosophic soft sets and bipolar neutrosophic soft topological spaces. *Neutrosophic Sets & Systems, 30*, 31–42.

24. Kharal, A., & Ahmad, B. (2011). Mappings on soft classes. *New Mathematics and Natural Computation, 7*(03), 471–481.

25. Alkhazaleh, S., & Marei, E. (2014). Mappings on neutrosophic soft classes. *Neutrosophic Sets and Systems, 2*, 3–8.

Multiple Attribute Decision-Making Based on Uncertain Linguistic Operators in Neutrosophic Environment

Chiranjibe Jana and Madhumangal Pal

1 Introduction

Neutrosophic set (NS) was initiated by Smarandache in his tremendous paper [1, 2]. Three membership functions characterize his proposed approach, namely, truth, indeterminacy, and falsity, which an extension of the fuzzy set (FS) founded by Zadeh [3]. The FS was expanded by Atanassov [4] in the form of the intuitionistic fuzzy set (IFS). Although (FS) and (IFS) are compelling sets, in some cases, these sets are not sufficient to master indeterminate and inconsistent information practice in real-world problems. Therefore, NS has influential approval to develop models displaying indeterminate and inconsistent data. To defeat difficulties in these regions, Wang et al. [5] defined the knowledge of single-valued neutrosophic set (SVNs). INS [6] is a generalization of NS, which is used in engineering and scientific applications.

Here, the measure of indeterminate and inconsistent information under interval neutrosophic environment as well as its linguistic information is a problem for decision-makers. The use of Dombi operator makes it easy for its measure of flexibility to count its parameter.

In the previous works made in the neutrosophic linguistic environment [7–10], they use aggregation operators to evaluate favourable one. In our present study, due to the use of Dombi operators to the same environment make us flexible for evaluation of the process. Whereas, existing studies have no such advantage to solve the problem. So, we focus our research on finding the best alternative based on the due method. In the next paragraph, we review some literatures related to the proposed environment.

C. Jana (✉) · M. Pal
Department of Applied Mathematics with Oceanology and Computer Programming, Vidyasagar University, Midnapore, West Bengal, India

© The Author(s), under exclusive license to Springer Nature Switzerland AG 2021
F. Smarandache, M. Abdel-Basset (eds.), *Neutrosophic Operational Research*,
https://doi.org/10.1007/978-3-030-57197-9_16

The aggregation operators (AOs) in knowledge retrieval are essential decision analysis areas. In 1988, the order weighted average (OWA) operator and its results are mentioned by Yager [11, 12]. Later, OWA operator has been utilized in (IFS) and (IVIFS) areas, and multi-criteria and multi-attribute decision-making approach has been constructed for these operators; for more information the readers are referred to [13–15]. Researchers have studied various decision-making problems [16–25] in different fuzzy uncertain environment. Since the concept of neutrosophic set (NS) is a more useful tool for modeling real-life problems. Abdel-Basset et al. [26] have studied linear time-cost trade-offs model in NS environment. In the same environment, Abdel-Basset et al. [27] utilized resource leveling for studying construction projects. The personal selection-based MCDM problems have been studied by Abdel-Basset et al. [28] using bipolar neutrosophic arguments. Abdel-Basset et al. [29] used the best-worst technique for solving supply chain problems based on the novel Plithogenic model. The development of different AOs for aggregating neutrosophic numbers is an essential research area. Ye [30] developed weighted operators-based decision-making in the neutrosophic environment. Peng et al. [31] have developed new operators using the concept in [30]. Thereafter, Ye [10] studied AOs in the linguistic environment in connection with INS. Zhang et al. in [32], Bausys, and Zavadskas used VIKOR approach-based MCDM with INS argument. Jana et al. [33] consider Hamacher AOs in their study to develop an MADM problem in single-valued trapezoidal neutrosophic environment. Broumi and Smarandache [34] followed the MADM methodology to decide aggregating information related to neutrosophic trapezoidal linguistic arguments. Ji et al. [35] focused on Frank operations of SVNNs and constructed the SVN-prioritized Bonferroni mean (SVNFNPBM) operator under Frank aggregation function. Zhang et al. [36] introduced a normal cloud method on neutrosophic set and construct an MADM approach under SVN environments. Sahin and Liu [37] derived correlation coefficient between two SVN-hesitant fuzzy numbers. Nancy and Garg [38] defined the operations of SVNNs based on Frank norm operations, and they proposed a decision-making method after identifying weighted aggregation operators. Biswas et al. [39, 40] utilized the TOPSIS method for MCDM problems under SVN environment. In [41], Liu et al. constructed a neutrosophic Bonferroni weighted geometric mean operator based on multi-valued functions. Jana et al. [42] studied weighted aggregation functions in the trapezoidal neutrosophic environment. Lu and Ye [43] proposed hybrid weighted arithmetic and geometric aggregation functions under SVN information and utilized these operators to develop decision-making problems. Broumi et al. [44] introduced an algorithm to solve a neutrosophic shortest problems from the source node to the destination node. Wu et al. [8] defined the technique of SVN 2-tuple linguistic element and its operational rules. They also developed some SVN2TL weighted arithmetic and geometric Hamacher aggregation operators under SVN2TL environment. Furthermore, they developed a MAGDM method based on these new operations. Abdel-Basset et al. [45] studied a MCGDM based on neutrosophic hierarchy method. Abdel-Basset et al. [46] proposed strategic planning and decision-making based on neutrosophic AHP-SWOT analysis. Pramanik et al. [47] studied cross entropy based on MAGDM in the environment of SVN numbers. Ye [48] analyzed the study of correlation coefficient of SVNNs and utilized it to develop an approach toward decision-making problems.

However, in many situations in real-world problems where problems are too complicated or too ill-defined to be malleable for the description of the vagueness of people's intelligent and the complex information in decision science. Some decision-making problems cannot always be conveyed in terms of crisp numbers or fuzzy numbers. Still, it may be comfortable to describe in linguistic words, especially for information as people's morality and vehicles performances. Then, decision-making results can be expressed in linguistic terms by the decision-makers such as "poor," "medium," and "good." The concept of linguistic variables was introduced by Zadeh [49] and concern in fuzzy reasoning. After that, a consensus decision-making approach with linguistic argument was mentioned by Herrera and Herrera-Viedma [50]. Xu [51] developed the MADM method by using goal programming with linguistic information. Ye [10] proposed a MADM problem in IN environment with linguistic information. Ye [9] submitted AOs in neutrosophic linguistic numbers environment and classified NLNWA, NLNWG operators. Tan et al. [7] introduced three generalized SVN linguistic operators, namely, GSVNLWA, GSVNLOWA, and GSVNLHA operators and set up a decision-making approach. The concept of intuitionistic linguistic fuzzy aggregation operators was proposed by Li [52]. To avoid the loss of information of an object, then researchers have introduced the continuous linguistic term. It is also observed in many areas that linguistic details may not be similar to any of the original linguistic labels, and may be placed between any two of them. In such a situation, Xu [53–56] introduced ULVs and provided some operational rules of them. Liu and Jin [57] introduced some UL aggregation operators on IFS and gave an application of it. Meng et al. [58] proposed some Choquet aggregation operators using ULIVIFS arguments, and applied operations, score, and accuracy functions for IVIULNs; introduced IVIFUL Choquet averaging (IVIULCA) operator and IVIFUL Choquet geometric (IVIULCG) operator; and utilized these operators to set up a MADM problem with IVIULVs. The main objective of this research is that INULNs have a constructive capacity to determine uncertain data that appears in real-world problems. Therefore, MADM methods were developed by the use of different fuzzy Dombi operations and uncertain linguistic information that made a significant interest to improve our propose study. In this research on MADM problems with INUL information is in its birth, and INULVs more definitely express fuzzy information than LVs. Therefore, based on the interval neutrosophic linguistic set proposed by Ye [10] and based on the concept of IVIULS proposed by Meng et al. [58], in this paper, we offer an INULS and develop MADM problems in which both the attribute weights take the form of real numbers, and attribute values take the form of INULVs. First, operational laws, score values, and accuracy of INULS are defined. Then, we constructed INULWA operator, INULWG operator, INULDWA operator, and INULDWG operator are introduced and investigate desirable properties of these operators. Finally, a specific application of a practical example is given to this paper.

The rest of the article is formulated as follows: In Sect. 2, some elementary definitions and operations on INNs and ULVs are considered. In Sect. 3, the definition of INULS set is defined and some operations of INULNs are proposed. In Sect. 4, the two operators, namely, interval neutrosophic uncertain linguistic weighted averaging (INULWA) operators and interval neutrosophic uncertain linguistic weighted geometric (INULWG) operators, are proposed; and some of their properties are highlighted. In Sect. 5, interval neutrosophic uncertain linguistic Dombi averaging (INULDWA) operator and interval neutrosophic uncertain linguistic Dombi geometric (INULDWG) operator have been introduced. In Sect. 6, a MADM model based on these operators is constructed. In Sect. 7, a practical example for the evaluation of the mutual fund selection is given. A comparative study of the proposed operator is established with the existing results in Sect. 8. In Sect. 9, concluding remarks are provided.

2 Preliminaries

Here, briefly recall some basic concepts related to NS [1] and INS [59], which are necessary for this work.

2.1 Some Concept of Interval Neurotrophic Set

Definition of neutrosophic set is defined [1] as follows:

[1] Let Z be a fixed set, with a collective element in Z followed by z. A NS q in Z is

$$q = \{\langle T_q(z), I_q(z), F_q(z)\rangle \,|z \in Z\}, \tag{1}$$

where $T_q(z)$, $I_q(z)$, and $F_q(z)$ are, respectively, truth-, indeterminacy-, and falsity-membership measure functions. $T_q(z)$, $I_q(z)$ and $F_q(z)$ are given standard or non-standard subsets of $]0^-, 1^+[$ follows as $T_q(z) : Z \rightarrow \;]0^-, 1^+[, I_q(z) : Z \rightarrow \;]0^-, 1^+[$ and $F_q(z) : Z \rightarrow \;]0^-, 1^+[$. The of $T_q(z)$, $I_q(z)$ and $F_q(z)$ have no restriction, and so $0^- \leq T_q(z) + I_q(z) + F_q(z) \leq 3^+$.

Smarandache [60] and Wang et al. and others [5] defined the concept of SVNS as follows.

[5] Let Z be a fixed set, with a collective element in Z denoted by z. Then SVNS is given as follows:

$$q = \{\langle T_q(z), I_q(z), F_q(z)\rangle \,|z \in Z\}, \tag{2}$$

where $T_q(z) : Z \rightarrow [0, 1], I_q(z) : Z \rightarrow [0, 1]$ and $F_q(z) : Z \rightarrow [0, 1]$ are, respectively, truth-, indeterminacy-, and falsity-membership measure functions of z to q with the condition $0 \leq T_q(z) + I_q(z) + F_q(z) \leq 3$.

An interval neutrosophic set (INS) [6] is a generalization of NS, which is used in engineering and scientific applications. In the next definition, [5] introduced the concept of INS.

[59] Let Z be a fixed set, with a collective element in Z denoted by z. INS specify following functions, namely, truth $T_q(z)$, indeterminacy $I_q(z)$, and falsity $F_q(z)$ where $T_q(z) = \left[T_q^l(z), T_q^u(z) \right]$, $I_q(z) = \left[I_q^l(z), I_q^u(z) \right]$, and $F_q(z) = \left[F_q^l(z), F_q^u(z) \right]$ with $0 \leq T_q^u(z) + I_q^u(z) + F_q^u(z) \leq 3$, for $z \in Z$.

[59] Let $q = \left\langle \left[T_q^l(z), T_q^u(z) \right], \left[I_q^l(z), I_q^u(z) \right], \left[F_q^l(z), F_q^u(z) \right] \right\rangle$ and $r = \left\langle \left[T_r^l(z), T_r^u(z) \right], \left[I_r^l(z), I_r^u(z) \right], \left[F_r^l(z), F_r^u(z) \right] \right\rangle$ be two INNs, then following operational rules are hold on INNs

1.
$$q + r = \left\langle \left[T_q^l(z) + T_r^l(z) - T_q^l(z)T_r^l(z), T_q^u(z) + T_r^u(z) \right. \right.$$
$$\left. \left. -T_q^u(z)T_r^u(z), I_q^l(z)I_r^l(z), I_q^u(z)I_r^u(z), F_q^l(z)F_r^l(z), F_q^u(z)F_r^u(z) \right] \right\rangle$$

$$q.r = \left\langle \left[T_q^l(z)T_r^l(z), T_q^u(z)T_r^u(z) \right], \right.$$

2.
$$\left[I_q^l(z) + I_r^l(z) - I_q^l(z)I_r^l(z), I_q^u(z) + I_r^u(z) - I_q^u(n)I_r^u(z) \right],$$

$$\left. \left[F_q^l(z) + F_r^l(z) - F_q^l(z)F_r^l(z), F_q^u(z) + F_r^u(z) - F_q^u(z)F_r^u(z) \right] \right\rangle$$

3. $\lambda q = \left\langle \left[1 - \left(1 - T^l{}_q(z)\right)^\lambda, 1 - \left(1 - T^u{}_q(z)\right)^\lambda, I_q^{l\lambda}(z), I_q^{u\lambda}(z), F_q^{l\lambda}(z), F_q^{u\lambda}(z) \right] \right\rangle$

4. $q^\lambda = \left\langle \left[T^{l\lambda}{}_q(z), T^{u\lambda}{}_q(z) \right], \left[1 - \left(1 - I_q^l(z)\right)^\lambda, 1 - \left(1 - I_q^u(z)\right)^\lambda \right], \left[1 - \left(1 - F_q^l(z)\right)^\lambda, 1 - \left(1 - F_q^u(z)\right)^\lambda \right] \right\rangle.$

2.2 Some Concept of Uncertain Linguistic Variables

In this part, considered some concepts and operational rules in which introduce qualitative as well as linguistic aspects by means of LVs [49, 50, 54, 55, 61–64]. Let $S = \{s_t | t = 1, 2, \ldots, p\}$ be a LTS with odd cardinality. In any stage, s_t constitutes a value for a linguistic variable, and shows the following properties:

(i) The $s_i \geq s_j$ if $i \geq j$ is an ordered set.
(ii) The $neg(s_i) = s_j$ such that $j = t - i$ is called negation operator.
(iii) $\max(s_i, s_j) = s_i$ if $s_i \geq s_j$ introduce as max operator.
(iv) $\min(s_i, s_j) = s_i$ if $s_i \leq s_j$ called min operator. For example [56] can be provided as:

$$S = \{s_0 = \text{extremely poor}, s_1 = \text{very poor}, s_2 = \text{poor}, s_3 = \text{medium}, s_4 = \text{good},$$

$$s_5 = \text{very good}, s_6 = \text{extremely good}\}.$$

To avoid the information loss, we enlarged the discrete term set S to a continuous term set $S = \{s_t | s_0 \le s_t \le s_p, t \in [1,p]\}$, where p is a sufficiently large positive integer. If $s_t \in S$, then it is called an original linguistic term (LT) or called virtual LT. To find alternatives and attributes, decision-makers generally used original LTS and virtual LTS only used to calculation [53–56, 65].

It is always seen in many real-world problems that the input LTS may not match any of the original linguistic labels, and it may be located between any two of them. In such situation, Xu [53–56] introduced uncertain linguistic variables (ULT) and provided some operational rules of them. [56] Let $s = [s_l, s_m]$, where $s_l, s_m \in S$, and s_l, s_m are the respectively lower and upper bounds of the LVs s. Also, let $\widetilde{S}$ be the set of all ULTs. Let $s = [s_l, s_m]$, $s_1 = \left[s_{l_1}, s_{m_1}\right]$ and $s_2 = \left[s_{l_2}, s_{m_2}\right]$ be three ULVs, where $s, s_1, s_2 \in \widetilde{S}$ and $\lambda \in [0, 1]$, then operational laws of them defined as follows:

$$s_1 \oplus s_2 = \left[s_{l_1}, s_{m_1}\right] \oplus \left[s_{l_2}, s_{m_2}\right] = \left[s_{l_1} \oplus s_{l_2}, s_{m_1} \oplus s_{m_2}\right] = \left[s_{l_1+l_2}, s_{m_1+m_2}\right]$$

$$s_1 \otimes s_2 = \left[s_{l_1}, s_{m_1}\right] \otimes \left[s_{l_2}, s_{m_2}\right] = \left[s_{l_1} \otimes s_{l_2}, s_{m_1} \otimes s_{m_2}\right] = \left[s_{l_1 l_2}, s_{m_1 m_2}\right]$$

$$\lambda s = \lambda [s_l, s_m] = [\lambda s_l, \lambda s_m] = [s_{\lambda l}, s_{\lambda m}]$$

$$(s)^\lambda = ([s_l, s_m])^\lambda = \left[(s_l)^\lambda, (s_m)^\lambda\right] = \left[s_{l^\lambda}, s_{m^\lambda}\right].$$

3 Interval Neutrosophic Uncertain Linguistic Set (IULN)

Based on the concepts of INS, ULS, and INLS, we introduce the INLS and ULS to define INULS and INULN. On the basis operational rules and ranking order of INULN is provided in this section.

Let Z be a universe of discourse, with a collective element in Z denoted by z. INULS ϱ in Z is defined by

$$\varrho = \left\{\langle z, s_{\phi(z)}, T_\varrho(z), I_\varrho(z), F_\varrho(z)\rangle | z \in Z\right\}, \tag{3}$$

where $s_{\phi(z)} = [s_{\eta(z)}, s_{\theta(z)}] \in S$, $T_\varrho(z) = \left[T_\varrho^l(z), T_\varrho^u(z)\right] \subseteq [0, 1]$, $I_\varrho(z) = \left[I_\varrho^l(z), I_\varrho^u(z)\right] \subseteq [0, 1]$, and $F_\varrho(z) = \left[F_\varrho^l(z), F_\varrho^u(z)\right] \subseteq [0, 1]$ with the condition $0 \le T_\varrho^u(z) + I_\varrho^u(z) + F_\varrho^u(z) \le 3$. The functions $T_\varrho(z), I_\varrho(z)$, and $F_\varrho(z)$ are, respectively, measured truth-, indeterminacy-, and falsity-membership values in an interval of the element z to the set Z to the ULVs $s_{\phi(z)} = [s_{\eta(z)}, s_{\theta(z)}]$. For convenience, $\varrho = \langle z, [s_{\eta(\varrho)}, s_{\theta(\varrho)}], [T^l(\varrho), T^u(\varrho)], [I^l(\varrho), I^u(\varrho)], [F^l(\varrho), F^u(\varrho)]\rangle$ is the eight tuples called INULNs. We defined some new operations on

INULNs:Let $\varrho = \langle [s_{\eta(\varrho)}, s_{\theta(\varrho)}], [T^l(\varrho), T^u(\varrho)], [I^l(\varrho), I^u(\varrho)], [F^l(\varrho), F^u(\varrho)] \rangle$ and $\delta = \langle [s_{\eta(\delta)}, s_{\theta(\delta)}], [T^l(\delta), T^u(\delta)], [I^l(\delta), I^u(\delta)], [F^l(\delta), F^u(\delta)] \rangle$ be any two INULNs, some operations of ϱ and δ defined for any real number $\lambda \in [0, 1]$

- $\varrho \oplus \delta = \langle [s_{\eta(\varrho) + \eta(\delta)}, s_{\theta(\varrho) + \theta(\delta)}], [T^l(\varrho) + T^l(\delta) - T^l(\varrho)T^l(\delta), T^u(\varrho) + T^u(\delta) - T^u(\varrho)T^u(\delta)], [I^l(\varrho)I^l(\delta), I^u(\varrho)I^u(\delta), F^l(\varrho)F^l(\delta), F^u(\varrho)F^u(\delta)] \rangle$
- $\varrho \otimes \delta = [s_{\eta(\varrho) \times \eta(\delta)}, s_{\theta(\varrho) \times \theta(\delta)}], [T^l(\varrho)T^l(\delta), T^u(\varrho)T^u(\delta))], [I^l(\varrho)+I^l(\delta)-I^l(\varrho)I^l(\delta), I^u(\varrho)+I^u(\delta)-I^u(\varrho)I^u(\delta)], [F^l(\varrho)+F^l(\delta)-F^l(\varrho)F^l(\delta), F^u(\varrho) + F^u(\delta) - F^u(\varrho)F^u(\delta)] \rangle$
- $\lambda\varrho = \langle [s_{\lambda\eta(\varrho)}, s_{\lambda\theta(\varrho)}], [1-(1-T^l(\varrho))^\lambda, 1-(1-T^u(\varrho))^\lambda, I^{l\lambda}(\varrho), I^{u\lambda}(\varrho), F^{l\lambda}(\varrho), F^{u\lambda}(\varrho)] \rangle$
- $\varrho^\lambda = \left\langle \left[s_{\eta(\varrho)^\lambda}, s_{\theta(\varrho)^\lambda} \right], T^{u\lambda}(\varrho), 1 - \left(1 - I^l(\varrho)\right)^\lambda, 1 - (1 - I^u(\varrho))^\lambda, 1 - \left(1 - F^l(\varrho)\right)^\lambda, 1 - \left(1 - F^u(\varrho)\right)^\lambda \right] \right\rangle$.

Let ϱ and δ be any two INULNs, then [(1)] $\varrho + \delta = \delta + \varrho$ [(2)] $\varrho . \delta = \varrho . \delta$ [(3)] $\lambda(\varrho + \delta) = \lambda\varrho + \lambda\delta$, for $\lambda \in [0, 1]$ [(4)] $(\varrho . \delta)^\lambda = \varrho^\lambda + \delta^\lambda$, for $\lambda \in [0, 1]$ [(5)] $\lambda_1\varrho + \lambda_2\varrho = (\lambda_1 + \lambda_2)\varrho$, for $\lambda_1, \lambda_2 \in [0, 1]$ [(6)] $\varrho^{\lambda_1} . \varrho^{\lambda_2} = \varrho^{\lambda_1 + \lambda_2}$, for $\lambda_1, \lambda_2 \in [0, 1]$ [(7)] $(\varrho + \delta) + \beta = \varrho + (\delta + \beta)$ [(8)] $(\varrho . \delta) . \beta = \varrho . (\delta . \beta)$. Based on the definition of score and accuracy function in [58] defined on interval-valued intuitionistic uncertain linguistic (IVIULNs) numbers, we defined score and accuracy on an interval neutrosophic uncertain linguistic information defined below.

Let $\varrho = \langle [s_{\eta(\varrho)}, s_{\theta(\varrho)}], [T^l(\varrho), T^u(\varrho)], [I^l(\varrho), I^u(\varrho)], [F^l(\varrho), F^u(\varrho)] \rangle$ be any INULN. Then, defined score function of ϱ is $\uplus(\varrho)$ by

$$\uplus(\varrho) = s_{\frac{(\eta(\varrho)+\theta(\varrho))\left(2+T^l(\varrho)+T^u(\varrho)-I^l(\varrho)-I^u(\varrho)-F^l(\varrho)-F^u(\varrho)\right)}{4}}, \uplus(\varrho) \in [0, 1] \qquad (4)$$

The accuracy function of ϱ is $\Im(\varrho)$ by

$$\Im(\varrho) = s_{\frac{(\eta(\varrho)+\theta(\varrho))\left(I^l(\varrho)+I^u(\varrho)+F^l(\varrho)+F^u(\varrho)\right)}{4}}, \Im(\varrho) \in [0, 1]. \qquad (5)$$

Based on the above design of score and accuracy, prioritized analysis between any two INULNs ϱ and n is defined as:

- If $\uplus(\varrho) < \uplus(\delta)$, imply $\varrho \prec \delta$
- If $\uplus(\varrho) > \uplus(\delta)$, imply $\varrho \succ \delta$
- If $\uplus(\varrho) = \uplus(\delta)$, then

  – If $\Im(\varrho) < I(\delta)$, imply $\varrho \prec \delta$.
  – If $\Im(\varrho) > I(\delta)$, imply $\varrho \succ \delta$.
  – If $\Im(\varrho) = \Im(\delta)$, imply $\varrho \sim \delta$.

4 Interval Neutrosophic Uncertain Linguistic Aggregation Operators

Here we defined INULWA operator and study some of its properties.

4.1 INULWA Operator

Let $\varrho_\xi = \left\langle \left[s_{\eta(\varrho_\xi)}, s_{\theta(\varrho_\xi)}, T^l(\varrho_\xi), T^u(\varrho_\xi), I^l(\varrho_\xi), I^u(\varrho_\xi), F^l(\varrho_\xi), F^u(\varrho_\xi) \right] \right\rangle$ be a set of INULNs for ($\xi = 1, 2, \ldots, \tau$). Then interval neutrosophic uncertain linguistic weighted average (INULWA) function $INULWA : \Theta^\tau \to \Theta$ is defined as follows:

$$INULWA_{\varpi}(\varrho_1, \varrho_2, \ldots, \varrho_\tau) = \overset{\tau}{\underset{\xi=1}{\oplus}} \left(\varpi_\xi \varrho_\xi \right) \tag{6}$$

where $\varpi = (\varpi_1, \varpi_2, \ldots, \varpi_\tau)^T$ is followed by the weight vector of $\varrho_\xi (\xi = 1, 2, \ldots, \tau)$, with $\varpi_\xi \in [0, 1]$ and $\sum_{\xi=1}^{\tau} \varpi_j = 1$.

By the operations on INULNs, we derive the following theorem.

Let $\varrho_\xi = \left\langle \left[s_{\eta(\varrho_\xi)}, s_{\theta(\varrho_\xi)}, T^l(\varrho_\xi), T^u(\varrho_\xi), I^l(\varrho_\xi), I^u(\varrho_\xi), F^l(\varrho_\xi), F^u(\varrho_\xi) \right] \right\rangle$ be a set of INULNs for ($j = 1, 2, \ldots, \tau$), then aggregating values of INULNs $\varrho_\xi (\xi = 1, 2, \ldots, \tau)$ are also an INULN, and further,

$$INULWA_{\varpi}(\varrho_1, \varrho_2, \ldots, \varrho_\tau) = \overset{\tau}{\underset{\xi=1}{\oplus}} \left(\varpi_\xi \varrho_\xi \right)$$

$$= \left\langle \left[s_{\sum_{\xi=1}^{\tau} \varpi_\xi \eta(\varrho_\xi)}, s_{\sum_{\xi=1}^{\tau} \varpi_j \theta(\varrho_\xi)}, 1 - \prod_{\xi=1}^{\tau} \left(1 - T^l(\varrho_\xi) \right)^{\varpi_\xi}, \right. \right.$$

$$1 - \prod_{\xi=1}^{\tau} \left(1 - T^u(\varrho_\xi) \right)^{\varpi_\xi}, \prod_{\xi=1}^{\tau} \left(I^l(\varrho_\xi) \right)^{\varpi_\xi}, \prod_{\xi=1}^{\tau} \left(I^l(\varrho_\xi) \right)^{\varpi_\xi} \right],$$

$$\left. \left[\prod_{\xi=1}^{\tau} \left(F^l(\varrho_\xi) \right)^{\varpi_\xi}, \prod_{\xi=1}^{\tau} \left(I^l(\varrho_\xi) \right)^{\varpi_\xi} \right] \right\rangle \tag{7}$$

where $\varpi = (\varpi_1, \varpi_2, \ldots, \varpi_\tau)^T$ is followed by the weight vector of $\varrho_\xi (\xi = 1, 2, \ldots, \tau)$, with $\varpi_\xi \in [0, 1]$, and $\sum_{\xi=1}^{\tau} \varpi_\xi = 1$.

Proof We prove Eq. (7) below using mathematical induction.

(i) when $\tau = 2$, we get

$$\left\langle \left[S_{\varpi_\xi \eta(\varrho_\xi)}, S_{\varpi_\xi \theta(\varrho_\xi)} \right], \left[1 - \left(1 - T^l \left(\varrho_\xi \right) \right)^{\varpi_\xi}, 1 - \left(1 - T^u \left(\varrho_\xi \right) \right)^{\varpi_\xi} \right], \right.$$

$$\left. \left[\left(I^l \left(\varrho_\xi \right) \right)^{\varpi_\xi}, \left(I^l \left(\varrho_\xi \right) \right)^{\varpi_\xi} \right], \left[\left(F^l \left(\varrho_\xi \right) \right)^{\varpi_\xi}, \left(I^l \left(\varrho_\xi \right) \right)^{\varpi_\xi} \right] \right\rangle \text{ for } \xi = 1, 2.$$

Then,

$$INULWA_\varpi \left(\varrho_1, \varrho_2 \right) = \overset{2}{\underset{\xi=1}{\oplus}} \varpi_\xi \varrho_\xi = \left\langle \left[S_{\sum_{\xi=1}^{2} \varpi_\xi \eta(\varrho_\xi)}, S_{\sum_{\xi=1}^{2} \varpi_\xi \theta(\varrho_\xi)} \right], \right.$$

$$\left[1 - \prod_{\xi=1}^{2} \left(1 - T^l \left(\varrho_\xi \right) \right)^{\varpi_\xi}, 1 - \prod_{\xi=1}^{2} \left(1 - T^u \left(\varrho_\xi \right) \right)^{\varpi_\xi} \right]$$

$$\left[\prod_{\xi=1}^{2} \left(I^l \left(\varrho_\xi \right) \right)^{\varpi_\xi}, \prod_{\xi=1}^{2} \left(I^l \left(\varrho_\xi \right) \right)^{\varpi_\xi} \right],$$

$$\left. \left[\prod_{\xi=1}^{2} \left(F^l \left(\varrho_\xi \right) \right)^{\varpi_\xi}, \prod_{\xi=1}^{2} \left(I^l \left(\varrho_\xi \right) \right)^{\varpi_\xi} \right] \right\rangle \tag{8}$$

(ii) Hypothesis, Eq. (7) holds for $\tau = k$ $(k \geq 2)$, then

$$INULWA_\varpi \left(\varrho_1, \varrho_2, \ldots, \varrho_k \right) = \overset{k}{\underset{\xi=1}{\oplus}} \left(\varpi_\xi \varrho_\xi \right) = \left\langle \left[S_{\sum_{\xi=1}^{k} \varpi_\xi \eta(\varrho_\xi)}, S_{\sum_{\xi=1}^{k} \varpi_\xi \theta(\varrho_\xi)} \right], \right.$$

$$\left[1 - \prod_{\xi=1}^{k} \left(1 - T^l \left(\varrho_\xi \right) \right)^{\varpi_\xi}, 1 - \prod_{\xi=1}^{k} \left(1 - T^u \left(\varrho_\xi \right) \right)^{\varpi_\xi} \right]$$

$$\left[\prod_{\xi=1}^{k} \left(I^l \left(\varrho_\xi \right) \right)^{\varpi_\xi}, \prod_{\xi=1}^{k} \left(I^l \left(\varrho_\xi \right) \right)^{\varpi_\xi} \right],$$

$$\left. \left[\prod_{\xi=1}^{k} \left(F^l \left(\varrho_\xi \right) \right)^{\varpi_\xi}, \prod_{\xi=1}^{k} \left(I^l \left(\varrho_\xi \right) \right)^{\varpi_\xi} \right] \right\rangle \tag{9}$$

When $\tau = k + 1$, we get $INULWA_\varpi \left(\varrho_1, \varrho_2, \ldots, \varrho_{k+1}, \varrho_k \right) = \overset{k}{\underset{\xi=1}{\oplus}} \left(\varpi_\xi \varrho_\xi \right) =$

$$\left\langle \left[S_{\sum_{\xi=1}^{k} \varpi_\xi \eta(\varrho_\xi)}, S_{\sum_{\xi=1}^{k} \varpi_\xi \theta(\varrho_\xi)} \right], \right.$$

$$\left[1 - \prod_{\xi=1}^{k} \left(1 - T^l \left(\varrho_\xi \right) \right)^{\varpi_\xi}, 1 - \prod_{\xi=1}^{k} \left(1 - T^u \left(\varrho_\xi \right) \right)^{\varpi_\xi} \right]$$

$$\left. \left[\prod_{\xi=1}^{k} \left(I^l \left(\varrho_\xi \right) \right)^{\varpi_\xi}, \prod_{\xi=1}^{k} \left(I^l \left(\varrho_\xi \right) \right)^{\varpi_\xi} \right], \left[\prod_{\xi=1}^{k} \left(F^l \left(\varrho_\xi \right) \right)^{\varpi_\xi}, \prod_{\xi=1}^{k} \left(I^l \left(\varrho_\xi \right) \right)^{\varpi_\xi} \right] \right\rangle$$

$$\oplus \left\langle \left[S_{\varpi_{k+1} \eta(\varrho_{k+1})}, S_{\varpi_{k+1} \theta(\varrho_{k+1})} \right], \left[1 - \left(1 - T^l \left(\varrho_{k+1} \right) \right)^{\varpi_{k+1}}, 1 - \left(1 - T^u \left(\varrho_{k+1} \right) \right)^{\varpi_{k+1}} \right], \right.$$

$$\left[\left(\varrho_{k+1} \right)^{\varpi_{k+1}}, \left(I^l \left(\varrho_{k+1} \right) \right)^{\varpi_{k+1}} \right], \left[\left(F^l \left(\varrho_{k+1} \right) \right)^{\varpi_{k+1}}, \left(I^l \left(\varrho_{k+1} \right) \right)^{\varpi_{k+1}} \right] \right\rangle$$

$$\left\langle \left[s_{\sum_{\xi=1}^{k+1} \varpi_\xi \eta(\varrho_\xi)}, s_{\sum_{\xi=1}^{k+1} \varpi_\xi \theta(\varrho_\xi)} \right], \right.$$

$$\left[1 - \prod_{\xi=1}^{k+1} \left(1 - T^l \left(\varrho_\xi \right) \right)^{\varpi_\xi}, 1 - \prod_{\xi=1}^{k+1} \left(1 - T^u \left(\varrho_\xi \right) \right)^{\varpi_\xi} \right],$$

$$\left[\prod_{\xi=1}^{k+1} \left(I^l \left(\varrho_\xi \right) \right)^{\varpi_\xi} \prod_{\xi=1}^{k+1} \left(I^l \left(\varrho_\xi \right) \right)^{\varpi_\xi} \right] \left[\prod_{\xi=1}^{k+1} \left(F^l \left(\varrho_\xi \right) \right)^{\varpi_\xi} \prod_{\xi=1}^{k+1} \left(I^l \left(\varrho_\xi \right) \right)^{\varpi_\xi} \right] \right\rangle$$

$$\tag{10}$$

Thus, for $\tau = k + 1$, Eq. (7) holds, and results is obtained.

(Idempotent property)

Let $\varrho_\xi = \left\langle \left[s_{\eta(\varrho_\xi)}, s_{\theta(\varrho_\xi)} \right], \left[T^l \left(\varrho_\xi \right), T^u \left(\varrho_\xi \right) \right], \left[I^l \left(\varrho_\xi \right), I^u \left(\varrho_\xi \right) \right], \right.$ $\left[F^l \left(\varrho_\xi \right), F^u \left(\varrho_\xi \right) \right] \right\rangle$ be a set of INULNs for ($\xi = 1, 2, \ldots, \tau$) are equal, i.e., $\varrho_\xi = \varrho$ for all ξ. Then

$$INULWA_\varpi \left(\varrho_1, \varrho_2, \ldots, \varrho_\tau \right) = \varrho. \tag{11}$$

(Boundedness property)

Let $\varrho_\xi = \left\langle \left[s_{\eta(\varrho_\xi)}, s_{\theta(\varrho_\xi)} \right], \left[T^l \left(\varrho_\xi \right), T^u \left(\varrho_\xi \right) \right], \left[I^l \left(\varrho_\xi \right), I^u \left(\varrho_\xi \right) \right], \right.$ $\left[F^l \left(\varrho_\xi \right), F^u \left(\varrho_\xi \right) \right] \right\rangle$ be a set of INULNs for ($\xi = 1, 2, \ldots, \tau$).

Let $s_\eta^- = \min_{1 \leq \xi \leq \tau} \left\{ s_{\eta(\varrho_\xi)} \middle| \left[s_{\eta(\varrho_\xi)}, s_{\theta(\varrho_\xi)} \right] \in \varrho_\xi \right\},$

$s_\eta^+ = \max_{1 \leq \xi \leq \tau} \left\{ s_{\eta(\varrho_\xi)} \middle| \left[s_{\eta(\varrho_\xi)}, s_{\theta(\varrho_\xi)} \right] \in \varrho_\xi \right\}, \quad s_\theta^- = \min_{1 \leq \xi \leq \tau} \left\{ s_{\theta(\varrho_\xi)} \middle| \left[s_{\eta(\varrho_\xi)}, s_{\theta(\varrho_\xi)} \right] \in \varrho_\xi \right\}$

and $s_\theta^+ = \max_{1 \leq \xi \leq \tau} \left\{ s_{\theta(\varrho_\xi)} \middle| \left[s_{\eta(\varrho_\xi)}, s_{\theta(\varrho_\xi)} \right] \in \varrho_\xi \right\}.$

Let $T^{l-} = \min_{1 \leq \xi \leq \tau} \left\{ T_j^l \middle| \left[T_\xi^l, T_\xi^u \right] \in \varrho_\xi \right\}$, and $T^{u-} = \min_{1 \leq \xi \leq \tau} \left\{ T^u \left(\varrho_\xi \right) \middle| \right.$ $\left[T^l \left(\varrho_\xi \right), T^u \left(\varrho_\xi \right) \right] \in \varrho_\xi \right\}$ and $T^{l+} = \max_{1 \leq \xi \leq \tau} \left\{ T^l \left(\varrho_\xi \right) \middle| \left[T^l \left(\varrho_\xi \right), T^u \left(\varrho_\xi \right) \right] \in \varrho_\xi \right\},$ and $T^{u+} = \max_{1 \leq \xi \leq \tau} \left\{ T^u \left(\varrho_\xi \right) \middle| \left[T^l \left(\varrho_\xi \right), T^u \left(\varrho_\xi \right) \right] \in \varrho_\xi \right\}$. Let $I^{l-} = \min_{1 \leq \xi \leq \tau} \left\{ I^l \left(\varrho_\xi \right) \middle| \right.$ $\left[I^l \left(\varrho_\xi \right), I^u \left(\varrho_\xi \right) \right] \in \varrho_\xi \right\}$, and $I^{u-} = \min_{1 \leq \xi \leq \tau} \left\{ I^u \left(\varrho_\xi \right) \middle| \left[I_\xi^l, I^u \left(\varrho_\xi \right) \right] \in \varrho_\xi \right\}$ and $I^{l+} = \max_{1 \leq \xi \leq \tau} \left\{ I^l \left(\varrho_\xi \right) \middle| \left[I^l \left(\varrho_\xi \right), I^u \left(\varrho_\xi \right) \right] \in \varrho_\xi \right\}$, and $I^{u+} = \max_{1 \leq \xi \leq \tau} \left\{ I^u \left(\varrho_\xi \right) \middle| \right.$ $\left[I_\xi^l \left(\varrho_\xi \right), I^u \left(\varrho_\xi \right) \right] \in \varrho_\xi \right\}$. Let $F^{l-} = \min_{1 \leq \xi \leq \tau} \left\{ F_\xi^l \middle| \left[F_\xi^l \left(\varrho_\xi \right), F^u \left(\varrho_\xi \right) \right] \in \varrho_\xi \right\}$, and $F^{u-} = \min_{1 \leq \xi \leq \tau} \left\{ F^u \left(\varrho_\xi \right) \middle| \left[F_\xi^l \left(\varrho_\xi \right), F^u \left(\varrho_\xi \right) \right] \in \varrho_\xi \right\}$ and $F^{l+} = \max_{1 \leq \xi \leq \tau} \left\{ F^l \left(\varrho_\xi \right) \middle| \right.$ $\left[F^l \left(\varrho_\xi \right), F^u \left(\varrho_\xi \right) \right] \in \varrho_\xi \right\}$, and $F^{u+} = \max_{1 \leq \xi \leq \tau} \left\{ F^u \left(\varrho_\xi \right) \middle| \left[F_\xi^l \left(\varrho_\xi \right), F^u \left(\varrho_\xi \right) \right] \in \varrho_\xi \right\},$ for all j, then we have

$$\left\{\left[s_\eta^-, s_\theta^-\right], \left[T^{l-}, T^{u-}\right], \left[I^{l-}, I^{u-}\right]\left[F^{l-}, F^{u-}\right]\right\} \leq INULWA_\varpi (\varrho_1, \varrho_2, \ldots, \varrho_\tau)$$

$$\leq \left\{\left[s_\eta^+, s_\theta^+\right], \left[T^{l+}, T^{u+}\right], \left[I^{l+}, I^{u+}\right], \left[F^{l+}, F^{u+}\right]\right\}.$$

(Monotonicity property)

Let $\varrho_\xi = \left\langle\left[s_{\eta(\varrho_\xi)}, s_{\theta(\varrho_\xi)}\right], \left[T^l(\varrho_\xi), T^u(\varrho_\xi)\right], \left[I^l(\varrho_\xi), I^u(\varrho_\xi)\right],\right.$
$\left.\left[F^l(\varrho_\xi), F^u(\varrho_\xi)\right]\right\rangle$ and $\varrho'_\xi \left\langle\left[s'_{\eta(\varrho'_\xi)}, s'_{\theta(\varrho'_\xi)}\right], \left[T'^l(\varrho'_\xi), T'^u(\varrho'_\xi)\right],\right.$
$\left.\left[I'^l(\varrho'_\xi), I'^u(\varrho'_\xi)\right], \left[F'^l(\varrho'_\xi), F'^u(\varrho'_\xi)\right]\right\rangle$ be two sets of INULNs for
$(\xi = 1, 2, \ldots, \tau)$. If $\varrho_\xi \leq \varrho'_\xi$ for all j, then

$$INULWA_\varpi (\varrho_1, \varrho_2 \ldots, \varrho_\tau) \leq INULWA_\varpi \left(\varrho'_1, \varrho'_2, \ldots, \varrho'_\tau\right). \tag{12}$$

□

4.2 INULWG Operator

Now, we will introduce interval neutrosophic uncertain linguistic weighted geometric (INULWG) operator and its properties:

Let $\varrho_\xi = \left\langle\left[s_{\eta(\varrho_\xi)}, s_{\theta(\varrho_\xi)}\right], \left[T^l(\varrho_\xi), T^u(\varrho_\xi)\right], \left[I^l(\varrho_\xi), I^u(\varrho_\xi)\right],\right.$
$\left.\left[F^l(\varrho_\xi), F^u(\varrho_\xi)\right]\right\rangle$ be a set of INULNs for $(\xi = 1, 2, \ldots, \tau)$. Then interval neutrosophic uncertain linguistic weighted geometric (INULWG) function $INULWG : \Theta^\tau \to \Theta$ is defined as follows:

$$INULWG_\varpi (m_1, m_2, \ldots, m_\tau) = \overset{\tau}{\underset{j=1}{\otimes}} (\varrho_\xi)^{\varpi_j} \tag{13}$$

where $\varpi = (\varpi_1, \varpi_2, \ldots, \varpi_\tau)^T$ is followed by the weight vector of $\varrho_\xi (\xi = 1, 2, \ldots, \tau)$, with $\varpi_\xi \in [0, 1]$ and $\sum_{\xi=1}^\tau \varpi_j = 1$.

By the operations on INULNs, we derive the following theorem:

Let $\varrho_\xi = \left\langle\left[s_{\eta(\varrho_\xi)}, s_{\theta(\varrho_\xi)}\right], \left[T^l(\varrho_\xi), T^u(\varrho_\xi)\right], \left[I^l(\varrho_\xi), I^u(\varrho_\xi)\right],\right.$
$\left.\left[F^l(\varrho_\xi), F^u(\varrho_\xi)\right]\right\rangle$ be a set of INULNs for $j = 1, 2, \ldots, \tau$, then aggregating values of INULNs ϱ_ξ for $\xi = 1, 2, \ldots, \tau$ using INULWG operator are also an INULN, and further,

$$INULWG_\varpi (\varrho_1, \varrho_2, \ldots, \varrho_\tau) = \overset{\tau}{\underset{\xi=1}{\otimes}} (\varrho_\xi)^{\varpi_\xi} = \left\langle\left[S_{\prod_{\xi=1}^\tau (\eta(\varrho_\xi))^{\varpi_\xi}}, S_{\prod_{\xi=1}^\tau (\theta(\varrho_\xi))^{\varpi_\xi}}\right],\right.$$

$$\left[\prod_{\xi=1}^{\tau}\left(T^{l}\left(\varrho_{\xi}\right)\right)^{\varpi_{\xi}},\prod_{\xi=1}^{\tau}\left(T^{u}\left(\varrho_{\xi}\right)\right)^{\varpi_{\xi}}\right]\left[1-\prod_{\xi=1}^{\tau}\left(1-I^{l}\left(\varrho_{\xi}\right)\right)^{\varpi_{\xi}},\right.$$

$$1-\prod_{\xi=1}^{\tau}\left(1-I^{u}\left(\varrho_{\xi}\right)\right)^{\varpi_{\xi}}\right],\left[1-\prod_{\xi=1}^{\tau}\left(1-F^{l}\left(\varrho_{\xi}\right)\right)^{\varpi_{\xi}},\right.$$

$$\left.1-\prod_{\xi=1}^{\tau}\left(1-F^{u}\left(\varrho_{\xi}\right)\right)^{\varpi_{\xi}}\right]\right\rangle \tag{14}$$

where $\varpi = (\varpi_1, \varpi_2, \ldots, \varpi_\tau)^T$ is followed by the weight vector of $\varrho_\xi(\xi = 1, 2, \ldots, \tau)$, with $\varpi_\xi \in [0, 1]$, and $\sum_{\xi=1}^{\tau}\varpi_\xi = 1$.

(Idempotent property)

Let $\varrho_\xi = \left\langle\left[s_{\eta(\varrho_\xi)}, s_{\theta(\varrho_\xi)}\right], \left[T^l\left(\varrho_\xi\right), T^u\left(\varrho_\xi\right)\right], \left[I^l\left(\varrho_\xi\right), I^u\left(\varrho_\xi\right)\right],\right.$
$\left[F^l\left(\varrho_\xi\right), F^u\left(\varrho_\xi\right)\right]\right\rangle$ be a set of INULNs for $(\xi = 1, 2, \ldots, \tau)$ are equal, i.e., $\varrho_\xi = \varrho$ for all ξ. Then

$$INULW\,G_\varpi\,(\varrho_1, \varrho_2, \ldots, \varrho_\tau) = \varrho. \tag{15}$$

(Boundedness property)

Let $\varrho_\xi = \left\langle\left[s_{\eta(\varrho_\xi)}, s_{\theta(\varrho_\xi)}\right], \left[T^l\left(\varrho_\xi\right), T^u\left(\varrho_\xi\right)\right], \left[I^l\left(\varrho_\xi\right), I^u\left(\varrho_\xi\right)\right],\right.$
$\left[F^l\left(\varrho_\xi\right), F^u\left(\varrho_\xi\right)\right]\right\rangle$ be a set of INULNs for $(j = 1, 2, \ldots, \tau)$.

Let $s_\eta^- = \min_{1 \leq \xi \leq \tau}\left\{s_{\eta(\varrho_\xi)}|\left[s_{\eta(\varrho_\xi)}, s_{\theta(\varrho_\xi)}\right] \in \varrho_\xi\right\}$ and $s_\eta^+ = \max_{1 \leq \xi \leq \tau}\left\{s_{\eta(\varrho_\xi)}|\right.$
$\left.\left[s_{\eta(\varrho_\xi)}, s_{\theta(\varrho_\xi)}\right] \in \varrho_\xi\right\}, s_\theta^- = \min_{1 \leq \xi \leq \tau}\left\{s_{\theta(\varrho_\xi)}|\left[s_{\eta(\varrho_\xi)}, s_{\theta(\varrho_\xi)}\right] \in \varrho_\xi\right\}$ and $s_\theta^+ = \max_{1 \leq \xi \leq \tau}\left\{s_{\theta(\varrho_\xi)}|\left[s_{\eta(\varrho_\xi)}, s_{\theta(\varrho_\xi)}\right] \in \varrho_\xi\right\}$. Let $T^{l-} = \min_{1 \leq \xi \leq \tau}\left\{T_\xi^l|\left[T^l\left(\varrho_\xi\right), T^u\left(\varrho_\xi\right)\right] \in \varrho_\xi\right\}$, and $T^{u-} = \min_{1 \leq \xi \leq \tau}\left\{T^u\left(\varrho_\xi\right)|\left[T^l\left(\varrho_\xi\right), T^u\left(\varrho_\xi\right)\right] \in \varrho_\xi\right\}$ and $T^{l+} = \max_{1 \leq \xi \leq \tau}\left\{T^l\left(\varrho_\xi\right)|\right.$
$\left.\left[T^l\left(\varrho_\xi\right), T^u\left(\varrho_\xi\right)\right] \in \varrho_\xi\right\}$, and $T^{u+} = \max_{1 \leq \xi \leq \tau}\left\{T^u\left(\varrho_\xi\right)|\left[T^l\left(\varrho_\xi\right), T^u\left(\varrho_\xi\right)\right] \in \varrho_\xi\right\}$.
Let $I^{l-} = \min_{1 \leq j \leq \tau}\left\{I^l\left(\varrho_\xi\right)|\left[I^l\left(\varrho_\xi\right), I^u\left(\varrho_\xi\right)\right] \in \varrho_\xi\right\}$, and $I^{u-} = \min_{1 \leq \xi \leq \tau}\left\{I^u\left(\varrho_\xi\right)|\right.$
$\left.\left[I_\xi^l\left(\varrho_\xi\right), I^u\left(\varrho_\xi\right)\right] \in \varrho_\xi\right\}$ and $I^{l+} = \max_{1 \leq j \leq \tau}\left\{I^l\left(\varrho_\xi\right)|\left[I^l\left(\varrho_\xi\right), I^u\left(\varrho_\xi\right)\right] \in \varrho_\xi\right\}$, and
$I^{u+} = \max_{1 \leq \xi \leq \tau}\left\{I^u\left(\varrho_\xi\right)|\left[I^l\left(\varrho_\xi\right), I^u\left(\varrho_\xi\right)\right] \in \varrho_\xi\right\}$. Let $F^{l-} = \min_{1 \leq \xi \leq \tau}\left\{F_\xi^l|\left[F^l\left(\varrho_\xi\right),\right.\right.$
$\left.\left.F^u\left(\varrho_\xi\right)\right] \in \varrho_\xi\right\}$, and $F^{u-} = \min_{1 \leq \xi \leq \tau}\left\{F^u\left(\varrho_\xi\right)|\left[F^l\left(\varrho_\xi\right), F^u\left(\varrho_\xi\right)\right] \in \varrho_\xi\right\}$ and
$F^{l+} = \max_{1 \leq \xi \leq \tau}\left\{F^l\left(\varrho_\xi\right)|\left[F^l\left(\varrho_\xi\right), F^u\left(\varrho_\xi\right)\right] \in \varrho_\xi\right\}$, and $F^{u+} = \max_{1 \leq \xi \leq \tau}\left\{F^u\left(\varrho_\xi\right)|\right.$
$\left.\left[F^l\left(\varrho_\xi\right), F^u\left(\varrho_\xi\right)\right] \in \varrho_\xi\right\}$, for all j, then we have

$$\left\{\left[s_\eta^-, s_\theta^-\right], \left[T^{l-}, T^{u-}\right], \left[I^{l-}, I^{u-}\right], \left[F^{l-}, F^{u-}\right]\right\} \leq INULWG_\varpi\,(\varrho_1, \varrho_2, \ldots, \varrho_\tau)$$

$$\leq \left\{\left[S_\eta^+, S_\theta^+\right], \left[T^{l+}, T^{u+}\right], \left[I^{l+}, I^{u+}\right], \left[F^{l+}, F^{u+}\right]\right\}.$$

(Monotonicity property)

Let $\varrho_\xi = \left\langle \left[s_{\eta(\varrho_\xi)}, s_{\theta(\varrho_\xi)} \right], \left[T^l(\varrho_\xi), T^u(\varrho_\xi) \right], \left[I^l(\varrho_\xi), I^u(\varrho_\xi) \right], \right.$
$\left. \left[F^l(\varrho_\xi), F^u(\varrho_\xi) \right] \right\rangle$ and $\varrho'_\xi \left\langle \left[s'_{\eta(\varrho'_\xi)}, s'_{\theta(\varrho'_\xi)} \right], \left[T'^l(\varrho'_\xi), T'^u(\varrho'_\xi) \right], \right.$
$\left. \left[I'^l(\varrho'_\xi), I'^u(\varrho'_\xi) \right], \left[F'^l(\varrho'_\xi), F'^u(\varrho'_\xi) \right] \right\rangle$ be two sets of INULNs for $(\xi = 1, 2, \dots, \tau)$. If $\varrho_\xi \leq \varrho'_\xi$ for all ξ, then

$$INULWG_\varpi(\varrho_1, \varrho_2 \dots, \varrho_\tau) \leq INULWG_\varpi(\varrho'_1, \varrho'_2, \dots, \varrho'_\tau). \tag{16}$$

5 Interval Neutrosophic Uncertain Linguistic Dombi Aggregation Operators

5.1 INULDWA Operator

Let $\varrho_\xi = \left\langle \left[s_{\eta(\varrho_\xi)}, s_{\theta(\varrho_\xi)} \right], \left[T^l(\varrho_\xi), T^u(\varrho_\xi) \right], \left[I^l(\varrho_\xi), I^u(\varrho_\xi) \right], \left[F^l(\varrho_\xi), F^u(\varrho_\xi) \right] \right\rangle$
be a set of INULNs for $(\xi = 1, 2, \dots, \tau)$. The INULDWA function $INULDWA : \Theta^\tau \to \Theta$ is defined as follows:

$$INULDWA_\varpi(\varrho_1, \varrho_2, \dots, \varrho_\tau) = \overset{\tau}{\underset{\xi=1}{\oplus}} \left(\varpi_\xi \varrho_\xi \right) \tag{17}$$

where $\varpi = (\varpi_1, \varpi_2, \dots, \varpi_\tau)^T$ is followed by the weight vector of $\varrho_\xi (\xi = 1, 2, \dots, \tau)$, with $\varpi_\xi \in [0, 1]$, and $\sum_{\xi=1}^\tau \varpi_\xi = 1$.

By the operations on INULNs, we derive the following theorem:

Let $\varrho_\xi = \left\langle \left[s_{\eta(\varrho_\xi)}, s_{\theta(\varrho_\xi)} \right], \left[T^l(\varrho_\xi), T^u(\varrho_\xi) \right], \left[I^l(\varrho_\xi), I^u(\varrho_\xi) \right], \right.$
$\left. \left[F^l(\varrho_\xi), F^u(\varrho_\xi) \right] \right\rangle$ be a set of INULNs for $(\xi = 1, 2, \dots, \tau)$, then aggregating values using INULDWA opearator ϱ_ξ $(\xi = 1, 2, \dots, \tau)$ are also an INULN, and further,

$$INULDWA_\varpi(\varrho_1, \varrho_2, \dots, \varrho_\tau) = \overset{\tau}{\underset{\xi=1}{\oplus}} \left(\varpi_\xi \varrho_\xi \right) = \left\langle \left[s_{\sum_{\xi=1}^\tau \varpi_\xi \eta(\varrho_\xi)}, s_{\sum_{\xi=1}^\tau \varpi_j \theta(\varrho_\xi)} \right] \right.$$

$$= \left[1 - \frac{1}{1 + \left\{ \sum_{\xi=1}^\tau \varpi_\xi \left(\frac{T^l(\varrho_\xi)}{1 - T^l(\varrho_\xi)} \right)^\sigma \right\}^{1/\sigma}}, 1 - \frac{1}{1 + \left\{ \sum_{\xi=1}^\tau \varpi_\xi \left(\frac{T^u(\varrho_\xi)}{1 - T^u(\varrho_\xi)} \right)^\sigma \right\}^{1/\sigma}} \right],$$

$$\left[\frac{1}{1 + \left\{ \sum_{\xi=1}^\tau \varpi_\xi \left(\frac{1 - I^l(\varrho_\xi)}{I^l(\varrho_\xi)} \right)^\sigma \right\}^{1/\sigma}}, \frac{1}{1 + \left\{ \sum_{\xi=1}^\tau \varpi_\xi \left(\frac{1 - I^u(\varrho_\xi)}{I^u(\varrho_\xi)} \right)^\sigma \right\}^{1/\sigma}} \right],$$

$$\left. \left[\frac{1}{1 + \left\{ \sum_{\xi=1}^\tau \varpi_\xi \left(\frac{1 - F^l(\varrho_\xi)}{F^l(\varrho_\xi)} \right)^\sigma \right\}^{1/\sigma}}, \frac{1}{1 + \left\{ \sum_{\xi=1}^\tau \varpi_\xi \left(\frac{1 - F^u(\varrho_\xi)}{F^u(\varrho_\xi)} \right)^\sigma \right\}^{1/\sigma}} \right] \right\rangle$$

where $\varpi = (\varpi_1, \varpi_2, \ldots, \varpi_\tau)^T$ is followed by the weight vector of $\varrho_\xi (\xi = 1, 2, \ldots, \tau)$, with $\varpi_\xi \in [0, 1]$, and $\sum_{\xi=1}^{\tau} \varpi_\xi = 1$.

Proof We prove the Eq. (18) below using mathematical induction.

(i) when $\tau = 2$, we get

$$\left\langle \left[S_{\varpi_\xi \eta(\varrho_\xi)}, S_{\varpi_\xi \theta(\varrho_\xi)} \right], \left[1 - \frac{1}{1 + \left\{ \varpi_\xi \left(\frac{T^l(\varrho_\xi)}{1 - T^l(\varrho_\xi)} \right)^\sigma \right\}^{1/\sigma}}, 1 - \frac{1}{1 + \left\{ \varpi_\xi \left(\frac{T^u(\varrho_\xi)}{1 - T^u(\varrho_\xi)} \right)^\sigma \right\}^{1/\sigma}} \right], \right.$$

$$\left[\frac{1}{1 + \left\{ \varpi_\xi \left(\frac{1 - I^l(\varrho_\xi)}{I^l(\varrho_\xi)} \right)^\sigma \right\}^{1/\sigma}}, \frac{1}{1 + \left\{ \varpi_\xi \left(\frac{1 - I^u(\varrho_\xi)}{I^u(\varrho_\xi)} \right)^\sigma \right\}^{1/\sigma}} \right],$$

$$\left. \left[\frac{1}{1 + \left\{ \varpi_\xi \left(\frac{1 - F^l(\varrho_\xi)}{F^l(\varrho_\xi)} \right)^\sigma \right\}^{1/\sigma}}, 1 - \frac{1}{1 + \left\{ \varpi_\xi \left(\frac{1 - F^u(\varrho_\xi)}{F^u(\varrho_\xi)} \right)^\sigma \right\}^{1/\sigma}} \right] \right\rangle$$

for $\xi = 1, 2$.

Then,

$$INULDWA_\varpi (\varrho_1, \varrho_2) = \bigoplus_{\xi=1}^{2} (\varpi_\xi \varrho_\xi) = \left\langle \left[S_{\sum_{\xi=1}^{2} \varpi_\xi \eta(\varrho_\xi)}, S_{\sum_{\xi=1}^{2} \varpi_j \theta(\varrho_\xi)} \right], \right.$$

$$\left[1 - \frac{1}{1 + \left\{ \sum_{\xi=1}^{2} \varpi_\xi \left(\frac{T^l(\varrho_\xi)}{1 - T^l(\varrho_\xi)} \right)^\sigma \right\}^{1/\sigma}}, 1 - \frac{1}{1 + \left\{ \sum_{\xi=1}^{2} \varpi_\xi \left(\frac{T^u(\varrho_\xi)}{1 - T^u(\varrho_\xi)} \right)^\sigma \right\}^{1/\sigma}} \right],$$

$$\left[\frac{1}{1 + \left\{ \sum_{\xi=1}^{2} \varpi_\xi \left(\frac{1 - I^l(\varrho_\xi)}{I^l(\varrho_\xi)} \right)^\sigma \right\}^{1/\sigma}}, \frac{1}{1 + \left\{ \sum_{\xi=1}^{2} \varpi_\xi \left(\frac{1 - I^u(\varrho_\xi)}{I^u(\varrho_\xi)} \right)^\sigma \right\}^{1/\sigma}} \right],$$

$$\left. \left[\frac{1}{1 + \left\{ \sum_{\xi=1}^{2} \varpi_\xi \left(\frac{1 - F^l(\varrho_\xi)}{F^l(\varrho_\xi)} \right)^\sigma \right\}^{1/\sigma}}, \frac{1}{1 + \left\{ \sum_{\xi=1}^{2} \varpi_\xi \left(\frac{1 - F^u(\varrho_\xi)}{F^u(\varrho_\xi)} \right)^\sigma \right\}^{1/\sigma}} \right] \right\rangle$$

(ii) Hypothesis, Eq. (18) holds for $\tau = k$ ($k \geq 2$), then $INULDWA_\varpi$ $(\varrho_1, \varrho_2, \ldots, \varrho_k) = \bigoplus_{\xi=1}^{k} (\varpi_\xi \varrho_\xi)$:

$$= \left\langle \left[S_{\sum_{\xi=1}^{k} \varpi_\xi \eta(\varrho_\xi)}, S_{\sum_{\xi=1}^{k} \varpi_\xi \theta(\varrho_\xi)} \right], \right.$$

$$\left[1 - \frac{1}{1 + \left\{ \sum_{\xi=1}^{k} \varpi_\xi \left(\frac{T^l(\varrho_\xi)}{1 - T^l(\varrho_\xi)} \right)^\sigma \right\}^{1/\sigma}}, 1 - \frac{1}{1 + \left\{ \sum_{\xi=1}^{k} \varpi_\xi \left(\frac{T^u(\varrho_\xi)}{1 - T^u(\varrho_\xi)} \right)^\sigma \right\}^{1/\sigma}} \right],$$

$$\left[\frac{1}{1 + \left\{ \sum_{\xi=1}^{k} \varpi_\xi \left(\frac{1 - I^l(\varrho_\xi)}{I^l(\varrho_\xi)} \right)^\sigma \right\}^{1/\sigma}}, \frac{1}{1 + \left\{ \sum_{\xi=1}^{k} \varpi_\xi \left(\frac{1 - I^u(\varrho_\xi)}{I^u(\varrho_\xi)} \right)^\sigma \right\}^{1/\sigma}} \right],$$

$$\left. \left[\frac{1}{1 + \left\{ \sum_{\xi=1}^{k} \varpi_\xi \left(\frac{1 - F^l(\varrho_\xi)}{F^l(\varrho_\xi)} \right)^\sigma \right\}^{1/\sigma}}, \frac{1}{1 + \left\{ \sum_{\xi=1}^{k} \varpi_\xi \left(\frac{1 - F^u(\varrho_\xi)}{F^u(\varrho_\xi)} \right)^\sigma \right\}^{1/\sigma}} \right] \right\rangle.$$

When $\tau = k + 1$, we get $INULDWA_\varpi (\varrho_1, \varrho_2, \ldots, \varrho_{k+1}, \varrho_k) = \bigoplus_{\xi=1}^{k} \left(\varpi_\xi \varrho_\xi \right) \oplus (\varpi_{k+1} \varrho_{k+1})$:

$$= \left\langle \left[S_{\sum_{\xi=1}^{k} \varpi_\xi \eta(\varrho_\xi)}, S_{\sum_{\xi=1}^{k} \varpi_\xi \theta(\varrho_\xi)} \right], \right.$$

$$\left[1 - \frac{1}{1 + \left\{ \sum_{\xi=1}^{k} \varpi_\xi \left(\frac{T^l(\varrho_\xi)}{1 - T^l(\varrho_\xi)} \right)^\sigma \right\}^{1/\sigma}}, 1 - \frac{1}{1 + \left\{ \sum_{\xi=1}^{k} \varpi_\xi \left(\frac{T^u(\varrho_\xi)}{1 - T^u(\varrho_\xi)} \right)^\sigma \right\}^{1/\sigma}} \right],$$

$$\left[\frac{1}{1 + \left\{ \sum_{\xi=1}^{k} \varpi_\xi \left(\frac{1 - I^l(\varrho_\xi)}{I^l(\varrho_\xi)} \right)^\sigma \right\}^{1/\sigma}}, \frac{1}{1 + \left\{ \sum_{\xi=1}^{k} \varpi_\xi \left(\frac{1 - I^u(\varrho_\xi)}{I^u(\varrho_\xi)} \right)^\sigma \right\}^{1/\sigma}} \right],$$

$$\left. \left[\frac{1}{1 + \left\{ \sum_{\xi=1}^{k} \varpi_\xi \left(\frac{1 - F^l(\varrho_\xi)}{F^l(\varrho_\xi)} \right)^\sigma \right\}^{1/\sigma}}, \frac{1}{1 + \left\{ \sum_{\xi=1}^{k} \varpi_\xi \left(\frac{1 - F^u(\varrho_\xi)}{F^u(\varrho_\xi)} \right)^\sigma \right\}^{1/\sigma}} \right] \right\rangle$$

$$\oplus \left\langle \left[S_{\varpi_{k+1} \eta(\varrho_{k+1})}, S_{\varpi_{k+1} \theta(\varrho_{k+1})} \right], \right.$$

$$\left[1 - \frac{1}{1 + \left\{ \varpi_{k+1} \left(\frac{T^l(\varrho_{k+1})}{1 - T^l(\varrho_{k+1})} \right)^\sigma \right\}^{1/\sigma}}, 1 - \frac{1}{1 + \left\{ \varpi_{k+1} \left(\frac{T^u(\varrho_{k+1})}{1 - T^u(\varrho_{k+1})} \right)^\sigma \right\}^{1/\sigma}} \right],$$

$$\left[\frac{1}{1 + \left\{ \varpi_{k+1} \left(\frac{1 - I^l(\varrho_{k+1})}{I^l(\varrho_{k+1})} \right)^\sigma \right\}^{1/\sigma}}, \frac{1}{1 + \left\{ \varpi_{k+1} \left(\frac{1 - I^u(\varrho_{k+1})}{I^u(\varrho_{k+1})} \right)^\sigma \right\}^{1/\sigma}} \right],$$

$$\left. \left[\frac{1}{1 + \left\{ \varpi_{k+1} \left(\frac{1 - F^l(\varrho_{k+1})}{F^l(\varrho_{k+1})} \right)^\sigma \right\}^{1/\sigma}}, 1 - \frac{1}{1 + \left\{ \varpi_{k+1} \left(\frac{1 - F^u(\varrho_{k+1})}{F^u(\varrho_{k+1})} \right)^\sigma \right\}^{1/\sigma}} \right] \right\rangle$$

$$= \left\langle \left[s_{\sum_{\xi=1}^{k+1} \varpi_\xi \eta(\varrho_\xi)}, s_{\sum_{\xi=1}^{k+1} \varpi_\xi \theta(\varrho_\xi)} \right], \left[1 - \frac{1}{\left\{ 1 + \left\{ \sum_{\xi=1}^{k+1} \varpi_\xi \left(\frac{T^l(\varrho_\xi)}{1-T^l(\varrho_\xi)} \right)^\sigma \right\}^{1/\sigma} \right.}, \right. \right.$$

$$1 - \frac{1}{\left\{ 1 + \left\{ \sum_{\xi=1}^{k+1} \varpi_\xi \left(\frac{T^u(\varrho_\xi)}{1-T^u(\varrho_\xi)} \right)^\sigma \right\}^{1/\sigma} \right.}$$

$$\left[\frac{1}{\left\{ 1 + \left\{ \sum_{\xi=1}^{k+1} \varpi_\xi \left(\frac{1-I^l(\varrho_\xi)}{I^l(\varrho_\xi)} \right)^\sigma \right\}^{1/\sigma} \right.}, \frac{1}{\left\{ 1 + \left\{ \sum_{\xi=1}^{k+1} \varpi_\xi \left(\frac{1-I^u(\varrho_\xi)}{I^u(\varrho_\xi)} \right)^\sigma \right\}^{1/\sigma} \right.} \right]$$

$$\left. \left[\frac{1}{\left\{ 1 + \left\{ \sum_{\xi=1}^{k+1} \varpi_\xi \left(\frac{1-F^l(\varrho_\xi)}{F^l(\varrho_\xi)} \right)^\sigma \right\}^{1/\sigma} \right.}, \frac{1}{\left\{ 1 + \left\{ \sum_{\xi=1}^{k+1} \varpi_\xi \left(\frac{1-F^u(\varrho_\xi)}{F^u(\varrho_\xi)} \right)^\sigma \right\}^{1/\sigma} \right.} \right] \right\rangle$$

Thus, for $\tau = k + 1$, Eq. (18) holds, and results is obtained. $\quad\square$

5.2   INULDWG Operator

Let $\varrho_\xi = \left\langle \left[s_{\eta(\varrho_\xi)}, s_{\theta(\varrho_\xi)} \right], \left[T^l(\varrho_\xi), T^u(\varrho_\xi) \right], \left[I^l(\varrho_\xi), I^u(\varrho_\xi) \right], \left[F^l(\varrho_\xi), \right. \right.$ $\left. F^u(\varrho_\xi) \right] \right\rangle$ be a set of INULNs for $(\xi = 1, 2, \ldots, \tau)$. Then interval neutrosophic uncertain linguistic Dombi weighted average (INULDWG) function $INULDWG : \Theta^\tau \to \Theta$ is defined as follows:

$$INULDWG_\varpi (\varrho_1, \varrho_2, \ldots, \varrho_\tau) = \overset{\tau}{\underset{\xi=1}{\otimes}} (\varrho_\xi)^{\varpi_\xi} \qquad (18)$$

where $\varpi = (\varpi_1, \varpi_2, \ldots, \varpi_\tau)^T$ is followed by the weight vector of $\varrho_\xi (\xi = 1, 2, \ldots, \tau)$, with $\varpi_\xi \in [0, 1]$, and $\sum_{\xi=1}^\tau \varpi_\xi = 1$.

In view of Dombi operation on INULNs, we derive the following theorem:

Let $\varrho_\xi = \left\langle \left[s_{\eta(\varrho_\xi)}, s_{\theta(\varrho_\xi)} \right], \left[T^l(\varrho_\xi), T^u(\varrho_\xi) \right], \left[I^l(\varrho_\xi), I^u(\varrho_\xi) \right], \left[F^l(\varrho_\xi), \right. \right.$ $\left. F^u(\varrho_\xi) \right] \right\rangle$ be a set of INULNs for $(\xi = 1, 2, \ldots, \tau)$, then aggregating values of INULNs ϱ_ξ $(\xi = 1, 2, \ldots, \tau)$ are also an INULN, and further,

$$INULDWG_\varpi (\varrho_1, \varrho_2, \ldots, \varrho_\tau) = \overset{\tau}{\underset{\xi=1}{\otimes}} (\varpi_\xi \varrho_\xi)$$

$$= \left\langle \left[S_{\prod_{\xi=1}^{\tau} (\eta(\varrho_{\xi}))^{\varpi_{\xi}}}, S_{\prod_{\xi=1}^{\tau} (\theta(\varrho_{\xi}))^{\varpi_{\xi}}} \right], \right.$$

$$\left[\frac{1}{1+\left\{ \sum_{\xi=1}^{\tau} \varpi_{\xi} \left(\frac{1-T^l(\varrho_{\xi})}{T^l(\varrho_{\xi})} \right)^{\sigma} \right\}^{1/\sigma}}, \frac{1}{1+\left\{ \sum_{\xi=1}^{\tau} \varpi_{\xi} \left(\frac{1-T^u(\varrho_{\xi})}{T^u(\varrho_{\xi})} \right)^{\sigma} \right\}^{1/\sigma}} \right],$$

$$\left[1 - \frac{1}{1+\left\{ \sum_{\xi=1}^{\tau} \varpi_{\xi} \left(\frac{I^l(\varrho_{\xi})}{1-I^l(\varrho_{\xi})} \right)^{\sigma} \right\}^{1/\sigma}}, 1 - \frac{1}{1+\left\{ \sum_{\xi=1}^{\tau} \varpi_{\xi} \left(\frac{I^u(\varrho_{\xi})}{1-I^u(\varrho_{\xi})} \right)^{\sigma} \right\}^{1/\sigma}} \right],$$

$$\left. \left[1 - \frac{1}{1+\left\{ \sum_{\xi=1}^{\tau} \varpi_{\xi} \left(\frac{F^l(\varrho_{\xi})}{1-F^l(\varrho_{\xi})} \right)^{\sigma} \right\}^{1/\sigma}}, 1 - \frac{1}{1+\left\{ \sum_{\xi=1}^{\tau} \varpi_{\xi} \left(\frac{F^u(\varrho_{\xi})}{1-F^u(\varrho_{\xi})} \right)^{\sigma} \right\}^{1/\sigma}} \right] \right\rangle .$$

where $\varpi = (\varpi_1, \varpi_2, \ldots, \varpi_\tau)^T$ is followed by the weight vector of $\varrho_\xi (\xi = 1, 2, \ldots, \tau)$, with $\varpi_\xi \in [0, 1]$, and $\sum_{\xi=1}^{\tau} \varpi = 1$.

Proof This theorem can be proved easily. $\square$

6 Model for MADM Method with INUL Information

In this study, we propose MADM method using INUL aggregation operators where the weights of the attributes are real numbers under INUL information. Here MADM method is used to develop the usefulness for the evaluation of rural development index selection under interval neutrosophic uncertain linguistic information. Let $A = \{A_1, A_2, \ldots, A_\zeta\}$ be a finite set of alternatives, and $b = \{b_1, b_2, \ldots, b_\tau\}$ be a set of attributes. Let $\varpi = (\varpi_1, \varpi_2, \ldots, \varpi_\zeta)^T$ be the weight vector for the attribute $\beta_\xi (\xi = 1, 2, \ldots, \tau)$ that is known such that $\varpi_\xi \in [0, 1]$, where $\sum_{\xi=1}^{\tau} \varpi_\xi = 1$. Suppose that $A = (a_{\rho\xi})_{\zeta \times \tau}$ is the INUL decision matrix, where $\varrho_{\rho\xi} = \left(\left[S_{\eta(\varrho_{\rho\xi})}, S_{\theta(\varrho_{\rho\xi})} \right], \left[T^l (\varrho_{\rho\xi}), T^u (\varrho_{\rho\xi}) \right], \left[I^l (\varrho_{\rho\xi}), I^u (\varrho_{\rho\xi}) \right], \left[F^l (\varrho_{\rho\xi}), F^u (\varrho_{\rho\xi}) \right] \right)$ is the INULN for the alternative $\varrho_{\rho\xi} \in M$ w.r.t. the attribute $b_\xi \in B$.

The algorithm follows a method to interpret MADM problem under INUL information using INULWA and INULWG operators.

Algorithm
Input: To the selection of desirable alternatives.
Output: Best alternative.

Case 1
Step 1. We utilize the the decision information provided in matrix A and by using the INULWA operator

$$INULWA_{\varpi}\left(\varrho_{11}, \varrho_{12}, \ldots, \varrho_{1\tau}\right) = \overset{\tau}{\underset{\xi=1}{\oplus}} \left(\varpi_{\xi}\varrho_{\rho\xi}\right)\beta_{\rho} =$$

$$\left\langle \left[S_{\sum_{\xi=1}^{\tau} \varpi_{\xi}\eta(\varrho_{\rho\xi})}, S_{\sum_{\xi=1}^{\tau} \varpi_{\xi}\theta(\varrho_{\rho\xi})}\right],\right.$$

$$\left[1 - \prod_{\xi=1}^{\tau}\left(1 - T^{l}\left(\varrho_{\rho\xi}\right)\right)^{\varpi_{\xi}}, 1 - \prod_{\xi=1}^{\tau}\left(1 - T^{u}\left(\varrho_{\rho\xi}\right)\right)^{\varpi_{\xi}}\right], \left[\prod_{\xi=1}^{\tau}\left(I^{l}\left(\varrho_{\rho\xi}\right)\right)^{\varpi_{\xi}},\right.$$

$$\left.\left.\prod_{\xi=1}^{\tau}\left(I^{l}\left(\varrho_{\rho\xi}\right)\right)^{\varpi_{\xi}}\right], \left[\prod_{\xi=1}^{\tau}\left(F^{l}\left(\varrho_{\rho\xi}\right)\right)^{\varpi_{\xi}}, \prod_{\xi=1}^{\tau}\left(I^{l}\left(\varrho_{\rho\xi}\right)\right)^{\varpi_{\xi}}\right]\right\rangle \qquad (19)$$

or

$$INULWG_{\varpi}\left(\varrho_{11}, \varrho_{12}, \ldots, \varrho_{1\tau}\right) = \overset{\tau}{\underset{\xi=1}{\otimes}} \left(\varrho_{\rho\xi}\right)^{\varpi_{\xi}}\beta_{\rho}$$

$$= \left\langle \left[S_{\prod_{\xi=1}^{\tau}(\eta(\varrho_{\varpi\xi}))^{\varpi_{\xi}}}, S_{\prod_{\xi=1}^{\tau}(\theta(\varrho_{\rho\xi}))^{\varpi_{\xi}}}\right],\right.$$

$$\left[\prod_{j=1}^{\tau}\left(T^{l}\left(\varpi_{\xi}\right)\right)^{\varpi_{\xi}}, \prod_{\xi=1}^{\tau}\left(T^{u}\left(\varrho_{\rho\xi}\right)\right)^{\varpi_{\xi}}\right], \left[1 - \prod_{\xi=1}^{\tau}\left(1 - I^{l}\left(\varrho_{\rho\xi}\right)\right)^{\varpi_{\xi}},\right.$$

$$1 - \prod_{\xi=1}^{\tau}\left(1 - I^{u}\left(\varrho_{\rho\xi}\right)\right)^{\varpi_{\xi}}\right], \left[1 - \prod_{\xi=1}^{\tau}\left(1 - F^{l}\left(\varrho_{\rho\xi}\right)\right)^{\varpi_{\xi}},\right.$$

$$\left.\left.1 - \prod_{\xi=1}^{\tau}\left(1 - F^{u}\left(\varrho_{\rho\xi}\right)\right)^{\varpi_{j}}\right]\right\rangle \qquad (20)$$

Case 2 If we apply INULDWA (INULDWG) operator, then we get the scheme as follows:

$$INULDWA_{\varpi}\left(\varrho_{1}, \varrho_{2}, \ldots, \varrho_{\tau}\right) = \overset{\tau}{\underset{\xi=1}{\oplus}} \left(\varpi_{\xi}\varrho_{\xi}\right) = \left\langle \left[S_{\sum_{\xi=1}^{\tau} \varpi_{\xi}\eta(\varrho_{\xi})}, S_{\sum_{\xi=1}^{\tau} \varpi_{\xi}\theta(\varrho_{\xi})}\right]\right.$$

$$= \left[1 - \frac{1}{1 + \left\{\sum_{\xi=1}^{\tau}\varpi_{\xi}\left(\frac{T^{l}(\varrho_{\xi})}{1-T^{l}(\varrho_{\xi})}\right)^{\sigma}\right\}^{1/\sigma}}, 1 - \frac{1}{1 + \left\{\sum_{\xi=1}^{\tau}\varpi_{\xi}\left(\frac{T^{u}(\varrho_{\xi})}{1-T^{u}(\varrho_{\xi})}\right)^{\sigma}\right\}^{1/\sigma}}\right],$$

$$\left[\frac{1}{1 + \left\{\sum_{\xi=1}^{\tau}\varpi_{\xi}\left(\frac{1-I^{l}(\varrho_{\xi})}{I^{l}(\varrho_{\xi})}\right)^{\sigma}\right\}^{1/\sigma}}, \frac{1}{1 + \left\{\sum_{\xi=1}^{\tau}\varpi_{\xi}\left(\frac{1-I^{u}(\varrho_{\xi})}{I^{u}(\varrho_{\xi})}\right)^{\sigma}\right\}^{1/\sigma}}\right],$$

$$\left.\left[\frac{1}{1 + \left\{\sum_{\xi=1}^{\tau}\varpi_{\xi}\left(\frac{1-F^{l}(\varrho_{\xi})}{F^{l}(\varrho_{\xi})}\right)^{\sigma}\right\}^{1/\sigma}}, \frac{1}{1 + \left\{\sum_{\xi=1}^{\tau}\varpi_{\xi}\left(\frac{1-F^{u}(\varrho_{\xi})}{F^{u}(\varrho_{\xi})}\right)^{\sigma}\right\}^{1/\sigma}}\right]\right\rangle$$

or

$$INULDWG_{\varpi}(\varrho_1, \varrho_2, \ldots, \varrho_\tau) = \bigotimes_{\xi=1}^{\tau} (\varrho_\xi)^{\varpi_\xi} \left\langle \left[S_{\prod_{\xi=1}^{\tau} (\eta(\varrho_\xi))^{\varpi_\xi}}, S_{\prod_{\xi=1}^{\tau} (\theta(\varrho_\xi))^{\varpi_\xi}} \right], \right.$$

$$\left[\frac{1}{1 + \left\{ \sum_{\xi=1}^{\tau} \varpi_\xi \left(\frac{1-T^l(\varrho_\xi)}{T^l(\varrho_\xi)} \right)^\sigma \right\}^{1/\sigma}}, \frac{1}{1 + \left\{ \sum_{\xi=1}^{\tau} \varpi_\xi \left(\frac{1-T^u(\varrho_\xi)}{T^u(\varrho_\xi)} \right)^\sigma \right\}^{1/\sigma}} \right],$$

$$\left[1 - \frac{1}{1 + \left\{ \sum_{\xi=1}^{\tau} \varpi_\xi \left(\frac{I^l(\varrho_\xi)}{1-I^l(\varrho_\xi)} \right)^\sigma \right\}^{1/\sigma}}, \frac{1}{1 + \left\{ \sum_{\xi=1}^{\tau} \varpi_\xi \left(\frac{I^u(\varrho_\xi)}{1-I^u(\varrho_\xi)} \right)^\sigma \right\}^{1/\sigma}} \right],$$

$$\left. \left[1 - \frac{1}{1 + \left\{ \sum_{\xi=1}^{\tau} \varpi_\xi \left(\frac{F^l(\varrho_\xi)}{1-F^l(\varrho_\xi)} \right)^\sigma \right\}^{1/\sigma}}, \frac{1}{1 + \left\{ \sum_{\xi=1}^{\tau} \varpi_\xi \left(\frac{F^u(\varrho_\xi)}{1-F^u(\varrho_\xi)} \right)^\sigma \right\}^{1/\sigma}} \right] \right\rangle$$

to obtained the overall values β_ρ ($\rho = 1, 2, \ldots, \zeta$) of the alternative ϱ_ξ.

Step 2. Evaluation of the score $\uplus(\beta_\rho)$ ($\rho = 1, 2, \ldots, \zeta$) based on overall INUL information β_ρ to determine the ranking of all the alternatives ϱ_ξ ($\xi = 1, 2, \ldots, \tau$) to obtain desirable choice A_ρ. If the values of $\uplus(\beta_\rho)$ and $\uplus(\beta_\xi)$ are same, then we next proceed to evaluate degrees of accuracy $\Im(\beta_\rho)$ and $\Im(\beta_\xi)$ rest on overall INUL information of β_i and β_ξ, and rank the alternative A_ρ depending with the accuracy function of $\Im(\beta_\rho)$ and $\Im(\beta_\xi)$.

Step 3. Rank all the alternatives A_ρ ($\rho = 1, 2, \ldots, \zeta$) in order to choose the best one(s) in accordance with $\uplus(\beta_\rho)$ ($\rho = 1, 2, \ldots, m$).

Step 4. Stop.

7 Numerical Example

7.1 Application

In the following, decision-making problem has been executed with a numerical example concerning investment selection to fitness of the proposed MADM problems. An investor wants to invest money in a mutual fund company. Before investment, an investor seeks an advise of an expert team/ Consider five mutual funds companies as alternatives such as:

(A_1): Large cap fund
(A_2): Liquid fund
(A_3): Blue chip fund
(A_4): Hybrid fund.

The expert team analyzed the mutual funds (alternatives) under the five characteristic given below and give their suggestion:

(b_1): Short term
(b_2): Mid-term
(b_3): Long term
(b_4): Risk of the funds
(b_5): wealth of the fund

Then collect the data and form the benefit rating information for four mutual funds with respect to experts team using the linguistic terms set $S = \{s_1 = $ extremely poor benefit, $s_2 = $ very poor benefit, $s_3 = $ poor benefit, $s_4 = $ medium benefit, $s_5 = $ good benefit, $s_6 = $ very good benefit, $s_7 = $ extremely good benefit$\}$ of the above five attributes and weight vector of them is $\psi = (0.4, 0.2, 0.1, 0.12, 0.18)^T$, and alternatives A_1, A_2, A_3, and A_4 evaluated with INULNs by the decision-makers have same dominance degree. Evaluation of decision-makers is given in Table 1.

Case 1:

Step 1. We aggregate INUL information $\beta_{\rho\xi}$ for $i = 1, 2, 3, 4$; $\xi = 1, 2, 3, 4, 5$ by using INULWA operator to obtain the overall preference values β_ξ for ($\rho = 1, 2, 3, 4$) representing the alternative A_ρ which is given in the following table:

Step 2. The score values of the alternatives A_ρ ($\rho = 1, 2, 3, 4$) are shown below by using aggregated values of the alternatives given in Table 2. Then, $\uplus(\beta_1) = s_{3.945}$, $\uplus(\beta_2) = s_{3.754}$, $\uplus(\beta_3) = s_{4.453}$ and $\uplus(\beta_4) = s_{4.066}$

Step 3. Based on the values of the score function, we design the ranking order of the alternatives as follows: we obtain $A_3 \succ A_4 \succ A_1 \succ A_2$. Blue chip fund A_3 is the best mutual funds for investment.

Table 1 Evaluations of decision-makers

	A_1	A_2
b_1	$\langle([s_4, s_5], [0.3, 0.5], [0.2, 0.3], [0.3, 0.5])\rangle$	$\langle([s_4, s_5], [0.4, 0.7], [0.2, 0.3], [0.5, 0.6])\rangle$
b_2	$\langle([s_5, s_5], [0.4, 0.6], [0.1, 0.2], [0.3, 0.4])\rangle$	$\langle([s_5, s_5], [0.5, 0.6], [0.3, 0.4], [0.3, 0.4])\rangle$
b_3	$\langle([s_3, s_4], [0.6, 0.7], [0.2, 0.3], [0.2, 0.4])\rangle$	$\langle([s_4, s_4], [0.6, 0.7], [0.1, 0.2], [0.4, 0.6])\rangle$
b_4	$\langle([s_6, s_6], [0.4, 0.5], [0.1, 0.3], [0.5, 0.6])\rangle$	$\langle([s_5, s_6], [0.3, 0.4], [0.2, 0.3], [0.3, 0.4])\rangle$
b_5	$\langle([s_3, s_4], [0.5, 0.6], [0.2, 0.4], [0.3, 0.4])\rangle$	$\langle([s_4, s_5], [0.4, 0.5], [0.3, 0.4], [0.4, 0.5])\rangle$
	A_3	A_4
	$\langle([s_5, s_5], [0.4, 0.5], [0.1, 0.2], [0.4, 0.5])\rangle$	$\langle([s_4, s_5], [0.6, 0.7], [0.1, 0.2], [0.4, 0.5])\rangle$
	$\langle([s_4, s_4], [0.3, 0.6], [0.2, 0.3], [0.5, 0.6])\rangle$	$\langle([s_2, s_3], [0.6, 0.7], [0.2, 0.3], [0.3, 0.4])\rangle$
	$\langle([s_4, s_5], [0.6, 0.8], [0.2, 0.4], [0.3, 0.4])\rangle$	$\langle([s_3, s_6], [0.6, 0.6], [0.1, 0.3], [0.2, 0.4])\rangle$
	$\langle([s_6, s_6], [0.7, 0.9], [0.1, 0.3], [0.4, 0.5])\rangle$	$\langle([s_4, s_5], [0.7, 0.8], [0.2, 0.3], [0.3, 0.4])\rangle$
	$\langle([s_3, s_4], [0.6, 0.7], [0.1, 0.2], [0.2, 0.3])\rangle$	$\langle([s_4, s_4], [0.4, 0.6], [0.1, 0.3], [0.4, 0.5])\rangle$

Table 2 Aggregated values of the alternatives using INULWA operators

Alternative (A_ρ)	INULWA
A_1	$\langle([s_{4.16}, s_{4.84}], [0.40698, 0.56353], [0.16021, 0.29133], [0.30629, 0.45915])\rangle$
A_2	$\langle([s_{4.32}, s_{5.02}], [0.43410, 0.62141], [0.21769, 0.32136], [0.39888, 0.50998])\rangle$
A_3	$\langle([s_{4.46}, s_{4.74}], [0.49171, 0.67192], [0.12311, 0.24405], [0.35873, 0.46258])\rangle$
A_4	$\langle([s_{3.5}, s_{4.52}], [0.58432, 0.59132], [0.12483, 0.25508], [0.34039, 0.42755])\rangle$

Table 3 Aggregated values of the alternatives using INULWG operators

Alternative (A_ρ)	INULWG
A_1	$\langle([s_{4.05}, s_{4.80}], [0.38649, 0.55421], [0.16927, 0.30073], [0.31866, 0.46869])\rangle$
A_2	$\langle([s_{4.29}, s_{4.99}], [0.42078, 0.59737], [0.23057, 0.33095], [0.41399, 0.52592])\rangle$
A_3	$\langle([s_{4.36}, s_{4.70}], [0.45242, 0.61966], [0.13125, 0.25520], [0.38127, 0.48262])\rangle$
A_4	$\langle([s_{3.38}, s_{4.42}], [0.56818, 0.68126], [0.13329, 0.26159], [0.35126, 0.46021])\rangle$

Table 4 Aggregated values of the alternatives using INULDWA operators

Alternative (A_ρ)	INULDWA
A_1	$\langle([s_{4.16}, s_{4.84}], [0.4168, 0.5696], [0.1515, 0.2844], [0.2994, 0.4545])\rangle$
A_2	$\langle([s_{4.32}, s_{5.02}], [0.4407, 0.6333], [0.2055, 0.3141], [0.3896, 0.5017])\rangle$
A_3	$\langle([s_{4.46}, s_{4.74}], [0.5128, 0.7222], [0.1176, 0.2372], [0.3409, 0.4498])\rangle$
A_4	$\langle([s_{3.5}, s_{4.52}], [0.5918, 0.6970], [0.1190, 0.2500], [0.3315, 0.4525])\rangle$

If using INULWG operator instead of INULWA operator, then find the following results:

Step 1. we aggregate INUL information $\beta_{\rho\xi}$ for $\rho = 1, 2, 3, 4$; $\xi = 1, 2, 3, 4, 5$ by using INULWG operator to obtain overall values of β_ρ ($\rho = 1, 2, 3, 4$) for A_ρ which is given in Table 3:

Step 2. The score for A_ρ ($\rho = 1, 2, 3, 4$) is shown below by using INULWG operator given in Table 3. Then, $\uplus(\beta_1) = s_{3.724}$, $\uplus(\beta_2) = s_{3.519}$, $\uplus(\beta_3) = s_{4.126}$ and $\uplus(\beta_4) = s_{3.984}$

Step 3. The score values of A_ρ ($\rho = 1, 2, 3, 4$) ; we design the ranking order of the alternatives as follows: $A_3 \succ A_4 \succ A_1 \succ A_2$. Hence, ($A_3$) is still the best mutual fund among all the funds.

Thus, the scores of all alternatives for operators INULWA (INULWG) are different but ranking order for A_ρ, ($\rho = 1, 2, 3, 4$) is still same, and best choice for both the operators is the alternative A_3.

Case 2:

Step 1. We aggregate INUL information $\beta_{\rho\zeta}$ for $\rho - 1, 2, 3, 4$; $\xi = 1, 2, 3, 4, 5$ by using INULDWA operator to obtain the accumulated values of β_ξ for ($\rho = 1, 2, 3, 4$) for A_ρ which is given in Table 4:

Step 2. The score of A_ρ ($\rho = 1, 2, 3, 4$) is shown below by using accumulated values of the alternatives given in Table 4. Then, $\uplus(\beta_1) = s_{3.042}$, $\uplus(\beta_2) = s_{3.883}$, $\uplus(\beta_3) = s_{4.806}$ and $\uplus(\beta_4) = s_{4.282}$

Step 3. Computed values of $\uplus(\beta_\rho)$, we design the ranking order of the alternatives as follows: $A_3 \succ A_4 \succ A_1 \succ A_2$. A_3 is the best choice.

If we use INULDWG operator instead of INULDWA operator, then we find the following results.

Step 1. We aggregate INUL data $\beta_{\rho\xi}$ for $\rho = 1, 2, 3, 4$; $\xi = 1, 2, 3, 4, 5$ by using INULDWG operator to obtain accumulated values of β_ρ ($\rho = 1, 2, 3, 4$) for A_ρ which is follows in table below:

Table 5 Aggregated values of the alternatives using INULDWG operators

Alternative (A_ρ)	INULDWG
A_1	$\langle([s_{4.05}, s_{4.80}], [0.3759, 0.5506], [0.1705, 0.3035], [0.3240, 0.4737])\rangle$
A_2	$\langle([s_{4.29}, s_{4.99}], [0.4138, 0.5856], [0.2331, 0.3339], [0.4199, 0.5334])\rangle$
A_3	$\langle([s_{4.36}, s_{4.70}], [0.4339, 0.6065], [0.1325, 0.2586], [0.3882, 0.4908])\rangle$
A_4	$\langle([s_{3.38}, s_{4.42}], [0.5593, 0.6785], [0.1346, 0.2632], [0.3543, 0.4624])\rangle$

Table 6 Evaluations of decision-makers

	A_1	A_2
b_1	$\langle(s_4, [0.3, 0.5], [0.2, 0.3], [0.3, 0.5])\rangle$	$\langle(s_4, [0.4, 0.7], [0.2, 0.3], [0.5, 0.6])\rangle$
b_2	$\langle(s_5, [0.4, 0.6], [0.1, 0.2], [0.3, 0.4])\rangle$	$\langle(s_5, [0.5, 0.6], [0.3, 0.4], [0.3, 0.4])\rangle$
b_3	$\langle(s_3, [0.6, 0.7], [0.2, 0.3], [0.2, 0.4])\rangle$	$\langle(s_4, [0.6, 0.7], [0.1, 0.2], [0.4, 0.6])\rangle$
b_4	$\langle(s_6, [0.4, 0.5], [0.1, 0.3], [0.5, 0.6])\rangle$	$\langle(s_5, [0.3, 0.4], [0.2, 0.3], [0.3, 0.4])\rangle$
b_5	$\langle(s_3, [0.5, 0.6], [0.2, 0.4], [0.3, 0.4])\rangle$	$\langle(s_4, [0.4, 0.5], [0.3, 0.4], [0.4, 0.5])\rangle$
	A_3	A_4
	$\langle(s_5, [0.4, 0.5], [0.1, 0.2], [0.4, 0.5])\rangle$	$\langle(s_4, [0.6, 0.7], [0.1, 0.2], [0.4, 0.5])\rangle$
	$\langle(s_4, [0.3, 0.6], [0.2, 0.3], [0.5, 0.6])\rangle$	$\langle(s_2, [0.6, 0.7], [0.2, 0.3], [0.3, 0.4])\rangle$
	$\langle(s_4, [0.6, 0.8], [0.2, 0.4], [0.3, 0.4])\rangle$	$\langle(s_3, [0.6, 0.6], [0.1, 0.3], [0.2, 0.4])\rangle$
	$\langle(s_6, [0.7, 0.9], [0.1, 0.3], [0.4, 0.5])\rangle$	$\langle(s_4, [0.7, 0.8], [0.2, 0.3], [0.3, 0.4])\rangle$
	$\langle(s_3, [0.6, 0.7], [0.1, 0.2], [0.2, 0.3])\rangle$	$\langle(s_4, [0.4, 0.6], [0.1, 0.3], [0.4, 0.5])\rangle$

Step 2. The score of A_ρ ($\rho = 1, 2, 3, 4$) is shown below using INULDWG operator given in Table 5. Then, $\uplus(\beta_1) = s_{3.662}$, $\uplus(\beta_2) = s_{3.432}$, $\uplus(\beta_3) = s_{4.01}$ and $\uplus(\beta_4) = s_{3.945}$

Step 3. Based on the values of the score, the ranking order of the alternatives is as follows: $A_3 \succ A_4 \succ A_1 \succ A_2$. Hence, ($A_3$) is still the best choice.

From the above calculations, the score values of the two operators INULDWA (INULDWG) are different but the ranking order of the alternatives A_ρ, ($\rho = 1, 2, 3, 4$) is still same, and best choice for both the operators is the alternative A_3. Therefore, proposed approach is stable for the decision-making frame.

8 Comparative Study

In order to compare propose study with the present existing problems was introduced by Ye [10] in an interval neutrosophic linguistic environment. In this study, he has imposed interval neutrosophic linguistic weighted average (INLWA) operator and interval neutrosophic linguistic weighted geometric (INLWG) operator. In this comparison study, the evaluation matrix in the form of interval neutrosophic linguistic numbers composed from proposed decision matrix (Table 6).

Table 7 Aggregated values of the alternatives using INLWA operators

Alternative (A_ρ)	INLWA
A_1	$\langle(s_{4.16}, [0.4070,0.5635], [0.1602,0.2913], [0.3063,0.4690])\rangle$
A_2	$\langle(s_{4.32}, [0.4341,0.6214], [0.2177,0.3214], [0.3989,0.5100])\rangle$
A_3	$\langle(s_{4.46}, [0.4917,0.6719], [0.1231,0.2441], [0.3587,0.4626])\rangle$
A_4	$\langle(s_{3.5}, [0.5843,0.5913], [0.1248,0.2551], [0.3404,0.4276])\rangle$

Table 8 Aggregated values of the alternatives using INLWG operators

Alternative (A_ρ)	INLWG
A_1	$\langle(s_{4.05}, [0.3865,0.5542], [0.1693,0.3007], [0.3187,0.4687])\rangle$
A_2	$\langle(s_{4.29}, [0.4208,0.5974], [0.2306,0.3310], [0.4140,0.5259])\rangle$
A_3	$\langle(s_{4.36}, [0.4524,0.6197], [0.1313,0.2552], [0.3813,0.4826])\rangle$
A_4	$\langle(s_{3.38}, [0.5682,0.6813], [0.1333,0.2616], [0.3513,0.4602])\rangle$

Step 1. we aggregate INL information $\beta_{\rho\xi}$ for $i = 1, 2, 3, 4;$ $\xi = 1, 2, 3, 4, 5$ by using INLWA operator to obtain accumulated preference values β_ξ for $(\rho = 1, 2, 3, 4)$ representing A_ρ which is given in Table 7:

Step 2. The score of A_ρ $(\rho = 1, 2, 3, 4)$ is shown below by using aggregate of the alternatives given in Table 7. Then, $\uplus(\beta_1) = s_{1.876}$, $\uplus(\beta_2) = s_{1.736}$, $\uplus(\beta_3) = s_{2.202}$ and $\uplus(\beta_4) = s_{1.774}$

Step 3. Based on the values of the score function, we design the ranking order of the alternatives as follows: we obtain $A_3 \succ A_1 \succ A_4 \succ A_2$. Thus, A_3 is the favourable fund.

If we use INLWG operator instead of INLWA operator, then we find the following results.

Step 1. we aggregate INL information $\beta_{\rho\xi}$ for $\rho = 1, 2, 3, 4;$ $\xi = 1, 2, 3, 4, 5$ by using INLWG operator to obtain the overall preference values of β_ρ $(\rho = 1, 2, 3, 4)$ for the alternatives A_ρ which is given in the following table:

Step 2. The score of A_ρ $(\rho = 1, 2, 3, 4)$ is shown below by using INLWG operator given in Table 8. Then, $\uplus(\beta_1) = s_{1.704}$, $\uplus(\beta_2) - s_{1.627}$, $\uplus(\beta_3) = s_{1.998}$ and $\uplus(\beta_4) = s_{1.726}$

Step 3. Based on the values of the score function, we design the ranking order of the alternatives as follows: we obtain $A_3 \succ A_4 \succ A_1 \succ A_2$. Hence, (A_3) is still the best choice.

From the above calculations, the score values of INLWA (INLWG) operators are different but the ranking order of the alternatives A_ρ, $(\rho = 1, 2, 3, 4)$ is still same, and best choice for both the operators is the alternative A_3.

9 Conclusion

In this article, we proposed a technique for solving a MADM problem with INULNs. We introduced IN uncertain averaging (INULWA) operator and IN uncertain linguistic geometric (INULWG) operator. Further, we have studied IN uncertain linguistic Dombi averaging (INULDWA) operator and uncertain linguistic Dombi geometric (INULDWG) operator. We have established the properties of these proposed operators. Next, We develop an approach to solve a MADM problem using these operators. Lastly, an illustrative example for the evaluation of mutual fund for investment is given for the proposed approach. In a further study, the proposed model can be applied in decision support,cognitive measure and different linguistic research, and other domains containing uncertainties.

10 Complaince with Ethical Standards

Funding: There is no funding of this research.

Conflict of Interest: There is no conflict of interest between the authors and the institute where the work has been carried out.

Ethical Approval: The article does not contain any studies with human participants or animals performed by any of the authors.

Acknowledgments We would like to thank the anonymous reviewers for their insightful and constructive comments and suggestions that have been helpful for providing a better version of the present work.

References

1. Smarandache, F. (1999). *A unifying field in logics. Neutrosophy: Neutrosophic probability, set and logic*. Rehoboth: American Research Press.
2. Smarandache, F. (2005). Neutrosophic set—A generalization of the intuitionistic fuzzy set. *International Journal of Pure and Applied Mathematics, 24*(3), 287–297.
3. Zadeh, L. A. (1965). Fuzzy sets. *Information and Control, 8,* 338–353.
4. Atanassov, K. T. (1999). *On intuitionistic fuzzy sets theory. Studies in fuzziness and soft computing* (p. 283). Berlin Heidelberg: Springer.
5. Wang, H., Smarandache, F., Zhang, Y. Q., & Sunderraman, R. (2010). Single valued neutrosophic sets. *Multispace and Multistructure, 4,* 410–413.
6. Ye, J. (2014). Similarity measures between interval neutrosophic sets and their applications in multicriteria decision-making. *Journal of Intelligent Fuzzy Systems, 26,* 165–172.
7. Tan, R., Zhang, W., & Chen, S. (2017). Some generalized single-valued neutrosophic linguistic operators and their application to multiple Attribute Group Decision Making. *Journal of Systems Science and Information, 5*(2), 148–162.
8. Wu, Q., Wu, P., Zhou, L., Chen, H., & Guan, X. (2018). Some new Hamacher aggregation operators under single-valued neutrosophic 2-tuple linguistic environment and their applications to multiattribute group decision making. *Computers and Industrial Engineering, 116,* 144–162.

9. Ye, J. (2016). Aggregation operators of neutrosophic linguistic numbers for multiple attribute group decision making. *Springerplus, 5*, 1–11.

10. Ye, J. (2014). Some aggregation operators of interval neutrosophic linguistic numbers for multiple attribute decision making. *Journal of Intelligent Fuzzy Systems, 27*, 2231–2241.

11. Yager, R. R. (1998). On ordered weighted averaging aggregation operators in multicriteria decision making. *IEEE Transactions on Systems, Man, and Cybernetics, 18*(1), 183–190.

12. Yager, R. R., & Kacprzyk, J. (1997). *The ordered weighted averaging operators: Theory and applications*. Boston, MA: Kluwer.

13. Beliakov, G., Pradera, A., & Calvo, T. (2007). *Aggregation functions: A guide for practitioners*. Heidelberg, Berlin, New York: Springer.

14. Deschrijver, G., Cornelis, C., & Kerre, E. E. (2004). On the representation of intuitionistic fuzzy *t*-norms and *t*-conorms. *IEEE Transactions on Fuzzy Systems, 12*, 45–61.

15. Xu, Z. S. (2007). Intuitionistic fuzzy aggregation operators. *IEEE Transactions on Fuzzy Systems, 15*(6), 1179–1187.

16. Abdel-Basset, M., Mohamed, R., Elhoseny, M., & Chang, V. (2020). Evaluation framework for smart disaster response systems in uncertainty environment. *Mechanical Systems and Signal Processing, 145*, 106941.

17. Jana, C., Pal, M., & Wang, J. Q. (2019). Bipolar fuzzy Dombi aggregation operators and its application in multiple attribute decision making process. *Journal of Ambient Intelligence and Humanized Computing, 10*, 3533–3549.

18. Jana, C., Senapati, T., Pal, M., & Yager, R. R. (2019). Picture fuzzy Dombi aggregation operators: Application to MADM process. *Applied Soft Computing, 74*(1), 99–109.

19. Jana, C., & Pal, M. (2019). Assessment of enterprise performance based on picture fuzzy Hamacher aggregation operators. *Symmetry, 11*(1), 75.

20. Jana, C., & Pal, M. (2019). A robust single-valued neutrosophic soft aggregation operators in multi-criteria decision making. *Symmetry, 11*(1), 110.

21. Jana, C., Senapati, T., & Pal, M. (2019). Pythagorean fuzzy Dombi aggregation operators and its applications in multiple attribut decision-making. *International Journal of Intelligence Systems, 34*(9), 2019–2038.

22. Jana, C., Pal, M., & Wang, J. Q. (2020). Bipolar fuzzy Dombi prioritized aggregation operators in multiple attribute decision making. *Soft Computing, 24*, 3631–3646.

23. Rani, D., & Garg, H. (2018). Complex intuitionistic fuzzy power aggregation operators and their applications in multicriteria decision-making. *Expert Systems, 35*, e12325. https://doi.org/10.1111/exsy.12325.

24. Wei, G. W., & Zhang, Z. (2019). Some single-valued neutrosophic Bonferroni power aggregation operators in multiple attribute decision making. *Journal of Ambient Intelligence and Humanized Computing, 10*, 863–882.

25. Zhang, H. Y., Wang, J. Q., & Chen, X. H. (2014). Interval neutrosophic sets and their application in multi criteria decision making problems. *The Scientific World Journal, 2014*, 645953, 15 pages.

26. Abdel-Basset, M., Ali, M., & Atef, A. (2020). Uncertainty assessments of linear time-cost tradeoffs using neutrosophic set. *Computers and Industrial Engineering, 141*, 106286.

27. Abdel-Basset, M., Ali, M., & Atef, A. (2020). Resource levelling problem in construction projects under neutrosophic environment. *The Journal of Supercomputing, 76*(2), 964–988.

28. Abdel-Basset, M., Gamal, A., Son, L. H., & Smarandache, F. (2020). A bipolar neutrosophic multi criteria decision making framework for professional selection. *Applied Sciences, 10*(4), 1202.

29. Abdel-Basset, M., Mohamed, R., Zaied, A. E. N. H., Gamal, A., & Smarandache, F. (2020). Solving the supply chain problem using the best-worst method based on a novel Plithogenic model. In F. Smarandache & M. Abdel-Basset (Eds.), *Optimization theory based on neutrosophic and plithogenic sets* (pp. 1–19). London: Academic Press.

30. Ye, J. (2014). A multicriteria decision-making method using aggregation operators for simplified neutrosophic sets. *Journal of Intelligent Fuzzy Systems, 26*, 2459–2466.

31. Peng, J., Wang, J. Q., & Chen, H. (2014). Simplified neutrosophic sets and their applications in multi-citeria group decision making problems. *International Journal of Systems Science, 47*(10), 2342–2358.

32. Bausys, R., & Zavadskas, E. K. (2015). Multicriteria decision making approach by VIKOR under interval neutrosophic set environment. *Economic Computation and Economic Cybernetics Studies and Research, 4,* 33–48.

33. Jana, C., Muhiuddin, G., & Pal, M. (2019). Multiple-attribute decision making problems based on SVTNH methods. *Journal of Ambient Intelligence and Humanized Computing, 11,* 3117–3733. https://doi.org/10.1007/s12652-019-01568-9.

34. Broumi, S., & Smarandache, F. (2014). Single valued neutrosophic trapezoid linguistic aggregation operators based multi-attribute decision making. *Bulletin of Pure & Applied Sciences, 33*(2), 135–155.

35. Ji, P., Wang, J. Q., & Zhang, H. Y. (2018). Frank prioritized Bonferroni mean operator with single-valued neutrosophic sets and its application in selecting third-party logistics providers. *Neural Computing and Applications, 30*(3), 799–823.

36. Zhang, H. Y., Ji, P., Wang, J. Q., & Chen, X. H. (2016). A neutrosophic normal cloud and its application in decision-making. *Cognitive Computation, 8*(4), 649–669.

37. Sahin, R., & Liu, P. (2017). Correlation coefficient of single-valued neutrosophic hesitant fuzzy sets and its applications in decision making. *Neural Computing and Applications, 28*(6), 1387–1395.

38. Nancy, & Garg, H. (2016). Novel single-valued neutrosophic aggregated operators under frank norm operation and its application to decision-making process. *International Journal for Uncertainty Quantifcation, 6*(4), 361–375.

39. Biswas, P., Pramanik, S., & Giri, B. C. (2014). A new methodology for neutrosophic multi-attribute decisionmaking with unknown weight information. *Neutrosophic Sets and Systems, 3,* 42–50.

40. Biswas, P., Pramanik, S., & Giri, B. C. (2016). TOPSIS method for multi-attribute group decision-making under single valued neutrosophic environment. *Neural Computing and Applications, 27*(3), 727–737.

41. Liu, P., Zhang, L., Liu, X., & Wang, P. (2016). Multi-valued neutrosophic number Bonferroni mean operators with their applications in multiple attribute group decision making. *International Journal of Information Technology and Decision Making, 15,* 1–28.

42. Jana, C., Pal, M., Karaaslan, F., & Wang, J. Q. (2018). Trapezoidal neutrosophic aggregation operators and its application in multiple attribute decision-making process. *Scientia Iranica E, 27,* 1655–1673. https://doi.org/10.24200/sci.2018.51136.2024.

43. Lu, Z., & Ye, J. (2014). Single-valued neutrosophic hybrid arithmetic and geometric aggregation operators and their decision-making method. *Information, 8,* 84.

44. Broumi, S., Bakali, A., Talea, M., Smarandache, F., Krishnan Kishore, K. P., & Sahin, R. (2018). Shortest path problem under interval valued neutrosophic setting. *Journal of Fundamental and Applied Sciences, 10*(4S), 168–174.

45. Abdel-Basset, M., Mohamed, M., Zhou, Y., & Hezam, I. (2017). Multi-criteria group decision making based on neutrosophic analytic hierarchy process. *Journal of Intelligent Fuzzy Systems, 33*(6), 4055–4066.

46. Abdel-Basset, M., Mohamed, M., & Smarandache, F. (2018). An extension of neutrosophic AHP-SWOT analysis for strategic planning and decision-making. *Symmetry, 10*(4), 116.

47. Pramanik, S., Dalapati, S., Alam, S., Smarandache, F., & Roy, T. K. (2018). NS-cross entropy based MAGDM under single valued neutrosophic set environment. *Information, 9*(2), 37.

48. Ye, J. (2013). Multicriteria decision-making method using the correlation coefficient under single-valued neutrosophic environment. *International Journal of General Systems, 42*(4), 386–394.

49. Zadeh, L. A. (1975–1976). The concept of a linguistic variable and its application to approximate reasoning. Part 1, 2 and 3, *Information Sciences, 8,* 199–249, 301–357; 9, 43–80.

50. Herrera, F., Herrera-Viedma, E., & Verdegay, J. L. (1996). A model of consensus in group decision making under linguistic assessments. *Fuzzy Sets and Systems, 78*, 73–87.
51. Xu, Z. S. (2006). Goal programming models for multiple attribute decision making under linguistic setting. *Journal of Management Sciences in China, 9*(2), 9–17.
52. Wang, J. Q., & Li, H. B. (2010). Multi-criteria decision making based on aggregation operators for intuitionistic linguistic fuzzy numbers. *Control and Decision, 25*(10), 1571–1574.
53. Xu, Z. S. (2004). *Uncertain multiple attribute decision making: Methods and applications.* Beijing: Tsinghua, University Press.
54. Xu, Z. S. (2006). An approach based on the uncertain LOWG and induced uncertain LOWG operators to group decision making with uncertain multiplicative linguistic preference relations. *Decision Support Systems, 41*, 488–499.
55. Xu, Z. S. (2006). Induced uncertain linguistic OWA operators applied to group decision making. *Information Fusion, 7*, 231–238.
56. Xu, Z. S. (2004). Uncertain linguistic aggregation operators based approach to multiple attribute group decision making under uncertain linguistic environment. *Information Sciences, 168*, 171–184.
57. Liu, P., & Jin, F. (2012). Methods for aggregating intuitionistic uncertain linguistic variables and their application to group decision making. *Information Sciences, 205*, 58–71.
58. Meng, F., Chen, X., & Zhang, Q. (2014). Some interval-valued intuitionistic uncertain linguistic Choquet operators and their application to multi-attribute group decision making. *Applied Mathematical Modelling, 38*, 2543–2557.
59. Wang, H., Smarandache, F., Zhang, Y. Q., & Sunderraman, R. (2005). *Interval neutrosophic sets and logic: Theory and applications in computing.* Phoenix, AZ: Hexis.
60. Smarandache, F. (1998). *Neutrosophy. neutrosophic probability, set, and logic* (p. 105). Ann Arbor, Michigan: ProQuest Information and Learning.
61. Herrera, F., Herrera-Viedma, E., & Martinez, L. (2000). A fusion approach for managing multigranularity linguistic term sets in decision making. *Fuzzy Sets and Systems, 114*, 43–58.
62. Kacprzyk, J. (1986). Group decision making with a fuzzy linguistic majority. *Fuzzy Sets and Systems, 18*, 105–118.
63. Liu, P., Liu, Z., & Zhang, X. (2014). Some intuitionistic uncertain linguistic Heronian mean operators and their application to group decision making. *Applied Mathematics and Computation, 230*, 570–586.
64. Zadeh, L. A. (1983). A computational approach to fuzzy quantifiers in natural languages. *Computers & Mathematcs with Applications, 9*, 149–184.
65. Xu, Z. S., & Da, Q. L. (2002). The uncertain OWA operator. *International Journal of Intelligent Systems, 17*(6), 569–575.

Neutrosophic Probabilistic Expert System for Decision-Making Support in Supply Chain Risk Management

Rafael Rojas-Gualdron, Florentin Smarandache, and Carlos Diaz-Bohorquez

1 Introduction

We can affirm that risk is a factor that must be taken into account in practically all aspects of our daily life; however, when it comes to business, these are more vulnerable to its adverse effects due to changing market trends, globalization, and the complexity and competitiveness of companies [1]. Changes in demand, uncertain supply, cost savings, and the implementation of lean production increase the probability of the appearance of risks [2]. Due to the existence of these risks, the company's supply chain (SC) is exposed to numerous interruptions, which, if not addressed in a timely manner, affect the company's performance [3]. For this reason, caused in part by the economic uncertainty created by the 2008 global financial crisis, supply chain risk management at local, national, or global levels has increasingly attracted the attention of researchers and professionals in recent years [4]. Supply chain risk management (SCRM), which has emerged as an area of knowledge since the early 2000s, has now become more than just the overlapping of directly related areas, such as business risk management and supply chain management [5]. Baryannis et al. [6] affirm that SCRM "encompasses the collaborative and coordinating efforts of all parties involved in a supply chain to identify, assess, mitigate and monitor risks with the aim of reducing vulnerability and increasing the strength and resilience of the supply chain, ensuring profitability and continuity." The large number of actions and decisions to be taken in SCRM

R. Rojas-Gualdron (✉) · C. Diaz-Bohorquez
Universidad Industrial de Santander, Bucaramanga, Colombia
e-mail: rafael2188797@correo.uis.edu.co; cediazbo@uis.edu.co

F. Smarandache
Department of Mathematics and Science, University of New Mexico, Gallup, NM, USA
e-mail: smarand@unm.edu

© The Author(s), under exclusive license to Springer Nature Switzerland AG 2021
F. Smarandache, M. Abdel-Basset (eds.), *Neutrosophic Operational Research*,
https://doi.org/10.1007/978-3-030-57197-9_17

has led to a large number of solutions proposed by researchers, which can be broadly classified into three categories: multiple criteria decision analysis techniques (MCDM), mathematical modeling and optimization, and, finally, artificial intelligence techniques. This research seeks to take aspects of each of these three categories to propose a new alternative solution that supports the decision-making process in SCRM. This paper is organized as follows: Sect. 2 includes a short conceptualization about topics that will be treated during the investigation, Sect. 3 explains the methodology proposed in this investigation, Sect. 4 shows the results of the simulation, and finally Sect. 5 presents the conclusions of this work.

2 Conceptualization

2.1 SCRM and the SCOR Model

The SCOR model (supply chain operations reference model) was proposed by the Council of Supply Chain Management Professionals [7] to link business processes, best practices, performance metrics, people, and technology in a unified structure. The SCOR model allows professionals to identify the characteristics that contribute to customer satisfaction by describing how processes interact throughout a supply chain and how relationships are established from a supplier's supplier to a customer's customer [8]. It has been recognized worldwide and widely applied in the industry, as it is a reference model that allows companies to communicate using the same terminology within and between organizations [9]. The SCOR model is an excellent tool for supporting tactical and operational management for decision-making in strategic planning [10]. Table 1 shows the five attributes suggested by the SCOR model for managing supply chain performance and that should be adopted for better strategic direction. Reliability, responsiveness, and agility are customer-centric and address the effectiveness of the company's external processes. The

Table 1 Performance attributes suggested by the SCOR model

Attribute	Description
Reliability	"The ability to perform tasks as expected. Reliability focuses on the predictability of the outcome of a process."
Responsiveness	"The speed at which tasks are performed. The speed at which a supply chain provides products to the customer."
Agility	"The ability to respond to external influences, the ability to respond to marketplace changes to gain or maintain competitive advantage."
Assets	"The ability to efficiently utilize assets. Asset management strategies in a supply chain include inventory reduction and in-sourcing vs. outsourcing."
Costs	"The costs of operating the supply chain processes. This includes labor costs, material costs, management, and transportation costs."

Source: Council of Supply Chain Management Professionals [7]

attributes of assets and costs are focused on the internal processes of the company and measure its efficiency [11].

2.2   Neutrosophic AHP-TOPSIS Technique

In the early 1980s, a multiple criteria decision-making technique became widespread known as the analytic hierarchy process or AHP [12], which became an indispensable method for managers to make decisions on a broad domain of complex problems involving technology choice, product design and planning, forecasting, and risk and opportunity modeling through decomposition of a complex subproblem problem to simplify it, using pairwise comparison judgments to calculate the weight of each criterion in the problem. Recently, Abdel-Basset et al. [13], proposed to integrate the AHP method proposed by Saaty with the neutrosophical set theory proposed by Smarandache [14]. In neutrosophical AHP, a neutrosophic ratio scale is used, which can be seen in Table 2, to indicate the relative importance of the factors of the corresponding criteria. Neutrosophical numbers are used to indicate the relative importance of criteria and alternatives, and a scoring function is later used to convert neutrosophical numbers to crisp numbers.

In the same way, Abdel-Basset et al. [13] integrate the technique for order preference by similarity to ideal solution (TOPSIS) with the theory of neutrosophic sets [14]. TOPSIS was developed to solve multiple criteria decision-making problems (MCDM) by Yoon and Hwang [15] where the selected alternative is the one with the shortest distance to the ideal positive solution and the longest distance to the ideal negative solution. TOPSIS was applied in various applications, such as supplier selection, location selection, carrier alternative classification, and company evaluation. In neutrosophical TOPSIS, the values are neutrosophical values, and the process includes neutrosophical mathematical techniques.

Table 2 The triangular neutrosophic scale of neutrosophic AHP

Saaty scale	Explanation	Neutrosophic triangular scale
1	Equally significant	$\tilde{1} = \langle (1, 1, 1) \rangle$
3	Slightly significant	$\tilde{3} = \langle (2, 3, 4) \rangle$
5	Strongly significant	$\tilde{5} = \langle (4, 5, 6) \rangle$
7	Very strongly significant	$\tilde{7} = \langle (6, 7, 8) \rangle$
9	Absolutely significant	$\tilde{9} = \langle (9, 9, 9) \rangle$
2	Sporadic values among two close scales	$\tilde{2} = \langle (1, 2, 3) \rangle$
4		$\tilde{4} = \langle (3, 4, 5) \rangle$
6		$\tilde{6} = \langle (5, 6, 7) \rangle$
8		$\tilde{8} = \langle (7, 8, 9) \rangle$

Source: Abdel-Basset et al. [13]

2.3 Probabilistic Expert Systems

Because deterministic rule-based expert systems do not address uncertainty and because in most practical applications uncertainty is present, the emergence of expert systems to address uncertain situations is necessary [16]. We can use medical diagnosis as an example, in which the doctor must make a diagnosis based on facts that are not necessarily accurate (subjectivity, imprecision, lack of information, errors, missing data, etc.) and relationships between symptoms and diseases that can lead to the same set of symptoms to more than one possible disease. In this case, the diagnosis does not seek to know in first instance exactly what disease the patient suffers from, but rather what would be the most probable. A probabilistic expert system (PES) is typically specified as a causal probabilistic network that relates the properties of conditional independence between variables to a directed graph with the set of vertices and is an efficient method of specifying and managing the joint probability distribution of a finite collection of random variables [17] which are compiled to produce a new structure known as the junction tree [18] which makes them susceptible to the application of simple algorithms [19].

3 Proposed Methodology

The methodology proposed in this research is divided into eight stages. In the first stage, a selection of the risks that may affect the supply chain is made. In the second stage, strategies are selected for managing these risks. In the third stage, performance indicators are selected to serve as a metric for quantifying risks. In the third stage, through the opinion of decision makers and the use of the neutrosophic AHP technique, we will give a weight to each attribute according to the importance of each of the performance metrics of the supply chain. In the fourth stage, the selection of the experts from whom we are going to obtain the information is made. The fifth stage allows us to use the concept of decision makers and a mathematical function to assign the probability of occurrence of each risk under normal operating conditions. In the sixth stage, again using the experts' concept and the neutrosophical TOPSIS technique, we will assign a weight to each risk on the supply chain, taking into account the impact on each of the performance metrics with their respective weight obtained in stage four. We then multiply this weight by the probability of occurrence obtained in stage five to obtain the quantified value of the risk. In the seventh stage, we again apply the neutrosophical TOPSIS technique to give a performance score to each strategy according to the effectiveness to manage each of the risks and the weights previously obtained, and we select the strategies with the highest performance score to suggest their application in SCRM. Finally, in stage eight, we proceed to create an algorithm that stores the information extracted from the opinion of the experts in a knowledge base that will feed our expert system.

Table 3 Risks in supply chain management

Risk code	Risk
R1	Lack of knowledge and experience of the project team
R2	Lack of clear goals
R3	Lack of coordination and cooperation among the members of the organization
R4	Delays in the timely supply of raw materials
R5	Transportation inappropriate

Table 4 Resilient supply chain strategies for risk mitigation

Code	Strategy	Relevant literature
S1	Fine-tune supply chain design	Chiang, Chhajed and Hess [22] and Qi, Shen and Snyder [23]
S2	Supply base strengthening	Sturgeon and Lester [24] and Nam, Vitton and Kurata [25]
S3	Inventory flexibility	Kemahliouglu-Ziya and Bartholdi III [26] and Sarker [27]
S4	Logistics flexibility	Atasoy [28] and Bruzzone and Longo [29]
S5	Process standardization	Ma, Wang and Xu [30] and Costinot, Vogel and Wang [31]

3.1   Selection of Risks That May Affect the Supply Chain

The methodology proposes to carry out a survey among the experts to know first-hand which are the risks that they consider may affect their supply chain. For the present paper and for illustrative purposes, five risks are selected from those proposed by Parsa and Torfi [20] which can be seen in Table 3.

3.2   Selection of Strategies for Supply Chain Risk Management

The methodology proposes to carry out a survey among the experts to know first-hand which are the strategies that they consider can be used to manage the risks of their supply chain. For the present paper, for illustrative purposes, five strategies are selected from those proposed by Rajesh [21], which can be found in Table 4.

3.3   Selection of Supply Chain Performance Metrics

The methodology of this research proposes the use of the metrics suggested by the SCOR model. Depending on the detail with which you want to measure the key performance indicators, level 1 metrics; level 1 and 2 metrics; or level 1, 2, and 3 metrics will be used. For the scope of this paper for illustrative purposes, the attributes of reliability, responsiveness, and costs will be used as metrics.

3.4   Selection of Experts

The methodology of this research proposes to inquire experts and decision makers about the information that will be used to estimate the probability of occurrence and the expected impact of each of the risks, as well as the effectiveness of each of the strategies used for its management. For the scope of this paper and in an illustrative way to check the results of the expert system, the authors will be the ones to answer the questions simulating the answers of three experts with different approaches: customer satisfaction, cost reduction, and increased profitability, which reflect the position of some decision makers from different areas within the same supply chain.

3.5   Estimation of Weights Using the Neutrosophical AHP Technique

The first step is to assign a weight to each of the three attributes selected from the SCOR model according to the importance that each of the decision makers gives to each of them within the supply chain. For this we are going to use the neutrosophic AHP technique proposed by Abdel-Basset et al. [13] using the scale proposed in Table 2. Each decision maker will be consulted at the end of the interviews, according to their knowledge, experience, and the topics consulted, what probability do they think they have given correct answers, what probability he sees that he was wrong, and what degree of indeterminacy, either due to lack of knowledge, ambiguity of the issues, or any other reason, he considers his answers to have. These values will be in the range between 0 and 1 and will be assigned as values of truth (T), falsehood (F), and indeterminacy (I). The values obtained from the pairwise comparison are shown in Table 5.

Using the score function proposed by the neutrosophical AHP technique [13], which can be seen in Eqs. (1) and (2), we convert the values to crisp numbers:

$$S\left(a_{ij}\right) = \frac{La_i + Ma_{ij} + Ua_{ij}}{3} + \left(Ta_{ij} - Ia_{ij} - Fa_{ij}\right) \tag{1}$$

$$S\left(a_{ji}\right) = \frac{1}{\left(\frac{La_i + Ma_{ij} + Ua_{ij}}{3} + \left(Ta_{ij} - Ia_{ij} - Fa_{ij}\right)\right)} \tag{2}$$

Once the values have been converted to crisp numbers, we proceed to normalize the decision matrix using Eq. (3):

Table 5 Results from the pairwise comparison

	Expert	Reliability	Responsiveness	Costs
Reliability	1	[(1,1);0.60, 0.40, 0.40]	[(2,3,4);0.60, 0.40, 0.40]	[(6,7,8);0.60, 0.40, 0.40]
	2	[(1,1);0.65, 0.40, 0.30]	[(1,2,3);0.65, 0.40, 0.30]	
	3	[(1,1);0.70, 0.25, 0.30]		
Responsiveness	1		[(1,1);0.60, 0.40, 0.40]	[(3,4,5);0.60, 0.40, 0.40]
	2		[(1,1);0.65, 0.40, 0.30]	
	3	[(2,3,4);0.70, 0.25, 0.30]	[(1,1);0.70, 0.25, 0.30]	
Costs	1		[(4,5,6);0.65, 0.40, 0.30]	[(1,1);0.60, 0.40, 0.40]
	2	[(3,4,5);0.65, 0.40, 0.30]		[(1,1);0.65, 0.40, 0.30]
	3	[(4,5,6);0.70, 0.25, 0.30]	[(2,3,4);0.70, 0.25, 0.30]	[(1,1);0.70, 0.25, 0.30]

$$\hat{S}_{ij} = \frac{S_{ij}}{\sum S_{.j}} \tag{3}$$

where $\sum S_j$ is the sum of all the values in column j. The next step is to find the value of the weights of each attribute using Eq. (4):

$$w_i = \frac{\sum \hat{S}_{i.}}{3} \tag{4}$$

where $\sum \hat{S}_{i.}$ is the sum of each element in row i in the normalized decision matrix. Now we perform the procedure described for each of the three decision makers. The de-neutrosophicated values, the decision matrices, and the weights of each attribute can be seen in Tables 6, 7, and 8.

The next step is to verify the consistency of the information obtained, using the consistency ratio for this. To calculate it, we must multiply the de-neutralized values by the weight corresponding to each attribute to obtain the weighted matrix, then add the rows and divide this value by the weight of the attribute. The obtained values are averaged to obtain λ_{max} of each expert as observed in Table 9.

Now we use Eq. (5) to calculate the consistency ratio (CR), which is the ratio between the consistency index (CI) and the random index (RI):

$$CR = \frac{CI}{RI}, \text{ where } CI = \frac{\lambda_{max} - n}{n - 1} \tag{5}$$

where n is the number of attributes, in this case $n = 3$, λ_{max} we find it in Table 9, the RI value for $n = 3$ is 0.58 and for each of the decision matrices to be considered consistent, the consistency ratio (CR) for 3 by 3 matrices must be less than 0.05. As we can see in Table 10, the three decision matrices are consistent since they have CR values less than 0.05, so the calculated weights are acceptable.

In Table 11 we can see the weights obtained by each expert for each of the attributes that we will use as metrics and we calculate the average weight for each attribute, which we will use later in this paper.

3.6 Estimation of the Probability of Occurrence of Each Risk

This methodology proposes to obtain from the experts the estimated probability values of each risk and then, using the neutrosophical theory [14], include an indeterminacy value according to the information collected. For this we propose the use of a neutrosophical triangle of indeterminacy such as the one shown in Fig. 1, in which we affirm that the further away a degree of truth is from the absolute values, 0 and 1, the greater is the presence of indeterminacy in their membership function.

Table 6 De-neutrosophicated values, normalized decision matrix, and attribute's weights for expert 1

De-neutrosophicated values				Normalized values				
	Reliability	Responsiveness	Costs		Reliability	Responsiveness	Costs	Weight
Reliability	1	3	7	Reliability	0.677	0.706	0.583	**0.656**
Responsiveness	0.333	1	4	Responsiveness	0.226	0.235	0.333	**0.265**
Costs	0.143	0.25	1	Costs	0.097	0.059	0.083	**0.080**

Table 7 De-neutrosophicated values, normalized decision matrix, and attribute's weights for expert 2

De-neutrosophicated values				Normalized values				
	Reliability	Responsiveness	Costs		Reliability	Responsiveness	Costs	Weight
Reliability	1	2	0.250	Reliability	0.182	0.250	0.172	**0.201**
Responsiveness	0.500	1	0.2	Responsiveness	0.091	0.125	0.138	**0.118**
Costs	4	5	1	Costs	0.727	0.625	0.690	**0.681**

Table 8 De-neutrosophicated values, normalized decision matrix, and attribute's weights for expert 3

De-neutrosophicated values				Normalized values				
	Reliability	Responsiveness	Costs		Reliability	Responsiveness	Costs	Weight
Reliability	1	0.333	0.2	Reliability	0.111	0.077	0.130	**0.106**
Responsiveness	3	1	0.333	Responsiveness	0.333	0.231	0.217	**0.260**
Costs	5	3	1	Costs	0.556	0.692	0.652	**0.633**

Table 9 Weighted decision matrix and λ_{max}

	Expert 1					Expert 2					Expert 3				
	Rel	Res	Cos	Sum		Rel	Res	Cos	Sum		Rel	Res	Cos	Sum	
Rel	0.66	0.79	0.56	2.01	3.06	0.20	0.24	0.17	0.61	3.02	0.11	0.09	0.13	0.32	3.01
Res	0.22	0.26	0.32	0.80	3.03	0.10	0.12	0.14	0.35	3.01	0.32	0.26	0.21	0.79	3.03
Cos	0.09	0.07	0.08	0.24	3.01	0.81	0.59	0.68	2.08	3.05	0.53	0.78	0.63	1.95	3.07
λ_{max}					3.03					3.02					3.04

Table 10 Consistency assessment of decision matrix for the three experts

	λ_{max}	CI	RI	CR	Consistent
Expert 1	3.0325	0.0163	0.58	0.0280	Yes
Expert 2	3.0247	0.0124	0.58	0.0213	Yes
Expert 3	3.0387	0.0194	0.58	0.0334	Yes

Table 11 Attribute's mean weights

	Expert 1	Expert 2	Expert 3	Mean weight
Reliability	0.656	0.201	0.106	0.321
Responsiveness	0.265	0.118	0.260	0.214
Costs	0.080	0.681	0.633	0.465

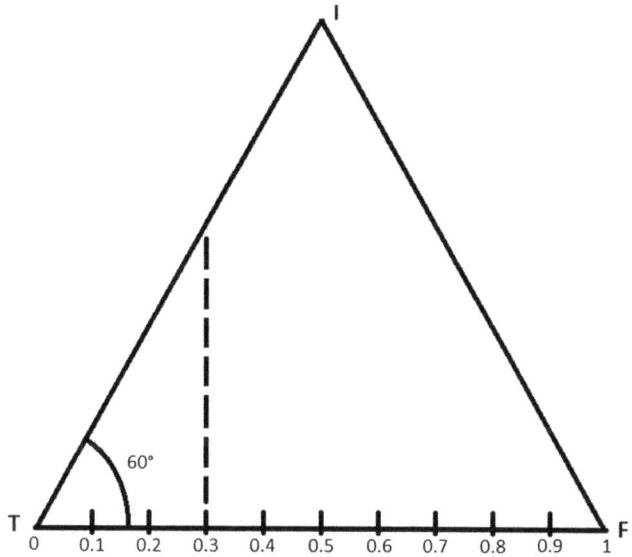

Fig. 1 Neutrosophical indeterminacy triangle

Given a probability of occurrence of a risk T_i, we can calculate the degrees of falsehood F_i and indeterminacy I_i with Eqs. (6), (7), and (8):

$$F_i = 1 - T_i \tag{6}$$

$$I_i = \frac{T_i}{\cos 60^\circ} \text{ for } T_i \leq 0.5 \tag{7}$$

$$I_i = \frac{1 - T_i}{\cos 60^\circ} \text{ for } T_i > 0.5 \tag{8}$$

Table 12 Normalized neutrosophic values for probability of occurrence of each risk

	Expert 1				Expert 2				Expert 3			
	Answer	Neut. number			Answer	Neut. number			Answer	Neut. number		
Risk	P	T	I	F	P	T	I	F	P	T	I	F
R1	0.15	0.12	0.23	0.65	0.20	0.14	0.29	0.57	0.10	0.08	0.17	0.75
R2	0.20	0.14	0.29	0.57	0.15	0.12	0.23	0.65	0.10	0.08	0.17	0.75
R3	0.25	0.17	0.33	0.50	0.20	0.14	0.29	0.57	0.15	0.12	0.23	0.65
R4	0.15	0.12	0.23	0.65	0.15	0.12	0.23	0.65	0.10	0.08	0.17	0.75
R5	0.10	0.08	0.17	0.75	0.15	0.12	0.23	0.65	0.10	0.08	0.17	0.75

Table 13 Normalized neutrosophic mean and crisp values for risk probability of occurrence

	Neutrosophic mean			Estimated probability
Risk	T	I	F	
R1	0.11	0.23	0.66	0.22
R2	0.11	0.23	0.66	0.22
R3	0.14	0.28	0.58	0.27
R4	0.10	0.21	0.69	0.20
R5	0.09	0.19	0.72	0.18

Once the values of T_i, I_i, F_i have been calculated, we normalize them and then calculate the neutrosophic mean [32] for each risk. Table 12 shows the values obtained from the experts of the probability of occurrence of each risk and its normalized neutrosophic equivalent. Now to de-neutrosophicate the risk probability obtained by the neutrosophical mean, we propose that a percentage of the value of indeterminacy has a probability k of becoming true. For our case we are going to use a value of $k = 0.46$ and we will convert the probability of occurrence of each risk to crisp numbers using Eq. (9). Table 13 shows the neutrosophical mean and the estimated probability as a crisp value.

$$P_i = T_i + k \cdot I_i \qquad (9)$$

3.7 Neutrosophic TOPSIS for the Estimation of Expected Impact for Each Risk

This methodology proposes to request the opinion of decision makers on the expected impact, in linguistic terms, of a risk on each of the metrics we selected in Table 2. For this, we will make use of the linguistic values conversion table on a triangular neutrosophical scale proposed by [33], which can be seen in Table 14. The impact values of each of the risks in Table 3 on the attributes of reliability (Rel), responsiveness (Res), and costs (Cos) obtained from the experts can be seen in the Table 15.

Then, we calculate the neutrosophical mean of the values obtained from the experts using Eq. (10), and the result can be seen in Table 16.

$$ll\overline{R_{ij}} = \left[\frac{(a1 + b1 + c1, a2 + b2 + c2, a3 + b3 + c3)}{3} \right.;$$

$$\left. \times \alpha_a \cap \alpha_b \cap \alpha_c, \theta_a \cup \theta_b \cup \theta_c, \beta_a \cup \beta_b \cup \beta_c \right] \tag{10}$$

The next step is to convert the values in Table 16 to crisp numbers using Eq. (11). Then, the decision matrix must be normalized using Eq. (12). The result can be seen in Table 17:

$$\text{Impact}_{ij} = \frac{1}{8} (a + b + c) * (2 + \alpha - \theta - \beta) \tag{11}$$

$$r_{ij} = \frac{x_{ij}}{\sqrt{\sum x_{.j}^2}} \text{ where } x_{.j} \text{ are all elements in column } j \tag{12}$$

The next step is to multiply the normalized decision matrix by the respective weight of each attribute calculated with the neutrosophic AHP technique. Then we must select the ideal best (V_j^+) and the ideal worst (V_j^-) for each attribute using Eqs. (13) and (14):

Table 14 Triangular neutrosophic scale for linguistic variables

Importance linguistic variable	Triangular neutrosophic scale	Rating linguistic variable
Very weakly important (VWI)	((0.10, 0.30, 0.35), 0.1, 0.2, 0.15)	Nothing (N)
Weakly important (WI)	((0.15, 0.25, 0.10), 0.6, 0.2, 0.3)	Very low (VL)
Partially important (PI)	((0.40, 0.35, 0.50), 0.6, 0.1, 0.2)	Low (L)
Equal important (EI)	((0.65, 0.60, 0.70), 0.8, 0.1, 0.1)	Medium (M)
Strong important (SI)	((0.70, 0.65, 0.80), 0.9, 0.2, 0.1)	High (H)
Very strongly important (VSI)	((0.90, 0.85, 0.90), 0.7, 0.2, 0.2)	Very high (VH)
Absolutely important (AI)	((0.95, 0.90, 0.95), 0.9, 0.1, 0.1)	Absolute (A)

Source: Abdel-Basset and Mohamed [33]

Table 15 Expert responses on the impact of a risk on each attribute

	Expert 1			Expert 2			Expert 3		
Risk	Rel	Res	Cos	Rel	Res	Cos	Rel	Res	Cos
R1	H	H	H	M	M	H	H	H	H
R2	H	M	L	M	M	M	H	M	M
R3	H	VH	M	H	H	M	VH	VH	H
R4	VH	VH	M	H	H	VH	H	VH	H
R5	VH	H	H	H	H	VH	H	VH	H

Table 16 Neutrosophic mean for impact of each risk in each attribute

Risk	Reliability	Responsiveness	Costs
R1	[(0.68,0.63,0.77); 0.80,0.20,0.10]	[(0.68,0.63,0.77); 0.80,0.20,0.10]	[(0.70,0.65,0.80); 0.90,0.20,0.10]
R2	[(0.68,0.63,0.77); 0.80,0.20,0.10]	[(0.65,0.60,0.70); 0.80,0.10,0.10]	[(0.57,0.52,0.63); 0.60,0.10,0.20]
R3	[(0.77,0.72,0.83); 0.70,0.20,0.20]	[(0.83,0.78,0.87); 0.70,0.20,0.20]	[(0.67,0.62,0.73); 0.80,0.20,0.10]
R4	[(0.77,0.72,0.83); 0.70,0.20,0.20]	[(0.83,0.78,0.87); 0.70,0.20,0.20]	[(0.75,0.70,0.80); 0.70,0.20,0.20]
R5	[(0.77,0.72,0.83); 0.70,0.20,0.20]	[(0.77,0.72,0.83); 0.70,0.20,0.20]	[(0.77,0.72,0.83); 0.70,0.20,0.20]

Table 17 De-neutrosophicated decision matrix and normalized decision matrix

De-neutrosophicated decision matrix				Normalized decision matrix			
				w	0.321	0.214	0.465
Risk	Reliability	Responsiveness	Costs	Risk	Reliability	Responsiveness	Costs
R1	0.6510	0.6510	0.6988	R1	0.4411	0.4304	0.4952
R2	0.6510	0.6338	0.4935	R2	0.4411	0.4189	0.3498
R3	0.6660	0.7140	0.6302	R3	0.4513	0.4719	0.4466
R4	0.6660	0.7140	0.6469	R4	0.4513	0.4719	0.4584
R5	0.6660	0.6660	0.6660	R5	0.4513	0.4403	0.4720

Table 18 Weighted decision matrix with performance scores

Risk	Reliability	Responsiveness	Costs	S+	S−	PS
R1	0.1416	0.0923	0.2300	0.00949564	0.06760296	0.87683771
R2	0.1416	0.0898	0.1625	0.06858599	0	0
R3	0.1449	0.1012	0.2075	0.02256523	0.04652183	0.67337979
R4	0.1449	0.1012	0.2130	0.01707824	0.0518473	0.75222185
R5	0.1449	0.0944	0.2193	0.01273102	0.05706778	0.81760399
V+	0.1449	0.1012	0.2300			
V−	0.1416	0.0898	0.1625			

$$V_j^+ = \max \left(r_{.j} \right) \tag{13}$$

$$V_j^- = \min \left(r_{.j} \right) \tag{14}$$

Then the Euclidean distance of each performance value must be calculated with respect to its ideal best (S_i^+) and its ideal worst (S_i^-) using Eqs. (15) and (16):

$$S_i^+ = \sqrt{\sum_j \left(r_{ij} - V_j^+ \right)^2} \tag{15}$$

$$S_i^- = \sqrt{\sum_j \left(r_{ij} - V_j^- \right)^2} \tag{16}$$

Finally, the performance scores are calculated (PS_i) using Eq. (17). The results can be seen in Table 18:

$$PS_i = \frac{S_i^-}{S_i^+ + S_i^-} \tag{17}$$

Table 19 Expert responses on the effectiveness of each strategy to manage each risk

Strategy	Expert 1				Expert 2				Expert 3			
	R1	R3	R4	R5	R1	R3	R4	R5	R1	R3	R4	R5
S1	L	A	VH	H	N	VH	VH	VH	M	A	VH	VH
S2	N	L	VH	L	N	VL	A	M	N	H	VH	L
S3	N	VL	H	VH	N	M	H	H	L	M	M	M
S4	N	M	H	VH	N	M	M	A	N	M	L	A
S5	L	L	L	VL	VL	L	L	L	M	M	L	L

Table 20 Neutrosophic mean for effectiveness of each strategy managing each risk

Strategy	Risk 1	Risk 3
S1	[(0.38,0.42,0.52); 0.10,0.20,0.20]	[(0.93,0.88,0.93); 0.70,0.20,0.20]
S2	[(0.10,0.30,0.35); 0.10,0.20,0.15]	[(0.42,0.42,0.47); 0.60,0.20,0.30]
S3	[(0.20,0.32,0.40); 0.10,0.20,0.20]	[(0.48,0.48,0.50); 0.60,0.20,0.30]
S4	[(0.10,0.30,0.35); 0.10,0.20,0.15]	[(0.65,0.60,0.70); 0.80,0.10,0.10]
S5	[(0.40,0.40,0.43); 0.60,0.20,0.30]	[(0.48,0.43,0.57); 0.60,0.10,0.20]
Strategy	*Risk 4*	*Risk 5*
S1	[(0.90,0.85,0.90); 0.70,0.20,0.20]	[(0.83,0.78,0.87); 0.70,0.20,0.20]
S2	[(0.92,0.87,0.92); 0.70,0.20,0.20]	[(0.48,0.43,0.57); 0.60,0.10,0.20]
S3	[(0.68,0.63,0.77); 0.80,0.20,0.10]	[(0.75,0.70,0.80); 0.70,0.20,0.20]
S4	[(0.58,0.53,0.67); 0.60,0.20,0.20]	[(0.93,0.88,0.93); 0.70,0.20,0.20]
S5	[(0.40,0.35,0.50); 0.60,0.10,0.20]	[(0.32,0.32,0.37); 0.60,0.20,0.30]

3.8 Neutrosophic TOPSIS for Strategies Selection in Supply Chain Risk Management

This methodology proposes to inquire the opinion of decision makers on the effectiveness, in linguistic terms, of a strategy to manage each of the risks. For this, we will make use of the table of conversion of linguistic values to a triangular neutrosophical scale proposed by [33], which can be seen in Table 14. The effectiveness values of each of the strategies on the risks obtained from the experts can be seen in Table 19. For the example discussed in this paper, risk 2 will not be considered since its weight on the attributes was zero.

Now we proceed to calculate the neutrosophical mean of the values obtained from the experts using Eq. (10) and the result can be seen in Table 20.

Knowing that the quantitative value of a risk can be estimated, as observed in Eq. (18), as the product of its expected impact (estimated in Sect. 3.7) by its probability of occurrence (estimated in Sect. 3.6), we proceed to calculate the weights that each risk will have in the decision matrix. The results can be seen in Table 21.

$$w_i = P_i * I_i \tag{18}$$

Table 21 Estimation of risks weights

Risk	Probability (P)	Impact (I)	Weight	Normalized weight
R1	0.2186	0.8768	0.1917	0.2846
R3	0.2719	0.6734	0.1831	0.2718
R4	0.2010	0.7522	0.1512	0.2245
R5	0.1805	0.8176	0.1476	0.2191

Table 22 De-neutrosophicated decision matrix and normalized decision matrix

De-neutrosophicated decision matrix					Normalized decision matrix				
					w	0.2846	0.2718	0.2245	0.2191
Strat	R1	R3	R4	R5	Strat	R1	R3	R4	R5
S1	0.0933	0.2635	0.2540	0.2380	S1	0.5337	0.6514	0.5419	0.5316
S2	0.0547	0.1138	0.2588	0.1422	S2	0.3129	0.2811	0.5521	0.3175
S3	0.0649	0.1283	0.2170	0.2156	S3	0.3715	0.3172	0.4631	0.4816
S4	0.0547	0.2113	0.1635	0.2635	S4	0.3129	0.5221	0.3488	0.5886
S5	0.1079	0.1422	0.1198	0.0875	S5	0.6175	0.3513	0.2556	0.1954

Table 23 Weighted decision matrix with performance scores

Strat	R1	R3	R4	R5	S+	S−	PS
S1	0.1519	0.1771	0.1217	0.1165	0.02703568	0.15371786	0.85042795
S2	0.0890	0.0764	0.1239	0.0696	0.14550146	0.07173703	0.33022245
S3	0.1057	0.0862	0.1040	0.1055	0.1187517	0.0804648	0.40390632
S4	0.0890	0.1419	0.0783	0.1290	0.10406584	0.11023329	0.51438983
S5	0.1757	0.0955	0.0574	0.0428	0.13603347	0.08874884	0.39482129
V+	0.1757	0.1771	0.1239	0.1290			
V−	0.0890	0.0764	0.0574	0.0428			

The next step is to convert the values in Table 20 to crisp numbers using Eq. (11). Then, the decision matrix must be normalized using Eq. (12). The result can be seen in Table 22.

The next step is to multiply the normalized decision matrix by the respective weight of each risk calculated in Table 21. Then the ideal best (V_j^+) and the ideal worst (V_j^-) must be selected in order to calculate the Euclidean distance of each strategy from its best ideal (S_i^+) and from its ideal worst (S_i^-). Then we calculate the performance score (PS_i) using Eqs. (13), (14), (15), (16), and (17). The result can be seen in Table 23.

3.9 Neutrosophic Probabilistic Expert System Programming

The next step is to carry out the programming of the knowledge base based on the information collected from the experts and calculated in the previous sections. The

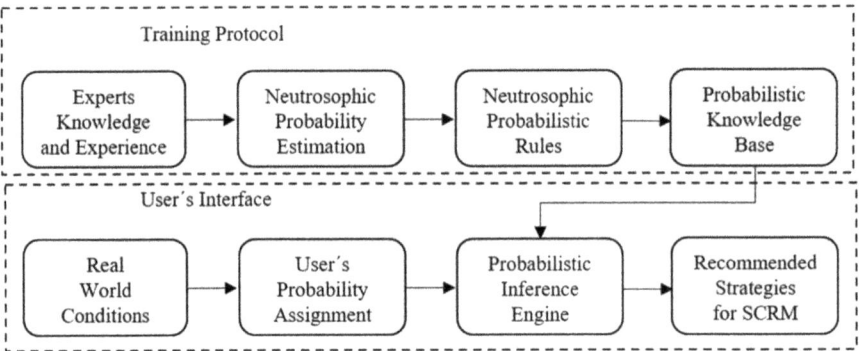

Fig. 2 Structure of the neutrosophic probabilistic expert system

strategies, risks, and metrics of the supply chain will form networks which will be connected by means of the probability values calculated previously. An algorithm that works as the inference engine of our expert system is used, which will ask the users if they want to work with the probability values with which the expert system was trained, if they want to change them according to their reality, since they may have the perception that at the time and place of the query a risk presents a greater probability of occurrence or is already occurring and seeks to mitigate its impact, for which the probability of occurrence of a risk that is affecting the supply chain is equal to 1. With the probability of occurrence values selected, the inference engine will query the knowledge base that was programmed with the values obtained with the methodology proposed in this research. Once the knowledge base has been created and the inference algorithm has been programmed, the expert system will be considered trained and will be ready to support decision-making in supply chain risk management. A diagram of the structure of the expert system can be seen in Fig. 2.

4   Simulation and Results

Let us consider the probability of occurrence of the risks calculated from the information extracted from the experts as the stable state of the supply chain. Considering the above, the performance score values obtained in Sect. 3.7 suggest the importance of the five strategies queried in a supply chain under normal conditions. Table 24 lists the five strategies in the order of importance to be applied with the probability of occurrence of the risks calculated in Sect. 3.5.

As we can see, strategy 1 has a significantly higher performance score than the other strategies, which is consistent, because a well-designed supply chain is of vital importance for all links in the supply chain. Now, we are going to query the expert system changing the probability of occurrence of risk 4 to 1; that is, we are going

Table 24 Strategies for supply chain risk management in the order of importance

Code	Strategy	Performance score
S1	Fine-tune supply chain design	0.8504
S4	Logistics flexibility	0.5143
S3	Inventory flexibility	0.4039
S5	Process standardization	0.3948
S2	Supply base strengthening	0.3302

Table 25 Strategies for supply chain risk management in the order of importance with $P(R_4) = 1$

Code	Strategy	Performance score
S1	Fine-tune supply chain design	0.9226
S2	Supply base strengthening	0.6954
S3	Inventory flexibility	0.6105
S4	Logistics flexibility	0.3794
S5	Process standardization	0.2014

Table 26 Strategies for supply chain risk management in the order of importance with $P(R_5) = 1$

Code	Strategy	Performance score
S1	Fine-tune supply chain design	0.8542
S4	Logistics flexibility	0.8224
S3	Inventory flexibility	0.6680
S2	Supply base strengthening	0.3139
S5	Process standardization	0.1536

to tell the expert system that the risk is present and affecting our chain. A change in the order of importance of the strategies would be expected, mainly giving greater weight to strategy 2 located in the last place, bearing in mind that strategy 1 serves to manage this risk. The new order of importance for this second scenario can be seen in Table 25.

As predicted, there was a change in the order of importance of the strategies. Strategy 1, which we had said was of high importance in risk management 4, increased its performance score; and also strategy 2, which we anticipated should be increased, considering it the most important from the authors' point of view to manage risk 4, also increased its performance score and was ranked as the second most important strategy in this scenario. We are going to propose a new change of scenario in which the probability of occurrence of risk 5 is 1; that is, the risk is present and affecting the supply chain. It is expected that strategy 4 associated with logistics and therefore with transport will increase its performance score. The results can be seen in Table 26.

As expected, strategy 4 increased its performance score and moved to second place of importance. This shows that the expert system is capable of making decisions as if it had the knowledge of not of one, but in this case three experts, which will be of great help to decision makers in risk management in the supply chain who must face problems that can include many more variables of strategies, risks, and performance metrics.

5 Conclusions

Through the research carried out and expressed in this paper, the effectiveness of the application of neutrosophical theory for the treatment of uncertainty in supply chain risk management [34] is verified. It is proposed to researchers interested in decision-making not only applied to SCRM, but to different fields of business research, to create their own neutrosophical probabilistic expert systems based on the methodology proposed in this paper and to improve it to be shared with decision makers around the world.

It is also verified once again that the use of expert systems allows transferring to a machine all the experience and knowledge of people who have spent years studying or learning empirically on a specific subject with the processing power of a computer, which can become an ally when making difficult decisions.

The field of neutrosophical theory is expected to continue to flourish to use advances in its theoretical, mathematical, and logical concepts in applications such as the one proposed in this research to continue advancing toward an increasingly competitive world with a lower proportion of errors and waste that allows a better quality of life for all human beings.

Acknowledgments The authors would like to express thank to the Universidad Industrial de Santander for the facilities and resources granted for the development of this research.

References

1. Blackhurst, J., Craighead, C. W., & Elkins, D. (2005). An empirically derived agenda of critical research issues for managing supply-chain disruptions. *International Journal of Production Research, 43*(19), 4067–4081. https://doi.org/10.1080/00207540500151549.
2. Lavastre, O., Gunasekaran, A., & Spalanzani, A. (2012). Supply chain risk management in French companies. *Decision Support Systems, 52*(4), 828–838. https://doi.org/10.1016/j.dss.2011.11.017.
3. Junaid, M., Xue, Y., Syed, M. W., Li, J. Z., & Ziaullah, M. (2020). A neutrosophic ahp and topsis framework for supply chain risk assessment in automotive industry of Pakistan. *Sustainability (Switzerland), 12*(1). https://doi.org/10.3390/SU12010154.
4. Baryannis, G., Dani, S., & Antoniou, G. (2019). Predicting supply chain risks using machine learning: The trade-off between performance and interpretability. *Future Generation Computer Systems, 101*, 993–1004. https://doi.org/10.1016/j.future.2019.07.059.
5. Sodhi, M. S., Son, B.-G., & Tang, C. S. (2012). Researchers' perspectives on supply chain risk management. *Production and Operations Management, 21*(1), 1–13. https://doi.org/10.1111/j.1937-5956.2011.01251.x.
6. Baryannis, G., Validi, S., Dani, S., & Antoniou, G. (2019a). Supply chain risk management and artificial intelligence: state of the art and future research directions. *International Journal of Production Research, 57*(7), 2179–2202. https://doi.org/10.1080/00207543.2018.1530476.
7. Council of Supply Chain Management Professionals. (2012). Supply Chain Operations Reference Model (SCOR) revision 11.0, *The Supply Chain Council, SCOR: The Supply Chain Reference.*

8. Ntabe, E. N., LeBel, L., Munson, A. D., & Santa-Eulalia, L. A. (2015). A systematic literature review of the supply chain operations reference (SCOR) model application with special attention to environmental issues. *International Journal of Production Economics, 169,* 310–332. https://doi.org/10.1016/j.ijpe.2015.08.008.

9. Akkawuttiwanich, P., & Yenradee, P. (2018). Fuzzy QFD approach for managing SCOR performance indicators. *Computers and Industrial Engineering, 122*(May), 189–201. https://doi.org/10.1016/j.cie.2018.05.044.

10. Estampe, D., Lamouri, S., Paris, J.-L., & Brahim-Djellou, S. (2013). A framework for analysing supply chain performance evaluation models. *International Journal of Production Economics, 142*(2), 247–258. https://doi.org/10.1016/j.ijpe.2010.11.024.

11. Chorfi, Z., Benabbou, L., & Berrado, A. (2018). An integrated performance measurement framework for enhancing public health care supply chains. *Supply Chain Forum: An International Journal, 19,* 191–203.

12. Saaty, T. L., & Vargas, L. G. (1979). Estimating technological coefficients by the analytic hierarchy process. *Socio-Economic Planning Sciences, 13*(6), 333–336.

13. Abdel-Basset, M., Gunasekaran, M., Mohamed, M., & Chilamkurti, N. (2019). A framework for risk assessment, management and evaluation: Economic tool for quantifying risks in supply chain. *Future Generation Computer Systems—The International Journal of Escience, 90,* 489–502. https://doi.org/10.1016/j.future.2018.08.035.

14. Smarandache, F. (1999). *A unifying field in logics. neutrosophy: Neutrosophic probability, set and logic.* Rehoboth: American Research Press.

15. Yoon, K., & Hwang, C. L. (1981). TOPSIS (technique for order preference by similarity to ideal solution)—a multiple attribute decision making, w: Multiple attribute decision making—Methods and applications, a state-of-the-at survey, *A state-of-the-at survey* (pp. 128–140).

16. Castillo, E., Gutiérrez, J. M., & Hadi, A. S. (1997). *Expert systems and probabilistic network models.* New York, NY: Springer.

17. Lauritzen, S. L., & Spiegelhalter, D. J. (1988). Local computations with probabilities on graphical structures and their application to expert systems. *Journal of the Royal Statistical Society: Series B: Methodological, 50*(2), 157–194. https://doi.org/10.1111/j.2517-6161.1988.tb01721.x.

18. Jensen, F. V. (1988). *Junction trees and decomposable hypergraphs.* Aalborg, Denmark: Judex Datasystemer, Tech. Rep.

19. Dawid, A. P. (1992). Applications of a general propagation algorithm for probabilistic expert systems. *Statistics and Computing, 2*(1), 25–36. https://doi.org/10.1007/BF01890546.

20. Parsa, K., & Torfi, F. (2017). Designing evaluation and ranking model for supply chain risk with multi-criteria decision approach. *Revista QUID, 1,* 532–541.

21. Rajesh, R. (2020). A grey-layered ANP based decision support model for analyzing strategies of resilience in electronic supply chains. *Engineering Applications of Artificial Intelligence, 87,* 103,338. https://doi.org/10.1016/j.engappai.2019.103338.

22. Chiang, W. K., Chhajed, D., & Hess, J. D. (2003). Direct marketing, indirect profits: A strategic analysis of dual-channel supply-chain design. *Management Science, 49*(1), 1–20.

23. Qi, L., Shen, Z.-J. M., & Snyder, L. V. (2010). The effect of supply disruptions on supply chain design decisions. *Transportation Science, 44*(2), 274–289.

24. Sturgeon, T., & Lester, R. K. (2004). The new global supply base: new challenges for local suppliers in East Asia, *Global production networking and technological change in East Asia* (pp. 35–87). Oxford: Oxford University Press.

25. Nam, S.-H., Vitton, J., & Kurata, H. (2011). Robust supply base management: Determining the optimal number of suppliers utilized by contractors. *International Journal of Production Economics, 134*(2), 333–343.

26. Kemahliouglu-Ziya, E., & Bartholdi, J. J., III. (2011). Centralizing inventory in supply chains by using Shapley value to allocate the profits. *Manufacturing & Service Operations Management, 13*(2), 146–162.

27. Sarker, B. R. (2014). Consignment stocking policy models for supply chain systems: A critical review and comparative perspectives'. *International Journal of Production Economics, 155,* 52–67.

28. Atasoy, B. (2013). *Integrated supply-demand models for the optimization of flexible transportation systems* . (No. THESIS). EPFL.

29. Bruzzone, A., & Longo, F. (2014). An application methodology for logistics and transportation scenarios analysis and comparison within the retail supply chain. *European Journal of Industrial Engineering, 8*(1), 112–142.

30. Ma, J., Wang, K., & Xu, L. (2011). Modelling and analysis of workflow for lean supply chains. *Enterprise Information Systems, 5*(4), 423–447.

31. Costinot, A., Vogel, J., & Wang, S. (2013). An elementary theory of global supply chains. *Review of Economic Studies, 80*(1), 109–144.

32. Salama, A. A., Smarandache, F., & Eisa, M. (2014). *Introduction to image processing via neutrosophic techniques.* Infinite Study.

33. Abdel-Basset, M., & Mohamed, R. (2020). A novel plithogenic TOPSIS-CRITIC model for sustainable supply chain risk management. *Journal of Cleaner Production, 247,* 119,586. https://doi.org/10.1016/j.jclepro.2019.119586.

34. Rojas-Gualdron, R., Smarandache, F., & Diaz-Bohorquez, C. (2019). Aplicación de la teoría neutrosófica para el tratamiento de la incertidumbre en la gestión del riesgo en la cadena de suministro. *Aglala, 10*(2), 1–19. https://doi.org/10.22519/22157360.1429.

Cloud Computing Technology Selection Using a Novel Neutrosophic Extension of the MULTIMOORA Method Based on the Use of Interval-Valued and Triangular-Valued Neutrosophic Numbers

Dragisa Stanujkic, Darjan Karabasevic, Gabrijela Popovic,
Edmundas Kazimieras Zavadskas, and Maja Stanujkic

1 Introduction

In recent decades, technological innovations, i.e., information and communication technologies (ICT), have significantly changed our society in general [1]. The advances toward higher socialization of ICT, the emergence of the Internet of Things, and Cloud computing have made the concept of digital business as a much more relevant concept [2]. The emergence of cloud computing represents a fundamental change in the way that ICT is shaped and developed [3].

Cloud computing can be seen as a concept based on earlier distributed services models. However, cloud computing stands out from traditional Internet services by its dynamic and flexible infrastructure [4]. Cloud computing can be seen as an area of computing in which highly scalable computing capabilities are provided to numerous external clients in the form of services delivered via the Internet [5]. As stated by Rehman et al. [6], the "cloud services have several attributes all of which are the criteria that have to be taken into account when making a service selection decision." One of the available tools that can help to overcome the problem of

D. Stanujkic · M. Stanujkic
Technical Faculty in Bor, University of Belgrade, Bor, Serbia
e-mail: dstanujkic@tfbor.bg.ac.rs; maja.stanujkic@gmail.com

D. Karabasevic (✉) · G. Popovic
Faculty of Applied Management, Economics and Finance, University Business Academy in Novi
Sad, Belgrade, Serbia
e-mail: darjan.karabasevic@mef.edu.rs; gabrijela.popovic@mef.edu.rs

E. K. Zavadskas
Institute of Sustainable Construction, Vilnius Gediminas Technical University, Vilnius, Lithuania
e-mail: edmundas.zavadskas@vgtu.lt

cloud technology selection is to apply multiple-criteria decision-making (MCDM) methods.

MCDM is one of the most important branches of operational research. Growing use of the MCDM methods for solving complex real-world problems has caused rapid development of the field [7, 8]. MCDM methods greatly facilitate decision-making process, and so far many prominent methods have been proposed, such as SAW, AHP, TOPSIS, PROMETHEE, ELECTRE, VIKOR, COPRAS, MOORA, and so forth. Also, many problems are associated with uncertainties, vagueness, and predictions; therefore, most of the MCDM methods have their extensions based on fuzzy, grey, or neutrosophic numbers [9, 10].

Neutrosophic sets (NSs) have been proposed by Smarandache [11, 12], as a generalization of numerous set theories, such as classical sets, fuzzy sets [13], intuitionistic fuzzy sets [14], bipolar fuzzy sets [15], and Pythagorean fuzzy sets [16].

In the fuzzy set theory, Zadeh considers using only one function, the membership function, for handling uncertainties and unreliability of information used for decision-making. Atanassov later [17] extends Zadeh's approach by introducing the non-membership function to the set, thus providing an opportunity to solve much more complex decision-making problems. In Attanassian intuitionistic fuzzy sets, the sum of membership and non-membership functions should be less than or equal to zero, which can limit their usage for solving some classes of real-world decision-making problems. This is why Yager introduces Pythagorean fuzzy sets that are somewhat more flexible about the previously discussed constraint.

In NSs theory, Smarandache extended previously mentioned fuzzy sets by introducing using of three independent memberships named the truth-membership $T_{A(x)}$, the falsity-membership $F_{A(x)}$, and the indeterminacy-membership $I_{A(x)}$ functions. Smarandache [12] and Wang et al. [18] further proposed a single-valued neutrosophic set, by modifying the conditions $T_{A(x)}$, $I_{A(x)}$, and $F_{A(x)} \in [0, 1]$ and $0 \leq T_{A(x)} + I_{A(x)} + F_{A(x)} \leq 3$, which are more suitable for solving scientific and engineering problems [19].

Compared with the fuzzy set and its extensions, the neutrosophic set can be identified as more flexible, for which reason an extension of the MULTIMORA method adapted for the purpose of using the interval-valued and triangular-valued neutrosophic numbers is proposed in this approach.

Therefore, the rest of this paper is organized as follows: In Sect. 2, some basic definitions related to the single-valued neutrosophic set are given. In Sect. 3, the ordinary MULTIMOORA method is presented, whereas in Sec. 4, the single-valued neutrosophic extension of the MULTIMOORA method is proposed as well as adoption of the neutrosophic extension of the MULTIMOORA method. In Sect. 5, an example is considered with the aim to explain in detail the proposed methodology. The conclusions are presented in the final section.

2   The Neutrosophic Sets, Concepts, and Operators

Definition 1 [12] Let X be the universe of discourse, with a generic element in X denoted by x. Then, the neutrosophic set (NS) A in X is as follows:

$$A = \left\{ x < T_{A(x)}, I_{A(x)}, F_{A(x)} > x \in X \right\}, \tag{1}$$

where $T_{A(x)}$, $I_{A(x)}$, and $F_{A(x)}$ are the truth-membership function, the indeterminacy-membership function, and the falsity-membership function, respectively, T_A, I_A, $F_A : X \to [^-0, 1^+]$ and $^-0 \leq T_{A(x)} + I_{A(x)} + F_{A(x)} \leq 3^+$.

2.1   Single-Valued Neutrosophic Sets

Definition 2 [12, 18] Let X be the universe of discourse. A single-valued neutrosophic set (SVNS) A over X is an object having the following form:

$$A = \left\{ x < T_{A(x)}, I_{A(x)}, F_{A(x)} > | x \in X \right\}, \tag{2}$$

where $T_A(x)$, $I_A(x)$, and $F_A(x)$ are the truth-membership function, the intermediacy-membership function, and the falsity-membership function, respectively, $T_{A(x)}$, $I_{A(x)}$ and $F_{A(x)} \in [0, 1]$ and $0 \leq T_{A(x)} + I_{A(x)} + F_{A(x)} \leq 3$.

Definition 3 [12] For an SVNS A in X, the triple $<t_A, i_A, f_A>$ is called the single-valued neutrosophic number (SVNN).

Definition 4 Let $x_1 = <t_1, i_1, f_1>$ and $x_2 = <t_2, i_2, f_2>$ be two SVNNs and $\lambda > 0$; then the basic operations are defined as follows:

$$x_1 + x_2 = < t_1 + t_2 - t_1 t_2, i_1 i_2, f_1 f_2 > . \tag{3}$$

$$x_1 \cdot x_2 = < t_1 t_2, i_1 + i_2 - i_1 i_2 f_1 + f_2 - f_1 f_2 > . \tag{4}$$

$$\lambda x_1 = < 1 - (1 - t_1)^\lambda, i_1^\lambda, f_1^\lambda > . \tag{5}$$

$$x_1^\lambda = < t_1^\lambda, i_1^\lambda, 1 - (1 - f_1)^\lambda > . \tag{6}$$

Definition 5 [20] Let $x = <t_x, i_x, f_x>$ be an SVNN; then the score function s_x of x can be as follows:

$$s_x = \frac{1 + t - 2i - f}{2}, \tag{7}$$

where $s_x \in [-1, 1]$.

Definition 6 Let $x_1 = <t_1, i_1, f_1>$ and $x_2 = <t_2, i_2, f_2>$ be two SVNNs. Then the distance between x_1 and x_2 is as follows:

$$d(x_1, x_2) = \frac{|t_1 - t_2| + |i_1 - i_2| + |f1 - f_2|}{3}. \tag{8}$$

Definition 7 [20] Let $A_j = <t_j, i_j, f_j>$ be a collection of SVNNs and $W = (w_1, w_2, \ldots, w_n)^T$ be an associated weighting vector. Then the single-valued neutrosophic weighted average (SVNWA) operator of A_j is as follows:

SVNWA $(A_1, A_2, \ldots, A_n)$
$$= \sum_{j=1}^{n} w_j A_j = \left\langle 1 - \prod_{j=1}^{n} \left(1 - t_j\right)^{w_j}, \prod_{j=1}^{n} \left(i_j\right)^{w_j}, \prod_{j=1}^{n} \left(f_j\right)^{w_j} \right\rangle, \tag{9}$$

where w_j is the element j of the weighting vector, $w_j \in [0, 1]$ and $\sum_{j=1}^{n} w_j = 1$.

Definition 8 [20] Let $A_j = <t_j, i_j, f_j>$ be a collection of SVNNs and $W = (w_1, w_2, \ldots, w_n)^T$ be an associated weighting vector. Then the single-valued neutrosophic weighted geometric (SVNWG) operator of A_j is as follows:

SVNWG $(A_1, A_2, \ldots, A_n)$
$$= \prod_{j=1}^{n} \left(A_j\right)^{w_j} = \left\langle \begin{array}{c} \prod_{j=1}^{n} \left(t_j\right)^{w_j}, \\ 1 - \prod_{j=1}^{n} \left(1 - i_j\right)^{w_j}, 1 - \prod_{j=1}^{n} \left(1 - f_j\right)^{w_j} \end{array} \right\rangle \tag{10}$$

where w_j is the element j of the weighting vector, $w_j \in [0, 1]$ and $\sum_{j=1}^{n} w_j = 1$.

2.2 Interval-Valued Neutrosophic Sets

Definition 9 [21] Let X be a space of points (objects) with generic elements in X, denoted by x. An interval-valued neutrosophic set (IVNS) A in X is characterized by an interval truth-membership function.

$T_{A(x)} = \left[T_A^l, T_A^l\right]$, an interval indeterminacy-membership function $I_{A(x)} = \left[I_A^l, I_A^l\right]$ and an interval falsity-membership function $F_{A(x)} = \left[F_A^l, F_A^l\right]$. For each point x in X, $T_{A(x)}$, $I_{A(x)}$ and $F_{A(x)} \in [0, 1]$ and $0 \leq T_A^u + I_A^u + F_A^u \leq 3$.

$$A = \left\{x < T_{A(x)}, I_{A(x)}, F_{A(x)} > | x \in X\right\}. \tag{11}$$

An IVNS A can be written as $A = <\left[T_A^l, T_A^u\right], \left[I_A^l, I_A^u\right], \left[F_A^l, F_A^u\right]>$.

Definition 10 For an IVNS A in X, the triple $<\left[t_a^l, t_a^u\right], \left[i_a^l, i_a^u\right], \left[f_a^l, f_a^u\right]>$ is called the interval-valued neutrosophic number (IVNN).

Definition 11 Let $x_1 = < \left[t_1^l, t_1^u\right], \left[i_1^l, i_1^u\right], \left[f_1^l, f_1^u\right] >$ and $x_2 = < \left[t_2^l, t_2^u\right], \left[i_2^l, i_2^u\right], \left[f_2^l, f_2^u\right] >$ be two IVNNs and $\lambda > 0$; then the basic operations on IVNNs are as follows:

$$\overline{x}_1 + \overline{x}_2 = \left\langle \left[t_1^l + t_2^l - t_1^l t_2^l, t_1^u + t_2^u - t_1^u t_2^u\right], \left[i_1^l i_2^l, i_1^u i_2^u\right], \left[f_1^l f_2^l, f_1^u f_2^u\right] \right\rangle. \tag{12}$$

$$\overline{x}_1 \cdot \overline{x}_2 = \left\langle \begin{array}{c} \left[t_1^l t_2^l, t_1^u t_2^u\right], \\ \left[i_1^l + i_2^l - i_1^l i_2^l, i_1^u / i_1^u i_2^u\right], \\ \left[f_1^l + f_2^l - f_1^l f_2^l, f_1^u - f_1^u f_2^u\right] \end{array} \right\rangle. \tag{13}$$

$$\lambda \overline{x} = \left\langle \left[1 - \left(1 - t_1^l\right)^\lambda, 1 - \left(1 - t_2^u\right)^\lambda\right], \left[\left(i_1^l\right)^\lambda, \left(i_1^u\right)^\lambda\right], \left[\left(f_1^l\right)^\lambda, \left(f_1^u\right)^\lambda\right] \right\rangle. \tag{14}$$

$$\overline{x}_\lambda = \left\langle \left[\left(t_1^l\right)^\lambda, \left(t_1^u\right)^\lambda\right], \right.$$
$$\left. \left[1 - \left(1 - i_1^l\right)^\lambda, 1 - \left(1 - i_2^u\right)^\lambda\right], \left[1 - \left(1 - f_1^l\right)^\lambda, 1 - \left(1 - f_2^u\right)^\lambda\right] \right\rangle. \tag{15}$$

Definition 12 [20] Let $x = <t_x, i_x, f_x>$ be an IVNN; then the score function s_x of x is as follows:

$$s_{\overline{x}} = \frac{2 + t^l + t^u - 2i^l - 2i^u - f^i - f^u}{4}. \tag{16}$$

Definition 13 Let $x_1 = < \left[t_1^l, t_1^u\right], \left[i_1^l, i_1^u\right], \left[f_1^l, f_1^u\right] >$ and $x_2 = < \left[t_2^l, t_2^u\right], \left[i_2^l, i_2^u\right], \left[f_2^l, f_2^u\right] >$ be two IVNNs. Then the distance between x_1 and x_2 is as follows:

$$d(x_1, x_2) = \frac{1}{3} \left(\begin{array}{c} \left|t_1^l - t_2^l\right| + \left|t_1^u - t_2^u\right| + \left|i_1^l + i_2^l\right| + \\ \left|i_1^u - i_2^u\right| + \left|f_1^l - f_2^l\right| + \left|f_1^u - f_2^u\right| \end{array} \right). \tag{17}$$

Definition 14 Let $\overline{A}_j = < \left[t_j^l, t_j^u\right], \left[i_j^l, i_j^u\right], \left[f_j^l, f_j^u\right] >$ be a collection of IVNNs and $W = (w_1, w_2, \ldots, w_n)^T$ be an associated weighting vector. Then a interval-valued neutrosophic weighted average (IVNWA) operator of A_j is as follows:

$$IVNWG\left(\tilde{A}_1, \tilde{A}_2, \ldots, \tilde{A}_n\right)$$

$$= \sum_{j=1}^{n} w_j \tilde{A}_j = \left\langle \begin{matrix} \left[1 - \prod_{j=1}^{n}\left(1 - t_j^l\right)^{w_j}, 1 - \prod_{j=1}^{n}\left(1 - t_j^u\right)^{w_j}\right], \\ \left[\prod_{j=1}^{n}\left(i_j^l\right)^{w_j}, \prod_{j=1}^{n}\left(i_j^u\right)^{w_j}\right], \\ \left[\prod_{j=1}^{n}\left(f_j^l\right)^{w_j}, \prod_{j=1}^{n}\left(f_j^u\right)^{w_j}\right] \end{matrix} \right\rangle \tag{18}$$

where w_j is the element j of the weighting vector, $w_j \in [0, 1]$ and $\sum_{j=1}^{n} w_j = 1$.

Definition 15 Let $\overline{A}_j = < \left[t_j^l, t_j^u\right], \left[i_j^l, i_j^u\right], \left[f_j^l, f_j^u\right] >$ be a collection of IVNNs and $W = (w_1, w_2, \ldots, w_n)^T$ be an associated weighting vector. Then a triangular-valued neutrosophic weighted geometric (IVNWG) operator of A_j is as follows:

$$IVNWG\left(\tilde{A}_1, \tilde{A}_2, \ldots, \tilde{A}_n\right)$$

$$= \prod_{j=1}^{n}(A_j)^{w_j} = \left\langle \begin{matrix} \left[\prod_{j=1}^{n}\left(t_j^l\right)^{w_j}, \prod_{j=1}^{n}\left(t_j^u\right)^{w_j}\right], \\ \left[1 - \prod_{j=1}^{n}\left(1 - i_j^l\right)^{w_j}, 1 - \prod_{j=1}^{n}\left(1 - i_j^u\right)^{w_j}\right], \\ \left[1 - \prod_{j=1}^{n}\left(1 - f_j^l\right)^{w_j}, 1 - \prod_{j=1}^{n}\left(1 - f_j^u\right)^{w_j}\right] \end{matrix} \right\rangle. \tag{19}$$

where w_j is the element j of the weighting vector, $w_j \in [0, 1]$ and $\sum_{j=1}^{n} w_j = 1$.

2.3 Triangular Fuzzy Neutrosophic Sets

Definition 16 [22] Let X be a space of points (objects) with generic elements in X, denoted by x. A triangular-valued neutrosophic set (TVNS) A in X is characterized by a triangular truth-membership function $\tilde{T}_{A(x)} = \left(T_A^l, T_A^m, T_A^u\right)$, a triangular indeterminacy-membership function $\tilde{I}_{A(x)} = \left(I_A^l, I_A^m, I_A^u\right)$, and a triangular falsity-membership function $\tilde{F}_{A(x)} = \left(F_A^l, F_A^m, F_A^u\right)$. For each point x in X, $\tilde{T}_{A(x)}$, $\tilde{I}_{A(x)}$, and $\tilde{F}_{A(x)} \in [0, 1]$ and $0 \leq T_A^u + I_A^u + F_A^u \leq 3$:

$$A = \left\{x < T_{A(x)}, I_{A(x)}, F_{A(x)} > | x \in X\right\}. \tag{20}$$

An TVNS A is also defined as $A = < \left(T_A^l, T_A^m, T_A^u\right), \left(I_A^l, I_A^m, I_A^u\right), \left(F_A^l, F_A^m, F_A^u\right) >$.

Definition 17 For a TVNS a in X, the triple $< \left(t_a^l, t_a^m, t_a^u\right), \left(i_a^l, i_a^m, i_a^u\right),$
$\left(f_a^l, f_a^m, f_a^u\right) >$ is called the triangular-valued neutrosophic number (TVNN).

Definition 18 Let $\tilde{x}_1 =< \left(t_1^l, t_1^m, t_1^u\right), \left(i_1^l, i_1^m, i_1^u\right), \left(f_1^l, f_1^m, f_1^u\right) >$ and $\tilde{x}_2 =<$
$\left(t_2^l, t_2^m, t_2^u\right), \left(i_2^l, i_2^m, i_2^u\right), \left(f_2^l, f_2^m, f_2^u\right) >$ be two TVNNs and $\lambda > 0$; then the basic
operations on TVNNs are as follows:

$$\tilde{x}_1 + \tilde{x}_2 = \left\langle \begin{array}{l} \left(t_1^l + t_2^l - t_1^l t_2^l, t_1^m + t_2^m - t_1^m t_2^m, t_1^u + t_2^u - t_1^u t_2^u\right), \\ \left(i_1^l i_2^l, i_1^m i_2^m, i_1^u i_2^u\right), \\ \left(f_1^l f_2^l, f_1^m f_2^m, f_1^u f_2^u\right) \end{array} \right\rangle. \tag{21}$$

$$\tilde{x}_1 \cdot \tilde{x}_2 = \left\langle \begin{array}{l} \left(t_1^l t_2^l, t_1^m t_2^m, t_1^u t_2^u\right), \\ \left(i_1^l + i_2^l - i_1^l i_2^l, i_1^m + i_2^m - i_1^m i_2^{mu}, i_1^u + i_2^u - i_1^u i_2^u\right), \\ \left(f_1^l + f_2^l - f_1^l f_2^l, f_1^m + f_2^m - f_1^m f_2^m, f_1^u + f_2^u - f_1^u f_2^u\right) \end{array} \right\rangle. \tag{22}$$

$$\lambda \tilde{x}_1 = \left\langle \begin{array}{l} \left(1 - \left(1 - t_1^l\right)^\lambda, 1 - \left(1 - t_2^m\right)^\lambda, 1 - \left(1 - t_2^u\right)^\lambda\right), \\ \left(\left(i_1^l\right)^\lambda, \left(i_1^m\right)^\lambda, \left(i_1^u\right)^\lambda\right), \\ \left(\left(f_1^l\right)^\lambda, \left(f_1^m\right)^\lambda, \left(f_1^u\right)^\lambda\right) \end{array} \right\rangle. \tag{23}$$

$$\tilde{x}_1^\lambda = \left\langle \begin{array}{l} \left(\left(t_1^l\right)^\lambda, \left(t_1^m\right)^\lambda, \left(t_1^u\right)^\lambda\right), \\ \left(1 - \left(1 - i_1^l\right)^\lambda, 1 - \left(1 - i_2^m\right)^\lambda, 1 - \left(1 - i_2^u\right)^\lambda\right), \\ \left(1 - \left(1 - f_1^l\right)^\lambda, 1 - \left(1 - f_2^m\right)^\lambda, 1 - \left(1 - f_2^u\right)^\lambda\right) \end{array} \right\rangle. \tag{24}$$

Definition 19 [23] Let $x = <(t^l, t^m, t^u), (i^l, i^m, i^u), (f^l, f^m, f^u)>$ be a TVNN; then the
score function s_x of x is as follows:

$$s_{\tilde{x}} = \frac{1}{3} \left(2 + \frac{t^l + 2t^m + t^u}{4} - \frac{i^l + 2i^m + i^u}{4} - \frac{f^l + 2f^m + f^u}{4}\right). \tag{25}$$

Definition 20 Let $\tilde{x}_1 =< \left(t_1^l, t_1^m, t_1^u\right), \left(i_1^l, i_1^m, i_1^u\right), \left(f_1^l, f_1^m, f_1^u\right) >$ and $\tilde{x}_2 =<$
$\left(t_2^l, t_2^m, t_2^u\right), \left(i_2^l, i_2^m, i_2^u\right), \left(f_2^l, f_2^m, f_2^u\right) >$ be two TVNNs. Then the distance
between x_1 and x_2 is as follows:

$$d\left(x_1, x_2\right) = \frac{1}{3} \left(\begin{array}{l} \mid t_1^l - t_2^l \mid + \mid t_1^m - t_2^m \mid + \mid t_1^u - t_2^u \mid + \\ \mid i_1^l - i_2^l \mid + \mid i_1^m - i_2^m \mid + \mid i_1^u - i_2^u \mid + \\ \mid f_1^l - f_2^l \mid + \mid f_1^m - f_2^m \mid + \mid f_1^u - f_2^u \mid \end{array} \right). \tag{26}$$

Definition 21 [23] Let $\tilde{A}_j =< \left(t^l_j, t^m_j, t^u_j\right), \left(i^l_j, i^m_j, i^u_j\right), \left(f^l_j, f^m_j, f^u_j\right) >$ be a collection of TVNNs and $W = (w_1, w_2, \ldots, w_n)^T$ be an associated weighting vector. Then a triangular-valued neutrosophic weighted average (TVNWA) operator of A_j is as follows:

$$
TVNWA\left(\tilde{A}_1, \tilde{A}_2, \ldots, \tilde{A}_n\right) = \sum_{j=1}^n w_j \tilde{A}_j
$$

$$
= \left\langle \begin{array}{c} \left(1 - \prod_{j=1}^n \left(1 - t^l_j\right)^{w_j}, 1 - \prod_{j=1}^n \left(1 - t^m_j\right)^{w_j}, 1 - \prod_{j=1}^n \left(1 - t^u_j\right)^{w_j}\right), \\ \left(\prod_{j=1}^n \left(i^l_j\right)^{w_j}, \prod_{j=1}^n \left(i^m_j\right)^{w_j}, \prod_{j=1}^n \left(i^u_j\right)^{w_j}\right), \\ \left(\prod_{j=1}^n \left(f^l_j\right)^{w_j}, \prod_{j=1}^n \left(f^m_j\right)^{w_j}, \prod_{j=1}^n \left(f^u_j\right)^{w_j}\right) \end{array} \right\rangle
$$

(27)

where w_j is the element j of the weighting vector, $w_j \in [0, 1]$ and $\sum_{j=1}^n w_j = 1$.

Definition 22 [23] Let $\tilde{A}_j =< \left(t^l_j, t^m_j, t^u_j\right), \left(i^l_j, i^m_j, i^u_j\right), \left(f^l_j, f^m_j, f^u_j\right) >$ be a collection of TVNNs and $W = (w_1, w_2, \ldots, w_n)^T$ be an associated weighting vector. Then a triangular-valued neutrosophic weighted geometric (TVNWG) operator of A_j is as follows:

$$
TVNWG\left(\tilde{A}_1, \tilde{A}_2, \ldots, \tilde{A}_n\right) = \prod_{j=1}^n (A_j)^{w_j}
$$

$$
= \left\langle \begin{array}{c} \left(\prod_{j=1}^n \left(t^l_j\right)^{w_j}, \prod_{j=1}^n \left(t^m_j\right)^{w_j}, \prod_{j=1}^n \left(t^u_j\right)^{w_j}\right), \\ \left(1 - \prod_{j=1}^n \left(1 - i^l_j\right)^{w_j}, 1 - \prod_{j=1}^n \left(1 - i^m_j\right)^{w_j}, 1 - \prod_{j=1}^n \left(1 - i^u_j\right)^{w_j}\right), \\ \left(1 - \prod_{j=1}^n \left(1 - f^l_j\right)^{w_j}, 1 - \prod_{j=1}^n \left(1 - f^m_j\right)^{w_j}, 1 - \prod_{j=1}^n \left(1 - f^u_j\right)^{w_j}\right) \end{array} \right\rangle
$$

(28)

where w_j is the element j of the weighting vector, $w_j \in [0, 1]$ and $\sum_{j=1}^n w_j = 1$.

3 The MULTIMOORA Method

The MULTIMOORA method consists of the three approaches named as follows: the ratio system (RS) approach, the reference point (RP) approach, and the full multiplicative form (FMF) [24].

The considered alternatives are ranked based on all the three approaches, and the final ranking order and the final decision is made based on the theory of dominance. In other words, the alternative with the highest number of appearances in the first positions on all ranking lists is the best-ranked alternative.

The process of determining the most appropriate alternative begins with the initial decision matrix D and the corresponding weighing vector W:

$$D = \left[x_{ij}\right]_{m \times n}, \tag{29}$$

$$W = \left[w_j\right], \tag{30}$$

where x_{ij} denotes a performance rating of alternative i to the criterion j; w_j denotes a weight of criterion j, $i = 1, 2, \ldots, m$; m is a number of alternatives, $j = 1, 2, \ldots, n$; n is a number of criteria.

Based on Brauers and Zavadskas [25], the computational procedure of the MULTIMOORA method can be concisely described by the following steps:

Step 1. Construct the normalized decision matrix. The normalized performance ratings are calculated as follows:

$$r_{ij} = \frac{x_{ij}}{\sqrt{\sum_{i=1}^{n} x_{ij}^2}}, \tag{31}$$

where r_{ij} denotes the normalized performance of the alternative i with respect to the criterion j.

3.1 The Ratio System Approach

Step 2.1. Calculate the overall importance of the each alternative. The overall importance of the alternative i is calculated as follows:

$$y_i = \sum_{j \in \Omega_{\max}} w_j r_{ij} - \sum_{j \in \Omega_{\min}} w_j r_{ij}, \tag{32}$$

where y_i denotes the overall importance of the alternative i, $\Omega_{\max}$ and $\Omega_{\min}$ denote the sets of the benefit cost criteria, respectively.

Step 2.2. Rank the alternatives based on the ratio system approach. In this approach the compared alternatives are ranked based on y_i in descending order and the alternative with the highest value of y_i is considered to be the most appropriate.

3.2 The Reference Point Approach

Step 3.1. Calculate the maximum distance of each alternative to the reference point. The maximum distance of alternative i to the reference point is calculated as follows:

$$d_i^{\max} = \max_j \left(w_j | r_j^* - r_{ij} | \right), \tag{33}$$

where $d_i^{\max}$ denotes the distance of the alternative i to the reference point and r_j^* denotes the coordinate j of the reference point, determined as follows:

$$r_j^* = \begin{cases} \max_i r_{ij}; \; j \in \Omega_{\max} \\ \min_i r_{ij}; \; j \in \Omega_{\min} \end{cases}. \tag{34}$$

Step 3.2. Rank the alternatives based on the reference point approach. In this approach the compared alternatives are ranked based on $d_i^{\max}$ in ascending order and the alternative with the lowest value of $d_i^{\max}$ is considered as the most appropriate.

3.3 The Full Multiplicative Form

Step 4.1. Calculate the overall utility of each alternative. The overall utility of the alternative i can be determined in following manner:

$$u_i = \frac{\prod\limits_{j \in \Omega_{\max}} \left(r_{ij} \right)^{w_j}}{\prod\limits_{j \in \Omega_{\min}} \left(r_{ij} \right)^{w_j}}, \tag{35}$$

where u_i denotes the overall utility of the alternative i.

Step 4.2. Rank the alternatives based on the full multiplicative form. In this approach the compared alternatives are ranked based on their u_i in descending order and the alternative with the highest value of u_i is considered as the most appropriate.

3.4 The Final Selection of the Most Appropriate Alternative

Step 5. The final ranking of alternatives based on the MULTIMOORA method.
As a result of evaluation by applying the MULTIMOORA method, three ranking lists of the considered alternatives are obtained. Based on Brauers and Zavadskas [26], the final ranking order of the alternatives is determined based on the theory of dominance.

4 An Extension of the MULTIMOORA Method Based on Single-Valued Neutrosophic Numbers

4.1 A Single-Valued Neutrosophic Extension of the MULTIMOORA Method

Stanujkic et al. [27] proposed an extension of the MULTIMOORA method adopted for using SVNNs.

The following is a simplified version of the above-mentioned extension adapted for solving decision-making problems that not include non-beneficial criteria. SVNNs belong to the interval [0,1], and therefore, they do not have to be normalized before applying the other steps of the MULTIMOORA method.

4.1.1 The Ratio System Approach

Step 1.1. Calculate the SVN overall importance of the each alternative. The SVN overall importance of alternative i is determined by using SVNWA operator, as follows:

$$Y_i = \left(1 - \prod \left(1 - t_j \right)^{w_j}, \prod \left(i_j \right)^{w_j}, \prod \left(f_j \right)^{w_j} \right), \tag{36}$$

where Y_i denotes the importance of the alternative i.

Step 1.2. Calculate the overall importance for each alternative. Before ranking, SVN values should be transformed into crisp numbers. One way for such conversion is by using the score function, as shown:

$$y_i = s\left(Y_i \right). \tag{37}$$

Step 1.3. Rank the alternatives and select the best one. The ranking of the alternatives can be performed in the same way as in the ratio system approach of the ordinary MULTIMOORA method.

4.1.2 The Reference Point Approach

The ranking of the alternatives and the selection of the best one, based on the reference point approach, can be expressed through the following sub-steps:

Step 2.1. Determine the reference point. In this approach, each coordinate of the reference point $r^* = \{r_1^*, r_2^*, \ldots, r_n^*\}$ is an SVNN, $r_j^* = < t_j^*, i_j^*, f_j^* >$, whose values are determined as follows:

$$r_j^* = \left\langle \max_i t_{ij}, \min_i i_{ij}, \min_i f_{ij} \right\rangle, \tag{38}$$

where r_j^* denotes the coordinate j of the reference point.

Step 2.2. Determine the maximum distance from each alternative to all the coordinates of the reference point as follows:

$$d_{ij}^{\max} = d\left(r_{ij}, r_j^*\right) w_j, \tag{39}$$

where $d_{ij}^{\max}$ denotes the distance of the alternative i obtained based on the criterion j determined by Eq. (8).

Step 2.3. Determine the maximum distance of each alternative, as follows:

$$d_i^{\max} = \max_j d_{ij}^{\max}. \tag{40}$$

Step 2.4. Rank the alternatives and select the best one. At this step, the ranking of the alternatives can be done in the same way as in the RPA of the ordinary MULTIMOORA method.

4.1.3 The Full Multiplicative Form

The ranking of the alternatives based on the full multiplicative form and SVN information can be expressed as follows:

Step 3.1. Calculate overall utility of each alternative, as follows:

$$U_i = \left(\prod \left(t_j\right)^{w_j}, 1 - \prod \left(1 - i_j\right)^{w_j}, 1 - \prod \left(1 - f_j\right)^{w_j} \right), \tag{41}$$

where U_i denotes the overall utility of the alternative i.

Step 3.2. Determine the overall utility for each alternative. As in the case of reference point approach, SVN overall utility of each alternative has to be transformed into crisp values, as follows:

$$u_i = s\left(U_i\right). \tag{42}$$

Step 3.3. Rank the alternatives and select the best one. The ranking of the alternatives has to be performed in the same way as in the full multiplicative form of the ordinary MULTIMOORA method.

4.1.4   The Final Selection of the Most Appropriate Alternative

Step 4. Determine the final ranking order of the alternatives. The final ranking order of the alternatives can be determined as in the case of the ordinary MULTI-MOORA method, i.e., based on the dominance theory.

4.2   An Adoption of the Neutrosophic Extension of the MULTIMOORA Method

It is known that numerous decision-making problems require the involvement of multiple decision-makers, domain experts, or respondents. In such cases, the individual ratings obtained from respondents should be transformed into group ratings, before the use of an MCDM method.

Numerous procedures for collecting respondents' attitudes, usually based on the use of linguistic variables, and their transformation into a group decision matrix are considered in a number of articles. In this article, an approach based on the use of a Likert 1-5 scale is proposed for evaluating alternatives, because such scales are simple for use and easy for understanding, especially when it is necessary to evaluate each alternative according to each criterion using three parameters that represent truth, intermediacy, and falsity.

Probably the simplest way for transforming individual ratings into group ratings is based on their arithmetic, or geometric, mean. However, in such cases, much of the meaningful information obtained on the basis of group decision-making can be irretrievably lost, and thus the benefits that the group decision-making approach provides can be minimized.

It is also known that respondents' attitudes can be transformed into different shapes of fuzzy, intuitionistic fuzzy, bipolar fuzzy, and similar numbers. However, the question remains on how to determine the boundaries, that is, the points of the selected shape.

Therefore, three cases are considered in this approach. In the first case, the ratings are represented by their median. In the second case, the ratings are represented by the interval, whose boundaries are determined on the basis of the median of all ratings. Finally, in the case approach, all ratings are represented by a triangle.

Suppose that an alternative is evaluated on the basis of the criterion j by K respondents, where x_{ij}^{k} denotes the rating of alternative i in relation to the criterion j obtained from respondent k. In cases when NS numbers are used for evaluation, we need three such ratings, for truth, intermediacy, and falsity membership.

In the first case the group rating of alternative i in relation to the criterion j is determined as follows:

$$x_{ij} = <t_{ij}, i_{ij}, f_{ij}> = \begin{cases} \text{med}\left(t_{ij}^k\right) \\ \text{med}\left(i_{ij}^k\right) \\ \text{med}\left(f_{ij}^k\right) \end{cases}, \tag{43}$$

where $\text{med}\left(t_{ij}^k\right)$ denotes median of t_{ij}^k, $\text{med}\left(i_{ij}^k\right)$ denotes median of i_{ij}^k, $\text{med}\left(f_{ij}^k\right)$ denotes median of f_{ij}^k, and $k = 1, 2, \ldots K$.

In the second case the group ratings are determined as follows:

$$\overline{x}_{ij} = <\left[t_{ij}^l, t_{ij}^u\right], \left[i_{ij}^l, i_{ij}^u\right], \left[f_{ij}^l, f_{ij}^u\right]>$$
$$= \begin{cases} t_{ij}^l = \text{avg}\left(t_{ij}^k \left| t_{ij}^k \le \text{med}\left(t_{ij}\right)\right.\right), & t_{ij}^u = \text{avg}\left(t_{ij}^k \left| t_{ij}^k \ge \text{med}\left(t_{ij}\right)\right.\right) \\ i_{ij}^l = \text{avg}\left(i_{ij}^k \left| i_{ij}^k \le \text{med}\left(i_{ij}\right)\right.\right), & i_{ij}^u = \text{avg}\left(i_{ij}^k \left| i_{ij}^k \ge \text{med}\left(i_{ij}\right)\right.\right) \\ f_{ij}^l = \text{avg}\left(f_{ij}^k \left| f_{ij}^k \le \text{med}\left(f_{ij}\right)\right.\right), & f_{ij}^u = \text{avg}\left(f_{ij}^k \left| f_{ij}^k \ge \text{med}\left(f_{ij}\right)\right.\right) \end{cases}, \tag{44}$$

where $\text{avg}\left(x_{ij}^k\right)$ denotes the mean of ratings that meet a certain criterion.

Finally, in the third case the group ratings are determined as follows:

$$\overline{x}_{ij} = <\left(t_{ij}^l, t_{ij}^m, t_{ij}^u\right), \left(i_{ij}^l, i_{ij}^m, i_{ij}^u\right), \left(f_{ij}^l, f_{ij}^m, f_{ij}^u\right)>$$
$$= \begin{cases} t_{ij}^l = \text{avg}\left(t_{ij}^k \left| t_{ij}^k \le \text{med}\left(t_{ij}\right)\right.\right), & t_{ij}^m = \text{med}\left(t_{ij}\right), & t_{ij}^u = \text{avg}\left(t_{ij}^k \left| t_{ij}^k \ge \text{med}\left(t_{ij}\right)\right.\right) \\ i_{ij}^l = \text{avg}\left(i_{ij}^k \left| i_{ij}^k \le \text{med}\left(i_{ij}\right)\right.\right), & i_{ij}^m = \text{med}\left(i_{ij}\right), & i_{ij}^u = \text{avg}\left(i_{ij}^k \left| i_{ij}^k \ge \text{med}\left(i_{ij}\right)\right.\right) \\ f_{ij}^l = \text{avg}\left(f_{ij}^k \left| f_{ij}^k \le \text{med}\left(f_{ij}\right)\right.\right), & f_{ij}^m = \text{med}\left(f_{ij}\right), & f_{ij}^u = \text{avg}\left(f_{ij}^k \left| f_{ij}^k \ge \text{med}\left(f_{ij}\right)\right.\right) \end{cases} \tag{45}$$

5 A Numerical Illustration

In order to demonstrate the usability and efficiency of the proposed approach a research based on Büyüközkan et al. [28] has been completed. In order to briefly demonstrate the advantages of the proposed methodology, this example has been slightly modified. It is also important to note that for the sake of simplicity all evaluation criteria are given the same importance. In order to emphasize the benefits that can be obtained using NS, the evaluation is performed based on the main set of criteria (sub-criteria are not considered), as presented below:

- C_1, Acquisition and transaction cost
- C_2, Availability
- C_3, Storage capacity
- C_4, CPU

Table 1 The ratings obtained from the first of the ten domain specialists

	C_1	C_2	C_3	C_4	C_5	C_6
A_1	<1.0,0.0,0.0>	<1.0,0.2,0.0>	<1.0,0.0,0.0>	<0.7,0.3,0.0>	<0.8,0.2,0.2>	<0.9,0.1,0.1>
A_2	<1.0,0.0,0.0>	<1.0,0.0,0.0>	<1.0,0.0,0.0>	<0.6,0.0,0.2>	<1.0,0.0,0.0>	<0.7,0.0,0.0>
A_3	<0.9,0.0,0.0>	<0.9,0.0,0.0>	<0.7,0.2,0.3>	<0.5,0.0,0.0>	<0.9,0.0,0.0>	<0.7,2.0,2.0>
A_4	<0.7,0.0,0.3>	<0.7,0.3,0.3>	<0.6,0.4,0.2>	<0.4,0.0,0.0>	<0.9,0.0,0.0>	<0.5,0.1,0.2>

Table 2 The ratings obtained from the second of the ten domain specialists

	C_1	C_2	C_3	C_4	C_5	C_6
A_1	<1.0,0.0,0.0>	<1.0,0.2,0.0>	<1.0,0.0,0.0>	<0.7,0.3,0.0>	<0.8,0.2,0.2>	<0.9,0.1,0.1>
A_2	<1.0,0.0,0.0>	<1.0,0.0,0.0>	<1.0,0.0,0.0>	<0.6,0.0,0.2>	<1.0,0.0,0.0>	<0.7,0.0,0.0>
A_3	<0.9,0.0,0.0>	<0.9,0.0,0.0>	<0.7,0.2,0.3>	<0.5,0.0,0.0>	<0.9,0.0,0.0>	<0.7,2.0,2.0>
A_4	<0.7,0.0,0.3>	<0.7,0.3,0.3>	<0.6,0.4,0.2>	<0.4,0.0,0.0>	<0.9,0.0,0.0>	<0.5,0.1,0.2>

Table 3 The ratings obtained from the third of the ten domain specialists

	C_1	C_2	C_3	C_4	C_5	C_6
A_1	<1.0,0.0,0.0>	<1.0,0.2,0.0>	<1.0,0.0,0.0>	<0.7,0.3,0.0>	<0.8,0.2,0.2>	<0.9,0.1,0.1>
A_2	<1.0,0.0,0.0>	<1.0,0.0,0.0>	<1.0,0.0,0.0>	<0.6,0.0,0.2>	<1.0,0.0,0.0>	<0.7,0.0,0.0>
A_3	<0.9,0.0,0.0>	<0.9,0.0,0.0>	<0.7,0.2,0.3>	<0.5,0.0,0.0>	<0.9,0.0,0.0>	<0.7,2.0,2.0>
A_4	<0.7,0.0,0.3>	<0.7,0.3,0.3>	<0.6,0.4,0.2>	<0.4,0.0,0.0>	<0.9,0.0,0.0>	<0.5,0.1,0.2>

- C_5, Performance
- C_6, Security

In this research, four cloud computing technologies were evaluated based on the opinions of ten domain specialists. The evaluated alternatives are shown below and the attitudes of three of the ten domain specialists examined are shown in Tables 1, 2, 3. It should be noted that the designation of alternatives is not in accordance with their appearance below, because the aim of this paper is not to declare any of the alternatives superior to the others.

The computing technologies evaluated in this research are:

- A_1, Azure Cloud
- A_2, Amazon Cloud
- A_3, Google Cloud
- A_4, IBM Cloud

5.1 Ranking Alternatives Using the Neutrosophic Extension of the MULTIMOORA Method

In the approach proposed by Stanujkic et al. [27], the group decision matrix is formed using SVNWA operator that is using Eq. (9). The group decision matrix formed using Eq. (9) is shown in Table 4.

Table 4 The group decision matrix

	C_1	C_2	C_3	C_4	C_5	C_6
A_1	<1.00, 0.00, 0.00>	<1.00, 0.00, 0.00>	<1.00, 0.00, 0.00>	<0.70, 0.30, 0.00>	<1.00, 0.00, 0.00>	<0.88, 0.00, 0.10>
A_2	<1.00, 0.00, 0.00>	<1.00, 0.00, 0.00>	<1.00, 0.00, 0.00>	<0.57, 0.00, 0.20>	<1.00, 0.00, 0.00>	<1.00, 0.00, 0.00>
A_3	<0.79, 0.00, 0.00>	<0.90, 0.00, 0.00>	<0.65, 0.20, 0.30>	<0.50, 0.00, 0.00>	<0.90, 0.00, 0.00>	<0.70, 0.00, 0.00>
A_4	<0.67, 0.00, 0.30>	<0.65, 0.30, 0.33>	<0.55, 0.40, 0.20>	<0.40, 0.00, 0.00>	<0.90, 0.00, 0.00>	<0.57, 0.00, 0.20>

Table 5 The ranking orders of the alternatives obtained on the basis of the RS approach

	Y_i	y_i	Rank
A_1	<1.00, 0.00, 0.00>	1.00	1
A_2	<1.00, 0.00, 0.00>	1.00	1
A_3	<0.79, 0.00, 0.00>	0.89	3
A_4	<0.68, 0.00, 0.00>	0.84	4

The ranking based on the RSA. The ranking results and the ranking order of the alternatives obtained based on the RS approach, i.e., by applying Eqs. (36) and (37), are accounted for in Table 5.

The ranking based on the RPA. The ranking of the alternatives based on the RP approach begins by determining the reference point, as it is shown in Table 6.

After that, the distances of alternatives with respect to coordinates of the reference point are calculated.The maximum distances of considered alternatives to the reference point, obtained using Eqs. (8) and (33), are presented in Table 7. The ranking order of the alternatives is also presented in Table 7.

The ranking based on the FMF. The ranking results and the ranking order of the alternatives obtained on the basis of the FMF approach, i.e., by applying Eqs. (41) and (42), are shown in Table 8.

The final ranking order of the alternatives which summarizes the three different ranks provided by the respective parts of the MULTIMOORA method is shown in Table 9.

As it can be seen from Table 9, all the three approaches, integrated in the MULTIMOORA, have resulted in different ranking orders, for which reason the final ranking order is determined based on the dominance theory.

5.2 Ranking Alternatives Using the Neutrosophic Extension of the MULTIMOORA Method and Modified Group Decision Matrix

As stated previously, the group decision matrix in the previous calculation was formed using the SVNWA operator. In the calculation shown below, the group ratings were calculated as the median of ratings obtained from the respondents, for each function separately (t, i and f).

Compared to Table 4, it can be seen that there are some differences in the values of SVNNs in Table 10.

The ranking based on the RSA. The calculation details and ranking order of alternatives obtained on the basis of data from Table 10 are shown in Table 11.

As can be seen from Table 12, the new approach used for forming the group decision matrix did not affect the ranking order of the alternatives.

The details of the calculations achieved by applying the PRA and FMF are shown in Tables 13 and 14, while comparisons of the two approaches are presented in

Table 6 The reference point

	C_1	C_2	C_3	C_4	C_5	C_6
r_j^*	<1.00, 0.00, 0.00>	<1.00, 0.00, 0.00>	<1.00, 0.00, 0.00>	<0.70, 0.00, 0.00>	<1.00, 0.00, 0.00>	<1.00, 0.00, 0.00>

Table 7 The ranking order of the alternatives obtained based on the RP approach

	d_i^{max}	Rank
A_1	0.02	1
A_2	0.03	2
A_3	0.06	3
A_4	0.08	4

Table 8 The ranking order of the alternatives obtained based on the FMF

	U_i	u_i	Rank
A_1	<0.92, 0.06, 0.02>	0.89	2
A_2	<0.91, 0.00, 0.04>	0.94	1
A_3	<0.72, 0.04, 0.06>	0.79	3
A_4	<0.60, 0.14, 0.19>	0.57	4

Table 9 The final ranking order of the alternatives according to the MULTIMOORA method

	RS	RP	FMF	Rank
A_1	1	1	2	1
A_2	1	2	1	1
A_3	3	3	3	3
A_4	4	4	4	4

Tables 15 and 16. The final ranking order of alternatives obtained on the basis of data from Table 10 is summarized in Table 17.

As can be seen from Table 12, the new approach used for forming the group decision matrix was only reflected in the ranking orders obtained by RP approach.

As can be inferred from the comparisons of Tables 9 and 17, the modifications performed in the group decision matrix, previously shown in Table 10, did not significantly affect the ranking order of considered alternatives.

5.3   Ranking Alternatives Using the Neutrosophic Extension of the MULTIMOORA Method and IVNNs

The details of calculation based on the MULTIMOORA method and IVNNs are shown below. Based on the data presented in Table 18, the overall importance, maximum distance, and overall utility of each alternative were calculated, as it is shown in Tables 19, 20, 21.

The calculation details of ranking alternatives based on the RS approach are shown in columns I and II of Table 19, that is, based on the left and right bounds of IVNNs. The ranking results obtained on the basis of the score function, that is, Eq. (16), are shown in column III of Table 19. As can be seen from Table 19, the alternative denoted as A_2 was first-ranked in all cases, indicating that the use of IVNNs did not have a significant influence on the ranking orders of alternatives, in this case.

Table 10 The group decision matrix

	C_1	C_2	C_3	C_4	C_5	C_6
A_1	<0.90, 0.20, 0.20>	<1.00, 0.00, 0.00>	<1.00, 0.00, 0.00>	<0.70, 0.30, 0.20>	<1.00, 0.00, 0.00>	<0.90, 0.10, 0.10>
A_2	<1.00, 0.00, 0.00>	<1.00, 0.00, 0.00>	<1.00, 0.00, 0.00>	<0.60, 0.00, 0.20>	<1.00, 0.00, 0.00>	<1.00, 0.10, 0.10>
A_3	<0.70, 0.30, 0.20>	<0.90, 0.00, 0.00>	<0.70, 0.20, 0.00>	<0.50, 0.00, 0.00>	<0.90, 0.00, 0.00>	<0.70, 0.20, 0.20>
A_4	<0.70, 0.00, 0.30>	<0.70, 0.30, 0.30>	<0.60, 0.40, 0.30>	<0.40, 0.00, 0.00>	<0.90, 0.00, 0.00>	<0.50, 0.00, 0.20>

Table 11 The ranking orders of the alternatives obtained on the basis of the RS approach

	Y_i	y_i	Rank
A_1	<1.00, 0.00, 0.00>	1.00	1
A_2	<1.00, 0.00, 0.00>	1.00	1
A_3	<0.78, 0.00, 0.00>	0.89	3
A_4	<0.69, 0.00, 0.00>	0.84	4

Table 12 The comparison of ranking orders obtained on the basis of the RS approach

	I		II	
	y_i	Rank	y_i	Rank
A_1	1.00	1	1.00	1
A_2	1.00	1	1.00	1
A_3	0.89	3	0.89	3
A_4	0.84	4	0.84	4

Table 13 The ranking order of the alternatives obtained based on the RP approach

	$d_i^{\max}$	Rank
A_1	0.03	2
A_2	0.02	1
A_3	0.05	3
A_4	0.06	4

Table 14 The ranking orders of the alternatives obtained on the basis of the FMF approach

	U_i	U_i	Rank
A_1	<0.91, 0.00, 0.00>	0.95	2
A_2	<0.92, 0.00, 0.00>	0.96	1
A_3	<0.73, 0.00, 0.00>	0.86	3
A_4	<0.64, 0.00, 0.00>	0.82	4

Table 15 The comparison of ranking orders obtained based on the RP approach

	I		II	
	$d_i^{\max}$	Rank	$d_i^{\max}$	Rank
A_1	0.02	1	0.03	2
A_2	0.03	2	0.02	1
A_3	0.06	3	0.05	3
A_4	0.08	4	0.06	4

Table 16 The comparison of ranking orders obtained on the basis of the FMF approach

	I		II	
	u_i	Rank	u_i	Rank
A_1	0.89	2	0.95	2
A_2	0.94	1	0.96	1
A_3	0.79	3	0.86	3
A_4	0.57	4	0.82	4

The calculation details of ranking alternatives based on the FMF are shown in Table 21. In columns I and II of Table 21 are shown ranking results based on left and right bounds of IVNNs, while in the column III are shown results obtained based on Eq. (16).

Table 17 The final ranking
order of the alternatives
according to the
MULTIMOORA method and
modified group decision
matrix

	RS	RP	FMF	Rank
A_1	1	2	2	2
A_2	1	1	1	1
A_3	3	3	3	3
A_4	4	4	4	4

The final ranking of alternatives based on the IVN MULTIMOORA are presented in Table 22.

5.4 Ranking Alternatives Using the Neutrosophic Extension of the MULTIMOORA Method and TVNNs

The calculation details on the MULTIMOORA method and TVNNs are presented below. Based on the data presented in Table 23, the overall importance, maximum distance, and overall utility of each alternative were calculated, as it is shown in Tables 24, 25, 26, 27.

As can be seen from Tables 24, 25, 26, the use of TVNNs did not influence the rank of the most appropriate alternative but it has an impact on the ranking order of the alternatives. The final ranking of alternatives obtained from the MULTIMOORA method and TVNNs are summarized in Table 27.

5.5 Final Considerations of the Use of MULTIMORA with Different Approaches

The final ranking orders of alternatives obtained using the intuitionistic extension of the MULTIMOORA method and four approaches for calculating group decision matrix, in group environment, are summarized in Table 28.

From Table 28, it can be concluded that the alternative denoted as $A2$ is selected as the most appropriate alternative by applying all the approaches, but it should be noticed that there is a certain difference in the ranking order of alternatives. However, what is important to note is that IVNNs and TVNNS contain significantly more information that can be used for analyzing different scenarios.

6 Conclusion

At the present time, cloud computing represents the evolution of Internet service that accelerates innovations in the computer industry. The proposed methodology has

Table 18 The IVNNs group decision matrix

	C_1	C_2	C_3	C_4	C_5	C_6
A_1	<[0.9, 1.0],[0.1, 0.6],[0.1, 0.6]>	<[1.0, 1.0],[0.0, 0.1],[0.0, 0.1]>	<[0.9, 1.0],[0.0, 0.1],[0.0, 0.1]>	<[0.7, 0.7],[0.3, 0.3],[0.1, 0.2]>	<[0.9, 1.0],[0.0, 0.1],[0.0, 0.1]>	<[0.9, 0.9],[0.1, 0.1],[0.1, 0.1]>
A_2	<[1.0, 1.0],[0.0, 0.0],[0.0, 0.0]>	<[1.0, 1.0],[0.0, 0.0],[0.0, 0.0]>	<[1.0, 1.0],[0.0, 0.0],[0.0, 0.0]>	<[0.6, 0.6],[0.0, 0.0],[0.2, 0.2]>	<[1.0, 1.0],[0.0, 0.0],[0.0, 0.0]>	<[0.9, 1.0],[0.0, 0.1],[0.1, 0.1]>
A_3	<[0.7, 0.8],[0.2, 0.3],[0.1, 0.2]>	<[0.9, 0.9],[0.0, 0.0],[0.0, 0.0]>	<[0.6, 0.7],[0.2, 0.2],[0.0, 0.0]>	<[0.5, 0.5],[0.0, 0.1],[0.0, 0.1]>	<[0.9, 0.9],[0.0, 0.1],[0.0, 0.1]>	<[0.7, 0.7],[0.1, 1.0],[0.1, 1.0]>
A_4	<[0.7, 0.7],[0.0, 0.0],[0.3, 0.3]>	<[0.6, 0.7],[0.3, 0.3],[0.3, 0.3]>	<[0.5, 0.6],[0.4, 0.4],[0.3, 0.3]>	<[0.4, 0.4],[0.0, 0.0],[0.0, 0.0]>	<[0.9, 0.9],[0.0, 0.1],[0.0, 0.1]>	<[0.5, 0.6],[0.0, 0.0],[0.2, 0.2]>

Table 19 The ranking orders of the alternatives obtained on the basis of the RS approach and IVNNs

		I		II		III	
		l	Rank	u	Rank	y_i	Rank
A_1	<[0.9, 1.0],[0.0, 0.1],[0.0, 0.1]>	0.95	2	0.80	4	0.87	3
A_2	<[1.0, 1.0],[0.0, 0.0],[0.0, 0.0]>	1.00	1	1.00	1	1.00	1
A_3	<[0.8, 0.8],[0.0, 0.0],[0.0, 0.0]>	0.88	3	0.90	2	0.89	2
A_4	<[0.7, 0.7],[0.0, 0.0],[0.0, 0.0]>	0.83	4	0.85	3	0.84	4

Table 20 The ranking order of the alternatives obtained based on the RP approach and IVNNs

	d_i^{max}	Rank
A_1	0.04	2
A_2	0.02	1
A_3	0.07	4
A_4	0.06	3

Table 21 The ranking orders of the alternatives obtained on the basis of the RP approach and IVNNs

		I		II		III	
		l	Rank	u	Rank	d_i^{max}	Rank
A_1	<[0.9, 0.9],[0.0, 0.1],[0.0, 0.1]>	0.93	2	0.76	4	0.84	3
A_2	<[0.9, 0.9],[0.0, 0.0],[0.0, 0.0]>	0.94	1	0.96	1	0.95	1
A_3	<[0.7, 0.7],[0.0, 0.0],[0.0, 0.0]>	0.85	3	0.87	2	0.86	2
A_4	<[0.6, 0.6],[0.0, 0.0],[0.0, 0.0]>	0.79	4	0.81	3	0.80	4

Table 22 The final ranking order of the alternatives according to the IVN MULTIMOORA method

	RS	RP	FMF	Rank
A_1	3	2	3	3
A_2	1	1	1	1
A_3	2	4	2	2
A_4	4	3	4	4

been able to fully respond to the task when it comes to cloud computing technology selection.

The MULTIMOORA method so far has been proven in solving different decision-making problems. In order to enable its usage, a numerous extensions have been also proposed.

In this paper, a possible improvement of the neutrosophic extension of the MULTIMOORA method based on interval-valued and triangular-valued neutrosophic numbers is considered. The proposed improvement considers the transformation of individual decision matrices into the group decision matrix whose elements are interval or triangular neutrosophic numbers, and its usage for conducting various analyzes.

The usage of interval-valued and triangular-valued neutrosophic numbers, instead of single-valued, is demonstrated on a numerical illustration of cloud computing technology selection. A conducted numerical illustration

Table 23 The TVNNs group decision matrix

	C_1	C_2	C_3	C_4	C_5	C_6
A_1	⟨0.9, 0.9, 1.0⟩(0.1, 0.2, 0.6)(0.1, 0.2, 0.6)⟩	⟨1.0, 1.0, 1.0⟩(0.0, 0.0, 0.1)(0.0, 0.0, 0.1)⟩	⟨0.9, 1.0, 1.0⟩(0.0, 0.0, 0.1)(0.0, 0.0, 0.1)⟩	⟨0.7, 0.7, 0.7⟩(0.3, 0.3, 0.3)(0.1, 0.2, 0.2)⟩	⟨0.9, 1.0, 1.0⟩(0.0, 0.0, 0.1)(0.0, 0.0, 0.1)⟩	⟨0.9, 0.9, 0.9⟩(0.1, 0.1, 0.1)(0.1, 0.1, 0.1)⟩
A_2	⟨1.0, 1.0, 1.0⟩(0.0, 0.0, 0.0)(0.0, 0.0, 0.0)⟩	⟨1.0, 1.0, 1.0⟩(0.0, 0.0, 0.0)(0.0, 0.0, 0.0)⟩	⟨1.0, 1.0, 1.0⟩(0.0, 0.0, 0.0)(0.0, 0.0, 0.0)⟩	⟨0.6, 0.6, 0.6⟩(0.0, 0.0, 0.0)(0.2, 0.2, 0.2)⟩	⟨1.0, 1.0, 1.0⟩(0.0, 0.0, 0.0)(0.0, 0.0, 0.0)⟩	⟨0.9, 1.0, 1.0⟩(0.1, 0.1, 0.1)(0.1, 0.1, 0.1)⟩
A_3	⟨0.7, 0.7, 0.8⟩(0.2, 0.3, 0.3)(0.1, 0.2, 0.2)⟩	⟨0.9, 0.9, 0.9⟩(0.0, 0.0, 0.0)(0.0, 0.0, 0.0)⟩	⟨0.6, 0.7, 0.7⟩(0.2, 0.2, 0.2)(0.0, 0.0, 0.0)⟩	⟨0.5, 0.5, 0.5⟩(0.0, 0.0, 0.1)(0.0, 0.0, 0.1)⟩	⟨0.9, 0.9, 0.9⟩(0.0, 0.0, 0.1)(0.0, 0.0, 0.1)⟩	⟨0.7, 0.7, 0.7⟩(0.1, 0.2, 1.0)(0.1, 0.2, 1.0)⟩
A_4	⟨0.7, 0.7, 0.7⟩(0.0, 0.0, 0.0)(0.3, 0.3, 0.3)⟩	⟨0.6, 0.7, 0.7⟩(0.3, 0.3, 0.3)(0.5, 0.3, 0.3)⟩	⟨0.5, 0.6, 0.6⟩(0.4, 0.4, 0.4)(0.3, 0.3, 0.3)⟩	⟨0.4, 0.4, 0.4⟩(0.0, 0.0, 0.0)(0.0, 0.0, 0.0)⟩	⟨0.9, 0.9, 0.9⟩(0.0, 0.0, 0.1)(0.0, 0.0, 0.1)⟩	⟨0.5, 0.5, 0.6⟩(0.0, 0.0, 0.0)(0.2, 0.2, 0.2)⟩

Table 24 The ranking orders of the alternatives obtained on TVNNs and the RS approach

		I		II		III		IV	
		l	Rank	m	Rank	u	Rank	y_i	Rank
A_1	<(0.9, 1.0, 1.0),(0.0, 0.0, 0.1),(0.0, 0.0, 0.1)>	0.95	2	1.00	1	0.80	4	0.97	2
A_2	<(1.0, 1.0, 1.0),(0.0, 0.0, 0.0),(0.0, 0.0, 0.0)>	1.00	1	1.00	1	1.00	1	1.00	1
A_3	<(0.8, 0.8, 0.8),(0.0, 0.0, 0.0),(0.0, 0.0, 0.0)>	0.88	3	0.89	3	0.90	2	0.93	3
A_4	<(0.7, 0.7, 0.7),(0.0, 0.0, 0.0),(0.0, 0.0, 0.0)>	0.83	4	0.84	4	0.85	3	0.89	4

Table 25 The ranking order of the alternatives obtained based on TVNNs and the RP approach

	d_i^{max}	Rank
A_1	0.12	2
A_2	0.04	1
A_3	0.21	3
A_4	0.26	4

Table 26 The ranking orders of the alternatives obtained on TVNNs and FMF

		I		II		III		IV	
		l	Rank	m	Rank	u	Rank	y_i	Rank
A_1	<(0.9, 0.9, 0.9),(0.1, 0.1, 0.2),(0.1, 0.1, 0.2)>	0.82	2	0.80	2	0.61	2	0.97	2
A_2	<(0.9, 0.9, 0.9),(0.0, 0.0, 0.0),(0.0, 0.1, 0.1)>	0.91	1	0.91	1	0.91	1	1.00	1
A_3	<(0.7, 0.7, 0.7),(0.1, 0.1, 0.5),(0.0, 0.1, 0.5)>	0.74	3	0.69	3	0.10	4	0.93	3
A_4	<(0.6, 0.6, 0.6),(0.1, 0.1, 0.2),(0.2, 0.2, 0.2)>	0.56	4	0.57	4	0.54	3	0.89	4

Table 27 The final ranking order of the alternatives according to the TVNNs and MULTIMOORA method

	RS	RP	FMF	Rank
A_1	2	2	2	2
A_2	1	1	1	1
A_3	3	3	4	3
A_4	4	4	3	4

Table 28 The final ranking order of the alternatives according to the MULTIMOORA method and modified group decision matrix

Approach	I	II	III	IV
A_1	1	2	3	2
A_2	1	1	1	1
A_3	3	3	2	3
A_4	4	4	4	4

demonstrates the possibilities and opportunities of the proposed extension. It is important to highlight that the interval-valued and triangular-valued neutrosophic numbers contain significantly more information that could be used for different scenarios. Therefore, justification of the proposed neutrosophic extension of the MULTIMOORA method based on interval-valued and triangular-valued neutrosophic numbers is verified.

References

1. Hankel, A., Heimeriks, G., & Lago, P. (2019). Green ICT adoption using a maturity model. *Sustainability, 11*(24), 7163.
2. Nachira, F., Dini, P., Nicolai, A., Le Louarn, M., & Rivera Lèon, L. (2007). *Digital business ecosystems: The results and the perspectives of the digital business ecosystem research and development activities in FP6*. Luxembourg: Office for Official Publications of the European Community.
3. Rajkumar, B., Broberg, J., & Goscinski, A. (2019). *Cloud computing principles and paradigms*. New York: John Wiley & Sons.
4. Stanoevska-Slabeva, K., & Wozniak, T. (2010). Cloud basics–an introduction to cloud computing. In *Grid and cloud computing* (pp. 47–61). Berlin, Heidelberg: Springer.
5. Plummer, D. C., Bittman, T. J., Austin, T., Cearley, D. W., & Smith, D. M. (2008). *Cloud computing: Defining and describing an emerging phenomenon*. Gartner, June, 17.
6. Rehman, Z., Hussain, O. K., & Hussain, F. K. (2012). Iaas cloud selection using MCDM methods. In *2012 IEEE Ninth international conference on e-business engineering* (pp. 246–251). IEEE.
7. Karabašević, D., Popović, G., Stanujkić, D., Maksimović, M., & Sava, C. (2019). An approach for hotel type selection based on the single-valued intuitionistic fuzzy numbers. *International Review, 1–2*, 7–14.
8. Naeini, A. B., Mosayebi, A., & Mohajerani, N. (2019). A hybrid model of competitive advantage based on Bourdieu capital theory and competitive intelligence using fuzzy Delphi and ism-gray Dematel (study of Iranian food industry). *International Review, 1–2*, 21–35.
9. Petrovic, I., & Kankaras, M. (2020). A hybridized IT2FS-DEMATEL-AHP-TOPSIS multicriteria decision making approach: Case study of selection and evaluation of criteria for determination of air traffic control radar position. *Decision Making: Applications in Management and Engineering, 3*(1), 146–164.
10. Stanujkić, D., & Karabašević, D. (2018). An extension of the WASPAS method for decision-making problems with intuitionistic fuzzy numbers: a case of website evaluation. *Operational Research in Engineering Sciences: Theory and Applications, 1*(1), 29–39.
11. Smarandache, F. (1998). *Neutrosophy probability set and logic*. Rehoboth: American Research Press.
12. Smarandache, F. (1999). *A unifying field in logics. Neutrosophy: Neutrosophic probability, set and logic*. Rehoboth: American Research Press.
13. Zadeh, L. A. (1965). Fuzzy sets. *Information and Control, 8*(3), 338–353.
14. Atanassov, K. T. (1986). Intuitionistic fuzzy sets. *Fuzzy Sets and Systems, 20*(1), 87–96.
15. Lee K. M. (2000). Bipolar-valued fuzzy sets and their basic operations. In Proceeding international conference, Bangkok, Thailand (pp. 307–317).
16. Yager, R. R. (2013). Pythagorean membership grades in multicriteria decision making. *IEEE Transactions on Fuzzy Systems, 22*(4), 958–965.
17. Atanassov, K., & Gargov, G. (1989). Interval valued intuitionistic fuzzy sets. *Fuzzy Sets and Systems, 31*(3), 343–349.
18. Wang, H., Smarandache, F., Zhang, Y. Q., & Sunderraman, R. (2010). Single valued neutrosophic sets. *Multispace and Multistructure, 4*, 410–413.
19. Li, Y., Liu, P., & Chen, Y. (2016). Some single valued neutrosophic number heronian mean operators and their application in multiple attribute group decision making. *Informatica, 27*(1), 85–110.
20. Sahin, R. (2014). Multi-criteria neutrosophic decision making method based on score and accuracy functions under neutrosophic environment. *arXiv preprint arXiv*:1412.5202.
21. Wang, H., Smarandache, F., Zhang, Y. Q., & Sunderraman, R. (2005). *Interval neutrosophic sets and logic: Theory and applications in computing*. Arizona: Hexis.
22. Biswas, P., Pramanik, S., & Giri, B. C. (2016). Aggregation of triangular fuzzy neutrosophic set information and its application to multi-attribute decision making. *Neutrosophic Sets and Systems, 12*, 20–40.

23. Ye, J. (2015). Trapezoidal neutrosophic set and its application to multiple attribute decision-making. *Neural Computing and Applications, 26*(5), 1157–1166.
24. Stanujkic, D., Zavadskas, E. K., Brauers, W. K. M., & Karabasevic, D. (2015). An extension of the MULTIMOORA method for solving complex decision-making problems based on the use of interval-valued triangular fuzzy numbers. *Transformations in Business and Economics, 14*(2B(35B)), 355–377.
25. Brauers, W. K. M., & Zavadskas, E. K. (2010). Project management by MULTIMOORA as an instrument for transition economies. *Technological and Economic Development of Economy, 16*(1), 5–24.
26. Brauers, W. K. M., & Zavadskas, E. K. (2011). MULTIMOORA optimization used to decide on a bank loan to buy property. *Technological and Economic Development of Economy, 17*(1), 174–188.
27. Stanujkic, D., Zavadskas, E. K., Smarandache, F., Brauers, W. K., & Karabasevic, D. (2017). A neutrosophic extension of the MULTIMOORA method. *Informatica, 28*(1), 181–192.
28. Büyüközkan, G., Göçer, F., & Feyzioğlu, O. (2018). Cloud computing technology selection based on interval-valued intuitionistic fuzzy MCDM methods. *Soft Computing, 22*(15), 5091–5114.

Neutrosophic Linear Differential Equation with a New Concept of Neutrosophic Derivative

Sandip Moi ⓘ, Suvankar Biswas ⓘ, and Smita Pal (Sarkar)

1 Introduction

The concept of fuzzy set was first introduced by L.A Zadeh [1, 2]. After that, many types of generalization of fuzzy sets have been done by various authors:

Types of fuzzy sets	Properties
Type 2 fuzzy sets [2]	Its membership function ranges over fuzzy sets of type 1
Type n fuzzy sets [2]	Its membership function ranges over fuzzy sets of type n-1
Intuitionistic fuzzy sets [3, 4]	It is characterized by membership function and non-membership function
Rough fuzzy set [5]	It is characterized by possible and certain membership function
Neutrosophic set [6–8]	It is characterized by truth-, indeterminacy-, and falsity-membership function

S. Moi · S. Pal (Sarkar)
Department of Mathematics, Indian Institute of Engineering Science and Technology, Shibpur, Howrah, India

S. Biswas (✉)
Department of Mathematics, Sonarpur Mahavidyalaya, Sahid Biswanath Sarani, Kolkata, India
e-mail: suvo180591@gmail.com

1.1 Neutrosophic Set

In the context of fuzzy mathematics, the concept of neutrosophic set which is the generalization of fuzzy set was first introduced by Smarandache [6–8]. Neutrosophic set was described independently by truth-membership function, indeterminacy-membership function, and falsity-membership function. Now, some work has been done [9–11], and many researchers are working on the neutrosophic set to develop the neutrosophic mathematics.

1.2 Neutrosophic Differential Equation

Many works and development on fuzzy differential equations have been done by many authors [12–17] where they have considered only the value of membership function. Like fuzzy, there are lots of scope to do generalization and development in neutrosophic differential equation by considering truth-membership function which is actually the membership function, indeterminacy-membership function, and falsity-membership function. But there are only few works done on neutrosophic differential equation. The work on neutrosophic differential equation was first introduced by Sumanthi et al. [18]. Then Sumanthi et al. [19] give a new approach on differential equation via trapezoidal neutrosophic number. Then Son et al. [20] solve the linear quadratic regulator problem governed by granular neutrosophic fractional differential equations.

After reading the above works, major drawbacks have been pointed out in the existing definition of neutrosophic derivative, and from there, we get the motivation to write this article. This article has been organized as follows. In Sect. 2, some preliminaries have been given which are related to our problem. Section 4 contains the definition of generalized neutrosophic derivative and the drawbacks of the definitions of neutrosophic derivative which are given in Son et al. [20] and Smarandache [21]. Section 4 contains the formulation of the problem. In Sect. 5, test examples have been given. Finally Sect. 6 contains a brief conclusion of this article.

2 Preliminaries

Definition 2.1 [22] Let U be a universe. A neutrosophic set X over U is defined by

$$X = \left\{ \left\langle x, \left(T_X(x), I_X(x), F_X(x) \right) \right\rangle : x \in U \right\}$$

where T_X, I_X, and F_X are called the truth-membership function, indeterminacy-membership function, and falsity-membership function, respectively. They are, respectively, defined by

$$T_X : U \to]0^-, 1^+[, \quad I_X : U \to]0^-, 1^+[, \quad F_X : U \to]0^-, 1^+[.$$

Definition 2.2 [22] Let U be a universe. A single-valued neutrosophic set (SVN-set) X over U is a neutrosophic set over U, but the truth, indeterminacy and falsity-membership function are, respectively, defined by $T_X : U \to]0^-, 1^+[, I_X : U \to]0^-, 1^+[, F_X : U \to]0^-, 1^+[$.

Definition 2.3 [23] Let U be a universal set. An interval neutrosophic set $X \in U$ is described by a truth-membership function $T_X(x)$, an indeterminacy-membership function $I_X(x)$, and falsity-membership function $F_X(x)$. For all $x \in U$, we have $T_X(x) = [\inf T_X(x), \mathrm{Sup} T_X(x)]$, $I_X(x) = [\inf I_X(x), \mathrm{Sup} I_X(x)]$, $F_X(x) = [\inf F_X(x), \mathrm{Sup} F_X(x)] \subseteq [0,1] \; \forall x \in U$. We only consider the sub-unitary interval of $[0, 1]$. It is the sub class of a neutrosophic set. Therefore, all interval neutrosophic sets are clearly neutrosophic sets.

Definition 2.4 [22] Let $\rho_X, \nu_X, \kappa_X \in [0, 1]$ be any real numbers and $a_i, b_i, c_i, d_i \in \mathbb{R}$ and $a_i \le b_i \le c_i \le d_i (i = 1, 2, 3)$. Then a single-valued neutrosophic number (SVN-number).
$X = \langle (a_1, b_1, c_1, d_1), \rho_X), ((a_2, b_2, c_2, d_2), \nu_X), ((a_3, b_3, c_3, d_3), \kappa_X) \rangle$ is a special neutrosophic set on the set of real numbers $\mathbb{R}$, whose truth-membership function T_X, indeterminacy-membership function I_X, and falsity-membership function F_X are, respectively, defined by

$$T_X : \mathbb{R} \to [0, \rho_X], T_X(x) = \begin{cases} f_T^l(x) & a_1 \le x < b_1 \\ \rho_X & b_1 \le x < c_1 \\ f_T^r(x) & c_1 \le x < d_1 \\ 0 & \text{Otherwise} \end{cases}$$

$$I_X : \mathbb{R} \to [\nu_X, 1], I_X(x) = \begin{cases} f_I^l(x) & a_2 \le x < b_2 \\ \nu_X & b_2 \le x < c_2 \\ f_I^r(x) & c_2 \le x < d_2 \\ 1 & \text{Otherwise} \end{cases}$$

$$F_X : \mathbb{R} \to [\kappa_X, 1], F_X(x) = \begin{cases} f_F^l(x) & a_3 \le x < b_3 \\ \kappa_X & b_3 \le x < c_3 \\ f_F^r(x) & c_3 \le x < d_3 \\ 1 & \text{Otherwise} \end{cases}$$

where the functions $f_T^l : [a_1, b_1] \to [0, \rho_X], f_I^r : [c_2, d_2] \to [\nu_X, 1], f_F^r : [c_3, d_3] \to [\kappa_X, 1]$ are continuous and non-decreasing, and satisfy the conditions: $f_T^l(a_1) = 0, f_T^l(b_1) = \rho_X, f_I^r(c_2) = \mu_X, f_I^r(d_2) = 1, f_F^r(c_3) = \kappa_X$ and $f_F^r(d_3) = 1$; then functions $f_T^r : [c_1, d_1] \to [0, \rho_X], f_I^l : [a_2, b_2] \to [\nu_X, 1], f_F^l : [a_3, b_3] \to [\kappa_X, 1]$ are continuous and non-increasing, and satisfy the conditions: $f_T^r(c_1) = \rho_X, f_T^r(d_1) = 0, f_I^l(a_2) = 1, f_I^l(b_2) = \nu_X, f_F^l(a_3) = 1$ and $f_F^l(b_3) = \kappa_X$. $[b_1, c_1], a_1$ and d_1 are called the mean interval and the lower and upper limits of the general neutrosophic number X for truth-membership function, respectively. $[b_2, c_2], a_2$ and d_2 are called the mean interval and the lower and upper limits of the general neutrosophic number X for the indeterminacy-membership function, respectively. $[b_3, c_3], a_3$ and d_3 are called the mean interval and the lower and upper limits of the general neutrosophic number X for the falsity-membership function, respectively. ρ_X, ν_X and κ_X are called the maximum truth-membership degree, minimum indeterminacy-membership degree and minimum falsity-membership degree, respectively.

Definition 2.5 [22] A single-valued triangular neutrosophic number (SVTN-number) $X = \langle(a, b, c); \rho_X, \nu_X, \kappa_X\rangle$ is a special neutrosophic set on the set of real numbers $\mathbb{R}$, whose truth- membership, indeterminacy-membership, and falsity-membership functions are, respectively, defined by

$$T_X(x) = \begin{cases} \left(\frac{x-a}{b-a}\right)\rho_X & \text{for } a \leq x \leq b \\ \left(\frac{c-x}{c-b}\right)\rho_X & \text{for } b \leq x \leq c \\ 0 & \text{Otherwise} \end{cases} \qquad I_X(x) = \begin{cases} \frac{(b-x+\nu_X(x-a))}{b-a} & \text{for } a \leq x \leq b \\ \frac{(x-b+\nu_X(c-x))}{c-b} & \nu\, b \leq x \leq c \\ 0 & \text{Otherwise} \end{cases}$$

$$F_X(x) = \begin{cases} \frac{(b-x+\kappa_X(x-a))}{b-a} & \text{for } a \leq x \leq b \\ \frac{(x-b+\kappa_X(c-x))}{c-b} & \text{for } b \leq x \leq c \\ 0 & \text{Otherwise} \end{cases}$$

Definition 2.6 [22] A single-valued trapezoidal neutrosophic number (SVTrN-number) $X = \langle(a, b, c); \rho_X, \nu_X, \kappa_X\rangle$ is a special neutrosophic set on the set of real numbers $\mathbb{R}$, whose truth-membership, indeterminacy-membership, and falsity-membership functions are, respectively, defined by

$$T_X(x) = \begin{cases} \left(\frac{x-a}{b-a}\right)\rho_X & \text{for } a \leq x \leq b \\ \rho_X & \text{for } b \leq x \leq c \\ \left(\frac{d-x}{d-c}\right)\rho_X & \text{for } c \leq x \leq d \\ 0 & \text{Otherwise} \end{cases} \qquad I_X(x) = \begin{cases} \frac{(b-x+\nu_X(x-a))}{b-a} & \text{for } a \leq x \leq b \\ \nu_X & \text{for } b \leq x \leq c \\ \frac{(x-c+\nu_X(d-x))}{d-c} & \text{for } c \leq x \leq d \\ 0 & \text{Otherwise} \end{cases}$$

$$F_X(x) = \begin{cases} \frac{(b-x+\kappa_X(x-a))}{b-a} & \text{for } a \leq x \leq b \\ \kappa_X & \text{for } b \leq x \leq c \\ \frac{(x-c+\kappa_X(d-x))}{d-c} & \text{for } c \leq x \leq d \\ 0 & \text{Otherwise} \end{cases}$$

Definition 2.7 [21] The definition of the neutrosophic derivative of the function $f_{Neu}(X)$ is defined by

$$f'_{Neu}(X) = \lim_{\sigma(H)\to 0} \frac{\langle\inf\, f\,(X + H) - \inf\, f(X), \sup\, f\,(X + H) - \sup\, f(X)\rangle}{H}$$

where $\langle a, b\rangle$ denote the open/closed/half-open-closed interval and $\sigma(H) = \max\{|\inf H|, |\sup H|\}$.

When H is an interval then the definition written as follows:

$$f'_{Neu}(X) = \lim_{[\inf H, \sup H]\to[0,0]} \frac{\langle\inf\, f\,(X + H) - \inf\, f(X), \sup\, f\,(X + H) - \sup\, f(X)\rangle}{[\inf H, \sup H]}$$

is the neutrosophic derivative of the function $f(X)$.

In a simplified way, one has

$$f'_{Neu}(X) = \lim_{h\to 0} \frac{[\inf\, f\,(X + H) - \inf\, f(X), \sup\, f\,(X + H) - \sup\, f(X)]}{h}$$

Both definitions are the generalizations of the classical derivative of a function, then for the crisp functions and for the crisp variables we have.

$[\inf H, \sup H] \equiv h$ and $\inf f(X + H) \equiv \sup f(X + H) \equiv f(x + h)$, $\inf f(X) \equiv \sup f(X) \equiv f(x)$.

Definition 2.8 [24] Let X and Y be two type-1 fuzzy numbers (T1FNs) whose horizontal membership functions [24] are $X^{gr}(\alpha, \mu_x)$ and $Y^{gr}(\alpha, \mu_y)$, respectively, and $\odot$ denote one of the four basic operations, i.e., addition, subtraction, multiplication, and division. Then $X \odot Y$ is a T1FN z such that $H(z) \triangleq X^{gr}(\alpha, \mu_x) \odot Y^{gr}(\alpha, \mu_y)$. The difference of X and Y is called granular difference (gr-difference).

Let $z = X \odot Y$. Then $[z]^{\alpha} = H^{-1}(X^{gr}(\alpha, \mu_x) \odot Y^{gr}(\alpha, \mu_y))$ always presents α-level sets of fuzzy number z.

Definition 2.9 [24] Let $g : [x, y] \subseteq \mathbb{R} \to E$ and $V_1, V_2. \ldots V_n$ are n distinct fuzzy numbers. Then the horizontal membership function of $g(x)$ at $x \in [x, y]$ is denoted by

$H(g(x)) \triangleq g^{gr}(x, \alpha, \mu_g)$ and defined as $g^{gr} : [x, y] \times [0, 1] \times \overbrace{[0, 1] \times \ldots \ldots [0, 1]}^{n} \to [p, q] \subseteq \mathbb{R}$ in which $\alpha_g \triangleq (\mu_{v_1, \ldots \ldots} \mu_{v_n})$ are the relative distance measure (RDM) ([24], p. 310–323) variables corresponding to the fuzzy numbers.

Definition 2.10 [20] Let $g : (x, y) \to E$ be a neutrosophic-valued function and $t_0 \in (x, y)$. Then g is granuler differentiable at the point t_0 if there is an element $\frac{d_{gr} g(t_0)}{dt} \in E$ such that $\lim_{h \to 0} \frac{g(t_0+h) \Theta^{gr} g(t_0)}{h} = \frac{d_{gr} g(t_0)}{dt}$ exists, where Θ is the gr-difference [24]. Then $\frac{d_{gr} g(t_0)}{dt}$ is called the granular derivative (gr-derivative) of the function g at t_0.

As a result, the function g is said to be gr-differentiable on the interval (x, y) if and only if the gr-derivative $\frac{d_{gr} g(t)}{dt}$ exists for all $t \in (x, y)$.

3   Generalized Neutrosophic Derivative

Definition 2.7 of the neutrosophic derivative of a function has some drawbacks. Let $f : (a, b) \to \xi$ be a neutrosophic-valued function defined as $f(x) = ng(x)$, where n be the neutrosophic number and $g : (0, 1) \to \mathbb{R}$ be a differentiable function such that $g(x)g'(x) < 0$, $\forall x \in (0, 1)$. Then neutrosophic derivative of the function $f(x)$ does not exist. Now, note that the derivative of the function $f(x)$ does not exist just because instead of taking n as crisp parameter, we have taken n as neutrosophic parameter. This is a serious drawback. For example, let us consider the function $f(x) = n\frac{1}{x}$, where $n = \langle (1, 2, 3); 0.7, 0.1, 0.1 \rangle$.

Now, we find the derivative of the function $f(x)$ for the truth-membership function. Then,

$$n\frac{1}{x} = \left[n_{T1}\frac{1}{x}, n_{T2}\frac{1}{x} \right] = \left[\left(1 + \frac{10\alpha}{7} \right) \frac{1}{x}, \left(3 - \frac{10\alpha}{7} \right) \frac{1}{x} \right] \qquad \text{(Now,)}$$

$$\lim_{[\inf H, \sup H] \to [0,0]} \frac{\langle \inf \ f(x+H) - \inf \ f(x), \sup \ f(x+H) - \sup \ f(x) \rangle}{[\inf \ H, \sup \ H]}$$

Then,

$$\left[\lim_{H \to 0} \frac{\left(1 + \frac{10\alpha}{7}\right)\left(\frac{1}{x+H} - \frac{1}{x}\right)}{H}, \ \lim_{H \to 0} \frac{\left(3 - \frac{10\alpha}{7}\right)\left(\frac{1}{x+H} - \frac{1}{x}\right)}{H} \right] = \left[-\left(1 + \frac{10\alpha}{7}\right)\frac{1}{x^2}, \right.$$

$$\left. -\left(3 - \frac{10\alpha}{7}\right)\frac{1}{x^2} \right].$$

But this is not an interval, because $-\left(1 + \frac{10\alpha}{7}\right)\frac{1}{x^2} > -\left(3 - \frac{10\alpha}{7}\right)\frac{1}{x^2}, \forall x \in$ (0, 1). In the similar manner we find the same result for the indeterministic and falsity membership function.

Now note that Definition 2.10 of the neutrosophic derivative of a function is better than Definition 2.7. Since this definition does not have the above drawback of Definition 2.7. But in Definition 2.10 of neutrosophic derivative has other drawbacks. If we find a series solution of a given differential equation as $y^{gr}(x, \alpha, \mu_y)$, $i.e$ $H^{-1}\left(y^{gr}(\alpha, \mu_y)\right) = [y]^{\alpha} =$

$$\left[\inf_{\beta \geq \alpha} \min_{\mu_y} y^{gr}(\beta, \mu_y), \sup_{\beta \geq \alpha} \max_{\mu_y} y^{gr}(\beta, \mu_y) \right], \quad \text{where} \quad \mu_y \triangleq (\mu_1, \mu_2, \dots, \mu_n),$$

then it is very difficult to find the infimum and supremum of $y^{gr}(x, \alpha, \mu_y)$ and as n increases the difficulty increases. In concern to the numerical solution of a differential equation, if the solution has been found in an infinite series form, then it is almost impossible to find the infimum and supremum in this case as there are infinitely many terms. Even if we take approximate solution, as a finite term of the infinite series solution, then also the complexity and difficulty for finding

$$H^{-1}\left(y^{gr}(\alpha, \mu_y)\right) = [y]^{\alpha} = \left[\inf_{\beta \geq \alpha} \min_{\mu_y} y^{gr}(\beta, \mu_y), \sup_{\beta \geq \alpha} \max_{\mu_y} y^{gr}(\beta, \mu_y) \right]$$

increases as we take better approximation by increasing the number of terms in the approximate solution from the infinite series solution. So, it is very difficult to find the α-level set of the vertical-membership function $y(x)$.

Therefore we are going to define a generalized neutrosophic derivative as follows.

Definition 3.1 Let $f : (a, b) \to \xi$ be a neutrosophic-valued function and $x_0 \in (a, b)$. Then the generalized neutrosophic derivative of $f(x)$ at x_0 is denoted by $\acute{f}(x_0)$ and defined by

1. $f'_{T\alpha} = \left[\min \left\{ f'_{T1}(x_0, \alpha), f'_{T2}(x_0, \alpha) \right\}, \right.$

   $\left. \max \left\{ f'_{T1}(x_0, \alpha), f'_{T2}(x_0, \alpha) \right\} \right], if \ f'_{T1}(x_0, \alpha), f'_{T2}(x_0, \alpha) \ exists.$

$$2. f'_{I\beta} = \left[\min\left\{f'_{I1}(x_0, \beta), f'_{I2}(x_0, \beta)\right\},\right.$$

$$\left.\max\left\{f'_{I1}(x_0, \beta), f'_{I2}(x_0, \beta)\right\}\right], \text{if } f'_{I1}(x_0, \beta), f'_{I2}(x_0, \beta) \text{ exists.}$$

$$3. f'_{F\gamma} = \left[\min\left\{f'_{F1}(x_0, \gamma), f'_{F2}(x_0, \gamma)\right\},\right.$$

$$\left.\max\left\{f'_{F1}(x_0, \gamma), f'_{F2}(x_0, \gamma)\right\}\right], \text{if } f'_{F1}(x_0, \gamma), f'_{F2}(x_0, \gamma) \text{ exists.}$$

$f'_{(x)}$, is said to be type-1 derivative at x_0 if $[f'(x_0)]_{(\alpha,\beta,\gamma)} = \langle[f'_{T1}(x_0, \alpha),$ $f'_{T2}(x_0, \alpha)], [f'_{I1}(x_0, \beta), f'_{I2}(x_0, \beta)], [f'_{F1}(x_0, \gamma), f'_{F2}(x_0, \gamma)]\rangle$ and type-2 derivative if $[f'(x_0)]_{(\alpha,\beta,\gamma)} = \langle[f'_{T2}(x_0, \alpha), f'_{T1}(x_0, \alpha)], [f'_{I2}(x_0, \beta), f'_{I1}(x_0, \beta)],$ $[f'_{F2}(x_0, \gamma), f'_{F1}(x_0, \gamma)]\rangle$. This definition of the neutrosophic derivative of a function has no drawbacks which occur in Definitions 2.7 and 2.10.

4   Formulation of the Problem

Let us consider the neutrosophic differential equation:

$$a_0(x)\frac{d^n y}{dx^n} + a_1(x)\frac{d^{n-1}y}{dx^{n-1}} + \cdots + a_{n-1}(x)\frac{dy}{dx} + a_n(x)y = f(x)$$

with the initial condition $y(x_0) = y_0$, where y_0 is a neutrosophic number and $a_0(x),$ $a_1(x), \ldots, a_n(x), f(x)$ are neutrosophic-valued function.

Taking (α, β, γ)-cut of the above equation, we find the linear systems of the form

$$\left.\begin{array}{l} a_{0T1}\dfrac{d^n y_{Ts_1}}{dx^n} + \cdots + a_{nT1}y_{Ts_n} = f_{T1}(x, \alpha) \\[2mm] a_{0T2}\dfrac{d^n y_{Ts'_1}}{dx^n} + \cdots + a_{nT2}y_{Ts'_n} = f_{T2}(x, \alpha) \end{array}\right\} \tag{4.1}$$

$$\left.\begin{array}{l} a_{0I1}\dfrac{d^n y_{Is_1}}{dx^n} + \cdots + a_{nI1}y_{Is_n} = f_{I1}(x, \beta) \\[2mm] a_{0I2}\dfrac{d^n y_{Ts'_1}}{dx^n} + \cdots + a_{nI2}y_{Is'_n} = f_{I2}(x, \beta) \end{array}\right\} \tag{4.2}$$

$$\left.\begin{array}{l} a_{0F1}\dfrac{d^n y_{Fs_1}}{dx^n} + \cdots + a_{nF1}y_{Fs_n} = f_{F1}(x, \gamma) \\[2mm] a_{0F2}\dfrac{d^n y_{Fs'_1}}{dx^n} + \cdots + a_{nF2}y_{Fs'_n} = f_{F2}(x, \gamma) \end{array}\right\} \tag{4.3}$$

where $s_i \in \{1, 2\}$ and $s'_i = \{1, 2\} \setminus \{s_i\}$.

The above equations can be written in the following form:

$$\left.\begin{array}{l} L_{T1}y_{T1} + L_{T2}y_{T2} = f_{T1}(x, \alpha) \\ L_{T3}y_{T1} + L_{T4}y_{T2} = f_{T2}(x, \alpha) \end{array}\right\} \tag{4.4}$$

$$L_{I1}y_{I1} + L_{I2}y_{I2} = f_{I1}(x, \beta) \left.\right\}$$
$$L_{I3}y_{I1} + L_{I4}y_{I2} = f_{I2}(x, \beta) \left.\right\} \qquad (4.5)$$

$$L_{F1}y_{F1} + L_{F2}y_{F2} = f_{F1}(x, \gamma) \left.\right\}$$
$$L_{F3}y_{F1} + L_{F4}y_{F2} = f_{F2}(x, \gamma) \left.\right\} \qquad (4.6)$$

where L_{K1}, L_{K2}, L_{K3}, and L_{K4} are linear differential operators with constant coefficients. That is L_{K1}, L_{K2}, L_{K3}, and L_{K4} are of the forms

$$L_{K1} \equiv c_{10}a_{0K1}D^n + \cdots + c_{1n}a_{nK1}, \, L_{K2} \equiv c'_{10}a_{0K1}D^n + \cdots + c'_{1n}a_{nK1}$$

$$L_{K3} \equiv c_{20}a_{0K2}D^n + \cdots + c_{2n}a_{nK2}, \, L_{K4} \equiv c'_{20}a_{0K2}D^n + \cdots + c'_{2n}a_{nK2}$$

where $K = T, I, F$. c_{ij} are either 0 or 1 and $c'_{ij} = \{0, 1\} \setminus c_{ij}, i = 1, 2 \, j = 0, 1, \ldots, n$.

Now we find the solution of the linear systems Eqs. (4.4), (4.5), and (4.6) by the operator method for linear systems with constant coefficients [25].

5 Example

In this section, we apply our given method on test problems. The results in Table 1 and Table 2 are calculated by Wolfram Mathematica 9.0 and the figures are drawn by MATLAB R2014a.

Example 1 Let us consider the neutrosophic differential equation:

$$\frac{d^2y}{dx^2} - y = ax \qquad (5.1)$$

with the initial conditions $y(0) = \langle(-1, 0, 1); 0.7, 0.6, 0.6\rangle$ and $y'(0) = \langle(-1, 0, 1); 0.5, 0.4, 0.4\rangle$ where $a = \langle(0, 1, 2); 0.6, 0.5, 0.4\rangle$.

Taking (α, β, γ)-cut of the Eq. (5.1), then we have

$$\frac{d^2y_{T1}}{dx^2} - y_{T2} = a_{T1}x \left.\right\}$$
$$\frac{d^2y_{T2}}{dx^2} - y_{T1} = a_{T2}x \left.\right\} \qquad (5.2)$$

$$\frac{d^2y_{I1}}{dx^2} - y_{I2} = a_{I1}x \left.\right\}$$
$$\frac{d^2y_{I2}}{dx^2} - y_{I1} = a_{I2}x \left.\right\} \qquad (5.3)$$

Table 1 Approximate solutions of $y_{T1}, y_{T2}, y_{I1}, y_{I1}, y_{F1}$, and y_{F2} at $x = 2$ for the different values of α, β, γ

$\alpha = (\alpha_1, \alpha_2, \alpha_3)$	y_{T1}	y_{T2}	$\beta = (\beta_1, \beta_2, \beta_3)$	y_{I1}	y_{I2}	$\gamma = (\gamma_1, \gamma_2, \gamma_3)$	$y_F 1$	$y_F 2$
(0,0,0)	0.0430072	3.21071	(0.5,0.6,0.4)	1.62686	1.62686	(0.4,0.6,0.4)	1.62686	1.62686
(0.2,0.2,0.2)	0.651395	2.60235	(0.8,0.8,0.8)	0.574314	2.67941	(0.8,0.8,0.8)	0.5016	2.75212
(0.6,0.7,0.5)	1.62686	1.62686	(1,1,1)	0.0430072	3.21071	(1,1,1)	0.0430072	3.21071

Table 2 Approximate solutions of $y_{T1}, y_{T2}, y_{I1}, y_{F1}$, and y_{F2} at $x = 2$ for the different values of α, β, γ

$\alpha = (\alpha_1, \alpha_2, \alpha_3)$	y_{T1}	y_{T2}	$\beta = (\beta_1, \beta_2, \beta_3)$	y_{I1}	y_{I2}	$\gamma = (\gamma_1, \gamma_2, \gamma_3)$	y_{F1}	y_{F2}
(0,0,0)	0.0430072	3.21071	(0.6,0.6,0.6)	1.62686	1.62686	(0.4,0.4,0.4)	1.62686	1.62686
(0.4,0.4,0.4)	0.83493	2.41878	(0.8,0.8,0.8)	0.83493	2.41878	(0.8,0.8,0.8)	0.57095	2.68276
(0.8,0.8,0.8)	1.62686	1.62686	(1,1,1)	0.0430072	3.21071	(1,1,1)	0.0430072	3.21071

$$\left.\begin{array}{l} \dfrac{d^2 y_{F1}}{dx^2} - y_{F2} = a_{F1}x \\ \dfrac{d^2 y_{F2}}{dx^2} - y_{F1} = a_{F2}x \end{array}\right\} \tag{5.4}$$

where $[a_{T1}(\alpha_1), a_{T2}(\alpha_1)] = \left[-\frac{5\alpha_1}{3}, 2 - \frac{5\alpha_1}{3}\right]$, $[a_{I1}(\beta_1), a_{I2}(\beta_1)] = [2(1 - \beta_1),$
$2\beta_2]$, and $[a_{F1}(\gamma_1), a_{F2}(\gamma_1)] = \left[-\frac{5(1-\gamma_1)}{3}, \frac{1+5\gamma_1}{3}\right]$ with the initial conditions

$$[y_{T1}(0, \alpha_2), y_{T2}(0, \alpha_2)] = \left[\frac{10\alpha_2}{7} - 1, 1 - \frac{10\alpha_2}{7}\right], [y_{I1}(0, \beta_2),$$

$$y_{I2}(0, \beta_2)] = \left[\frac{3}{2} - \frac{5\beta_2}{2}, \frac{5\beta_2}{2} - \frac{3}{2}\right],$$

$$[y_{F1}(0, \gamma_2), y_{F2}(0, \gamma_2)] = \left[\frac{3}{2} - \frac{5\gamma_2}{2}, \frac{\gamma_2}{2} - \frac{3}{2}\right],$$

$$\left[y'_{T1}(0, \alpha_3), y'_{T2}(0, \alpha_3)\right] = [2\alpha_3 - 1, 1 - 2\alpha_3],$$

$\left[y'_{I1}(0, \beta_3), y'_{I2}(0, \beta_3)\right] = \left[\frac{2}{3} - \frac{5\beta_3}{3}, \frac{5\beta_3}{3} - \frac{2}{3}\right]$ and $\left[y'_{F1}(0, \gamma_3), y'_{F2}(0, \gamma_3)\right] =$
$\left[\frac{2}{3} - \frac{5\gamma_3}{3}, \frac{5\gamma_3}{3} - \frac{2}{3}\right]$ where,
$\alpha_1 \in [0, 0.6], \alpha_2 \in [0, 0.7], \alpha_3 \in [0, 0.5], \beta_1 \in [0.5, 1], \beta_2 \in [0.6, 1], \beta_3 \in [0.4, 1],$
$\gamma_1 \in [0.6, 1], \gamma_2 \in [0.6, 1]$ and $\gamma_3 \in [0.4, 1]$.

Now we solve the system of Eq. 5.2 by the operator method for linear system with constant coefficients. Now we introduce operator notation and write the system Eq. (5.2) in the form

$$\left.\begin{array}{l} D^2 y_{T1} - y_{T2} = a_{T1}x \\ D^2 y_{T2} - y_{T1} = a_{T2}x \end{array}\right\} \tag{5.5}$$

Let $L_{T1} = D^2$ and $L_{T2} = -1$, the above system written as follows:

$$\left.\begin{array}{l} L_{T1}y_{T1} + L_{T2}y_{T2} = a_{T1}x \\ L_{T1}y_{T2} + L_{T2}y_{T1} = a_{T2}x \end{array}\right\} \tag{5.6}$$

Now we remove y_{T2} and we get the differential equation

$$\left(D^4 - 1\right) y_{T1} = \frac{3}{5}(2 - \alpha) x \tag{5.7}$$

The general solution of the differential Eq. (5.7) is

$$y_{T1}(x, \alpha) = c_1 \cos x + c_2 \sin x + c_3 e^x + c_4 e^{-x} + \left(\frac{5\alpha_1}{3} - 2\right) x$$

Now substituting the value of $y_{T1}(x, \alpha)$ into the first equation of the system Eq. (5.6), we get

$$y_{T1}(x, \alpha) = -c_1 \cos x - c_2 \sin x + c_3 e^x + c_4 e^{-x} - \left(\frac{5\alpha_1}{3} - 2\right) x$$

Now applying the given initial conditions, we obtain

$$c_1 + c_3 + c_4 = \frac{10\alpha_2}{7} - 1 \tag{5.8}$$

$$c_2 + c_3 - c_4 = 2\alpha_3 - \frac{5\alpha_1}{3} + 1 \tag{5.9}$$

$$c_1 - c_3 - c_4 = \frac{10\alpha_2}{7} - 1 \tag{5.10}$$

$$c_2 - c_3 + c_4 = 2\alpha_3 - \frac{5\alpha_1}{3} - 1 \tag{5.11}$$

Now we solve the Eqs. (5.8), (5.9), (5.10) and (5.11); then we find

$$c_1 = \frac{1}{7}(-7 + 10\alpha_2), \quad c_2 = \frac{1}{3}(-5\alpha_1 + 6\alpha_3), \quad c_3 = \frac{1}{2} \text{ and } c_4 = -\frac{1}{2}$$

Therefore the solution of the system Eq. (5.2) is

$$y_{T1}(x, \alpha) = \frac{1}{7}(-7 + 10\alpha_2) \cos x + \frac{1}{3}(-5\alpha_1 + 6\alpha_3) \sin x + \frac{1}{2} e^x - \frac{1}{2}e^{-x} + \left(\frac{5\alpha_1}{3} - 2\right) x$$
$$y_{T2}(x, \alpha) = -\frac{1}{7}(-7 + 10\alpha_2) \cos x - \frac{1}{3}(-5\alpha_1 + 6\alpha_3) \sin x + \frac{1}{2} e^x - \frac{1}{2}e^{-x} - \frac{5\alpha_1}{3}x$$

where $\alpha = (\alpha_1, \alpha_2, \alpha_3)$.

In the similar manner we solve the system of Eqs. 5.3 and 5.4. Then the solutions of the systems are

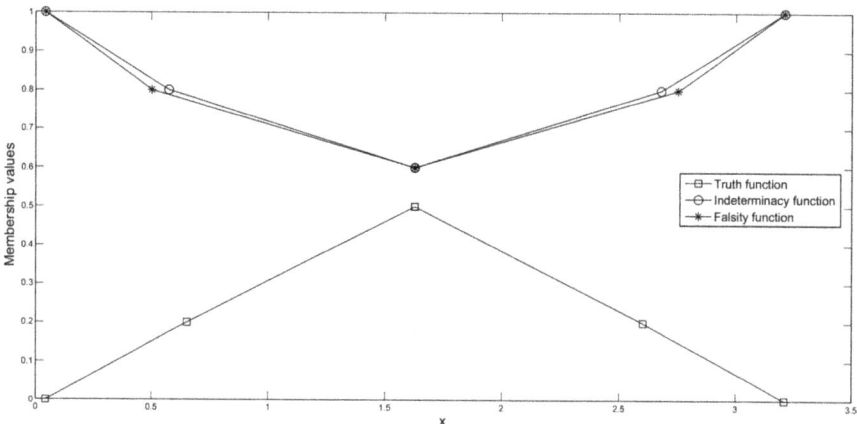

Fig. 1 The graph of y at $x = 2$ for left branch and right branch of truth-, indeterminacy-, and falsity-membership functions, respectively

$$y_{I1}(x, \beta) = \frac{1}{2}(3 - 5\beta_2)\cos x + \frac{1}{3}(6\beta_1 - 5\beta_3 - 1)\sin x + \frac{1}{2}e^x - \frac{1}{2}e^{-x} - 2\beta_1 x$$

$$y_{I2}(x, \beta) = -\frac{1}{2}(3 - 5\beta_2)\cos x - \frac{1}{3}(6\beta_1 - 5\beta_3 - 1)\sin x + \frac{1}{2}e^x - \frac{1}{2}e^{-x} - 2(1 - \beta_1)x$$

$$y_{F1}(x, \gamma) = \frac{1}{2}(3 - 5\gamma_2)\cos x + \frac{5}{3}(\gamma_1 - \gamma_3)\sin x + \frac{1}{2}e^x - \frac{1}{2}e^{-x} - \frac{1}{3}(1 + 5\gamma_1)$$

$$y_{F2}(x, \gamma) = -\frac{1}{2}(3 - 5\gamma_2)\cos x - \frac{5}{3}(\gamma_1 - \gamma_3)\sin x + \frac{1}{2}e^x - \frac{1}{2}e^{-x} - \frac{5}{3}(1 - \gamma_1)x$$

where $\beta = (\beta_1, \beta_2, \beta_3)$ and $\gamma = (\gamma_1, \gamma_2, \gamma_3)$.

In Table 1 it has been seen that when α increases, the solution of left branch for truth-membership function at $x = 2$ increases and the solution of right branch for truth-membership function decreases. For $\alpha = (\alpha_1, \alpha_2, \alpha_3) = (0.6, 0.7, 0.5)$, the left and right branch of truth-membership function gives the same solution. Similarly, when β and γ increases, the solution of left branch for indeterminacy- and falsity-membership function at $x = 2$ decreases and the solution of right branch increases. For $\beta = (\beta_1, \beta_2, \beta_3) = (0.5, 0.6, 0.4)$ and $\gamma = (\gamma_1, \gamma_2, \gamma_3) = (0.4, 0.6, 0.4)$, the left and right branch of indeterminacy- and falsity-membership function, respectively, gives the same solution. So, from Table 1 it has been seen that the solutions at $x = 2$

exists as neutrosophic number. In Fig. 1 it has been seen that the graph of y at $x = 2$ is a triangular neutrosophic number.

Example 2 We consider the same example (5.1) with different initial condition $y(0) = \langle(-1, 0, 1); 0.8, 0.6, 0.4\rangle$ and $y'(0) = \langle(-1, 0, 1); 0.8, 0.6, 0.4\rangle$ where $a = \langle(0, 1, 2); 0.8, 0.6, 0.4\rangle$.

Then the solution of the neutrosophic differential equation is

$$y_{T1}(x, \alpha) = \left(-1 + \tfrac{5\alpha_2}{4}\right) \cos x - \tfrac{5}{4}(\alpha_1 - \alpha_3) \sin x + \tfrac{1}{2} e^x - \tfrac{1}{2}e^{-x} + \left(\tfrac{5\alpha_1}{4} - 2\right) x$$
$$y_{T2}(x, \alpha) = -(-1 + 10\alpha_2) \cos x + \tfrac{5}{4}(\alpha_1 - \alpha_3) \sin x + \tfrac{1}{2} e^x - \tfrac{1}{2}e^{-x} - \tfrac{5\alpha_1}{4} x$$

$$y_{I1}(x, \beta) = \tfrac{1}{2}(3 - 5\beta_2) \cos x + \tfrac{5}{2}(\beta_1 - \beta_3) \sin x + \tfrac{1}{2} e^x - \tfrac{1}{2}e^{-x} - \tfrac{1}{2}(-1 + 5\beta_1) x$$
$$y_{I2}(x, \beta) = -\tfrac{1}{2}(3 - 5\beta_2) \cos x - \tfrac{5}{2}(\beta_1 - \beta_3) \sin x + \tfrac{1}{2} e^x - \tfrac{1}{2}e^{-x} - \tfrac{5}{2}(1 - \beta_1) x$$

$$y_{F1}(x, \gamma) = \tfrac{1}{3}(2 - 5\gamma_2) \cos x + \tfrac{5}{3}(\gamma_1 - \gamma_3) \sin x + \tfrac{1}{2} e^x - \tfrac{1}{2}e^{-x} - \tfrac{1}{3}(1 + 5\gamma_1) x$$

$$y_{F2}(x, \gamma) = -\tfrac{1}{3}(2 - 5\gamma_2) \cos x - \tfrac{5}{3}(\gamma_1 - \gamma_3) \sin x + \tfrac{1}{2} e^x - \tfrac{1}{2}e^{-x} - \tfrac{5}{3}(1 - \gamma_1) x$$

where $\alpha = (\alpha_1, \alpha_2, \alpha_3)$, $\beta = (\beta_1, \beta_2, \beta_3)$ and $\gamma = (\gamma_1, \gamma_2, \gamma_3)$.

From Table 2 also we can make some conclusion like Table 1. The solution exists as a neutrosophic number. In Fig. 2 also it has been seen that the graph of y at $x = 2$ is a triangular neutrosophic number for Example 2.

6   Conclusion

It has been shown in Sect. 3 that the existing Definition 2.7 of neutrosophic derivative does not exist for the function $f(x) = ng(x)$, where n be the neutrosophic number and $g : (0, 1) \to \mathbb{R}$ is a differentiable function such that $g(x)g'(x) < 0$, $\forall x \in (0, 1)$. It is not always possible to find the neutrosophic derivative (Definition 2.10) in practical even if it exists. It has also been shown that Definition 3.1 introduced in this article is better than the existing Definitions 2.7 and 2.10 in terms of the drawbacks understanding and finding the derivative. A method has been modified in the neutrosophic system to find the exact solution. From Sect. 4 it can be concluded that any nth order linear neutrosophic differential equation can be solved by the given method. This is the first time where a general *nth*-order linear differential equation has been solved in the neutrosophic system. From Table 1, Table 2 and Fig. 1, Fig. 2 it can be seen that the solution exists as

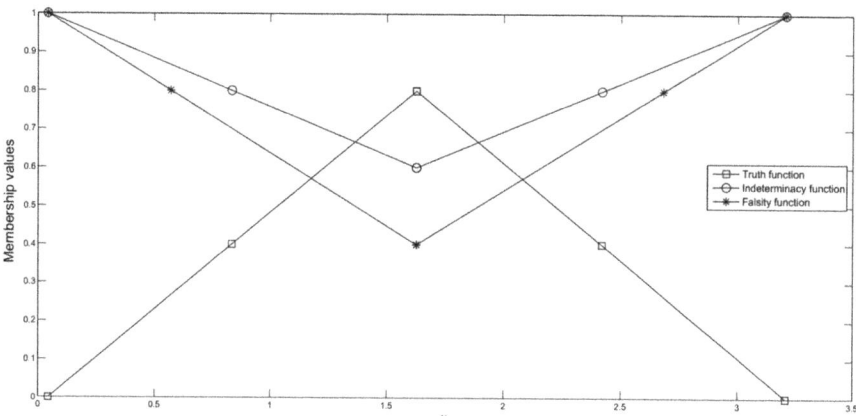

Fig. 2 The graph of y at $x = 2$ for left branch and right branch of truth-, indeterminacy- and falsity membership functions, respectively

a triangular neutrosophic number as the parameters were taken in the form of triangular neutrosophic number.

Acknowledgments In this article the study of Sandip Moi is funded by the Council of Scientific and Industrial Research (CSIR) government of India (File No.- 08/003(0135)/2019-EMR-I).

References

1. Chang, S. S & Zadeh, L. A. (1996). On fuzzy mapping and control. In Fuzzy sets, fuzzy logic, and fuzzy systems: selected papers by Lotfi A Zadeh, World Scientific, 180–184.
2. Zadeh, L. A. (1975). The concept of a linguistic variable and its application to approximate reasoning—I. *Information Sciences, 8*(3), 199–249.
3. Atanassov, K. T. (1986). Intuitionistic fuzzy sets. *Fuzzy Sets and Systems, 20*(1), 87–96.
4. Atanassov, K. T. (1999). Intuitionistic fuzzy sets. In *Intuitionistic fuzzy sets* (pp. 1–137). Springer.
5. Dubois, D., & Prade, H. (1990). Rough fuzzy sets and fuzzy rough sets*. *International Journal of General Systems, 17*(2–3), 191–209.
6. Smarandache, F. (2003). Proceedings of the first international conference on Neutrosophy, Neutrosophic Logic, Neutrosophic Set, Neutrosophic Porbability and Statistics.
7. Smarandache, F. (1999). A unifying field in logics: Neutrosophic logic.
8. Smarandache, F. (2005). Neutrosophic set-a generalization of the intuitionistic fuzzy set. *International Journal of Pure and Applied Mathematics, 24*(3), 287.
9. Haibin, W., Smarandache, F., Zhang, Y. & Sunderraman, R. (2010). Single valued neutrosophic sets. Infinite Study.
10. Haibin, W., Smarandache, Y., Sunderraman, R., & Zhang, Y. (2005). Interval neutrosophic sets and logic: theory and applications in computing: Theory and applications in computing. *Infinite Study, 5*.
11. Broumi, S., & Smarandache, F. (2013). Several similarity measures of neutrosophic sets. *Infinite Study*.

12. Kaleva, O. (1987). Fuzzy differential equations. *Fuzzy Sets and Systems, 24*(3), 301–317.
13. Seikkala, S. (1987). On the fuzzy initial value problem. *Fuzzy Sets and Systems, 24*(3), 319–330.
14. Buckley, J. J., & Feuring, T. (2000). Fuzzy differential equations. *Fuzzy Sets and Systems, 110*(1), 43–54.
15. Ahmad, M. Z., Hasan, M. K., & Baets, B. D. (2013). Analytical and numerical solutions of fuzzy differential equations. *Information Sciences, 236*, 156–167.
16. Biswas, S., & Roy, T. K. (2016). Adomian decomposition method for fuzzy differential equations with linear differential operator. *Journal of Information and Computing Science, 11*(4), 243–250.
17. Biswas, S., & Roy, T. K. (2019). A semianalytical method for fuzzy integro-differential equations under generalized seikkala derivative. *Soft Computing, 23*(17), 7959–7975.
18. Sumathi, I. R., & Priya, V. M. (2018). A new perspective on Neutrosophic differential equation. *Infinite Study*.
19. Sumathi, I. R., & Sweety, C. A. C. (2019). New approach on differential equation via trapezoidal neutrosophic number. *Complex & Intelligent Systems, pages*, 1–8.
20. Son, N. T. K., Dong, N. P., Long, H. V., & Khastan, A. (2019). Linear quadratic regulator problem governed by granular neutrosophic fractional differential equations. *ISA transactions*.
21. Smarandache, F. (2015). Neutrosophic Precalculus and Neutrosophic Calculus: Neutrosophic applications. *Infinite Study*.
22. Deli, I., & Şubaş, Y. (2017). A ranking method of single valued neutrosophic numbers and its applications to multi-attribute decision making problems. *International Journal of Machine Learning and Cybernetics, 8*(4), 1309–1322.
23. Sun, H., Yang, H., Wu, J., & Ouyang, Y. (2015). Interval neutrosophic numbers choquet integral operator for multi-criteria decision making. *Journal of Intelligent & Fuzzy Systems, 28*(6), 2443–2455.
24. Mazandarani, M., Pariz, N., & Kamyad, A. V. (2017). Granular differentiability of fuzzy-number-valued functions. *IEEE Transactions on Fuzzy Systems, 26*(1), 310–323.
25. Ross, S. L. (2007). *Differential Equation* (3rd ed.). Limited: Wiley India Pvt.

Linear and Nonlinear Programming Methods of Two-Person Interval-Valued Neutrosophic Games

İrfan Deli

1 Introduction

Since game theory [1–3] has a remarkable importance in the field of decision theory, many real problems can be modeled as a game in optimization. In real-life management problems, the information in games is rather uncertain due to their complexity and diversity, and it is useful to model the problems as games with fuzzy set [4], which is called fuzzy matrix game [5]; intuitionistic fuzzy set [6] is called intuitionistic fuzzy matrix game [7, 8], neutrosophic sets [9] is called neutrosophic matrix game [10], and so on. "Membership degrees of neutrosophic sets with truth-membership degrees of neutrosophic sets and non-membership degrees of intu-itionistic fuzzy sets with both indeterminacy-membership and falsity-membership of neutrosophic sets have many similar features. Moreover, the constructs of game with neutrosophic sets are remarkably similar to intuitionistic fuzzy games, and as a result, the models and methods of intuitionistic fuzzy games can be extended to neutrosophic games" [10]. Therefore, some authors gave the application of neutrosophic game theory in [11–13]. However, it may not be easy to identify exact values for the degree of truth-membership, the degree of indeterminacy-membership, and the falsity-membership degrees of neutrosophic sets due to the complexity and diversity of game (based on the comment of Li [7, 8] in the intuitionistic fuzzy games). As far as we know, there is no investigation on matrix games with payoffs of interval-valued neutrosophic sets. Therefore, by inspiring and/or generalizing the study in [1, 3, 5, 7, 8, 10], this chapter will focus on definition and solution of two-

İ. Deli (✉)
Muallim Rıfat Faculty of Education, Kilis 7 Aralık University, Kilis, Turkey
e-mail: irfandeli@kilis.edu.tr; irfandeli20@gmail.com

© The Author(s), under exclusive license to Springer Nature Switzerland AG 2021
F. Smarandache, M. Abdel-Basset (eds.), *Neutrosophic Operational Research*,
https://doi.org/10.1007/978-3-030-57197-9_20

person interval-valued neutrosophic matrix game based on linear programming and
nonlinear programming.

2 Preliminaries

We now recall some basic notions of neutrosophic set, interval neutrosophic set,
game theory, neutrosophic game theory, and so on. For more details, the reader may
refer to [1-19].

Definition 1 [14] Let X be a universe of discourse; then the neutrosophic set A is
an object having the form.

$A = \{< x: T_A (x) I_A (x), F_A (x) >, x \in X\}$ where the functions T_A, I_A,
$F_A: X \to [0,1]$ define, respectively, the degree of truth-membership, the degree of
indeteminacy-membership, and the degree of falsity-membership of the element x
$\in$ X to the set A with the condition.

$$0 \le T_A (x) + I_A (x) + F_A (x) \le 3$$

Definition 2 [15] A neutrosophic set A is contained in another neutrosophic set B,
denoted $A \subseteq B$ if $\forall x \in X$, $\mu_A(x) \le \mu_B(x)$, $\nu_A(x) \ge \nu_B(x)$, $\omega_A(x) \ge \omega_B(x)$.

Also [10], if $A \subseteq B$, then max{A,B} = B and min{A,B} = A.

Definition 3 [1] A matrix game is a two player game, denoted by G = (Player
$1 = P_1$, Player $2 = P_2$, S_1, S_2), such that

1. P_1 has a finite strategy set S_1 with m elements.
2. P_2 has a finite strategy set S_2 with n elements.
3. The payoffs of the players are functions $u_1 (s^1, s^2)$ and $u_2 (s^1, s^2)$ of the outcomes
   $(s^1, s^2) \in S_1 \times S_2 z$.

The matrix game is played as follows: At a certain time P_1 chooses a strategy
$s^1 \in S_1$ and simultaneously P_2 chooses a strategy $s^2 \in S_2$. Once this is done, each
P_i (i = 1, 2) receives the payoff $u_i (s^1, s^2)$ (i = 1, 2). If $S_1 = \{s_1^1, s_2^1, \dots, s_m^1\}$, $S_2 = \{s_1^2, s_2^2, \dots, s_m^2\}$ (S_1 and S_2 are called all pure strategy sets available for players P_1
and P_2, respectively.) and we put $a_{ij} = u_1 (s_i^1, s_j^2)$ and $b_{ij} = u_2 (s_i^1, s_j^2)$, then the
payoffs can be arranged in the form of the m $\times$ n matrix shown below:

$$[a_{ij}]_{m \times n} = \begin{pmatrix} (a_{11}, b_{11}) & (a_{12}, b_{12}) & \cdots & (a_{1n}, b_{1n}) \\ (a_{21}, b_{21}) & (a_{22}, b_{22}) & & (a_{2n}, b_{2n}) \\ \vdots & & \ddots & \vdots \\ (a_{m1}, b_{m1}) & (a_{m2}, b_{m2}) & \cdots & (a_{mn}, b_{mn}) \end{pmatrix}$$ which is called an m $\times$

n matrix game of the matrix game G.

Definition 4 [5] Let R^n be the n-dimensional Euclidean space and R_+^n be its nonnegative orthant. Let

M be an m $\times$ n real matrix, $e^T = (1,1,\dots, 1)$ be a vector of ones whose dimension
is specified as per the specific context. By a crisp two-person zero-sum matrix game

G, we mean the triplet $G = (S^m, S^m, M)$ where $S^m = \{x \in R_+^m, e^T x = 1\}$ and $S^n = \{y \in R_+^n, e^T y = 1\}$

In the terminology of the matrix game theory, S^m (respectively S^n) is called the strategy space for P_1 (respectively P_2) and M is called the payoff matrix. Also it is a convention to assume that P_1 is a maximizing player and P_2 is a minimizing player. Further for $x \in S^m, y \in S^n$, the scalar x^T Ay is the payoff to P_1 and the matrix game G is zero sum, the payoff to P_2 is $-x^T$ Ay. Here, $S^m = \{x \in R_+^m, e^T x = 1\}$ and $S^n = \{y \in R_+^n, e^T y = 1\}$ is called mixed strategy sets available for players P_1 and P_2, respectively.

Definition 5 [10] A two-person simplified neutrosophic game (TPSN-game) is a two player game, denoted by $N = (S_1, S_2, A)$, such that

1. P_1 has a finite strategy set S_1 with m elements.
2. P_2 has a finite strategy set S_2 with n elements
3. $\hat{A}$ is a simplified neutrosophic set over $S_1 \times S_2$ that is defined as follows:
$$\hat{A} = \{< (s^1, s^2), (T_{\hat{A}}(s^1, s^2), I_{\hat{A}}(s^1, s^2), F_{\hat{A}}(s^1, s^2))>: (s^1, s^2) \in S_1 \times S_2\}$$

The TPSN-game is played as follows: Simultaneously, P_1 chooses $s^1 \in S_1$ and P_2 chooses $s^2 \in S_2$, each unaware of the choice of the other, then the payoff for P_1 is expressed with $< T_{\hat{A}}(s^1, s^2), I_{\hat{A}}(s^1, s^2), F_{\hat{A}}(s^1, s^2)>$. In this word, P_1 gains the payoff expressed $< T_{\hat{A}}(s^1, s^2), I_{\hat{A}}(s^1, s^2), F_{\hat{A}}(s^1, s^2) >$ at the situation (s^1, s^2). The outcomes of P_2 on the situation (s^1, s^2) is negations of that of P_1.

If $S_1 = \{s_1^1, s_2^1, \ldots, s_m^1\}$, $S_2 = \{s_1^2, s_2^2, \ldots, s_m^2\}$ and we put $A_{ij} = \left(T_{\hat{A}}\left(s_i^1, s_j^2\right), I_{\hat{A}}\left(s_i^1, s_j^2\right), F_{\hat{A}}\left(s_i^1, s_j^2\right) \right)$, then the payoffs of TPSN-game can be arranged in the form of the m × n matrix shown below:

$$[A_{ij}]_{m \times n} - \begin{pmatrix} A_{11} & A_{12} & \cdots & A_{1n} \\ A_{21} & A_{22} & & A_{2n} \\ \vdots & & \ddots & \vdots \\ A_{m1} & A_{m2} & \cdots & A_{mn} \end{pmatrix}$$ which is called an m × n matrix game of

TPSN-game N.

Definition 6 [16] For $\hat{a} = [a_L, a_U]$

1. The minimization problem is defined as follows:

$$\min \{\hat{a}\}$$
$$s.t. \ \hat{a} \in \Omega_1,$$

or

$$\min\left\{a_U, m\left(\widehat{a}\right) = \frac{a_L + a_U}{2}\right\}$$
$$s.t.\ \widehat{a} \in \Omega_1,$$

The maximization problem is defined as follows:

$$\max\left\{\widehat{a}\right\}$$
$$s.t.\ \widehat{a} \in \Omega_2,$$

or

$$\max\left\{a_L, m\left(\widehat{a}\right) = \frac{a_L + a_U}{2}\right\}$$
$$s.t.\ \widehat{a} \in \Omega_2,$$

where Ω_1 and Ω_2 are the set of collection of $\widehat{a}$.

Definition 7 [17] Let X be a universe of discourse; then the interval-valued neutrosophic set (for short IVNS) A is an object having the form
$A = \{<x, [T_A^L(x), T_A^U(x)], [I_A^L(x), I_A^U(x)], [F_A^L(x), F_A^U(x)]> \mid x \in X\}$ where the functions $[T_A^L(x), T_A^U(x)], [I_A^L(x), I_A^U(x)], [F_A^L(x), F_A^U(x)] \subseteq [0,1]$ define, respectively, the degree of truth-membership, the degree of indeterminacy-membership, and the degree of falsity-membership of the element $x \in X$ to the set A with the condition $0 \leq T_A^U(x) + I_A^U(x) + F_A^U(x) \leq 3$.

Definition 8 [17, 18] Let
$A = \{<x, [T_A^L(x), T_A^U(x)], [I_A^L(x), I_A^U(x)], [F_A^L(x), F_A^U(x)]> \mid x \in X\}$ and
$B = \{<x, [T_B^L(x), T_B^U(x)], [I_B^L(x), I_B^U(x)], [F_B^L(x), F_B^U(x)]> \mid x \in X\}$ be two IVNSs and $\lambda > 0$. Then

1. $A \subseteq B$ if and only if $T_A^L(x) \leq T_B^L(x), T_A^U(x) \leq T_B^U(x), I_A^L(x) \geq I_B^L(x), I_A^U(x) \geq I_B^U(x), F_A^L(x) \geq F_B^L(x), F_A^U(x) \geq F_B^U(x)$.
   (Also, if $A \subseteq B$ then $\max\{A, B\} = B$ and $\min\{A, B\} = A$)
2. $A+B = \{<x, [T_A^L(x) + T_B^L(x) - T_A^L(x).T_B^L(x), T_A^U(x) + T_B^U(x) - T_A^U(x).T_B^U(x)], [I_A^L(x).I_B^U(x), I_A^L(x).I_B^U(x)], [F_A^L(x).F_B^U(x), F_A^L(x).F_B^U(x)]> \mid x \in X\}$
3. $A.B = \{<x, [T_A^L(x).T_B^L(x), T_A^U(x).T_B^U(x)],$

$$\left[I_A^L(x) + I_B^L(x) - I_A^L(x).I_B^L(x), I_A^U(x) + I_A^U(x) - I_A^U(x).I_A^U(x)\right],$$

$$\left[F_A^L(x) + F_B^L(x) - F_A^L(x).F_B^L(x), F_A^U(x) + F_A^U(x) - F_A^U(x).F_A^U(x)\right]> \qquad \mid$$
$x \in X\}$

4. $\lambda A = \{<x, [1 - \left(1 - T_A^L(x)\right)^\lambda, 1 - \left(1 - T_A^U(x)\right)^\lambda], \left[\left(I_A^L(x)\right)^\lambda, \left(I_A^U(x)\right)^\lambda\right],$
$\left[\left(F_A^L(x)\right)^\lambda, \left(F_A^U(x)\right)^\lambda\right]> \mid x \in X\}$

5. $A^\lambda = \{<x, [(T_A^L(x))^\lambda, (T_A^U(x))^\lambda], [1 - (1 - I_A^L(x))^\lambda, 1 - (1 - I_A^U(x))^\lambda],$
$[1 - (1 - F_A^L(x))^\lambda, 1 - (1 - F_A^U(x))^\lambda]> | x \in X\}$

3 Two-Person Interval-Valued Neutrosophic Matrix Game

In this section, we define two-person interval-valued neutrosophic matrix game with payoffs which is interval-valued neutrosophic sets. Some of it is quoted and/or inspired and/or generalized from [1, 3, 5, 7, 8, 10].

Definition 9 A two-person interval-valued neutrosophic matrix game (TPIVNM-game) is a two player game, denoted by $N = (S_1, S_2, A)$, such that

1. P_1 has a finite strategy set S_1 with m elements.
2. P_2 has a finite strategy set S_2 with n elements.
3. A is an interval-valued neutrosophic neutrosophic set over $S_1 \times S_2$ that is defined as follows:

$$A = \left\{< \left(s_i^{\ 1}, s_j^{\ 2}\right), ([T_{ij}^L, T_{ij}^U], [I_{ij}^L, I_{ij}^U], [F_{ij}^L, F_{ij}^U]) > \left(s_i^{\ 1}, s_j^{\ 2}\right) \in S_1 \times S_2\right\} \tag{1}$$

The TPIVNM-game N is played as follows:

Simultaneously, P_1 chooses $s_i^{\ 1} \in S_1$ and P_2 chooses $s_j^{\ 2} \in S_2$, each unaware of the choice of the other, then the payoff for P_1 is expressed with $< [T_{ij}^L, T_{ij}^U], [I_{ij}^L, I_{ij}^U], [F_{ij}^L, F_{ij}^U] >$. In this word, P_1 gains the payoff expressed $< [T_{ij}^L, T_{ij}^U], [I_{ij}^L, I_{ij}^U], [F_{ij}^L, F_{ij}^U] >$ at the situation $(s_i^{\ 1}, s_j^{\ 2})$. The outcomes of P_2 on the situation $(s_i^{\ 1}, s_j^{\ 2})$ is negations of that of P_1.

If $S_1 = \{s_1^1, s_2^1, \ldots, s_m^1\}$, $S_2 = \{s_1^2, s_2^2, \ldots, s_m^2\}$ and we put $l_{ij} = < \left([T_{ij}^L, T_{ij}^U], [I_{ij}^L, I_{ij}^U], [F_{ij}^L, F_{ij}^U]\right) >$, then the payoffs of TPIVNM-game N can be arranged in the form of the m $\times$ n matrix as follows:

$$A = \begin{pmatrix} l_{11} & l_{12} & \cdots & l_{1n} \\ l_{21} & l_{22} & & l_{2n} \\ \vdots & & \ddots & \vdots \\ l_{m1} & l_{m2} & \cdots & l_{mn} \end{pmatrix}$$ which is called an m $\times$ n matrix game of TPIVNM-

game N.

Definition 10 Let $X = \left\{x : x \in R_+^m, x^T e_m = 1\right\}$ and $Y = \left\{y : y \in R_+^n, y^T e_n = 1\right\}$ be the mixed strategy spaces of players P_1 and P_2, respectively, and $N = (S_1, S_2, A)$ be a TPIVNM-game. If player P_1 chooses any mixed strategy $X = (x_1, x_2, \ldots, x_m)^T$ and player P_2, chooses any mixed strategy $Y = (y_1, y_2, \ldots, y_n)^T$, then the expected payoff of player P_1 at the mixed strategy situation (x, y) is obtained as follows:

$$E(x, y) = x^T A y \tag{2}$$

According to the operations of Definition 8, Eq. (2) is computed as follows:

$$E(x, y) = \left\langle \left[1 - \prod_{j=1}^{n} \prod_{i=1}^{m} \left(1 - T_{ij}^{L} \right)^{x_i y_j}, 1 - \prod_{j=1}^{n} \prod_{i=1}^{m} \left(1 - T_{ij}^{U} \right)^{x_i y_j} \right], \left[\prod_{j=1}^{n} \prod_{i=1}^{m} \left(I_{ij}^{L} \right)^{x_i y_j}, \right. \right.$$
$$\left. \prod_{j=1}^{n} \prod_{i=1}^{m} \left(I_{ij}^{U} \right)^{x_i y_j} \right], \left[\prod_{j=1}^{n} \prod_{i=1}^{m} \left(F_{ij}^{L} \right)^{x_i y_j}, \prod_{j=1}^{n} \prod_{i=1}^{m} \left(F_{ij}^{U} \right)^{x_i y_j} \right] \right\rangle \tag{3}$$

Since players are rational there exists strategies $x_0 \in X$ and $y_0 \in Y$ in a TPIVNM-game as follows:

$$x_0^T A y_0 = \max_{x \in X} \min_{y \in Y} \left\{ x^T A y \right\} = \min_{y \in Y} \max_{x \in X} \left\{ x^T A y \right\} \tag{4}$$

then x_0 and y_0 are called optimal strategies((x_0, y_0) is saddle point) for players P_1 and P_2, respectively. $x_0^T A y_0$ is called a value of the TPIVNM-game N with payoffs of interval-valued neutrosophic sets.

If there is not x_0 and y_0 for Eq. (4), then $\max_{x \in X} \min_{y \in Y} \{x^T A y\}$ and $\min_{y \in Y} \max_{x \in X} \{x^T A y\}$ is programming problem as follows:

$$\xi = \left[1 - \prod_{j=1}^{n} \prod_{i=1}^{m} \left(1 - T_{ij}^{L} \right)^{x_i y_j}, 1 - \prod_{j=1}^{n} \prod_{i=1}^{m} \left(1 - T_{ij}^{U} \right)^{x_i y_j} \right] \tag{5}$$

$$\zeta = \left[\prod_{j=1}^{n} \prod_{i=1}^{m} \left(I_{ij}^{L} \right)^{x_i y_j}, \prod_{j=1}^{n} \prod_{i=1}^{m} \left(I_{ij}^{U} \right)^{x_i y_j} \right] \tag{6}$$

and

$$\varsigma = \left[\prod_{j=1}^{n} \prod_{i=1}^{m} \left(F_{ij}^{L} \right)^{x_i y_j}, \prod_{j=1}^{n} \prod_{i=1}^{m} \left(F_{ij}^{U} \right)^{x_i y_j} \right] \tag{7}$$

Definition 11 Let V_1 and V_2 be two interval-valued neutrosophic sets and $x^* \in X$ and $y^* \in Y$. If

$$x^{*T} A y \supseteq V_1$$

$$x^T A y^* \subseteq V_2$$

then (x^*, y^*, V_1, V_2) is a solution of two-person interval-valued neutrosophic matrix game. Also, all V_1 and V_2 are denoted by the sets V and W, respectively.

Definition 12 Assume that there is a solution $V_1^* \in V$ and $V_2^* \in W$ for players P_1 and P_2 in TPIVNM-games, respectively. If there do not exist any reasonable values $V_1 \in V \left(V_1 \neq V_1^* \right)$ and $V_2 \in W \left(V_2 \neq V_2^* \right)$ so that

$$V_1 \supseteq V_1^*$$

$$V_2 \subseteq V_2^*$$

then $\left(x^*, y^*, V_1^*, V_2^* \right)$ is a solution of of two-person interval-valued neutrosophic matrix game based on maximin-minimax strategy.

If P_1 chose mixed strategy $x \in X$ in a TPIVNM-game, then its payoff is $\min_{y \in Y}\{E(x, y)\}$ as follows:

$$\begin{cases} \sum\limits_{i=1}^{m}\left[\lambda \ln\left(1 - \mu_{ij}^{U}\right) + (1 - \lambda)\ln\left(v_{ij}^{L}\right) + \lambda \ln\left(1 - \mu_{ij}^{L}\right) + (1 - \lambda)\ln\left(v_{ij}^{U}\right)\right] \\ \breve{x}_i \leq \breve{u} \ (j = 1, 2, \ldots, n) \\ \sum\limits_{\hat{i}=1}^{m} \breve{x}_i = 1 \\ \breve{u} \leq 0, \breve{x}_i \geq 0 \ (i = 1, 2, \ldots, m) \end{cases} \qquad \text{(8 of Chap. 69)}$$

Also, player P_1 use a mixed strategy $x^* \in X$ based on maximize the interval-valued neutrosophic set, we have

$$\theta = < \left[T^{L*}, T^{U*}\right], \left[I^{L*}, I^{U*}\right], \left[F^{L*}, F^{U*}\right] >$$

$$= < \max_{x \in X}\min_{y \in Y}\left\{\left[1 - \prod_{j=1}^{n}\prod_{i=1}^{m}\left(1 - T_{ij}^{L}\right)^{x_i y_j}, 1 - \prod_{j=1}^{n}\prod_{i=1}^{m}\left(1 - T_{ij}^{U}\right)^{x_i y_j}\right]\right\},$$

$$\min_{x \in X}\max_{y \in Y}\left\{\left[\prod_{j=1}^{n}\prod_{i=1}^{m}\left(I_{ij}^{L}\right)^{x_i y_j}, \prod_{j=1}^{n}\prod_{i=1}^{m}\left(I_{ij}^{U}\right)^{x_i y_j}\right]\right\},$$

$$\min_{x \in X}\max_{y \in Y}\left\{\left[\prod_{j=1}^{n}\prod_{i=1}^{m}\left(F_{ij}^{L}\right)^{x_i y_j}, \prod_{j=1}^{n}\prod_{i=1}^{m}\left(F_{ij}^{U}\right)^{x_i y_j}\right]\right\} >$$

$$\tag{8}$$

Similarly, P_2 use mixed strategy $y \in Y$ in TPIVNM-game; then its expected loss's maximum is $\max_{x \in X}\{E(x, y)\}$ as follows:

$$\vartheta = < \left[\sigma^L, \sigma^U\right], \left[\rho^L, \rho^U\right], \left[\kappa^L, \kappa^U\right] >$$

$$= < \max_{x \in X} \left\{ \left[1 - \prod_{j=1}^{n}\prod_{i=1}^{m}\left(1 - T_{ij}^L\right)^{x_i y_j}, 1 - \prod_{j=1}^{n}\prod_{i=1}^{m}\left(1 - T_{ij}^U\right)^{x_i y_j}\right]\right\},$$

$$\min_{x \in X} \left\{ \left[\prod_{j=1}^{n}\prod_{i=1}^{m}\left(I_{ij}^L\right)^{x_i y_j}, \prod_{j=1}^{n}\prod_{i=1}^{m}\left(I_{ij}^U\right)^{x_i y_j}\right]\right\},$$

$$\min_{x \in X} \left\{ \left[\prod_{j=1}^{n}\prod_{i=1}^{m}\left(F_{ij}^L\right)^{x_i y_j}, \prod_{j=1}^{n}\prod_{i=1}^{m}\left(F_{ij}^U\right)^{x_i y_j}\right]\right\} >$$

Also, player P_2 should choose a mixed strategy $y^* \in Y$ to minimize the interval-valued neutrosophic set ϑ as follows:\stop

$$\vartheta = < \left[\sigma^{L*}, \sigma^{U*}\right], \left[\rho^{L*}, \rho^{U*}\right], \left[\kappa^{L*}, \kappa^{U*}\right] >$$
$$= < \min_{y \in Y}\max_{x \in X} \left\{ \left[1 - \prod_{j=1}^{n}\prod_{i=1}^{m}\left(1 - T_{ij}^L\right)^{x_i y_j}, 1 - \prod_{j=1}^{n}\prod_{i=1}^{m}\left(1 - T_{ij}^U\right)^{x_i y_j}\right]\right\},$$
$$\max_{y \in Y}\min_{x \in X} \left\{ \left[\prod_{j=1}^{n}\prod_{i=1}^{m}\left(I_{ij}^L\right)^{x_i y_j}, \prod_{j=1}^{n}\prod_{i=1}^{m}\left(I_{ij}^U\right)^{x_i y_j}\right]\right\},$$
$$\max_{y \in Y}\min_{x \in X} \left\{ \left[\prod_{j=1}^{n}\prod_{i=1}^{m}\left(F_{ij}^L\right)^{x_i y_j}, \prod_{j=1}^{n}\prod_{i=1}^{m}\left(F_{ij}^U\right)^{x_i y_j}\right]\right\} >$$

$$(9)$$

Theorem 13 θ^* and ϑ^* are interval-valued neutrosophic sets and $\theta^* \subseteq \vartheta^*$.
   Proof
We have from Eq. (3) that the expected payoff $E(x,y)$ of player P_1 is an interval-valued neutrosophic set, i.e.,

$$E(x, y) = < \left[1 - \prod_{j=1}^{n}\prod_{i=1}^{m}\left(1 - T_{ij}^L\right)^{x_i y_j}, 1 - \prod_{j=1}^{n}\prod_{i=1}^{m}\left(1 - T_{ij}^U\right)^{x_i y_j}\right],$$

$$\left[\prod_{j=1}^{n}\prod_{i=1}^{m}\left(I_{ij}^L\right)^{x_i y_j}, \prod_{j=1}^{n}\prod_{i=1}^{m}\left(I_{ij}^U\right)^{x_i y_j}\right],$$

$$\left[\prod_{j=1}^{n}\prod_{i=1}^{m}\left(F_{ij}^L\right)^{x_i y_j}, \prod_{j=1}^{n}\prod_{i=1}^{m}\left(F_{ij}^U\right)^{x_i y_j}\right] >$$

Then, we have

$$0 \leq 1 - \prod_{j=1}^{n} \prod_{i=1}^{m} \left(1 - T_{ij}^{L}\right)^{x_i y_j} \leq 1, 0 \leq 1 - \prod_{j=1}^{n} \prod_{i=1}^{m} \left(1 - T_{ij}^{U}\right)^{x_i y_j}$$

$$\leq 1, 0 \leq \prod_{j=1}^{n} \prod_{i=1}^{m} \left(I_{ij}^{L}\right)^{x_i y_j} \leq 1,$$

$$0 \leq \prod_{j=1}^{n} \prod_{i=1}^{m} \left(I_{ij}^{U}\right)^{x_i y_j} \leq 1, 0 \leq \prod_{j=1}^{n} \prod_{i=1}^{m} \left(F_{ij}^{L}\right)^{x_i y_j}$$

$$\leq 1, 0 \leq \prod_{j=1}^{n} \prod_{i=1}^{m} \left(F_{ij}^{U}\right)^{x_i y_j} \leq 1$$

and

$$0 \leq 1 - \prod_{j=1}^{n} \prod_{i=1}^{m} \left(1 - T_{ij}^{U}\right)^{x_i y_j} + \prod_{j=1}^{n} \prod_{i=1}^{m} \left(I_{ij}^{U}\right)^{x_i y_j} + \prod_{j=1}^{n} \prod_{i=1}^{m} \left(F_{ij}^{U}\right)^{x_i y_j} \leq 3,$$

$$(10)$$

which infer that

$$0 \leq \max_{x \in X} \min_{y \in Y} \left\{ 1 - \prod_{j=1}^{n} \prod_{i=1}^{m} \left(1 - T_{ij}^{L}\right)^{x_i y_j} \right\} \leq 1,$$

$$0 \leq \max_{x \in X} \min_{y \in Y} \left\{ 1 - \prod_{j=1}^{n} \prod_{i=1}^{m} \left(1 - T_{ij}^{U}\right)^{x_i y_j} \right\} \leq 1,$$

$$0 \leq \min_{x \in X} \max_{y \in Y} \left\{ \prod_{j=1}^{n} \prod_{i=1}^{m} \left(I_{ij}^{L}\right)^{x_i y_j} \right\} \leq 1,$$

$$0 \leq \min_{x \in X} \max_{y \in Y} \left\{ \prod_{j=1}^{n} \prod_{i=1}^{m} \left(I_{ij}^{U}\right)^{x_i y_j} \right\} < 1,$$

$$0 \leq \min_{x \in X} \max_{y \in Y} \left\{ \prod_{j=1}^{n} \prod_{i=1}^{m} \left(F_{ij}^{L}\right)^{x_i y_j} \right\} \leq 1, \text{ and}$$

$$0 \leq \min_{x \in X} \max_{y \in Y} \left\{ \prod_{j=1}^{n} \prod_{i=1}^{m} \left(F_{ij}^{U}\right)^{x_i y_j} \right\} \leq 1.$$

It is derived from Eq. (10) that

$$
0 \leq \left[1 - \prod_{j=1}^{n} \prod_{i=1}^{m} \left(1 - T_{ij}^{U} \right)^{x_i y_j} \right] + \min_{y \in Y} \left\{ \prod_{j=1}^{n} \prod_{i=1}^{m} \left(I_{ij}^{U} \right)^{x_i y_j} \right\}
$$
$$
+ \min_{y \in Y} \left\{ \prod_{j=1}^{n} \prod_{i=1}^{m} \left(F_{ij}^{U} \right)^{x_i y_j} \right\}
$$
$$
\leq \left[1 - \prod_{j=1}^{n} \prod_{i=1}^{m} \left(1 - T_{ij}^{U} \right)^{x_i y_j} \right]
$$
$$
+ \prod_{j=1}^{n} \prod_{i=1}^{m} \left(I_{ij}^{U} \right)^{x_i y_j} + \prod_{j=1}^{n} \prod_{i=1}^{m} \left(F_{ij}^{U} \right)^{x_i y_j}
$$
$$
\leq \left[1 - \prod_{j=1}^{n} \prod_{i=1}^{m} \left(1 - T_{ij}^{U} \right)^{x_i y_j} \right] + \max_{y \in Y} \left\{ \prod_{j=1}^{n} \prod_{i=1}^{m} \left(I_{ij}^{U} \right)^{x_i y_j} \right\}
$$
$$
+ \max_{y \in Y} \left\{ \prod_{j=1}^{n} \prod_{i=1}^{m} \left(F_{ij}^{U} \right)^{x_i y_j} \right\} \leq 3
$$

which infers that

$$
0 \leq \min_{y \in Y} \left\{ 1 - \prod_{j=1}^{n} \prod_{i=1}^{m} \left(1 - T_{ij}^{U} \right)^{x_i y_j} \right\} + \min_{y \in Y} \left\{ \prod_{j=1}^{n} \prod_{i=1}^{m} \left(I_{ij}^{U} \right)^{x_i y_j} \right\}
$$
$$
+ \min_{y \in Y} \left\{ \prod_{j=1}^{n} \prod_{i=1}^{m} \left(F_{ij}^{U} \right)^{x_i y_j} \right\}
$$
$$
\leq \min_{y \in Y} \left\{ 1 - \prod_{j=1}^{n} \prod_{i=1}^{m} \left(1 - T_{ij}^{U} \right)^{x_i y_j} \right\} + \max_{y \in Y} \left\{ \prod_{j=1}^{n} \prod_{i=1}^{m} \left(I_{ij}^{U} \right)^{x_i y_j} \right\}
$$
$$
+ \max_{y \in Y} \left\{ \prod_{j=1}^{n} \prod_{i=1}^{m} \left(F_{ij}^{U} \right)^{x_i y_j} \right\}
$$
$$
\leq \left[1 - \prod_{j=1}^{n} \prod_{i=1}^{m} \left(1 - T_{ij}^{U} \right)^{x_i y_j} \right] + \max_{y \in Y} \left\{ \prod_{j=1}^{n} \prod_{i=1}^{m} \left(I_{ij}^{U} \right)^{x_i y_j} \right\}
$$
$$
+ \max_{y \in Y} \left\{ \prod_{j=1}^{n} \prod_{i=1}^{m} \left(F_{ij}^{U} \right)^{x_i y_j} \right\} \leq 3
$$

Thus,

$$0 \leq \min_{y \in Y} \left\{ 1 - \prod_{j=1}^{n} \prod_{i=1}^{m} \left(1 - T_{ij}^{U} \right)^{x_i y_j} \right\} + \min_{x \in X} \min_{y \in Y} \left\{ \prod_{j=1}^{n} \prod_{i=1}^{m} \left(I_{ij}^{U} \right)^{x_i y_j} \right\}$$

$$+ \min_{x \in X} \min_{y \in Y} \left\{ \prod_{j=1}^{n} \prod_{i=1}^{m} \left(F_{ij}^{U} \right)^{x_i y_j} \right\}$$

$$\leq \min_{y \in Y} \left\{ 1 - \prod_{j=1}^{n} \prod_{i=1}^{m} \left(1 - T_{ij}^{U} \right)^{x_i y_j} \right\} + \min_{x \in X} \max_{y \in Y} \left\{ \prod_{j=1}^{n} \prod_{i=1}^{m} \left(I_{ij}^{U} \right)^{x_i y_j} \right\}$$

$$+ \min_{x \in X} \max_{y \in Y} \left\{ \prod_{j=1}^{n} \prod_{i=1}^{m} \left(F_{ij}^{U} \right)^{x_i y_j} \right\}$$

$$\leq \left[1 - \prod_{j=1}^{n} \prod_{i=1}^{m} \left(1 - T_{ij}^{U} \right)^{x_i y_j} \right] + \min_{x \in X} \max_{y \in Y} \left\{ \prod_{j=1}^{n} \prod_{i=1}^{m} \left(I_{ij}^{U} \right)^{x_i y_j} \right\}$$

$$+ \min_{x \in X} \max_{y \in Y} \left\{ \prod_{j=1}^{n} \prod_{i=1}^{m} \left(F_{ij}^{U} \right)^{x_i y_j} \right\} \leq 3$$

which infers that

$$0 \leq \max_{x \in X} \min_{y \in Y} \left\{ 1 - \prod_{j=1}^{n} \prod_{i=1}^{m} \left(1 - T_{ij}^{U} \right)^{x_i y_j} \right\} + \min_{x \in X} \min_{y \in Y} \left\{ \prod_{j=1}^{n} \prod_{i=1}^{m} \left(I_{ij}^{U} \right)^{x_i y_j} \right\}$$

$$+ \min_{x \in X} \min_{y \in Y} \left\{ \prod_{j=1}^{n} \prod_{i=1}^{m} \left(F_{ij}^{U} \right)^{x_i y_j} \right\}$$

$$\leq \max_{x \in X} \min_{y \in Y} \left\{ 1 - \prod_{j=1}^{n} \prod_{i=1}^{m} \left(1 - T_{ij}^{U} \right)^{x_i y_j} \right\} + \min_{x \in X} \max_{y \in Y} \left\{ \prod_{j=1}^{n} \prod_{i=1}^{m} \left(I_{ij}^{U} \right)^{x_i y_j} \right\}$$

$$+ \min_{x \in X} \max_{y \in Y} \left\{ \prod_{j=1}^{n} \prod_{i=1}^{m} \left(F_{ij}^{U} \right)^{x_i y_j} \right\}$$

$$\leq \max_{x \in X} \left\{ 1 - \prod_{j=1}^{n} \prod_{i=1}^{m} \left(1 - T_{ij}^{U} \right)^{x_i y_j} \right\} + \min_{x \in X} \max_{y \in Y} \left\{ \prod_{j=1}^{n} \prod_{i=1}^{m} \left(I_{ij}^{U} \right)^{x_i y_j} \right\}$$

$$+ \min_{x \in X} \max_{y \in Y} \left\{ \prod_{j=1}^{n} \prod_{i=1}^{m} \left(F_{ij}^{U} \right)^{x_i y_j} \right\} \leq 3$$

i.e.,

$$0 \leq \max_{x \in X} \min_{y \in Y} \left\{ 1 - \prod_{j=1}^{n} \prod_{i=1}^{m} \left(1 - T_{ij}^{U} \right)^{x_i y_j} \right\}$$

$$+ \min_{x \in X} \max_{y \in Y} \left\{ \prod_{j=1}^{n} \prod_{i=1}^{m} \left(I_{ij}^{U} \right)^{x_i y_j} \right\}$$

$$+ \min_{x \in X} \max_{y \in Y} \left\{ \prod_{j=1}^{n} \prod_{i=1}^{m} \left(F_{ij}^{U} \right)^{x_i y_j} \right\} \leq 3$$

Based on θ^* in Eq. (7), we have $0 \leq T_{ij}^{U*} + I_{ij}^{U*} + F_{ij}^{U*} \leq 3$. Finaly, θ^* is an interval-valued neutrosophic set. Similarly, for ϑ^*

$$\min_{y \in Y} \left\{ 1 - \prod_{j=1}^{n} \prod_{i=1}^{m} \left(1 - T_{ij}^{L} \right)^{x_i y_j} \right\} \leq 1 - \prod_{j=1}^{n} \prod_{i=1}^{m} \left(1 - T_{ij}^{L} \right)^{x_i y_j} \text{ and}$$

$$\max_{x \in X} \left\{ 1 - \prod_{j=1}^{n} \prod_{i=1}^{m} \left(1 - T_{ij}^{L} \right)^{x_i y_j} \right\} \geq 1 - \prod_{j=1}^{n} \prod_{i=1}^{m} \left(1 - T_{ij}^{L} \right)^{x_i y_j}$$

and

$$\min_{y \in Y} \left\{ 1 - \prod_{j=1}^{n} \prod_{i=1}^{m} \left(1 - T_{ij}^{L} \right)^{x_i y_j} \right\} \leq \max_{x \in X} \left\{ 1 - \prod_{j=1}^{n} \prod_{i=1}^{m} \left(1 - T_{ij}^{L} \right)^{x_i y_j} \right\}$$

Therefore, we have

$$\min_{y \in Y} \left\{ 1 - \prod_{j=1}^{n} \prod_{i=1}^{m} \left(1 - T_{ij}^{L} \right)^{x_i y_j} \right\} \leq \min_{y \in Y} \max_{x \in X} \left\{ 1 - \prod_{j=1}^{n} \prod_{i=1}^{m} \left(1 - T_{ij}^{L} \right)^{x_i y_j} \right\}$$

and

$$\max_{x \in X} \min_{y \in Y} \left\{ 1 - \prod_{j=1}^{n} \prod_{i=1}^{m} \left(1 - T_{ij}^{L} \right)^{x_i y_j} \right\} \leq \min_{y \in Y} \max_{x \in X} \left\{ 1 - \prod_{j=1}^{n} \prod_{i=1}^{m} \left(1 - T_{ij}^{L} \right)^{x_i y_j} \right\} \tag{11}$$

Similarly, we have

$$\max_{x \in X} \min_{y \in Y} \left\{ 1 - \prod_{j=1}^{n} \prod_{i=1}^{m} \left(1 - T_{ij}^{U} \right)^{x_i y_j} \right\} \leq \min_{y \in Y} \max_{x \in X} \left\{ 1 - \prod_{j=1}^{n} \prod_{i=1}^{m} \left(1 - T_{ij}^{U} \right)^{x_i y_j} \right\} \tag{12}$$

Also, it readily follows from Eqs. (11, 12) that

$$\max_{x \in X} \min_{y \in Y} \left\{ \left[1 - \prod_{j=1}^{n} \prod_{i=1}^{m} \left(1 - T_{ij}^{L} \right)^{x_i y_j}, 1 - \prod_{j=1}^{n} \prod_{i=1}^{m} \left(1 - T_{ij}^{U} \right)^{x_i y_j} \right] \right\}$$
$$\leq \min_{y \in Y} \max_{x \in X} \left\{ \left[1 - \prod_{j=1}^{n} \prod_{i=1}^{m} \left(1 - T_{ij}^{L} \right)^{x_i y_j}, 1 - \prod_{j=1}^{n} \prod_{i=1}^{m} \left(1 - T_{ij}^{U} \right)^{x_i y_j} \right] \right\} \tag{13}$$

On the other hand,

$$\max_{y \in Y} \left\{ \prod_{j=1}^{n} \prod_{i=1}^{m} \left(I_{ij}^{L} \right)^{x_i y_j} \right\} \geq \prod_{j=1}^{n} \prod_{i=1}^{m} \left(I_{ij}^{L} \right)^{x_i y_j} \text{ and } \min_{x \in X} \left\{ \prod_{j=1}^{n} \prod_{i=1}^{m} \left(I_{ij}^{L} \right)^{x_i y_j} \right\} \leq \prod_{j=1}^{n} \prod_{i=1}^{m} \left(I_{ij}^{L} \right)^{x_i y_j}$$

and

$$\max_{y \in Y} \left\{ \prod_{j=1}^{n} \prod_{i=1}^{m} \left(I_{ij}^{L}\right)^{x_i y_j} \right\} \geq \min_{x \in X} \left\{ \prod_{j=1}^{n} \prod_{i=1}^{m} \left(I_{ij}^{L}\right)^{x_i y_j} \right\}$$

Therefore, we have

$$\max_{y \in Y} \left\{ \prod_{j=1}^{n} \prod_{i=1}^{m} \left(I_{ij}^{L}\right)^{x_i y_j} \right\} \geq \max_{y \in Y} \min_{x \in X} \left\{ \prod_{j=1}^{n} \prod_{i=1}^{m} \left(I_{ij}^{L}\right)^{x_i y_j} \right\}$$

Hence, we have

$$\min_{x \in X} \max_{y \in Y} \left\{ \prod_{j=1}^{n} \prod_{i=1}^{m} \left(I_{ij}^{L}\right)^{x_i y_j} \right\} \geq \max_{y \in Y} \min_{x \in X} \left\{ \prod_{j=1}^{n} \prod_{i=1}^{m} \left(I_{ij}^{L}\right)^{x_i y_j} \right\} \tag{14}$$

and

$$\min_{x \in X} \max_{y \in Y} \left\{ \prod_{j=1}^{n} \prod_{i=1}^{m} \left(I_{ij}^{U}\right)^{x_i y_j} \right\} \geq \max_{y \in Y} \min_{x \in X} \left\{ \prod_{j=1}^{n} \prod_{i=1}^{m} \left(I_{ij}^{U}\right)^{x_i y_j} \right\} \tag{15}$$

Also, from Eqs. (14, 15), we have

$$\max_{x \in X} \min_{y \in Y} \left\{ \left[\prod_{j=1}^{n} \prod_{i=1}^{m} \left(I_{ij}^{L}\right)^{x_i y_j}, \prod_{j=1}^{n} \prod_{i=1}^{m} \left(I_{ij}^{U}\right)^{x_i y_j} \right] \right\} \leq$$
$$\min_{y \in Y} \max_{x \in X} \left\{ \left[1 \prod_{j=1}^{n} \prod_{i=1}^{m} \left(I_{ij}^{L}\right)^{x_i y_j}, \prod_{j=1}^{n} \prod_{i=1}^{m} \left(I_{ij}^{U}\right)^{x_i y_j} \right] \right\} \tag{16}$$

On the other hand, we have

$$\max_{y \in Y} \left\{ \prod_{j=1}^{n} \prod_{i=1}^{m} \left(F_{ij}^{L}\right)^{x_i y_j} \right\} \geq \prod_{j=1}^{n} \prod_{i=1}^{m} \left(F_{ij}^{L}\right)^{x_i y_j} and$$

$$\min_{x \in X} \left\{ \prod_{j=1}^{n} \prod_{i=1}^{m} \left(F_{ij}^{L}\right)^{x_i y_j} \right\} \leq \prod_{j=1}^{n} \prod_{i=1}^{m} \left(F_{ij}^{L}\right)^{x_i y_j}$$

and

$$\max_{y \in Y} \left\{ \prod_{j=1}^{n} \prod_{i=1}^{m} \left(F_{ij}^{L} \right)^{x_i y_j} \right\} \geq \min_{x \in X} \left\{ \prod_{j=1}^{n} \prod_{i=1}^{m} \left(F_{ij}^{L} \right)^{x_i y_j} \right\}$$

Also we have

$$\max_{y \in Y} \left\{ \prod_{j=1}^{n} \prod_{i=1}^{m} \left(F_{ij}^{L} \right)^{x_i y_j} \right\} \geq \max_{y \in Y} \min_{x \in X} \left\{ \prod_{j=1}^{n} \prod_{i=1}^{m} \left(F_{ij}^{L} \right)^{x_i y_j} \right\}$$

Hence, we have

$$\min_{x \in X} \max_{y \in Y} \left\{ \prod_{j=1}^{n} \prod_{i=1}^{m} \left(F_{ij}^{L} \right)^{x_i y_j} \right\} \geq \max_{y \in Y} \min_{x \in X} \left\{ \prod_{j=1}^{n} \prod_{i=1}^{m} \left(F_{ij}^{L} \right)^{x_i y_j} \right\} \tag{17}$$

Similarly, we have

$$\min_{x \in X} \max_{y \in Y} \left\{ \prod_{j=1}^{n} \prod_{i=1}^{m} \left(F_{ij}^{U} \right)^{x_i y_j} \right\} \geq \max_{y \in Y} \min_{x \in X} \left\{ \prod_{j=1}^{n} \prod_{i=1}^{m} \left(F_{ij}^{U} \right)^{x_i y_j} \right\} \tag{18}$$

We have from Eqs. (17, 18) that

$$\max_{x \in X} \min_{y \in Y} \left\{ \left[\prod_{j=1}^{n} \prod_{i=1}^{m} \left(F_{ij}^{L} \right)^{x_i y_j}, \prod_{j=1}^{n} \prod_{i=1}^{m} \left(F_{ij}^{U} \right)^{x_i y_j} \right] \right\} \leq$$
$$\min_{y \in Y} \max_{x \in X} \left\{ \left[\prod_{j=1}^{n} \prod_{i=1}^{m} \left(F_{ij}^{L} \right)^{x_i y_j}, \prod_{j=1}^{n} \prod_{i=1}^{m} \left(F_{ij}^{U} \right)^{x_i y_j} \right] \right\} \tag{19}$$

Finaly, by combining with Eqs. (13), (16), (19), we have

$$\theta^* = \max_{x \in X} \min_{y \in Y} \left\{ \left[1 - \prod_{j=1}^{n} \prod_{i=1}^{m} \left(1 - T_{ij}^{L} \right)^{x_i y_j}, 1 - \prod_{j=1}^{n} \prod_{i=1}^{m} \left(1 - T_{ij}^{U} \right)^{x_i y_j} \right], \right.$$
$$\left[\prod_{j=1}^{n} \prod_{i=1}^{m} \left(I_{ij}^{L} \right)^{x_i y_j}, \prod_{j=1}^{n} \prod_{i=1}^{m} \left(I_{ij}^{U} \right)^{x_i y_j} \right],$$
$$\left[\prod_{j=1}^{n} \prod_{i=1}^{m} \left(F_{ij}^{L} \right)^{x_i y_j}, \prod_{j=1}^{n} \prod_{i=1}^{m} \left(F_{ij}^{U} \right)^{x_i y_j} \right] \right\}$$
$$\leq \min_{y \in Y} \max_{x \in X} \left\{ \left[1 - \prod_{j=1}^{n} \prod_{i=1}^{m} \left(1 - T_{ij}^{L} \right)^{x_i y_j}, 1 - \prod_{j=1}^{n} \prod_{i=1}^{m} \left(1 - T_{ij}^{U} \right)^{x_i y_j} \right], \right.$$
$$\left[\prod_{j=1}^{n} \prod_{i=1}^{m} \left(I_{ij}^{L} \right)^{x_i y_j}, \prod_{j=1}^{n} \prod_{i=1}^{m} \left(I_{ij}^{U} \right)^{x_i y_j} \right], \left[\prod_{j=1}^{n} \prod_{i=1}^{m} \left(F_{ij}^{L} \right)^{x_i y_j}, \prod_{j=1}^{n} \prod_{i=1}^{m} \left(F_{ij}^{U} \right)^{x_i y_j} \right] \right\} = \vartheta^*$$

Finally, the proof is valid.

Lemma 14 [7] Let z_1 and z_2 are any nonpositive variables. Then,

$$\min\{z_1 + z_2\} = \min\{z_1\} + \min\{z_2\} \tag{20}$$

$$\min\{z_1 + z_2\} = \min\{z_1\} + \min\{z_2\} \tag{21}$$

is correct.

Proof

It is very obvious that

$$\min_{\Omega}\{z_1 + z_2\} \leq z_1 + z_2$$

which implies that

$$\min\{z_1 + z_2\} \leq \min\{z_1\} + \min\{z_2\}$$

On the other hand, it readily follows that

$$z_1 \geq \min\{z_1\} \quad and \quad z_2 \geq \min\{z_2\}$$

Hence, we have

$$z_1 + z_2 \geq \min\{z_1\} + \min\{z_2\}$$

which implies that

$$\min\{z_1 + z_2\} \geq \min\{z_1\} + \min\{z_2\}$$

Finally Eq. (20) is valid. Similarly, we can prove that (21) is valid.

A Mathematical Programming Model: In a TPIVNM-game, from Eq. (8), Definitions 11 and 12, the maximin strategy x^* and the gain-floor $\theta^* = <[T^{L*}, T^{U*}]$, $[I^{L*}, I^{U*}], [F^{L*}, F^{U*}]>$ of player P_1 can be generated through solving the following mathematical programming model:

$$\max\left\{\left[T^L, T^U\right]\right\}, \min\{\left[I^L, I^U\right], \min\{\left[F^L, F^U\right]\}$$

$$s.t. \begin{cases} \left[1 - \prod\limits_{j=1}^{n}\prod\limits_{i=1}^{m}\left(1-T_{ij}^L\right)^{x_i y_j}, 1 - \prod\limits_{j=1}^{n}\prod\limits_{i=1}^{m}\left(1-T_{ij}^U\right)^{x_i y_j}\right] \geq \left[T^L, T^U\right] \ (y \in Y) \\[3mm] \left[\prod\limits_{j=1}^{n}\prod\limits_{i=1}^{m}\left(I_{ij}^L\right)^{x_i y_j}, \prod\limits_{j=1}^{n}\prod\limits_{i=1}^{m}\left(I_{ij}^U\right)^{x_i y_j}\right] \leq \left[I^L, I^U\right] \ (y \in Y) \\[3mm] \left[\prod\limits_{j=1}^{n}\prod\limits_{i=1}^{m}\left(F_{ij}^L\right)^{x_i y_j}, \prod\limits_{j=1}^{n}\prod\limits_{i=1}^{m}\left(F_{ij}^U\right)^{x_i y_j}\right] \leq \left[F^L, F^U\right] \ (y \in Y) \\[3mm] 0 \leq T^U + I^U + F^U \leq 3 \\[2mm] \sum\limits_{i=1}^{m} x_i = 1 \\[2mm] T^L \geq 0, T^U \geq 0, I^L \geq 0, I^U \geq 0, F^L \geq 0, F^U \geq 0, x_i \geq 0 \ (i=1,2,\dots,m) \end{cases}$$

$$(22)$$

where

$$T^L = \min_{y \in Y}\left\{1 - \prod_{j=1}^{n}\prod_{i=1}^{m}\left(1 - T_{ij}^L\right)^{x_i y_j}\right\},$$

$$T^U = \min_{y \in Y}\left\{1 - \prod_{j=1}^{n}\prod_{i=1}^{m}\left(1 - T_{ij}^U\right)^{x_i y_j}\right\},$$

$$I^L = \max_{y \in Y}\left\{\prod_{j=1}^{n}\prod_{i=1}^{m}\left(I_{ij}^L\right)^{x_i y_j}\right\},$$

$$I^U = \max_{y \in Y}\left\{\prod_{j=1}^{n}\prod_{i=1}^{m}\left(I_{ij}^U\right)^{x_i y_j}\right\},$$

$$F^L = \max_{y \in Y}\left\{\prod_{j=1}^{n}\prod_{i=1}^{m}\left(F_{ij}^L\right)^{x_i y_j}\right\} \quad and$$

$$F^U = \max_{y \in Y}\left\{\prod_{j=1}^{n}\prod_{i=1}^{m}\left(F_{ij}^U\right)^{x_i y_j}\right\}$$

It is easy to see that $\max\{[T^L, T^U]\}$ is equivalent to $\min\{[1 - T^L, 1 - T^U]\}$; which further is equivalent to $\min\{[\ln(1 - T^L), \ln(1 - T^U)]\}$ for $0 \leq T^L \leq 1$ and $0 \leq T^U \leq 1$. Similarly, $\min\{[I^L, I^U]\}$ and $\min\{[F^L, F^U]\}$ is equivalent to $\min\{[\ln(I^L), \ln(I^U)]\}$ for $0 \leq I^L \leq 1$ and $0 \leq I^U \leq 1$ and $\min\{[\ln(F^L), \ln(F^U)]\}$ for $0 \leq F^L \leq 1$ and $0 \leq F^U \leq 1$, respectively. Thus, using the linear weighted averaging method of multiobjective programming, $\max\{[T^L, T^U]\}$, $\min\{[I^L, I^U]\}$ and $\min\{[F^L, F^U]\}$ in Eq. (22) may be aggregated as follows:

$$\min\{[\lambda \ln\left(1 - T^U\right) + (1 - \lambda)\ln\left(I^L\right) + (1 - \lambda)\ln\left(F^L\right),$$
$$\lambda \ln\left(1 - T^L\right) + (1 - \lambda)\ln\left(I^U\right) + (1 - \lambda)\ln\left(F^U\right)]\}$$

$$(23)$$

where $\lambda \in [0, 1]$ is a weight determined by players a priori.

Also,

$$\left[1 - \prod_{j=1}^{n}\prod_{i=1}^{m}\left(1 - T_{ij}^{L}\right)^{x_i y_j}, 1 - \prod_{j=1}^{n}\prod_{i=1}^{m}\left(1 - T_{ij}^{U}\right)^{x_i y_j}\right] \geq \left[T^{L}, T^{U}\right]$$

can be transformed into the two inequalities as follows:

$$\prod_{j=1}^{n}\prod_{i=1}^{m}\left(1 - T_{ij}^{L}\right)^{x_i y_j}, \leq 1 - T^{L} \tag{24}$$

$$\prod_{j=1}^{n}\prod_{i=1}^{m}\left(1 - T_{ij}^{L}\right)^{x_i y_j}, \leq 1 - T^{L} \tag{25}$$

which are

$$\sum_{j=1}^{n}\sum_{i=1}^{m} x_i y_j \ln\left(1 - T_{ij}^{L}\right)^{x_i y_j} \leq \ln\left(1 - T^{L}\right) \tag{26}$$

$$\sum_{j=1}^{n}\sum_{i=1}^{m} x_i y_j \ln\left(1 - T_{ij}^{L}\right)^{x_i y_j} \leq \ln\left(1 - T^{L}\right) \tag{27}$$

where $0 \leq T^{L} \leq 1, 0 \leq T^{U} \leq 1, 0 \leq T_{ij}^{L} \leq 1$ and $0 \leq T_{ij}^{U} \leq 1 (i = 1, 2, \ldots, m; j = 1, 2, \ldots, n)$

Similarly,

$$\left[\prod_{j=1}^{n}\prod_{i=1}^{m}\left(I_{ij}^{L}\right)^{x_i y_j}, \prod_{j=1}^{n}\prod_{i=1}^{m}\left(I_{ij}^{U}\right)^{x_i y_j}\right] \leq \left[I^{L}, I^{U}\right]$$

can be rewritten as follows:

$$\sum_{j=1}^{n}\sum_{i=1}^{m} x_i y_j \ln\left(I_{ij}^{L}\right)^{x_i y_j} \leq \ln\left(I^{L}\right) \tag{28}$$

$$\sum_{j=1}^{n}\sum_{i=1}^{m} x_i y_j \ln\left(I_{ij}^{U}\right)^{x_i y_j} \leq \ln\left(I^{U}\right) \tag{29}$$

where $0 \leq I^L \leq 1, 0 \leq I^U \leq 1, 0 \leq I_{ij}^L \leq 1$ and $0 \leq I_{ij}^U \leq 1 (i = 1,2,\ldots,m;$ $j = 1,2,\ldots,n)$

Similarly,

$$\left[\prod_{j=1}^{n} \prod_{i=1}^{m} \left(F_{ij}^L \right)^{x_i y_j}, \prod_{j=1}^{n} \prod_{i=1}^{m} \left(F_{ij}^U \right)^{x_i y_j} \right] \leq \left[F^L, F^U \right]$$

can be rewritten as follows:

$$\sum_{j=1}^{n} \sum_{i=1}^{m} x_i y_j \ln \left(F_{ij}^L \right)^{x_i y_j} \leq \ln \left(F^L \right) \tag{30}$$

$$\sum_{j=1}^{n} \sum_{i=1}^{m} x_i y_j \ln \left(F_{ij}^U \right)^{x_i y_j} \leq \ln \left(F^U \right) \tag{31}$$

where $0 \leq F^L \leq 1, 0 \leq F^U \leq 1, 0 \leq F_{ij}^L \leq 1$ and $0 \leq F_{ij}^U \leq 1 (i = 1,2,\ldots,m;$ $j = 1,2,\ldots,n)$

Thus, we have

$$\begin{cases} \sum_{j=1}^{n} \sum_{i=1}^{m} \left[\lambda \ln \left(1 - T_{ij}^U \right) + (1-\lambda) \ln \left(I_{ij}^L \right) + (1-\lambda) \ln \left(F_{ij}^L \right) \right] x_i y_j \leq \\ \qquad \lambda \ln \left(1 - T^U \right) + (1-\lambda) \ln \left(I^L \right) + (1-\lambda) \ln \left(F^L \right) \ (y \in Y) \\ \sum_{j=1}^{n} \sum_{i=1}^{m} \left[\lambda \ln \left(1 - T_{ij}^L \right) + (1-\lambda) \ln \left(I_{ij}^U \right) + (1-\lambda) \ln \left(F_{ij}^U \right) \right] x_i y_j \leq \\ \qquad \lambda \ln \left(1 - T^L \right) + (1-\lambda) \ln \left(T^U \right) + (1-\lambda) \ln \left(F^U \right) \ (y \in Y) \\ 0 \leq T^U + I^U + F^U \leq 3 \\ \sum_{i=1}^{m} x_i = 1 \\ T^L \geq 0, T^U \geq 0, I^L \geq 0, I^U \geq 0, F^L \geq 0, F^U \geq 0 \\ x_i \geq 0 \ (i = 1, 2, \ldots, m) \end{cases} \tag{32}$$

where $0 \leq T^L \leq 1, 0 \leq T^U \leq 1, 0 \leq I^L \leq 1, 0 \leq I^U \leq 1, 0 \leq F^L \leq 1, 0 \leq F^U \leq 1,$ $0 \leq T_{ij}^L \leq 1, 0 \leq T_{ij}^U \leq 1, 0 \leq I_{ij}^L \leq 1, 0 \leq I_{ij}^U \leq 1, 0 \leq F_{ij}^L \leq 1$ and $0 \leq F_{ij}^U \leq 1 (i = 1,2,\ldots,m; j = 1,2,\ldots,n)$. Then, from Eqs. (23) and (32), Eq. (22) we have the interval-valued mathematical programming model as follows:

$$\min \{[\, \lambda \ln \left(1 - T^U\right) + (1 - \lambda) \ln \left(I^L\right) + (1 - \lambda) \ln \left(F^L\right), \lambda \ln \left(1 - T^L\right) +$$
$$(1 - \lambda) \ln \left(I^U\right) + (1 - \lambda) \ln \left(F^U\right)\,]\}$$

$$s.t. \begin{cases} \sum\limits_{j=1}^{n} \sum\limits_{i=1}^{m} \left[\lambda \ln \left(1 - T_{ij}^U\right) + (1 - \lambda) \ln \left(I_{ij}^L\right) + (1 - \lambda) \ln \left(F_{ij}^L\right)\right] x_i y_j \leq \\ \qquad \lambda \ln \left(1 - T^U\right) + (1 - \lambda) \ln \left(I^L\right) + (1 - \lambda) \ln \left(F^L\right) \ (y \in Y) \\ \sum\limits_{j=1}^{n} \sum\limits_{i=1}^{m} \left[\lambda \ln \left(1 - T_{ij}^U\right) + (1 - \lambda) \ln \left(I_{ij}^U\right) + (1 - \lambda) \ln \left(F_{ij}^U\right)\right] x_i y_j \leq \\ \qquad \lambda \ln \left(1 - T^L\right) + (1 - \lambda) \ln \left(I^U\right) + (1 - \lambda) \ln \left(F^U\right) \ (y \in Y) \\ 0 \leq T^U + I^U + F^U \leq 3 \\ \sum\limits_{i=1}^{m} x_i = 1 \\ T^L \geq 0, T^U \geq 0, I^L \geq 0, I^U \geq 0, F^L \geq 0, F^U \geq 0 \\ x_i \geq 0 \ (i = 1, 2, \ldots, m) \end{cases}$$

$$(33)$$

If $\tau^L = \lambda \ln (1 - T^U) + (1 - \lambda) \ln (I^L) + (1 - \lambda) \ln (F^L)$ and $\tau^U = \lambda \ln (1 - T^L)$ $+ (1 - \lambda) \ln (I^U) + (1 - \lambda) \ln (F^U)$ \qquad Then, it is obvious that $\tau^L \leq 0$ and $\tau^U \leq 0$ due to $\lambda \in [0, 1], 0 \leq T^L \leq 1, 0 \leq T^U \leq 1, 0 \leq I^L \leq 1, 0 \leq I^U \leq 1,$ $0 \leq F^L \leq 1, 0 \leq F^U \leq 1$. Thus, Eq. (33) may be rewritten as follows:

$$\min \left\{\left[\tau^L, \tau^U\right]\right\}$$

$$s.t. \begin{cases} \sum\limits_{j=1}^{n} \sum\limits_{i=1}^{m} \left[\lambda \ln \left(1 - T_{ij}^U\right) + (1 - \lambda) \ln \left(I_{ij}^L\right) + (1 - \lambda) \ln \left(F_{ij}^L\right)\right] x_i y_j \leq \tau^L \ (y \in Y) \\ \sum\limits_{j=1}^{n} \sum\limits_{i=1}^{m} \left[\lambda \ln \left(1 - T_{ij}^L\right) + (1 - \lambda) \ln \left(I_{ij}^U\right) + (1 - \lambda) \ln \left(F_{ij}^U\right)\right] x_i y_j \leq \tau^U \ (y \in Y) \\ \sum\limits_{i=1}^{m} x_i = 1 \\ \tau^L \leq 0, \tau^U \leq 0, x_i \geq 0, (i = 1, 2, \ldots, m) \end{cases}$$

$$(34)$$

Also, Eq. (34) can rewritten as follows:

$$\min \{[\tau^L, \tau^U]\}$$

$$s.t. \begin{cases} \sum\limits_{i=1}^{m} \left[\lambda \ln \left(1 - T_{ij}^U\right) + (1 - \lambda) \ln \left(I_{ij}^L\right) + (1 - \lambda) \ln \left(F_{ij}^L\right)\right] x_i \leq \tau^L \ (j = 1, 2, \ldots, n) \\ \sum\limits_{i=1}^{m} \left[\lambda \ln \left(1 - T_{ij}^L\right) + (1 - \lambda) \ln \left(I_{ij}^U\right) + (1 - \lambda) \ln \left(F_{ij}^U\right)\right] x_i \leq \tau^U \ (j = 1, 2, \ldots, n) \\ \sum\limits_{i=1}^{m} x_i = 1, \tau^L \leq 0, \tau^U \leq 0, x_i \geq 0, (i = 1, 2, \ldots, m) \end{cases}$$

$$(35)$$

Then we have from Eq. (35) the linear programming model as follows:

$$\min\left\{\tau^U\right\}\ \min\left\{\frac{\tau^L+\tau^U}{2}\right\}$$

$$s.t.\begin{cases}\sum_{i=1}^{m}\left[\lambda\ln\left(1-T_{ij}^U\right)+(1-\lambda)\ln\left(I_{ij}^L\right)+(1-\lambda)\ln\left(F_{ij}^L\right)\right]x_i\leq\tau^L\ (j=1,2,\ldots,n)\\[4pt]\sum_{i=1}^{m}\left[\lambda\ln\left(1-T_{ij}^L\right)+(1-\lambda)\ln\left(I_{ij}^U\right)+(1-\lambda)\ln\left(F_{ij}^U\right)\right]x_i\leq\tau^U\ (j=1,2,\ldots,n)\\[4pt]\sum_{i=1}^{m}x_i=1\\[4pt]\tau^L\leq0,\tau^U\leq0,x_i\geq0\\[2pt](i=1,2,\ldots,m)\end{cases}$$

$$(36)$$

Then, from Eq. (20), Eq. (36), we have

$$\min\left\{\tau^U\right\},\ \min\left\{\frac{\tau^U}{2}\right\}+\min\left\{\frac{\tau^L}{2}\right\}$$

$$s.t.\begin{cases}\sum_{i=1}^{m}\left[\lambda\ln\left(1-T_{ij}^U\right)+(1-\lambda)\ln\left(I_{ij}^L\right)+(1-\lambda)\ln\left(F_{ij}^L\right)\right]x_i\leq\tau^L\ (j=1,2,\ldots,n)\\[4pt]\sum_{i=1}^{m}\left[\lambda\ln\left(1-T_{ij}^L\right)+(1-\lambda)\ln\left(I_{ij}^U\right)+(1-\lambda)\ln\left(F_{ij}^U\right)\right]x_i\leq\tau^U\ (j=1,2,\ldots,n)\\[4pt]\sum_{i=1}^{m}x_i=1\\[4pt]\tau^L\leq0,\tau^U\leq0,x_i\geq0\ (i=1,2,\ldots,m)\end{cases}$$

$$(37)$$

And we have

$$\min\left\{\tau^L\right\},\ \min\left\{\tau^U\right\}$$

$$s.t.\begin{cases}\sum_{i=1}^{m}\left[\lambda\ln\left(1-T_{ij}^U\right)+(1-\lambda)\ln\left(I_{ij}^L\right)+(1-\lambda)\ln\left(F_{ij}^L\right)\right]x_i\leq\tau^L\ (j=1,2,\ldots,n)\\[4pt]\sum_{i=1}^{m}\left[\lambda\ln\left(1-T_{ij}^L\right)+(1-\lambda)\ln\left(I_{ij}^U\right)+(1-\lambda)\ln\left(F_{ij}^U\right)\right]x_i\leq\tau^U\ (j=1,2,\ldots,n)\\[4pt]\sum_{i=1}^{m}x_i=1\\[4pt]\tau^L\leq0,\tau^U\leq0,x_i\geq0\ (i=1,2,\ldots,m)\end{cases}$$

$$(38)$$

If τ^L and τ^U in Eq. (38) are of equal importance, then Eq. (38) can be written as follows:

$$\min \left\{ \frac{\tau^L + \tau^U}{2} \right\},$$

$$s.t. \begin{cases} \sum_{i=1}^{m} \left[\lambda \ln \left(1 - T_{ij}^U \right) + (1 - \lambda) \ln \left(I_{ij}^L \right) + (1 - \lambda) \ln \left(F_{ij}^L \right) + \lambda \ln \left(1 - T_{ij}^L \right) \\ + (1 - \lambda) \ln \left(I_{ij}^U \right) + (1 - \lambda) \ln \left(F_{ij}^U \right) \right] x_i \leq \tau^L + \tau^U \quad (j = 1, 2, \ldots, n) \\ \sum_{i=1}^{m} x_i = 1 \\ \tau^L \leq 0, \tau^U \leq 0, x_i \geq 0 \ (i = 1, 2, \ldots, m) \end{cases}$$

$$(39)$$

And Eq. (39) may be further simplified for $u = \tau^L + \tau^U$ as follows:

$$\min \left\{ \frac{u}{2} \right\},$$

$$s.t. \begin{cases} \sum_{i=1}^{m} \left[\lambda \ln \left(1 - T_{ij}^U \right) + (1 - \lambda) \ln \left(I_{ij}^L \right) + (1 - \lambda) \ln \left(F_{ij}^L \right) + \lambda \ln \left(1 - T_{ij}^L \right) \\ + (1 - \lambda) \ln \left(I_{ij}^U \right) + (1 - \lambda) \ln \left(F_{ij}^U \right) \right] x_i \leq u \quad (j = 1, 2, \ldots, n) \\ \sum_{i=1}^{m} x_i = 1, u \leq 0, x_i \geq 0 \ (i = 1, 2, \ldots, m) \end{cases}$$

$$(40)$$

According to Eq. (9) and Definitions 11–12, the minimax strategy y^* and the loss-ceiling $\vartheta^* = <[\sigma^{L*}, \sigma^{U*}], [\rho^{L*}, \rho^{U*}], [\kappa^{L*}, \kappa^{U*}]>$ of player P_2 can be generated through solving the following mathematical programming model:

$$\min \left\{ [\sigma^L, \sigma^U] \right\}, \max \{ [\rho^L, \rho^U] \}, \max \{ [\kappa^L, \kappa^U] \}$$

$$s.t. \begin{cases} \left[1 - \prod_{j=1}^{n} \prod_{i=1}^{m} \left(1 - T_{ij}^L \right)^{x_i y_j}, 1 - \prod_{j=1}^{n} \prod_{i=1}^{m} \left(1 - T_{ij}^U \right)^{x_i y_j} \right] \leq [\sigma^L, \sigma^U] \ (x \in X) \\ \left[\prod_{j=1}^{n} \prod_{i=1}^{m} \left(I_{ij}^L \right)^{x_i y_j}, \prod_{j=1}^{n} \prod_{i=1}^{m} \left(I_{ij}^U \right)^{x_i y_j} \right] \geq [\rho^L, \rho^U] \ (x \in X) \\ \left[\prod_{j=1}^{n} \prod_{i=1}^{m} \left(F_{ij}^L \right)^{x_i y_j}, \prod_{j=1}^{n} \prod_{i=1}^{m} \left(F_{ij}^U \right)^{x_i y_j} \right] \geq [\kappa^L, \kappa^U] \ (x \in X) \\ 0 \leq \sigma^U + \rho^U + \kappa^U \leq 3 \\ \sum_{j=1}^{n} y_j = 1 \\ \sigma^L \geq 0, \sigma^U \geq 0, \rho^L \geq 0, \rho^U \geq 0, \kappa^L \geq 0, \kappa^U \geq 0, y_j \geq 0 \ (j = 1, 2, \ldots, n) \end{cases}$$

$$(41)$$

where

$$\sigma^L = \max_{x \in X} \left\{ 1 - \prod_{j=1}^{n} \prod_{i=1}^{m} \left(1 - T_{ij}^L\right)^{x_i y_j} \right\},$$

$$\sigma^U = \max_{x \in X} \left\{ 1 - \prod_{j=1}^{n} \prod_{i=1}^{m} \left(1 - T_{ij}^U\right)^{x_i y_j} \right\},$$

$$\rho^L = \min_{x \in X} \left\{ \prod_{j=1}^{n} \prod_{i=1}^{m} \left(I_{ij}^L\right)^{x_i y_j} \right\},$$

$$\rho^U = \min_{x \in X} \left\{ \prod_{j=1}^{n} \prod_{i=1}^{m} \left(I_{ij}^U\right)^{x_i y_j} \right\},$$

$$\kappa^L = \min_{x \in X} \left\{ \prod_{j=1}^{n} \prod_{i=1}^{m} \left(F_{ij}^L\right)^{x_i y_j} \right\} \quad and$$

$$\kappa^U = \min_{x \in X} \left\{ \prod_{j=1}^{n} \prod_{i=1}^{m} \left(F_{ij}^U\right)^{x_i y_j} \right\}.$$

$\min\{[\sigma^L, \sigma^U]\}$, $\max\{[\rho^L, \rho^U]\}$ and $\max\{[\kappa^L, \kappa^U]\}$ is equivalent to $\max\{[\ln(1 - \sigma^L), \ln(1 - \sigma^U)]\}$, $\max\{[\ln(\rho^L), \ln(\rho^U)]\}$ and $\max\{[\ln(\kappa^L), \ln(\kappa^U)]\}$, respectively, where $0 \le \sigma^L \le 1$, $0 \le \sigma^U \le 1$, $0 \le \rho^L \le 1$, $0 \le \rho^U \le 1$, $0 \le \kappa^L \le 1$ and $0 \le \kappa^U \le 1$.

Thus, we have

$$\max \{[\lambda \ln\left(1 - \sigma^U\right) + (1 - \lambda) \ln\left(\rho^L\right) + (1 - \lambda) \ln\left(\kappa^L\right),$$
$$\lambda \ln\left(1 - \sigma^L\right) + (1 - \lambda) \ln\left(\rho^U\right) + (1 - \lambda) \ln\left(\kappa^U\right)]\} \tag{42}$$

where $0 \le \sigma^L \le 1$, $0 \le \sigma^U \le 1$, $0 \le \rho^L \le 1$, $0 \le \rho^U \le 1$, $0 \le \kappa^L \le 1$ and $0 \le \kappa^U \le 1$.

Then, from Eq. (41), we have

$$\left[1 - \prod_{j=1}^{n} \prod_{i=1}^{m} \left(1 - T_{ij}^L\right)^{x_i y_j}, 1 - \prod_{j=1}^{n} \prod_{i=1}^{m} \left(1 - T_{ij}^U\right)^{x_i y_j} \right] \le \left[\sigma^L, \sigma^U\right]$$

And also,

$$\prod_{j=1}^{n} \prod_{i=1}^{m} \left(1 - T_{ij}^L\right)^{x_i y_j} \ge 1 - \sigma^L \tag{43}$$

$$\prod_{j=1}^{n} \prod_{i=1}^{m} \left(1 - T_{ij}^U\right)^{x_i y_j} \ge 1 - \sigma^U \tag{44}$$

which are equivalent to

$$\sum_{j=1}^{n}\sum_{i=1}^{m} x_i y_j \ln\left(1 - T_{ij}^{L}\right) \geq \ln\left(1 - \sigma^{L}\right) \tag{45}$$

$$\sum_{j=1}^{n}\sum_{i=1}^{m} x_i y_j \ln\left(1 - T_{ij}^{U}\right) \geq \ln\left(1 - \sigma^{U}\right) \tag{46}$$

respectively, where $0 \leq \sigma^{L} \leq 1$, $0 \leq \sigma^{U} \leq 1$, $0 \leq T_{ij}^{L} \leq 1.0 \leq T_{ij}^{U} \leq 1$.
Similarly, fom (41), we have

$$\left[\prod_{j=1}^{n}\prod_{i=1}^{m}\left(I_{ij}^{L}\right)^{x_i y_j}, \prod_{j=1}^{n}\prod_{i=1}^{m}\left(I_{ij}^{U}\right)^{x_i y_j}\right] \geq \left[\rho^{L}, \rho^{U}\right]$$

And also

$$\sum_{j=1}^{n}\sum_{i=1}^{m} x_i y_j \ln\left(I_{ij}^{L}\right) \geq \ln\left(\rho^{L}\right) \tag{47}$$

$$\sum_{j=1}^{n}\sum_{i=1}^{m} x_i y_j \ln\left(I_{ij}^{U}\right) \geq \ln\left(\rho^{U}\right) \tag{48}$$

where $0 \leq \rho^{L} \leq 1$, $0 \leq \rho^{U} \leq 1$ $0 \leq I_{ij}^{L} \leq 1$, $0 \leq I_{ij}^{U} \leq 1$.
Similarly, from (47), we have

$$\left[\prod_{j=1}^{n}\prod_{i=1}^{m}\left(F_{ij}^{L}\right)^{x_i y_j}, \prod_{j=1}^{n}\prod_{i=1}^{m}\left(F_{ij}^{U}\right)^{x_i y_j}\right] \geq \left[\kappa^{L}, \kappa^{U}\right]$$

And also

$$\sum_{j=1}^{n}\sum_{i=1}^{m} x_i y_j \ln\left(F_{ij}^{L}\right) \geq \ln\left(\kappa^{L}\right) \tag{49}$$

$$\sum_{j=1}^{n}\sum_{i=1}^{m} x_i y_j \ln\left(F_{ij}^{U}\right) \geq \ln\left(\kappa^{U}\right) \tag{50}$$

where $0 \leq \kappa^{L} \leq 1$, $0 \leq \kappa^{U} \leq 1$, $0 \leq F_{ij}^{L} \leq 1$, $0 \leq F_{ij}^{U} \leq 1$.

Then, Eq. (41) may be written as follows:

$$
s.t. \begin{cases}
\sum_{j=1}^{n} \sum_{i=1}^{m} \dfrac{\left[\lambda \ln\left(1-\mu_{ij}^{U}\right) + (1-\lambda)\ln\left(v_{ij}^{L}\right) + (1-\lambda)\ln\left(v_{ij}^{L}\right)\right] x_i y_j \geq}{\lambda \ln\left(1-\sigma^{U}\right) + (1-\lambda)\ln\left(\rho^{L}\right) + (1-\lambda)\ln\left(\rho^{L}\right)} \quad (x \in X) \\[4mm]
\sum_{j=1}^{n} \sum_{i=1}^{m} \dfrac{\left[\lambda \ln\left(1-\mu_{ij}^{L}\right) + (1-\lambda)\ln\left(v_{ij}^{U}\right) + (1-\lambda)\ln\left(v_{ij}^{U}\right)\right] x_i y_j \geq}{\lambda \ln\left(1-\sigma^{L}\right) + (1-\lambda)\ln\left(\rho^{U}\right) + (1-\lambda)\ln\left(\rho^{U}\right)} \quad (x \in X) \\[4mm]
\sum_{j=1}^{n} y_i = 1, 0 \leq \sigma^{L} + \rho^{U} + \kappa^{U} \leq 3, \\[2mm]
\sigma^{L} \geq 0, \sigma^{U} \geq 0, \rho^{L} \geq 0, \rho^{U} \geq 0, \kappa^{L} \geq 0, \kappa^{U} \geq 0, y_i \geq 0 \quad (j = 1, 2, \ldots, n)
\end{cases}
\tag{51}
$$

where $0 \leq \sigma^{L} \leq 1$, $0 \leq \sigma^{U} \leq 1$, $0 \leq \rho^{L} \leq 1$, $0 \leq \rho^{U} \leq 1$, $0 \leq \kappa^{L} \leq 1$, $0 \leq \kappa^{U} \leq 1$, $0 \leq T_{ij}^{L} \leq 1$, $0 \leq T_{ij}^{U} \leq 1$, $0 \leq I_{ij}^{L} \leq 1$, $0 \leq I_{ij}^{U} \leq 1$, $0 \leq F_{ij}^{L} \leq 1$ and $0 \leq F_{ij}^{U} \leq 1$.

From Eqs. (42) and (51), Eq. (41), we have

$$
\max \{[\lambda \ln\left(1-\sigma^{U}\right) + (1-\lambda)\ln\left(\rho^{L}\right) + (1-\lambda)\ln\left(\kappa^{L}\right), \lambda \ln\left(1-\sigma^{L}\right) +
$$
$$
(1-\lambda)\ln\left(\rho^{U}\right) + (1-\lambda)\ln\left(\kappa^{U}\right)]\}
$$
$$
s.t. \begin{cases}
\sum_{j=1}^{n} \sum_{i=1}^{m} \left[\lambda \ln\left(1-T_{ij}^{U}\right) + (1-\lambda)\ln\left(I_{ij}^{L}\right) + (1-\lambda)\ln\left(F_{ij}^{L}\right)\right] x_i y_j \geq \lambda \ln\left(1-\sigma^{U}\right) + \\
\qquad\qquad\qquad\qquad (1-\lambda)\ln\left(\rho^{L}\right) + (1-\lambda)\ln\left(\kappa^{L}\right) \quad (x \in X) \\[3mm]
\sum_{j=1}^{n} \sum_{i=1}^{m} \left[\lambda \ln\left(1-T_{ij}^{L}\right) + (1-\lambda)\ln\left(I_{ij}^{U}\right) + (1-\lambda)\ln\left(F_{ij}^{U}\right)\right] x_i y_j \geq \lambda \ln\left(1-\sigma^{L}\right) + \\
\qquad\qquad\qquad\qquad (1-\lambda)\ln\left(\rho^{U}\right) + (1-\lambda)\ln\left(\kappa^{U}\right) \quad (x \in X) \\[3mm]
\sum_{j=1}^{n} y_i = 1, 0 \leq \sigma^{L} + \rho^{U} + \kappa^{U} \leq 3, \\[2mm]
\sigma^{L} \geq 0, \sigma^{U} \geq 0, \rho^{L} \geq 0, \rho^{U} \geq 0, \kappa^{L} \geq 0, \kappa^{U} \geq 0, y_i \geq 0 \quad (j = 1, 2, \ldots, n)
\end{cases}
\tag{52}
$$

Let $\delta^{L} = \lambda \ln(1-\sigma^{U}) + (1-\lambda)\ln(\rho^{L}) + (1-\lambda)\ln(\kappa^{L})$ and $\delta^{U} = \lambda \ln(1-\sigma^{L}) + (1-\lambda)\ln(\rho^{U}) + (1-\lambda)\ln(\kappa^{U})$ then, it is obvious that $\delta^{L} \leq 0$, $\delta^{U} \leq 0$, due to $\lambda \in [0,1]$, $0 \leq 1 - \sigma^{L} \leq 1$, $0 \leq 1 - \sigma^{U} \leq 1$, $0 \leq \rho^{L} \leq 1$, $0 \leq \rho^{U} \leq 1$, $0 \leq \kappa^{L} \leq 1$, $0 \leq \kappa^{U} \leq 1$. Therefore, Eq. (52) may be written as follows:

$$
\max \left\{\left[\delta^{L}, \delta^{U}\right]\right\}
$$
$$
s.t. \begin{cases}
\sum_{j=1}^{n} \sum_{i=1}^{m} \left[\lambda \ln\left(1-T_{ij}^{U}\right) + (1-\lambda)\ln\left(I_{ij}^{L}\right) + (1-\lambda)\ln\left(F_{ij}^{L}\right)\right] x_i y_j \geq \delta^{L} \quad (x \in X) \\[3mm]
\sum_{j=1}^{n} \sum_{i=1}^{m} \left[\lambda \ln\left(1-T_{ij}^{L}\right) + (1-\lambda)\ln\left(I_{ij}^{U}\right) + (1-\lambda)\ln\left(F_{ij}^{U}\right)\right] x_i y_j \geq \delta^{U} \quad (x \in X) \\[3mm]
\sum_{j=1}^{n} y_i = 1, \delta^{L} \leq 0, \delta^{U} \leq 0, y_i \geq 0 \quad (j = 1, 2, \ldots, n)
\end{cases}
\tag{53}
$$

Moreover Eq. (53) may be written as follows:

$$\max\left\{\left[\delta^L, \delta^U\right]\right\}$$

$$s.t.\begin{cases} \sum_{j=1}^{n}\left[\lambda\ln\left(1 - T_{ij}^{U}\right) + (1 - \lambda)\ln\left(I_{ij}^{L}\right) + (1 - \lambda)\ln\left(F_{ij}^{L}\right)\right]y_j \geq \delta^L \\ \quad (i = 1, 2, \ldots, m) \\ \sum_{j=1}^{n}\left[\lambda\ln\left(1 - T_{ij}^{L}\right) + (1 - \lambda)\ln\left(I_{ij}^{U}\right) + (1 - \lambda)\ln\left(F_{ij}^{U}\right)\right]y_j \geq \delta^U \\ \quad (i = 1, 2, \ldots, m) \\ \sum_{j=1}^{n} y_i = 1, \delta^L \leq 0, \delta^U \leq 0, y_i \geq 0 \;\; (j = 1, 2, \ldots, n) \end{cases}$$

$$(54)$$

According to Eq. (35) and Eq. (54), we have

$$\max\left\{\delta^L\right\}, \max\left\{\frac{\delta^L + \delta^U}{2}\right\}$$

$$s.t.\begin{cases} \sum_{j=1}^{n}\left[\lambda\ln\left(1 - T_{ij}^{U}\right) + (1 - \lambda)\ln\left(I_{ij}^{L}\right) + (1 - \lambda)\ln\left(F_{ij}^{L}\right)\right]y_j \geq \delta^L \\ \quad (i = 1, 2, \ldots, m) \\ \sum_{j=1}^{n}\left[\lambda\ln\left(1 - T_{ij}^{L}\right) + (1 - \lambda)\ln\left(I_{ij}^{U}\right) + (1 - \lambda)\ln\left(F_{ij}^{U}\right)\right]y_j \geq \delta^U \\ \quad (i = 1, 2, \ldots, m) \\ \sum_{j=1}^{n} y_i = 1, \delta^L \leq 0, \delta^U \leq 0, y_i \geq 0 \;\; (j = 1, 2, \ldots, n) \end{cases}$$

$$(55)$$

And, from Eq. (36) and Eq. (55), we have

$$\max\left\{\delta^L\right\}, \max\left\{\frac{\delta^L}{2}\right\} + \max\left\{\frac{\delta^U}{2}\right\}$$

$$s.t.\begin{cases} \sum_{j=1}^{n}\left[\lambda\ln\left(1 - T_{ij}^{U}\right) + (1 - \lambda)\ln\left(I_{ij}^{L}\right) + (1 - \lambda)\ln\left(F_{ij}^{L}\right)\right]y_j \geq \delta^L \;\; (i = 1, 2, \ldots, m) \\ \sum_{j=1}^{n}\left[\lambda\ln\left(1 - T_{ij}^{L}\right) + (1 - \lambda)\ln\left(I_{ij}^{U}\right) + (1 - \lambda)\ln\left(F_{ij}^{U}\right)\right]y_j \geq \delta^U \;\; (i = 1, 2, \ldots, m) \\ \sum_{j=1}^{n} y_i = 1, \delta^L \leq 0, \delta^U \leq 0, y_i \geq 0 \;\; (j = 1, 2, \ldots, n) \end{cases}$$

$$(56)$$

which is equivalent to

$$\max\left\{\delta^L\right\}, \max\left\{\delta^U\right\}$$

$$s.t. \begin{cases} \sum\limits_{j=1}^{n}\left[\lambda\ln\left(1-T_{ij}^U\right)+(1-\lambda)\ln\left(I_{ij}^L\right)+(1-\lambda)\ln\left(F_{ij}^L\right)\right]y_j \geq \delta^L \\ \quad (i=1,2,\dots,m) \\ \sum\limits_{j=1}^{n}\left[\lambda\ln\left(1-T_{ij}^L\right)+(1-\lambda)\ln\left(I_{ij}^U\right)+(1-\lambda)\ln\left(F_{ij}^U\right)\right]y_j \geq \delta^U \\ \quad (i=1,2,\dots,m) \\ \sum\limits_{j=1}^{n} y_i = 1, \delta^L \leq 0, \delta^U \leq 0, y_i \geq 0 \ (j=1,2,\dots,n) \end{cases}$$

$$(57)$$

Then, we have

$$\max\left\{\frac{\delta^L+\delta^U}{2}\right\}$$

$$s.t. \begin{cases} \sum\limits_{j=1}^{n}\left[\lambda\ln\left(1-T_{ij}^U\right)+(1-\lambda)\ln\left(I_{ij}^L\right)+(1-\lambda)\ln\left(F_{ij}^L\right)+\lambda\ln\left(1-T_{ij}^L\right)\right. \\ \quad \left. +(1-\lambda)\ln\left(I_{ij}^U\right)+(1-\lambda)\ln\left(F_{ij}^U\right)\right]y_j \geq \delta^L+\delta^U \ (i=1,2,\dots,m) \\ \sum\limits_{j=1}^{n} y_i = 1 \\ \delta^L \leq 0, \delta^U \leq 0, y_i \geq 0 \ (j=1,2,\dots,n) \end{cases}$$

$$(58)$$

If $v = \delta^L + \delta^U$ then, Eq. (58) can be rewritten as follows:

$$\max\left\{\frac{v}{2}\right\}$$

$$s.t. \begin{cases} \sum\limits_{j=1}^{n}\left[\lambda\ln\left(1-T_{ij}^U\right)+(1-\lambda)\ln\left(v_{ij}^L\right)+(1-\lambda)\ln\left(F_{ij}^L\right)+\lambda\ln\left(1-T_{ij}^L\right)\right. \\ \quad \left. +(1-\lambda)\ln\left(I_{ij}^U\right)+(1-\lambda)\ln\left(F_{ij}^U\right)\right]y_j \geq \delta^L+\delta^U \ (i=1,2,\dots,m) \\ \sum\limits_{j=1}^{n} y_i = 1 \\ \delta^L \leq 0, \delta^U \leq 0, y_i \geq 0 \ (j=1,2,\dots,n) \end{cases}$$

$$(59)$$

In here Eqs. (40) and (59) are linear programming models for any given value of the weight $\lambda\in[0,1]$ in a TPIVNM-game.

Theorem 15 For $\lambda \in [0, 1]$ a TPIVNM-game has a solution $(x^*, y^*, x^{*T}Ay^*)$.
Proof
By using Eqs. (40) and (59) for $\lambda \in [0, 1]$, we have

$$A' = \left(\lambda\ln\left(1-T_{ij}^U\right)+(1-\lambda)\ln\left(I_{ij}^L\right)+(1-\lambda)\ln\left(F_{ij}^L\right)+\lambda\ln\left(1-T_{ij}^L\right)\right.$$
$$\left. +(1-\lambda)\ln\left(I_{ij}^U\right)+(1-\lambda)\ln\left(F_{ij}^U\right)\right)_{m\times n}$$

From maximin theorem of matrix games [3], A' always has a solution and therefore Eqs. (40) and (59) always have optimal solutions, denoted by (x^*, u^*) and (y^*, v^*), respectively, where $u^* = v^*$.

Theorem 16 u and v are monotonic and nondecreasing functions of the weight $\lambda \in [0,1]$ in Definition 15.

Proof: Let

$$\tau^L = \lambda \ln\left(1 - T^U\right) + (1 - \lambda) \ln\left(I^L\right) + (1 - \lambda) \ln\left(F^L\right),$$
$$\tau^U = \lambda \ln\left(1 - T^L\right) + (1 - \lambda) \ln\left(I^U\right) + (1 - \lambda) \ln\left(F^U\right) \rightarrow u = \tau^L + \tau^U$$

and

$$A' = \left(\lambda \ln\left(1 - T^U\right) + (1 - \lambda) \ln\left(I^L\right) + (1 - \lambda) \ln\left(F^L\right)\right.$$
$$\left. + \lambda \ln\left(1 - T^L\right) + (1 - \lambda) \ln\left(I^U\right) + (1 - \lambda) \ln\left(F^U\right)\right)_{m \times n}$$

where $0 \le T^L \le 1$, $0 \le T^U \le 1$, $0 \le I^L \le 1$, $0 \le I^U \le 1$, $0 \le F^L \le 1$, $0 \le F^U \le 1$.

Now; we can obtain the derivative of u with respect to λ as follows:

$$\frac{du}{d\lambda} = \ln\left(1 - T^U\right) - \ln\left(I^L\right) - \ln\left(F^L\right) + \ln\left(1 - T^L\right) - \ln\left(I^U\right) - \ln\left(F^U\right)$$

(60)

Noticing that $[T^L, T^U]$, $[I^L, I^U]$ and $[F^L, F^U]$ are three subintervals on the unit interval $[0,1]$. Thus, $0 \le T^L \le T^U \le 1$, $0 \le I^L \le I^U \le 1$, $0 \le F^L \le F^U \le 1$. Therefore, we can obtain

$$0 < 1 - T^U \le 1 - T^L \le 1$$

Hence, we have $\ln(1 - T^U) \le \ln(1 - T^L) \le 0$ and $\ln(I^L) \le \ln(I^U) \le 0$ and $\ln(F^L) \le \ln(F^U) \le 0$.

By combining with Eq. (60), we have

$$\frac{du}{d\lambda} \ge 2\left[\ln\left(1 - T^U\right) - \ln\left(I^U\right) - \ln\left(F^U\right)\right] = 2\ln\frac{1 - T^U}{I^U.F^U}$$

(61)

Since $1 - T^U \ge I^U.F^U$ and $0 \le I^U, F^U \le 1$, we have $\frac{1-T^U}{I^U.F^U} \ge 1 \rightarrow \ln\left(\frac{1-T^U}{I^U.F^U}\right) \ge 0 \rightarrow \frac{du}{d\lambda} \ge 0$

Similarly, we have

$$\delta^L = \lambda \ln\left(1 - \sigma^U\right) + (1 - \lambda) \ln\left(\rho^L\right) + (1 - \lambda) \ln\left(\kappa^L\right), \delta^U$$
$$= \lambda \ln\left(1 - \sigma^L\right) + (1 - \lambda) \ln\left(\rho^U\right) + (1 - \lambda) \ln\left(\kappa^U\right) \rightarrow v = \delta^L + \delta^U$$

and

$$v = \lambda \ln \left(1 - \sigma^U\right) + (1 - \lambda) \ln \left(\rho^L\right) + (1 - \lambda) \ln \left(\kappa^L\right)$$
$$+ \lambda \ln \left(1 - \sigma^L\right) + (1 - \lambda) \ln \left(\rho^U\right) + (1 - \lambda) \ln \left(\kappa^U\right)$$

where $0 \leq T^L \leq T^U \leq 1,\ 0 \leq I^L \leq I^U \leq 1,\ 0 \leq F^L \leq F^U \leq 1$. Then, from the derivative of the function v with respect to λ we have

$$\frac{dv}{d\lambda} = \ln \left(1 - \sigma^U\right) + \ln \left(1 - \sigma^L\right) - \ln \left(\rho^L\right) - \ln \left(\rho^L\right) - \ln \left(\kappa^L\right) - \ln \left(\kappa^L\right) \tag{62}$$

It is obvious that $0 \leq \sigma^L \leq \sigma^U < 1,\ 0 \leq \rho^L \leq \rho^U < 1$ *and* $0 \leq \kappa^L \leq \kappa^U < 1$ since $[\sigma^L, \sigma^U],\ [\rho^L, \rho^U]$ *and* $[\kappa^L, \kappa^U]$ are the subintervals on the unit interval $[0,1]$. Then, we have

$$0 < 1 - \sigma^U \leq 1 - \sigma^L \leq 1 \to \ln \left(1 - \sigma^U\right)$$
$$\leq \ln \left(1 - \sigma^L\right) \leq 0 \to \ln \left(\rho^L\right) \leq \ln \left(\rho^U\right) \leq 0.$$

By combining with Eq. (62), we have

$$\frac{dv}{d\lambda} \geq 2 \left[\ln \left(1 - \sigma^U\right) - \ln \left(\rho^U\right) - \ln \left(\kappa^U\right)\right] = 2 \ln \frac{1 - \sigma^U}{\rho^U . \kappa^U} \tag{63}$$

Noticing that $<[\sigma^L, \sigma^U], [\rho^L, \rho^U], [\kappa^L, \kappa^U]>$ is an interval-valued neutrosophic set. Then, it readily follows from Definition 8 that

$$\sigma^U + \rho^U \leq 1 \to 1 - \sigma^U \geq \rho^U \to \frac{1 - \sigma^U}{\rho^U} \geq 1 \to \ln \left(\frac{1 - \sigma^U}{\rho^U}\right) \geq 0 \to \frac{dv}{d\lambda} \geq 0.$$

Finally, the proof is valid.

The relations between optimal solutions of Eqs. (40) and (59) and the solution of any a TPIVNM-game N with payoffs of interval-valued neutrosophic sets are summarized as in the following Theorem 17.

Theorem 17 For any given value of the weight $\lambda \in (0,1)$, assume that (x^*, u^*) and (y^*, v^*) are optimal solutions of Eqs. (40) and (59), respectively. Then, (x^*, u^*) and (y^*, v^*) are noninferior solutions of Eqs. (22) and (41), respectively, where $\theta^* = \ <[T^L, T^U], [I^L, I^U], [F^L, F^U]>$ and $\vartheta^* = \ <[\sigma^L, \sigma^U], [\rho^L, \rho^U], [\kappa^L, \kappa^U]>$ are interval-valued neutrosophic sets.

Proof: If

(x^*, θ^*) is not a solution of Eq. (22), then there is a solution $\left(\overset{\smile}{x}, \overset{\smile}{\underset{....}{\theta}}\right)$

$\left(\overset{\smile}{x} \in X, \overset{\smile}{\underset{....}{\theta}} = < \left[\overset{\smile}{\sigma}^L, \overset{\smile}{\sigma}^U\right], \left[\overset{\smile}{\rho}^L, \overset{\smile}{\rho}^U\right], \left[\overset{\smile}{\kappa}^L, \overset{\smile}{\kappa}^U\right] >\right)$ as follows:

$$
\begin{cases}
1- \prod_{j=1}^{n} \prod_{i=1}^{m} \left(1-T_{ij}^{L}\right)^{\breve{x}_i y_j} \geq \breve{T}^{L}, \quad 1- \prod_{j=1}^{n} \prod_{i=1}^{m} \left(1-T_{ij}^{U}\right)^{\breve{x}_i y_j} \geq \breve{T}^{U} \quad (y \in Y), \\
\prod_{j=1}^{n} \prod_{i=1}^{m} \left(I_{ij}^{L}\right)^{\breve{x}_i y_j} \leq \breve{I}^{L}, \quad \prod_{j=1}^{n} \prod_{i=1}^{m} \left(I_{ij}^{U}\right)^{\breve{x}_i y_j} \leq \breve{I}^{U} \quad (y \in Y) \\
\prod_{j=1}^{n} \prod_{i=1}^{m} \left(F_{ij}^{L}\right)^{\breve{x}_i y_j} \leq \breve{F}^{L}, \quad \prod_{j=1}^{n} \prod_{i=1}^{m} \left(F_{ij}^{U}\right)^{\breve{x}_i y_j} \leq \breve{F}^{U} \quad (y \in Y) \\
0 \leq \breve{T}^{U} + \breve{I}^{U} + \breve{F}^{U} \leq 3, \breve{T}^{L} \geq 0, \breve{T}^{U} \geq 0, \breve{I}^{L} \geq 0, \breve{I}^{U} \geq 0, \breve{F}^{L} \geq 0, \breve{F}^{U} \geq 0 \\
\sum_{i=1}^{m} \breve{x}_i = 1, \breve{x}_i \geq 0 \, (i=1,2,\ldots,m)
\end{cases}
$$

(64)

$$
\left[\breve{T}^{L}, \breve{T}^{U} \right] \geq \left[T^{L*}, T^{U*} \right]
$$

(65)

$$
\left[\breve{I}^{L}, \breve{I}^{U} \right] \leq \left[I^{L*}, I^{U*} \right]
$$

(66)

$$
\left[\breve{F}^{L}, \breve{F}^{U} \right] \leq \left[F^{L*}, F^{U*} \right]
$$

(67)

From Eqs. (64–67) and $0 < \lambda < 1$, we have

$$
\begin{cases}
\sum_{j=1}^{n} \sum_{i=1}^{m} \begin{bmatrix} \lambda \ln \left(1 - T_{ij}^{U}\right) + (1-\lambda) \ln \left(I_{ij}^{L}\right) + (1-\lambda) \ln \left(F_{ij}^{L}\right) \end{bmatrix} \breve{x}_i y_j \leq \\
\qquad \lambda \ln \left(1 - \breve{T}^{U}\right) + (1-\lambda) \ln \left(\breve{I}^{L}\right) + (1-\lambda) \ln \left(\breve{F}^{L}\right) \Big] \quad (y \in Y) \\
\sum_{j=1}^{n} \sum_{i=1}^{m} \begin{bmatrix} \lambda \ln \left(1 - T_{ij}^{L}\right) + (1-\lambda) \ln \left(I_{ij}^{U}\right) + (1-\lambda) \ln \left(F_{ij}^{U}\right) \end{bmatrix} \breve{x}_i y_j \leq \\
\qquad \lambda \ln \left(1 - \breve{T}^{L}\right) + (1-\lambda) \ln \left(\breve{I}^{U}\right) + (1-\lambda) \ln \left(\breve{F}^{U}\right) \Big] \quad (y \in Y) \\
0 \leq \breve{T}^{U} + \breve{I}^{U} + \breve{F}^{U} \leq 3, \breve{T}^{L} \geq 0, \breve{T}^{U} \geq 0, \breve{I}^{L} \geq 0, \breve{I}^{U} \geq 0, \breve{F}^{L} \geq 0, \breve{F}^{U} \geq 0 \\
\sum_{i=1}^{m} \breve{x}_i = 1, \breve{x}_i \geq 0 \, (i=1,2,\ldots,m)
\end{cases}
$$

(68)

$$
\lambda \ln \left(1 - \breve{T}^{U}\right) + (1-\lambda) \ln \left(\breve{I}^{L}\right) + (1-\lambda) \ln \left(\breve{F}^{L}\right)
$$
$$
\leq \lambda \ln \left(1 - T^{U*}\right) + (1-\lambda) \ln \left(I^{L*}\right) + (1-\lambda) \ln \left(F^{L*}\right)
$$

(69)

$$\lambda \ln\left(1 - \overset{\smile L}{T}\right) + (1 - \lambda) \ln\left(\overset{\smile U}{I}\right) + (1 - \lambda) \ln\left(\overset{\smile U}{F}\right)$$
$$\leq \lambda \ln\left(1 - T^{L*}\right) + (1 - \lambda) \ln\left(I^{U*}\right) + (1 - \lambda) \ln\left(F^{U*}\right) \tag{70}$$

If

$$\overset{\smile L}{\tau} = \lambda \ln\left(1 - \overset{\smile U}{T}\right) + (1 - \lambda) \ln\left(\overset{\smile L}{I}\right) + (1 - \lambda) \ln\left(\overset{\smile L}{F}\right) \tag{71}$$

and

$$\overset{\smile U}{\tau} = \lambda \ln\left(1 - \overset{\smile L}{T}\right) + (1 - \lambda) \ln\left(\overset{\smile U}{I}\right) + (1 - \lambda) \ln\left(\overset{\smile U}{F}\right) \tag{72}$$

Then, from Eqs. (69)–(72), we have

$$\begin{cases} \overset{\smile L}{\tau} \leq \tau^{L*} \\ \overset{\smile U}{\tau} \leq \tau^{U*} \end{cases} \tag{73}$$

Moreover, there is at least one inequality in Eq. (73) which is strictly valid. Therefore, we have

$$\overset{\smile L}{\tau} + \overset{\smile U}{\tau} \leq \tau^{L*} + \tau^{U*} \tag{74}$$

Then, Eq. (64) may be rewritten as follows:

$$\begin{cases} \sum\limits_{j=1}^{n}\sum\limits_{i=1}^{m} \left[\lambda \ln\left(1 - T_{ij}^{U}\right) + (1 - \lambda) \ln\left(I_{ij}^{L}\right) + (1 - \lambda) \ln\left(F_{ij}^{L}\right)\right] \overset{\smile}{x}_i y_j \leq \overset{\smile L}{\tau} \ (y \in Y) \\ \sum\limits_{j=1}^{n}\sum\limits_{i=1}^{m} \left[\lambda \ln\left(1 - T_{ij}^{L}\right) + (1 - \lambda) \ln\left(I_{ij}^{U}\right) + (1 - \lambda) \ln\left(F_{ij}^{U}\right)\right] \overset{\smile}{x}_i y_j \leq \overset{\smile U}{\tau} \ (y \in Y) \\ \sum\limits_{\hat{i}=1}^{m} \overset{\smile}{x}_i = 1 \\ \overset{\smile L}{\tau} \leq 0, \overset{\smile U}{\tau} \leq 0, \overset{\smile}{x}_i \geq 0 \ (i = 1, 2, \ldots, m) \end{cases} \tag{75}$$

Also, Eq. (75) can be written as follows:

$$
\begin{cases}
\sum_{i=1}^{m} \left[\lambda \ln \left(1 - T_{ij}^{U} \right) + (1 - \lambda) \ln \left(I_{ij}^{L} \right) + (1 - \lambda) \ln \left(F_{ij}^{L} \right) \right] \breve{x}_i \leq \breve{\tau}^{L} \quad (j = 1, 2, \ldots, n) \\
\sum_{i=1}^{m} \left[\lambda \ln \left(1 - T_{ij}^{L} \right) + (1 - \lambda) \ln \left(I_{ij}^{U} \right) + (1 - \lambda) \ln \left(F_{ij}^{U} \right) \right] \breve{x}_i \leq \breve{\tau}^{U} \quad (j = 1, 2, \ldots, n) \\
\sum_{\hat{i}=1}^{m} \breve{x}_i = 1 \\
\breve{\tau}^{L} \leq 0, \breve{\tau}^{U} \leq 0, \breve{x}_i \geq 0 \quad (i = 1, 2, \ldots, m)
\end{cases}
\tag{76}
$$

And then we have

$$
\begin{cases}
\sum_{i=1}^{m} \begin{bmatrix} \lambda \ln \left(1 - T_{ij}^{U} \right) + (1 - \lambda) \ln \left(I_{ij}^{L} \right) + (1 - \lambda) \ln \left(F_{ij}^{L} \right) + \lambda \ln \left(1 - T_{ij}^{L} \right) \\ + (1 - \lambda) \ln \left(I_{ij}^{U} \right) + (1 - \lambda) \ln \left(F_{ij}^{U} \right) \end{bmatrix} \breve{x}_i \leq \breve{\tau}^{L} + \breve{\tau}^{U} \quad (j = 1, 2, \ldots, n) \\
\sum_{\hat{i}=1}^{m} \breve{x}_i = 1 \\
\breve{\tau}^{L} \leq 0, \breve{\tau}^{U} \leq 0, \breve{x}_i \geq 0 \quad (i = 1, 2, \ldots, m)
\end{cases}
\tag{77}
$$

If $\breve{u} = \breve{\tau}^{L} + \breve{\tau}^{U}$, then Eq. (77) can be rewritten as follows:

$$
\begin{cases}
\sum_{i=1}^{m} \begin{bmatrix} \lambda \ln \left(1 - T_{ij}^{U} \right) + (1 - \lambda) \ln \left(I_{ij}^{L} \right) + (1 - \lambda) \ln \left(F_{ij}^{L} \right) + \\ \lambda \ln \left(1 - T_{ij}^{L} \right) + (1 - \lambda) \ln \left(I_{ij}^{U} \right) + (1 - \lambda) \ln \left(F_{ij}^{U} \right) \end{bmatrix} \breve{x}_i \leq \breve{u} \quad (j = 1, 2, \ldots, n) \\
\sum_{\hat{i}=1}^{m} \breve{x}_i = 1 \\
\breve{u} \leq 0, \breve{x}_i \geq 0 \quad (i = 1, 2, \ldots, m)
\end{cases}
\tag{78}
$$

which show that (x,u) is a feasible solution of Eq. (40) and we have from Eq. (74)

$$
\frac{\breve{u}}{2} \leq \frac{u^*}{2}
$$

Finally, (x^*, θ^*) is positively solution of Eq. (22).

Similarly, If (y^*, ϑ^*) is not a solution of Eq. (41), then there is a solution $(\breve{y}, \breve{\vartheta})$ of Eq. (41) ($\breve{y} \in Y$ and $\breve{\vartheta} = \left< \left[\breve{\sigma}_{\mathrm{L}}, \breve{\sigma}_{\mathrm{U}} \right], \left[\breve{\rho}_{\mathrm{L}}, \breve{\rho}_{\mathrm{U}} \right], \left[\breve{\kappa}_{\mathrm{L}}, \breve{\kappa}_{\mathrm{U}} \right] \right>$) as follows:

$$
\begin{cases}
1 - \prod_{j=1}^{n} \prod_{i=1}^{m} \left(1 - T_{ij}^{L}\right)^{x_i \breve{y}_j} \leq \breve{\sigma}^{L}, \;\; 1 - \prod_{j=1}^{n} \prod_{i=1}^{m} \left(1 - T_{ij}^{U}\right)^{x_i \breve{y}_j} \leq \breve{\sigma}^{U} \quad (x \in X) \\[2mm]
\prod_{j=1}^{n} \prod_{i=1}^{m} \left(I_{ij}^{L}\right)^{x_i \breve{y}_j} \geq \breve{\rho}^{L}, \;\; \prod_{j=1}^{n} \prod_{i=1}^{m} \left(I_{ij}^{U}\right)^{x_i \breve{y}_j} \geq \breve{\rho}^{U} \quad (x \in X) \\[2mm]
\prod_{j=1}^{n} \prod_{i=1}^{m} \left(F_{ij}^{L}\right)^{x_i \breve{y}_j} \geq \breve{\kappa}^{L}, \;\; \prod_{j=1}^{n} \prod_{i=1}^{m} \left(F_{ij}^{U}\right)^{x_i \breve{y}_j} \geq \breve{\kappa}^{U} \quad (x \in X) \\[2mm]
0 \leq \breve{\sigma}^{U} + \breve{\rho}^{U} + \breve{\kappa}^{U} \leq 3 \\[2mm]
\sum_{j=1}^{n} \breve{y}_j = 1, \breve{\sigma}^{L} \geq 0, \breve{\sigma}^{U} \geq 0, \breve{\rho}^{L} \geq 0, \breve{\rho}^{U} \geq 0, \breve{\kappa}^{L} \geq 0, \breve{\kappa}^{U} \geq 0, \breve{y}_i \geq 0 \, (i = 1, 2, \ldots, m)
\end{cases}
\tag{79}
$$

$$
\left[\breve{\sigma}^{L}, \breve{\sigma}^{U}\right] \leq \left[\sigma^{L*}, \sigma^{U*}\right]
\tag{80}
$$

$$
\left[\breve{\rho}^{L}, \breve{\rho}^{U}\right] \geq \left[\rho^{L*}, \rho^{U*}\right]
\tag{81}
$$

and

$$
\left[\breve{\kappa}^{L}, \breve{\kappa}^{U}\right] \geq \left[\kappa^{L*}, \kappa^{U*}\right]
\tag{82}
$$

Furtermore, from (79–82) we have

$$
\begin{cases}
\sum_{j=1}^{n} \sum_{i=1}^{m} \left[\lambda \ln\left(1 - T_{ij}^{U}\right) + (1 - \lambda) \ln\left(I_{ij}^{L}\right) + (1 - \lambda) \ln\left(F_{ij}^{L}\right)\right] x_i \breve{y}_j \geq \\
\qquad\qquad \lambda \ln\left(1 - \breve{\sigma}^{U}\right) + (1 - \lambda) \ln\left(\breve{\rho}^{L}\right) + (1 - \lambda) \ln\left(\breve{\kappa}^{L}\right) \quad (x \in X) \\[2mm]
\sum_{j=1}^{n} \sum_{i=1}^{m} \left[\lambda \ln\left(1 - T_{ij}^{L}\right) + (1 - \lambda) \ln\left(I_{ij}^{U}\right) + (1 - \lambda) \ln\left(F_{ij}^{U}\right)\right] x_i \breve{y}_j \geq \\
\qquad\qquad \lambda \ln\left(1 - \breve{\sigma}^{L}\right) + (1 - \lambda) \ln\left(\breve{\rho}^{U}\right) + (1 - \lambda) \ln\left(\breve{\kappa}^{U}\right) \quad (x \in X) \\[2mm]
\sum_{j=1}^{n} \breve{y}_i = 1, 0 \leq \breve{\sigma}^{U} + \breve{\rho}^{U} + \breve{\kappa}^{U} \leq 3 \\[2mm]
\breve{\sigma}^{U} \geq 0, \breve{\sigma}^{L} \geq 0, \breve{\rho}^{L} \geq 0, \breve{\rho}^{U} \geq 0, \breve{\kappa}^{L} \geq 0, \breve{\kappa}^{U} \geq 0, \breve{y}_i \geq 0 \; (j = 1, 2, \ldots, n)
\end{cases}
\tag{83}
$$

$$
\begin{aligned}
&\lambda \ln\left(1 - \breve{\sigma}^{U}\right) + (1 - \lambda) \ln\left(\breve{\rho}^{L}\right) + (1 - \lambda) \ln\left(\breve{\kappa}^{L}\right) \\
&\geq \lambda \ln\left(1 - \sigma^{U*}\right) + (1 - \lambda) \ln\left(\rho^{L*}\right) + (1 - \lambda) \ln\left(\kappa^{L*}\right)
\end{aligned}
\tag{84}
$$

$$\lambda \ln \left(1 - \breve{\sigma}^L\right) + (1 - \lambda) \ln \left(\breve{\rho}^U\right) + (1 - \lambda) \ln \left(\breve{\kappa}^U\right)$$
$$\geq \lambda \ln \left(1 - \sigma^{L*}\right) + (1 - \lambda) \ln \left(\rho^{U*}\right) + (1 - \lambda) \ln \left(\kappa^{U*}\right) \tag{85}$$

Also let

$$\breve{\delta}^L = \lambda \ln \left(1 - \breve{\sigma}^U\right) + (1 - \lambda) \ln \left(\breve{\rho}^L\right) + (1 - \lambda) \ln \left(\breve{\kappa}^L\right) \tag{86}$$

and

$$\breve{\delta}^U = \lambda \ln \left(1 - \breve{\sigma}^L\right) + (1 - \lambda) \ln \left(\breve{\rho}^U\right) + (1 - \lambda) \ln \left(\breve{\kappa}^U\right) \tag{87}$$

From Eqs. (84)–(87) we have

$$\begin{cases} \breve{\delta}^L \leq \delta^{L*} \\ \breve{\delta}^U \leq \delta^{U*} \end{cases} \tag{88}$$

and we have

$$\breve{\delta}^L + \breve{\delta}^U \leq \delta^{L*} + \delta^{U*} \tag{89}$$

By combining with Eqs. (86) and (87), Eq. (83), we have

$$\begin{cases} \displaystyle\sum_{j=1}^{n}\sum_{i=1}^{m} \left[\lambda \ln \left(1 - T_{ij}^U\right) + (1 - \lambda) \ln \left(I_{ij}^L\right) + (1 - \lambda) \ln \left(F_{ij}^L\right)\right] x_i \breve{y}_j \geq \breve{\delta}^L \quad (x \in X) \\ \displaystyle\sum_{j=1}^{n}\sum_{i=1}^{m} \left[\lambda \ln \left(1 - T_{ij}^L\right) + (1 - \lambda) \ln \left(I_{ij}^U\right) + (1 - \lambda) \ln \left(F_{ij}^U\right)\right] x_i \breve{y}_j \geq \breve{\delta}^U \quad (x \in X) \\ \displaystyle\sum_{j=1}^{n} \breve{y}_j = 1 \\ \breve{\delta}^L \leq 0, \ \breve{\delta}^U \leq 0, \ \breve{y}_j \geq 0 \ (j = 1, 2, \dots, n) \end{cases} \tag{90}$$

Also, from Eq. (90), we have

$$
\begin{cases}
\sum_{j=1}^{n} \left[\lambda \ln \left(1 - T_{ij}^{U}\right) + (1-\lambda) \ln \left(I_{ij}^{L}\right) + (1-\lambda) \ln \left(F_{ij}^{L}\right) \right] \breve{y}_j \geq \breve{\delta}^{L} \quad (i = 1, 2, \ldots, m) \\[2mm]
\sum_{j=1}^{n} \left[\lambda \ln \left(1 - T_{ij}^{L}\right) + (1-\lambda) \ln \left(I_{ij}^{U}\right) + (1-\lambda) \ln \left(F_{ij}^{U}\right) \right] \breve{y}_j \geq \breve{\delta}^{U} \quad (i = 1, 2, \ldots, m) \\[2mm]
\sum_{j=1}^{n} \breve{y}_j = 1 \\[2mm]
\breve{\delta}^{L} \leq 0, \; \breve{\delta}^{U} \leq 0, \; \breve{y}_j \geq 0 \quad (j = 1, 2, \ldots, n)
\end{cases}
\tag{91}
$$

And we have

$$
\begin{cases}
\sum_{j=1}^{n} \left[\lambda \ln \left(1 - T_{ij}^{U}\right) + (1-\lambda) \ln \left(I_{ij}^{L}\right) + (1-\lambda) \ln \left(F_{ij}^{L}\right) + \lambda \ln \left(1 - T_{ij}^{L}\right) + \right. \\
\qquad\qquad \left. (1-\lambda) \ln \left(I_{ij}^{U}\right) + (1-\lambda) \ln \left(F_{ij}^{U}\right) \right] \breve{y}_j \geq \breve{\delta}^{L} + \breve{\delta}^{U} \\[2mm]
\sum_{j=1}^{n} \breve{y}_j = 1 \\[2mm]
\breve{\delta}^{L} \leq 0, \; \breve{\delta}^{U} \leq 0, \; \breve{y}_j \geq 0 \quad (j = 1, 2, \ldots, n) \; (i = 1, 2, \ldots, m)
\end{cases}
\tag{92}
$$

If $\breve{v} = \breve{\delta}^{L} + \breve{\delta}^{U}$ then, Eq. (92) can be written as follows:

$$
\begin{cases}
\sum_{j=1}^{n} \left[\lambda \ln \left(1 - T_{ij}^{U}\right) + (1-\lambda) \ln \left(I_{ij}^{L}\right) + (1-\lambda) \ln \left(F_{ij}^{L}\right) + \right. \\
\qquad \left. \lambda \ln \left(1 - T_{ij}^{L}\right) + (1-\lambda) \ln \left(I_{ij}^{U}\right) + (1-\lambda) \ln \left(F_{ij}^{U}\right) \right] \breve{y}_j \geq \breve{v} \\[2mm]
\sum_{j=1}^{n} \breve{y}_j = 1 \\[2mm]
\breve{v} \leq 0, \; \breve{y}_j \geq 0 \quad (j = 1, 2, \ldots, n) \; (i = 1, 2, \ldots, m)
\end{cases}
\tag{93}
$$

Furtermore, from Eq. (89), we have

$$
\frac{\breve{v}}{2} \leq \frac{v^*}{2}.
$$

Therefore (y^*, v^*) is the optimal solution of Eq. (59) and (y^*, ϑ^*) is positively the solution of Eq. (41).

Nonlinear Programming Model: Based on Theorem 17 any TPIVNM-game N has a solution $(x^*, y^*, x^{*T} A y^*)$ for $\lambda \in [0, 1]$. Also Eqs. (40) and (59) are a linear programming models, which are easily solved by using the simplex method of linear programming; however, when some interval $\left[T_{ij}^{L}, T_{ij}^{U} \right] = [1, 1]$ or $\left[I_{ij}^{L}, I_{ij}^{U} \right] = [0, 0]$ or $\left[F_{ij}^{L}, F_{ij}^{U} \right] = [0, 0]$, i.e., the interval-valued neutrosophic set $< T_{ij}^L, T_{ij}^U\big], \big[I_{ij}^L, I_{ij}^U\big], \big[F_{ij}^L, F_{ij}^U\big] >$ degenerates to a real number, then $\ln(1-T_{ij}^L) \to -\infty$, $\ln(1-T_{ij}^U) \to -\infty$, $\ln(1-I_{ij}^L) \to -\infty$, $\ln(1-I_{ij}^L) \to -\infty$., $\ln(1-F_{ij}^L) \to -\infty$ and/or $\ln(1-F_{ij}^L) \to -\infty$. Therefore Eqs. (40) and (59) make no sense and Eqs. (40) and (59) should be formally rewritten as the following nonlinear programming models:

$$\min \left\{ \left(1-T^L\right)^\lambda \left(I^L\right)^{1-\lambda} \left(F^L\right)^{1-\lambda} \left(1-T^U\right)^\lambda \left(I^U\right)^{1-\lambda} \left(F^U\right)^{1-\lambda} \right\}$$

$$s.t. \begin{cases} \prod_{i=1}^{m} \left[\left(1-T_{ij}^L\right)^\lambda \left(I_{ij}^L\right)^{1-\lambda} \left(F_{ij}^L\right)^{1-\lambda} \left(1-T_{ij}^U\right)^\lambda \left(I_{ij}^U\right)^{1-\lambda} \left(F_{ij}^U\right)^{1-\lambda} \right]^{x_i} \leq \\ \qquad \left(1-T^L\right)^\lambda \left(I^L\right)^{1-\lambda} \left(F^L\right)^{1-\lambda} \left(1-T^U\right)^\lambda \left(I^U\right)^{1-\lambda} \left(F^U\right)^{1-\lambda} \\ \sum_{i=1}^{m} x_i = 1 \\ 0 \leq T^U + I^U + F^U \leq 3, (j=1,2,\ldots,n) \\ T^L \geq 0, T^U \geq 0, I^L \geq 0, I^U \geq 0, F^L \geq 0, F^U \geq 0, x_i \geq 0 \ (i=1,2,\ldots,m) \end{cases}$$
$$(94)$$

and

$$\max \left\{ \left(1-\sigma^L\right)^\lambda \left(\rho^L\right)^{1-\lambda} \left(\kappa^L\right)^{1-\lambda} \left(1-\sigma^U\right)^\lambda \left(\rho^U\right)^{1-\lambda} \left(\kappa^U\right)^{1-\lambda} \right\}$$

$$s.t. \begin{cases} \prod_{i=1}^{m} \left[\left(1-T_{ij}^L\right)^\lambda \left(I_{ij}^L\right)^{1-\lambda} \left(F_{ij}^L\right)^{1-\lambda} \left(1-T_{ij}^U\right)^\lambda \left(I_{ij}^U\right)^{1-\lambda} \left(F_{ij}^U\right)^{1-\lambda} \right]^{y_j} \geq \\ \qquad \left(1-\sigma^L\right)^\lambda \left(\rho^L\right)^{1-\lambda} \left(\kappa^L\right)^{1-\lambda} \left(1-\sigma^U\right)^\lambda \left(\rho^U\right)^{1-\lambda} \left(\kappa^U\right)^{1-\lambda} \\ \sum_{j=1}^{n} y_j = 1, 0 \leq \sigma^U + \rho^U + \kappa^U \leq 3, (i=1,2,\ldots,m) \\ \sigma^L \geq 0, \sigma^U \geq 0, \rho^L \geq 0, \rho^U \geq 0, \kappa^L \geq 0, \kappa^U \geq 0, y_j \geq 0 \ (j=1,2,\ldots,n) \end{cases}$$
$$(95)$$

respectively.

Let $p = (1-T^L)^\lambda (I^L)^{1-\lambda} (F^L)^{1-\lambda} (1-T^U)^\lambda (I^U)^{1-\lambda} (F^U)^{1-\lambda}$, then, it is obvious that $0 \leq p \leq 1$ due to $\lambda \in [0,1]$; $00 \leq T^L \leq 1$, $0 \leq T^U \leq 1$, $0 \leq I^L \leq 1$, $0 \leq I^U \leq 1$, $0 \leq F^L \leq 1$, $0 \leq F^U \leq 1$.

From Eq. (95) we have nonlinear programming model as follows:

$$\min \{p\}$$

$$s.t. \begin{cases} \prod_{i=1}^{m} \left[\left(1-T_{ij}^L\right)^\lambda \left(I_{ij}^L\right)^{1-\lambda} \left(F_{ij}^L\right)^{1-\lambda} \left(1-T_{ij}^U\right)^\lambda \left(I_{ij}^U\right)^{1-\lambda} \left(F_{ij}^U\right)^{1-\lambda} \right]^{x_i} \leq p \ (j=1,2,\ldots,n) \\ \sum_{i=1}^{m} x_i = 1, p \geq 0, x_i \geq 0 \ (i=1,2,\ldots,m) \end{cases}$$
$$(96)$$

Similarly, If $q = (1 - \sigma^L)^\lambda (\rho^L)^{1-\lambda} (\kappa^L)^{1-\lambda} (1 - \sigma^U)^\lambda (\rho^U)^{1-\lambda} (\kappa^U)^{1-\lambda}$, $0 \le q \le 1$ $(\lambda \in [0,1])$ $0 \le 1 - \sigma^L \le 1$, $0 \le 1 - \sigma^U \le 1$, $0 \le \rho^L \le 1$, $0 \le \rho^U \le 1$, $0 \le \kappa^L \le 1$, $0 \le \kappa^U \le 1$ then, Eq. (96) may be written as;

$$
\max\{q\}
$$
$$
s.t. \begin{cases} \prod_{i=1}^{m} \left[\left(1 - T_{ij}^L\right)^\lambda \left(I_{ij}^L\right)^{1-\lambda} \left(F_{ij}^L\right)^{1-\lambda} \left(1 - T_{ij}^U\right)^\lambda \left(I_{ij}^U\right)^{1-\lambda} \left(F_{ij}^U\right)^{1-\lambda} \right]^{y_j} \ge q \quad (i = 1, 2, \ldots, m) \\ \sum_{j=1}^{n} y_j = 1 \\ q \ge 0, y_j \ge 0 \quad (j = 1, 2, \ldots, n) \end{cases}
$$
$$(97)$$

Finally from Eqs. (40), (59), (96), and (97) we have $p^* = q^*$, $p^* = e^{u^*}$ and $q^* = e^{v^*}$, where (x^*, u^*) and (y^*, v^*) are the optimal solutions of Eqs. (40) and (59), and (x^*, p^*) and (y^*, q^*) are the optimal solutions of Eqs. (96) and (97), respectively.

References

1. Aliprantis, C. D., & Chakrabarti, S. K. (2000). Games and decision making. *Oxford University Press*.
2. Newmann, J. V., & Morgenstern, O. (1947). *Theory of games and economic behaviour*. Princeton, New Jersey: Princeton University press.
3. Owen, G. (1982). *Game theory* (2nd ed.). New York: Academic Press.
4. Zadeh, L. A. (1965). Fuzzy sets. *Information and Control, 8*, 338–353.
5. Bector, C. R. (2005). *Fuzzy mathematical programming and fuzzy matrix games*. Berlin Heidelberg: Springer-Verlag.
6. Atanassov, K. (1986). Intuitionistic fuzzy sets. *Fuzzy Sets and Systems, 20*, 87–96.
7. Li, D. F. (2014). *Decision and game theory in management with intuitionistic fuzzy Sets studies in fuzziness and soft computing volume 308*. Springer.
8. Li, D. F. (2010). Mathematical-programming approach to matrix games with payoffs represented by Atanassov's interval-valued intuitionistic sets. *IEEE Transactions on Fuzzy Systems, 18*(6), 1112–1128.
9. Smarandache, F. (1998). *A unifying field in logics. Neutrosophy: Neutrosophic Probability, Set and Logic*. Rehoboth: American Research Press.
10. Deli, İ. (2019). Matrix games with simplified Neutrosophic payoffs, fuzzy multicriteria decision making using Neutrosophic Sets. *Springer Nature Switzerland AG, 369*, 233–246.
11. Abu-Faty, H. G., El-Hefnawy, N. A., & Kafafy, A. (2017). Neutrosophic TOPSIS based game theory for solving MCGDM problems. *Australian Journal of Basic and Applied Sciences, 11*(13), 29–38.
12. Bhattacharya S., Smarandache, F., Khoshnevisan, M.: The Israel-Palestine question? A case for application of neutrosophic game theory, game theory and information (book / working paper) 2004.
13. Pramanik, S., & Roy, T. K. (2014). Neutrosophic game theoretic approach to indo-pak conflict overjammu-kashmir. *Neutrosophic Sets and Syst., 2*, 82–101.
14. Wang, H., Smarandache, F., Zhang, Y. Q., & Sunderraman, R. (2010). Single valued neutrosophic set. *Multispace and Multistructure, 4*, 410–413.

15. Peng, J. J., Wang, J. Q., Zhang, H. Y., & Chen, X. H. (2014). An outranking approach for multi-criteria decision-making problemswith simplified neutrosophic sets. *Applied Soft Computing, 25*, 336–346.

16. Ishibuchi, H., & Tanaka, H. (1990). Multiobjective programming in optimization of the interval objective function. *European Journal of Operational Research, 48*(2), 219–225.

17. Wang, H., Smarandache, F., Zhang, Y. Q., & Sunderraman, R. (2005). *Interval Neutrosophic Sets and logic: Theory and applications in computing.* Phoenix, AZ: Hexis.

18. Zhang, H. Y., Wang, J. Q., & Chen, X. H. (2014). Interval Neutrosophic Sets and Their Application in Multicriteria Decision Making Problems. *The Scientific World Journal*, 645953. https://doi.org/10.1155/2014/645953.

Neutrosophic Hypersoft Matrix Theory: Its Definition, Operators, and Application in Decision-Making of Personnel Selection Problem

Muhammad Saqlain, Muhammad Saeed, Rana Muhammad Zulqarnain, and Sana Moin

1 Introduction

Mathematicians and researchers are developing many algorithms and techniques to solve the issues that arise in optimal choice selection problem. But the foremost difficult problems were associated with the issues that were additional qualitative rather than quantitative in nature. Thus, to handle with uncertainty, vagueness, and impressions in real-life problems, researchers develop new methods and new theoretical tools. Zadeh [2] initiated fuzzy sets in 1965 which deals with uncertainty; later Atanassov [3] proposed the concept of intuitionistic sets which deals with the vagueness. In 1998, the concept of neutrosophic sets were proposed by Smarandache [4]; impressions and hesitation components were solved with this concept. A new concept to deal with MCDM problem which was based on parameters, [5] initiated soft set theory in 1999, aggregate operators were proposed by [6, 7]. In 2018 Smarandache [8, 9] extended the theory of soft set into hypersoft set and neutrosophic hypersoft set by altering the function F into a multi-attribute function and introduced the hybrids of crisp, fuzzy, intuitionistic fuzzy, neutrosophic, and plithogenic hypersoft set. Saqlain, et al. [10] proposed the aggregate operators of neutrosophic hypersoft set NHSS [11] along with applications.

M. Saqlain · S. Moin
Lahore Garrison University, DHA Phase-VI, Sector C, Lahore, Pakistan
e-mail: msaqlain@lgu.edu.pk

M. Saeed (✉)
School of Mathematics, University of Management and Technology, Lahore, Pakistan
e-mail: Muhammad.saeed@umt.edu.pk

R. M. Zulqarnain
School of Mathematics, Northwest University Xi'an, Xi'an, China

© The Author(s), under exclusive license to Springer Nature Switzerland AG 2021   449
F. Smarandache, M. Abdel-Basset (eds.), *Neutrosophic Operational Research*,
https://doi.org/10.1007/978-3-030-57197-9_21

Matrix notion was necessary to be defined because matrix theory plays a vital role in finding the best option in real-life issue. so due to this reason in this article, the focus is to develop matrix theory for neutrosophic hypersoft set NHSS along with NHSM algorithms proposed in Sect. 4, comparing with the results of Saqlain et al. [1] in which an algorithm was proposed to solve multi-criteria decision problems.

1.1 Literature Review

After the invention of fuzzy set, interval-valued fuzzy sets and intuitionistic fuzzy sets came into being to deal with the uncertainty for membership values. In comparison with intuitionistic fuzzy sets and interval-valued intuitionistic fuzzy sets, the indeterminacy is communicated in a neutrosophic set. A neutrosophic set has three essential components such as truth membership T, indeterminacy membership I, and falsity membership F, which are independent of each other. But a neutrosophic set is troublesome to apply in logical and engineering problems. In this manner, Wang et al. [12] and Wang et al. [13] display the concepts of single-valued neutrosophic set (SVNS). SVNSs give uncertainty, imprecise, conflicting, and incomplete data existing in real world. Moreover, it would be more reasonable to handle vague and conflicting information. Molodtsov presents a total new idea of soft set which helps in various real-life problems [6, 7, 14–18]. Maji [19] comes up with neutrosophic soft set depicted by truth, indeterminacy, and falsity membership values which are independent in nature. Neutrosophic soft set can deal with inadequate, uncertain, and inconsistence information, whereas intuitionistic fuzzy soft set and fuzzy soft set can just deal with partial information. Smarandache [8, 9] presented a new procedure to deal with uncertainty. He generalized the soft set to hypersoft set by changing the function into multi-decision function and Abdel-Baset et al. [20–23], Nabeeh et al. [24], and Smarandache [25] moreover talk about the different expansion of neutrosophic sets in TOPSIS and MCDM. Since we realize uncertainty is all over to defeat this issue another framework is modeled by joining neutrosophic set to hypersoft set which results in neutrosophic hypersoft set. This will overcome the uncertainty in a continuously correct way.

1.2 Objective/Motivation of the Study

Multi-criteria multi-attributive decision problems (MCMADP), in reality very complex to solve due to intricate dependence of attributes in which decision makers play a vital role. Hypersoft sets are very suitable to solve MCMADP due to their structure in which decision maker and attribute can be considered separately. Those criteria/attributes, which can be important in optimal choice, can be dealt easily. Therefore this structure can be easier to conclude best choice if we may construct their matrix since matrices are very important to solve MCDM problems. Thus,

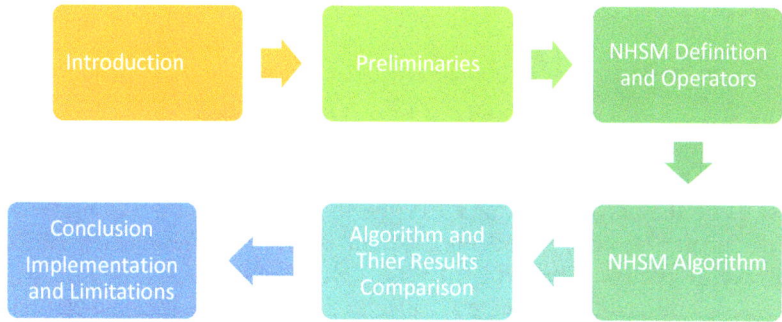

Fig. 1 The paper presentation in pictorial form

motivation of this paper is to propose some basic matrix operators of neutrosophic hypersoft sets (NHSSs). Finally, NHSM algorithm is designed to solve MCMADP and examples are illustrated.

1.3 Structure of the Article

Some basic definitions are mentioned in Sect. 2, which will be used in rest of this paper. In Sect. 3, first time neutrosophic hypersoft matrix theory and its operators are defined along with some necessary examples. Section 4 consists of NHSM algorithm along with suitable examples for the validity of proposed algorithm and also compared with the algorithm defined by Saqlain et al. [1]. In Sect. 5, results are discussed. Finally, the implementation and limitations are discussed along with the conclusion of the present work and have been stated in Sect. 6 (Fig. 1).

2 Preliminaries

This section consists of some basic definitions, which lead us to neutrosophic hypersoft sets and will be helpful in rest of the article.

Definition 2.1 Neutrosophic Soft Set

Let $\mathbb{U}$ be the universal set and $\mathbb{E}$ be the set of attributes with respect to $\mathbb{U}$. Let $\mathbb{P}(\mathbb{U})$ be the set of neutrosophic values of $\mathbb{U}$ and $\subseteq \mathbb{E}$. A pair $(\mathbb{F}, \mathbb{A})$ is called a neutrosophic soft set over $\mathbb{U}$ and its mapping is given as

$$\mathbb{F} : \mathbb{A} \rightarrow \mathbb{P}(\mathbb{U})$$

Definition 2.2 Hypersoft Set:

Let $\mathbb{U}$ be the universal set and $\mathbb{P}(\mathbb{U})$ be the power set of $\mathbb{U}$. Consider $l^1, l^2, l^3 \ldots l^n$ for $n \geq 1$, be n well-defined attributes, whose corresponding attributive values are, respectively, the set $L^1, L^2, L^3 \ldots L^n$ with $L^i \cap L^j = \varnothing$, for $i \neq j$ and i, j∈{1, 2, 3 ... n}, then the pair $(\mathbb{F}, L^1 \times L^2 \times L^3 \ldots L^n)$ is said to be hypersoft set over $\mathbb{U}$ where

$$\mathbb{F} : L^1 \times L^2 \times L^3 \ldots L^n \rightarrow \mathbb{P}(\mathbb{U})$$

Definition 2.3 Neutrosophic Hypersoft Set

Let $\mathbb{U}$ be the universal set and $\mathbb{P}(\mathbb{U})$be the power set of $\mathbb{U}$. Consider l^1, l^2, $l^3 \ldots l^n$ for $n \geq 1$, be n well-defined attributes, whose corresponding attributive values are, respectively, the set L^1, L^2, $L^3 \ldots L^n$ with $L^i \cap L^j = \varnothing$, for $i \neq j$ and i, j∈{1, 2, 3 ... n} and their relation $L^1 \times L^2 \times L^3 \ldots L^n = \$$, then the pair $(\mathbb{F}, \$)$ is said to be neutrosophic hypersoft set (NHSS) over $\mathbb{U}$ where

$$\mathbb{F} : L^1 \times L^2 \times L^3 \ldots L^n \rightarrow \mathbb{P}(\mathbb{U})$$

and

$\mathbb{F}(L^1 \times L^2 \times L^3 \ldots L^n) = \{< x, T(\mathbb{F}(\$)), I(\mathbb{F}(\$)), F(\mathbb{F}(\$)) >, x \in \mathbb{U}\}$ where T is the membership value of truthiness, I is the membership value of indeterminacy, and F is the membership value of falsity such that $T, I, F : \mathbb{U} \rightarrow [0, 1]$ also $0 \leq T(\mathbb{F}(\$)) + I(\mathbb{F}(\$)) + F(\mathbb{F}(\$)) \leq 3$.

3   Definition and Aggregate Operators of NHSM

Neutrosophic Hypersoft Matrix (NHSM).

In this section, we have introduced some definitions with suitable examples.

3.1   Definition: NHSM

Let $\mathbb{U} = \{u^1, u^2, \ldots u^a\}$ and $\mathbb{P}(\mathbb{U})$ be the universal set and power set of universal set, respectively; also consider $\mathbb{L}_1, \mathbb{L}_2, \ldots \mathbb{L}_{\ell}$ for $\ell \geq 1$, ℓ well-defined attributes, whose corresponding attributive values are, respectively, the set $\mathbb{L}_1^a, \mathbb{L}_2^b, \ldots \mathbb{L}_{\ell}^z$ and their relation $\mathbb{L}_1^a \times \mathbb{L}_2^b \times \ldots \mathbb{L}_{\ell}^z$ where $a, b, c, \ldots z = 1, 2, \ldots n$; then the pair $\left(\mathbb{F}, \mathbb{L}_1^a \times \mathbb{L}_2^b \times \ldots \mathbb{L}_{\ell}^z\right)$ is said to be neutrosophic hypersoft set over $\mathbb{U}$ where $\mathbb{F} : \left(\mathbb{L}_1^a \times \mathbb{L}_2^b \times \ldots \mathbb{L}_{\ell}^z\right) \rightarrow \mathbb{P}(\mathbb{U})$ and it is define as

$$\mathbb{F} : \left(\mathbb{L}_1^a \times \mathbb{L}_2^b \times \ldots \mathbb{L}_{\ell}^z\right) = \left\{< u, \mathbb{T}_{\ell}(u), \mathbb{I}_{\ell}(u), \mathbb{F}_{\ell}(u) > u \in \mathbb{U}, \ell \in \left(\mathbb{L}_1^a \times \mathbb{L}_2^b \times \ldots \mathbb{L}_{\ell}^z\right)\right\}$$

Let $\mathbb{R}_{\ell} = \left(\mathbb{L}_1^a \times \mathbb{L}_2^b \times \ldots \mathbb{L}_{\ell}^z\right)$ be the relation, and its characteristic function is $\mathcal{X}_{\mathbb{R}_{\ell}} : \left(\mathbb{L}_1^a \times \mathbb{L}_2^b \times \ldots \mathbb{L}_{\ell}^z\right) \rightarrow \mathbb{P}(\mathbb{U})$ and it is defined as $\mathcal{X}_{\mathbb{R}_{\ell}} =$

$\{< u, \mathbb{T}_\ell(u), \mathbb{I}_\ell(u), \mathbb{F}_\ell(u) > u \in \mathbb{U}, \ell \in (\mathbb{L}_1^a \times \mathbb{L}_2^b \times \ldots \mathbb{L}_{\&}^z)\}$. The tabular representation of $\mathbb{R}_\ell$ is given as.

	$\mathbb{L}_1^a$	$\mathbb{L}_1^b$	...	$\mathbb{L}_{\&}^z$
u^1	$\mathcal{X}_{\mathbb{R}_\ell}(u^1, \mathbb{L}_1^a)$	$\mathcal{X}_{\mathbb{R}_\ell}(u^1, \mathbb{L}_1^b)$	...	$\mathcal{X}_{\mathbb{R}_\ell}(u^1, \mathbb{L}_{\&}^z)$
u^2	$\mathcal{X}_{\mathbb{R}_\ell}(u^2, \mathbb{L}_1^a)$	$\mathcal{X}_{\mathbb{R}_\ell}(u^2, \mathbb{L}_1^b)$	...	$\mathcal{X}_{\mathbb{R}_\ell}(u^2, \mathbb{L}_{\&}^z)$
$\vdots$	$\vdots$	$\vdots$	$\ddots$	$\vdots$
u^a	$\mathcal{X}_{\mathbb{R}_\ell}(u^a, \mathbb{L}_1^a)$	$\mathcal{X}_{\mathbb{R}_\ell}(u^a, \mathbb{L}_1^b)$	...	$\mathcal{X}_{\mathbb{R}_\ell}(u^a, \mathbb{L}_{\&}^z)$

If $A_{ij} = \mathcal{X}_{\mathbb{R}_\ell}(u^i, \mathbb{L}_j^k)$, where $i = 1, 2, 3 \ldots a, j = 1, 2, 3, \ldots \&, k = a, b, c, \ldots z$, then a matrix is defined as

$$[A_{ij}]_{a \times \&} = \begin{pmatrix} A_{11} & A_{12} & \cdots & A_{1\&} \\ A_{21} & A_{22} & \cdots & A_{2\&} \\ \vdots & \vdots & \ddots & \vdots \\ A_{a1} & A_{a2} & \cdots & A_{a\&} \end{pmatrix}$$

where $A_{ij} = \left(\mathbb{T}_{\mathbb{L}_j^k}(u_i), \mathbb{I}_{\mathbb{L}_j^k}(u_i), \mathbb{F}_{\mathbb{L}_j^k}(u_i), u_i \in \mathbb{U}, \mathbb{L}_j^k \in (\mathbb{L}_1^a \times \mathbb{L}_2^b \times \ldots \mathbb{L}_{\&}^z)\right)$
$= \left(\mathbb{T}_{ijk}^A, \mathbb{I}_{ijk}^A, \mathbb{F}_{ijk}^A\right)$

Thus, we can represent any neutrosophic hypersoft set in term of neutrosophic hypersoft matrix (NHSM); it means that they are interchangeable. Its generalized form is given as.

$$[A_{ij}]_{a \times b} = \begin{bmatrix} \mathbb{T}_{\mathbb{L}_1^1}(a), \mathbb{I}_{\mathbb{L}_1^1}(a), \mathbb{F}_{\mathbb{L}_1^1}(a) & \mathbb{T}_{\mathbb{L}_1^2}(a), \mathbb{I}_{\mathbb{L}_1^2}(a), \mathbb{F}_{\mathbb{L}_1^2}(a) & \cdots & \mathbb{T}_{\mathbb{L}_1^{\&}}(a), \mathbb{I}_{\mathbb{L}_1^{\&}}(a), \mathbb{F}_{\mathbb{L}_1^{\&}}(a) \\ \mathbb{T}_{\mathbb{L}_2^1}(a), \mathbb{I}_{\mathbb{L}_2^1}(a), \mathbb{F}_{\mathbb{L}_2^1}(a) & \mathbb{T}_{\mathbb{L}_2^2}(a), \mathbb{I}_{\mathbb{L}_2^2}(a), \mathbb{F}_{\mathbb{L}_2^2}(a) & \cdots & \mathbb{T}_{\mathbb{L}_2^{\&}}(a), \mathbb{I}_{\mathbb{L}_2^{\&}}(a), \mathbb{F}_{\mathbb{L}_2^{\&}}(a) \\ \vdots & \vdots & \ddots & \vdots \\ \mathbb{T}_{\mathbb{L}_a^1}(a), \mathbb{I}_{\mathbb{L}_a^1}(a), \mathbb{F}_{\mathbb{L}_a^1}(a) & \mathbb{T}_{\mathbb{L}_a^2}(a), \mathbb{I}_{\mathbb{L}_a^2}(a), \mathbb{F}_{\mathbb{L}_a^2}(a) & \cdots & \mathbb{T}_{\mathbb{L}_a^{\&}}(a), \mathbb{I}_{\mathbb{L}_a^{\&}}(a), \mathbb{F}_{\mathbb{L}_a^{\&}}(a.) \end{bmatrix}$$

Example 1 Let $\mathbb{U}$ be the set of teachers selected for promotion to 18th grade:

$$\mathbb{U} = \left\{T^1, T^2, T^3, T^4, T^5\right\}$$

also consider the set of attributes as

$P_1 = $ Academic record, $P_2 = $ Manegement skills, $P_3 = $ Experiance, $P_4 = $ Awards

And their respective attributes are given as

$$P_1^a = \text{Academic record} = \{100\%, 90\%, 85\%\}$$

$$P_2^{\ell} = \text{Manegement skills} = \{\text{Good, Average, Poor}\}$$

$$P_3^c = \text{Experiance} = \{8\text{yr}, 10\text{yr}, 12\text{yr}\}$$

$$P_4^d = \text{Awards} = \{10+, 15+, 20+\}$$

Let the function be $\mathbb{F} : P_1^a \times P_2^{\ell} \times P_3^c \times P_4^d \to \mathbb{P}(\mathbb{U})$.

Below are the tables of their neutrosophic values from different decision makers:

P_1^a (Academic record)	T^1	T^2	T^3	T^4	T^5
100%	(0.9, 0.3, 0.1)	(0.5, 0.2, 0.1)	(0.8, 0.5, 0.2)	(0.8, 0.2, 0.1)	(0.7, 0.4, 0.2)
90%	(0.5, 0.3, 0.6)	(0.3, 0.2, 0.1)	(0.3, 0.6, 0.2)	(0.7, 0.3, 0.6)	(0.5, 0.4, 0.5)
85%	(0.8, 0.2, 0.4)	(0.9, 0.5, 0.3)	(0.9, 0.4, 0.1)	(0.6, 0.3, 0.2)	(0.6, 0.1, 0.2)

P_2^{ℓ} (Manegement skills)	T^1	T^2	T^3	T^4	T^5
Good	(0.3, 0.4, 0.7)	(0.6, 0.5, 0.3)	(0.5, 0.6, 0.8)	(0.6, 0.4, 0.8)	(0.3, 0.6, 0.7)
Average	(0.8, 0.2, 0.1)	(0.6, 0.4, 0.3)	(0.9, 0.4, 0.1)	(0.6, 0.2, 0.3)	(0.5, 0.3, 0.2)
Poor	(0.7, 0.2, 0.3)	(0.9, 0.3, 0.1)	(0.8, 0.3, 0.2)	(0.5, 0.4, 0.3)	(0.5, 0.2, 0.1)

P_3^c (Experiance)	T^1	T^2	T^3	T^4	T^5
8 year	(0.5, 0.3, 0.4)	(0.6, 0.3, 0.4)	(0.5, 0.7, 0.2)	(0.8, 0.4, 0.1)	(0.7, 0.4, 0.3)
10 year	(0.6, 0.4, 0.7)	(0.3, 0.6, 0.4)	(0.8, 0.2, 0.1)	(0.4, 0.5, 0.6)	(0.8, 0.4, 0.2)
12 year	(0.7, 0.2, 0.3)	(0.9, 0.3, 0.1)	(0.8, 0.3, 0.2)	(0.5, 0.4, 0.3)	(0.5, 0.2, 0.1)

P_4^d (Awards)	T^1	T^2	T^3	T^4	T^5
10+	(0.6, 0.4, 0.5)	(0.7, 0.5, 0.3)	(0.6, 0.4, 0.3)	(0.6, 0.2, 0.1)	(0.4, 0.5, 0.3)
15+	(0.8, 0.2, 0.4)	(0.7, 0.3, 0.2)	(0.8, 0.3, 0.1)	(0.3, 0.4, 0.5)	(0.3, 0.5, 0.8)
20+	(0.4, 0.9, 0.6)	(0.8, 0.4, 0.2)	(0.2, 0.6, 0.5)	(0.7, 0.5, 0.2)	(0.6, 0.4, 0.7)

Neutrosophic hypersoft set is defined as

$$\mathbb{F} : \left(P_1^a \times P_2^{\theta} \times P_3^c \times P_4^d \right) \to \mathbb{P}\,(\mathbb{U})$$

Let's assume $\mathbb{F}\left(\left(P_1^a \times P_2^{\theta} \times P_3^c \times P_4^d\right)\right) = \mathcal{F}\,(100\%, \text{Good}, 15+) = \{T^3, T^4, T^5\}$

Then neutrosophic hypersoft set of above assumed relation is

$$\mathbb{F}\left(\left(P_1^a \times P_2^{\theta} \times P_3^c \times P_4^d\right)\right) = \mathcal{F}\,(100\%, \text{Good}, 15+)$$

$$= \Big\{ \ll T^3, (100\%\,\{0.8, 0.5, 0.2\}, \text{Good}\,\{0.5, 0.6, 0.8\}, 15 + \{0.8, 0.3, 0.1\}) \gg,$$

$$\ll T^4, (100\%\,\{0.8, 0.2, 0.1\}, \text{Good}\,\{0.6, 0.4, 0.8\}, 15 + \{0.3, 0.4, 0.5\}) \gg,$$

$$\ll T^5 (100\%\,\{0.7, 0.4, 0.2\}, \text{Good}\,\{0.3, 0.6, 0.7\}, 15 + \{0.3, 0.5, 0.8\}) \gg, \Big\}$$

The tabular representation of characteristic function is given as

	A_1^a	A_2^{θ}	A_4^d
T^3	$(100\ \%\ \{0.8, 0.5, 0.2\})$	$\text{Good}\{0.5, 0.6, 0.8\}$	$15 + \{0.8, 0.3, 0.1\}$
T^4	$100\ \%\ \{0.8, 0.2, 0.1\}$	$\text{Good}\{0.6, 0.4, 0.8\}$	$15 + \{0.3, 0.4, 0.5\}$
T^5	$100\ \%\ \{0.7, 0.4, 0.2\}$	$\text{Good}\{0.3, 0.6, 0.7\}$	$15 + \{0.3, 0.5, 0.8\}$

And its matrix is defined as

$$[A]_{3 \times 3} = \begin{bmatrix} 100\%\,\{0.8, 0.5, 0.2\}\ Good\,\{0.5, 0.6, 0.8\}\ 15 + \{0.8, 0.3, 0.1\} \\ 100\%\,\{0.8, 0.2, 0.1\}\ Good\,\{0.6, 0.4, 0.8\}\ 15 + \{0.3, 0.4, 0.5\} \\ 100\%\,\{0.7, 0.4, 0.2\}\ Good\,\{0.3, 0.6, 0.7\}\ 15 + \{0.3, 0.5, 0.8\} \end{bmatrix}$$

3.2 Definition: Row NHSM

Let $A = \left[A_{ij}\right]$ be the NHSM of order $a \times \theta$, where $A_{ij} = \left(\mathbb{T}_{ijk}^A, \mathbb{I}_{ijk}^A, \mathbb{F}_{ijk}^A\right)$, then A is said to be row NHSM if $a = 1$. It means that if a NHSM contains only one attribute, i.e., $\mathbb{U} = \{u^1\}$, then it is a row NHSM.

Example If $\mathbb{U} = \{T^3\}$, then Example is said to be row NHSM and it is given as

$$[100\%\,\{0.8, 0.5, 0.2\}\ \ Good\,\{0.5, 0.6, 0.8\}\ \ 15 + \{0.8, 0.3, 0.1\}]$$

3.3   Definition: Column NHSM

Let $A = \left[A_{ij}\right]$ be the NHSM of order $a \times b$, where $A_{ij} = \left(\mathbb{T}_{ijk}^{A}, \mathbb{I}_{ijk}^{A}, \mathbb{F}_{ijk}^{A}\right)$, then A is said to be column NHSM if $b = 1$. It means that if a NHSM contains only one alternative, i.e., $\ell = \left\{\mathbb{L}_{1}^{a}\right\}$, then it is a column NHSM.

Example If $\mathbb{U} = \left\{T^{3}, T^{4}, T^{5}\right\}$ and P_{1}^{a} is the only attribute, then Example is said to be column NHSM and it is given as

$$\begin{bmatrix} 100\% \; \{0.8, 0.5, 0.2\} \\ 100\% \; \{0.8, 0.2, 0.1\} \\ 100\% \; \{0.7, 0.4, 0.2\} \end{bmatrix}$$

3.4   Definition: Zero NHSM

Let $A = \left[A_{ij}\right]$ be the NHSM of order $a \times b$, where $A_{ij} = \left(\mathbb{T}_{ijk}^{A}, \mathbb{I}_{ijk}^{A}, \mathbb{F}_{ijk}^{A}\right)$, then A is said to be zero NHSM if $A_{ij} = (0, 1, 1)$.

Example For Example, zero NHSM is given as

$$[A]_{3 \times 3} = \begin{bmatrix} 100\% \; \{0, 1, 1\} \; Good \; \{0, 1, 1\} \; 15 + \{0, 1, 1\} \\ 100\% \; \{0, 1, 1\} \; Good \; \{0, 1, 1\} \; 15 + \{0, 1, 1\} \\ 100\% \; \{0, 1, 1\} \; Good \; \{0, 1, 1\} \; 15 + \{0, 1, 1\} \end{bmatrix}$$

3.5   Definition: Universal NHSM

Let $A = \left[A_{ij}\right]$ be the NHSM of order $a \times b$, where $A_{ij} = \left(\mathbb{T}_{ijk}^{A}, \mathbb{I}_{ijk}^{A}, \mathbb{F}_{ijk}^{A}\right)$, then A is said to be universal NHSM if $A_{ij} = (1, 0, 0)$.

Example: For Example, universal NHSM is given as

$$[A]_{3 \times 3} = \begin{bmatrix} 100\% \; \{1, 0, 0\} \; Good \; \{1, 0, 0\} \; 15 + \{1, 0, 0\} \\ 100\% \; \{1, 0, 0\} \; Good \; \{1, 0, 0\} \; 15 + \{1, 0, 0\} \\ 100\% \; \{1, 0, 0\} \; Good \; \{1, 0, 0\} \; 15 + \{1, 0, 0\} \end{bmatrix}$$

3.6 Definition: Neutrosophic Hypersoft Submatrix (NHSSM)

Let $A = [A_{ij}]$ and $B = [B_{ij}]$ be the two NHSM of order $a \times b$, where $A_{ij} = \left(\mathbb{T}^A_{ijk}, \mathbb{I}^A_{ijk}, \mathbb{F}^A_{ijk}\right)$ and $B_{ij} = \left(\mathbb{T}^B_{ijk}, \mathbb{I}^B_{ijk}, \mathbb{F}^B_{ijk}\right)$, then $[A_{ij}]$ is said to be NHSSM of $[B_{ij}]$ if $\mathbb{T}^A_{ijk} \leq \mathbb{T}^B_{ijk}, \mathbb{I}^A_{ijk} \leq \mathbb{I}^B_{ijk}, \mathbb{F}^A_{ijk} \geq \mathbb{F}^B_{ijk}$.

3.7 Definition: Neutrosophic Hypersoft Equal Matrix (NHSEM)

Let $A = [A_{ij}]$ and $B = [B_{ij}]$ be the two NHSM of order $a \times b$, where $A_{ij} = \left(\mathbb{T}^A_{ijk}, \mathbb{I}^A_{ijk}, \mathbb{F}^A_{ijk}\right)$ and $B_{ij} = \left(\mathbb{T}^B_{ijk}, \mathbb{I}^B_{ijk}, \mathbb{F}^B_{ijk}\right)$, then $[A_{ij}]$ is said to be NHSEM of $[B_{ij}]$ if $\mathbb{T}^A_{ijk} = \mathbb{T}^B_{ijk}, \mathbb{I}^A_{ijk} = \mathbb{I}^B_{ijk}, \mathbb{F}^A_{ijk} = \mathbb{F}^B_{ijk}$.

3.8 Definition: AND Operation

Let $A = [A_{ij}]$ and $B = [B_{ij}]$ be the two NHSM of order $a \times b$, where $A_{ij} = \left(\mathbb{T}^A_{ijk}, \mathbb{I}^A_{ijk}, \mathbb{F}^A_{ijk}\right)$ and $B_{ij} = \left(\mathbb{T}^B_{ijk}, \mathbb{I}^B_{ijk}, \mathbb{F}^B_{ijk}\right)$ then $[A_{ij}] \wedge [B_{ij}]$ is given as

$$[A_{ij}] \wedge [B_{ij}] = \left(\min\left(\mathbb{T}\left([A_{ij}]\right), \mathbb{T}\left([B_{ij}]\right)\right), \frac{\left(\mathbb{I}\left([A_{ij}]\right), \mathbb{I}\left([B_{ij}]\right)\right)}{n},\right.$$

$$\left. \max\left(\mathbb{F}\left([A_{ij}]\right), \mathbb{F}\left([B_{ij}]\right)\right)\right)$$

Example Let $A = [A_{ij}]$ and $B = [B_{ij}]$ be the two NHSM of order 3×3

$$[A]_{3\times3} = \begin{bmatrix} 100\% \{0.8, 0.5, 0.2\} \; Good \; \{0.5, 0.6, 0.8\} \; 15 + \{0.8, 0.3, 0.1\} \\ 100\% \{0.8, 0.2, 0.1\} \; Good \; \{0.6, 0.4, 0.8\} \; 15 + \{0.3, 0.4, 0.5\} \\ 100\% \{0.7, 0.4, 0.2\} \; Good \; \{0.3, 0.6, 0.7\} \; 15 + \{0.3, 0.5, 0.8\} \end{bmatrix}$$

$$[B]_{3\times3} = \begin{bmatrix} 100\% \{0.9, 0.4, 0.2\} \; Good \; \{0.8, 0.5, 0.3\} \; 15 + \{0.4, 0.3, 0.7\} \\ 100\% \{0.6, 0.3, 0.2\} \; Good \; \{0.8, 0.2, 0.2\} \; 15 + \{0.7, 0.2, 0.3\} \\ 100\% \{0.7, 0.4, 0.1\} \; Good \; \{0.9, 0.2, 0.0\} \; 15 + \{0.8, 0.3, 0.1\} \end{bmatrix}$$

Then $[A] \wedge [B]$ is given as

$[C] = [A] \wedge [B]$

$$= \begin{bmatrix} 100\% \ \{0.0, 0.3, 0.8\} \ Good \ \{0.0, 0.3, 0.8\} \ 15 + \{0.0, 0.2, 0.8\} \\ 100\% \ \{0.0, 0.2, 0.8\} \ Good \ \{0.0, 0.3, 0.8\} \ 15 + \{0.0, 0.3, 0.8\} \\ 100\% \ \{0.0, 0.3, 0.8\} \ Good \ \{0.0, 0.2, 0.8\} \ 15 + \{0.0, 0.3, 0.8\} \end{bmatrix}$$

One calculation is provided for the convenience of reader:

$$C_{11} = ((100\% \ \{0.8, 0.5, 0.2\} \, , \, 100\% \ \{0.9, 0.4, 0.2\})) \, ,$$

$$((Good \ \{0.5, 0.6, 0.8\} \, , \, Good \ \{0.0, 0.0, 0.0\})) \, ,$$

$$((15 + \{0.8, 0.3, 0.1\} \, , \, 15 + \{0.0, 0.0, 0.0\}))$$

$$C_{11} = (((\{0.8, 0.45, 0.2\}) \, , \, (\{0.0, 0.3, 0.8\}) \, , \, (\{0, 0.15, 0.1\}))$$

$$C_{11} = 100\% \ \{0.0, 0.3, 0.8\}$$

3.9 Definition: OR Operation

Let $A = [A_{ij}]$ and $B = [B_{ij}]$ be the two NHSM of order $a \times b$, where $A_{ij} = \left(\mathbb{T}_{ijk}^A, \mathbb{I}_{ijk}^A, \mathbb{F}_{ijk}^A \right)$ and $B_{ij} = \left(\mathbb{T}_{ijk}^B, \mathbb{I}_{ijk}^B, \mathbb{F}_{ijk}^B \right)$ then $[A_{ij}] \vee [B_{ij}]$ is given as

$$[A_{ij}] \vee [B_{ij}] = \left(\max \left(\mathbb{T} \left([A_{ij}] \right), \mathbb{T} \left([B_{ij}] \right) \right), \frac{\left(\mathbb{I} \left([A_{ij}] \right), \mathbb{I} \left([B_{ij}] \right) \right)}{n}, \right.$$

$$\min \left(\mathbb{F} \left([A_{ij}] \right), \mathbb{F} \left([B_{ij}] \right) \right)$$

Example Let $A = [A_{ij}]$ and $B = [B_{ij}]$ be the two NHSM of order 3×3

$$[A]_{3 \times 3} = \begin{bmatrix} 100\% \ \{0.8, 0.5, 0.2\} \ Good \ \{0.5, 0.6, 0.8\} \ 15 + \{0.8, 0.3, 0.1\} \\ 100\% \ \{0.8, 0.2, 0.1\} \ Good \ \{0.6, 0.4, 0.8\} \ 15 + \{0.3, 0.4, 0.5\} \\ 100\% \ \{0.7, 0.4, 0.2\} \ Good \ \{0.3, 0.6, 0.7\} \ 15 + \{0.3, 0.5, 0.8\} \end{bmatrix}$$

$$[B]_{3 \times 3} = \begin{bmatrix} 100\% \ \{0.9, 0.4, 0.2\} \ Good \ \{0.8, 0.5, 0.3\} \ 15 + \{0.4, 0.3, 0.7\} \\ 100\% \ \{0.6, 0.3, 0.2\} \ Good \ \{0.8, 0.2, 0.2\} \ 15 + \{0.7, 0.2, 0.3\} \\ 100\% \ \{0.7, 0.4, 0.1\} \ Good \ \{0.9, 0.2, 0.0\} \ 15 + \{0.8, 0.3, 0.1\} \end{bmatrix}$$

Then $[A] \vee [B]$ is given as

$$[C] = [A] \vee [B]$$

$$= \begin{bmatrix} 100\% \{0.9, 0.3, 0.0\} \ Good \ \{0.8, 0.3, 0.0\} \ 15 + \{0.8, 0.2, 0.0\} \\ 100\% \{0.8, 0.2, 0.0\} \ Good \ \{0.8, 0.3, 0.0\} \ 15 + \{0.8, 0.3, 0.0\} \\ 100\% \{0.7, 0.3, 0.0\} \ Good \ \{0.9, 0.2, 0.0\} \ 15 + \{0.8, 0.3, 0.0\} \end{bmatrix}$$

One calculation is provided for the convenience of reader:

$$C_{11} = ((100\% \{0.8, 0.5, 0.2\}, 100\% \{0.9, 0.4, 0.2\})),$$

$$((Good \{0.5, 0.6, 0.8\}, Good \{0.0, 0.0, 0.0\})),$$

$$((15 + \{0.8, 0.3, 0.1\}, 15 + \{0.0, 0.0, 0.0\}))$$

$$C_{11} = ((\{0.9, 0.45, 0.2\}), (\{0.5, 0.3, 0.0\}), (\{0.8, 0.15, 0.0\}))$$

$$C_{11} = 100\% \{0.9, 0.3, 0.0\}$$

3.10 Definition: Product of NHSM

Let $A = [A_{ij}]$ and $B = [B_{j\ell}]$ be the two NHSM of order $a \times b$ and $b \times c$, where $A_{ij} = (\mathbb{T}_{ijk}^A, \mathbb{I}_{ijk}^A, \mathbb{F}_{ijk}^A)$ and $B_{j\ell} = (\mathbb{T}_{jk\ell}^B, \mathbb{I}_{jk\ell}^B, \mathbb{F}_{jk\ell}^B)$, then $[A_{ij}]_{a \times b} \times [B_{j\ell}]_{b \times c} = [C_{i\ell}]_{a \times c}$ is given as

$$[C_{i\ell}] = \left(\max_{jk} \min \left(\mathbb{T}_{ijk}^A, \mathbb{T}_{jk\ell}^B \right), \min_{jk} \max \left(\mathbb{I}_{ijk}^A, \mathbb{I}_{jk\ell}^B \right), \min_{jk} \max \left(\mathbb{F}_{ijk}^A, \mathbb{F}_{jk\ell}^B \right) \right).$$

Example Let A and B be the two NHSM of order 3×3:

$$[A]_{3 \times 3} = \begin{bmatrix} 100\% \{0.8, 0.5, 0.2\} \ Good \ \{0.5, 0.6, 0.8\} \ 15 + \{0.8, 0.3, 0.1\} \\ 100\% \{0.8, 0.2, 0.1\} \ Good \ \{0.6, 0.4, 0.8\} \ 15 + \{0.3, 0.4, 0.5\} \\ 100\% \{0.7, 0.4, 0.2\} \ Good \ \{0.3, 0.6, 0.7\} \ 15 + \{0.3, 0.5, 0.8\} \end{bmatrix}$$

$$[B]_{3 \times 3} = \begin{bmatrix} 100\% \{0.9, 0.4, 0.2\} \ Good \ \{0.8, 0.5, 0.3\} \ 15 + \{0.4, 0.3, 0.7\} \\ 100\% \{0.6, 0.3, 0.2\} \ Good \ \{0.8, 0.2, 0.2\} \ 15 + \{0.7, 0.2, 0.3\} \\ 100\% \{0.7, 0.4, 0.1\} \ Good \ \{0.9, 0.2, 0.0\} \ 15 + \{0.8, 0.3, 0.1\} \end{bmatrix}$$

Then their product is given as

$$[C] = \begin{bmatrix} 100\% \{0.8, 0.3, 0.1\} \ Good \ \{0.5, 0.3, 0.1\} \ 15 + \{0.4, 0.3, 0.2\} \\ 100\% \{0.6, 0.3, 0.2\} \ Good \ \{0.6, 0.2, 0.1\} \ 15 + \{0.3, 0.2, 0.1\} \\ 100\% \{0.7, 0.4, 0.2\} \ Good \ \{0.3, 0.4, 0.2\} \ 15 + \{0.3, 0.4, 0.2\} \end{bmatrix}$$

One calculation is provided for the convenience of reader:

$$C_{11} = ((100\% \{0.8, 0.5, 0.2\}, 100\% \{0.9, 0.4, 0.2\})),$$
$$((Good \{0.5, 0.6, 0.8\}, Good \{0.0, 0.0, 0.0\})),$$
$$((15 + \{0.8, 0.3, 0.1\}, 15 + \{0.0, 0.0, 0.0\}))$$

$$C_{11} = ((\{0.8, 0.5, 0.2\}), (\{0.0, 0.6, 0.8\}), (\{0.0, 0.3, 0.1\}))$$

$$C_{11} = 100\% \{0.8, 0.3, 0.1\}$$

3.11 Definition: Sum of NHSM

Let $A = [A_{ij}]$ and $B = [B_{ij}]$ be the two NHSM of order $a \times b$, where $A_{ij} = \left(\mathbb{T}^A_{ijk}, \mathbb{I}^A_{ijk}, \mathbb{F}^A_{ijk}\right)$ and $B_{ij} = \left(\mathbb{T}^B_{ijk}, \mathbb{I}^B_{ijk}, \mathbb{F}^B_{ijk}\right)$, then $[A_{ij}] + [B_{ij}] = [C_{ij}]$ is given as

$$\mathbb{T}\left([C_{ij}]\right) = \left(\frac{\mathbb{T}\left([A_{ij}]\right) + \mathbb{T}\left([B_{ij}]\right)}{n}\right)$$

$$\mathbb{I}\left([C_{ij}]\right) = \left(\frac{\left(\mathbb{I}\left([A_{ij}]\right) + \mathbb{I}\left([B_{ij}]\right)\right)}{n}\right)$$

$$\mathbb{F}\left([C_{ij}]\right) = \left(\frac{\mathbb{F}\left([A_{ij}]\right) + \mathbb{F}\left([B_{ij}]\right)}{n}\right)$$

Example Let A and B be the two NHSM of order 3×3:

$$[A]_{3\times3} = \begin{bmatrix} 100\% \{0.8, 0.5, 0.2\} & Good \{0.5, 0.6, 0.8\} & 15 + \{0.8, 0.3, 0.1\} \\ 100\% \{0.8, 0.2, 0.1\} & Good \{0.6, 0.4, 0.8\} & 15 + \{0.3, 0.4, 0.5\} \\ 100\% \{0.7, 0.4, 0.2\} & Good \{0.3, 0.6, 0.7\} & 15 + \{0.3, 0.5, 0.8\} \end{bmatrix}$$

$$[B]_{3\times3} = \begin{bmatrix} 100\% \{0.9, 0.4, 0.2\} & Good \{0.8, 0.5, 0.3\} & 15 + \{0.4, 0.3, 0.7\} \\ 100\% \{0.6, 0.3, 0.2\} & Good \{0.8, 0.2, 0.2\} & 15 + \{0.7, 0.2, 0.3\} \\ 100\% \{0.7, 0.4, 0.1\} & Good \{0.9, 0.2, 0.0\} & 15 + \{0.8, 0.3, 0.1\} \end{bmatrix}$$

Then their sum is given as

$$[C] = \begin{bmatrix} 100\% \{0.85, 0.45, 0.2\} & Good \{0.65, 0.55, 0.55\} & 15 + \{0.6, 0.3, 0.4\} \\ 100\% \{0.7, 0.25, 0.15\} & Good \{0.7, 0.3, 0.5\} & 15 + \{0.5, 0.3, 0.4\} \\ 100\% \{0.7, 0.4, 0.15\} & Good \{0.6, 0.4, 0.35\} & 15 + \{0.55, 0.4, 0.45\} \end{bmatrix}$$

One calculation is provided for the convenience of reader:

$$C_{11} = ((100\% \{0.8, 0.5, 0.2\}, 100\% \{0.9, 0.4, 0.2\}))$$

$$C_{11} = 100\% \{0.85, 0.45, 0.2\}$$

4 Neutrosophic Hypersoft Matrix Algorithm (NHSM Algorithm)

In this section NHSM algorithm is proposed and soundness is proved by the illustrating example. An algorithm proposed by [1] is also compared in this section. NHSM algorithm is most suitable in decision-making, since in this algorithm, individual matrix is constructed for each decision maker's choice, whereas this was not possible in algorithm [1].

4.1 NHSM Algorithm

4.2 Problem Description and Formulation

Teachers' recruitment problem (TRP) is the most complex and absurd task. There is no fix and fabricated design to know their subject knowledge or pedagogical skills. Therefore, decision makers find themselves in a blind alley. Consequently, based on their own knowledge and experience, they select a person which do not meet the institutional requirement. Thus TRP is typically a multi-attributive multi-criteria decision problem (*MAMCDP*).

Assumptions:

- Independent attributes are considered.
- Everyone attends the interview.
- Hesitant environment is not yet considered.

Parameters: Multi-attributive multi-criteria decision problem (MAMCDP) notations

NHSM algorithm	Algorithm proposed by [1]
Step: 1 construction of NHSM $$[A_{ij}]_{a\times b} = \left[\left(\mathbb{T}^A_{ijk}, \mathbb{I}^A_{ijk}, \mathbb{F}^A_{ijk}\right)\right]_{a\times b}$$ **Step: 2** construction of combined Choice matrices. $$[Y_{ij}]_{a\times b} \wedge [Z_{ij}]_{a\times b} = [P_{ij}]_{a\times b}$$ **Step: 3** calculate the product of each combined choice matrix. $[X_{ij}]_{a\times b} \otimes [P_{ij}]_{a\times b} = [U_{ij}]_{a\times b}$ **Step: 4** calculate the resultant R of NHSM $$[U_{ij}]_{a\times b} \oplus [V_{ij}]_{a\times b} \oplus [W_{ij}]_{a\times b} = [R_{ij}]_{a\times b}$$ **Step: 5** then calculate the weight $(\dot{W})$ of each object $(\dot{O})$ by adding truthness, indeterminacy, and falseness values of the attributes of ith row of resultant matrix $[R_{ij}]_{a\times b}$. $$[\dot{W}_{ij}]_{ax} = \begin{bmatrix} \mathbb{T}_{\mathbb{L}^1_1}(w), \mathbb{I}_{\mathbb{L}^1_1}(w), \mathbb{F}_{\mathbb{L}^1_1}(w) \\ \vdots \\ \mathbb{T}_{\mathbb{L}^1_a}(w), \mathbb{I}_{\mathbb{L}^1_a}(w), \mathbb{F}_{\mathbb{L}^1_a}(w) \end{bmatrix}$$ **Step: 6** the object having highest weight $(\dot{W})$ becomes an optimal choice of object. $$\left[< \left(\mathbb{T}^A_{ijk} > \mathbb{T}^B_{ijk}\right), \left(\mathbb{F}^A_{ijk} < \mathbb{F}^B_{ijk}\right), \left(\mathbb{I}^A_{ijk} < \mathbb{I}^B_{ijk}\right) > \right]$$	**Step 1:** Construct a matrix of multiple-valued Pq of attributes of order $m \times n$. $A = [pqr]m \times n, 1 \le q \le m, 1 \le r \le n$ **Step 2:** Fill the column values with zeros if multiple-valued attributes are less than equal to n to form a matrix of order $m \times n$ as defined in the below example. **Step 3:** Decision makers will assign fuzzy neutrosophic numbers (FNNs) to each multiple valued linguistic variable. **Step 4:** Selection of the subset of NHSS. **Step 5:** Conversion of fuzzy neutrosophic values of step: 4 into crisp numbers by using accuracy function $A(N) = [Pij]$. **Step 6:** Calculate the relative closeness by using the TOPSIS technique of MCDM. **Step 7:** Determine the rank of relative closeness by arranging in ascending order.

T_i=Universal set of teachers where $i = 1, 2, 3$
$F_i = $ Decision makers where $i = 1, 2, 3$
A_i=Attributes that are further categorized into the following

$$A_1 = \text{Qualification} = \{\text{M.Sc., M.phil., Ph.D.}\}$$

$$A_2 = \text{Experience in years} = \{2, 3, 4, 5\}$$

$$A_3 = \text{Reasearch publications} = \{1 - 5, 6 - 10, 11 - 15\} = \{P_1, P_2, P_3\}$$

Formulation of the Problem

Let us consider an institute wants to hire a teacher which is appropriate to its requirements; the institute received the following statistics-based CVs.

Firstly, we define a mapping by considering the attributes defined above and given by,

$$F : A_1 \times A_2 \times A_3 \to P(T_i)$$

Now consider there are three decisions makers F_1, F_2, F_3 having different selection criteria shown below, and having different choices to assign neutrosophic numbers:

$$F_1 (Qualification \geq M.Sc., Experience\ in\ years \geq 4,$$

$$Reasearch\ publication \geq 2) = \{T_1, T_2, T_4\}$$

$$F_2 (Qualification \geq M.Phil, Experience\ in\ years \geq 3,$$

$$Reasearch\ publication \geq 3) = \{T_1, T_2, T_4\}$$

$F_3(Qualification \geq Ph.\ D, Experience\ in\ years \geq 5, Research\ publication \geq 2) = \{T_1, T_2, T_4\}$

Secondly,

According to the decision makers' recruitment criteria, they will assign neutrosophic numbers against each attribute. All the other decision makers' will also follow the same stipulated formula.

F_1 *will assign neutrosophic numbers against his/her stipulated criteria*

̌T	A_1 = Qualification	A_2 = Experience	A_3 = Research publications
T_1	M.Sc.<(0.6, 0.4, 0.1)>	5<(0.4, 0.3, 0.1)>	P_2 <(0.6, 0.2, 0.3)>
T_2	M.Phil. <(0.6, 0.2, 0.1)>	4 <(0.2, 0.1, 0.1)>	P_3 < (0.7, 0.3, 0.2)>
T_4	Ph.D.<(0.9, 0.1, 0.3)>	7 < (0.7, 0.3, 0.2)>	P_1 < (0.4, 0.1, 0.2)>

F_2 *will assign neutrosophic numbers against each preferred attribute, shown below.*

̌T	A_1 = Qualification	A_2 = Experience	A_3 = Research publications
T_1	M.Sc.<(0.9, 0.4, 0.3)>	5<(0.4, 0.3, 0.1)>	P_2 <(0.6, 0.2, 0.3)>
T_2	M.Phil. <(0.6, 0.1, 0.2)>	4 <(0.2, 0.1, 0.1)>	P_3 < (0.7, 0.3, 0.2)>
T_4	Ph.D.<(0.8, 0.4, 0.3)>	7 < (0.7, 0.3, 0.2)>	P_1 < (0.4, 0.1, 0.2)>

F_3 *will assign neutrosophic numbers against each preferred attribute, shown below.*

Ť	$A_1=$ Qualification	$A_2=$ Experience	$A_3=$ Research publications
T_1	M.Sc.<(0.9, 0.3, 0.2)>	5<(0.4, 0.3, 0.1)>	P_2 <(0.6, 0.2, 0.3)>
T_2	M.Phil. <(0.7, 0.1, 0.3)>	4 <(0.2, 0.1, 0.1)>	P_3 < (0.7, 0.3, 0.2)>
T_4	Ph.D.<(0.5, 0.2, 0.1)>	7 < (0.7, 0.3, 0.2)>	P_1 < (0.4, 0.1, 0.2)>

4.3 Calculations Using NHSM Algorithm

Step 1: Construction of NHSM $\left[A_{ij} \right]_{a \times b} = \left[\left(\mathbb{T}^A_{ijk}, \mathbb{I}^A_{ijk}, \mathbb{F}^A_{ijk} \right) \right]_{a \times b}$ NFN assigned by the decision makers:

$$[X] = \begin{bmatrix} (0.6, 0.4, 0.1) & (0.4, 0.3, 0.1) & (0.6, 0.2, 0.3) \\ (0.6, 0.2, 0.1) & (0.2, 0.1, 0.1) & (0.7, 0.3, 0.2) \\ (0.9, 0.1, 0.3) & (0.7, 0.3, 0.2) & (0.4, 0.1, 0.2) \end{bmatrix} \tag{1}$$

$$[Y] = \begin{bmatrix} (0.9, 0.4, 0.3) & (0.4, 0.3, 0.1) & (0.6, 0.2, 0.3) \\ (0.6, 0.1, 0.2) & (0.2, 0.1, 0.1) & (0.7, 0.3, 0.2) \\ (0.8, 0.4, 0.3) & (0.7, 0.3, 0.2) & (0.4, 0.1, 0.2) \end{bmatrix} \tag{2}$$

$$[Z] = \begin{bmatrix} (0.9, 0.3, 0.2) & (0.4, 0.3, 0.1) & (0.6, 0.2, 0.3) \\ (0.7, 0.1, 0.3) & (0.2, 0.1, 0.1) & (0.7, 0.3, 0.2) \\ (0.5, 0.2, 0.1) & (0.7, 0.3, 0.2) & (0.4, 0.1, 0.2) \end{bmatrix} \tag{3}$$

Step 2: Construction of combined choice matrices, using $\left[Y_{ij} \right]_{a \times b} \wedge \left[Z_{ij} \right]_{a \times b} = \left[P_{ij} \right]_{a \times b}$:

$$[Y] \wedge [Z] = [P] = \begin{bmatrix} (0, 9, 0.1, 0.1) & (0.9, 0.1, 0.1) & (0.8, 0.2, 0.1) \\ (0, 9, 0.1, 0.1) & (0.7, 0.1, 0.1) & (0.7, 0.1, 0.1) \\ (0, 9, 0.1, 0.1) & (0.8, 0.1, 0.1) & (0.8, 0.1, 0.2) \end{bmatrix}$$

$$[Z] \wedge [X] = [Q] = \begin{bmatrix} (0, 9, 0.1, 0.1) & (0.9, 0.1, 0.1) & (0.9, 0.1, 0.1) \\ (0, 9, 0.1, 0.1) & (0.7, 0.1, 0.1) & (0.7, 0.1, 0.1) \\ (0, 9, 0.1, 0.1) & (0.7, 0.1, 0.1) & (0.7, 0.1, 0.1) \end{bmatrix}$$

$$[X] \wedge [Y] = [R] = \begin{bmatrix} (0, 9, 0.1, 0.1) & (0.7, 0.1, 0.1) & (0.7, 0.1, 0.1) \\ (0, 9, 0.1, 0.1) & (0.7, 0.1, 0.1) & (0.7, 0.1, 0.1) \\ (0, 9, 0.1, 0.1) & (0.9, 0.1, 0.1) & (0.9, 0.1, 0.2) \end{bmatrix}$$

Step 3: Calculate the product of each combined choice matrix:

$$[U] = \begin{bmatrix} (0,75,0.1,0.15) & (0.75,0.1,0.1) & (0.75,0.1,0.1) \\ (0,75,0.1,0.1) & (0.75,0.1,0.1) & (0.75,0.1,0.1) \\ (0,9,0.1,0.15) & (0.9,0.1,0.15) & (0.9,0.1,0.15) \end{bmatrix}$$

$$[V] = \begin{bmatrix} (0,9,0.1,0.1) & (0.9,0.1,0.1) & (0.9,0.1,0.1) \\ (0,8,0.1,0.1) & (0.75,0.1,0.15) & (0.75,0.1,0.1) \\ (0,85,0.1,0.15) & (0.85,0.1,0.15) & (0.85,0.1,0.2) \end{bmatrix}$$

$$[W] = \begin{bmatrix} (0,9,0.1,0.1) & (0.8,0.1,0.1) & (0.75,0.1,0.1) \\ (0,75,0.1,0.1) & (0.75,0.1,0.1) & (0.8,0.1,0.1) \\ (0,8,0.1,0.1) & (0.7,0.1,0.1) & (0.7,0.1,0.1) \end{bmatrix}$$

Step 4: Calculate the resultant R of NHSM, and converting into column matrices:

$$[U] = \begin{bmatrix} (0.75,0.1,0.1) \\ (0.75,0.1,0.1) \\ (0.9,0.1,0.15) \end{bmatrix}$$

$$[V] = \begin{bmatrix} (0.9,0.1,0.1) \\ (0.7667,0.1,0.1167) \\ (0.85,0.1,0.1167) \end{bmatrix}$$

$$[W] = \begin{bmatrix} (0.7667,0.1,0.1) \\ (0.7667,0.1,0.1) \\ (0.7334,0.1,0.1) \end{bmatrix}$$

Step 5: Calculation of the weight ($\dot{W}$) of each object ($\acute{O}$) by adding truthness, indeterminacy, and falseness values of the attributes of i-th row of resultant matrix $[R_{ij}]_{a \times b}$:

$$[U] \oplus [V] \oplus [W] = \begin{bmatrix} \widetilde{W} \end{bmatrix} = \begin{bmatrix} (0.805567,0.1,0.1) \\ (0.761134,0.1,0.105567) \\ (0.8278,0.1,0.122234) \end{bmatrix}$$

Step 6: Optimal choice selection, now consider the membership value of $\begin{bmatrix} \widetilde{W} \end{bmatrix}$:

$$T_1 = 0.805567$$

$$T_2 = 0.761134$$

$$T_4 = 0.8278$$

Hence, teacher T_4 is the optimal choice.

4.4 Algorithm Proposed by Saqlain et al. [1]

In this section we primarily use the defined algorithm to solve the above-mentioned TRP.

Step 1: Construct a matrix of multiple-valued attributes of order $m \times n$. Presented in the formulation of the problem:

$$\text{Linguistic matrix} = \begin{matrix} & A_1 \ A_2 \ A_3 \\ T_1 \\ T_2 \\ T_4 \end{matrix} \begin{bmatrix} M.Sc. & 5 & P_2 \\ M.Phil. & 4 & P_3 \\ Ph.D. & 7 & P_1 \end{bmatrix}$$

Step 2: In problem formulation, hesitant environment is not yet considered.

Step 3: Decision makers will assign fuzzy neutrosophic numbers (FNNs) to each multiple-valued linguistic variable individually which takes the form:

$$[X] = \begin{bmatrix} (0.6, 0.4, 0.1) & (0.4, 0.3, 0.1) & (0.6, 0.2, 0.3) \\ (0.6, 0.2, 0.1) & (0.2, 0.1, 0.1) & (0.7, 0.3, 0.2) \\ (0.9, 0.1, 0.3) & (0.7, 0.3, 0.2) & (0.4, 0.1, 0.2) \end{bmatrix} \qquad (4)$$

$$[Y] = \begin{bmatrix} (0.9, 0.4, 0.3) & (0.4, 0.3, 0.1) & (0.6, 0.2, 0.3) \\ (0.6, 0.1, 0.2) & (0.2, 0.1, 0.1) & (0.7, 0.3, 0.2) \\ (0.8, 0.4, 0.3) & (0.7, 0.3, 0.2) & (0.4, 0.1, 0.2) \end{bmatrix} \qquad (5)$$

$$[Z] = \begin{bmatrix} (0.9, 0.3, 0.2) & (0.4, 0.3, 0.1) & (0.6, 0.2, 0.3) \\ (0.7, 0.1, 0.3) & (0.2, 0.1, 0.1) & (0.7, 0.3, 0.2) \\ (0.5, 0.2, 0.1) & (0.7, 0.3, 0.2) & (0.4, 0.1, 0.2) \end{bmatrix} \qquad (6)$$

Step 4: Conversion of fuzzy neutrosophic values of step: 3 into crisp numbers by using accuracy function (N):

$$[X] = \begin{bmatrix} 0.366 & 0.266 & 0.366 \\ 0.30 & 0.133 & 0.40 \\ 0.433 & 0.40 & 0.233 \end{bmatrix} \tag{7}$$

$$[Y] = \begin{bmatrix} 0.533 & 0.266 & 0.366 \\ 0.30 & 0.133 & 0.40 \\ 0.50 & 0.40 & 0.233 \end{bmatrix} \tag{8}$$

$$[Z] = \begin{bmatrix} 0.466 & 0.266 & 0.366 \\ 0.366 & 0.133 & 0.40 \\ 0.266 & 0.40 & 0.233 \end{bmatrix} \tag{9}$$

Step 6: Calculate the relative closeness by using the TOPSIS technique of MCDM:

Decision makers' F_1 calculations of (1)

di+	di-	ci	Result – Rank	
0.337783604	0.567287	0.626788	2	T_1
0.312895149	0.816137	0.722864	1	T_2
0.753774271	0.312895	0.293338	3	T_4

Decision makers' F_2 calculations of (2)

di+	di-	ci	Result – Rank	
0.434441265	0.478704	0.524236	2	T_1
0.312895149	0.836817	0.727849	1	T_2
0.77611824	0.312895	0.28732	3	T_4

Decision makers' F_3 calculations of (3)

di+	di-	ci	Result – Rank	
0.506572438	0.401598	0.442205	2	T_1
0.312895149	0.736282	0.701771	1	T_2
0.666488903	0.312895	0.319482	3	T_4

Step 7: Relative closeness by arranging in ascending order shows that T_2 will be the best teacher to hire.

5   Result Comparison and Discussion

To check the validity and feasibility of proposed NHSM algorithm methods, a comparison analysis is conducted with the method proposed by Saqlain et al. [1]. It should be noted that this approach is useful where the attributes are independent (Table 1 and Fig 2).

The results are presented in Table 2; if the NHSM algorithm is used, the required alternative is T_4, and if other proposed method [1] is utilized, the best choice is T_2. Therefore, for the same teacher recruitment problem TRP and MAMCDP information, the results obtained by the proposed method in this paper are different with those obtained using the compared method. This represents the effectiveness and feasibility of the NHSM algorithm in MAMCDP environment.

Table 1 Teacher recruitment problem (TRP), initial formulation data

Ť	A_1 = Qualification	A_2 = Experience	A_3 = Research publications
T_1	M.Sc.	5	P_2
T_2	M.Phil.	4	P_3
T_3	B.Sc.	3	P_1
T_4	Ph.D.	7	P_1

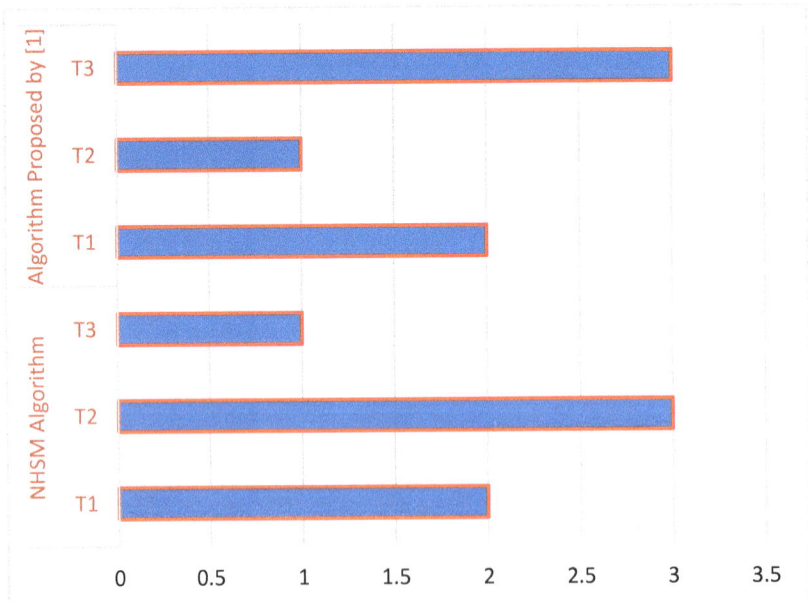

Fig. 2 Result comparison of TRP, using NHSM algorithm and algorithm proposed by Saqlain et al. [1]

Table 2 Result comparison of TRP, using NHSM algorithm and algorithm proposed by Saqlain et al. [1]

Method	Final ranking	Best alternative	Worst alternative
NHSM algorithm	$T_4 > T_1 > T_2$	T_4	T_2
Algorithm proposed by [1]	$T_2 > T_1 > T_4$	T_2	T_4

6 Conclusion

Firstly, in this paper, some basic definitions and notions regarding neutrosophic hypersoft matrices have been proposed. *Secondly,* using some of the defined operators, an algorithm, namely, NHSM algorithm is proposed, which is illustrated with the decision-making problem of teacher recruitment. *Next,* to prove the validity and the feasibility of the proposed method in decision-making in a real-life example was considered. The proposed algorithm is useful when the problem is MAMCDP. In future, applications in medical, engineering, and administration; and many other fields of knowledge of uncertainty, vague, and imprecise environment should be investigated, and decision-making can be done.

Acknowledgments Authors are thankful to the editor-in-chief and referees.

References

1. Saqlain, M., Saeed, M., Ahmad, M. R., & Smarandache, F. (2019). Generalization of TOPSIS for Neutrosophic Hypersoft set using accuracy function and its application. *Neutrosophic Sets and Systems (NSS), 27,* 131–137.
2. Zadeh, L. A. (1965). Fuzzy sets. *Information and Control, 8,* 338–353.
3. Atanassov, K. (1986). Intuitionistic fuzzy sets. *Fuzzy Sets and Systems, 20,* 87–96.
4. Smarandache, F. (2005). Neutrosophic set, a generalization of the intuitionistic fuzzy sets. *International Journal of Pure and Applied Mathematics, 24,* 287–297.
5. Molodtsov, D. (1999). Soft set theory - first results. *Computers and Mathematics with Applications, 37,* 19–31.
6. Maji, P. K., Roy, A. R., & Biswas, R. (2002a). An application of soft sets in a decision-making problem. *Computers and Mathematics with Applications, 44,* 1077–1083.
7. Maji, P. K., Roy, A. R., & Biswas, R. (2002b). An application of soft sets in a decision-making problem. *Computers and Mathematics with Applications, 44,* 1077–1083.
8. Smarandache, F. (2018a). Extension of soft set to Hypersoft set, and then to Plithogenic Hypersoft set. *Neutrosophic sets and system, 22,* 168–170.
9. Smarandache, F. (2018b). Extension of soft set to Hypersoft set, and then to plithogenic hypersoft set. *Neutrosophic Sets and System, 22,* 168–170.
10. Saqlain, M., Sana, M., Jafar, N., Saeed, M., & Said, B. (2020). Single and multi-valued Neutrosophic Hypersoft set and tangent similarity measure of single valued Neutrosophic Hypersoft sets. *Neutrosophic Sets and Systems (NSS), 32,* 317–329.
11. Saqlain, M., Sana, M., Jafar, N., Saeed, M., & Smarandache, F. (2020). Aggregate operators of Neutrosophic Hypersoft sets. *Neutrosophic Sets and Systems (NSS), 32,* 294–306.

12. Wang, H., Smarandache, F., Zhang, Y. Q., & Sunderraman, R. (2010). Single valued neutrosophic sets. *Multispace and Multistructure, 4,* 410–413.
13. Wang, H., Smarandache, F., Zhang, Y. Q., & Sunderraman. (2005). *Interval neutrosophic sets and logic: Theory and applications in computing.* Hexis, Phoenix, AZ: Neutrosophic Sets and System book series.
14. Ali, M. I., Feng, F., Liu, X., & Min, W. K. (2009). On some new operations in soft set theory. *Computers and Mathematics with Applications, 57*(9), 1547–1553.
15. Cagman, N., & Enginoglu, S. (2010a). Soft matrix theory and its decision making. *Computers and Mathematics with Applications, 59,* 3308–3314.
16. Cagman, N., & Enginoglu, S. (2010b). Soft set theory and uni-int decision making. *European Journal of Operations Research, 207,* 848–855.
17. Pawlak, Z. (1982). Rough sets. *International Journal of Information and Computer Science, 1,* 341–356.
18. Tripathy, B. K., & Arun, K. R. (2015). A new approach to soft sets, soft multisets and their properties. *International Journal of Reasoning Based Intelligent Systems, 7*(3/4), 244–253.
19. Maji, P. K. (2013). Neutrosophic soft set. *Ann. Fuzzy Math. Inform., 5*(1), 157–168.
20. Abdel-Baset, M., Chang, V., Gamal, A., & Smarandache, F. (2019). An integrated neutrosophic ANP and VIKOR method for achieving sustainable supplier selection: A case study in the importing field. *Computers in Industry, 106,* 94–110.
21. Abdel-Basset, M., Manogaran, G., Gamal, A., & Smarandache, F. (2019). A group decision-making framework based on the neutrosophic TOPSIS approach for smart medical device selection. *Journal of Medical Systems, 43*(2), 38–51.
22. Abdel-Basset, M., Saleh, M., Gamal, A., & Smarandache, F. (2019). An approach of the TOPSIS technique for developing supplier selection with group decision making under type-2 neutrosophic number. *Applied Soft Computing, 77,* 438–452.
23. Abdel-Basset, M., Manogaran, G., Gamal, A., & Smarandache, F. (2018). A hybrid approach of neutrosophic sets and DEMATEL method for developing supplier selection criteria. *Design Automation for Embedded Systems,* 1–22.
24. Nabeeh, N. A., Smarandache, F., Abdel-Basset, M., El-Ghareeb, H. A., & Aboelfetouh, A. (2019). An integrated Neutrosophic-TOPSIS approach and its application to personnel selection: A new trend in brain processing and analysis. *IEEE Access, 7,* 29734–29744.
25. Smarandache, F. (1998). Neutrosophy. In *Neutrosophic probability, set, and logic.* Ann Arbor, Michigan: ProQuest Information & Learning.

TOPSIS Method for Multi-Attribute Group Decision Making Based on Neutrality Aggregation Operator Under Single Valued Neutrosophic Environment: A Case Study of Airline Companies

Salih Berkan Aydemir and Tugba Kaya

1 Introduction

In a globalizing world, competition among companies is increasing rapidly. In order to be successful in the competition, it is necessary to focus on customer and customer satisfaction. Airline companies are among these important companies. Customers decide among many airline companies and choose the most suitable one for them. In this case, in many studies, it has been deemed appropriate to use a decision making process to choose the most suitable one. For example, sustainable energy decision making [27], Foreign and Domestic Turkish Banks [1], domestic airlines service quality, supplier evaluation and selection [9], credit risk assessment [6], and so on. The decision making is the process of finding the best option from all of the feasible alternatives. During the decision making process, the decision maker tries to determine the most appropriate and the most reliable alternative from the limited number of alternatives. But there must be a need to reliable data at the decision making process. The decision maker takes into account generally past experiences in decision process. The people's past experiences can affect the decision process. If these experiences may be analyzed properly, the process is more reliable and effective. Since the decision making process is based on the opinions of individuals, there are ambiguous and incomplete interpretations in solving multiple criteria decision making (MCDM) problems.

The human brain cannot examine an expression such as a computer only in black and white in clear steps. In addition, the expressions cannot be expected to be precise, as the human thinking process can verbally and transfer what they know to others with verbal expressions in the same way. Therefore, due to human

S. B. Aydemir (✉) · Tugba Kaya
Computer Engineering Department, Eskisehir Technical University, Eskisehir, Turkey
e-mail: sbaydemir@eskisehir.edu.tr; tugbaturkoglu@eskisehir.edu.tr

471

nature, she/he can make statements containing fuzzy and uncertainty from the decision making phase. For these reasons, the value of an attribute cannot always be expressed in crisp numbers in multi-criteria decision making problems. Therefore, the solution of multiple criteria problems gives more sensible results with fuzzy sets. Fuzzy sets (FSs) were proposed by L. Zadeh in 1965 [20]. Fuzzy sets have the concept of membership degree. Thus, imprecise information can be handled more clearly. Fuzzy sets have been applied to MCDM structures and successful results have been obtained [11]. On the other hand, Atanassov mentioned that there should be a degree of non-membership (v) in addition to membership function (μ) in fuzzy sets. Such sets are called intuitionistic fuzzy sets (IFSs). However, since the sum of the values of membership and non-membership in the IFSs should be less than 1, there are many generalized fuzzy sets such as Pythagorean fuzzy set which the main feature of Pythagorean fuzzy sets is that it is characterized by four parameters, namely membership degree, nonmembership degree, strength of commitment about membership, and direction of commitment., Fermatean fuzzy set which can handle uncertain information more easily in the process of decision making, q-rung orthopair fuzzy set [30], Fermatean fuzzy set [19], q-rung orthopair fuzzy set [31]. However, the widest of the fuzzy set mentioned are neutrosophic sets(NSs) [21]. NSs include three different independent membership functions. So, neutrosophic set (NS), which is the generalized version of methods such as fuzzy, intuitionistic fuzzy, q-rung orthopair fuzzy, etc. which takes into consideration the concepts of incomplete and indeterminacy, as well as the concepts of uncertainty and imprecise are used in this study because it can be used in real and engineering applications recommended by Wang et al. [26]. NS method consists of three membership functions such that the truth-membership, the indeterminacy-membership, and the falsity-membership. For example, let there be a voting system with ten voters. At the first time, three voters are "positive", four voters are "negative" and three voters are undecided. Neutroscopic expression is stated as (0.3, 0.4, 0.3). But in the second time, one voter "positive", four voters "negative", two voters give up and three voters are undecided then this situation is expressed as (0.1, 0.3, 0.4). Therefore, it can be said that the use of NS in the presence of such conditions gives successful results.

MCDM can be examined in two parts as traditional methods and functional aggregation methods. Traditional methods, such as technique for order preference by similarity to ideal solution (TOPSIS) [10], Promethee [3], VIKOR [15], ELEC-TRE [17]. On the other hand, NS has been tested on multi-attribute group decision making (MAGDM) and MCDM with various aggregation operators. Harish Garg has combined the prioritized operator with the Muirhed mean (MM) under the NS environment [8]. With the prioritized operator, the priorities of the criteria are taken into account. The MM operator takes into account the interrelationship between arguments. Since NSs are difficult to apply in real engineering problems and scientific applications, a subclass of NS has been proposed by Wang et al. [26]. These sets are called single valued neutrosophic sets. SVNSs are well suited for handling ambiguous, incomplete, imprecise information. The flexible parameter of the Dombi operator is proposed in the SVNS environment [4]. In addition, operators that take into account the interrelationship between arguments are expressed in

SVNS numbers such as Bonferroni mean [13], Heronian mean [12]. On the other hand, the neutrality aggregation operator has been proposed by Harish garg in the SVNS environment against the biases and subjective evaluations of decision makers [7]. In the study [7], the effect of the probabilistic sum and the coefficient of membership degrees among each other were emphasized. Traditional methods such as TOPSIS and VIKOR have been extended with SVNS [16]. SVNS and TOPSIS have been applied on multiple criteria problems with unknown criteria and decision maker weights [2]. However, single valued neutrosophic weighted averaging (SVNWA) aggregation operator was used in the study proposed by Biswas et al. But, decision makers did not take into account neutrality. The motivation of this study can be summarized as follows.

1. In the study, single valued neutrosophic set is used for the decision makers to evaluate it more broadly.
2. Neutrality is an important factor for the data set, which contains many decision makers in terms of application area. Hence the neutrality aggregation operator is combined with the TOPSIS method.
3. As a result of competition between airline companies with a real-life example, the best airline company and the most preferred criterion are investigated. This case study will create awareness for airline companies.
4. In addition, weights obtained from neutrality aggregation-based TOPSIS method are used in Promethee method. Thus, Promethee and TOPSIS methods are compared in SVNS environment.

The purpose of this study is to consist of decision mechanism according to passenger or customer's past experiences. The passengers are commenting verbally and numerically on these criteria that they are satisfied with or not satisfied with. If the passenger is satisfied with his/her travel, the ratings will be a plus for the company. However, if these assessments need to be properly analysed then companies can develop different strategies. The developing strategies by researcher should not be single-person oriented for every person who has different characteristic features. So, almost all travelled passengers are considered and their comments are assessed, and different attitudes can be created. The fact that each company takes into consideration the users' evaluations allows it to be moved to the top in the market sector. For example, as a result of the studies carried out, a feature missing from the firms has been found by the decision makers evaluations. Since improvement by these companies increase user satisfaction, it carries the company to the top in the market sector. Thus, the results obtained from the studies will have positive results for companies and customers. In the study, users are also mentioned as decision makers.

The parts of the study are planned as follows. In the first part, some traditional and functional methods are mentioned in the field of multiple criteria decision making problems. In the second section, basic concepts related to the methods proposed in the study are given. The method suggested in third section is explained that it is a neutrality operator based TOPSIS method. In fourth section, a case study for the proposed method is considered. The most preferred airline company and its criteria are investigated according to the airline companies and their criteria. In fifth section, the Promethee method is used on the airline's data. In the last section, the proposed method and the Promethee method are compared. In addition, the most preferred companies and criteria of the airline companies are interpreted in both methods. Finally, methods that can guide and improve for the future are mentioned.

2 Basic Concepts

In this section, the methods used in the study are summarized. Related concepts are explained step by step.

2.1 Neutrosophic Sets

Neutrosophic Sets (NS) provide a general structure due to its independent membership functions. It ensures that each membership function covers many fuzzy sets (IFS, PFS, Spherical Fuzzy Set, q-Rung Orthopair Fuzzy Set) [22, 23] by changing between 0 and 1.

Definition 1 (Neutrosophic Sets [21]) Let X be a universal space of points(objects) with generic elements in X denoted by x. A Neutrosophic Sets l(NS) in $A \subset X$ is characterized by truth-membership function $T_A(x)$, indeterminacy-membership function $I_A(x)$, and falsity-membership function $F_A(x)$. $T_A(x), I_A(x), F_A(x)$ subset of $[^-0, 1^+]$. Therefore, all membership functions are $T_A(x) \rightarrow [^-0, 1^+]$, $I_A(x) \rightarrow [^-0, 1^+]$ and $F_A(x) \rightarrow [^-0, 1^+]$

$I_A(x)$ represents ambiguous, incomplete, imprecise, etc. data. The total range of three independent membership values is as follows. $^-0 \leq T_A(x) + I_A(x) + F_A(x) \leq 3^+$.

Definition 2 (Single Valued Neutrosophic Sets [26]) Let X be a space of points(objects) with generic elements in X denoted by x. A Single Valued Neutrosophic Sets (SVNS) in $A \subset X$ is characterized by truth-membership function $T_A(x)$, indeterminacy-membership function $I_A(x)$, and falsity-membership function $F_A(x)$. $T_A(x), I_A(x), F_A(x) \in [0, 1]$ for all $x \in X$. Also, there are two different SVNS for discrete and continuous values. If X in continuous, a SVNS A can be following as $\int_x \langle T_A(x), I_A(x), F_A(x) \rangle$ for all $x \in X$, If X in discrete, a SVNS A can be following as $\sum_x \langle T_A(x), I_A(x), F_A(x) \rangle$ for all $x \in X$.

Mathematical set representation of SVNS can be written as
$A = \{(x_1 | \langle T_A(x_1), I_A(x_1), F_A(x_1) \rangle), \ldots, (x_n | \langle T_A(x_n), I_A(x_n), F_A(x_n) \rangle)\}$. For the sake of the shortness, it can be written as $A = \langle T_A(x), I_A(x), F_A(x) \rangle$ for all $x \in X$.

Definition 3 ([26]) Let $\mathcal{A} = \langle T_{\mathcal{A}}(x), I_{\mathcal{A}}(x), F_{\mathcal{A}}(x) \rangle$ and $\mathcal{B} = \langle T_{\mathcal{B}}(x), I_{\mathcal{B}}(x), F_{\mathcal{B}}(x) \rangle$ be any two SVNSs. Some set operations used for SVNSs are as follows.

1. $\mathcal{A} \subseteq \mathcal{B}$ if and only if $T_{\mathcal{A}}(x) \leq T_{\mathcal{B}}(x), I_{\mathcal{A}}(x) \geq I_{\mathcal{B}}(x), F_{\mathcal{A}}(x) \geq F_{\mathcal{B}}(x)$ for all $x \in X$.
2. $\mathcal{A} = \mathcal{B}$ if and only if $\mathcal{A} \subseteq \mathcal{B}$ and $\mathcal{B} \subseteq \mathcal{A}$ for all $x \in X$.
3. $\mathcal{A}^c = \{(x | \langle T_A(x), 1 - I_A(x), F_A(x) \rangle) | x \in X\}$ for all $x \in X$.
4. $\mathcal{A} \cup \mathcal{B} = \langle max(T_{\mathcal{A}}(x), T_{\mathcal{B}}(x)), min(I_{\mathcal{A}}(x), I_{\mathcal{B}}(x)), min(F_{\mathcal{A}}(x), F_{\mathcal{B}}(x)) \rangle$ for all $x \in X$.
5. $\mathcal{A} \cap \mathcal{B} = \langle min(T_{\mathcal{A}}(x), T_{\mathcal{B}}(x)), max(I_{\mathcal{A}}(x), I_{\mathcal{B}}(x)), max(F_{\mathcal{A}}(x), F_{\mathcal{B}}(x)) \rangle$ for all $x \in X$.

On the other hand, in addition to the above operational procedures, the following operators have been proposed by Liu and Wang [13].

Definition 4 Let $\mathcal{A} = \langle T_{\mathcal{A}}(x), I_{\mathcal{A}}(x), F_{\mathcal{A}}(x) \rangle$ and $\mathcal{B} = \langle T_{\mathcal{B}}(x), I_{\mathcal{B}}(x), F_{\mathcal{B}}(x) \rangle$ be any two SVNSs. Then,

1. $\mathcal{A} \oplus \mathcal{B} = \langle T_{\mathcal{A}}(x) + T_{\mathcal{B}}(x) - T_{\mathcal{A}}(x).T_{\mathcal{B}}(x), I_{\mathcal{A}}(x).I_{\mathcal{B}}(x), F_{\mathcal{A}}(x).F_{\mathcal{B}}(x) \rangle$
2. $\mathcal{A} \otimes \mathcal{B} = \langle T_{\mathcal{A}}(x).T_{\mathcal{B}}(x), I_{\mathcal{A}}(x) + I_{\mathcal{B}}(x) - I_{\mathcal{A}}(x).I_{\mathcal{B}}(x), F_{\mathcal{A}}(x) + F_{\mathcal{B}}(x) - F_{\mathcal{A}}(x).F_{\mathcal{B}}(x) \rangle$.

In the study proposed by Majumdar and Samanta, basic entropy and distance methods on SVNS were proposed [14]. In this study, generally used Euclidean distance is used as distance function.

Definition 5 Distance between two SVNSs
Let $\mathcal{A} = \{(x_1 | \langle T_{\mathcal{A}}(x_1), I_{\mathcal{A}}(x_1), F_{\mathcal{A}}(x_1) \rangle), \ldots, (x_n | \langle T_{\mathcal{A}}(x_n), I_{\mathcal{A}}(x_n), F_{\mathcal{A}}(x_n) \rangle)\}$ and $\mathcal{B} = \{(x_1 | \langle T_{\mathcal{B}}(x_1), I_{\mathcal{B}}(x_1), F_{\mathcal{B}}(x_1) \rangle), \ldots, (x_n | \langle T_{\mathcal{B}}(x_n), I_{\mathcal{B}}(x_n), F_{\mathcal{B}}(x_n) \rangle)\}$ be two SVNSs. Euclidean distance between sets $\mathcal{A}$ and $\mathcal{B}$ is defined as

$$\mathcal{D}_{Eucl}(\mathcal{A}, \mathcal{B}) = \sqrt{\sum_{i=1}^{n} (T_{\mathcal{A}}(x_i) - T_{\mathcal{B}}(x_i))^2 + (I_{\mathcal{A}}(x_i) - I_{\mathcal{B}}(x_i))^2 + (F_{\mathcal{A}}(x_i) - F_{\mathcal{B}}(x_i))^2}$$

$$(1)$$

In addition, the normalized Euclidean distance between $\mathcal{A}$ and $\mathcal{B}$ is calculated as follows [18]

$$\mathcal{D}_{Eucl}^{N}(\mathcal{A}, \mathcal{B}) = \sqrt{\frac{1}{3n} \sum_{i=1}^{n} (T_{\mathcal{A}}(x_i) - T_{\mathcal{B}}(x_i))^2 + (I_{\mathcal{A}}(x_i) - I_{\mathcal{B}}(x_i))^2 + (F_{\mathcal{A}}(x_i) - F_{\mathcal{B}}(x_i))^2}$$

$$(2)$$

2.2 Neutrality Aggregation

Neutrality aggregation (NA) operator takes into account neutral or fair decision factor of decision makers. NA deals with interactive coefficients between membership functions and probability sum. NA operator is proposed by Harish Garg on

SVNS [7]. The definition of the SVNWNA (Single valued Neutrosophic Weighted Neutrality Aggregation) operator is as follows:

Definition 6 ([7]) Let $\mathcal{A}_i$ be a collection of single valued neutrosophic numbers (SVNNs). The SVNWNA is a mapping defined on $\mathcal{A}_i$ by SVNWNA: $\varphi^n \rightarrow \varphi$. $SVNWNA(\mathcal{A}_1, \mathcal{A}_2, \ldots, \mathcal{A}_n) = \Theta_{i=1}^n w_i \mathcal{A}_i$ where $w_i > 0$ and $\sum w_i = 1$ be the weighting vector of $\mathcal{A}_i$. The aggregated value of the SVNNs is as follows:

For $\mathcal{A}_i = (\mathcal{T}_i, \mathcal{I}_i, \mathcal{F}_i)$,

$$SVNWNA(\mathcal{A}_1, \mathcal{A}_2, \ldots, \mathcal{A}_n)$$

$$= \left(\frac{3 \sum_{i=1}^n w_i T_i}{\sum_{i=1}^n w_i (T_i + I_i + F_i)} \left(1 - \prod_{i=1}^n \pi_i^{w_i} \right), \frac{3 \sum_{i=1}^n w_i I_i}{\sum_{i=1}^n w_i (T_i + I_i + F_i)} \left(1 - \prod_{i=1}^n \pi_i^{w_i} \right), \right.$$

$$\left. \frac{3 \sum_{i=1}^n w_i F_i}{\sum_{i=1}^n w_i (T_i + I_i + F_i)} \left(1 - \prod_{i=1}^n \pi_i^{w_i} \right) \right)$$

$$(3)$$

where $\pi_i = 1 - \frac{T_i + I_i + F_i}{3} \in [0, 1]$.

2.3 Promethee Method and GAIA Plane

In multi-criteria decision making, it is aimed to determine the most suitable among more than one alternative. There are many methods in the literature to achieve this goal. Therefore, Promethee method, one of the outranking models, is used in this study. Promethee method evaluates supplier alternatives with different preference functions. The peculiarity of this method is that it achieves both partial and full priorities of alternatives and allows for more detailed analysis. Promethee (Preference Ranking Organization Method for Enrichment Evaluations) method, which is one of the most used of multi-criteria decision making methods, has been developed based on the difficulties arising in the application of existing prioritization methods in the literature. The Promethee method is a multi-criteria decision making method developed by J. P. Brans in 1982, consisting of two stages: Promethee I (partial order) and Promethee II (full order) [3, 25]. Promethee method consists of 7 steps:

- Determining the alternatives, criteria, and criteria weights of the decision maker
- Determination of preference functions
- Determination of common preference functions and preference indices
- Determination of positive and negative superiority values
- Partial order of alternatives with Promethee I
- Calculation of net priority values
- Obtaining the exact sequence of alternatives with Promethee II

Presenting multi-criteria problems as a graphical visual enable the user to have a more positive effect in the decision making process. Consequently, GAIA tools have been developed that combine the visual and Promethee method that helps users.

3 TOPSIS Method for Multiple Attribute Group Decision Making with Neutrality Aggregation Operator Based Single-Neutrosophic Information

Consider a decision making process consisting of m alternative and n attributes. Let the set of alternatives be $A = \{A_1, A_2, \ldots A_m\}$. Let the set of attributes be $C = \{C_1, C_2, \ldots C_n\}$. Decision making process is defined as the rating given to C_j attribute for alternative A_i. Also, the weight values assigned to attributes by the decision makers are $W = \{w_1, w_2, \ldots w_n\}$. On the other hand, decision makers also have a weight vector. Let us assume that there are k decision makers and we can express them as $D = \{d_1, d_2, \ldots, d_k\}$. Let us show the weight vector of decision makers with $\Lambda = \{\lambda_1, \lambda_2, \ldots, \lambda_k\}$. The decision matrix used for the multiple attribute group decision making process is designed as follows:

$$
R = \langle r_{ij} \rangle_{m \times n}^k =
\begin{array}{c}
\\
A_1 \\
A_2 \\
\vdots \\
A_m
\end{array}
\begin{array}{c}
C_1 \quad C_2 \cdots \quad C_n \\
\left[
\begin{array}{cccc}
r_{11}^k & a_{12}^k & \cdots & a_{1n}^k \\
r_{21}^k & a_{22}^k & \cdots & a_{2n}^k \\
\vdots & \vdots & \ddots & \vdots \\
r_{m1}^k & a_{m2}^k & \cdots & a_{mn}^k
\end{array}
\right]
\end{array}
\tag{4}
$$

The decision matrix consists of m alternative and n attributes for the kth decision-maker. In the multiple group decision making process, the decision matrix is expressed in the SVNS environment. So three membership degrees are also taken into account. Truth membership degree, falsity-membership degree, and indeterminacy membership degree. Equation (4) is expressed by SVNS for multiple criteria problems as follows:

$$
R_{\mathcal{A}} = \langle r_{ij} \rangle_{m \times n}^k = \langle \mathcal{T}_{ij}^{(k)}, \mathcal{I}_{ij}^{(k)}, \mathcal{F}_{ij}^{(k)} \rangle_{m \times n}
$$

$$
=
\begin{array}{c}
A_1 \\
A_2 \\
\vdots \\
A_m
\end{array}
\begin{array}{c}
C_1 \qquad\qquad C_2 \qquad\qquad \cdots \qquad\qquad C_n \\
\left[
\begin{array}{cccc}
\langle \mathcal{T}_{11}^{(k)}, \mathcal{I}_{11}^{(k)}, \mathcal{F}_{11}^{(k)} \rangle & \langle \mathcal{T}_{12}^{(k)}, \mathcal{I}_{12}^{(k)}, \mathcal{F}_{12}^{(k)} \rangle & \cdots & \langle \mathcal{T}_{1n}^{(k)}, \mathcal{I}_{1n}^{(k)}, \mathcal{F}_{1n}^{(k)} \rangle \\
\langle \mathcal{T}_{21}^{(k)}, \mathcal{I}_{21}^{(k)}, \mathcal{F}_{21}^{(k)} \rangle & \langle \mathcal{T}_{22}^{(k)}, \mathcal{I}_{22}^{(k)}, \mathcal{F}_{22}^{(k)} \rangle & \cdots & \langle \mathcal{T}_{2n}^{(k)}, \mathcal{I}_{2n}^{(k)}, \mathcal{F}_{2n}^{(k)} \rangle \\
\vdots & \vdots & \ddots & \vdots \\
\langle \mathcal{T}_{m1}^{(k)}, \mathcal{I}_{m1}^{(k)}, \mathcal{F}_{m1}^{(k)} \rangle & \langle \mathcal{T}_{m2}^{(k)}, \mathcal{I}_{m2}^{(k)}, \mathcal{F}_{m2}^{(k)} \rangle & \cdots & \langle \mathcal{T}_{mn}^{(k)}, \mathcal{I}_{mn}^{(k)}, \mathcal{F}_{mn}^{(k)} \rangle
\end{array}
\right]
\end{array}
\tag{5}
$$

where $\mathcal{T}_{ij}$ is called the truth-membership degree, $\mathcal{I}_{ij}$ is the indeterminacy-membership degree, and $\mathcal{F}_{ij}$ is the falsity-membership degree. $\mathcal{A}$ is the represents neutrosophic sets.

Some basic cases and properties related to single valued neutrosophic sets are given below.

1. $0 \leq \mathcal{T}_{ij} \leq 1; 0 \leq \mathcal{I}_{ij} \leq 1, 0 \leq \mathcal{F}_{ij} \leq 1$.
2. $0 \leq \mathcal{T}_{ij} + \mathcal{I}_{ij} + \mathcal{F}_{ij} \leq 3$ for $i = 1, 2, \ldots m$ and $j = 1, 2, \ldots n$.

In the study [5], neutrosophic sets were shown with neutrosophic cube. The vertices of the cube are $(0, 0, 0)$, $(1, 0, 0)$, $(1, 0, 1)$, $(0, 0, 1)$, $(0, 1, 0)$, $(1, 1, 0)$, $(0, 1, 1)$, and $(1, 1, 1)$. These points can be considered as limit values. Depending on this situation, the rating values given in the neutrosophic sets environment are evaluated in three categories. These three categories are mentioned below.

1. **Highly acceptable ratings** It is possible with eight SVNSs that the votes cast by decision makers are highly acceptable. These points are $(0.5, 0, 0)$, $(1, 0, 0)$, $(1, 0, 0.5)$, $(0.5, 0, 0.5)$, $(0.5, 0, 0.5)$, $(1, 0, 0.5)$, $(1, 0.5, 0.5)$, and $(0.5, 0.5, 0.5)$. This set is represented as a subcube of the neutrosophic cube. It can also be defined as follows: $\mathcal{H} = \langle \mathcal{T}_{ij}, \mathcal{I}_{ij}, \mathcal{F}_{ij} \rangle$ where $0.5 < \mathcal{T}_{ij} < 1$, $0 < \mathcal{I}_{ij} < 0.5$, and $0 < \mathcal{F}_{ij} < 0.5$ for $i = 1, 2, \ldots m$ and $j = 1, 2, \ldots n$.
2. **Unacceptable ratings** If the truth-membership degree is 0%, indeterminacy-membership degree is 100%, and falsity-membership degree is 100%, the unacceptable ratings set is obtained. The set avoided by decision makers is as follows: $\mathcal{U} = \langle \mathcal{T}_{ij}, \mathcal{I}_{ij}, \mathcal{F}_{ij} \rangle$ where $\mathcal{T}_{ij} = 0$, $0 < \mathcal{I}_{ij} \leq 1$ and $0 < \mathcal{F}_{ij} \leq 1$ for $i = 1, 2, \ldots m$ and $j = 1, 2, \ldots n$.
3. **Tolerable ratings** The tolerable ratings are in the area where highly acceptable and unacceptable ratings are not included. The set expressed by tolerable ratings is as follows: $\mathcal{L} = \langle \mathcal{T}_{ij}, \mathcal{I}_{ij}, \mathcal{F}_{ij} \rangle$ where $0 < \mathcal{T}_{ij} < 0.5$, $0.5 < \mathcal{I}_{ij} < 1$, and $0.5 < \mathcal{F}_{ij} < 1$ for $i = 1, 2, \ldots m$ and $j = 1, 2, \ldots n$.

In light of all the information provided, we can define the TOPSIS method based on neutrality aggregation operator in the neutrosophic environment. The steps of the method are explained step by step.

- **Calculation of the weights of decision makers**
  The importance of each decision maker may be different. Suppose there are p people for the decision maker. So let the neutrosophic number of the kth decision maker be $\Lambda_k = \langle \mathcal{T}_k, \mathcal{I}_k, \mathcal{F}_k \rangle$. The weight of the kth decision maker is then calculated by

$$\lambda_k = \frac{1 - \sqrt{((1 - \mathcal{T}_k(x))^2 + (\mathcal{I}_k(x))^2 + (\mathcal{F}_k(x))^2)/3}}{\sum_{i=1}^{p}(1 - \sqrt{((1 - \mathcal{T}_k(x))^2 + (\mathcal{I}_k(x))^2 + (\mathcal{F}_k(x))^2)/3})} \tag{6}$$

  where $\sum_{i=1}^{p} \lambda_k = 1$
- **Creating an aggregated decision matrix based on decision makers using neutral aggregation**

Although the weight of the decision makers was determined in the previous step, the neutral operator is used in this step against the high number of decision makers and biased evaluations. Show the weight of the decision makers with $\Lambda = (\lambda_1, \lambda_2, \ldots, \lambda_p)^T$. Due to the aggregate operator, group evaluations will turn into a single decision matrix. The SVNWNA operator given in Eq. (3) and proposed by [7] is used. The aggregated matrix is as follows:

$$R^k = \langle r_{ij} \rangle^k_{m \times n} \text{ where, } r_{ij} = SVNWNA_\Lambda \left(r_{ij}^{(1)}, r_{ij}^{(2)}, \ldots, r_{ij}^{(p)} \right)$$

$$= \lambda_1 r_{ij}^{(1)} \oplus \lambda_2 r_{ij}^{(2)}, \ldots, \oplus \lambda_p r_{ij}^{(p)}, i = 1, 2, \ldots m, j = 1, 2, \ldots n$$

$$= \left(\frac{3 \sum_{k=1}^p \lambda_k T_{ij}^{(k)}}{\sum_{k=1}^p \lambda_k (T_{ij}^{(k)} + I_{ij}^{(k)} + F_{ij}^{(k)})} \left(1 - \prod_{k=1}^p (\pi_{ij}^{(k)})^{\lambda_k} \right), \right.$$

$$\frac{3 \sum_{k=1}^p \lambda_k I_{ij}^{(k)}}{\sum_{k=1}^p \lambda_k (T_{ij}^{(k)} + I_{ij}^{(k)} + F_{ij}^{(k)})} \left(1 - \prod_{k=1}^p (\pi_{ij}^{(k)})^{\lambda_k} \right),$$

$$\left. \frac{3 \sum_{k=1}^p \lambda_k F_{ij}^{(k)}}{\sum_{k=1}^p \lambda_k (T_{ij}^{(k)} + I_{ij}^{(k)} + F_{ij}^{(k)})} \left(1 - \prod_{k=1}^p (\pi_{ij}^{(k)})^{\lambda_k} \right) \right) \tag{7}$$

The aggregated neutrosophic decision matrix is as follows:

$$R = \langle r_{ij} \rangle_{m \times n} = \langle \mathcal{T}_{ij}, \mathcal{I}_{ij}, \mathcal{F}_{ij} \rangle_{m \times n}$$

$$= \begin{array}{c} \\ A_1 \\ A_2 \\ \vdots \\ A_m \end{array} \begin{bmatrix} \langle \mathcal{T}_{11}, \mathcal{I}_{11}, \mathcal{F}_{11} \rangle & \langle \mathcal{T}_{12}, \mathcal{I}_{12}, \mathcal{F}_{12} \rangle & \cdots & \langle \mathcal{T}_{1n}, \mathcal{I}_{1n}, \mathcal{F}_{1n} \rangle \\ \langle \mathcal{T}_{21}, \mathcal{I}_{21}, \mathcal{F}_{21} \rangle & \langle \mathcal{T}_{22}, \mathcal{I}_{22}, \mathcal{F}_{22} \rangle & \cdots & \langle \mathcal{T}_{2n}, \mathcal{I}_{2n}, \mathcal{F}_{2n} \rangle \\ \vdots & \vdots & \ddots & \vdots \\ \langle \mathcal{T}_{m1}, \mathcal{I}_{m1}, \mathcal{F}_{m1} \rangle & \langle \mathcal{T}_{m2}, \mathcal{I}_{m2}, \mathcal{F}_{m2} \rangle & \cdots & \langle \mathcal{T}_{mn}, \mathcal{I}_{mn}, \mathcal{F}_{mn} \rangle \end{bmatrix}$$

$$\begin{array}{cccc} C_1 & C_2 & \cdots & C_n \end{array}$$

$$\tag{8}$$

- **Calculation of attribute weights**

  Weights given by decision makers to attributes may depend on their personal thoughts. Therefore, in this step, neutral aggregation operator is used when calculating attribute weights. Let the weight given to the C_j attribute by the kth decision maker be $w_k^j = w_j^{(1)}, w_j^{(2)}, \ldots, w_j^{(p)}$. There are n weights ($W = \{w_1, w_2, \ldots, w_n\}$) for n attributes.

$$R^k = \langle r_{ij} \rangle_{m \times n}^k \text{ where } w_j = SVNWNA_\Lambda \left(w_j^{(1)}, w_j^{(2)}, \ldots, w_j^{(p)} \right)$$

$$= \lambda_1 w_j^{(1)} \oplus \lambda_2 w_j^{(2)}, \ldots, \oplus \lambda_p w_j^{(p)}, j = 1, 2, \ldots n$$

$$= \left(\frac{3 \sum_{k=1}^{p} \lambda_k T_j^{(k)}}{\sum_{k=1}^{p} \lambda_k (T_j^{(k)} + I_j^{(k)} + F_j^{(k)})} \left(1 - \prod_{k=1}^{p} (\pi_j^{(k)})^{\lambda_k} \right), \right. \tag{9}$$

$$\frac{3 \sum_{k=1}^{p} \lambda_k I_j^{(k)}}{\sum_{k=1}^{p} \lambda_k (T_j^{(k)} + I_j^{(k)} + F_j^{(k)})} \left(1 - \prod_{k=1}^{p} (\pi_j^{(k)})^{\lambda_k} \right),$$

$$\left. \frac{3 \sum_{k=1}^{p} \lambda_k F_j^{(k)}}{\sum_{k=1}^{p} \lambda_k (T_j^{(k)} + I_j^{(k)} + F_j^{(k)})} \left(1 - \prod_{k=1}^{p} (\pi_j^{(k)})^{\lambda_k} \right) \right)$$

- **Aggregation of the weighted neutrosophic decision matrix**
  In this step, the aggreted decision matrix and weights are multiplied. So aggregated weighted neutrosophic decision matrix is obtained.

$$R \otimes W = R^W = \langle r_{ij}^{w_j} \rangle_{m \times n} = \langle \mathcal{T}_{ij}^{w_j}, \mathcal{I}_{ij}^{w_j}, \mathcal{F}_{ij}^{w_j} \rangle_{m \times n} \tag{10}$$

$$= \begin{array}{c} \\ A_1 \\ A_2 \\ \vdots \\ A_m \end{array} \overset{\displaystyle \begin{array}{cccc} C_1 & C_2 & \cdots & C_n \end{array}}{\left[\begin{array}{cccc} \langle \mathcal{T}_{11}^{w_1}, \mathcal{I}_{11}^{w_1}, \mathcal{F}_{11}^{w_1} \rangle & \langle \mathcal{T}_{12}^{w_2}, \mathcal{I}_{12}^{w_2}, \mathcal{F}_{12}^{w_2} \rangle & \cdots & \langle \mathcal{T}_{1n}^{w_n}, \mathcal{I}_{1n}^{w_n}, \mathcal{F}_{1n}^{w_n} \rangle \\ \langle \mathcal{T}_{21}^{w_1}, \mathcal{I}_{21}^{w_1}, \mathcal{F}_{21}^{w_1} \rangle & \langle \mathcal{T}_{22}^{w_2}, \mathcal{I}_{22}^{w_2}, \mathcal{F}_{22}^{w_2} \rangle & \cdots & \langle \mathcal{T}_{2n}^{w_n}, \mathcal{I}_{2n}^{w_n}, \mathcal{F}_{2n}^{w_n} \rangle \\ \vdots & \vdots & \ddots & \vdots \\ \langle \mathcal{T}_{m1}^{w_1}, \mathcal{I}_{m1}^{w_1}, \mathcal{F}_{m1}^{w_1} \rangle & \langle \mathcal{T}_{m2}^{w_2}, \mathcal{I}_{m2}^{w_2}, \mathcal{F}_{m2}^{w_2} \rangle & \cdots & \langle \mathcal{T}_{mn}^{w_n}, \mathcal{I}_{mn}^{w_n}, \mathcal{F}_{mn}^{w_n} \rangle \end{array} \right]} \tag{11}$$

where $\langle r_{ij}^{w_j} \rangle$ is aggregated weighted neutrosophic decision matrix
- **Determination of positive and negative ideal solution for single valued neutrosophic sets**
  Let us examine the attribute types for the determination of positive and negative ideal solutions. Let J_1 and J_2 be benefit type and cost type attributes, respectively. $S_{\mathcal{A}}^{+}$ is the neutrosophic positive ideal solution and $S_{\mathcal{A}}^{-}$ is the neutrosophic negative ideal solution. Positive and negative ideal solutions are defined as follows: $S_{\mathcal{A}}^{+} = [r_1^{w+}, r_2^{w+}, \ldots, r_n^{w+}]$. $\mathcal{A}$ is the represents neutrosophic sets. $r_j^{w+} = \langle \mathcal{T}_j^{w+}, \mathcal{I}_j^{w+}, \mathcal{F}_j^{w+} \rangle$ for $j = 1, 2, \ldots, n$

$$\mathcal{T}_j^{w+} = \left\{ \left(\max_i \mathcal{T}_{ij}^{w_j} | j \in J_1 \right), \left(\min_i \mathcal{T}_{ij}^{w_j} | j \in J_2 \right) \right\} \tag{12}$$

$$\mathcal{I}_j^{w+} = \left\{ \left(\min_i \mathcal{I}_{ij}^{w_j} | j \in J_1 \right), \left(\max_i \mathcal{I}_{ij}^{w_j} | j \in J_2 \right) \right\} \tag{13}$$

$$\mathcal{F}_j^{w+} = \left\{ \left(\min_i \mathcal{F}_{ij}^{w\,j} | j \in J_1 \right), \left(\max_i \mathcal{F}_{ij}^{w\,j} | j \in J_2 \right) \right\} \tag{14}$$

$S_{\mathcal{A}}^- = [r_1^{w-}, r_2^{w-}, \dots, r_n^{w-}]$. $\mathcal{A}$ is the represents neutrosophic sets. $r_j^{w-} = \langle \mathcal{T}_j^{w-}, \mathcal{I}_j^{w-}, \mathcal{F}_j^{w-} \rangle$ for $j = 1, 2, \dots, n$.

$$\mathcal{T}_j^{w-} = \left\{ \left(\min_i \mathcal{T}_{ij}^{w\,j} | j \in J_1 \right), \left(\max_i \mathcal{T}_{ij}^{w\,j} | j \in J_2 \right) \right\} \tag{15}$$

$$\mathcal{I}_j^{w-} = \left\{ \left(\max_i \mathcal{I}_{ij}^{w\,j} | j \in J_1 \right), \left(\min_i \mathcal{I}_{ij}^{w\,j} | j \in J_2 \right) \right\} \tag{16}$$

$$\mathcal{F}_j^{w-} = \left\{ \left(\max_i \mathcal{F}_{ij}^{w\,j} | j \in J_1 \right), \left(\min_i \mathcal{F}_{ij}^{w\,j} | j \in J_2 \right) \right\} \tag{17}$$

- **Calculation of distances between alternatives and ideal solutions.**
  Equation (2) was used for distance calculations. The distance between each alternative $\langle \mathcal{T}_{ij}^{w\,j}, \mathcal{I}_{ij}^{w\,j}, \mathcal{F}_{ij}^{w\,j} \rangle$ and the positive ideal solution $\langle \mathcal{T}_j^{w+}, \mathcal{I}_j^{w+}, \mathcal{F}_j^{w+} \rangle$ is calculated as follows:

$$\mathcal{D}_{Eucl}^{i+}(r_{ij}^{w\,j}, r_j^{w+}) = \sqrt{\frac{1}{3n} \sum_{j=1}^{n} \left\{ \begin{array}{l} (\mathcal{T}_{ij}^{w\,j}(x_j) - \mathcal{T}_j^{w+}(x_j))^2 \\ +(\mathcal{I}_{ij}^{w\,j}(x_j) - \mathcal{I}_j^{w+}(x_j))^2 \\ +(\mathcal{F}_{ij}^{w\,j}(x_j) - \mathcal{F}_j^{w+}(x_j))^2 \end{array} \right\}} \tag{18}$$

The distance between each alternative $\langle \mathcal{T}_{ij}^{w\,j}, \mathcal{I}_{ij}^{w\,j}, \mathcal{F}_{ij}^{w\,j} \rangle$ and the positive ideal solution $\langle \mathcal{T}_j^{w-}, \mathcal{I}_j^{w-}, \mathcal{F}_j^{w-} \rangle$ is calculated as follows:

$$\mathcal{D}_{Eucl}^{i-}(r_{ij}^{w\,j}, r_j^{w-}) = \sqrt{\frac{1}{3n} \sum_{j=1}^{n} \left\{ \begin{array}{l} (\mathcal{T}_{ij}^{w\,j}(x_j) - \mathcal{T}_j^{w-}(x_j))^2 \\ +(\mathcal{I}_{ij}^{w\,j}(x_j) - \mathcal{I}_j^{w-}(x_j))^2 \\ +(\mathcal{F}_{ij}^{w\,j}(x_j) - \mathcal{F}_j^{w-}(x_j))^2 \end{array} \right\}} \tag{19}$$

- **Determination of the closeness coefficient to neutrosophic ideal solution for single valued neutrosophic numbers]**
  In this step, closeness coefficient is calculated according to positive and negative ideal solutions for neutrosophic sets.

$$C_i^* = \frac{\mathcal{D}_{Eucl}^{i-}(r_{ij}^{w\,j}, r_j^{w-})}{\mathcal{D}_{Eucl}^{i+}(r_{ij}^{w\,j}, r_j^{w+}) + \mathcal{D}_{Eucl}^{i-}(r_{ij}^{w\,j}, r_j^{w-})} \tag{20}$$

where C_i^* is between 0 and 1.
- **Ranking the alternatives** Alternative rankings are determined according to closeness coefficient values (C_i^*).

4   Numerical Example: Case Study of Airline Companies

We use two different datasets for this study. For the first data set, we select the subset of all of user reviews about airlines which Kaya used in her work [24]. These are costumer review data given from 1 January 2017 to 31 December 2017 (DS2).In addition to this data set, passenger of 2018 year data has also been added. Therefore, this study uses passenger data between 1 January 2017 and 31 December 2018. There are five airlines and it consists of 2232 users (decision makers) and 18 attributes. For us, numerical ratings are important so, data set consists of six subratings, and an overall rating. These sub ratings are value of money (VM), seat comfort (SC), cabin staff service (SS), food and beverages (Catering), inflight entertainment (Ent) and ground service (GS). While subratings are shown with stars from 1 to 5, overall rating is shown with a bar from 1 to 10.

1. **Value for money:** It is all about that costs paid by passengers. For example, "How much were you charged for your air fare?", "Were there additional costs such as booking fees, Baggage fees, Seat fees, Food and Beverage fees, Wi-Fi or entertainment fees?", "Even if there were fees do you think they were still pretty good value or over the top expensive?".
2. **Seat comfort:** The criterion is related to the safety and comfort of airline companies seat for passenger not dissatisfaction. These are the evaluations given according to the layout between the front and rear seats and the side-by-side seats so as not to disturb the passengers. Especially in long-term journeys, this criterion is at a level that can satisfy the journeys.
3. **Cabin staff service:** The criterion recognizes the highest all-round performance of cabin staff, for hard service (e.g. techniques, efficiency, attention etc.), and soft service characteristics (e.g. staff enthusiasm, attitude, friendliness and hospitality). The criterion is about behaviour and attitude of staff towards passengers.
4. **Food and Beverages:** It is about airlines decide what passengers will eat and drink on board. The criterion is very important especially for long-term flights. The quality of the products distributed by the staff is very important. In families with children, there are positive evaluations for the quality of this criterion.
5. **Inflight entertainment:** The criteria recognizes airlines delivering the best choice and quality of inflight entertainment. From the choice and currency of movies, TV programmes, music, and games through to the availability and functionality of on board Wi-Fi, this criterion recognizes those airlines that seek to deliver a fantastic experience to their passengers.

6. **Ground service:** Ground Service is responsible for all ground operations in the entire airline network and ensures that the company's approved standards are met and the operation is safe, cost-effective, and efficient.

At this section, it is aimed to examine the passengers who have experience in airline companies and to show the direction in which firms have improved over the years. In this study, our aim is that intuitionistic fuzzy set will be applied on the two data sets obtained from passengers. There are six criteria for this data set: C_1 : Seat Comfort (SM), C_2 : Staff Service (SS), C_3 : Ground Service (GS), C_4 : Value for Money (VM), C_5 : Food and Beverages (Catering), C_6 : Inflight Entertainment (Ent.).

4.1 Converting to Neutrosophic Numbers of Users' Rating of the Criteria

In this step, users' ratings ranging from 1 to 5 are converted to neutrosophic environment. But, the data set contains inconsistent data. Both overall rating of the criteria given to ratings and the ratings of the criteria given to overall rating should be consistent with each other. For example, in the case where all criteria are 1, assigning overall rating to 10 causes inconsistency in the data. In the data used in the study, no adverse effects were observed. But, in two cases, the following method is considered.

Overall Rating Rearrange When the criteria are considered, although the ratings given to the criteria are low, if the overall rating is high, then relevant criteria may not be included in the dataset. Conversely, although the ratings given to the criteria are high, if the overall rating is low, the criterion causing the negativity is not included in the data set. In the light of existing criteria, it is necessary to make adjustments to the data set. In this study, it focuses on the difference between the average of the ratings given to the criteria and the overall rating. Considering the ratings of the decision makers to the criteria and the overall rating, the following situations are discussed. U is a set of users, R is a set of related criteria, and O is a set of overall. $U_R = (u_i, r_j) | u_i \in U, r_j \in R$ and $U_O = (u_i, o) | u_i \in U, o \in O, i = 1, 2, \ldots, M$ and $j = 1, 2, \ldots, N$. M is the number of users and N is the number of criteria (except for overall criteria). The limit values created above are obtained as a result of the experiments. If these limit values are changed, the integrity of the data is impaired. Table 1 shows the number of data discarded from each data set.

Airline company/year	2017	2018
Air China (A1)	1	2
ANA (A2)	1	2
Lufthansa (A3)	2	7
THY (A4)	5	7
United (A5)	4	9

Table 1 Inconsistent data extracted on the data set according to years

Table 2 Linguistic terms for rating of decision makers and attributes

Stars on data	Linguistic terms	SVNNs
5-Star	Very good/very important (VG/VI)	$\langle 0.90, 0.10, 0.10 \rangle$
4-Star	Good/important (G/I)	$\langle 0.80, 0.20, 0.15 \rangle$
3-Star	Fair/medium (F/M)	$\langle 0.50, 0.40, 0.45 \rangle$
2-Star	Bad/unimportant (B/UI)	$\langle 0.35, 0.60, 0.70 \rangle$
1-Star	Very bad/very unimportant (VB/VUI)	$\langle 0.10, 0.80, 0.90 \rangle$

The preprocessing process on the data is as follows (Preprocessing of inconsistent data on the dataset):

Program Code

```
[VI_user, I_user, M_user]=Function Preprocess (data)
InputRating values about m users n criteria m × n
OutputProcessed dataset Load m × n matrix
Ifmean(data(u_i, r) − data(u_i, o)) ≤ 1) VI_user = data(u_i, r)
Else If 1 < mean(data(u_i, r) − data(u_i, o)) ≤ 2) I_user = data(u_i, r)
Else If 2 < mean(data(u_i, r) − data(u_i, o)) ≤ 5) M_user = data(u_i, r)
Else Discard_user
```

4.2 Case Study of Airline Companies

In this section, the neutrosophic TOPSIS structure which mentions in section-3 is applied to the airline companies database. Due to the high number of decision makers in the data set, neutral operator and neutrosophic TOPSIS are used together.

Step 1: Determine the weights of decision makers: In this study, the decision makers are accepted as passengers/decision makers who gained travel experience. The average of the values given by each decision makers to the sub-criteria in each airline company is taken and it is converted which neutrosophic numbers corresponds to Table 2. Then the weight calculation is made for each decision makers using Eq. (6). Let the neutrosophic number of the kth decision maker be $\Lambda_k = \langle \mathcal{T}_k, \mathcal{I}_k, \mathcal{F}_k \rangle$. For example, the weights of the decision makers for the Airchina

Table 3 Aggregated neutrosophic decision matrix

Airline company	C_1	C_2	C_3
A_1	⟨0.5651, 0.3997, 0.4303⟩	⟨0.6099, 0.3679, 0.3934⟩	⟨0.4856, 0.4755, 0.5195⟩
A_2	⟨0.7829, 0.2074, 0.2023⟩	⟨0.8551, 0.1435, 0.1463⟩	⟨0.8442, 0.1518, 0.1469⟩
A_3	⟨0.6925, 0.2865, 0.2896⟩	⟨0.7780, 0.2153, 0.2182⟩	⟨0.6808, 0.3064, 0.3186⟩
A_4	⟨0.6555, 0.3230, 0.3368⟩	⟨0.7052, 0.2793, 0.2917⟩	⟨0.5551, 0.4149, 0.4488⟩
A_5	⟨0.4717, 0.4879, 0.5358⟩	⟨0.5483, 0.4212, 0.4590⟩	⟨0.4574, 0.5020, 0.5534⟩
	C_4	C_5	C_6
A_1	⟨0.6189, 0.3632, 0.3955⟩	⟨0.5019, 0.4557, 0.5011⟩	⟨0.4015, 0.5432, 0.6037⟩
A_2	⟨0.8348, 0.1619, 0.1541⟩	⟨0.7587, 0.2315, 0.2400⟩	⟨0.6706, 0.3125, 0.3266⟩
A_3	⟨0.6911, 0.2938, 0.3025⟩	⟨0.6805, 0.3013, 0.3116⟩	⟨0.4740, 0.4879, 0.5339⟩
A_4	⟨0.6581, 0.3233, 0.3481⟩	⟨0.6928, 0.2885, 0.3042⟩	⟨0.6585, 0.3212, 0.3393⟩
A_5	⟨0.4408, 0.5178, 0.5754⟩	⟨0.4092, 0.5419, 0.5998⟩	⟨0.3853, 0.5635, 0.6246⟩

Table 4 Aggregated weighted neutrosophic decision matrix

Airline company	C_1	C_2	C_3
A_1	⟨0.4178, 0.8435, 0.8515⟩	⟨0.4640, 0.8488, 0.8549⟩	⟨0.3048, 0.8048, 0.8211⟩
A_2	⟨0.5788, 0.7933, 0.7920⟩	⟨0.6506, 0.7952, 0.7958⟩	⟨0.5299, 0.6842, 0.6824⟩
A_3	⟨0.5119, 0.8140, 0.8148⟩	⟨0.5920, 0.8123, 0.8130⟩	⟨0.4273, 0.7418, 0.7463⟩
A_4	⟨0.4846, 0.8235, 0.8271⟩	⟨0.5366, 0.8277, 0.8306⟩	⟨0.3484, 0.7822, 0.7948⟩
A_5	⟨0.3487, 0.8665, 0.8790⟩	⟨0.4171, 0.8616, 0.8706⟩	⟨0.2871, 0.8146, 0.8337⟩
	C_4	C_5	C_6
A_1	⟨0.4168, 0.7921, 0.8026⟩	⟨0.3231, 0.8061, 0.8223⟩	⟨0.2142, 0.7869, 0.8151⟩
A_2	⟨0.5622, 0.7263, 0.7238⟩	⟨0.4885, 0.7263, 0.7293⟩	⟨0.3577, 0.6792, 0.6858⟩
A_3	⟨0.4654, 0.7694, 0.7723⟩	⟨0.4382, 0.7512, 0.7548⟩	⟨0.2528, 0.7611, 0.7825⟩
A_4	⟨0.4432, 0.7790, 0.7872⟩	⟨0.4461, 0.7466, 0.7522⟩	⟨0.3513, 0.6833, 0.6917⟩
A_5	⟨0.2968, 0.8425, 0.8614⟩	⟨0.2634, 0.8368, 0.8575⟩	⟨0.2055, 0.7964, 0.8248⟩

airline company are given following: λ_{DM-1}: 0.0017, λ_{DM-2}: 0.0102, λ_{DM-3}: 0.0102, λ_{DM-4}: 0.0113, λ_{DM-5}: 0.0017, and λ_{DM-195}: 0.0069.

Step 2: Determine aggregated neutrosophic decision matrix based on the opinions of decision makers: Each DM has an neutrosophic decision matrix. Let us fusion all individual idea to express group idea. Can be done using aggregated neutrosophic decision matrix. Single valued neutrosophic weighted neutrality aggregation (SVNWNA) is used in neutrosophic matrix operation [7]. For the creation of SVNWNA matrix, multiplied by the ith weight and matrix of the ith decision maker. Then all decision makers combined with SVNWNA operator. The aggregated neutrosophic decision matrix based on aggregation of decision makers' opinions is constructed as Table 3 using Eq. (7)

Step 3: Determine the weights of the criteria: Neutrosophic numbers transformations of numerical ratings given by the decision makers to the criteria are shown in Table 2. Opinions of decision makers on criteria are aggregated using Eq. (9) to determine the weight of each criterion for dataset

Table 5 Negative and positive neutrosophic ideal solutions

Criteria	$\mathcal{D}_{\mathcal{A}}^{+}$	$\mathcal{D}_{\mathcal{A}}^{-}$
C_1	$\langle 0.5788, 0.7933, 0.7920 \rangle$	$\langle 0.3487, 0.8665, 0.8790 \rangle$
C_2	$\langle 0.6506, 0.7952, 0.7958 \rangle$	$\langle 0.4171, 0.8616, 0.8706 \rangle$
C_3	$\langle 0.5299, 0.6842, 0.6824 \rangle$	$\langle 0.2871, 0.8146, 0.8337 \rangle$
C_4	$\langle 0.5622, 0.7263, 0.7238 \rangle$	$\langle 0.2968, 0.8425, 0.8614 \rangle$
C_5	$\langle 0.4885, 0.7263, 0.7293 \rangle$	$\langle 0.2634, 0.8368, 0.8575 \rangle$
C_6	$\langle 0.3577, 0.6792, 0.6858 \rangle$	$\langle 0.2055, 0.7964, 0.8248 \rangle$

Table 6 Separation measures and the relative closeness coefficient of each alternative

Criteria	$\mathcal{D}_{\mathcal{A}}^{+}$	$\mathcal{D}_{\mathcal{A}}^{-}$	$C_{\mathcal{A}}^{*}$
A_1	0.1368	0.0487	0.2627
A_2	0	0.1766	**1**
A_3	0.0692	0.1148	0.6240
A_4	0.0838	0.1100	0.5678
A_5	0.1766	0	0

$$
W_{criteria} = \begin{bmatrix} 0.7393 & 0.3873 & 0.4064 \\ 0.7609 & 0.3737 & 0.3989 \\ 0.6277 & 0.4879 & 0.5297 \\ 0.6735 & 0.4496 & 0.4874 \\ 0.6438 & 0.4716 & 0.5113 \\ 0.5334 & 0.5644 & 0.6199 \end{bmatrix} \tag{21}
$$

Step 4: Determine aggregated weighted neutrosophic decision matrix: Weighted neutrosophic decision matrix is generated using Eqs. (10) and (11). (Table 4)

Step 5: Obtain neutrosophic positive and negative ideal solution: We obtain negative and positive neutrosophic ideal solution using Eq. (12)–(17)

Table 5 contains all ideal solutions according to the criteria.

Step 6: Calculate the separation measures and rank the alternatives: The normalized Euclidean distance is used for measure separation between alternatives on neutrosophic set [18]. According to the distance measure, positive and negative ideal solutions are calculated.

The largest relative closeness coefficient expressed in bold indicates that A_2 is the most preferred alternative. Thus, it is seen that the most preferred airline company "ANA" is based on the given criteria. Since the high number of decision makers causes prejudice in decision making mechanism, it is seen that it gives more consistent results with the use of neutral aggregation operator.

Step 7: Ranking the alternatives: According to the relative closeness coefficient values given in Table 6, ranking order of five alternatives is $A_2 \succ A_3 \succ A_4 \succ A_1 \succ A_5$ Thus, the airline companies are indicated with $ANA \succ Lufthansa \succ THY \succ AirChina \succ United$ from the most preferred to the least preferred (Table 6).

Table 7 Evaluations of each alternative based on criteria

Criteria	VM	SC	SS	Cat	Ent.	GS
Air China	2.45	2.38	2.42	2.09	1.86	1.94
ANA	4.43	4.08	4.53	4.12	3.98	4.54
Lufthansa	3.42	3.47	3.55	3.51	3.57	3.46
THY	3.19	3.27	4	3.59	3.71	2.73
United	1.83	2.04	2.33	2.01	2.05	1.94
Weights	0.67	0.74	0.76	0.64	0.53	0.63

Table 8 Statistical values for each criterion

Criteria	VM	SC	SS	Cat	Ent.	GS
Min	1.83	2.38	2.42	2.09	1.86	1.94
Max	4.43	4.08	4.53	4.12	3.98	4.54
Mean	3.06	3.05	3.37	3.06	3.04	2.92
S.Dev.	0.88	0.74	0.87	0.85	0.89	0.99

5 Comparative Analysis with Promethee Method

The hybrid method proposed in this study is compared with the Promethee method which is state of the art. Promethee method has been applied on airline companies data.

5.1 Base Method: Promethee

In this section, Promethee method is used to determine the order of importance between the criteria/alternatives. The weight values obtained for each criterion are used in the first experiment to determine the criterion weights, the first stage of the Promethee method. Tables 7 and 8 are shown the weight values, criteria values for each alternative and statistical values of this criteria, respectively.

In the Fig. 1, the order of each alternative and criteria is given according to the phi values.

The parts shown in blue and brown in Fig. 1 represent the passenger numbers in the alternatives. When the importance order of each alternative is examined in terms of criteria, it is seen that the SC criterion is in the first place in ANA and THY companies and this criterion is in the last place in United firm. While the entertainment criterion is the last place in general, it ranks first in the United company. It can be said that the importance of GS and Catering criteria varies according to the companies. Generally, seat comfort, staff service, and value for money have high level importance, there is vice versa for entertainment and ground service.

Figure 2 shows the phi value changes in the alternatives as a result of the changes made in the criteria weights. In this example, a change regarding SS is given. When an increase in SS weight value is increased, THY and Air China increase, while

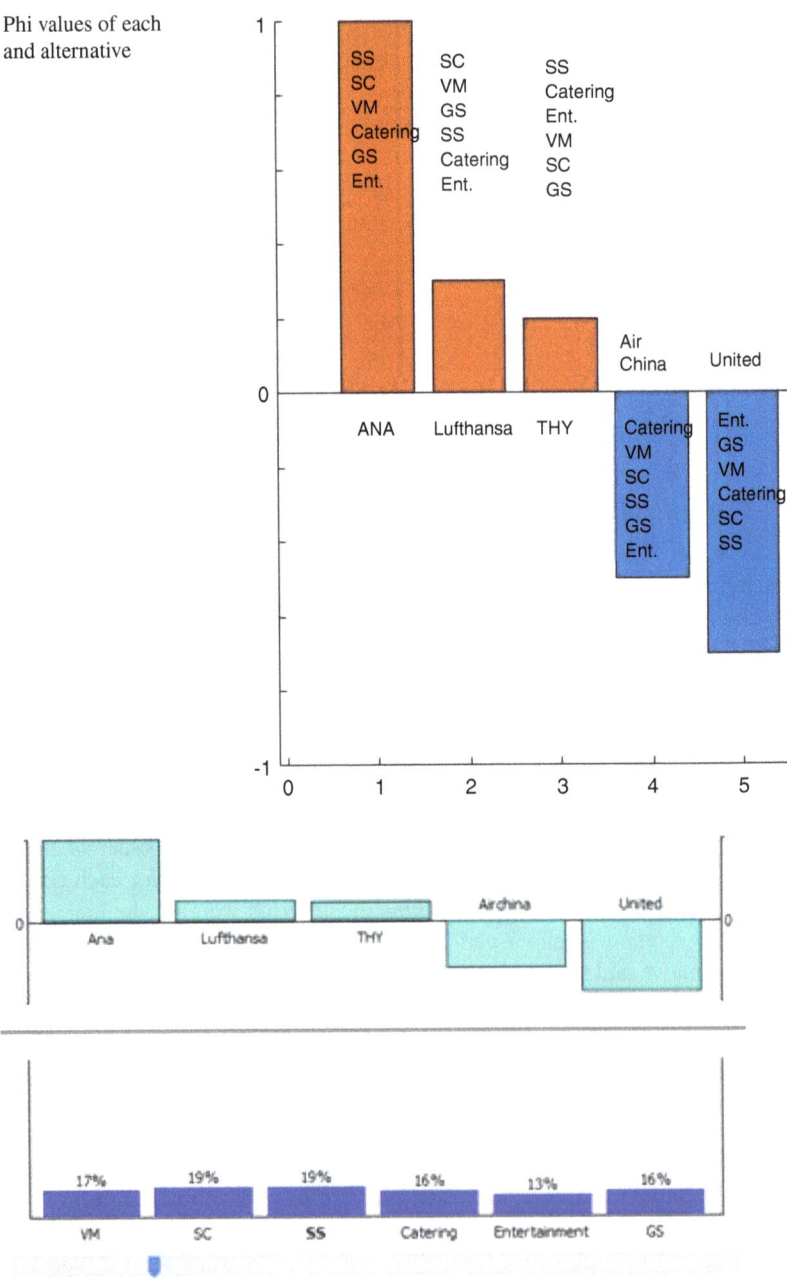

Fig. 1 Phi values of each criteria and alternative

Fig. 2 Phi values of alternatives according to changing criteria weights

Fig. 3 Phi values of
alternatives according to
changing criteria weights

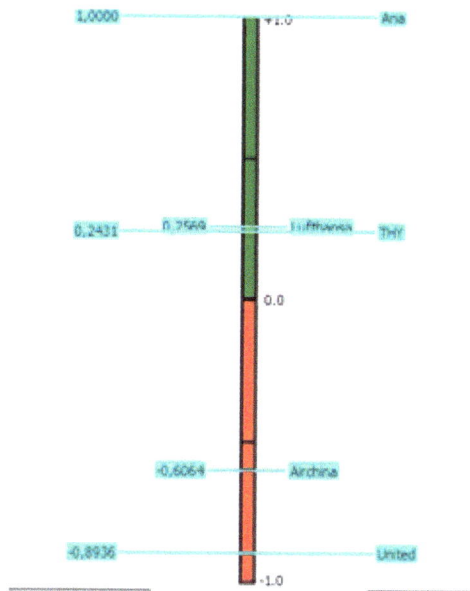

Lufthansa and United have the opposite situation. According to these results, the reason why a change in the weight of the criteria affects the alternatives positively or negatively is due to the importance ordering of the criteria in the alternatives.

After determining the order of importance of the criteria, it is seen that both figures are included in the order of alternatives. When looking at the order of the alternatives in Fig. 1, it can be said that ANA, Lufthansa, and THY, which have positive phi values in the first three places. It is seen that there is an approximately 1% difference between Lufthansa and THY, while ANA has the top value in Fig. 3. The last row is shown in Fig. 3 with Air China and United airline companies.

6   Conclusion

The development of the Internet has made user satisfaction an important criterion in the competitive environment. In the past, negative/positive opinions about a product or any situation experienced by users may not be known by people in different regions. But nowadays, person that experiences are related to any product or situation can present their opinions in written form on the Internet. The provided information is light to those who will experience the same situation. As well as being able to express written thoughts, one can also evaluate the important situation/product for himself/herself in terms of criteria. The determined of user characteristics into the evaluation process yields more consistent, appropriate results for subsequent users. In this study, the reviews of passengers who have experience of airline companies are examined. Different approaches have been applied by evaluating the overall performance of the 5 airline companies that the passengers travel to, such as value for money, seat comfort, staff service, food and beverages (Catering), inflight entertainment, and ground service. In this study, the rating values are first converted to neutrosophic numbers and then the study is compared with the other study in the literature. In this study, each passenger is determined as the decision maker. So there are decision makers in Air China 195, ANA 102, Lufthansa 450, THY 436, and united 1049. After determination of the weight values for each decision maker, aggregated neutrosophic decision matrix is formed using these values (See Table 3). When the results of the approach applied using the neutrosophic set are examined, it is seen that Ana, Lufthansa, and THY companies ranked first three in the ranking among the alternatives. The last two places are Air China and United. When the examination is made in terms of criteria, it is seen that the staff service and seat comfort criterion of ANA company, which ranks first, are quite pleased with the passengers. On the contrary, it can be said that catering and entertainment criteria are missing for this company. When the criteria of United firm, which is at the end of the ranking, are examined, staff service and seat comfort are important as other companies, while food service and entertainment are at the top. Considering the overall rankings of all companies, it can be concluded that the passengers are satisfied with the seat comfort and staff service of the first two rows. In line with these results, companies with low satisfaction level of passengers can develop based on other high value companies. In the other study, Promethee II method, the weight values required for the criteria are used as the weight values obtained from the first approach. Looking at the results of the approach, it is seen that the same results are obtained with the first approach in the ranking among the alternatives. While the ranking of the criteria in terms of alternatives is staff service and seat comfort for ANA, the last place is the ground services and entertainment criteria. While this result is similar to the first approach, there is the opposite situation for Air China. While the most important criterion in this company is catering and value for money, the worst criteria are ground services and entertainment. As the result of the evaluations made by the decision makers, United, the worst company, entertainment which is the worst criterion in general is the most important criterion in this company. When the results of both approaches are examined, it was seen that anomalous results are obtained. The rankings of the alternatives are similar in both approaches. Considering the ranking of the criteria according to the general weight values in the first approach, it can be concluded that the same ranking takes place in ANA, which is in the top position in the second approach. In general, in the first three criteria, VM, SC, and SS results are seen. General assessment, it can be concluded that United and Air China, one of the worst airlines in the rankings, should take into account passenger satisfaction in order to move to the top of the competition. According to experimental results, a firm improvement can be made by considering the criteria rankings in the top ranked firms. As a result, in this study, it can be concluded that in order to improve the customer satisfaction for airline companies, it is necessary to make improvements in terms of criteria. At the same time, the high number of decision makers causes the decision makers to influence the decision making process with their bias. Therefore, in the method proposed in the study, neutrosophic information and neutral aggregation-based TOPSIS method were used. This method was also compared to the Promethee method, which is frequently used in decision making problems. In future studies, with neutrality aggregation operator induced aggregation [28], prioritized aggregation [29] operators can be combined. In addition, neutrality aggregation operator can be transformed into structures containing parameters with T-norm structures. Such as Dombi, Einstein, and Hamacher T-norm.

Declaration of Competing Interest The authors declare that they have no known competing financial interests or personal relationships that could have appeared to influence the work reported in this paper.

References

1. Bayyurt, N. (2013). Ownership effect on bank's performance: Multi criteria decision making approaches on foreign and domestic Turkish banks. *Procedia-Social and Behavioral Sciences 99*, 919–928.
2. Biswas, P., Pramanik, S., & Giri, B.C. (2016). TOPSIS method for multi-attribute group decision-making under single-valued neutrosophic environment. *Neural computing and Applications 27*(3), 727–737.
3. Brans, J.P., Vincke, P., & Mareschal, B. (1986). How to select and how to rank projects: The Promethee method. *European journal of operational research 24*(2), 228–238.
4. Chen, J. & Ye, J. (2017). Some single-valued neutrosophic Dombi weighted aggregation operators for multiple attribute decision-making. *Symmetry 9*(6), 82.
5. Dezert, J. (2002). Open questions in neutrosophic inferences. *Multiple-Valued Logic/An International Journal 8*(3), 439–472.
6. Doumpos, M., Kosmidou, K., Baourakis, G., & Zopounidis, C. (2002). Credit risk assessment using a multicriteria hierarchical discrimination approach: A comparative analysis. *European Journal of Operational Research 138*(2), 392–412.
7. Garg, H. (2020). Novel neutrality aggregation operator-based multiattribute group decision-making method for single-valued neutrosophic numbers. *Soft Computing, 24*(14), 10327–10349.
8. Garg, H. et al. (2018). Multi-criteria decision-making method based on prioritized muirhead mean aggregation operator under neutrosophic set environment. *Symmetry 10*(7), 280.
9. Ho, W., Xu, X., & Dey, P. K. (2010). Multi-criteria decision making approaches for supplier evaluation and selection: A literature review. *European Journal of operational research 202*(1), 16–24.
10. Hwang, C. L. & Yoon, K. (1981). *Multiple attribute decision making–methods and applications springer-verlag berlin Heidelberg*. New York: Springer.
11. Kahraman, C. (2008). *Fuzzy multi-criteria decision making: Theory and applications with recent developments*, vol. 16. Berlin: Springer.
12. Li, Y., Liu, P., & Chen, Y. (2016). Some single valued neutrosophic number Heronian mean operators and their application in multiple attribute group decision making. *Informatica 27*(1), 85–110.
13. Liu, P. & Wang, Y. (2014). Multiple attribute decision-making method based on single-valued neutrosophic normalized weighted Bonferroni mean. *Neural Computing and Applications 25*(7-8), 2001–2010.

14. Majumdar, P. & Samanta, S. K. (2014). On similarity and entropy of neutrosophic sets. *Journal of Intelligent and Fuzzy Systems 26*(3), 1245–1252

15. Opricovic, S., & Tzeng, G. H. (2004). Compromise solution by MCDM methods: A comparative analysis of VIKOR and TOPSIS. *European Journal of Operational Research 156*(2), 445–455.

16. Pouresmaeil, H., Shivanian, E., Khorram, E., & Fathabadi, H. S. (2017). An extended method using TOPSIS and VIKOR for multiple attribute decision making with multiple decision makers and single valued neutrosophic numbers. *Advanced Applied Statistical, 50*, 261–292.

17. Roy, B. (1991). The outranking approach and the foundations of ELECTRE methods. *Theory and Decisions 31*(1), 49–73.

18. Salama, A., Abdelfattah, M., & Eisa, M. (2014). Distances, hesitancy degree and flexible querying via neutrosophic sets. *International Journal of Computer Applications 101*(10), 7–12.

19. Senapati, T. & Yager, R. R. (2019). Fermatean fuzzy weighted averaging/geometric operators and its application in multi-criteria decision-making methods. *Engineering Applications of Artificial Intelligence 85*, 112–121.

20. Sets, F. (1965). L. zadeh. *Information and Control.–NY 8*(3), 338–353

21. Smarandache, F. (1999). A unifying field in logics: Neutrosophic logic. In *Philosophy* (pp. 1–141). Santa Fe: American Research Press.

22. Smarandache, F. (2019). Neutrosophic Set is a Generalization of Intuitionistic Fuzzy Set, Inconsistent Intuitionistic Fuzzy Set (Picture Fuzzy Set, Ternary Fuzzy Set), Pythagorean Fuzzy Set (Atanassov's Intuitionistic Fuzzy Set of second type), q-Rung Orthopair Fuzzy Set, Spherical Fuzzy Set, and n-HyperSpherical Fuzzy Set, while Neutrosophication is a Generalization of Regret Theory, Grey System Theory, and Three-Ways Decision (revisited). Infinite Study.

23. Smarandache, F. et al. (2010). Neutrosophic set–a generalization of the intuitionistic fuzzy set. *Journal of Defense Resources Management (JoDRM) 1*(1), 107–116.

24. Türkoğlu, T. (2016). Çoklu ölçüt oy değerleri üzerinden veri madenciliği, Master's thesis. Anadolu: Anadolu Üniversitesi.

25. Vincke, J. & Brans, P. (1985). A preference ranking organization method. the promethee method for MCDM. *Management Science 31*(6), 647–656.

26. Wang, H., Smarandache, F., Zhang, Y., & Sunderraman, R.: *Single valued neutrosophic sets*. Infinite study (2010).

27. Wang, J. J., Jing, Y. Y., Zhang, C. F., & Zhao, J. H. (2009). Review on multi-criteria decision analysis aid in sustainable energy decision-making. *Renewable and sustainable energy reviews 13*(9), 2263–2278.

28. Yager, R. R. (2003). Induced aggregation operators. *Fuzzy sets and systems 137*(1), 59–69.

29. Yager, R. R. (2008). Prioritized aggregation operators. *International Journal of Approximate Reasoning 48*(1), 263–274.

30. Yager, R. R. (2013). Pythagorean fuzzy subsets. In *Proceedings of the 2013 Joint IFSA World Congress and NAFIPS Annual Meeting (IFSA/NAFIPS)* (pp. 57–61). New York: IEEE.

31. Yager, R. R. (2016). Generalized orthopair fuzzy sets. *IEEE Transactions on Fuzzy Systems 25*(5), 1222–1230.

Aggregate, Arithmetic, and Geometric Operators of Octagonal Neutrosophic Numbers and Its Application in Multi-Criteria Decision-Making Problems

Muhammad Saqlain, Muhammad Saeed, Rana Muhammad Zulqarnain, and Ali Hamza

1 Introduction

Uncertainty theory plays a vital role in numerous fields of life such as modeling, medical, and engineering. However, a general question is raised that in mathematical modeling how we can express and use the uncertainty concept. A lot of researchers in the whole world proposed and recommended different approaches to use uncertainty theory. First of all, Zadeh [1] presented the notion of fuzzy sets to solve those problems which contain uncertainty and vagueness. Some cases where the conclusion is offensive and the decision-maker assertion is ambiguous cannot be handled by fuzzy sets. To overcome such situations, the idea of the interval-valued fuzzy set was developed by Turksen [2]. But these theories only deal with the insufficient data considering both membership and non-membership values; these theories cannot handle the incompatible and imprecise information. The idea of the neutrosophic set (NS) was developed by Smarandache [3] to deal with such incompatible and imprecise data. Recently he developed a new approach known as hypersoft set [4].

M. Saqlain
Lahore Garrison University, DHA Phase-VI, Sector C, Lahore, Pakistan
e-mail: msaqlain@lgu.edu.pk

M. Saeed (✉)
School of Mathematics, University of Management and Technology, Lahore, Pakistan
e-mail: Muhammad.saeed@umt.edu.pk

R. M. Zulqarnain
School of Mathematics, Northwest University Xi'an, Xi'an, China

A. Hamza
Department of Mathematics, Lahore Garrison University, Lahore, Pakistan

Many researchers from all over the world worked on hypersoft set and developed different decision-making method to deal with the problems which contain uncertainty. Saqlain, Moin, Jafar, Saeed, & Smarandache [5] defined the neutrosophic hypersoft set and proposed different basic operations and their properties with numerical examples.

Nowadays many groups of mathematicians and scientists developed decision-making methods by using the recently developed neutrosophic method. However, still many viewing platforms about neutrosophic numbers (NNs) and their de-impreciseness in several forms are very necessary. Deli & Şubaş [6] proposed the generalized concept of fuzzy numbers and intuitionistic fuzzy numbers known as single-valued neutrosophic numbers (SVNNs). They also presented to solve MCDM problems with SVNNs in Deli & Şubaş [7] and constructed the concept of cut sets of SVNNs. Broumi, et al. [8] extended the idea of SVNNs and proposed single-valued 2n and 2n + 1-sided polygonal NNs and its matrix form with basic properties. In Biswas, Pramanik, & Giri [9], the authors defined the trapezoidal fuzzy neutrosophic number with properties. Şahin, Kargın, & Smarandache [10] transformed the SVNNs into single-valued trapezoidal neutrosophic numbers (SVTNNs) and proposed the generalized concept of SVTNNs. Karaaslan [11] gave the idea of Gaussian SVNNs with some arithmetic operations and developed a decision-making method based on newly developed operators and used the proposed method in medical diagnosis. Mo & Huang [12] introduced the λ-cutting matrix of SVNNs.

Chakraborty, et al. [13] introduced the linear and non-linear generalized triangular NNs and developed the concept of de-neutrosophication to convert the NNs into crisp form. They also extended the idea of pentagonal fuzzy numbers and developed a new technique known as interval-valued fuzzy numbers [14, 15] and defuzzified the pentagonal fuzzy numbers by different methods.

An advanced technique of NNs known as type 2 NNs with different properties and operations was developed by Abdel-Basset, Saleh, Gamal, & Smarandache [16]. In Liu, Chu, Li, & Chen [17], the authors constructed some operations on neutrosophic numbers with properties by using Hamacher operations.

Liu, Zhang, Liu, & Wang [18] combined the concept of Bonferroni mean and multi-valued neutrosophic numbers and introduced different operators on the developed method. Pramanik, Dalapati, Alam, & Roy [19] gave the idea of score and accuracy functions with their some properties for neutrosophic cubic numbers and developed a new approach to solve multi-attribute group decision-making problems (MAGDM). They also extended the TOPSIS under the neutrosophic cubic environment and developed the Euclidean distance between two neutrosophic cubic numbers in Pramanik, Dey, Giri, & Smarandache [20].

Pramanik & Mallick [21] extended the approaches of VIKOR and MAGDM under SVTNN environment and presented their numerical examples. The idea of linguistic neutrosophic cubic numbers with different operators and properties was developed by Ye [22]; he also developed a decision-making method by using developed operators under neutrosophic cubic environment. Fahmi, Amin, Khan, & Smarandache [23] proposed the triangular neutrosophic cubic fuzzy numbers with their score and accuracy functions. The idea of pentagonal neutrosophic

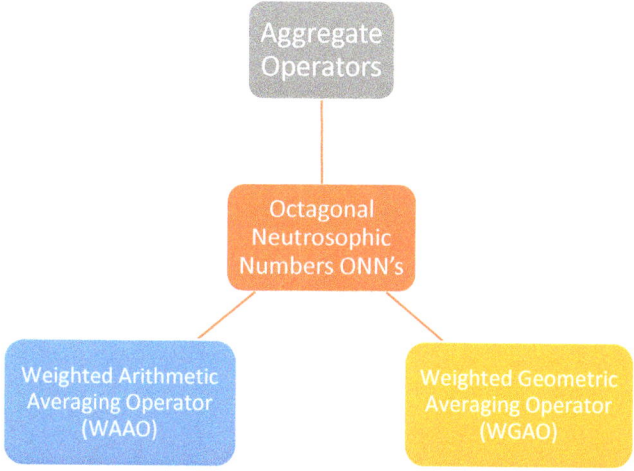

Fig. 1 Pictorial view for the paper presentation

number was constructed by Chakraborty, Broumi, & Singh [14] with its logical score and accuracy function. Selvakumari & Lavanya [24] introduced the octagonal neutrosophic numbers (ONNs) with α, β, and γ cuts and some other related operators; they also developed a method on proposed numbers by using the heavy ordered weighted averaging operator.

In Saqlain, Hamza, & Farooq [25], the authors extended the concept of ONNs and introduced some types of ONNs such as linear, non-linear, symmetric, and asymmetric with their α-cuts and mathematical notions.

1.1 Motivation

From the literature, it is found that fundamentals of octagonal neutrosophic numbers (ONNs) have not yet been studied, hence the motivation of the present study.

1.2 The Paper Presentation

In this paper, the fundamentals of octagonal neutrosophic numbers (ONNs) are presented.

Aggregate operators (sum, difference, product, power rules and relations, etc.)

Weighted arithmetic averaging operator (WAAO) and weighted geometric averaging operator WGAO

A case study of personal selection (Fig. 1)

2 Preliminaries

Definition 2.1 Let $\acute{X}$ be a universe of discourse; octagonal neutrosophic set $\check{N}$ in $\acute{X}$ is defined as:

$$\check{N} = \left\{ \acute{X}, \acute{T}_{\check{N}}\left(\acute{X}\right), \acute{I}_{\check{N}}\left(\acute{X}\right), \acute{F}_{\check{N}}\left(\acute{X}\right) \mid \acute{x} \in \acute{X} \right\},$$

where $\acute{T}_{\check{N}}\left(\acute{X}\right) \subseteq [0, 1]$, $\acute{I}_{\check{N}}\left(\acute{X}\right) \subseteq [0, 1]$ and $\acute{F}_{\check{N}}\left(\acute{X}\right) \subseteq [0, 1]$ are three octagonal fuzzy numbers $\acute{T}_{\check{N}}\left(\acute{X}\right) = \left(\acute{t}^1_{\check{N}}, \acute{t}^2_{\check{N}}, \acute{t}^3_{\check{N}}, \acute{t}^4_{\check{N}}, \acute{t}^5_{\check{N}}, \acute{t}^6_{\check{N}}, \acute{t}^7_{\check{N}}, \acute{t}^8_{\check{N}}\right) : \check{X} \longmapsto [0, 1]$,

$\acute{I}_{\check{N}}\left(\acute{X}\right) = \left(\acute{i}^1_{\check{N}}, \acute{i}^2_{\check{N}}, \acute{i}^3_{\check{N}}, \acute{i}^4_{\check{N}}, \acute{i}^5_{\check{N}}, \acute{i}^6_{\check{N}}, \acute{i}^7_{\check{N}}, \acute{i}^8_{\check{N}}\right) :$

$\acute{X} \longmapsto [0, 1]$ and $\acute{F}_{\check{N}}\left(\acute{X}\right) = \left(\acute{f}^1_{\check{N}}, \acute{f}^2_{\check{N}}, \acute{f}^3_{\check{N}}, \acute{f}^4_{\check{N}}, \acute{f}^5_{\check{N}}, \acute{f}^6_{\check{N}}, \acute{f}^7_{\check{N}}, \acute{f}^8_{\check{N}}\right) : \check{X} \longmapsto$

$[0, 1]$ with the condition $0 \leq \acute{T}_{\check{N}}\left(\acute{X}\right) + \acute{I}_{\check{N}}\left(\acute{X}\right) + \acute{F}_{\check{N}}\left(\acute{X}\right) \leq 3$, $\acute{x} \in$ $\acute{X}$. For a better experience, the Octagonal neutrosophic number is denoted by $\acute{T}_{\check{N}}(\acute{x}) = \left(\acute{a}, \acute{b}, \acute{c}, \acute{d}, \acute{e}, \acute{f}, \acute{g}, \acute{h}\right)$, $\acute{I}_{\check{N}}(\acute{x}) = \left(\acute{i}, \acute{j}, \acute{k}, \acute{l}, \acute{m}, \acute{n}, \acute{o}, \acute{p}\right)$ and $\acute{F}_{\check{N}}(\acute{x}) = \left(\acute{q}, \acute{r}, \acute{s}, \acute{t}, \acute{u}, \acute{v}, \acute{w}, \acute{x}\right)$. Then the octagonal neutrosophic number is denoted by: $\hat{n} = \left\langle \left(\acute{a}, \acute{b}, \acute{c}, \acute{d}, \acute{e}, \acute{f}, \acute{g}, \acute{h}\right), \left(\acute{i}, \acute{j}, \acute{k}, \acute{l}, \acute{m}, \acute{n}, \acute{o}, \acute{p}\right), \left(\acute{q}, \acute{r}, \acute{s}, \acute{t}, \acute{u}, \acute{v}, \acute{w}, \acute{x}\right) \right\rangle$, and it is a basic element in the octagonal neutrosophic set.

2.1 Deductions

1. **Pentagonal**
   If $\acute{f} = \acute{g} = \acute{h} = 0$, $\acute{n} = \acute{o} = \acute{p} = 0$ and $\acute{v} = \acute{w} = \acute{x} = 0$, then octagonal neutrosophic number reduces into pentagonal neutrosophic number.
2. **Trapezoidal**
   If $\acute{e} = \acute{f} = \acute{g} = \acute{h} = 0$, $\acute{m} = \acute{n} = \acute{o} = \acute{p} = 0$ and $\acute{u} = \acute{v} = \acute{w} = \acute{x} = 0$, then octagonal neutrosophic number reduces into pentagonal neutrosophic number.
3. **Triangular**
   If $\acute{d} = \acute{e} = \acute{f} = \acute{g} = \acute{h} = 0$, $\acute{l} = \acute{m} = \acute{n} = \acute{o} = \acute{p} = 0$ and $\acute{t} = \acute{u} = \acute{v} = \acute{w} = \acute{x} = 0$, then octagonal neutrosophic number reduces into pentagonal neutrosophic number.

3 Calculations

Let $\hat{\bar{n}}_1 = \left\langle \left(\acute{a}_1, \acute{b}_1, \acute{c}_1, \acute{d}_1, \acute{e}_1, \acute{f}_1, \acute{g}_1, \acute{h}_1\right), \left(\acute{i}_1, \acute{j}_1, \acute{k}_1, \acute{l}_1, \acute{m}_1, \acute{n}_1, \acute{o}_1, \acute{p}_1\right),\right.$
$\left. \left(\acute{q}_1, \acute{r}_1, \acute{s}_1, \acute{t}_1, \acute{u}_1, \acute{v}_1, \acute{w}_1, \acute{x}_1\right)\right\rangle$ and $\hat{\bar{n}}_2 = \left\langle \left(\acute{a}_2, \acute{b}_2, \acute{c}_2, \acute{d}_2, \acute{e}_2, \acute{f}_2, \acute{g}_2, \acute{h}_2\right),\right.$
$\left(\acute{i}_2, \acute{j}_2, \acute{k}_2, \acute{l}_2, \acute{m}_2, \acute{n}_2, \acute{o}_2, \acute{p}_2\right), \left(\acute{q}_2, \acute{r}_2, \acute{s}_2, \acute{t}_2, \acute{u}_2, \acute{v}_2, \acute{w}_2, \acute{x}_2\right)\right\rangle,$

which can be considered as two octagonal neutrosophic numbers.
Then these following rules will be obtained:

3.1 Sum of ONNs' Sum

Let $\acute{\bar{n}}_1$ and $\acute{\bar{n}}_2$ be two ONNs; then the sum can be defined as:

$$
\begin{aligned}
\hat{\bar{n}}_1 + \hat{\bar{n}}_2 = \Big\{ &\left(\acute{a}_1 + \acute{a}_2 - \acute{a}_1\acute{a}_2, \acute{b}_1 + \acute{b}_2 - \acute{b}_1\acute{b}_2, \acute{c}_1 + \acute{c}_1 - \acute{c}_1\acute{c}_1, \acute{d}_1 + \acute{d}_2 - \acute{d}_1\acute{d}_2,\right. \\
&\left. \acute{e}_1 + \acute{e}_2 - \acute{e}_1\acute{e}_2, \acute{f}_1 + \acute{f}_2 - \acute{f}_1\acute{f}_2, \acute{g}_1 + \acute{g}_2 - \acute{g}_1\acute{g}_2, \acute{h}_1 + \acute{h}_2 - \acute{h}_1\acute{h}_2\right), \\
&\left(\acute{i}_1\acute{i}_2, \acute{j}_1\acute{j}_2, \acute{k}_1\acute{k}_2, \acute{l}_1\acute{l}_2, \acute{m}_1\acute{m}_2, \acute{n}_1\acute{n}_2, \acute{o}_1\acute{o}_2, \acute{p}_1\acute{p}_2\right), \\
&(\acute{q}_1\acute{q}_2, \acute{r}_1\acute{r}_2, \acute{s}_1\acute{s}_2, \acute{t}_1\acute{t}_2, \acute{u}_1\acute{u}_2, \acute{v}_1\acute{v}_2, \acute{w}_1\acute{w}_2, \acute{x}_1x_2) \Big\}
\end{aligned}
$$

Example 1 Consider $\acute{\bar{n}}_1$ and $\acute{\bar{n}}_2$ be two ONNs; then the sum will be:

$$\acute{\bar{n}}_1 = \langle (0.0, 0.1, 0.1, 0.2, 0.2, 0.2, 0.2, 0.2), (0.0, 0.1, 0.1, 0.2, 0.2, 0.3, 0.3, 0.4),$$
$$(0.0, 0.1, 0.2, 0.2, 0.2, 0.2, 0.3, 0.4)\rangle$$

$$\acute{\bar{n}}_2 = \langle (0.0, 0.1, 0.2, 0.2, 0.2, 0.3, 0.4, 0.4), (0.0, 0.1, 0.2, 0.2, 0.2, 0.4, 0.5, 0.5),$$
$$(0.1, 0.1, 0.2, 0.2, 0.2, 0.4, 0.5, 0.5)\rangle$$

$$\acute{\bar{n}}_1 + \acute{\bar{n}}_2 = \{(0, 0.19, 0.28, 0.36, 0.36, 0.44, 0.52), (0, 0.01, 0.02, 0.04, 0.04,$$
$$0.12, 0.15, 0.2), (0, 0.01, 0.04, 0.04, 0.04, 0.08, 0.15, 0.2)\}$$

3.2 Product of Two ONNs

Let $\acute{\tilde{n}}_1$ and $\acute{\tilde{n}}_2$ be two ONNs; then the product can be defined as:

$$\acute{\tilde{n}}_1 * \acute{\tilde{n}}_2 = \left\{ \left(\acute{a}_1\acute{a}_2, \acute{b}_1\acute{b}_2, \acute{c}_1\acute{c}_2, \acute{d}_1\acute{d}_2, \acute{e}_1\acute{e}_2, \acute{f}_1\acute{f}_2, \acute{g}_1\acute{g}_2, \acute{h}_1\acute{h}_2\right) \left(\acute{i}_1 + \acute{i}_2 - \acute{i}_1\acute{i}_2,\right.\right.$$

$$\acute{j}_1 + \acute{j}_2 - \acute{j}_1\acute{j}_2$$

$$\acute{k}_1 + \acute{k}_2 - \acute{k}_1\acute{k}_2, \acute{l}_1 + \acute{l}_2 - \acute{l}_1\acute{l}_2, \acute{m}_1 + \acute{m}_2 - \acute{m}_1\acute{m}_2, \acute{n}_1 + \acute{n}_2 - \acute{n}_1\acute{n}_2,$$

$$\acute{o}_1 + \acute{o}_2 - \acute{o}_1\acute{o}_2, \acute{p}_1 + \acute{p}_2 - \acute{p}_1\acute{p}_2), \left(\acute{q}_1 + \acute{q}_2 - \acute{q}_1\acute{q}_2, \acute{r}_1 + \acute{r}_2 - \acute{r}_1\acute{r}_2, \acute{s}_1 + \acute{s}_2 - \acute{s}_1\acute{s}_2,\right.$$

$$\left.\left.\acute{t}_1 + \acute{t}_2 - \acute{t}_1\acute{t}_2, \acute{u}_1 + \acute{u}_2 - \acute{u}_1\acute{u}_2, \acute{v}_1 + \acute{v}_2 - \acute{v}_1\acute{v}_2, \acute{w}_1 + \acute{w}_2 - \acute{w}_1\acute{w}_2, \acute{x}_1 + \acute{x}_2 - \acute{x}_1\acute{x}_2\right) \right\}$$

Example 2 Consider $\acute{\tilde{n}}_1$ and $\acute{\tilde{n}}_2$ are two ONNs; then the product will be:

$$\acute{\tilde{n}}_1 = \langle (0.0, 0.1, 0.1, 0.2, 0.2, 0.2, 0.2, 0.2), (0.0, 0.1, 0.1, 0.2, 0.2, 0.3, 0.3, 0.4),$$
$$(0.0, 0.1, 0.2, 0.2, 0.2, 0.2, 0.3, 0.4) \rangle$$

$$\acute{\tilde{n}}_2 = \langle (0.0, 0.1, 0.2, 0.2, 0.2, 0.3, 0.4, 0.4), (0.0, 0.1, 0.2, 0.2, 0.2, 0.4, 0.5, 0.5),$$
$$(0.1, 0.1, 0.2, 0.2, 0.2, 0.4, 0.5, 0.5) \rangle$$

$$\acute{\tilde{n}}_1 * \acute{\tilde{n}}_2 = \{(0, 0.01, 0.02, 0.04, 0.04, 0.06, 0.08, 0.08),$$
$$(0, 0.19, 0.28, 0.36, 0.36, 0.58, 0.65, 0.7),$$
$$(0.1, 0.19, 0.36, 0.36, 0.36, 0.52, 0.65, 0.7)\}$$

3.3 Scalar Product of Two ONNs

Let $\acute{\tilde{n}}_1$ be an ONN and λ be any scalar number; then the scalar product can be defined as:

$$\lambda \acute{\tilde{n}} = \left\{ \left(1 \dot{-} (1 \dot{-} \acute{a}_1)^\lambda, \left(1 - \left(1 \dot{-} \acute{b}_1\right)^\lambda, \left(1 - \left(1 \dot{-} \acute{c}_1\right)^\lambda, \left(1 - \left(1 \dot{-} \acute{d}_1\right)^\lambda,\right.\right.\right.\right.$$

$$\left(1 - \left(1 \dot{-} \acute{e}_1\right)^\lambda, \left(1 - \left(1 \dot{-} \acute{f}_1\right)^\lambda, \left(1 - \left(1 \dot{-} \acute{g}_1\right)^\lambda, \left(1 - \left(1 \dot{-} \acute{h}_1\right)^\lambda,\right.$$

$$\left.\left(\acute{i}_1^\lambda, \acute{j}_1^\lambda, \acute{k}_1^\lambda, \acute{l}_1^\lambda, \acute{m}_1^\lambda, \acute{n}_1^\lambda, \acute{o}_1^\lambda, \acute{p}_1^\lambda\right), \left(\acute{q}_1^\lambda, \acute{r}_1^\lambda, \acute{s}_1^\lambda, \acute{t}_1^\lambda, \acute{u}_1^\lambda, \acute{v}_1^\lambda, \acute{w}_1^\lambda, \acute{x}_1^\lambda\right) \right\} \quad \lambda > 0$$

Example 3 Consider, $\acute{\tilde{n}}_1$ is an ONN and λ is any scalar number; then the scalar product can be defined as:

$$\acute{\tilde{n}}_1 = \langle (0.0, 0.1, 0.1, 0.2, 0.2, 0.2, 0.2, 0.2), (0.0, 0.1, 0.1, 0.2, 0.2, 0.3, 0.3, 0.4),$$
$$(0.0, 0.1, 0.2, 0.2, 0.2, 0.2, 0.3, 0.4) \rangle$$

$$\lambda \acute{\tilde{n}}_1 = \left(0^\lambda, 0.1^\lambda, 0.1^\lambda, 0.2^\lambda, 0.2^\lambda, 0.2^\lambda, 0.2^\lambda, 0.2^\lambda \right), \left(0^\lambda, 0.1^\lambda, 0.1^\lambda, 0.2^\lambda, 0.2^\lambda, \right.$$
$$\left. 0.3^\lambda, 0.3^\lambda, 0.4^\lambda \right), 0^\lambda, 0.1^\lambda, 0.2^\lambda, 0.2^\lambda, 0.2^\lambda, 0.2^\lambda, 0.3^\lambda, 0.4^\lambda \right)$$

3.4 Power Rule of ONNs

Let $\acute{\tilde{n}}_1$ be an ONN and λ be any scalar number; then the scalar product can be defined as:

$$\acute{\tilde{n}}_1^\lambda = \{ (\acute{a}_1^\lambda, \acute{b}_1^\lambda, \acute{c}_1^\lambda, \acute{d}_1^\lambda, \acute{e}_1^\lambda, \acute{f}_1^\lambda, \acute{g}_1^\lambda, \acute{h}_1^\lambda), \left(1 - \left(1 - \acute{i}_1\right)^\lambda, \left(1 - \left(1 - \acute{j}_1\right)^\lambda, \right. \right.$$
$$\left(1 - \left(1 - \acute{k}_1\right)^\lambda, \left(1 - \left(1 - \acute{l}_1\right)^\lambda, \right.$$
$$\left(1 - \left(1 - \acute{m}_1\right)^\lambda, \left(1 - \left(1 - \acute{n}_1\right)^\lambda, \left(1 - \left(1 - \acute{o}_1\right)^\lambda, \left(1 - \left(1 - \acute{p}_1\right)^\lambda\right), \right.\right.$$
$$\left(\left(1 - \left(1 - \acute{q}_1\right)^\lambda, \left(1 - \left(1 - \acute{r}_1\right)^\lambda, \right.\right.$$
$$\left(1 - \left(1 - \acute{s}_1\right)^\lambda, \left(1 - \left(1 - \acute{t}_1\right)^\lambda, \left(1 - \left(1 - \acute{u}_1\right)^\lambda, \left(1 - \left(1 - \acute{v}_1\right)^\lambda, \right.\right.$$
$$\left(1 - \left(1 - \acute{w}_1\right)^\lambda\right), \left(1 - \left(1 - \acute{x}_1\right)^\lambda\right) \}$$

$$\tilde{\lambda} > 0$$

Example 4 Consider $\acute{\tilde{n}}_1$ is an ONN and λ be any scalar number (taken as power); then the scalar product can be defined as:

$$\acute{\tilde{n}}_1 = \langle (0.0, 0.1, 0.1, 0.2, 0.2, 0.2, 0.2, 0.2), (0.0, 0.1, 0.1, 0.2, 0.2, 0.3, 0.3, 0.4),$$
$$(0.0, 0.1, 0.2, 0.2, 0.2, 0.2, 0.3, 0.4) \rangle$$

$$\acute{\tilde{n}}_1^\lambda = \left(0^\lambda, 0.1^\lambda, 0.1^\lambda, 0.2^\lambda, 0.2^\lambda, 0.2^\lambda, 0.2^\lambda, 0.2^\lambda\right), \ \left(0^\lambda, 0.1^\lambda, 0.1^\lambda, 0.2^\lambda, 0.2^\lambda,\right.$$
$$\left.0.3^\lambda, 0.3^\lambda, 0.4^\lambda\right), \left(0^\lambda, 0.1^\lambda, 0.2^\lambda, 0.2^\lambda, 0.2^\lambda, 0.2^\lambda, 0.3^\lambda, 0.4^\lambda\right)$$

3.5 Score Function of ONNs

Let $\acute{\tilde{n}}_1$ be an ONN; then the score function $\acute{\tilde{S}}(\tilde{n})$ can be defined as:

Let $\acute{\tilde{n}} = \left\langle \left(\acute{a}, \acute{b}, \acute{c}, \acute{d}, \acute{e}, \acute{f}, \acute{g}, \acute{h}\right), \left(\acute{i}, \acute{j}, \acute{k}, \acute{l}, m, \acute{n}, \acute{o}, \acute{p}\right), \left(\acute{q}, \acute{r}, \acute{s}, \acute{t}, u, \acute{v}, \acute{w}, \acute{x}\right)\right\rangle$
be an octagonal neutrosophic number; then score function of octagonal neutrosophic number is given as:

$$\acute{\tilde{S}}\left(\tilde{n}\right) = \frac{1}{3}\left(2 + \frac{\acute{a} + \acute{b} + \acute{c} + \acute{d} + \acute{e} + \acute{f} + \acute{g} + \acute{h}}{8} - \frac{\acute{i} + \acute{j} + \acute{k} + \acute{l} + \acute{m} + \acute{n} + \acute{o} + \acute{p}}{8}\right.$$
$$\left. - \frac{\acute{q} + \acute{r} + \acute{s} + \acute{t} + \acute{u} + \acute{v} + \acute{w} + \acute{x}}{8}\right), \acute{\tilde{S}}\left(\acute{\tilde{n}}\right) \in [0, 1],$$

The bigger value of $\acute{\tilde{S}}\left(\acute{n}\right)$, the bigger the octagonal neutrosophic number $\tilde{n}$. As well as when If $\acute{b} = \acute{c} = \acute{d}, \acute{e} = \acute{f} = \acute{g}, \ \acute{j} = \acute{k} = \acute{l}, \ \acute{m} = \acute{n} = \acute{o}$ and $\acute{r} = \acute{s} = \acute{t}, \ \acute{u} = \acute{v} = \acute{w}$, then

$$\acute{\tilde{S}}\left(\acute{n}\right) = \frac{1}{3}\left(2 + \frac{\acute{a} + 3\acute{b} + 3\acute{e} + \acute{h}}{8} - \frac{\acute{i} + 3\acute{j} + 3\acute{m} + \acute{p}}{8} - \frac{\acute{q} + 3\acute{r} + 3\acute{u} + \acute{x}}{8}\right),$$

$$\acute{\tilde{S}}\left(\acute{n}\right) \in [0, 1].$$

3.6 Accuracy Function of ONNs

Let $\acute{\tilde{n}}_1$ be an ONN; then the score function $\acute{H}(\tilde{n})$ can be defined as:

Let $\acute{\tilde{n}} = \left\langle \left(\acute{a}, \acute{b}, \acute{c}, \acute{d}, \acute{e}, \acute{f}, \acute{g}, \acute{h}\right), \left(\acute{i}, \acute{j}, \acute{k}, \acute{l}, \acute{m}, \acute{n}, \acute{o}, \acute{p}\right), \left(\acute{q}, \acute{r}, \acute{s}, \acute{t}, \acute{u}, \acute{v}, \acute{w}, \acute{x}\right)\right\rangle$
be an octagonal neutrosophic number; then accuracy function of octagonal neutrosophic number is given as:

$$\acute{H}\left(\tilde{n}\right) = \frac{1}{3}\left(2 + \frac{\acute{a}+\acute{b}+\acute{c}+\acute{d}+\acute{e}+\acute{f}+\acute{g}+\acute{h}}{8} - \frac{\acute{i}+\acute{j}+\acute{k}+\acute{l}+\acute{m}+\acute{n}+\acute{o}+\acute{p}}{8}\right.$$

$$\left. - \frac{\acute{q}+\acute{r}+\acute{s}+\acute{t}+\acute{u}+\acute{v}+\acute{w}+\acute{x}}{8}\right) \acute{H}\left(\acute{n}\right) \in [-1, 1],$$

The bigger value of $\acute{H}\left(\tilde{n}\right)$, the bigger the accuracy degree of octagonal neutrosophic number $\tilde{n}$. As well as, if $\acute{b} = \acute{c} = \acute{d}$, $\mathrm{e} = \acute{f} = \acute{g}$, $\acute{j} = \acute{k} = \acute{l}$, $\acute{m} = \acute{n} = \acute{o}$ and $\acute{r} = \acute{s} = \acute{t}$, $\acute{u} = \acute{v} = \acute{w}$, then,

$$\acute{H}\left(\acute{n}\right) = \frac{1}{3}\left(2 + \frac{\acute{a}+3\acute{b}+3\acute{e}+\acute{h}}{8} - \frac{\acute{i}+3\acute{j}+3\acute{m}+\acute{p}}{8} - \frac{\acute{q}+3\acute{r}+3\acute{u}+\acute{x}}{8}\right),$$

$$\acute{H}\left(\acute{n}\right) \in [-1, 1].$$

3.7   Metric Relations between Two ONNs

Let
$$\tilde{n}_1 = \left\langle \left(\acute{a}_1, \acute{b}_1, \acute{c}_1, \acute{d}_1, \acute{e}_1, \acute{f}_1, \acute{g}_1, \acute{h}_1\right), \left(\acute{i}_1, \acute{j}_1, \acute{k}_1, \acute{l}_1, \acute{m}_1, \acute{n}_1, \acute{o}_1, \acute{p}_1\right), \right.$$
$$\left. \left(\acute{q}_1, \acute{r}_1, \acute{s}_1, \acute{t}_1, \acute{u}_1, \acute{v}_1, \acute{w}_1, \acute{x}_1\right) \right\rangle$$
and
$$\tilde{n}_2 = \left\langle \left(\acute{a}_2, \acute{b}_2, \acute{c}_2, \acute{d}_2, \acute{e}_2, \acute{f}_2, \acute{g}_2, \acute{h}_2\right), \left(\acute{i}_2, \acute{j}_2, \acute{k}_2, \acute{l}_2, \acute{m}_2, \acute{n}_2, \acute{o}_2, \acute{p}_2\right), \right.$$
$$\left. \left(\acute{q}_2, \acute{r}_2, \acute{s}_2, \acute{t}_2, \acute{u}_2, \acute{v}_2, \acute{w}_2, \acute{x}_2\right) \right\rangle,$$

which can be considered as two octagonal neutrosophic numbers. Thus $\acute{S}(\tilde{n}_1)$ and $\acute{S}(\tilde{n}_2)$ are two scores of $\tilde{n}_1$ and $\tilde{n}_2$ and $\acute{H}\left(\tilde{n}_1\right)$ and $\acute{H}(\tilde{n}_2)$ are two accuracy degree of $\tilde{n}_1$ and $\tilde{n}_2$; then the relation between two octagonal neutrosophic number is defined as:

1. If $\acute{S}\left(\tilde{n}_1\right) > \acute{S}\left(\tilde{n}_2\right)$, then $\tilde{n}_1 > \tilde{n}_2$;
2. If $\acute{S}\left(\tilde{n}_1\right) = \acute{S}(\tilde{n}_2)$, and

   - If $\acute{H}\left(\tilde{n}_1\right) = \acute{H}(\tilde{n}_2)$, then $\tilde{n}_1 = \tilde{n}_2$;
   - If $\acute{H}\left(\tilde{n}_1\right) > \acute{h}\left(\tilde{n}_2\right)$, then $\tilde{n}_1 > \tilde{n}_2$;

4 Weighted Operators

4.1 Weighted Arithmetic Averaging Operator

Definition 4.1 Let $\hat{n}_j = \left\langle \left(\acute{a}_j, \acute{b}_j, \acute{c}_j, \acute{d}_j, \acute{e}_j, \acute{f}_j, \acute{g}_j, \acute{h}_j \right), \right.$

$\left. \left(\acute{i}_j, \acute{j}_j, \acute{k}_j, \acute{l}_j, \acute{m}_j, \acute{n}_j, \acute{o}_j, \acute{p}_j \right) \left(\acute{q}_j, \acute{r}_j, \acute{s}_j, \acute{t}_j, \acute{u}_j, \acute{v}_j, \acute{w}_j, \acute{x}_j \right) \right\rangle (j = 1, 2, \ldots, n)$

be the collection of octagonal neutrosophic numbers. Then octagonal neutrosophic number weighted arithmetic averaging operator (ONNWAA = P) is defined as:

$$\text{P}(\hat{n}_1, \hat{n}_2, \ldots, \hat{n}_n) = \acute{w}_1 \hat{n}_1 + \acute{w}_2 \hat{n}_2 + \ldots + \acute{w}_n \hat{n}_n = +_{j=1}^{n}(\acute{w}_j \hat{n}_j),$$

where $\acute{w}_j$ ($j = 1, 2, 3, \ldots, n$) is the weight of jth octagonal neutrosophic number $\hat{n}_j \left(j = 1, 2, 3, \ldots, \hat{n}_{.} \right)$ with $\acute{w}_j \in [0, 1]$ and $\sum_{j=1}^{n} \acute{w}_j = 1$.

Proof

$$\acute{w}_1 \hat{n}_1 = \left\{ \left(1 \dot{-} (1 \dot{-} \acute{a}_1)^{\acute{w}_1} \right), \left(1 - \left(1 \dot{-} \acute{b}_1 \right)^{\acute{w}_1} \right), \left(1 - \left(1 \dot{-} \acute{c}_1 \right)^{\acute{w}_1} \right), \right.$$

$$\left(1 - \left(1 \dot{-} \acute{d}_1 \right)^{\acute{w}_1} \right), \left(1 - \left(1 \dot{-} \acute{e}_1 \right)^{\acute{w}_1} \right)$$

$$\left(1 - \left(1 \dot{-} \acute{f}_1 \right)^{\acute{w}_1} \right), \left(1 - \left(1 \dot{-} \acute{g}_1 \right)^{\acute{w}_1} \right), \left(1 - \left(1 \dot{-} \acute{h}_1 \right)^{\acute{w}_1} \right),$$

$$\left(\acute{i}_1^{\acute{w}_1}, \acute{j}_1^{\acute{w}_1}, \acute{k}_1^{\acute{w}_1}, \acute{l}_1^{\acute{w}_1}, \acute{m}_1^{\acute{w}_1}, \acute{n}_1^{\acute{w}_1}, \acute{o}_1^{\acute{w}_1}, \acute{p}_1^{\acute{w}_1} \right),$$

$$\left. \left(\acute{q}_1^{\acute{w}_1}, \acute{r}_1^{\acute{w}_1}, \acute{s}_1^{\acute{w}_1}, \acute{t}_1^{\acute{w}_1}, \acute{u}_1^{\acute{w}_1}, \acute{v}_1^{\acute{w}_1}, \acute{w}_1^{\acute{w}_1}, \acute{x}_1^{\acute{w}_1} \right) \right\}$$

$$\acute{w}_2 \tilde{n} = \left\{ \left(1 \dot{-} (1 \dot{-} \acute{a}_2)^{\acute{w}_2} \right), \left(1 - \left(1 \dot{-} \acute{b}_2 \right)^{\acute{w}_2} \right), \left(1 - \left(1 \dot{-} \acute{c}_2 \right)^{\acute{w}_2} \right), \right.$$

$$\left(1 - \left(1 \dot{-} \acute{d}_2 \right)^{\acute{w}_2} \right), \left(1 - \left(1 \dot{-} \acute{e}_2 \right)^{\acute{w}_2} \right),$$

$$\left(1 - \left(1 \dot{-} \acute{f}_2 \right)^{\acute{w}_2} \right), \left(1 - \left(1 \dot{-} \acute{g}_2 \right)^{\acute{w}_2} \right), \left(1 - \left(1 \dot{-} \acute{h}_2 \right)^{\acute{w}_2} \right),$$

$$\left(\acute{i}_2^{\acute{w}_2}, \acute{j}_2^{\acute{w}_2}, \acute{k}_2^{\acute{w}_2}, \acute{l}_2^{\acute{w}_2}, \acute{m}_2^{\acute{w}_2}, \acute{n}_2^{\acute{w}_2}, \acute{o}_2^{\acute{w}_2}, \acute{p}_2^{\acute{w}_2} \right),$$

$$\left. \left(\acute{q}_2^{\acute{w}_2}, \acute{r}_2^{\acute{w}_2}, \acute{s}_2^{\acute{w}_2}, \acute{t}_2^{\acute{w}_2}, \acute{u}_2^{\acute{w}_2}, \acute{v}_2^{\acute{w}_2}, \acute{w}_2^{\acute{w}_2}, \acute{x}_2^{\acute{w}_2} \right) \right\}$$

Thus, $P\left(\hat{n}_1, \hat{n}_2\right)$ given as:

$$
\acute{w}_1\hat{n}_1 + \acute{w}_2\hat{n}_2 = \left\{\left(\left(1 - (1\dot{-}\acute{a}_2)^{\acute{w}_1}\right) + \left(1 - (1\dot{-}\acute{a}_2)^{\acute{w}_2}\right) - \left(1 - \left(1\dot{-}\acute{a}_1\right)^{\acute{w}_1}\right)\right.\right.
$$
$$
\left(1 - \left(1\dot{-}\acute{a}_2\right)^{\acute{w}_2}\right)\Bigg),
$$

$$
\left(\left(1 - \left(1\dot{-}\acute{b}_1\right)^{\acute{w}_1}\right) + \left(1 - \left(1\dot{-}\acute{b}_2\right)^{\acute{w}_2}\right) - \left(1 - \left(1\dot{-}\acute{b}_1\right)^{\acute{w}_1}\right)\right.
$$
$$
\left(1 - \left(1\dot{-}\acute{b}_2\right)^{\acute{w}_2}\right)\Bigg),
$$

$$
\left(\left(1 - \left(1\dot{-}\acute{c}_1\right)^{\acute{w}_1}\right) + \left(1 - \left(1\dot{-}\acute{c}_2\right)^{\acute{w}_2}\right) - \left(1 - \left(1\dot{-}\acute{c}_1\right)^{\acute{w}_1}\right)\right.
$$
$$
\left(1 - \left(1\dot{-}\acute{c}_2\right)^{\acute{w}_2}\right)\Bigg),
$$

$$
\left(\left(1 - \left(1\dot{-}\acute{d}_1\right)^{\acute{w}_1}\right) + \left(1 - \left(1\dot{-}\acute{d}_2\right)^{\acute{w}_2}\right) - \left(1 - \left(1\dot{-}\acute{d}_1\right)^{\acute{w}_1}\right)\right.
$$
$$
\left(1 - \left(1\dot{-}\acute{d}_2\right)^{\acute{w}_2}\right)\Bigg),
$$

$$
\left(\left(1 - \left(1\dot{-}\acute{e}_1\right)^{\acute{w}_1}\right) + \left(1 - \left(1\dot{-}\acute{e}_2\right)^{\acute{w}_2}\right) - \left(1 - \left(1\dot{-}\acute{e}_1\right)^{\acute{w}_1}\right)\right.
$$
$$
\left(1 - \left(1\dot{-}\acute{e}_2\right)^{\acute{w}_2}\right)\Bigg),
$$

$$
\left(\left(1 - \left(1\dot{-}\acute{f}_1\right)^{\acute{w}_1}\right) + \left(1 - \left(1\dot{-}\acute{f}_2\right)^{\acute{w}_2}\right) - \left(1 - \left(1\dot{-}\acute{f}_1\right)^{\acute{w}_1}\right)\right.
$$
$$
\left(1 - \left(1\dot{-}\acute{f}_2\right)^{\acute{w}_2}\right)\Bigg),
$$

$$
\left(\left(1 - \left(1\dot{-}\acute{g}_1\right)^{\acute{w}_1}\right) + \left(1 - \left(1\dot{-}\acute{g}_2\right)^{\acute{w}_2}\right) - \left(1 - \left(1\dot{-}\acute{g}_1\right)^{\acute{w}_1}\right)\right.
$$
$$
\left(1 - \left(1\dot{-}\acute{g}_2\right)^{\acute{w}_2}\right)\Bigg),
$$

$$
\left(\left(1 - \left(1\dot{-}\acute{h}_1\right)^{\acute{w}_1}\right) + \left(1 - \left(1\dot{-}\acute{h}_2\right)^{\acute{w}_2}\right) - \left(1 - \left(1\dot{-}\acute{h}_1\right)^{\acute{w}_1}\right)\right.
$$
$$
\left(1 - \left(1\dot{-}\acute{h}_2\right)^{\acute{w}_2}\right)\Bigg),
$$

$$\left(\acute{i}_1^{\acute{w}_1}\acute{i}_2^{\acute{w}_2}, \acute{j}_1^{\acute{w}_1}\acute{j}_2^{\acute{w}_2}, \acute{k}_1^{\acute{w}_1}\acute{k}_2^{\acute{w}_2}, \acute{l}_1^{\acute{w}_1}\acute{l}_2^{\acute{w}_2}, \acute{m}_1^{\acute{w}_1}\acute{m}_2^{\acute{w}_2}, \acute{n}_1^{\acute{w}_1}\acute{n}_2^{\acute{w}_2}, \acute{o}_1^{\acute{w}_1}\acute{o}_2^{\acute{w}_2}, \acute{p}_1^{\acute{w}_1}\acute{p}_2^{\acute{w}_2}\right),$$

$$\left(\acute{q}_1^{\acute{w}_1}\acute{q}_2^{\acute{w}_2}, \acute{r}_1^{\acute{w}_1}\acute{r}_2^{\acute{w}_2}, \acute{s}_1^{\acute{w}_1}\acute{s}_2^{\acute{w}_2}, \acute{t}_1^{\acute{w}_1}\acute{t}_2^{\acute{w}_2}, \acute{u}_1^{\acute{w}_1}\acute{u}_2^{\acute{w}_2}, \acute{v}_1^{\acute{w}_1}\acute{v}_2^{\acute{w}_2}, \acute{w}_1^{\acute{w}_1}\acute{w}_2^{\acute{w}_2}, \acute{x}_1^{\acute{w}_1}\acute{x}_2^{\acute{w}_2}\right)\right\}$$

$$= \left(1-\left(1\dot{-}\acute{a}_1\right)^{\acute{w}_1}\right)1-\left(1\dot{-}\acute{a}_2\right)^{\acute{w}_2}, \left(1-\left(1\dot{-}\acute{b}_1\right)^{\acute{w}_1}\right)\left(1-\left(1\dot{-}\acute{b}_2\right)^{\acute{w}_2}\right),$$

$$\left(1-\left(1\dot{-}\acute{c}_1\right)^{\acute{w}_1}\right)\left(1-\left(1\dot{-}\acute{c}_2\right)^{\acute{w}_2}\right), \left(1-\left(1\dot{-}\acute{d}_1\right)^{\acute{w}_1}, \left(1-\left(1\dot{-}\acute{d}_2\right)^{\acute{w}_2}\right)\right),$$

$$\left(1-\left(1\dot{-}\acute{e}_1\right)^{\acute{w}_1}\right)\left(1-\left(1\dot{-}\acute{e}_2\right)^{\acute{w}_2}\right),$$

$$\left(1-\left(1\dot{-}\acute{f}_1\right)^{\acute{w}_1}\right)\left(1-\left(1\dot{-}\acute{f}_2\right)^{\acute{w}_2}\right)\left(1-\left(1\dot{-}\acute{g}_1\right)^{\acute{w}_1}\right)\left(1-\left(1\dot{-}\acute{g}_2\right)^{\acute{w}_2}\right),$$

$$\left(1-\left(1\dot{-}\acute{h}_1\right)^{\acute{w}_1}\right)\left(1-\left(1\dot{-}\acute{h}_2\right)^{\acute{w}_2}\right)$$

$$\left(\prod_{j=1}^{2}\acute{i}_j^{\acute{w}_j}, \prod_{j=1}^{2}\acute{j}_j^{\acute{w}_j}, \prod_{j=1}^{2}\acute{k}_j^{\acute{w}_j}, \prod_{j=1}^{2}\acute{l}_j^{\acute{w}_j}, \prod_{j=1}^{2}\acute{m}_j^{\acute{w}_j}, \prod_{j=1}^{2}\acute{n}_j^{\acute{w}_j},\right.$$

$$\left.\prod_{j=1}^{2}\acute{o}_j^{\acute{w}_j}, \prod_{j=1}^{2}\acute{p}_j^{\acute{w}_j}\right)$$

$$\left(\prod_{j=1}^{2}\acute{q}_j^{\acute{w}_j}, \prod_{j=1}^{2}\acute{r}_j^{\acute{w}_j}, \prod_{j=1}^{2}\acute{s}_j^{\acute{w}_j}, \prod_{j=1}^{2}\acute{t}_j^{\acute{w}_j}, \prod_{j=1}^{2}\acute{u}_j^{\acute{w}_j}, \prod_{j=1}^{2}\acute{v}_j^{\acute{w}_j},\right.$$

$$\left.\prod_{j=1}^{2}\acute{w}_j^{\acute{w}_j}, \prod_{j=1}^{2}\acute{x}_j^{\acute{w}_j}\right)$$

Theorem 1 Let $\tilde{n}_j = \left\langle\left(\acute{a}_j, \acute{b}_j, \acute{c}_j, \acute{d}_j, \acute{e}_j, \acute{f}_j, \acute{g}_j, \acute{h}_j\right)\left(\acute{i}_j, \acute{j}_j, \acute{k}_j, \acute{l}_j, \acute{m}_j, \acute{n}_j, \acute{o}_j, \acute{p}_j\right),\right.$ $\left.\left(\acute{q}_j, \acute{r}_j, \acute{s}_j, \acute{t}_j, \acute{u}_j, \acute{v}_j, \acute{w}_j, \acute{x}_j\right)\right\rangle$ $(j = 1, 2, \ldots, n)$ be the collection of octagonal neutrosophic number, and the aggregate value by the help of 'Ɐ' operator is also an octagonal neutrosophic number; then

$$Ɐ\left(\tilde{n}_1, \tilde{n}_2, \ldots, \tilde{n}_n\right) = \acute{w}_1\tilde{n}_1\,\acute{w}_n\tilde{n}_n \ldots \acute{w}_n\tilde{n}_n = +_{j=1}^{\acute{n}}\left(\ddot{w}_j\tilde{n}_j\right)$$

$$= \left\{\left(1-\prod_{j=1}^{n}\left(1-\acute{a}_j\right)^{\acute{w}_j}\right), 1-\prod_{j=1}^{n}\left(1-\acute{b}_j\right)^{\acute{w}_j}\right),$$

$$1-\prod_{j=1}^{n}\left(1-\acute{c}_j\right)^{\acute{w}_j}\right), 1-\prod_{j=1}^{n}\left(1-\acute{d}_j\right)^{\acute{w}_j}\right)$$

$$1 - \prod_{j=1}^{n}\left(1 - \acute{e}_j\right)^{\acute{w}_j}\Big), 1 - \prod_{j=1}^{n}\left(1 - \acute{f}_j\right)^{\acute{w}_j}, 1 - \prod_{j=1}^{n}\left(1 - \acute{g}_j\right)^{\acute{w}_j} 0, 1$$

$$- \prod_{j=1}^{n}\left(1 - \acute{h}_j\right)^{\acute{w}_j}\Big),$$

$$* \left(\prod_{j=1}^{n}\acute{i}_j^{\acute{w}_j}, \prod_{j=1}^{n}\acute{j}_j^{\acute{w}_j}, \prod_{j=1}^{n}\acute{k}_j^{\acute{w}_j}, \prod_{j=1}^{n}\acute{l}_j^{\acute{w}_j}, \prod_{j=1}^{n}\acute{m}_j^{\acute{w}_j}, \prod_{j=1}^{n}\acute{n}_j^{\acute{w}_j}, \right.$$

$$\left. \prod_{j=1}^{n}\acute{o}_j^{\acute{w}_j}, \prod_{j=1}^{n}\acute{p}_j^{\acute{w}_j}\right),$$

$$\left(\prod_{j=1}^{n}\acute{q}_j^{\acute{w}_j}, \prod_{j=1}^{n}\acute{r}_j^{\acute{w}_j}, \prod_{j=1}^{n}\acute{s}_j^{\acute{w}_j}, \prod_{j=1}^{n}\acute{t}_j^{\acute{w}_j}, \prod_{j=1}^{n}\acute{u}_j^{\acute{w}_j}, \prod_{j=1}^{n}\acute{v}_j^{\acute{w}_j}\right.$$

$$\left. \prod_{j=1}^{n}\acute{w}_j^{\acute{w}_j}, \prod_{j=1}^{n}\acute{x}_j^{\acute{w}_j}\right)\Big\}$$

If n = k, then equation becomes:

$$\mathsf{P}\left(\tilde{n}_1, \tilde{n}_2, \ldots, \tilde{n}_k\right) = \ddot{w}_1\tilde{n}_1 \quad \ddot{w}_2\tilde{n}_2 \ldots \ldots \ddot{w}_k\tilde{n}_k = +_{j=1}^{\tilde{n}}\left(\ddot{w}_j\tilde{n}_j\right)$$

$$= \left\{\left(1 - \prod_{j=1}^{k}\left(1 - \acute{a}_j\right)^{\acute{w}_j}\right), \ 1 - \prod_{j=1}^{k}\left(1 - \acute{b}_j\right)^{\acute{w}_j}\right),$$

$$1 - \prod_{j=1}^{k}\left(1 - \acute{c}_j\right)^{\acute{w}_j}, 1 - \prod_{j=1}^{k}\left(1 - \acute{d}_j\right)^{\acute{w}_j}\right)$$

$$1 - \prod_{j=1}^{k}\left(1 - \acute{e}_j\right)^{\acute{w}_j}\Big), 1 - \prod_{j=1}^{k}\left(1 - \acute{f}_j\right)^{\acute{w}_j}\Big),$$

$$1 - \prod_{j=1}^{k}\left(1 - \acute{g}_j\right)^{\acute{w}_j}\Big), 1 - \prod_{j=1}^{k}\left(1 - \acute{h}_j\right)^{\acute{w}_j}\Big),$$

$$* \left(\prod_{j=1}^{n}\acute{i}_j^{\acute{w}_j}, \prod_{j=1}^{n}\acute{j}_j^{\acute{w}_j}, \prod_{j=1}^{n}\acute{k}_j^{\acute{w}_j}, \prod_{j=1}^{n}\acute{l}_j^{\acute{w}_j}, \right.$$

$$\left. \prod_{j=1}^{n}\acute{m}_j^{\acute{w}_j}, \prod_{j=1}^{n}\acute{n}_j^{\acute{w}_j}, \prod_{j=1}^{n}\acute{o}_j^{\acute{w}_j}, \prod_{j=1}^{n}\acute{p}_j^{\acute{w}_j}\right),$$

$$\left(\prod_{j=1}^{n}\acute{q}_j^{\acute{w}_j}, \prod_{j=1}^{n}\acute{r}_j^{\acute{w}_j}, \prod_{j=1}^{n}\acute{s}_j^{\acute{w}_j}, \prod_{j=1}^{n}\acute{t}_j^{\acute{w}_j}, \prod_{j=1}^{n}\acute{u}_j^{\acute{w}_j}, \right.$$

$$\left. \prod_{j=1}^{n}\acute{v}_j^{\acute{w}_j} \prod_{j=1}^{n}\acute{w}_j^{\acute{w}_j}, \prod_{j=1}^{n}\acute{x}_j^{\acute{w}_j}\right)\Big\}$$

When n = k + 1, then equation becomes:
$$\mathsf{P}\left(\tilde{n}_1, \tilde{n}_2, \ldots, \tilde{n}_k + 1\right) \text{ given as:}$$

$$= \left\{ \left(1 - \prod_{j=1}^{k} \cdot \left(1 \dot{-} \acute{a}_j\right)^{\acute{w}_j} \right) + \left(1 - \left(1 \dot{-} \acute{a}_{k+1}\right)^{w_{k+1}'} \right) \right.$$

$$- \left(1 - \prod_{j=1}^{k} \cdot \left(1 \dot{-} \acute{a}_j\right)^{\acute{w}_j} \right) \left(1 - \left(1 \dot{-} a_{k+1}'\right)^{w_{k+1}'} \right),$$

$$\left(\left(1 - \prod_{j=1}^{k} \cdot \left(1 \dot{-} \acute{b}_j\right)^{\acute{w}_j} \right) + \left(1 - \left(1 \dot{-} \acute{b}_{k+1}\right)^{w_{k+1}'} \right) \right.$$

$$- \left(1 - \prod_{j=1}^{k} \cdot \left(1 \dot{-} \acute{b}_j\right)^{\acute{w}_j} \right) \left(1 - \left(1 \dot{-} b_{k+1}'\right)^{w_{k+1}'} \right),$$

$$\left(\left(1 - \prod_{j=1}^{k} \cdot \left(1 \dot{-} \acute{c}_j\right)^{\acute{w}_j} \right) + \left(1 - \left(1 \dot{-} c_{k+1}'\right)^{w_{k+1}'} \right) \right.$$

$$- \left(1 - \prod_{j=1}^{k} \cdot \left(1 \dot{-} \acute{c}_j\right)^{\acute{w}_j} \right) \left(1 - \left(1 \dot{-} c_{k+1}'\right)^{w_{k+1}'} \right),$$

$$\left(\left(1 - \prod_{j=1}^{k} \cdot \left(1 \dot{-} \acute{d}_j\right)^{\acute{w}_j} \right) + \left(1 - \left(1 \dot{-} d_{k+1}'\right)^{w_{k+1}'} \right) \right.$$

$$- \left(1 - \prod_{j=1}^{k} \cdot \left(1 \dot{-} \acute{d}_j\right)^{\acute{w}_j} \right) \left(1 - \left(1 \dot{-} d_{k+1}'\right)^{w_{k+1}} \right),$$

$$\left(\left(1 - \prod_{j=1}^{k} \cdot \left(1 \dot{-} \acute{e}_j\right)^{\acute{w}_j} \right) + \left(1 - \left(1 \dot{-} e_{k+1}'\right)^{w_{k+1}'} \right) \right.$$

$$- \left(1 - \prod_{j=1}^{k} \cdot \left(1 \dot{-} \acute{e}_j\right)^{\acute{w}_j} \right) \left(1 - \left(1 \dot{-} e_{k+1}'\right)^{w_{k+1}'} \right),$$

$$\left(\left(1 - \prod_{j=1}^{k} \cdot \left(1 \dot{-} \acute{f}_j\right)^{\acute{w}_j} \right) + \left(1 - \left(1 \dot{-} f_{k+1}'\right)^{w_{k+1}'} \right) \right.$$

$$- \left(1 - \prod_{j=1}^{k} \cdot \left(1 \dot{-} \acute{f}_j\right)^{\acute{w}_j} \right) \left(1 - \left(1 \dot{-} f_{k+1}'\right)^{w_{k+1}'} \right),$$

$$\left(\left(1 - \prod_{j=1}^{k} \cdot \left(1 \dot{-} \acute{g}_j\right)^{\acute{w}_j} \right) + \left(1 - \left(1 \dot{-} g_{k+1}'\right)^{w_{k+1}'} \right) \right.$$

$$- \left(1 - \prod_{j=1}^{k} \cdot \left(1 \dot{-} \acute{g}_j\right)^{\acute{w}_j} \right) \left(1 - \left(1 \dot{-} g_{k+1}'\right)^{w_{k+1}'} \right),$$

$$\left(\left(\left(1-\prod\nolimits_{j=1}^{k}\cdot\left(1\dot{-}\acute{h_j}\right)^{\acute{w}_j}\right)+\left(1-\left(1\dot{-}\acute{h}_{k+1}\right)^{\acute{w}_{k+1}}\right)\right.$$

$$\left.-\left(1-\prod\nolimits_{j=1}^{k}\cdot\left(1\dot{-}\acute{h_j}\right)^{\acute{w}_j}\right)\left(1-\left(1\dot{-}h_{k+1}^{\prime}\right)^{w_{k+1}^{\prime}}\right),\right.$$

$$\left(\prod\nolimits_{j=1}^{k+1}\acute{i}_j^{\acute{w}_j},\prod\nolimits_{j=1}^{k+1}\acute{j}_j^{\acute{w}_j},\prod\nolimits_{j=1}^{k+1}\acute{k}_j^{\acute{w}_j},\prod\nolimits_{j=1}^{k+1}\acute{l}_j^{\acute{w}_j},\right.$$

$$\left.\prod\nolimits_{j=1}^{k+1}\acute{m}_j^{\acute{w}_j},\prod\nolimits_{j=1}^{k+1}\acute{n}_j^{\acute{w}_j},\prod\nolimits_{j=1}^{k+1}\acute{o}_j^{\acute{w}_j},\prod\nolimits_{j=1}^{k+1}\acute{p}_j^{\acute{w}_j}\right),$$

$$\left(\prod\nolimits_{j=1}^{k+1}\acute{q}_j^{\acute{w}_j},\prod\nolimits_{j=1}^{k+1}\acute{r}_j^{\acute{w}_j},\prod\nolimits_{j=1}^{k+1}\acute{s}_j^{\acute{w}_j},\prod\nolimits_{j=1}^{k+1}\acute{t}_j^{\acute{w}_j},\right.$$

$$\left.\prod\nolimits_{j=1}^{k+1}\acute{u}_j^{\acute{w}_j},\prod\nolimits_{j=1}^{k+1}\acute{v}_j^{\acute{w}_j}\prod\nolimits_{j=1}^{k+1}\acute{w}_j^{\acute{w}_j},\prod\nolimits_{j=1}^{k+1}\acute{x}_j^{\acute{w}_j}\right)$$

$$=\left\{\left(1-\prod\nolimits_{j=1}^{k}\cdot\left(1\dot{-}\acute{a}_j\right)^{\acute{w}_j},1-\prod\nolimits_{j=1}^{k}\cdot\left(1\dot{-}\acute{b}_j\right)^{\acute{w}_j},\ ,1-\prod\nolimits_{j=1}^{k}\cdot\left(1\dot{-}\acute{c}_j\right)^{\acute{w}_j},\right.\right.$$

$$1-\prod\nolimits_{j=1}^{k}\cdot\left(1\dot{-}\acute{d}_j\right)^{\acute{w}_j},1-\prod\nolimits_{j=1}^{k}\cdot\left(1\dot{-}\acute{e}_j\right)^{\acute{w}_j},$$

$$\left.1-\prod\nolimits_{j=1}^{k}\cdot\left(1\dot{-}\acute{f}_j\right)^{\acute{w}_j},1-\prod\nolimits_{j=1}^{k}\cdot\left(1\dot{-}\acute{g}_j\right)^{\acute{w}_j},1-\prod\nolimits_{j=1}^{k}\cdot\left(1\dot{-}\acute{h}_j\right)^{\acute{w}_j}\right),$$

$$\left(\prod\nolimits_{j=1}^{k+1}\acute{i}_j^{\acute{w}_j},\prod\nolimits_{j=1}^{k+1}\acute{j}_j^{\acute{w}_j},\prod\nolimits_{j=1}^{k+1}\acute{k}_j^{\acute{w}_j},\prod\nolimits_{j=1}^{k+1}\acute{l}_j^{\acute{w}_j},\prod\nolimits_{j=1}^{k+1}\acute{m}_j^{\acute{w}_j},\prod\nolimits_{j=1}^{k+1}\acute{n}_j^{\acute{w}_j},\right.$$

$$\left.\prod\nolimits_{j=1}^{k+1}\acute{o}_j^{\acute{w}_j},\prod\nolimits_{j=1}^{k+1}\acute{p}_j^{\acute{w}_j}\right),$$

$$\left(\prod\nolimits_{j=1}^{k+1}\acute{q}_j^{\acute{w}_j},\prod\nolimits_{j=1}^{k+1}\acute{r}_j^{\acute{w}_j},\prod\nolimits_{j=1}^{k+1}\acute{s}_j^{\acute{w}_j},\prod\nolimits_{j=1}^{k+1}\acute{t}_j^{\acute{w}_j},\prod\nolimits_{j=1}^{k+1}\acute{u}_j^{\acute{w}_j},\right.$$

$$\left.\prod\nolimits_{j=1}^{k+1}\acute{v}_j^{\acute{w}_j}\prod\nolimits_{j=1}^{k+1}\acute{w}_j^{\acute{w}_j},\prod\nolimits_{j=1}^{k+1}\acute{x}_j^{\acute{w}_j}\right)\right\}$$

Especially, when $\acute{W}=(1/n,1/n,\ldots,1/n)^{\acute{T}}$, then '$\mathbb{P}$' operator reduces to octagonal neutrosophic number arithmetic averaging operator.

So, the 'Ҏ' operator has these following properties (Ҏ1)–(Ҏ3):

- **Idempotency** (Ҏ1): Let $\acute{\tilde{n}}_j = \left\langle \left(\acute{a}, \grave{b}, \acute{c}, \acute{d}, \acute{e}, \acute{f}, \acute{g}, \grave{h}\right), \left(\acute{i}, \acute{j}, \acute{k}, \acute{l}, \acute{m}, \acute{n}, \acute{o}, \acute{p}\right), \right.$
  $\left(\acute{q}, \acute{r}, \acute{s}, \acute{t}, \acute{u}, \acute{v}, \acute{w}, \acute{x}\right) \right\rangle$ (j = 1, 2, ..., n) be the collection of octagonal neutro-
  sophic number, and if each $\acute{\tilde{n}}_j$ (j = 1, 2, ..., n) is equal to ń, i.e., $\acute{\tilde{n}}_j = \acute{\tilde{n}}$ for
  J = 1, 2, ... , ń, then Ҏ $\left(\acute{\tilde{n}}_1, \acute{\tilde{n}}_2, .., \acute{\tilde{n}}_n\right) = \acute{\tilde{n}}$.

- **Boundedness** (Ҏ2): Let$\acute{\tilde{n}}_j = \left\langle \left(\acute{a}_j, \grave{b}_j, \acute{c}_j, \acute{d}_j, \acute{e}_j, \acute{f}_j, \acute{g}_j, \grave{h}_j\right), \right.$
  $\left(\acute{i}_j, \acute{j}_j, \acute{k}_j, \acute{l}_j, \acute{m}_j, \acute{n}_j, \acute{o}_j, \acute{p}_j\right), \left(\acute{q}_j, \acute{r}_j, \acute{s}_j, \acute{t}_j, \acute{u}_j, \acute{v}_j, \acute{w}_j, \acute{x}_j.\right)\right\rangle$
  (j = 1, 2, ..., n) be the collection of octagonal neutrosophic number. Let

$$n^- = \left\{ \left({}_j{}^{\min} \acute{a}_j, {}_j{}^{\min} \grave{b}_j, {}_j{}^{\min} \grave{b}_j, {}_j{}^{\min} \acute{c}_j, {}_j{}^{\min} \acute{d}_j, {}_j{}^{\min} \grave{e}, {}_j{}^{\min} \acute{f}_j, {}_j{}^{\min} \acute{g}_j, {}_j{}^{\min} \grave{h}_j\right), \right.$$

$$\left({}_j{}^{\max} \acute{i}_j, {}_j{}^{\max} \acute{j}_j, {}_j{}^{\max} \acute{k}_j, {}_j{}^{\max} \acute{l}_j, {}_j{}^{\max} \acute{m}_j, {}_j{}^{\max} \acute{n}_j\right), \left({}_j{}^{\max} \acute{o}_j, {}_j{}^{\max} \acute{p}_j\right),$$

$$\left({}_j{}^{\max} \acute{q}_j, {}_j{}^{\max} \acute{r}_j, {}_j{}^{\max} \acute{s}_j, {}_j{}^{\max} \acute{t}_j, {}_j{}^{\max} \acute{u}_j, {}_j{}^{\max} \acute{v}_j, {}_j{}^{\max} \acute{w}_j, {}_j{}^{\max} \grave{b}_j, \right.$$

$$ {}_j{}^{\max} \grave{b}_j, {}_j{}^{\max} \acute{c}_j, {}_j{}^{\max} \acute{d}_j, {}_j{}^{\max} \grave{e}, {}_j{}^{\max} \acute{f}_j, {}_j{}^{\max} \acute{g}_j,$$

$$ {}_j{}^{\max} \grave{h}_j), ({}_j{}^{\min} \acute{i}_j, {}_j{}^{\min} \acute{j}_j, {}_j{}^{\min} \acute{k}_j, {}_j{}^{\min} \acute{l}_j, {}_j{}^{\min} \acute{m}_j, {}_j{}^{\min} \acute{n}_j,$$
$$ {}_j{}^{\min} \acute{o}_j, {}_j{}^{\min} \acute{p}_j) ,$$

$$\left({}_j{}^{\min} \grave{q}_j, {}_j{}^{\min} \grave{r}_j, {}_j{}^{\min} \grave{s}_j, {}_j{}^{\min} \grave{t}_j, {}_j{}^{\min} \grave{u}_j, {}_j{}^{\min} \grave{v}_j, {}_j{}^{\min} \grave{w}_j, {}_j{}^{\min} \acute{x}_j\right) \right\}.$$

$$n^- \le Ҏ \left(\acute{\tilde{n}}_1, \acute{\tilde{n}}_2, \ldots, \acute{\tilde{n}}_n \le n^+. \right)$$

- **Monotonicity** (Ҏ3): Let $\acute{\tilde{n}}_j$ (j = 1, 2, ..., n) and $\tilde{n}_j^*$ (j = 1, 2, , ..., n) be two
  collections of octagonal neutrosophic numbers. If $\acute{\tilde{n}}_j \le \tilde{n}_j^*$ for j= 1, 2, ..., n,
  then ONNWAA$(\acute{\tilde{n}}_1, \acute{\tilde{n}}_2, \ldots, \acute{\tilde{n}}_n) \le Ҏ \left(\tilde{n}_1^*, \tilde{n}_2^*, \ldots, \tilde{n}_n^*\right).$

Proof of Ҏ1 Consider, $\tilde{n}_j = \acute{\tilde{n}}$ for j = 1, 2, ..., n, then, Ҏ $\left(\acute{\tilde{n}}_1, \acute{\tilde{n}}_2, \ldots, \acute{\tilde{n}}_n\right) =$
$\left(\acute{\tilde{n}}_1, \acute{\tilde{n}}_2, \ldots, \acute{\tilde{n}}_k\right) = \acute{w}_1 \acute{\tilde{n}}_1 \square \; \acute{w}_2 \acute{\tilde{n}}_2 \square ..... \square \acute{w}_n \acute{\tilde{n}}_k = +_{j=1}^{\acute{n}} \left(\acute{w}_j \acute{\tilde{n}}_j\right) =$
$\{(\left(1 - \prod_{j=1}^k \cdot (1 \dot{-} \acute{a}_j)^{\acute{w}_j}\right), \left(1 - \prod_{j=1}^k \cdot \left(1 \dot{-} \grave{b}_j\right)^{\acute{w}_j}\right), \left(1 - \prod_{j=1}^k \cdot (1 \dot{-} \acute{c}_j)^{\acute{w}_j}\right),$
$\left(1 - \prod_{j=1}^k \cdot \left(1 \dot{-} \acute{d}_j\right)^{\acute{w}_j}\right), \left(1 - \prod_{j=1}^k \cdot (1 \dot{-} \acute{e}_j)^{\acute{w}_j}\right), \left(1 - \prod_{j=1}^k \cdot \left(1 \dot{-} \acute{f}_j\right)^{\acute{w}_j}\right),$
$\left(1 - \prod_{j=1}^k \cdot \left(1 \dot{-} \acute{g}_j\right)^{\acute{w}_j}\right), \left(1 - \prod_{j=1}^k \cdot \left(1 \dot{-} \acute{h}_j\right)^{\acute{w}_j}\right))) \quad * \quad \left(\prod_{j=1}^n i_j^{\acute{w}_j}, \prod_{j=1}^n j_j^{\acute{w}_j}, \right.$
$\prod_{j=1}^n k_j^{\acute{w}_j}, \prod_{j=1}^n l_j^{\acute{w}_j}, \prod_{j=1}^n m_j^{\acute{w}_j}, \prod_{j=1}^n n_j^{\acute{w}_j}, \prod_{j=1}^n o_j^{\acute{w}_j}, \prod_{j=1}^n p_j^{\acute{w}_j}\right), \left(\prod_{j=1}^n q_j^{\acute{w}_j}, \right.$

$$\left. \prod\nolimits_{j=1}^{n} r_j^{\acute{w}_j}, \prod\nolimits_{j=1}^{n} s_j^{\acute{w}_j}, \prod\nolimits_{j=1}^{n} t_j^{\acute{w}_j}, \prod\nolimits_{j=1}^{n} u_j^{\acute{w}_j}, \prod\nolimits_{j=1}^{n} v_j^{\acute{w}_j} \prod\nolimits_{j=1}^{n} w_j^{\acute{w}_j}, \prod\nolimits_{j=1}^{n} x_j^{\acute{w}_j} \right) \right\}$$

$$= \left\{ \left(\left(1 - (1-\acute{a}) \sum\nolimits_{j=1}^{n} \acute{w}_j \right), \left(1 - \left(1 - \acute{b} \right) \sum\nolimits_{j=1}^{n} \acute{w}_j \right), \right. \right.$$

$$\left(1 - (1-\acute{c}) \sum\nolimits_{j=1}^{n} \acute{w}_j \right), \left(1 - \left(1 - \acute{d} \right) \sum\nolimits_{j=1}^{n} \acute{w}_j \right), \left(1 - (1-\acute{e}) \sum\nolimits_{j=1}^{n} \acute{w}_j \right),$$

$$\left(1 - \left(1 - \acute{f} \right) \sum\nolimits_{j=1}^{n} \acute{w}_j \right), \left(1 - (1-\acute{g}) \sum\nolimits_{j=1}^{n} \acute{w}_j \right), \left(1 - \left(1 - \acute{h} \right) \sum\nolimits_{j=1}^{n} \acute{w}_j \right) \right)$$

$$* \left(\acute{i} \sum\nolimits_{j=1}^{n} \acute{w}_j, \acute{j} \sum\nolimits_{j=1}^{n} \acute{w}_j, \acute{k} \sum\nolimits_{j=1}^{n} \acute{w}_j, \acute{l} \sum\nolimits_{j=1}^{n} \acute{w}_j, \acute{m} \sum\nolimits_{j=1}^{n} \acute{w}_j, \acute{n} \sum\nolimits_{j=1}^{n} \acute{w}_j, \right.$$

$$\left. \acute{o} \sum\nolimits_{j=1}^{n} \acute{w}_j, \acute{p} \sum\nolimits_{j=1}^{n} \acute{w}_j \right), \left(\acute{q} \sum\nolimits_{j=1}^{n} \acute{w}_j, \acute{r} \sum\nolimits_{j=1}^{n} \acute{w}_j, \acute{s} \sum\nolimits_{j=1}^{n} \acute{w}_j, \acute{t} \sum\nolimits_{j=1}^{n} \acute{w}_j, \right.$$

$$left. \acute{u} \sum\nolimits_{j=1}^{n} \acute{w}_j, \acute{v} \sum\nolimits_{j=1}^{n} \acute{w}_j, \acute{w} \sum\nolimits_{j=1}^{n} \acute{w}_j, \acute{x} \sum\nolimits_{j=1}^{n} \acute{w}_j \right) \right\}$$

$$= \left\{ \left(\acute{a}, \acute{b}, \acute{c}, \acute{d}, \acute{e}, \acute{f}, \acute{g}, \acute{h} \right), \left(\acute{i}, \acute{j}, \acute{k}, \acute{l}, \acute{m}, \acute{n}, \acute{o}, \acute{p} \right), \left(\acute{q}, \acute{r}, \acute{s}, \acute{t}, \acute{u}, \acute{v}, \acute{w}, \acute{x} \right) \right\} = \tilde{\acute{n}}.$$

Proof of P2 since $n^- \le \acute{n}_j^+ \le n^+$ for $(j = 1, 2, \ldots, n)$ then

$$\sum\nolimits_{j=1}^{n} \acute{w}_j n^- \le \sum\nolimits_{j=1}^{n} \acute{w}_j n_j \le \sum\nolimits_{j=1}^{n} \acute{w}_j n^+ . \text{Then } \sum\nolimits_{j=1}^{n} \acute{w}_j n_j \le \grave{n^+} \text{ by } (\square 1),$$

i.e., $\acute{n^-} \le \square \left(\tilde{\acute{n}}_1, \tilde{\acute{n}}_2, \ldots, \tilde{\acute{n}}_n \right) \le n^+ \mathbb{P} \left(\tilde{\acute{n}}_1, \tilde{\acute{n}}_2, \ldots, \tilde{\acute{n}}_n \right) \le n^+ .$

Proof of P2 When $\tilde{\acute{n}}_j \le \tilde{n}_j^*$ for $(j = 1, 2, \ldots, n)$, there is $\sum\nolimits_{j=1}^{n} \acute{w}_j n_j \le$
$\sum\nolimits_{j=1}^{n} \acute{w}_j \tilde{n}_j^* \sum\nolimits_{j=1}^{n} \acute{w}_j n_j \le \sum\nolimits_{j=1}^{n} \acute{w}_j \tilde{n}_j^*,$ i.e., $\mathbb{P} \left(\tilde{\acute{n}}_1, \tilde{\acute{n}}_2, \ldots, \tilde{\acute{n}}_n \right) \le \mathbb{P}$
$\left(\tilde{n}_1^*, \tilde{n}_2^*, \ldots, \tilde{n}_n^* \right).$ Hence it is the proof of these properties.

4.2 Weighted Geometric Averaging Operator (WGAO)

Definition 4.2 Let $\tilde{\acute{n}}_j = \left\langle \left(\acute{a}_j, \acute{b}_j, \acute{c}_j, \acute{d}_j, \acute{e}_j, \acute{f}_j, \acute{g}_j, \acute{h}_j \right), \left(\acute{i}_j, \acute{j}_j, \acute{k}_j, \acute{l}_j, \acute{m}_j, \acute{n}_j, \acute{o}_j, \right. \right.$
$\acute{p}_j \big), \left(\acute{q}_j, \acute{r}_j, \acute{s}_j, \acute{t}_j, \acute{u}_j, \acute{v}_j, \acute{w}_j, \acute{x}_j \right) \big\rangle$ $(j = 1, 2, \ldots, n)$
Be the collection of octagonal neutrosophic numbers. Then octagonal neutrosophic number weighted geometric averaging (ONNWGA $= \mho$) operator is defined as:

$$\mho \left(\tilde{\acute{n}}_1, \tilde{\acute{n}}_2, \ldots, \tilde{\acute{n}}_n \right) = \tilde{\acute{n}}_1^{\acute{w}_1} \square \; \tilde{\acute{n}}_2^{\acute{w}_2} \square \ldots \square \tilde{\acute{n}}_n^{\acute{w}_n} = +_{j=1}^{n} \left(\tilde{\acute{n}}_j^{\acute{w}_j} \right)$$

where $\ddot{w}_j$ $(j = 1, 2, 3, \ldots, n)$is the weight of jth octagonal neutrosophic number $\acute{\hat{n}}_j \left(j = 1, 2, 3, \ldots, \acute{\hat{n}}.\right)$ with $\acute{w}_j \in [0, 1]$ and $\sum_{j=1}^n \acute{w}_j = 1$.

Theorem 2 Let $\acute{\hat{n}}_j = \langle\langle \acute{a}_j, \acute{b}_j, \acute{c}_j, \acute{d}_j, \acute{e}_j, \acute{f}_j, \acute{g}_j, \hat{h} \rangle, \left(\acute{i}_j, \acute{j}_j, \acute{k}_j, \acute{l}_j, \acute{m}_j, \acute{n}_j, \acute{o}_j, \acute{p}_j\right),$
$\left(\acute{q}_j, \acute{r}_j, \acute{s}_j, \acute{t}_j, \acute{u}_j, \acute{v}_j, \acute{w}_j, \acute{x}_j\right)\rangle$ $(j = 1, 2, \ldots, n)$ be the collection of octagonal neutrosophic number, and the aggregate value by the help of Ᵽ operator is also an octagonal neutrosophic number, then:

$$\text{Ᵽ}\left(\acute{\hat{n}}_1, \acute{\hat{n}}_2, \ldots, \acute{\hat{n}}_n\right) = \acute{\hat{n}}_1^{\acute{w}_1} \square \ \acute{\hat{n}}_2^{\acute{w}_2} \square \ldots \square \acute{\hat{n}}_n^{\acute{w}_n} = +_{j=1}^{\hat{n}} \left(\acute{\hat{n}}_j^{\acute{w}_j}\right)$$

$$= \left(\left(\left(1 - \prod_{j=1}^k \cdot (1 \dot{-} i)^{\acute{w}_j}\right), \left(1 - \prod_{j=1}^k \cdot \left(1 \dot{-} j_j\right)^{\acute{w}_j}\right),\right.\right.$$

$$\left(1 - \prod_{j=1}^k \cdot \left(1 \dot{-} k_j\right)^{\acute{w}_j}\right), \left(1 - \prod_{j=1}^k \cdot \left(1 \dot{-} l_j\right)^{\acute{w}_j}\right),$$

$$\left(1 - \prod_{j=1}^k \cdot \left(1 \dot{-} m_j\right)^{\acute{w}_j}\right), \left(1 - \prod_{j=1}^k \cdot \left(1 \dot{-} n_j\right)^{\acute{w}_j}\right),$$

$$\left.\left(1 - \prod_{j=1}^k \cdot \left(1 \dot{-} o_j\right)^{\acute{w}_j}\right), \left(1 - \prod_{j=1}^k \cdot (1 \dot{-} p_j)^{\acute{w}_j}\right)\right),$$

$$\left\{\left(\left(1 - \prod_{j=1}^k \cdot (1 \dot{-} q_j)^{\acute{w}_j}\right), \left(1 - \prod_{j=1}^k \cdot \left(1 \dot{-} r_j\right)^{\acute{w}_j}\right),\right.\right.$$

$$\left(1 - \prod_{j=1}^k \cdot \left(1 \dot{-} s_j\right)^{\acute{w}_j}\right), \left(1 - \prod_{j=1}^k \cdot \left(1 \dot{-} t_j\right)^{\acute{w}_j}\right),$$

$$\left(1 - \prod_{j=1}^k \cdot \left(1 \dot{-} u_j\right)^{\acute{w}_j}\right), \left(1 - \prod_{j=1}^k \cdot \left(1 \dot{-} v_j\right)^{\acute{w}_j}\right),$$

$$\left.\left(1 - \prod_{j=1}^k \cdot \left(1 \dot{-} w_j\right)^{\acute{w}_j}\right), \left(1 - \prod_{j=1}^k \cdot \left(1 \dot{-} x_j\right)^{\acute{w}_j}\right)\right)\right\}$$

where $\ddot{w}_j$ $(j = 1, 2, 3, \ldots, n)$ is the weight of jth octagonal neutrosophic number $\acute{\hat{n}}_j \left(j = 1, 2, 3, \ldots, \acute{\hat{n}}.\right)$ with $\acute{w}_j \in [0, 1]$ and $\sum_{j=1}^n \acute{w}_j = 1$.

Proof is similar to Theorem 1.

Especially, when $\acute{W} = (1/n, 1/n, \ldots, 1/n)^{\acute{T}}$, then 'Ᵽ' operator reduces to octagonal neutrosophic number arithmetic averaging operator.

So, the 'Ᵽ' operator has these following properties (Ᵽ1)–(Ᵽ3):

Idempotency Ᵽ1: Let $\acute{\tilde{n}}_j = \left\langle \left(\acute{a}_j, \grave{b}_j, \acute{c}_j, \grave{d}_j, \acute{e}_j, \grave{f}_j, \grave{g}_j, \grave{h}_j \right), \left(\grave{i}_j, \grave{j}_j, \grave{k}_j, \acute{l}_j, \acute{m}_j, \acute{n}_j, \acute{o}_j, \grave{p}_j \right), \left(\acute{q}_j, \grave{r}_j, \acute{s}_j, \grave{t}_j, \acute{u}_j, \acute{v}_j, \grave{w}_j, \acute{x}_j \right) \right\rangle$ (j = 1, 2, ..., n) be the collection of octagonal neutrosophic number, and if each $\acute{\tilde{n}}_j$ (j = 1, 2, ..., n) is equal to $\acute{n}$, i.e., $\acute{\tilde{n}}_j = \acute{\tilde{n}}$ for j = 1, 2, ..., $\acute{n}$, then

$$\text{Ᵽ}\left(\acute{\tilde{n}}_1, \acute{\tilde{n}}_2, .., \acute{\tilde{n}}_n \right) = \acute{\tilde{n}}.$$

Boundedness Ᵽ2 : Let $\acute{\tilde{n}}_j = \left\langle \left(\acute{a}_j, \grave{b}_j, \acute{c}_j, \grave{d}_j, \acute{e}_j, \grave{f}_j, \grave{g}_j, \grave{h}_j \right), \left(\grave{i}_j, \grave{j}_j, \grave{k}_j, \acute{l}_j, \acute{m}_j, \acute{n}_j, \acute{o}_j, \grave{p}_j \right), \left(\acute{q}_j, \grave{r}_j, \acute{s}_j, \grave{t}_j, \acute{u}_j, \acute{v}_j, \grave{w}_j, \acute{x}_j \right) \right\rangle$ (j = 1, 2, ..., n) be the collection of octagonal neutrosophic number.

Let $n^- = {}_j^{\min} \acute{a}_j, {}_j^{\min} \grave{b}_j, {}_j^{\min} \grave{b}_j, {}_j^{\min} \acute{c}_j, {}_j^{\min} \grave{d}_j, {}_j^{\min} \acute{e}, {}_j^{\min} \grave{f}_j, {}_j^{\min} \grave{g}_j, {}_j^{\min} \grave{h}_j)$,

$\left({}_j^{\max} \grave{i}_j, {}_j^{\max} \grave{j}_j, {}_j^{\max} \grave{k}_j, {}_j^{\max} \acute{l}_j, {}_j^{\max} \acute{m}_j, {}_j^{\max} \acute{n}_j, {}_j^{\max} \acute{o}_j, {}_j^{\max} \grave{p}_j \right)$,

$\left({}_j^{\max} \acute{q}_j, {}_j^{\max} \grave{r}_j, {}_j^{\max} \acute{s}_j, {}_j^{\max} \grave{t}_j, {}_j^{\max} \acute{u}_j, {}_j^{\max} \acute{v}_j, {}_j^{\max} \grave{w}_j, {}_j^{\max} \acute{x}_j \right)$ and n^+

$= \{ \left({}_j^{\max} \acute{a}_j, {}_j^{\max} \grave{b}_j, {}_j^{\max} \grave{b}_j, {}_j^{\max} \acute{c}_j, {}_j^{\max} \grave{d}_j, {}_j^{\max} \acute{e}, {}_j^{\max} \grave{f}_j, {}_j^{\max} \grave{g}_j, {}_j^{\max} \grave{h}_j \right)$,

$\left({}_j^{\min} \grave{i}_j, {}_j^{\min} \grave{j}_j, {}_j^{\min} \grave{k}_j, {}_j^{\min} \acute{l}_j, {}_j^{\min} \acute{m}_j, {}_j^{\min} \grave{n}_j, {}_j^{\min} \acute{o}_j, {}_j^{\min} \grave{p}_j \right)$,

$\left({}_j^{\min} \grave{q}_{jj}{}^{\min} \grave{r}_j, {}_j^{\min} \acute{s}_j, {}_j^{\min} \grave{t}_j, {}_j^{\min} \acute{u}_j, {}_j^{\min} \acute{v}_j, {}_j^{\min} \grave{w}_j, {}_j^{\min} \acute{x}_j) \}$

$$\text{Ᵽ}n^- \leq \left(\acute{\tilde{n}}_1, \acute{\tilde{n}}_2, ..., \acute{\tilde{n}}_n \right) \leq n^+.$$

Monotonically Ᵽ3: Let $\acute{\tilde{n}}_j$ (j = $\dot{1}, \ddot{2}, ..., n$) and $\tilde{n}_j^*$ (j = $\dot{1}, \ddot{2}, ..., n$) be two collections of octagonal neutrosophic numbers. If $\acute{\tilde{n}}_j \leq \tilde{n}_j^*$ for j $\overset{.}{=}$ 1, $\ddot{2}, ..., n$, then,

$$\text{Ᵽ}\left(\acute{\tilde{n}}_1, \acute{\tilde{n}}_2, ..., \acute{\tilde{n}}_n \right) \leq \text{Ᵽ}\left(\tilde{n}_1^*, \tilde{n}_2^*, ..., \tilde{n}_n^* \right)$$

4.3 ONNWAA(Ᵽ) and ONNWGA(℧) Operators-Based Algorithm

Now we will develop an approach based on Ᵽ and ℧ operators as well as the score and accuracy functions to handle multiple-attribute decision-making problems with octagonal neutrosophic information. If we talk about multiple-attribute decision-making problems with octagonal neutrosophic information, here we have a set of alternatives $\acute{\tilde{A}} = \left\{ \acute{\tilde{A}}_1, \acute{\tilde{A}}_2, ..., \acute{\tilde{A}}_{\acute{m}} \right\}$, which can satisfy a set of attributes $\tilde{\mathcal{C}} = \{ \mathcal{C}_1, \mathcal{C}_2, ..., \mathcal{C}_{\acute{n}} \}$. By decision maker, an alternative on attributes is evaluated and these values are represented by the form of octagonal neutrosophic number. So, we have an octagonal neutrosophic decision matrix:

$$\acute{D} = \left(\tilde{\acute{d}}_{ij}\right)_{m\acute{X}n} = \left(\left(\acute{a}_{ij}, \acute{b}_{ij}, \acute{c}_{ij}, \acute{d}_{ij}, \acute{e}_{ij}, \acute{f}_{ij}, \acute{g}_{ij}, \acute{h}_{ij}\right), \left(\acute{i}_{ij}, \acute{j}_{ij}, \acute{k}_{ij}, \acute{l}_{ij}, \acute{m}_{ij}, \acute{n}_{ij},\right.\right.$$

$$\left.\left.\acute{o}_{ij}, \acute{p}_{ij}\right), \left(\acute{q}_{ij}, \acute{r}_{ij}, \acute{s}_{ij}, \acute{t}_{ij}, \acute{u}_{ij}, \acute{v}_{ij}, \acute{w}_{ij}, \acute{x}_{ij}\right)\right)_{m\acute{X}n}$$

where $\left(\acute{a}_{ij}, \acute{b}_{ij}, \acute{c}_{ij}, \acute{d}_{ij}, \acute{e}_{ij}, \acute{f}_{ij}, \acute{g}_{ij}, \acute{h}_{ij}\right) \subset [0, 1]$ shows the degree that alternative $\acute{A}_i$ satisfies the attribute C_j, and $\left(\acute{i}_{ij}, \acute{j}_{ij}, \acute{k}_{ij}, \acute{l}_{ij}, \acute{m}_{ij}, \acute{n}_{ij}, \acute{o}_{ij}, \acute{p}_{ij}\right) \subset \left[0, 1\right]$ shows the degree that alternative $\acute{A}_i$ is uncertain the attribute C_j, and $\left(\acute{q}_{ij}, \acute{r}_{ij}, \acute{s}_{ij}, \acute{t}_{ij}, \acute{u}_{ij}, \acute{v}_{ij}, \acute{w}_{ij}, \acute{x}_{ij}\right) \subset [0, 1]$ shows the degree that alternative $\acute{A}_i$ does not satisfy the attribute the C_j with $0 \leq \acute{d}_{ij} + \acute{h}_{ij} + \acute{p}_{ij} \leq 3$ for $\acute{i} = 1, 2, \ldots, $ m, and $\acute{j} = 1, 2, \ldots, $ n.

On this, we will apply the Ᵽ and ℧ operators, the score and accuracy function to multiple attribute decision-making problem with octagonal neutrosophic information, which can be given as follows.

Step 1:

Insert the Ᵽ operator $\tilde{\acute{d}}_i = \left(\left(\acute{a}_i, \acute{b}_i, \acute{c}_i, \acute{d}_i, \acute{e}_i, \acute{f}_i, \acute{g}_i, \acute{h}_i\right), \left(\acute{i}_i, \acute{j}_i, \acute{k}_i, \acute{l}_i, \acute{m}_i, \acute{n}_i, \acute{o}_i, \acute{p}_i\right),\right.$ $\left.\left(\acute{q}_i, \acute{r}_i, \acute{s}_i, \acute{t}_i, \acute{u}_i, \acute{v}_i, \acute{w}_i, \acute{x}_i\right)\right) = $ Ᵽ $\left(\tilde{\acute{d}}_{i1}, \tilde{\acute{d}}_{i2}, \ldots, \tilde{\acute{d}}_{in}\right)$ or by the ℧ operator $\tilde{\acute{d}}_i = \left(\left(\acute{a}_i, \acute{b}_i, \acute{c}_i, \acute{d}_i, \acute{e}_i, \acute{f}_i, \acute{g}_i, \acute{h}_i\right), \left(\acute{i}_i, \acute{j}_i, \acute{k}_i, \acute{l}_i, \acute{m}_i, \acute{n}_i, \acute{o}_i, \acute{p}_i\right), \left(\acute{q}_i, \acute{r}_i, \acute{s}_i, \acute{t}_i, \acute{u}_i, \acute{v}_i, \acute{w}_i, \acute{x}_i\right)\right)$ $= ℧ (\tilde{\acute{d}}_{i1}, \tilde{\acute{d}}_{i2}, \ldots, \tilde{\acute{d}}_{in})$ $(i = 1, 2, \ldots, m)$ to get the combination of the overall octagonal neutrosophic number of $\tilde{\acute{d}}_i \left(i = 1, 2, \ldots, m\right)$ for every alternative $\acute{A}_i \left(i = 1, 2, \ldots, m\right)$.

Step 2:

Utilize the score $\acute{S}\left(\tilde{\acute{d}}_i\right) \left(i = 1, 2, \ldots, m\right)$ of the combination of the overall octagonal neutrosophic number of $(\tilde{\acute{d}}_i) \; i = 1, 2, \ldots, m\right)$ for rank the alternatives of $\acute{A}_i\left(i = 1, 2, \ldots, m\right)$.

Suppose there is similar to scores $\acute{S}\left(\tilde{\acute{d}}_i\right)$ and $\acute{S}\left(\tilde{\acute{d}}_j\right)$, then we have to calculate the accuracy degrees $\acute{H}\left(\acute{d}_i\right)$ and $\acute{H}\left(\acute{d}_j\right)$ of the combination of overall octagonal neutrosophic number, for the ranking of alternative A_i and $\acute{A}_j$ according to accuracy degree $\acute{H}\left(\tilde{\acute{d}}_i\right)$ and $\acute{H}\left(\tilde{\acute{d}}_J\right)$.

Alternatives				
A_1	A_2	A_3	A_4	A_4
Mr. A	Mr. B	Mr. C	Mr. D	Mr. E

4.4.3 Assumptions

Following linguistic values will be assigned:

$$D = \begin{bmatrix} \langle(0.2, 0.3, 0.4, 0.5, 0.6, 0.7, 0.8, 0.9), (0.0, 0.1, 0.2, 0.3, 0.4, 0.5, 0.6, 0.7), (0.1, 0.1, 0.1, 0.1, 0.1, 0.1, 0.1, 0.1)\rangle \\ \langle(0.9, 0.9, 0.9, 0.9, 0.9, 0.9, 0.9, 0.9), (0.2, 0.3, 0.4, 0.5, 0.6, 0.7, 0.8, 0.9), (0.1, 0.1, 0.1, 0.1, 0.1, 0.1, 0.2, 0.2)\rangle \\ \langle(0.1, 0.1, 0.1, 0.1, 0.1, 0.1, 0.1, 0.1), (0.0, 0.1, 0.1, 0.2, 0.2, 0.2, 0.2, 0.2), (0.1, 0.1, 0.1, 0.1, 0.2, 0.2, 0.3, 0.4)\rangle \\ \langle(0.2, 0.3, 0.4, 0.5, 0.6, 0.7, 0.8, 0.9), (0.1, 0.2, 0.3, 0.3, 0.3, 0.3, 0.3, 0.3), (0.1, 0.1, 0.1, 0.1, 0.1, 0.2, 0.3, 0.3)\rangle \\ \langle(0.2, 0.3, 0.4, 0.5, 0.6, 0.7, 0.8, 0.9), (0.1, 0.1, 0.1, 0.1, 0.1, 0.1, 0.2, 0.2), (0.2, 0.2, 0.4, 0.5, 0.7, 0.7, 0.8, 0.9)\rangle \end{bmatrix}$$

$\langle(0.0, 0.1, 0.1, 0.2, 0.2, 0.2, 0.2, 0.2), (0.0, 0.1, 0.1, 0.2, 0.2, 0.3, 0.3, 0.4), (0.0, 0.1, 0.2, 0.2, 0.2, 0.2, 0.3, 0.4)\rangle$

$\langle(0.0, 0.1, 0.2, 0.2, 0.2, 0.3, 0.4, 0.4), (0.0, 0.1, 0.2, 0.2, 0.2, 0.4, 0.5, 0.5), (0.1, 0.1, 0.2, 0.2, 0.2, 0.4, 0.5, 0.5)\rangle$

$\langle(0.0, 0.1, 0.1, 0.2, 0.2, 0.3, 0.3, 0.4), (0.1, 0.1, 0.2, 0.2, 0.2, 0.3, 0.4, 0.4), (0.1, 0.1, 0.1, 0.1, 0.1, 0.1, 0.1, 0.1)\rangle$

$\langle(0.1, 0.1, 0.2, 0.2, 0.2, 0.3, 0.4, 0.5), (0.1, 0.2, 0.2, 0.2, 0.2, 0.3, 0.3, 0.4), (0.1, 0.1, 0.2, 0.2, 0.2, 0.3, 0.4, 0.4)\rangle$

$\langle(0.1, 0.1, 0.1, 0.1, 0.2, 0.2, 0.4, 0.4), (0.1, 0.1, 0.1, 0.1, 0.2, 0.2, 0.4, 0.5), (0.1, 0.1, 0.1, 0.1, 0.2, 0.2, 0.4, 0.4)\rangle$

$\langle(0.9, 0.9, 0.9, 0.9, 0.9, 0.9, 0.9, 0.9), (0.1, 0.1, 0.1, 0.2, 0.2, 0.3, 0.4, 0.4), (0.1, 0.2, 0.3, 0.3, 0.3, 0.3, 0.3, 0.3)\rangle$

$\langle(0.1, 0.1, 0.1, 0.1, 0.2, 0.3, 0.4, 0.4), (0.0, 0.1, 0.1, 0.1, 0.1, 0.1, 0.4, 0.5), (0.0, 0.1, 0.4, 0.4, 0.4, 0.4, 0.4, 0.5)\rangle$

$\langle(0.0, 0.1, 0.1, 0.1, 0.1, 0.1, 0.4, 0.4), (0.0, 0.1, 0.1, 0.1, 0.3, 0.3, 0.4, 0.5), (0.0, 0.1, 0.1, 0.1, 0.2, 0.3, 0.4, 0.5)\rangle$

$\langle(0.0, 0.1, 0.1, 0.1, 0.1, 0.1, 0.4, 0.8), (0.0, 0.1, 0.1, 0.1, 0.6, 0.7, 0.8, 0.9), (0.0, 0.1, 0.1, 0.1, 0.2, 0.3, 0.4, 0.5)\rangle$

$\langle(0.0, 0.1, 0.1, 0.1, 0.1, 0.1, 0.4, 0.9), (0.0, 0.1, 0.1, 0.1, 0.5, 0.5, 0.5, 0.5), (0.0, 0.1, 0.1, 0.2, 0.2, 0.2, 0.4, 0.5)\rangle$

$$\begin{bmatrix} (0.1, 0.1, 0.1, 0.1, 0.1, 0.1, 0.1, 0.1), (0.1, 0.1, 0.1, 0.1, 0.1, 0.2, 0.4, 0.8), (0.1, 0.1, 0.1, 0.1, 0.5, 0.6, 0.7, 0.9) \\ (0.1, 0.1, 0.2, 0.2, 0.2, 0.2, 0.4, 0.8), (0.1, 0.1, 0.1, 0.2, 0.2, 0.3, 0.4, 0.9), (0.1, 0.1, 0.1, 0.2, 0.6, 0.7, 0.8, 0.9) \\ (0.0, 0.1, 0.2, 0.2, 0.2, 0.2, 0.2, 0.5), (0.0, 0.1, 0.2, 0.2, 0.3, 0.3, 0.3, 0.4), (0.2, 0.2, 0.2, 0.2, 0.2, 0.2, 0.2, 0.4) \\ (0.1, 0.2, 0.3, 0.4, 0.5, 0.6, 0.7, 0.8), (0.1, 0.2, 0.3, 0.3, 0.5, 0.6, 0.7, 0.7), (0.1, 0.2, 0.3, 0.3, 0.5, 0.7, 0.8, 0.9) \\ (0.1, 0.2, 0.6, 0.8, 0.7, 0.8, 0.9, 0.9), (0.1, 0.2, 0.5, 0.8, 0.8, 0.9, 0.9, 0.9), (0.1, 0.2, 0.8, 0.8, 0.8, 0.8, 0.8, 0.9) \end{bmatrix}$$

Now apply the proposed algorithm for the selection of the best one:

Step 1:

$$\acute{C}_1 \qquad \acute{C}_4 \qquad \acute{C}_2 \qquad \acute{C}_3$$

$$\begin{bmatrix} Á_1 \\ Á_2 \\ Á_3 \\ Á_4 \\ Á_5 \end{bmatrix} \begin{bmatrix} (0.5), (0.3), (0.1) & (0.1), (0.2), (0.2) & (0.7), (0.2), (0.2) & (0.1), (0.2), (0.3) \\ (0.9), (0.4), (0.1) & (0.2), (0.2), (0.2) & (0.2), (0.1), (0.3) & (0.2), (0.2), (0.4) \\ (0.9), (0.1), (0.1) & (0.2), (0.2), (0.1) & (0.1), (0.4), (0.2) & (0.2), (0.2), (0.2) \\ (0.5), (0.2), (0.1) & (0.2), (0.2), (0.2) & (0.2), (0.4), (0.2) & (0.4), (0.4), (0.5) \\ (0.8), (0.1), (0.5) & (0.2), (0.2), (0.2) & (0.2), (0.3), (0.2) & (0.6), (0.6), (0.6) \end{bmatrix}$$

$$\acute{C}_1 \qquad \acute{C}_4 \qquad \acute{C}_2 \qquad \acute{C}_3$$

$$\begin{bmatrix} Á_1 \\ Á_2 \\ Á_3 \\ Á_4 \\ Á_5 \end{bmatrix} \begin{bmatrix} (0.5), (0.3), (0.1) & (0.1), (0.2), (0.2) & (0.7), (0.2), (0.2) & (0.1), (0.2), (0.3) \\ (0.9), (0.4), (0.1) & (0.2), (0.2), (0.2) & (0.2), (0.1), (0.3) & (0.2), (0.2), (0.4) \\ (0.9), (0.1), (0.1) & (0.2), (0.2), (0.1) & (0.1), (0.4), (0.2) & (0.2), (0.2), (0.2) \\ (0.5), (0.2), (0.1) & (0.2), (0.2), (0.2) & (0.2), (0.4), (0.2) & (0.4), (0.4), (0.5) \\ (0.8), (0.1), (0.5) & (0.2), (0.2), (0.2) & (0.2), (0.3), (0.2) & (0.6), (0.6), (0.6) \end{bmatrix} \begin{bmatrix} \acute{W}_1 \\ \acute{W}_2 \\ \acute{W}_3 \\ \acute{W}_4 \end{bmatrix}$$

$$\acute{C}_1 \qquad \acute{C}_4 \qquad \acute{C}_2 \qquad \acute{C}_3$$

$$\begin{bmatrix} Á_1 \\ Á_2 \\ Á_3 \\ Á_4 \\ Á_5 \end{bmatrix} \begin{bmatrix} (0.5), (0.3), (0.1) & (0.1), (0.2), (0.2) & (0.7), (0.2), (0.2) & (0.1), (0.2), (0.3) \\ (0.9), (0.4), (0.1) & (0.2), (0.2), (0.2) & (0.2), (0.1), (0.3) & (0.2), (0.2), (0.4) \\ (0.9), (0.1), (0.1) & (0.2), (0.2), (0.1) & (0.1), (0.4), (0.2) & (0.2), (0.2), (0.2) \\ (0.5), (0.2), (0.1) & (0.2), (0.2), (0.2) & (0.2), (0.4), (0.2) & (0.4), (0.4), (0.5) \\ (0.8), (0.1), (0.5) & (0.2), (0.2), (0.2) & (0.2), (0.3), (0.2) & (0.6), (0.6), (0.6) \end{bmatrix} \begin{bmatrix} 0.25 \\ 0.25 \\ 0.2 \\ 0.3 \end{bmatrix}$$

Apply the Ɵ operator to get the collection of a complete octagonal neutrosophic number of $\acute{d}_i \left(\acute{i} = 1, 2, 3, 4, 5 \right)$ for the selection $Á_i \left(\acute{i} = 1, 2, 3, 4, 5 \right)$ as follows:

$$\acute{d}_1 = \{(0.1, 0, 0.2, 0), (0.7, 0.7, 0.7, 0.6), (0.5, 0.7, 0.7, 0.7)\}$$

$$\acute{d}_2 = \{(0.4, 0.1, 0, 0.1), (0.8, 0.7, 0.6, 0.6), (0.6, 0.7, 0.8, 0.8)\}$$

$$\acute{d}_3 = \{(0.4, 0.1, 0, 0.1), (0.6, 0.7, 0.8, 0.6), (0.6, 0.6, 0.7, 0.6)\}$$

$$\acute{d}_4 = \{(0.2, 0.1, 0, 0.1), (0.7, 0.7, 0.8, 0.8), (0.7, 0.7, 0.7, 0.9)\}$$

$$\acute{d}_5 = \{(0.3, 0.1, 0, 0.2), (0.6, 0.7, 0.8, 0.9), (0.8, 0.7, 0.7, 0.9)\}$$

Apply the Ʊ operator to get the collection of a complete octagonal neutrosophic number of $\acute{d}_i \left(\acute{i} = 1, 2, 3, 4, 5 \right)$ for the selection $Á_i \left(\acute{i} = 1, 2, 3, 4, 5 \right)$ as follows:

$$\acute{d}_1 = \{(0.8, 0.6, 0.9, 0.5), (0.9, 0.1, 0, 0.1), (0, 0.1, 0, 0.1)\}$$

$$\acute{d}_2 = \{(1, 0.7, 0.7, 0.6), (0.2, 0.1, 0, 0.1), (0, 0.1, 0.1, 0.1)\}$$

$$\acute{\tilde{d}}_3 = \{(1, 0.7, 0.6, 0.6), (0, 0, 0.1, 0.1), (0, 0, 0, 0.1)\}$$

$$\acute{\tilde{d}}_4 = \{(0.8, 0.7, 0.7, 0.8), (0.1, 0.1, 0.1, 0.1), (0, 0.1, 0, 0.2)\}$$

$$\acute{\tilde{d}}_5 = \{(0.9, 0.7, 0.7, 0.9), (0, 0.1, 0.1, 0.2), (0.2, 0.1, 0, 0.2)\}$$

Step 2: Find the score values of $\acute{\tilde{S}}(\acute{\tilde{d}}_i)(\acute{i} = 1, 2, 3, 4, 5)$ for the collective overall octagonal neutrosophic numbers of $(\acute{\tilde{d}}_i)(\acute{i} = 1, 2, 3, 4, 5)$, as mentioned below:

Alternative $\acute{A}_i$	Score value ONNWAA	Score value ONNWGA
$\acute{A}_1$	0.64	0.67
$\acute{A}_2$	0.67	0.86
$\acute{A}_3$	0.68	0.88
$\acute{A}_4$	0.65	0.84
$\acute{A}_5$	0.66	0.85

Step 3: Rank all the persons of $(\acute{\tilde{A}}_i)(\acute{i} = 1, 2, 3, 4, 5)$ on the base of score value, and remember that "$\succeq$ "means "preferred to". So, we are now on the conclusion that two types of ranking are enough to select the best one.

The ranking order of the alternative Enrollment in local colleges, 2005:

Aggregation operation	Ranking order
ONNWAA ₽	$\tilde{A}_3 \succeq \tilde{A}_2 \succeq \tilde{A}_5 \succeq \tilde{A}_4 \succeq \tilde{A}_1$
ONNWGA Ն	$\tilde{A}_3 \succeq \tilde{A}_2 \succeq \tilde{A}_5 \succeq \tilde{A}_4 \succeq \tilde{A}_1$

5 Conclusion

Generally, neutrosophic numbers are used in decision-making where uncertain, hesitant, and ambiguous data of a real-life is involved. This chapter presents the fundamentals of octagonal Neutrosophic Numbers (ONNs), consisting of metric operations, weighted averaging, and geometric averaging operators. Further the proposed operators are applied in the real-life application of multi-criteria decision-making. In future, more algorithms can be developed by using octagonal neutrosophic numbers in various fields like multi-criteria decision-making problems, image processing problems, and pattern recognition.

In forthcoming work, the focus will be on matrix theory of octagonal neutrosophic numbers ONNs and its applications in decision-making problems.

References

1. Zadeh, L. A. (1965). Fuzzy sets. *Information and Control, 8*(3), 338–353.
2. Turksen, I. B. (1986). Interval valued fuzzy sets based on normal forms. *Fuzzy Sets and Systems, 20*(2), 191–210.
3. Smarandache, F. (2005). Neutrosophic set-a generalization of the intuitionistic fuzzy set. *International journal of pure and applied mathematics, 24*(3), 287–297.
4. Smarandache, F. (2018). Extension of soft set to hypersoft set, and then to plithogenic hypersoft set. *Neutrosophic Sets and Systems, 22*, 168–170.
5. Saqlain, M., Moin, S., Jafar, M. N., Saeed, M., & Smarandache, F. (2020b). Aggregate operators of Neutrosophic Hypersoft set. *Neutrosophic Sets and Systems, 32*, 294–306.
6. Deli, I., & Şubaş, Y. (2014). Single valued neutrosophic numbers and their applications to multicriteria decision making problem. *Neutrosophic Sets and Systems, 2*(1), 1–13.
7. Deli, I., & Şubaş, Y. (2017). A ranking method of single valued neutrosophic numbers and its applications to multi-attribute decision making problems. *International Journal of Machine Learning and Cybernetics, 8*(4), 1309–1322.
8. Broumi, S., Murugappan, M., Talea, M., Bakali, A., Smarandache, F., Singh, P. K., & Dey, A. (2018). Single valued (2N+ 1) sided polygonal neutrosophic numbers and single valued (2N) sided polygonal neutrosophic numbers. *Neutrosophic Sets and Systems, 25*, 54–65.
9. Biswas, P., Pramanik, S., & Giri, B. C. (2015). Cosine similarity measure based multi-attribute decision-making with trapezoidal fuzzy neutrosophic numbers. *Neutrosophic Sets and Systems, 8*, 46–56.
10. Şahin, M., Kargın, A., & Smarandache, F. (2018). Generalized single valued triangular Neutrosophic numbers and aggregation operators for application to multi-attribute group decision making. *New Trends in Neutrosophic Theory and Applications, 2*, 51–84.
11. Karaaslan, F. (2018). Gaussian single-valued neutrosophic numbers and its application in multi-attribute decision making. *Neutrosophic Sets and Systems, 22*, 101–117.
12. Mo, J., & Huang, H.-L. (2019). (T, S)-based single-valued Neutrosophic number equivalence matrix and clustering method. *Mathematics, 7*(1), 36.
13. Chakraborty, A., Mondal, S. P., Ahmadian, A., Senu, N., Alam, S., & Salahshour, S. (2018). Different forms of triangular neutrosophic numbers, de-neutrosophication techniques, and their applications. *Symmetry, 10*(8), 1–28.
14. Chakraborty, A., Broumi, S., & Singh, P. K. (2019a). Some properties of pentagonal neutrosophic numbers and its applications in transportation problem environment. *Neutrosophic Sets and Systems, 28*, 200–215.
15. Chakraborty, A., Mondal, S. P., Alam, S., Ahmadian, A., Senu, N., De, D., & Salahshour, S. (2019b). The pentagonal fuzzy number: Its different representations, properties, ranking, Defuzzification and application in game problems. *Symmetry, 11*(2), 248.
16. Abdel-Basset, M., Saleh, M., Gamal, A., & Smarandache, F. (2019). An approach of TOPSIS technique for developing supplier selection with group decision making under type-2 neutrosophic number. *Applied Soft Computing, 77*, 438–452.
17. Liu, P., Chu, Y., Li, Y., & Chen, Y. (2014). Some generalized neutrosophic number Hamacher aggregation operators and their application to group decision making. *International Journal of Fuzzy Systems, 16*(2), 242–255.
18. Liu, P., Zhang, L., Liu, X., & Wang, P. (2016). Multi-valued neutrosophic number Bonferroni mean operators with their applications in multiple attribute group decision making. *International Journal of Information Technology & Decision Making, 15*(5), 1181–1210.

19. Pramanik, S., Dalapati, S., Alam, S., & Roy, T. K. (2017a). NC-TODIM-based MAGDM under a neutrosophic cubic set environment. *Information, 8*(4), 1–21.
20. Pramanik, S., Dey, P. P., Giri, B. C., & Smarandache, F. (2017b). An extended TOPSIS for multi-attribute decision making problems with neutrosophic cubic information. *Neutrosophic Sets and Systems, 17*, 20–28.
21. Pramanik, S., & Mallick, R. (2018). VIKOR based MAGDM strategy with trapezoidal neutrosophic numbers. *Neutrosophic Sets and Systems, 22*, 118–129.
22. Ye, J. (2017). Linguistic neutrosophic cubic numbers and their multiple attribute decision-making method. *Information, 8*(3), 110.
23. Fahmi, A., Amin, F., Khan, M., & Smarandache, F. (2019). Group decision making based on triangular neutrosophic cubic fuzzy einstein hybrid weighted averaging operators. *Symmetry, 11*(2), 180.
24. Selvakumari, K., & Lavanya, S. (2018). Solving fuzzy game problem in octagonal NEUTROSOPHIC numbers using heavy OWA operator. *International Journal of Engineering & Technology, 7*, 497–499.
25. Saqlain, M., Hamza, A., & Farooq, S. (2020a). Linear and non-linear octagonal Neutrosophic numbers: Its representation, α-cut and applications. *International Journal of Neutrosophic Science, 3*(1), 29–43.

A Study of Neutrosophic Cubic Finite State Machines, Subsystems, and Applications

Muhammad Gulistan, Ismat Beg, and Mateen Javed Abbasi

1 Introduction

In 1965 first time the fuzzy set (FS) was introduced by Zadeh [31] which is a new addition in the mathematical logic. There are so many extensions of FS, the interval valued fuzzy set (IVFS) was suggested by Turksons [29], Intutionistic fuzzy set (IFS) was presented by Atanasov [3]. IFS was the generalization of FS and is provably corresponding to IVFS, where the bound at lower position is called membership degree and bound at upper position is known as non-membership degree. Vague set whose idea is given by Gau and Buehrer [4], and the idea of bipolar fuzzy set (BFS) was described by Zhang [32]. In (2012) the concept of cubic set (CS) was proposed by Jun et al. [17]. CS is the combination of IVFS and FS in the form of ordered pair. These all are mathematical tools to determine the complications in our daily life. Smarandache [28] gave the concept of neutrosophic set (NS) which is the extension of FS, IVFS, IFS. In NS we deal with its three components, that is, truthfulness, indeterminate, and untruthfulness; these three functions are independent completely. Neutrosophy gives us a support for a whole family of new mathematical theories with the abstraction of both classical and fuzzy counterparts. In real life and in scientific problems to apply neutrosophic set Wang et al. [30] introduced the new idea of single valued neutrosophic set (SVNS) and interval neutrosophic set (INS). These are subclasses of NS, in which truthfulness, indeterminate, and untruthfulness were taken in closed interval $[0, 1]$, see also

M. Gulistan (✉) · M. J. Abbasi
Department of Mathematics and Statistics, Hazara University, Mansehra, Pakistan
e-mail: gulistanmath@hu.edu.pk

I. Beg
Lahore School of Economics, Lahore, Pakistan
e-mail: ibeg@lahoreschool.edu.pk

519

[15]. Jun et al. [18] gave the idea of neutrosophic cubic set. For application of neutrosophic cubic sets we refer to [5–11, 33]. Ali et al. [2] introduced the theory of neutrosophic cubic sets and their application in pattern recognition. For more application we refer [12, 19–22]. Malik et al. [24–26] proposed fuzzy finite state machines (FFSM), product of fuzzy finite state machine, subsystem of FFSM, and submachine of FFSM and also consider some properties related to these FSM. In (2002) Kumbhojkar and Chaudhari [23] gave the several ideas of covering of FFSM. Sato and Kuroki [27] in the same year described fuzzy finite switchboard state machine. In (2005) Jun [13] theorized the work of Malik et al. and described the new idea of intutionistic fuzzy finite state machine (IFFSM) and submachine of IFFSM. In (2006) Jun [14] presented the idea of intutionistic fuzzy finite switchboard state machine, commutative IFFSM, and strong homomorphism. Jun and Kavikumar [16] gave the idea of bipolar fuzzy finite state machine (BFFSM). Recently in (2018) Abughazalah and Yaqoob [1] gave the idea of subsystems of cubic FSM.

This chapter is organized as follows:

In Sect. 2, we discuss basic helping material from the existing literature. In Sect. 3, we describe NCFSM and some related results. In Sect. 4 we described proof of some results which are based on neutrosophic cubic subsystems of FSM. We also described some results related to internal and external neutrosophic cubic subsystems of neutrosophic cubic FSM. In Sect. 5, we define some operation on subsystems of neutrosophic cubic FSM. At the last in Sect. 6, we accommodate the applications of neutrosophic cubic FSM, and test the use of our presented model.

2 Preliminaries

In this section, we give some helping material from the existing literature.

Definition 2.1 ([30]) A neutrosophic set is a structure

$$N = \{(x; \eta_{NT}(x), \eta_{NI}(x), \eta_{NF}(x) \mid x \in X)\}$$

in X. Here $(\eta_{NT}(x), \eta_{NI}(x), \eta_{NF}(x) \in [0, 1])$ are called truth, indeterminacy, and falsity functions, respectively.

Definition 2.2 ([30]) Let X be a space of points (objects), by means of interval valued neutrosophic set (IVNS) in X is express in X by a truthfull membership function ζ_{NT}, indeterminate membership function ζ_{NI} and a untruthfull membership function ζ_{NF} and is given as

$$\zeta = \left\{ \left(x, \left[\zeta_{NT}(x)^L, \zeta_{NT}(x)^U \right], \left[\zeta_{NI}(x)^L, \zeta_{NI}(x)^U \right], \left[\zeta_{NF}(x)^L, \zeta_{NF}(x)^U \right] \right) / x \in X \right\}$$

for each point x in X, $\zeta_{NT} : X \rightarrow B]0, 1^+[, \zeta_{NI} : X \rightarrow B]0, 1^+[, \zeta_{NF} : X \rightarrow B]0, 1^+[$ and can be written as

$$\widetilde{\zeta} = \left(x, \left(\widetilde{\zeta}_{NT}\left(x \right), \widetilde{\zeta}_{NI}\left(x \right), \widetilde{\zeta}_{NF}\left(x \right) \right) / x \in X \right).$$

Definition 2.3 ([18]) Let X be a space of points (objects), by means of neutrosophic cubic set (NCS) in X. And it is given as $NCS = \left\{ \left(x, \widetilde{\zeta}_{NT}\left(x \right), \widetilde{\zeta}_{NI}\left(x \right), \widetilde{\zeta}_{NF}\left(x \right), \eta_{NT}\left(x \right), \eta_{NI}\left(x \right), \eta_{NF}\left(x \right) \right) / x \in X \right\}$ in which $\widetilde{\zeta}_{NT}, \widetilde{\zeta}_{NI}$ and $\widetilde{\zeta}_{NF}$ shows that the interval valued neutrosophic set in X and η_{NT}, η_{NT} and η_{NT} is a simple neutrosophic set in X.

Definition 2.4 ([26]) A fuzzy finite state machine is based on three components, that is, $FM = (T, P, F)$, where T shows the non-empty finite set of states and P shows the non-empty finite set of input symbols, respectively, and F is a fuzzy set in $T \times P \times T$, that is, $F : T \times P \times T \rightarrow [0, 1]$. Let P^* denote the set of all words of elements of P of finite length. Let λ denote the empty word in P^* and $|p|$ denote the length of P for every $p \in P^*$.

Definition 2.5 ([13]) An intutionistic finite state machine is based on three components, that is, $IM = (T, P, I)$, where T shows the non-empty finite set of states and P shows the non-empty finite set of input symbols, respectively, and $I = \langle \eta_T, \eta_F \rangle$ is a intutionistic function in $T \times P \times T$, that is, $I : T \times P \times T \rightarrow B [0, 1]$. Let P^* denote the set of all words of elements of P of finite length. Let λ denote the empty word in P^* and $|p|$ denote the length of P for every $p \in P^*$.

Definition 2.6 ([1]) A cubic finite state machine is based on three components, that is, $CM = (T, P, C)$, where T shows the non-empty finite set of states and P shows the non-empty finite set of input symbols, respectively, and $C = \left(\widetilde{\zeta}_A, \eta_A \right)$ is a cubic set in $T \times P \times T$, that is, $C : T \times P \times T \rightarrow [0, 1]$. Let P^* denote the set of all words of elements of P of finite length. Let λ denote the empty word in P^* and $|p|$ denote the length of P for every $p \in P^*$.

3   Neutrosophic Cubic Finite State Machine (NCFSM)

In this section we discuss NCFSM (neutrosophic cubic finite state machine) and some related results are also provided.

Definition 3.1 A neutrosophic cubic finite state machine shortly we denote (NeutrosophiccubicFSM) is a triplet $NCM = (T, P, NC)$, where T is the set of non-empty finite states and P is the set of non-empty finite inputs symbols, respectively, and $NC = \left(\widetilde{\zeta}_{NTx}, \widetilde{\zeta}_{,NIx}, \widetilde{\zeta}_{NFx}, \eta_{NTx}, \eta_{NIx}, \eta_{NFx} \right)$ is a neutrosophic cubic set in $T \times P \times T$. Let P^* denote the set of all words of elements of P of finite length. Let λ denote the empty word in P^* and $|p|$ denote the length of P for every $p \in P^*$.

Definition 3.2 Let $NCM = (T, P, NC)$ be a neutrosophic cubic finite state machine. The neutrosophic cubic set can be define as $NC^* = \left(\widetilde{\zeta}_{NTx}, \widetilde{\zeta}_{,NIx}, \widetilde{\zeta}_{NFx}, \eta_{NTx}, \eta_{NIx}, \eta_{NFx} \right)$ in $T \times P^* \times T$. For truthness and indeterminacy

$$\tilde{\zeta}^*_{NT}(m,\lambda,n) = \begin{cases} [1,1] \text{ if } m = n \\ [0,0] \text{ if } m \neq n \end{cases}, \quad \eta^*_{NT}(m,\lambda,n) = \begin{cases} 0 \text{ if } m = n \\ 1 \text{ if } m \neq n \end{cases},$$

$$\tilde{\zeta}^*_{NI}(m,\lambda,n) = \begin{cases} [1,1] \text{ if } m = n \\ [0,0] \text{ if } m \neq n \end{cases}, \quad \eta^*_{NI}(m,\lambda,n) = \begin{cases} 0 \text{ if } m = n \\ 1 \text{ if } m \neq n \end{cases},$$

and for falsity we can define

$$\tilde{\zeta}^*_{NF}(m,\lambda,n) = \begin{cases} [0,0] \text{ if } m = n \\ [1,1] \text{ if } m \neq n \end{cases}, \quad \eta^*_{NF}(m,\lambda,n) = \begin{cases} 1 \text{ if } m = n \\ 0 \text{ if } m \neq n \end{cases}.$$

Also for truthness and indeterminacy

$$\tilde{\zeta}^*{}_{NCT}(m, pa, n) = \vee_{r \in t} \left[\tilde{\zeta}^*{}_{NCT}(m, p, r) \wedge \tilde{\zeta}_{NCT}(r, a, n) \right],$$

$$\eta^*_{NCT}(m, pa, n) = \wedge_{r \in t} \left[\eta^*_{NCT}(m, p, r) \vee \eta_{NCT}(r, a, n) \right]$$

$$\tilde{\zeta}^*{}_{NCI}(m, pa, n) = \vee_{r \in t} \left[\tilde{\zeta}^*{}_{NCI}(m, p, r) \wedge \tilde{\zeta}_{NCI}(r, a, n) \right],$$

$$\eta^*_{NCI}(m, pa, n) = \wedge_{r \in t} \left[\eta^*_{NCI}(m, p, r) \vee \eta_{NCI}(r, a, n) \right]$$

for all $m, n \in T, p \in P^*$ and $a \in P$. For falsity

$$\tilde{\zeta}^*{}_{NCF}(m, pa, n) = \wedge_{r \in t} \left[\tilde{\zeta}^*{}_{NCF}(m, p, r) \vee \tilde{\zeta}_{NCF}(r, a, n) \right],$$

$$\eta^*_{NCF}(m, pa, n) = \vee_{r \in t} \left[\eta^*_{NCF}(m, p, r) \wedge \eta_{NCF}(r, a, n) \right]$$

for all $m, n \in T, p \in P^*$, and $a \in P$.

Lemma 3.3 *Let* $NCM = (T, P, NC)$ *be a neutrosophic cubic FSM, then show that*

$$\tilde{\zeta}^*{}_{NCT}(m, pq, n) = \vee_{r \in t} \left[\tilde{\zeta}^*{}_{NCT}(m, p, r) \wedge \tilde{\zeta}^*{}_{NCT}(r, q, n) \right],$$

$$\eta^*_{NCT}(m, pq, n) = \wedge_{r \in t} \left[\eta^*_{NCT}(m, p, r) \vee \eta_{NCT}(r, q, n) \right]$$

$$\tilde{\zeta}^*{}_{NCI}(m, pq, n) = \vee_{r \in t} \left[\tilde{\zeta}^*{}_{NCI}(m, p, r) \wedge \tilde{\zeta}^*{}_{NCI}(r, q, n) \right],$$

$$\eta^*_{NCI}(m, pq, n) = \wedge_{r \in t} \left[\eta^*_{NCI}(m, p, r) \vee \eta_{NCI}(r, q, n) \right]$$

$$\tilde{\zeta}^*{}_{NCF}(m, pq, n) = \wedge_{r \in t} \left[\tilde{\zeta}^*{}_{NCF}(m, p, r) \vee \tilde{\zeta}^*{}_{NCF}(r, q, n) \right],$$

$$\eta^*_{NCF}(m, pq, n) = \vee_{r \in t} \left[\eta^*_{NCF}(m, p, r) \wedge \eta_{NCF}(r, q, n) \right]$$

for all $m, n \in T$ *and* $p, q \in P^*$.

Proof Let $m, n \in T$ and $p, q \in P^*$ suppose $|P| = n$ we prove the result by induction.

If $n = 0$, then $q = \lambda$ and so $pq = p\lambda = p$. Now

$$\vee_{r \in t} \left[\tilde{\zeta}^*_{NCT}(m, p, r) \wedge \tilde{\zeta}^*_{NCT}(r, q, n) \right] = \vee_{r \in t} \left[\tilde{\zeta}^*_{NCT}(m, p, r) \wedge \tilde{\zeta}^*_{NCT}(r, \lambda, n) \right].$$

Hence

$$\tilde{\zeta}^*_{NCT}(m, pa, n) = \tilde{\zeta}^*_{NCT}(m, pq, n).$$

Now

$$\wedge_{r \in t} \left[\eta^*_{NCT}(m, p, r) \vee \eta_{NCT}(r, q, n) \right] = \wedge_{r \in t} \left[\eta^*_{NCT}(m, p, r) \vee \eta_{NCT}(r, \lambda, n). \right]$$

Hence

$$\eta^*_{NCT}(m, p, n) = \eta^*_{NCT}(m, pq, n).$$

Hence the result is true for $n = 0$. Let us consider that the result is true for all $c \in P^*$ such that $|c| = n - 1, n > 0$ then $q = cd$, where $c \in P^*$ and $d \in p$, $|c| = n - 1, n > 0$ then

$$
\begin{aligned}
\tilde{\zeta}^*_{NCT}(m, pq, n) &= \tilde{\zeta}^*_{NCT}(m, pcd, n) \\
&= \vee_{r \in t} \left[\tilde{\zeta}^*_{NCT}(m, pc, r) \wedge \tilde{\zeta}^*_{NCT}(r, d, n) \right] \\
&= \vee_{r \in t} \left[\vee_{s \in S} \left(\tilde{\zeta}^*_{NCT}(m, p, s) \wedge \tilde{\zeta}^*_{NCT}(s, c, r) \right) \wedge \tilde{\zeta}^*_{NCT}(r, d, n) \right] \\
&= \vee_{r, s \in t} \left[\tilde{\zeta}^*_{NCT}(m, p, s) \wedge \tilde{\zeta}^*_{NCT}(s, c, r) \wedge \tilde{\zeta}^*_{NCT}(r, d, n) \right] \\
&= \vee_{s \in t} \left[\tilde{\zeta}^*_{NCT}(m, p, s) \wedge \left(\vee_{r \in t} \left(\tilde{\zeta}^*_{NCT}(s, c, r) \wedge \tilde{\zeta}^*_{NCT}(r, d, n) \right) \right) \right] \\
&= \vee_{s \in t} \left[\tilde{\zeta}^*_{NCT}(m, p, s) \wedge \tilde{\zeta}^*_{NCT}(s, cd, n) \right] \\
&= \vee_{s \in t} \left[\tilde{\zeta}^*_{NCT}(m, p, s) \wedge \tilde{\zeta}^*_{NCT}(s, q, n) \right] \\
&= \tilde{\zeta}^*_{NCT}(m, pq, n)
\end{aligned}
$$

and

$$
\begin{aligned}
\eta^*_{NCT}(m, pq, n) &= \eta^*_{NCT}(m, pcd, n) \\
&= \wedge_{r \in t} \left[\eta^*_{NCT}(m, pc, r) \vee \eta^*_{NCT}(r, d, n) \right] \\
&= \wedge_{r \in t} \left[\wedge_{s \in t} \left(\eta^*_{NCT}(m, p, s) \vee \eta^*_{NCT}(s, c, r) \vee \eta^*_{NCT}(r, d, n) \right) \right] \\
&= \wedge_{r, s \in t} \left[\eta^*_{NCT}(m, p, s) \vee \eta^*_{NCT}(s, c, r) \vee \eta^*_{NCT}(r, d, n) \right] \\
&= \wedge_{s \in t} \left[\eta^*_{NCT}(m, p, s) \vee \left(\wedge_{r \in t} \left(\eta^*_{NCT}(s, c, r) \vee \eta^*_{NCT}(r, d, n) \right) \right) \right] \\
&= \wedge_{s \in t} \left[\eta^*_{NCT}(m, p, s) \vee \eta^*_{NCT}(s, cd, n) \right]
\end{aligned}
$$

$$= \wedge_{s \in t} \left[\eta^*_{NCT} \left(m, p, s \right) \vee \eta^*_{NCT} \left(s, q, n \right) \right]$$

$$== \eta^*_{NCT} \left(m, pq, n \right)$$

therefore the result is true for $|b| = n, n > 0$. Now the case of indeterminacy is same as that of truthness. For falsity, let $m, n \in T$ and $p, q \in P^*$ suppose $|P| = n$ we prove the result by induction if $n = 0$, then $q = \lambda$ and so $pq = p\lambda = p$. Now

$$\wedge_{r \in t} \left[\tilde{\zeta}^*_{NCF} \left(m, p, r \right) \vee \tilde{\zeta}^*_{NCF} \left(r, q, n \right) \right] = \wedge_{r \in t} \left[\tilde{\zeta}^*_{NCF} \left(m, p, r \right) \vee \tilde{\zeta}^*_{NCF} \left(r, \lambda, n \right) \right]$$

hence

$$\tilde{\zeta}^*_{NCF} \left(m, pa, n \right) = \tilde{\zeta}^*_{NCF} \left(m, pq, n \right),$$

and

$$\vee_{r \in t} \left[\eta^*_{NCF} \left(m, p, r \right) \wedge \eta_{NCF} \left(r, q, n \right) \right] = \vee_{r \in t} \left[\eta^*_{NCF} \left(m, p, r \right) \wedge \eta_{NCF} \left(r, \lambda, n \right) \right]$$

hence

$$\eta^*_{NCF} \left(m, p, n \right) = \eta^*_{NCF} \left(m, pq, n \right).$$

Hence the result is true for $n = 0$.

Let us consider that the result is true for all $c \in P^*$ such that $|c| = n - 1, n > 0$ then $q = cd$, where $c \in P^*$ and $d \in p, |c| = n - 1, n > 0$ then

$$\tilde{\zeta}^*_{NCF} \left(m, pq, n \right) = \tilde{\zeta}^*_{NCF} \left(m, pcd, n \right) = \wedge_{r \in t} \left[\tilde{\zeta}^*_{NCF} \left(m, pc, r \right) \vee \tilde{\zeta}^*_{NCF} \left(r, d, n \right) \right]$$

$$= \wedge_{r, s \in t} \left[\tilde{\zeta}^*_{NCF} \left(m, p, s \right) \vee \tilde{\zeta}^*_{NCF} \left(s, c, r \right) \vee \tilde{\zeta}^*_{NCF} \left(r, d, n \right) \right]$$

$$= \wedge_{r \in t} \left[\tilde{\zeta}^*_{NCF} \left(m, p, s \right) \vee \left(\wedge_{r \in t} \left(\tilde{\zeta}^*_{NCF} \left(s, c, r \right) \vee \tilde{\zeta}^*_{NCF} \left(r, d, n \right) \right) \right) \right]$$

$$= \wedge_{r \in t} \left[\tilde{\zeta}^*_{NCF} \left(m, p, s \right) \vee \tilde{\zeta}^*_{NCF} \left(s, cd, n \right) \right]$$

$$= \wedge_{r \in t} \left[\tilde{\zeta}^*_{NCF} \left(m, p, s \right) \vee \tilde{\zeta}^*_{NCF} \left(s, cd, n \right) \right]$$

$$= \wedge_{r \in t} \left[\tilde{\zeta}^*_{NCF} \left(m, p, s \right) \vee \tilde{\zeta}^*_{NCF} \left(s, q, n \right) \right]$$

$$== \tilde{\zeta}^*_{NCF} \left(m, pq, n \right)$$

and

$$\eta^*_{NCF} \left(m, pq, n \right) = \eta^*_{NCF} \left(m, pcd, n \right)$$

$$= \vee_{r \in t} \left[\eta^*_{NCF} \left(m, pc, r \right) \wedge \eta^*_{NCF} \left(r, d, n \right) \right]$$

$$= \vee_{r \in t} \left[\wedge_{s \in t} \left(\eta^*_{NCF} \left(m, p, s \right) \wedge \eta^*_{NCF} \left(s, c, r \right) \wedge \eta^*_{NCF} \left(r, d, n \right) \right) \right]$$

$$= \vee_{r, s \in t} \left[\eta^*_{NCF} \left(m, p, s \right) \wedge \eta^*_{NCF} \left(s, c, r \right) \wedge \eta^*_{NCF} \left(r, d, n \right) \right]$$

$$= \vee_{s \in t} \left[\eta^*_{NCF} \left(m, p, s \right) \wedge \left(\vee_{r \in t} \left(\eta^*_{NCF} \left(s, c, r \right) \wedge \eta^*_{NCF} \left(r, d, n \right) \right) \right) \right]$$

$$= \vee_{s \in t} \left[\eta^*_{NCF} (m, p, s) \wedge \eta^*_{NCF} (s, cd, n) \right]$$

$$= \vee_{s \in t} \left[\eta^*_{NCF} (m, p, s) \wedge \eta^*_{NCF} (s, q, n) \right]$$

$$= \eta^*_{NCF} (m, pq, n)$$

therefore the result is true for $|b| = n, n > 0$.

4 Subsystems of Neutrosophic Cubic Finite State Machines

In this section we study the subsystems of neutrosophic cubic finite state machine and some related results under our discussion and also discuss internal and external neutrosophic cubic subsystems with related examples.

Definition 4.1 Let NCM $=$ (T, P, NC) be a neutrosophic cubic finite state machine. Let

$$\widehat{NC} = \left(\tilde{\zeta}_{\widehat{NT}}, \tilde{\zeta}_{\widehat{NI}}, \tilde{\zeta}_{\widehat{NF}}, \eta_{\widehat{NT}}, \eta_{\widehat{NI}}, \eta_{\widehat{NF}} \right)$$

be a neutrosophic cubic subset in T. Then $\left(T, \widehat{NC}, P, NC \right)$ is called subsystems of $\widehat{NC}$ if and only if

$$\eta_{\widehat{NC}_T} (n_2) \geq r \min \left\{ \tilde{\zeta}_{\widehat{NC}_T} (n_1), \tilde{\zeta}_{\widehat{NC}_T} (n_1, a, n_2) \right\}, \eta_{\widehat{NC}_T} (n_2) \leq \max \left\{ \eta_{\widehat{NC}_T} (n_2), \eta_{\widehat{NC}_T} (n_1, a, n_2) \right\}$$

$$\eta_{\widehat{NC}_I} (n_2) \geq r \min \left\{ \tilde{\zeta}_{\widehat{NC}_I} (n_1), \tilde{\zeta}_{\widehat{NC}_I} (n_1, a, n_2) \right\}, \eta_{\widehat{NC}_I} (n_2) \leq \max \left\{ \eta_{\widehat{NC}_I} (n_2), \eta_{\widehat{NC}_I} (n_1, a, n_2) \right\}$$

$$\tilde{\zeta}_{\widehat{NC}_F} (n_2) \leq r \max \left\{ \tilde{\zeta}_{\widehat{NC}_F} (n_1), \tilde{\zeta}_{\widehat{NC}_F} (n_1, a, n_2) \right\}, \eta_{\widehat{NC}_F} (n_2) \geq \min \left\{ \eta_{\widehat{NC}_F} (n_2), \eta_{\widehat{NC}_F} (n_1, a, n_2) \right\}$$

for all $n_1, n_2 \in T$ and $a \in P$. If the neutrosophic cubic subsystem of NCM is $\left(T, \widehat{NC}, P, NC \right)$, then simply we can write $\widehat{NC}$ for $\left(T, \widehat{NC}, P, NC \right)$.

Example Let $T = \{ n_1, n_2, n_3 \}$ and $P = \{ a, b \}$. Let $NC = \left(\tilde{\zeta}_{\widehat{NT}}, \tilde{\zeta}_{\widehat{NI}}, \tilde{\zeta}_{\widehat{NF}}, \eta_{\widehat{NT}}, \eta_{\widehat{NI}}, \eta_{\widehat{NF}} \right)$ be a neutrosophic cubic subset in $T \times P \times T$ define by the table

$T \times P \times T$	$\tilde{\zeta}_{\widehat{NC}_T}$	$\eta_{\widehat{NC}_T}$
(n_1, a, n_2)	[0.3, 0.6], [0.4, 0.6], [0.3, 0.1]	0.7, 0.8, 0.3
(n_2, a, n_3)	[0.1, 0.5], [0.2, 0.6], [0.2, 0.4]	0.9, 0.9, 0.2
(n_3, a, n_1)	[0.2, 0.4], [0.3, 0.5], [0.6, 0.7]	0.7, 0.8, 0.1
(n_1, b, n_3)	[0.1, 0.6], [0.2, 0.6], [0.4, 0.6]	0.5, 0.5, 0.4
(n_3, b, n_2)	[0.3, 0.5], [0.2, 0.4], [0.6, 0.7]	0.7, 0.6, 0.4
(n_2, b, n_2)	[0.3, 0.4], [0.2, 0.5], [0.3, 0.5]	0.8, 0.7, 0.7

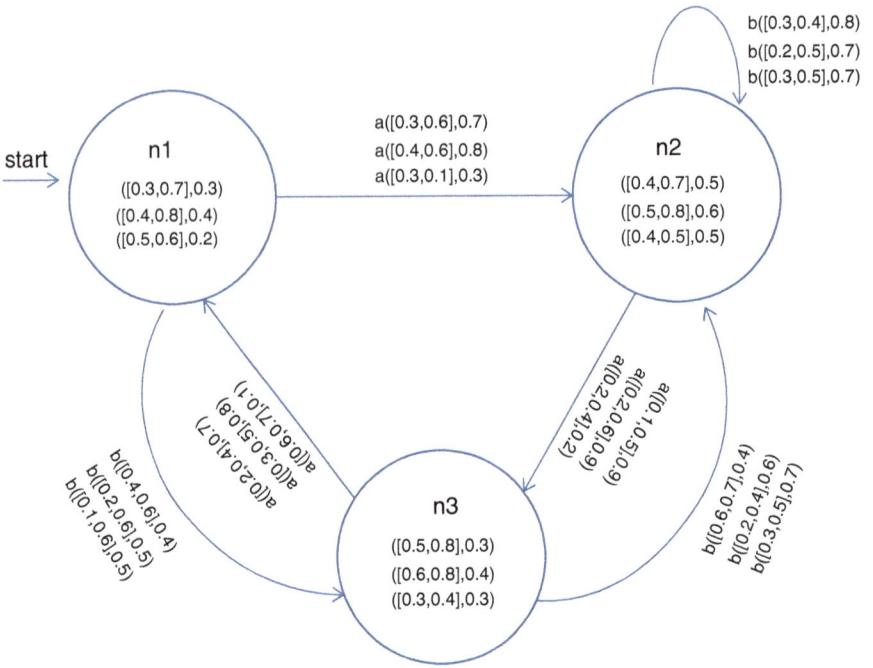

Fig. 1 Neutrosophic cubic subsystem $\widehat{NC}$

Thus NCM $=$ (T, P, NC) is a neutrosophic cubic FSM. Let NC $=$ $\left(\tilde{\zeta}_{\widehat{NT}}, \tilde{\zeta}_{\widehat{NI}}, \tilde{\zeta}_{\widehat{NF}}, \eta_{\widehat{NT}}, \eta_{\widehat{NI}}, \eta_{\widehat{NF}}\right)$ be a neutrosophic cubic subset in T defined by the table

T	$\tilde{\zeta}_{\widehat{NC}_{T,I,F}}$	$\eta_{\widehat{NC}_{T,I,F}}$
n_1	$[0.3, 0.7]$, $[0.4, 0.8]$, $[0.5, 0.6]$	$0.3, 0.4, 0.2$
n_2	$[0.4, 0.7]$, $[0.5, 0.8]$, $[0.4, 0.5]$	$0.5, 0.6, 0.5$
n_3	$[0.5, 0.8]$, $[0.6, 0.8]$, $[0.3, 0.4]$	$0.3.0.4, 0.3$

then the transition diagram is shown in Fig. 1;

Theorem 4.2 *Let* NCM $=$ (T, P, NC) *be a neutrosophic cubic FSM. Let* $\widehat{NC}^* = \left(\tilde{\zeta}_{\widehat{NT}}, \tilde{\zeta}_{\widehat{NI}}, \tilde{\zeta}_{\widehat{NF}}, \eta_{\widehat{NT}}, \eta_{\widehat{NI}}, \eta_{\widehat{NF}}\right)$ *be a neutrosophic cubic subset in T. Then* $\left(T, \widehat{NC}, P, NC\right)$ *is a neutrosophic cubic subsystem of NCM if and only if*

$$\tilde{\zeta}_{\widehat{NC}_T}(n_2) \geq r \min\left\{\tilde{\zeta}_{\widehat{NC}_T}(n_1), \tilde{\zeta}^*_{\widehat{NC}_T}(n_1, p, n_2)\right\},$$

$$\eta_{\widehat{NC}_T}(n_2) \leq \max\left\{\eta_{\widehat{NC}_T}(n_1), \eta^*_{\widehat{NC}_T}(n_1, p, n_2)\right\}$$

$$\tilde{\zeta}_{\widehat{NC}_I}(n_2) \geq r \min\left\{\tilde{\zeta}_{\widehat{NC}_I}(n_1), \tilde{\zeta}^*_{\widehat{NC}_I}(n_1, p, n_2)\right\},$$

$$\eta_{\widehat{NC}_I}(n_2) \leq \max\left\{\eta_{\widehat{NC}_I}(n_2), \eta^*_{\widehat{NC}_I}(n_1, p, n_2)\right\}$$

$$\tilde{\zeta}_{\widehat{NC}_F}(n_2) \leq r \max\left\{\tilde{\zeta}_{\widehat{NC}_F}(n_1), \tilde{\zeta}^*_{\widehat{NC}_F}(n_1, p, n_2)\right\},$$

$$\eta_{\widehat{NC}_F}(n_2) \geq \min\left\{\eta_{\widehat{NC}_F}(n_2), \eta^*_{\widehat{NC}_F}(n_1, p, n_2)\right\}$$

for all $n_1, n_2 \in T$ and $p \in P^*$.

Definition 4.3 Let $NCM = (T, P, NC)$ is a neutrosophic cubic FSM. Let $m, n \in T$. Then the immediate successor m of n is defined as, if $\exists a \in p$ such that

$$\tilde{\zeta}_{NC_T}(n, a, m) \succ [0, 0] \quad \text{and} \quad \eta_{NC_T}(n, a, m) \prec 1$$

$$\tilde{\zeta}_{NC_I}(n, a, m) \succ [0, 0] \quad \text{and} \quad \eta_{NC_I}(n, a, m) \prec 1$$

$$\tilde{\zeta}_{NC_F}(n, a, m) \prec [1, 1] \quad \text{and} \quad \eta_{NC_F}(n, a, m) \succ 0,$$

where m is called the successor of n if $\exists p \in P^*$ such that $\tilde{\zeta}^*_{\widehat{NC}_T}(n, p, m) \succ [0, 0]$ and $\eta^*_{\widehat{NC}_T} \prec 1$. Let $n \in T$ then the set of neutrosophic cubic successor of n is denoted by $NCT(n)$. If u is contained in T, then we define $NCT(U) = \cup\{NCT(n) / n \in U\}$.

Definition 4.4 Let $NCM = (T, P, NC)$ is a neutrosophic cubic FSM. Let $\hat{\rho} = \left(\zeta_{\hat{\rho}T}, \zeta_{\hat{\rho}I}, \zeta_{\hat{\rho}F}, \eta_{\hat{\rho}T}, \eta_{\hat{\rho}I}, \eta_{\hat{\rho}F}\right)$ be a cubic subset of T for all $p \in P^*$, define the neutrosophic cubic subset $\hat{\rho}p$ of T by

$$\left(\tilde{\zeta}_{\hat{\rho}T} p\right)(n) = r \sup_{m \in T}\left\{r \min\left\{\tilde{\zeta}_{\hat{\rho}T}(m), \tilde{\zeta}_{NCM_T}(m, p, n)\right\}\right\},$$

$$\left(\eta_{\hat{\rho}I} p\right)(n) - \inf_{m \subset T}\left\{\max\left\{\eta_{\hat{\rho}I}(m), \eta_{NCM_I}(m, p, n)\right\}\right\}.$$

$$\left(\tilde{\zeta}_{\hat{\rho}I} p\right)(n) = r \sup_{m \in I}\left\{r \min\left\{\tilde{\zeta}_{\hat{\rho}I}(m), \tilde{\zeta}_{NCM_I}(m, p, n)\right\}\right\},$$

$$\left(\eta_{\hat{\rho}I} p\right)(n) = \inf_{m \in I}\left\{\max\left\{\eta_{\hat{\rho}I}(m), \eta_{NCM_I}(m, p, n)\right\}\right\}.$$

$$\left(\tilde{\zeta}_{\hat{\rho}F} p\right)(n) = r \inf_{m \in T}\left\{r \max\left\{\tilde{\zeta}_{\hat{\rho}F}(m), \tilde{\zeta}_{NCM_F}(m, p, n)\right\}\right\},$$

$$\left(\eta_{\hat{\rho}F} p\right)(n) = \sup_{m \in T}\left\{\min\left\{\eta_{\hat{\rho}F}(m), \eta_{NCM_F}(m, p, n)\right\}\right\}.$$

Proposition 4.5 *Let $NCM = (T, P, NC)$ be a neutrosophic cubic FSM. Then for all neutrosophic cubic subsets $\widehat{\rho} = \left(\zeta_{\widehat{\rho}T}, \zeta_{\widehat{\rho}I}, \zeta_{\widehat{\rho}F}, \eta_{\widehat{\rho}T}, \eta_{\widehat{\rho}I}, \eta_{\widehat{\rho}F}\right)$ of T and for all $p, q \in P^*$ we have*

$$\left(\widetilde{\zeta_{\widehat{\rho}T}} p\right) q = \widetilde{\zeta_{\widehat{\rho}T}} (pq) \quad \text{and} \quad \left(\eta_{\widehat{\rho}T} p\right) q = \eta_{\widehat{\rho}T} (pq),$$

$$\left(\widetilde{\zeta_{\widehat{\rho}I}} p\right) q = \widetilde{\zeta_{\widehat{\rho}I}} (pq) \quad \text{and} \quad \left(\eta_{\widehat{\rho}I} p\right) q = \eta_{\widehat{\rho}I} (pq),$$

$$\left(\widetilde{\zeta_{\widehat{\rho}F}} p\right) q = \widetilde{\zeta_{\widehat{\rho}F}} (pq) \quad \text{and} \quad \left(\eta_{\widehat{\rho}F} p\right) q = \eta_{\widehat{\rho}F} (pq).$$

Proof Consider neutrosophic cubic subset $\widehat{\rho} = \left(\zeta_{\widehat{\rho}T}, \eta_{\widehat{\rho}T}\right)$ of T for truthness and let $p, q \in P^*$ if $n = 0$, then $q = \lambda$. Let $n \in T$ then

$$\left(\left(\zeta_{\widehat{\rho}T} p\right) \lambda\right) (n) = r \sup_{m \in T} r \min \left\{\left(\zeta_{\widehat{\rho}T} p\right) (m), \widetilde{\zeta}^*_{NCT} (m, \lambda, n)\right\}$$

$$= \left(\widetilde{\zeta_{\widehat{\rho}T}} p\right) (n)$$

and

$$\left(\left(\eta_{\widehat{\rho}T} p\right) \lambda\right) (n) = \inf_{m \in T} \left\{\max \left\{\left(\eta_{\widehat{\rho}T} p\right) (m), \eta^*_{NCT} (m, \lambda, n)\right\}\right\}$$

$$= \left(\eta_{\widehat{\rho}T} p\right) (n)$$

hence $\left(\widetilde{\zeta_{\widehat{\rho}T}} p\right) \lambda = \widetilde{\zeta_{\widehat{\rho}T}} p = \widetilde{\zeta_{\widehat{\rho}T}} (p\lambda)$ and $\left(\eta_{\widehat{\rho}T} p\right) \lambda = \eta_{\widehat{\rho}T} p = \eta_{\widehat{\rho}T} (p\lambda)$. Now we assume that the result is true for all $u \in P^*$ such that $|u| = n - 1, n \succ 0$ and for all $\widehat{\rho}$. Let $q = ua$, where $a \in P^*$ and $|u| = n - 1$. Let $n \in T$. Then

$$\left(\widetilde{\zeta_{\widehat{\rho}T}} (pq)\right) (n) = \left(\widetilde{\zeta_{\widehat{\rho}T}} (pua)\right) (n) = \left(\widetilde{\zeta_{\widehat{\rho}T}} ((pu) a)\right) (n)$$

$$= r \sup_{r \in T} \left\{r \min \left\{\left(\widetilde{\zeta_{\widehat{\rho}T}} (pu) (r), \widetilde{\zeta}^*_{NCT} (r, a, n)\right)\right\}\right\}$$

$$= r \sup_{r \in T} \left\{r \min \left\{r \sup_{t \in T} \left\{r \min \left\{\left(\widetilde{\zeta_{\widehat{\rho}T}} (p)\right) (t), \widetilde{\zeta}^*_{NCT} (t, u, r)\right\}\right\}, \widetilde{\zeta}^*_{NCT} (r, a, n)\right\}\right\}$$

$$= r \sup_{t \in T} \left\{r \min \left\{\left(\widetilde{\zeta_{\widehat{\rho}T}} (p)\right) (t), r \sup_{r \in T} \left\{r \min \left\{\widetilde{\zeta}^*_{NCT} (t, u, r), \widetilde{\zeta}^*_{NCT} (r, a, n)\right\}\right\}\right\}\right\}$$

$$= r \sup_{t \in T} \left\{r \min \left\{\left(\widetilde{\zeta_{\widehat{\rho}T}} (p)\right) (t), \widetilde{\zeta}^*_{NCT} (t, ua, n)\right\}\right\}$$

$$= \left(\widetilde{\zeta_{\widehat{\rho}T}} (p) ua\right) (n)$$

$$= \left(\widetilde{\zeta_{\widehat{\rho}T}} (p) q\right) (n)$$

and

$$\left(\eta_{\widehat{\rho}T} (pq)\right) (n) = \left(\eta_{\widehat{\rho}T} (pua)\right) (n) = \left(\eta_{\widehat{\rho}T} ((pu) a)\right) (n)$$

$$= \inf_{r \in T} \left\{\max \left\{\left(\eta_{\widehat{\rho}T} (pu) (r), \eta^*_{NCT} (r, a, n)\right)\right\}\right\}$$

$$= \inf_{r \in T} \left\{ \max \left\{ \inf_{t \in T} \left\{ \max \left\{ \left(\eta_{\widehat{\rho}T}\,(p) \right)(t)\,, \eta^*_{NCT}\,(t, u, r) \right\} \right\}, \eta^*_{NCT}\,(r, a, n) \right\} \right\}$$

$$= \inf_{t \in T} \left\{ \max \left\{ \left(\eta_{\widehat{\rho}T}\,(p) \right)(t)\,, \inf_{r \in T} \left\{ \max \left\{ \eta^*_{NCT}\,(t, u, r)\,, \eta^*_{NCT}\,(r, a, n) \right\} \right\} \right\} \right\}$$

$$= \inf_{t \in T} \left\{ \max \left\{ \left(\eta_{\widehat{\rho}T}\,(p) \right)(t)\,, \eta^*_{NCT}\,(t, ua, n) \right\} \right\}$$

$$= \left(\eta_{\widehat{\rho}T}\,(p)\,ua \right)(n)$$

$$= \left(\eta_{\widehat{\rho}T}\,(p)\,q \right)(n)$$

hence $\left(\widetilde{\zeta_{\widehat{\rho}T}}\,p \right)q = \widetilde{\zeta_{\widehat{\rho}T}}\,(pq)$ and $\left(\eta_{\widehat{\rho}T}\,p \right)q = \eta_{\widehat{\rho}T}\,(pq)$. Now for the case of indeterminacy one can prove in the similar way as above. And for falsity let $n \in T$ now

$$\left(\left(\zeta_{\widehat{\rho}F}\,p \right) \lambda \right)(n) = r\inf_{m \in T} \left\{ r\max \left\{ \left(\zeta_{\widehat{\rho}F}\,p \right)(m)\,, \widetilde{\zeta}^*_{NCF}\,(m, \lambda, n) \right\} \right\} = \left(\widetilde{\zeta_{\widehat{\rho}F}}\,p \right)(n)$$

and

$$\left(\left(\eta_{\widehat{\rho}F}\,p \right) \lambda \right)(n) = \sup_{m \in T} \left\{ \min \left\{ \left(\eta_{\widehat{\rho}F}\,p \right)(m)\,, \eta^*_{NCF}\,(m, \lambda, n) \right\} \right\} = \left(\eta_{\widehat{\rho}F}\,p \right)(n)$$

hence

$$\left(\widetilde{\zeta_{\widehat{\rho}F}}\,p \right) \lambda = \widetilde{\zeta_{\widehat{\rho}F}}\,p = \widetilde{\zeta_{\widehat{\rho}F}}\,(p\lambda) \text{ and } \left(\eta_{\widehat{\rho}F}\,p \right) \lambda = \eta_{\widehat{\rho}F}\,p = \eta_{\widehat{\rho}F}\,(p\lambda).$$

Now we assume that the result is true for all $u \in P^*$ such that $|u| = n - 1, n \succ 0$ and for all $\widehat{\rho}$. Let $q = ua$, where $a \in P^*$ and $|u| = n - 1$. Let $n \in T$. Then

$$\left(\widetilde{\zeta_{\widehat{\rho}F}}\,(pq) \right)(n) = \left(\widetilde{\zeta_{\widehat{\rho}F}}\,(pua) \right)(n) = \left(\widetilde{\zeta_{\widehat{\rho}F}}\,((pu)\,a) \right)(n)$$

$$= r\inf_{r \in T} \left\{ r\max \left\{ \widetilde{\zeta_{\widehat{\rho}F}}\,(pu)\,(r)\,, \widetilde{\zeta}^*_{NCF}\,(r, a, n) \right\} \right\}$$

$$= r\inf_{r \in T} \left\{ r\max \left\{ r\inf_{t \in T} \left\{ r\max \left\{ \widetilde{\zeta_{\widehat{\rho}F}}\,(p)\,(t)\,, \widetilde{\zeta}^*_{NCF}\,(t, u, r) \right\} \right\}, \widetilde{\zeta}^*_{NCF}\,(r, a, n) \right\} \right\}$$

$$= r\inf_{t \in T} \left\{ r\max \left\{ \widetilde{\zeta_{\widehat{\rho}F}}\,(p)\,(t)\,, r\inf_{r \in T} \left\{ r\max \left\{ \widetilde{\zeta}^*_{NCF}\,(t, u, r)\,, \widetilde{\zeta}^*_{NCF}\,(r, a, n) \right\} \right\} \right\} \right\}$$

$$= r\inf_{t \in T} \left\{ r\max \left\{ \left(\widetilde{\zeta_{\widehat{\rho}F}}\,(p) \right)(t)\,, \widetilde{\zeta}^*_{NCF}\,(t, ua, n) \right\} \right\}$$

$$= \left(\widetilde{\zeta_{\widehat{\rho}F}}\,(p)\,ua \right)(n) = \left(\widetilde{\zeta_{\widehat{\rho}F}}\,(p)\,q \right)(n)$$

and

$$\left(\eta_{\widehat{\rho}F}\,(pq) \right)(n) = \left(\eta_{\widehat{\rho}F}\,(pua) \right)(n) = \left(\eta_{\widehat{\rho}F}\,((pu)\,a) \right)(n)$$

$$= r\sup_{r \in T} \left\{ \min \left\{ \left(\eta_{\widehat{\rho}F}\,(pu)\,(r)\,, \eta^*_{NCF}\,(r, a, n) \right) \right\} \right\}$$

$$= r \sup_{r \in T} \left\{ \min \left\{ r \sup_{t \in T} \left\{ \min \left\{ (\eta \widehat{\rho}_F \, (p)) \, (t) \, , \eta^*_{NCF} \, (t, u, r) \right\} \right\} , \eta^*_{NCF} \, (r, a, n) \right\} \right\}$$

$$= r \sup_{t \in T} \left\{ \min \left\{ (\eta_{\widehat{\rho} F} \, (p)) \, (t) \, , r \sup_{r \in T} \left\{ \min \left\{ \eta^*_{NCF} \, (t, u, r) \, , \eta^*_{NCF} \, (r, a, n) \right\} \right\} \right\} \right\}$$

$$= r \sup_{t \in T} \left\{ \min \left\{ (\eta_{\widehat{\rho} F} \, (p)) \, (t) \, , \eta^*_{NCF} \, (t, ua, n) \right\} \right\}$$

$$= \left(\eta_{\widehat{\rho} F} \, (p) \, ua \right) (n) = \left(\eta_{\widehat{\rho} F} \, (p) \, q \right) (n)$$

hence

$$\left(\widetilde{\zeta_{\widehat{\rho}_F}} \, p \right) q = \widetilde{\zeta_{\widehat{\rho}_F}} \, (pq) \quad \text{and} \quad \left(\eta_{\widehat{\rho}_F} \, p \right) q = \eta_{\widehat{\rho}_F} \, (pq) \, .$$

Theorem 4.6 *Let $NCM = (T, P, NC)$ be a neutrosophic cubic FSM. Then for all neutrosophic cubic subsets $\widehat{\rho} = \left(\zeta_{\widehat{\rho} T}, \zeta_{\widehat{\rho} I}, \zeta_{\widehat{\rho} F}, \eta_{\widehat{\rho} T}, \eta_{\widehat{\rho} I}, \eta_{\widehat{\rho} F} \right)$ of T. Then $\widehat{\rho}$ subsystem of NCM if and only if*

(1) $\widetilde{\zeta_{\widehat{\rho} T}} \, p \subseteq \widetilde{\zeta_{\widehat{\rho} T}}$ and $\eta_{\widehat{\rho} T} \, p \supseteq \eta_{\widehat{\rho} T}$ for all $p \in P^*$, (2) $\widetilde{\zeta_{\widehat{\rho} I}} \, p \subseteq \widetilde{\zeta_{\widehat{\rho} I}}$ and $\eta_{\widehat{\rho} I} \, p \supseteq \eta_{\widehat{\rho} I}$ for all $p \in P^*$,

(3) $\widetilde{\zeta_{\widehat{\rho} F}} \, p \supseteq \widetilde{\zeta_{\widehat{\rho} F}}$ and $\eta_{\widehat{\rho} F} \, p \subseteq \eta_{\widehat{\rho} F}$ for all $p \in P^*$.

Proof

(1) Let $\widehat{\rho}$ be a subsystem of NCM and let $p \in P^*$ and $n \in T$ then

$$\left(\widetilde{\zeta_{\widehat{\rho} T}} \, p \right) (n) = r \sup_{m \in T} \left\{ r \min \left\{ \widetilde{\zeta_{\widehat{\rho} T}} \, (m) \, , \widetilde{\zeta}^*_{NCT} \, (m, p, n) \right\} \right\} \preceq \widetilde{\zeta_{\widehat{\rho} T}} n,$$

and

$$\left(\eta_{\widehat{\rho} T} \, p \right) (n) = \inf_{m \in T} \left\{ \max \left\{ \eta_{\widehat{\rho} T} \, (m) \, , \eta^*_{NCT} \, (m, p, n) \right\} \right\} \geq \eta_{\widehat{\rho} T} \, (n) \, .$$

Hence

$$\widetilde{\zeta_{\widehat{\rho} T}} \, p \subseteq \widetilde{\zeta_{\widehat{\rho} T}} \quad \text{and} \quad \eta_{\widehat{\rho} T} \, p \supseteq \eta_{\widehat{\rho} T}.$$

Conversely, let us suppose that $\widetilde{\zeta_{\widehat{\rho} T}} \, p \subseteq \widetilde{\zeta_{\widehat{\rho} T}}$ and $\eta_{\widehat{\rho} T} \, p \supseteq \eta_{\widehat{\rho} T}$ for all $p \in P^*$. Let $n \in T$ and $p \in P^*$. Now

$$\widetilde{\zeta_{\widehat{\rho} T}} \, (n) \succeq \left(\widetilde{\zeta_{\widehat{\rho} T}} \, p \right) (n) = r \sup_{m \in T} \left\{ r \min \left\{ \widetilde{\zeta_{\widehat{\rho} T}} \, (m) \, , \widetilde{\zeta}^*_{NCT} \, (m, p, n) \right\} \right\}$$

$$\succeq r \min \left\{ \widetilde{\zeta_{\widehat{\rho} T}} \, (n) \, , \widetilde{\zeta}^*_{NCT} \, (m, p, n) \right\} ,$$

and

$$\eta_{\widehat{\rho} T} \, (n) \leq \left(\eta_{\widehat{\rho} T} \, p \right) (n) = \inf_{m \in T} \left\{ \max \left\{ \eta_{\widehat{\rho} T} \, (m) \, , \eta^*_{NCT} \, (m, p, n) \right\} \right\}$$

$$\leq \max \left\{ \eta_{\widehat{\rho}T}^{}(m), \eta_{NCT}^{*}(m, p, n) \right\}.$$

For indeterminacy case we can use the above case. And for falsity, let $\widehat{\rho}$ be a subsystem of NCM and let $p \in P^*$ and $n \in T$ then

$$\left(\widetilde{\zeta}_{\widehat{\rho}F}\, p \right)(n) = r \inf_{m \in T} \left\{ r \max \left\{ \widetilde{\zeta}_{\widehat{\rho}F}(m), \widetilde{\zeta}_{NCF}^{*}(m, p, n) \right\} \right\} \geq \widetilde{\zeta}_{\widehat{\rho}F}(n),$$

and

$$\left(\eta_{\widehat{\rho}F}\, p \right)(n) = \sup_{m \in T} \left\{ \min \left\{ \eta_{\widehat{\rho}F}(m), \eta_{NCF}^{*}(m, p, n) \right\} \right\} \leq \eta_{\widehat{\rho}F}(n).$$

Hence $\widetilde{\zeta}_{\widehat{\rho}F}\, p \supseteq \widetilde{\zeta}_{\widehat{\rho}F}$ and $\eta_{\widehat{\rho}F}\, p \subseteq \eta_{\widehat{\rho}F}$. Conversely,let us suppose that $\widetilde{\zeta}_{\widehat{\rho}F}\, p \supseteq \widetilde{\zeta}_{\widehat{\rho}F}$ and $\eta_{\widehat{\rho}F}\, p \subseteq \eta_{\widehat{\rho}F}$ for all $p \in P^*$. Let $n \in T$ and $p \in P^*$. Now

$$\widetilde{\zeta}_{\widehat{\rho}F}(n) \leq \left(\widetilde{\zeta}_{\widehat{\rho}F}\, p \right)(n) = r \inf_{m \in T} \left\{ r \max \left\{ \widetilde{\zeta}_{\widehat{\rho}F}(m), \widetilde{\zeta}_{NCF}^{*}(m, p, n) \right\} \right\}$$

$$\leq r \max \left\{ \widetilde{\zeta}_{\widehat{\rho}F}(n), \widetilde{\zeta}_{NCF}^{*}(m, p, n) \right\},$$

and

$$\eta_{\widehat{\rho}F}(n) \geq \left(\eta_{\widehat{\rho}F}\, p \right)(n) = \sup_{m \in T} \left\{ \min \left\{ \eta_{\widehat{\rho}F}(m), \eta_{NCF}^{*}(m, p, n) \right\} \right\}$$

$$\geq \min \left\{ \eta_{\widehat{\rho}F}(m), \eta_{NCF}^{*}(m, p, n) \right\}.$$

Hence $\widehat{\rho}$ is a subsystem of NCM.

Definition 4.7 Let $NCM = (T, P, NC)$ be a neutrosophic cubic FSM. Let $h = ([i, j], k) \in D(0, 1] \times (0, 1]$ and $n \in T$. Define the neutrosophic cubic subset $n_h P = \langle n[i, j]P, n_k P \rangle$ of T by

$$(n[i, j]P)(n) = r \sup_{a \in P} \left\{ r \min \left\{ [i, j], \widetilde{\zeta}_{NCT}(n, a, m) \right\} \right\} \text{ and}$$

$$(n_k P)(n) = \inf_{a \in P} \left\{ \max \left\{ k, \eta_{NCT}(n, a, m) \right\} \right\}$$

$$(n[i, j]P)(n) = r \sup_{a \in P} \left\{ r \min \left\{ [i, j], \widetilde{\zeta}_{NCI}(n, a, m) \right\} \right\} \text{ and}$$

$$(n_k P)(n) = \inf_{a \in P} \left\{ \max \left\{ k, \eta_{NCI}(n, a, m) \right\} \right\}$$

$$(n[i, j]P)(n) = r \inf_{a \in P} \left\{ r \max \left\{ [i, j], \widetilde{\zeta}_{NCF}(n, a, m) \right\} \right\} \text{ and}$$

$$(n_k P)(n) = \sup_{a \in P} \left\{ \min \left\{ k, \eta_{NCF}(n, a, m) \right\} \right\}$$

for all $m \in T$.

Definition 4.8 Let $NCM = (T, P, NC)$ is a neutrosophic cubic FSM. Let $h = ([i, j], k) \in D(0, 1] \times (0, 1]$ and $n \in T$. Define the neutrosophic cubic subset $n_h P^* = \langle n[i, j] P^*, n_k P^* \rangle$ of T by

$$\left(n[i, j] P^*\right)(n) = r \sup_{a \in p^*} \left\{r \min \left\{[i, j], \tilde{\zeta}_{NCT}(n, a, m)\right\}\right\} \quad \text{and}$$

$$\left(n_k P^*\right)(n) = \inf_{a \in P^*} \left\{\max \left\{k, \eta_{NCT}(n, a, m)\right\}\right\}$$

$$\left(n[i, j] P^*\right)(n) = r \sup_{a \in p^*} \left\{r \min \left\{[i, j], \tilde{\zeta}_{NCI}(n, a, m)\right\}\right\} \quad \text{and}$$

$$\left(n_k P^*\right)(n) = \inf_{a \in P^*} \left\{\max \left\{k, \eta_{NCI}(n, a, m)\right\}\right\}$$

$$\left(n[i, j] P^*\right)(n) = r \inf_{a \in P^*} \left\{r \max \left\{[i, j], \tilde{\zeta}_{NCF}(n, a, m)\right\}\right\} \quad \text{and}$$

$$\left(n_k P^*\right)(n) = r \sup_{a \in P^*} \left\{\min \left\{k, \eta_{NCF}(n, a, m)\right\}\right\}$$

for all $m \in T$.

Theorem 4.9 *Let $NCM = (T, P, NC)$ be a neutrosophic cubic FSM. Let $h = ([i, j], k) \in D(0, 1] \times (0, 1]$ and $n \in T$. Then the following assertions hold*
(1) $n_h P^* = \langle n[i, j] P^*, n_k P^* \rangle$ is a subsystem of NCM. (2) $supp(n_h P^*) = T(n)$.

Proof For truthness. Let $b \in T$ and $p \in P^*$ now

$$\left(\langle n[i, j] P^* \rangle p\right)(b) = r \sup_{r \in T} \left\{r \min \left\{\langle n[i, j] P^* \rangle(r), \tilde{\zeta}^*_{NCT}(r, p, b)\right\}\right\}$$

$$= r \sup_{r \in T} \left\{r \min \left\{r \sup_{q \in P^*} \left\{r \min p \left\{[i, j], \tilde{\zeta}^*_{NCT}(n, q, r)\right\}\right\}, \tilde{\zeta}^*_{NCT}(r, p, b)\right\}\right\}$$

$$= r \sup_{r \in T, q \in P^*} \left\{r \min \left\{[i, j], \tilde{\zeta}^*_{NCT}(n, q, r), \tilde{\zeta}^*_{NCT}(r, p, b)\right\}\right\}$$

$$= r \sup_{q \in P^*} \left\{r \min \left\{[i, j], \tilde{\zeta}^*_{NCT}(n, qp, b)\right\}\right\}$$

$$\leq r \sup_{u \in P^*} \left\{r \min \left\{[i, j], \tilde{\zeta}^*_{NCT}(n, u, b)\right\}\right\} = \left(n[i, j] P^*\right)(b),$$

and

$$\left(\langle nk P^* \rangle p\right)(b) = \inf_{r \in T} \left\{\max \left\{\langle nk P^* \rangle(r), \eta^*_{NCT}(r, p, b)\right\}\right\}$$

$$= \inf_{r \in T} \left\{\max \left\{\inf_{q \in P^*} \left\{max \left\{k, \eta^*_{NCT}(n, q, r)\right\}\right\}, \eta^*_{NCT}(r, p, b)\right\}\right\}$$

$$= \inf_{r \in T, q \in P^*} \left\{\max \left\{k, \eta^*_{NCT}(n, q, r), \eta^*_{NCT}(r, p, b)\right\}\right\}$$

$$= \inf_{q \in P^*} \left\{\max \left\{k, \eta^*_{NCT}(n, qp, b)\right\}\right\}$$

$$\succeq \inf_{u\in P^*} \left\{\max\left\{k, \eta^*_{NCT}(n, u, b)\right\}\right\} = \left(nkP^*\right)(b).$$

Hence

$$\left\langle n\,[i, j]\,P^*\right\rangle p \subseteq n\,[i, j]\,P^* \text{ and } \left\langle nkP^*\right\rangle p \supseteq \left(nkP^*\right)$$

thus $n_h P^* = \left\langle n\,[i, j]\,P^*, n_k P^*\right\rangle$ is a subsystem of NCM according to this statement $\tilde{\zeta}_{\widehat{\rho}T}\,p \subseteq \tilde{\zeta}_{\widehat{\rho}T}$ and $\eta_{\widehat{\rho}T}\,p \supseteq \eta_{\widehat{\rho}T}$ for all $p \in P^*$. Now (2) $\mathrm{supp}(n_h P^*) = T(n)$ is straightforward. Now the case of indeterminacy is similar as above case 1. And for falsity, let $b \in T$ and $p \in P^*$ now

$$\left(\left\langle n\,[i, j]\,P^*\right\rangle p\right)(b) = r \inf_{r\in T}\left\{r\max\left\{\left\langle n\,[i, j]\,P^*\right\rangle(r), \tilde{\zeta}^*_{NCF}(r, p, b)\right\}\right\}$$

$$= r \inf_{r\in T}\left\{r\max\left\{r \inf_{q\in P^*}\left\{r\max\left\{[i, j], \tilde{\zeta}^*_{NCF}(n, q, r)\right\}\right\}, \tilde{\zeta}^*_{NCF}(r, p, b)\right\}\right\}$$

$$= r \inf_{r\in T, q\in P^*}\left\{r\max\left\{[i, j], \tilde{\zeta}^*_{NCF}(n, q, r), \tilde{\zeta}^*_{NCF}(r, p, b)\right\}\right\}$$

$$= r \inf_{q\in P^*}\left\{r\max\left\{[i, j], \tilde{\zeta}^*_{NCF}(n, qp, b)\right\}\right\}$$

$$=\succeq r \inf_{u\in P^*}\left\{r\max\left\{[i, j], \tilde{\zeta}^*_{NCF}(n, u, b)\right\}\right\} = \left(n\,[i, j]\,P^*\right)(b)$$

and

$$\left(\left\langle nkP^*\right\rangle p\right)(b) = \sup_{r\in T}\left\{\min\left\{\left\langle nkP^*\right\rangle(r), \eta^*_{NCF}(r, p, b)\right\}\right\}$$

$$= \sup_{r\in T}\left\{\min\left\{\sup_{q\in P^*}\left\{\min\left\{k, \eta^*_{NCF}(n, q, r)\right\}\right\}, \eta^*_{NCF}(r, p, b)\right\}\right\}$$

$$= \sup_{r\in T, q\in P^*}\left\{\min\left\{k, \eta^*_{NCF}(n, q, r), \eta^*_{NCF}(r, p, b)\right\}\right\}$$

$$= \sup_{q\in P^*}\left\{\min\left\{k, \eta^*_{NCF}(n, qp, b)\right\}\right\}$$

$$=\succeq \sup_{u\in P^*}\left\{\min\left\{k, \eta^*_{NCF}(n, u, b)\right\}\right\}$$

$$= \left(nkP^*\right)(b).$$

Hence

$$\left\langle n\,[i, j]\,P^*\right\rangle p \supseteq n\,[i, j]\,P^* \text{ and } \left\langle nkP^*\right\rangle p \subseteq \left(nkP^*\right).$$

Thus $n_h P^* = \left\langle n\,[i, j]\,P^*, n_k P^*\right\rangle$ is a subsystem of NCM according to this statement $\tilde{\zeta}_{\widehat{\rho}F}\,p \subseteq \tilde{\zeta}_{\widehat{\rho}F}$ and $\eta_{\widehat{\rho}F}\,p \supseteq \eta_{\widehat{\rho}F}$ for all $p \in P^*$. Now (2)$\mathrm{supp}(n_h P^*) = T(n)$ is straightforward.

Definition 4.10 A subsystem $\widehat{NC^I} = \left\langle \tilde{\zeta}_{\widehat{NC_T^I}}, \tilde{\zeta}_{\widehat{NC_I^I}}, \tilde{\zeta}_{\widehat{NC_F^I}}, \eta_{\widehat{NC_T^I}}, \eta_{\widehat{NC_I^I}}, \eta_{\widehat{NC_F^I}} \right\rangle$ of a neutrosophic cubic subsystems of FSM, $NCM = (T, P, NC)$ is said to be internal neutrosophic cubic subsystem (INC − subsystem) if (1) $=$

$$
\begin{cases}
\zeta^-_{\widehat{NC_T}} \le \eta_{\widehat{NC_T}} \le \zeta^+_{\widehat{NC_T}} \\
\zeta^-_{\widehat{NC_I}} \le \eta_{\widehat{NC_I}} \le \zeta^+_{\widehat{NC_I}} \\
\zeta^-_{\widehat{NC_F}} \le \eta_{\widehat{NC_F}} \le \zeta^+_{\widehat{NC_F}}
\end{cases},
$$

$$
(2) = \begin{cases}
\zeta_{NC_T^-}(n, a, m) \le \eta_{NC_T}(n, a, m) \le \zeta_{NC_T^+}(n, a, m) \\
\zeta_{NC_I^-}(n, a, m) \le \eta_{NC_I}(n, a, m) \le \zeta_{NC_I^+}(n, a, m) \quad \text{for all } n, m \in T \text{ and } a \in P . \\
\zeta_{NC_F^-}(n, a, m) \le \eta_{NC_F}(n, a, m) \le \zeta_{NC_F^+}(n, a, m)
\end{cases}
$$

Definition 4.11 A subsystem $\widehat{NC^E} = \left\langle \tilde{\zeta}_{\widehat{NC_T^E}}, \tilde{\zeta}_{\widehat{NC_I^E}}, \tilde{\zeta}_{\widehat{NC_F^E}}, \eta_{\widehat{NC_T^E}}, \eta_{\widehat{NC_I^E}}, \eta_{\widehat{NC_F^E}} \right\rangle$ of a neutrosophic cubic subsystems of FSM, $NCM = (T, P, NC)$ is said to be external neutrosophic cubic subsystem (ENC − subsystem) if (1) $=$

$$
\begin{cases}
\eta_{\widehat{NC_T}}(n) \notin \left(\zeta^-_{\widehat{NC_T}}(n), \zeta^+_{\widehat{NC_T}}(n) \right) \\
\eta_{\widehat{NC_I}}(n) \notin \left(\zeta^-_{\widehat{NC_I}}(n), \zeta^+_{\widehat{NC_I}}(n) \right) \\
\eta_{\widehat{NC_F}}(n) \notin \left(\zeta^-_{\widehat{NC_F}}(n), \zeta^+_{\widehat{NC_F}}(n) \right)
\end{cases},
$$

$$
(2) = \begin{cases}
\eta_{NC_T}(n, a, m) \notin \left(\zeta_{NC_T^-}(n, a, m), \zeta_{NC_T^+}(n, a, m) \right) \\
\eta_{NC_I}(n, a, m) \notin \left(\zeta_{NC_I^-}(n, a, m), \zeta_{NC_I^+}(n, a, m) \right) \quad \text{for all } n, m \in T \text{ and } a \in P . \\
\eta_{NC_F}(n, a, m) \notin \left(\zeta_{NC_F^-}(n, a, m), \zeta_{NC_F^+}(n, a, m) \right)
\end{cases}
$$

Example The neutrosophic cubic $FSMs$

$$
NCM^I = \left(T, \widehat{NC^I}, P, NC \right) \text{ and } NCM^E = \left(T, \widehat{NC^E}, P, NC \right)
$$

are internal and external neutrosophic cubic subsystems, respectively, as shown in Figs. 2 and 3;

Theorem 4.12 *Let $NCM = (T, P, NC)$ be a neutrosophic cubic FSM. Then*

$$
\widehat{NC^I} = \left\{ \tilde{\zeta}_{\widehat{NC^I}_T}, \tilde{\zeta}_{\widehat{NC^I}_I}, \tilde{\zeta}_{\widehat{NC^I}_F}, \eta_{\widehat{NC^I}_T}, \eta_{\widehat{NC^I}_I}, \eta_{\widehat{NC^I}_F} \right\}
$$

is an internal neutrosophic cubic subsystem of NCM if

$$
\begin{cases}
\zeta^-_{\widehat{NC_T}} \le \eta_{\widehat{NC_T}} \le \zeta^+_{\widehat{NC_T}} \\
\zeta^-_{\widehat{NC_I}} \le \eta_{\widehat{NC_I}} \le \zeta^+_{\widehat{NC_I}} \\
\zeta^-_{\widehat{NC_F}} \le \eta_{\widehat{NC_F}} \le \zeta^+_{\widehat{NC_F}}
\end{cases}
$$

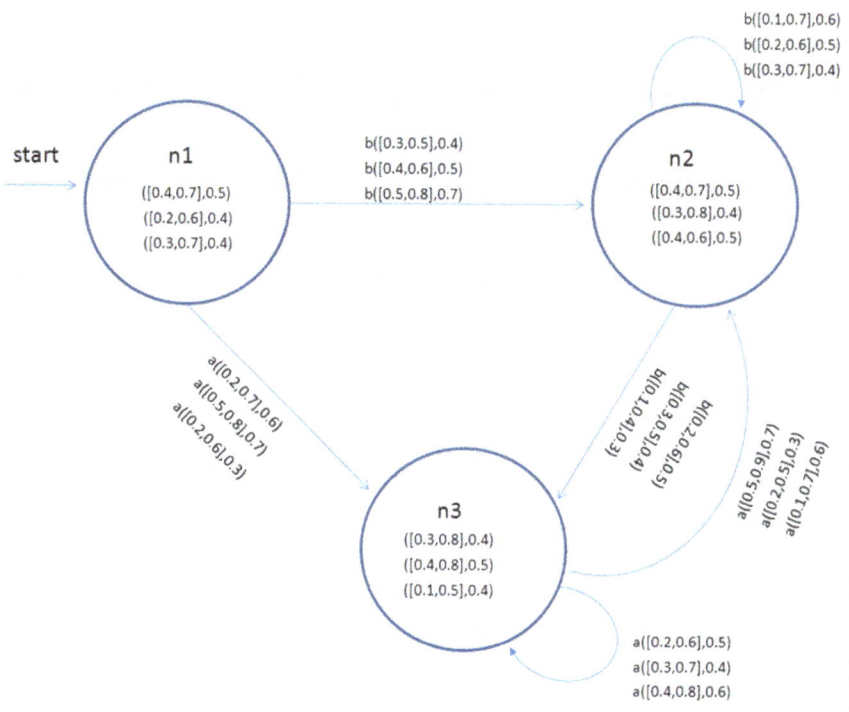

Fig. 2 INC-subsystem, for example, (4)

$$\text{and} \quad \begin{cases} \zeta^*_{NC_T^-}(n, a, m) \le \eta^*_{NC_T}(n, a, m) \le \zeta^*_{NC_T^+}(n, a, m) \\ \zeta^*_{NC_I^-}(n, a, m) \le \eta^*_{NC_I}(n, a, m) \le \zeta^*_{NC_I^+}(n, a, m) \; for\, all\, n, m \in T\, and\, a \in P^* \\ \zeta^*_{NC_F^-}(n, a, m) \le \eta^*_{NC_F}(n, a, m) \le \zeta^*_{NC_F^+}(n, a, m) \end{cases}$$

Proof As it is given that

$$\begin{cases} \zeta^-_{\widehat{NC_T}} \le \eta_{\widehat{NC_T}} \le \zeta^+_{\widehat{NC_T}} \\ \zeta^-_{\widehat{NC_I}} \le \eta_{\widehat{NC_I}} \le \zeta^+_{\widehat{NC_I}} \\ \zeta^+_{\widehat{NC_F}} \le \eta_{\widehat{NC_F}} \le \zeta^+_{\widehat{NC_F}} \end{cases}$$

$$\text{and} \begin{cases} \zeta_{NC_T^-}(n, a, m) \le \eta_{NC_T}(n, a, m) \le \zeta_{NC_T^+}(n, a, m) \\ \zeta_{NC_I^-}(n, a, m) \le \eta_{NC_I}(n, a, m) \le \zeta_{NC_I^+}(n, a, m) \; for\, all\, n, m \in T\, and\, a \in P. \\ \zeta_{NC_F^-}(n, a, m) \le \eta_{NC_F}(n, a, m) \le \zeta_{NC_F^+}(n, a, m) \end{cases}$$

Fig. 3 ENC-subsystem, for example, (4)

This implies that NC and $\widehat{NC^I} = \left\{\widetilde{\zeta}_{\widehat{NC^I}_T}, \widetilde{\zeta}_{\widehat{NC^I}_I}, \widetilde{\zeta}_{\widehat{NC^I}_F}, \eta_{\widehat{NC^I}_T}, \eta_{\widehat{NC^I}_I}, \eta_{\widehat{NC^I}_F}\right\}$ are internal neutrosophic cubic subsets of $T \times P \times T$ and T, respectively. So $\widehat{NC^I} = \left\{\widetilde{\zeta}_{\widehat{NC^I}_T}, \widetilde{\zeta}_{\widehat{NC^I}_I}, \widetilde{\zeta}_{\widehat{NC^I}_F}, \eta_{\widehat{NC^I}_T}, \eta_{\widehat{NC^I}_I}, \eta_{\widehat{NC^I}_F}\right\}$ is a neutrosophic cubic subsystem of NCM. Thus, $\widehat{NC^I} = \left\{\widetilde{\zeta}_{\widehat{NC^I}_T}, \widetilde{\zeta}_{\widehat{NC^I}_I}, \widetilde{\zeta}_{\widehat{NC^I}_F}, \eta_{\widehat{NC^I}_T}, \eta_{\widehat{NC^I}_I}, \eta_{\widehat{NC^I}_F}\right\}$ is an internal neutrosophic cubic subsystem of NCM. This completes the required proof.

Theorem 4.13 *Let* $NCM = (T, P, NC)$ *be a neutrosophic cubic FSM. Then*

$$\widehat{NC^E} = \left\{\widetilde{\zeta}_{\widehat{NC^E}_T}, \widetilde{\zeta}_{\widehat{NC^E}_I}, \widetilde{\zeta}_{\widehat{NC^E}_F}, \eta_{\widehat{NC^E}_T}, \eta_{\widehat{NC^E}_I}, \eta_{\widehat{NC^E}_F}\right\}$$

is an external neutrosophic cubic subsystem of NCM if

$$(1) = \begin{cases} \eta_{\widehat{NC}_T}(n) \notin \left(\zeta^-_{\widehat{NC}_T}(n), \zeta^+_{\widehat{NC}_T}(n)\right) \\ \eta_{\widehat{NC}_I}(n) \notin \left(\zeta^-_{\widehat{NC}_I}(n), \zeta^+_{\widehat{NC}_I}(n)\right) \\ \eta_{\widehat{NC}_F}(n) \notin \left(\zeta^-_{NC_F}(n), \zeta^+_{NC_F}(n)\right) \end{cases}$$

$$(2) = \begin{cases} \eta^*_{NC_T} (n, a, m) \notin \left(\zeta^*_{NC_T^-} (n, a, m), \zeta^*_{NC_T^+} (n, a, m) \right) \\ \eta^*_{NC_I} (n, a, m) \notin \left(\zeta^*_{NC_I^-} (n, a, m), \zeta^*_{NC_I^+} (n, a, m) \right) \ for\ all\ n, m \in T\ and\ a \in P^* . \\ \eta^*_{NC_F} (n, a, m) \notin \left(\zeta^*_{NC_F^-} (n, a, m), \zeta^*_{NC_F^+} (n, a, m) \right) \end{cases}$$

Proof The proof of this theorem is straightforward.

Theorem 4.14 $Let \left\{ \widehat{NC_i^I} = \left\langle \tilde{\zeta}_{\widehat{NC_i^I}_T}, \tilde{\zeta}_{\widehat{NC_i^I}_I}, \tilde{\zeta}_{\widehat{NC_i^I}_F}, \eta_{\widehat{NC_i^I}_T}, \eta_{\widehat{NC_i^I}_I}, \eta_{\widehat{NC_i^I}_F} \right\rangle / i \in \wedge \right\}$ be a family of $INC-subsystem$ Of neutrosophic cubic FSMs, $NCMi = (T_i, P_i, NC_i)$. Then $\cup_{i \in \wedge} p\widehat{NC_i^I}$ is an $INC-subsystem$ Of NCM.

Proof Since $\widehat{NC_i^I}$ is an $INC-$subsystem, we have

$$\begin{cases} \zeta^-_{\widehat{NC_T^I}} (n) \leq \eta_{\widehat{NC_T^I}} (n) \leq \zeta^+_{\widehat{NC_T^I}} (n) \\ \zeta^-_{\widehat{NC_I^I}} (n) \leq \eta_{\widehat{NC_I^I}} (n) \leq \zeta^+_{\widehat{NC_I^I}} (n) \\ \zeta^-_{\widehat{NC_F^I}} (n) \leq \eta_{\widehat{NC_F^I}} (n) \leq \zeta^+_{\widehat{NC_F^I}} (n) \end{cases}$$

$$, and \begin{cases} \zeta_{NC_T^-} (n, a, m) \leq \eta_{NC_T} (n, a, m) \leq \zeta_{NC_T^+} (n, a, m) \\ \zeta_{NC_I^-} (n, a, m) \leq \eta_{NC_I} (n, a, m) \leq \zeta_{NC_I^+} (n, a, m)\ for\ i \in \wedge . \\ \zeta_{NC_F^-} (n, a, m) \leq \eta_{NC_F} (n, a, m) \leq \zeta_{NC_F^+} (n, a, m) \end{cases}$$

This implies that

$$\left(\cup_{i \in \wedge} \zeta^-_{\widehat{NC_T^I}} \right) (n) \leq \left(\vee_{i \in \wedge} \eta_{\widehat{NC_T^I}} \right) (n) \leq \left(\cup_{i \in \wedge} \zeta^+_{\widehat{NC_T^I}} \right) (n) ,$$

$$\left(\cup_{i \in \wedge} \zeta^-_{\widehat{NC_I^I}} \right) (n) \leq \left(\vee_{i \in \wedge} \eta_{\widehat{NC_I^I}} \right) (n) \leq \left(\cup_{i \in \wedge} \zeta^+_{\widehat{NC_I^I}} \right) (n)$$

$$\left(\cup_{i \in \wedge} \zeta^-_{\widehat{NC_F^I}} \right) (n) \leq \left(\vee_{i \in \wedge} \eta_{\widehat{NC_F^I}} \right) (n) \leq \left(\cup_{i \in \wedge} \zeta^+_{\widehat{NC_F^I}} \right) (n) ,$$

and

$$\left(\cup_{i \in \wedge} \zeta^-_{NC_T} \right) (n, a, m) \leq \left(\vee_{i \in \wedge} \eta_{NC_T} \right) (n, a, m) \leq \left(\cup_{i \in \wedge} \zeta^+_{NC_T} \right) (n, a, m)$$

$$\left(\cup_{i \in \wedge} \zeta^-_{NC_I} \right) (n, a, m) \leq \left(\vee_{i \in \wedge} \eta_{NC_I} \right) (n, a, m) \leq \left(\cup_{i \in \wedge} \zeta^+_{NC_I} \right) (n, a, m)$$

$$\left(\cup_{i \in \wedge} \zeta^-_{NC_F} \right) (n, a, m) \leq \left(\vee_{i \in \wedge} \eta_{NC_F} \right) (n, a, m) \leq \left(\cup_{i \in \wedge} \zeta^+_{NC_F} \right) (n, a, m) .$$

Hence $\cup_{i \in \wedge} p\widehat{NC_i^I}$ is an $INC-$subsystem of NCM.

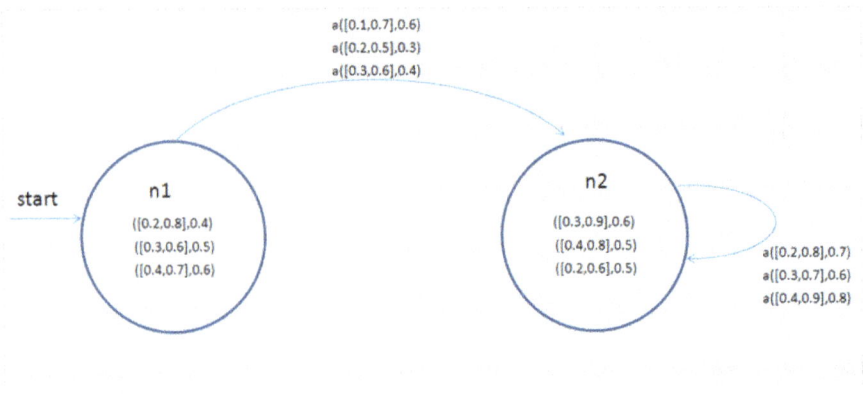

Fig. 4 First INC-subsystem

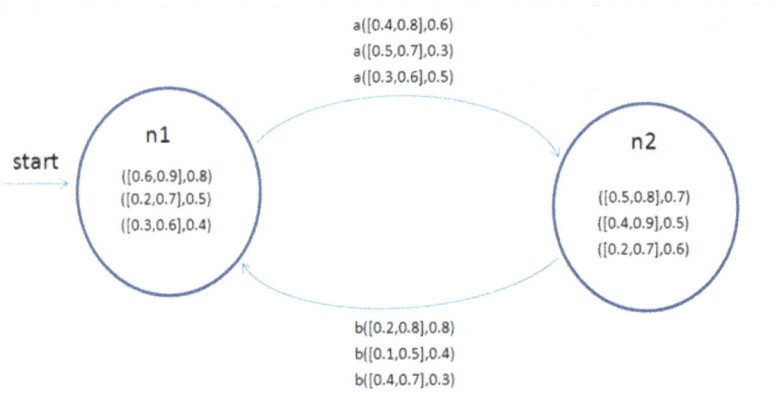

Fig. 5 Second INC-subsystem

The following theorem shows that the $R-$union of INC-subsystem need not to be an INC-subsystem(ENC − subsystem) (Figs. 4 and 5).

Example We have the following two INC-subsystem;
Then we have (Fig. 6)

Hence we provide the condition for the $R-$union of two INC-subsystems to be an INC-subsystem.

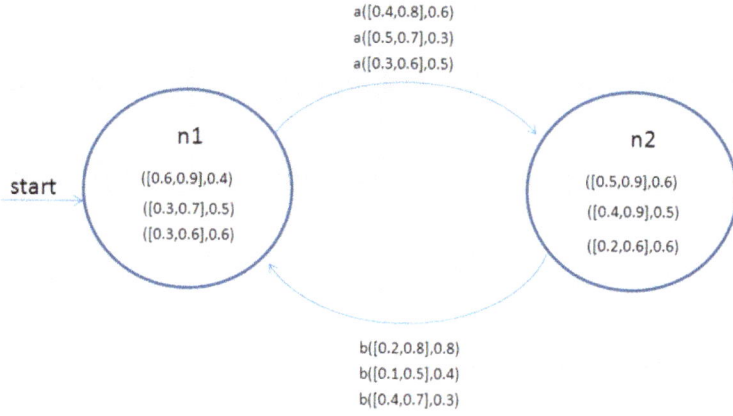

a([0.4,0.8],0.6)
a([0.5,0.7],0.3)
a([0.3,0.6],0.5)

n1

start

([0.6,0.9],0.4)
([0.3,0.7],0.5)
([0.3,0.6],0.6)

n2

([0.5,0.9],0.6)
([0.4,0.9],0.5)
([0.2,0.6],0.6)

b([0.2,0.8],0.8)
b([0.1,0.5],0.4)
b([0.4,0.7],0.3)

Fig. 6 Union of two INC-subsystems, for example, (4)

Theorem 4.15 *Let* $\widehat{NC_1^I}$ *and* $\widehat{NC_1^I}$ *be two INC-subsystems of neutrosophic cubic FSMs,* $NCM_1 = (T_1, P_1, NC_1)$ *and FSMs* $NCM_2 = (T_2, P_2, NC_2)$, *respectively, such that*

$$\max\left\{\zeta^-_{\widehat{NC_1^I}}(n_1), \zeta^+_{\widehat{NC_2^I}}(n_2)\right\} \le \min\left\{\eta_{\widehat{NC_1^I}}(n_1), \eta_{\widehat{NC_2^I}}(n_2)\right\}$$

and $\max\left\{\zeta^-_{NC_1}(n_1, a_1, m_1), \zeta^-_{NC_2}(n_2, a_2, m_2)\right\} \le \min\left\{\eta_{NC_1}(n_1, a_1, m_1),\right.$

$$\left.\eta_{NC_2}(n_2, a_2, m_2)\right\}$$

for all $n_1, m_1 \in T_1, n_2, m_2 \in T_2, a_1 \in P_1$ *and* $a_2 \in P_2$. *Then the R−union of two INC-subsystems to be an INC-subsystem.*

Proof Let $\widehat{NC_1^I}$ and $\widehat{NC_1^I}$ be a INC-subsystem which satisfy the conditions

$$\max\left\{\zeta^-_{\widehat{NC_1^I}}(n_1), \zeta^+_{\widehat{NC_2^I}}(n_2)\right\} \le \min\left\{\eta_{\widehat{NC_1^I}}(n_1), \eta_{\widehat{NC_2^I}}(n_2)\right\}$$

and $\max\left\{\zeta^-_{NC_1}(n_1, a_1, m_1), \zeta^-_{NC_2}(n_2, a_2, m_2)\right\} \le \min\left\{\eta_{NC_1}(n_1, a_1, m_1),\right.$

$$\left.\eta_{NC_2}(n_2, a_2, m_2)\right\}$$

for all $n_1, m_1 \in T_1, n_2, m_2 \in T_2, a_1 \in P_1$ and $a_2 \in P_2$. Since

$$\eta_{\widehat{NC_1^I}}(n_1) \in \left\{\zeta^-_{\widehat{NC_1^I}}(n_1), \zeta^+_{\widehat{NC_1^I}}(n_1)\right\}, \eta_{NC_1}(n_1, a_1, m_1) \in \left\{\zeta^-_{NC_1}(n_1, a_1, m_1),\right.$$

$$\left.\zeta^+_{NC_2}(n_2, a_2, m_2)\right\}$$

and

$$\eta_{\widehat{NC_2^I}}(n_2) \in \left\{\zeta^-_{\widehat{NC_2^I}}(n_2), \zeta^+_{\widehat{NC_2^I}}(n_2)\right\}, \eta_{NC_2}(n_2, a_2, m_2) \in \left\{\zeta^-_{NC_2}(n_2, a_2, m_2),\right.$$

$$\left.\zeta^+_{NC_2}(n_2, a_2, m_2)\right\}.$$

This implies that

$$\min\left\{\eta_{\widehat{NC_1^I}}(n_1), \eta_{\widehat{NC_2^I}}(n_2)\right\} \le \max\left\{\zeta^+_{\widehat{NC_1^I}}(n_1), \zeta^+_{\widehat{NC_2^I}}(n_2)\right\}$$

and

$$\min\left\{\eta_{NC_1}(n_1, a_1, m_1), \eta_{NC_2}(n_2, a_2, m_2)\right\} \le \max\left\{\zeta^+_{NC_1}(n_1, a_1, m_1),\right.$$

$$\left.\zeta^+_{NC_2}(n_2, a_2, m_2)\right\}.$$

Thus from the given condition we get

$$\max\left\{\zeta^-_{\widehat{NC_1^I}}(n_1), \zeta^-_{\widehat{NC_2^I}}(n_2)\right\} \le \min\left\{\eta_{\widehat{NC_1^I}}(n_1), \eta_{\widehat{NC_2^I}}(n_2)\right\}$$

$$\le \max\left\{\zeta^+_{\widehat{NC_1^I}}(n_1), \zeta^+_{\widehat{NC_2^I}}(n_2)\right\} \max\left\{\zeta^-_{NC_1}(n_1, a_1, m_1),\right.$$

$$\left.\zeta^-_{NC_2}(n_2, a_2, m_2)\right\}$$

$$\le \min\left\{\eta_{NC_1}(n_1, a_1, m_1), \eta_{NC_2}(n_2, a_2, m_2)\right\}$$

$$\le \max\left\{\left\{\zeta^+_{NC_1}(n_1, a_1, m_1), \zeta^+_{NC_2}(n_2, a_2, m_2)\right\}\right\}.$$

This shows that $\widehat{NC_1^I} \cup_R \widehat{NC_2^I}$ is an INC-subsystem.

Remark With the help of an example, it is very easy to show that the P−union and R−union of ENC-subsystems need not to be an ENC-subsystem(INC − subsystem). We provide the condition for the P−union and R−union of two ENC-subsystems to be an ENC-subsystem.

Step 3:

Rank every alternative of $(A_i \left(\acute{i} = 1, 2, \ldots, m \right)$ according to $\acute{s} \left(\tilde{\tilde{d}}_i \right), H \left(\left(\tilde{\tilde{d}}_i \right) \right.$ $\left(\acute{i} = 1, 2, \ldots, m \right)$ and select the very best one.

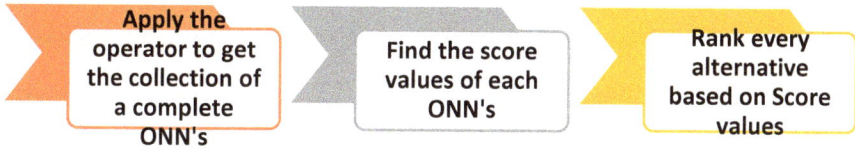

4.4 Case Study

Under discussion:

- Qualification and experience.
- Reliability.

  The MCDM problem, in this case, the selection of a well-groomed teacher.

4.4.1 Problem Formulation

A company requires an experienced and impressive person to proceed with a task; hence it is a problem of multiple-attribute decision-making. The selection will be based on different criteria. We will use neutrosophic set for the selection of effective volunteers.

4.4.2 Parameters

The selection will be based on high-end specs and the criteria are:

Criteria			
$\acute{C}_1$	$\acute{C}_2$	$\acute{C}_3$	C_4
Qualification	Experience	Conviviality	Publications

Theorem 4.16 *Let* $\widehat{NC_1^E}$ *and* $\widehat{NC_1^E}$ *be a ENC-subsystems of neutrosophic cubic FSMs,* $NCM_1 = (T_1, P_1, NC_1)$ *and FSMs* $NCM_2 = (T_2, P_2, NC_2)$, *respectively, such that*

$$
\min \left\{ \begin{array}{l} \max \left\{ \zeta^+_{\widehat{NC_1^E}}(n_1), \zeta^-_{\widehat{NC_2^E}}(n_2) \right\}, \\ \max \left\{ \zeta^-_{\widehat{NC_1^E}}(n_1), \zeta^+_{\widehat{NC_2^E}}(n_2) \right\} \end{array} \right\} > \max \left\{ \eta_{\widehat{NC_1^E}}(n_1), \eta_{\widehat{NC_2^E}}(n_2) \right\}
$$

$$
\geq \max \left\{ \begin{array}{l} \min \left\{ \zeta^+_{\widehat{NC_1^E}}(n_1), \zeta^-_{\widehat{NC_2^E}}(n_2) \right\}, \\ \min \left\{ \zeta^-_{\widehat{NC_1^E}}(n_1), \zeta^+_{\widehat{NC_2^E}}(n_2) \right\} \end{array} \right\}
$$

and

$$
\min \left\{ \begin{array}{l} \max \left\{ \zeta^+_{NC_1}(n_1, a_1, m_1), \zeta^-_{NC_2}(n_2, a_2, m_2) \right\}, \\ \max \left\{ \zeta^+_{NC_1}(n_1, a_1, m_1), \zeta^-_{NC_2}(n_2, a_2, m_2) \right\} \end{array} \right\}
$$

$$
> \max \left\{ \eta_{NC_1}(n_1, a_1, m_1), \eta_{NC_2}(n_2, a_2, m_2) \right\}
$$

$$
\geq \max \left\{ \begin{array}{l} \min \left\{ \zeta^+_{NC_1}(n_1, a_1, m_1), \zeta^-_{NC_2}(n_2, a_2, m_2) \right\}, \\ \min \left\{ \zeta^+_{NC_1}(n_1, a_1, m_1), \zeta^-_{NC_2}(n_2, a_2, m_2) \right\} \end{array} \right\}
$$

for all $n_1, m_1 \in T_1, n_2, m_2 \in T_2, a_1 \in P_1$ *and* $a_2 \in P_2$. *Then the* $P-$*union of two ENC-subsystems to be an ENC-subsystem.*

Theorem 4.17 *Let* $\widehat{NC_1^E}$ *and* $\widehat{NC_2^E}$ *be a ENC-subsystems of neutrosophic cubic FSMs,* $NCM_1 = (T_1, P_1, NC_1)$ *and FSMs* $NCM_2 = (T_2, P_2, NC_2)$, *respectively, such that*

$$
\min \left\{ \begin{array}{l} \max \left\{ \zeta^+_{\widehat{NC_1^E}}(n_1), \zeta^-_{\widehat{NC_2^E}}(n_2) \right\}, \\ \max \left\{ \zeta^-_{\widehat{NC_1^E}}(n_1), \zeta^+_{\widehat{NC_2^E}}(n_2) \right\} \end{array} \right\} > \min \left\{ \eta_{\widehat{NC_1^E}}(n_1), \eta_{\widehat{NC_2^E}}(n_2) \right\}
$$

$$
\geq \max \left\{ \begin{array}{l} \min \left\{ \zeta^+_{\widehat{NC_1^E}}(n_1), \zeta^-_{\widehat{NC_2^E}}(n_2) \right\}, \\ \min \left\{ \zeta^-_{\widehat{NC_1^E}}(n_1), \zeta^+_{\widehat{NC_2^E}}(n_2) \right\} \end{array} \right\}
$$

and

$$\min \left\{ \begin{array}{l} \max \left\{ \zeta^+_{NC_1} (n_1, a_1, m_1), \zeta^-_{NC_2} (n_2, a_2, m_2) \right\}, \\ \max \left\{ \zeta^+_{NC_1} (n_1, a_1, m_1), \zeta^-_{NC_2} (n_2, a_2, m_2) \right\} \end{array} \right\}$$

$$> \max \left\{ \eta_{NC_1} (n_1, a_1, m_1), \eta_{NC_2} (n_2, a_2, m_2) \right\}$$

$$\geq \max \left\{ \begin{array}{l} \min \left\{ \zeta^+_{NC_1} (n_1, a_1, m_1), \zeta^-_{NC_2} (n_2, a_2, m_2) \right\}, \\ \min \left\{ \zeta^+_{NC_1} (n_1, a_1, m_1), \zeta^-_{NC_2} (n_2, a_2, m_2) \right\} \end{array} \right\}$$

for all $n_1, m_1 \in T_1, n_2, m_2 \in T_2, a_1 \in P_1$ and $a_2 \in P_2$. Then the $R-$union of two ENC-subsystems to be an ENC-subsystem.

Theorem 4.18 Let $\widehat{NC} = \left(\tilde{\zeta}_{\widehat{NT}}, \tilde{\zeta}_{\widehat{NI}}, \tilde{\zeta}_{\widehat{NF}}, \eta_{\widehat{NT}}, \eta_{\widehat{NI}}, \eta_{\widehat{NF}} \right)$ be a subsystem of neutrosophic cubic FSM, $NCM = (T, P, NC)$. If $\widehat{NC}$ is both an INC-subsystem and an ENC-subsystem. Then $\eta_{\widehat{NC}} (m_i) \in U \left(\tilde{\zeta}_{\widehat{NC}} \right) \cup L \left(\tilde{\zeta}_{\widehat{NC}} \right)$ and $\eta_{NC} (n_i, p_i, m_i) \in U \left(\tilde{\zeta}_{NC} \right) \cup L \left(\tilde{\zeta}_{NC} \right)$ for all $m_i, n_i \in T$ and $p_i \in P$. Where

$$U \left(\tilde{\zeta}_{\widehat{NC}} \right) = \left\{ \zeta^+_{\widehat{NC}} (m_i) / m_i \in T \right\}, L \left(\tilde{\zeta}_{\widehat{NC}} \right) = \left\{ \zeta^-_{\widehat{NC}} (m_i) / m_i \in T \right\}$$

and

$$U \left(\tilde{\zeta}_{NC} \right) = \left\{ \zeta^+_{NC} (n_i, p_i, m_i) / (n_i, p_i, m_i) \in T \times P \times T \right\}, L \left(\tilde{\zeta}_{NC} \right)$$

$$= \left\{ \zeta^-_{NC} (n_i, p_i, m_i) / (n_i, p_i, m_i) \in T \times P \times T \right\}.$$

Proof Assume that $\widehat{NC} = \left(\tilde{\zeta}_{\widehat{NT}}, \tilde{\zeta}_{\widehat{NI}}, \tilde{\zeta}_{\widehat{NF}}, \eta_{\widehat{NT}}, \eta_{\widehat{NI}}, \eta_{\widehat{NF}} \right)$ is both an INC-subsystem and an ENC-subsystem. Then by definition we have

$$\eta_{\widehat{NC}} (m_i) \in \left[\zeta^-_{\widehat{NC}} (m_i), \zeta^+_{\widehat{NC}} (m_i) \right], \eta_{NC} (n_i, p_i, m_i) \in \left[\zeta^-_{NC} (n_i, p_i, m_i), \right.$$

$$\left. \zeta^+_{NC} (n_i, p_i, m_i) \right]$$

and

$$\eta_{\widehat{NC}} (m_i) \notin \left(\zeta^-_{\widehat{NC}} (m_i), \zeta^+_{\widehat{NC}} (m_i) \right), \eta_{NC} (n_i, p_i, m_i) \notin \left[\zeta^-_{NC} (n_i, p_i, m_i), \right.$$

$$\left. \zeta^+_{NC} (n_i, p_i, m_i) \right].$$

Thus

$$\eta_{\widehat{NC}} (m_i) = \zeta^-_{\widehat{NC}} (m_i) \, or \, \eta_{\widehat{NC}} (m_i) = \zeta^+_{\widehat{NC}} (m_i)$$

and

$$\eta_{NC}\left(n_i, p_i, m_i\right) = \zeta_{NC}^{-}\left(n_i, p_i, m_i\right)$$

$$\text{or } \eta_{NC}\left(n_i, p_i, m_i\right) = \zeta_{NC}^{+}\left(n_i, p_i, m_i\right).$$

Hence

$$\eta_{\widehat{NC}}\left(m_i\right) \in U\left(\tilde{\zeta}_{\widehat{NC}}\right) \cup L\left(\tilde{\zeta}_{\widehat{NC}}\right)$$

and

$$\eta_{NC}\left(n_i, p_i, m_i\right) \in U\left(\tilde{\zeta}_{NC}\right) \cup L\left(\tilde{\zeta}_{NC}\right)$$

for all $m_i, n_i \in T$ and $p_i \in P$.

5 Operations on Subsystems of Neutrosophic Cubic FSMs (Finite State Machines)

In this part we discuss few operations on subsystems of neutrosophic cubic FSMs and provided some related results.

Definition 5.1 Let $\widehat{NC}_1$ and $\widehat{NC}_2$ be the two neutrosophic cubic subsystems of neutrosophic cubic FSMs. As

$$\widehat{NC}_1 = \left\{\tilde{\zeta}_{\widehat{NC}_{T1}}, \tilde{\zeta}_{\widehat{NC}_{I1}}, \tilde{\zeta}_{\widehat{NC}_{F1}}, \eta_{\widehat{NC}_{T1}}, \eta_{\widehat{NC}_{I1}}, \eta_{\widehat{NC}_{F1}}\right\}$$

$$\text{and } \widehat{NC}_2 = \left\{\tilde{\zeta}_{\widehat{NC}_{T2}}, \tilde{\zeta}_{\widehat{NC}_{I2}}, \tilde{\zeta}_{\widehat{NC}_{F2}}, \eta_{\widehat{NC}_{T2}}, \eta_{\widehat{NC}_{I2}}, \eta_{\widehat{NC}_{F2}}\right\}$$

such that $NCM_1 = (T_1, P_1, NC_1)$ and $NCM_2 = (T_2, P_2, NC_2)$, respectively, and let $P_1 \cap P_2 = \phi$. The cartesian composition of $\widehat{NC}_1$ and $\widehat{NC}_2$ is defined by

$$\widehat{NC} \circ \widehat{NC}_2 = \left\langle T_1 \times T_2, \widehat{NC} \circ \widehat{NC}_2, P_1 \cup P_2, NC_2 \circ NC_2\right\rangle$$

and is defined as follows:

For truthness (a) $\begin{cases} \left(\tilde{\zeta}_{\widehat{NC}_{T1}} \circ \tilde{\zeta}_{\widehat{NC}_{T2}}\right)(n_1, n_2) = \text{rmin}\left\{\tilde{\zeta}_{\widehat{NC}_{T1}}(n_1), \tilde{\zeta}_{\widehat{NC}_{T2}}(n_2)\right\} \\ \left(\eta_{\widehat{NC}_{T1}} \circ \eta_{\widehat{NC}_{T2}}\right)(n_1, n_2) = \text{max}\left\{\eta_{\widehat{NC}_{T1}}(n_1), \eta_{\widehat{NC}_{T2}}(n_2)\right\} \end{cases}$

for all $(n_1, n_2) \in T_1 \times T$.

(b) $\begin{cases} \left(\tilde{\zeta}_{NC_{T1}} \circ \tilde{\zeta}_{NC_{T2}}\right)((m_1, r), a, (n_1, r)) = \tilde{\zeta}_{NC_{T1}}(m_1, a, n_1) \\ \left(\eta_{NC_{T1}} \circ \eta_{NC_{T2}}\right)((m_1, r), a, (n_1, r)) = \eta_{NC_{T1}}(m_1, a, n_1) \end{cases}$ for all $m_1, n_1 \in$

$T_1, r \in T_2$ and $a \in P_1 \cup P_2$.

(c) $\begin{cases} \left(\tilde{\zeta}_{NCT1} \circ \tilde{\zeta}_{NCT2}\right)((r, m_2), a, (r, n_2)) = \tilde{\zeta}_{NCT2}(m_2, a, n_2) \\ \left(\eta_{NCT1} \circ \eta_{NCT2}\right)((r, m_2), a, (r, n_2)) = \eta_{NCT2}(m_2, a, n_2) \end{cases}$ for all $m_2, n_2 \in$

$T_2, r \in T_1$ and $a \in P_1 \cup P_2$. For the case of indeterminacy conditions are same as that of truthness.

For falsity, (a) $\begin{cases} \left(\tilde{\zeta}_{\widehat{NC}F1} \circ \tilde{\zeta}_{\widehat{NC}F2}\right)(n_1, n_2) = rmax\left\{\tilde{\zeta}_{\widehat{NC}F1}(n_1), \tilde{\zeta}_{\widehat{NC}F2}(n_2)\right\} \\ \left(\eta_{\widehat{NC}F1} \circ \eta_{\widehat{NC}F2}\right)(n_1, n_2) = min\left\{\eta_{\widehat{NC}F1}(n_1), \eta_{\widehat{NC}F2}(n_2)\right\} \end{cases}$

for all $(n_1, n_2) \in T_1 \times T_2,$.

(b) $\begin{cases} \left(\tilde{\zeta}_{NCF1} \circ \tilde{\zeta}_{NCF2}\right)((m_1, r), a, (n_1, r)) = \tilde{\zeta}_{NCF1}(m_1, a, n_1) \\ \left(\eta_{NCF1} \circ \eta_{NCF2}\right)((m_1, r), a, (n_1, r)) = \eta_{NCF1}(m_1, a, n_1) \end{cases}$ for all $m_1, n_1 \in$

$T_1, r \in T_2$ and $a \in P_1 \cap P_2$.

(c) $\begin{cases} \left(\tilde{\zeta}_{NCF1} \circ \tilde{\zeta}_{NCF2}\right)((r, m_2), a, (r, n_2)) = \tilde{\zeta}_{NCF2}(m_2, a, n_2) \\ \left(\eta_{NCF1} \circ \eta_{NCF2}\right)((r, m_2), a, (r, n_2)) = \eta_{NCF2}(m_2, a, n_2) \end{cases}$ for all $m_2, n_2 \in$

$T_2, r \in T_1$ and $a \in P_1 \cap P_2$.

Example Let $\widehat{NC}_1$ and $\widehat{NC}_2$ be the two neutrosophic cubic subsystems of neutrosophic cubic FSMs. As

$$\widehat{NC}_1 = \left\{\tilde{\zeta}_{\widehat{NC}T1}, \tilde{\zeta}_{\widehat{NC}I1}, \tilde{\zeta}_{\widehat{NC}F1}, \eta_{\widehat{NC}T1}, \eta_{\widehat{NC}I1}, \eta_{\widehat{NC}F1}\right\}$$

and $$\widehat{NC}_2 = \left\{\tilde{\zeta}_{\widehat{NC}T2}, \tilde{\zeta}_{\widehat{NC}I2}, \tilde{\zeta}_{\widehat{NC}F2}, \eta_{\widehat{NC}T2}, \eta_{\widehat{NC}I2}, \eta_{\widehat{NC}F2}\right\}$$

such that $NCM_1 = (T_1, P_1, NC_1)$ and $NCM_2 = (T_2, P_2, NC_2)$, respectively, as shown in Figs. 7 and 8;

Then, their corresponding cartesian composition $\widehat{NC}_1 \circ \widehat{NC}_2$ is shown in Fig. 9:

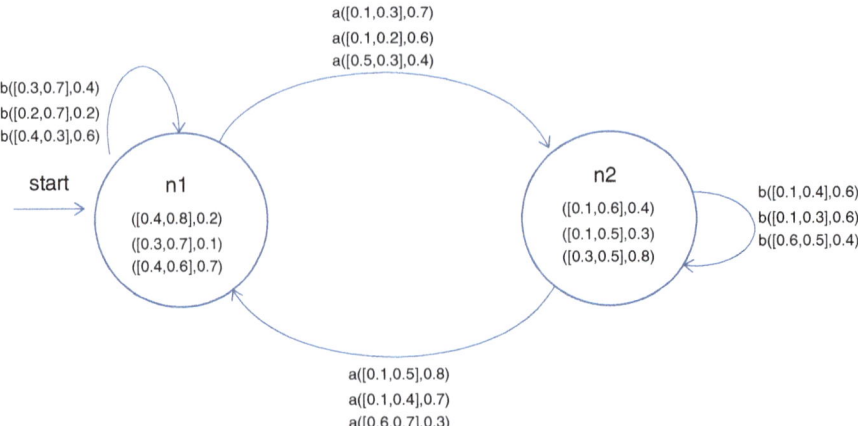

Fig. 7 Neutrosophic cubic subsystems $\widehat{NC}_1$

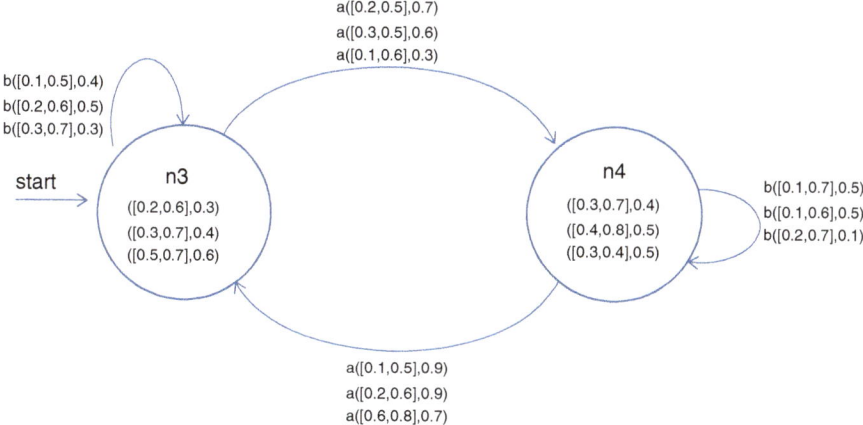

Fig. 8 Neutrosophic cubic subsystems $\widehat{NC}_2$

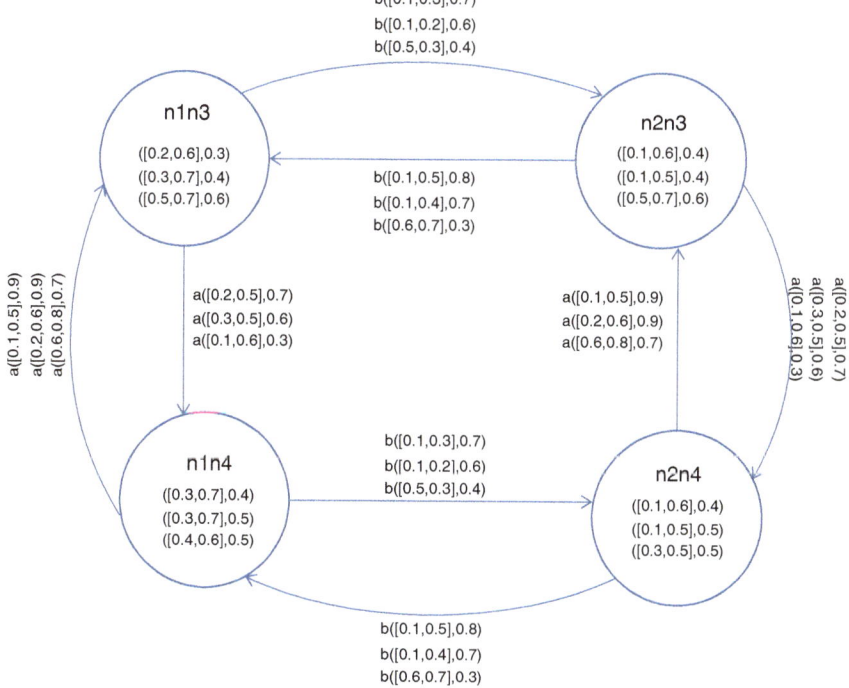

Fig. 9 Cartesian composition $\widehat{NC}_1 \circ \widehat{NC}_2$

Proposition 5.2 *The cartesian composition of two neutrosophic cubic subsystems is a neutrosophic cubic subsystem.*

Proof For truthness. The condition (a) of Definition 5.1 is obvious, therefore, we verify only condition (b) and (c). Let $n_1 \in T_1$ and $m_2, n_2 \in T_2$. Then

$$r \min \left\{ \left(\tilde{\zeta}_{\widehat{NC}_{T1}} \circ \tilde{\zeta}_{\widehat{NC}_{T2}} \right)(n_1, m_2), \left(\tilde{\zeta}_{NC_{T1}} \circ \tilde{\zeta}_{NC_{T2}} \right)((n_1, m_2), a, (n_1, n_2)) \right\}$$

$$= r \min \left\{ r \min \left\{ \tilde{\zeta}_{\widehat{NC}_{T1}}(n_1), \tilde{\zeta}_{\widehat{NC}_{T2}}(m_2) \right\}, \tilde{\zeta}_{NC_{T2}}(m_2, a, n_2) \right\}$$

$$= r \min \left\{ \tilde{\zeta}_{\widehat{NC}_{T1}}(n_1), r \min \left\{ \tilde{\zeta}_{\widehat{NC}_{T2}}(m_2), \tilde{\zeta}_{NC_{T2}}(m_2, a, n_2) \right\} \right\}$$

$$\preceq r \min \left\{ \tilde{\zeta}_{\widehat{NC}_{T1}}(n_1), \tilde{\zeta}_{NC_{T2}}(n_2) \right\}$$

$$= \left(\tilde{\zeta}_{\widehat{NC}_{T1}} \circ \tilde{\zeta}_{\widehat{NC}_{T2}} \right)(n_1, n_2),$$

and

$$\max \left\{ \left(\eta_{\widehat{NC}_{T1}} \circ \eta_{\widehat{NC}_{T2}} \right)(n_1, m_2), \left(\eta_{NC_{T1}} \circ \eta_{NC_{T2}} \right)((n_1, m_2), a, (n_1, n_2)) \right\}$$

$$= \max \left\{ \max \left\{ \eta_{\widehat{NC}_{T1}}(n_1), \eta_{\widehat{NC}_{T2}}(m_2) \right\}, \eta_{NC_{T2}}(m_2, a, n_2) \right\}$$

$$= \max \left\{ \eta_{\widehat{NC}_{T1}}(n_1), \max \left\{ \eta_{\widehat{NC}_{T2}}(m_2), \eta_{NC_{T2}}(m_2, a, n_2) \right\} \right\}$$

$$\geq \max \left\{ \eta_{\widehat{NC}_{T1}}(n_1), \eta_{\widehat{NC}_{T2}}(n_2) \right\} = \left(\eta_{\widehat{NC}_{T1}} \circ \eta_{\widehat{NC}_{T2}} \right)(n_1, n_2).$$

Similarly, we can prove condition (c) for $m_1, n_1 \in T_1$ and $m_2 \in T_2$. The case of indeterminacy is straightforward. And for falsity, the condition (a) of definition (5.1) is obvious, therefore, we verify only condition (b) and (c). Let $n_1 \in T_1$ and $m_2, n_2 \in T_2$. Then

$$r \max \left\{ \left(\tilde{\zeta}_{\widehat{NC}_{F1}} \circ \tilde{\zeta}_{\widehat{NC}_{F2}} \right)(n_1, m_2), \left(\tilde{\zeta}_{NC_{F1}} \circ \tilde{\zeta}_{NC_{F2}} \right)((n_1, m_2), a, (n_1, n_2)) \right\}$$

$$= r \max \left\{ r \max \left\{ \tilde{\zeta}_{\widehat{NC}_{F1}}(n_1), \tilde{\zeta}_{\widehat{NC}_{F2}}(m_2) \right\}, \tilde{\zeta}_{NC_{F2}}(m_2, a, n_2) \right\}$$

$$= r \max \left\{ \tilde{\zeta}_{\widehat{NC}_{F1}}(n_1), r \max \left\{ \tilde{\zeta}_{\widehat{NC}_{F2}}(m_2), \tilde{\zeta}_{NC_{F2}}(m_2, a, n_2) \right\} \right\}$$

$$\preceq r \max \left\{ \tilde{\zeta}_{\widehat{NC}_{F1}}(n_1), \tilde{\zeta}_{NC_{F2}}(n_2) \right\}$$

$$= \left(\tilde{\zeta}_{\widehat{NC}_{F1}} \circ \tilde{\zeta}_{\widehat{NC}_{F2}} \right)(n_1, n_2),$$

and

$$\min \left\{ \left(\eta_{\widehat{NC}_{F1}} \circ \eta_{\widehat{NC}_{F2}} \right)(n_1, m_2), \left(\eta_{NC_{F1}} \circ \eta_{NC_{F2}} \right)((n_1, m_2), a, (n_1, n_2)) \right\}$$

$$= \min \left\{ \min \left\{ \eta_{\widehat{NC}_{F1}}(n_1), \eta_{\widehat{NC}_{F2}}(m_2) \right\}, \eta_{NC_{F2}}(m_2, a, n_2) \right\}$$

$$= \min \left\{ \eta_{\widehat{NC}_{F1}}(n_1), \min \left\{ \eta_{\widehat{NC}_{F2}}(m_2), \eta_{NC_{F2}}(m_2, a, n_2) \right\} \right\}$$

$$\geq \min \left\{ \eta_{\widehat{NC}_{F1}}(n_1), \eta_{\widehat{NC}_{F2}}(n_2) \right\} = \left(\eta_{\widehat{NC}_{F1}} \circ \eta_{\widehat{NC}_{F2}} \right)(n_1, n_2).$$

Similarly, we can prove condition (c) for $m_1, n_1 \in T_1$ and $m_2 \in T_2$.

Definition 5.3 Let $\widehat{NC}_1$ and $\widehat{NC}_2$ be the two neutrosophic cubic subsystems of neutrosophic cubic FSMs. As

$$\widehat{NC}_1 = \left\{ \tilde{\zeta}_{\widehat{NC}_{T1}}, \tilde{\zeta}_{\widehat{NC}_{I1}}, \tilde{\zeta}_{\widehat{NC}_{F1}}, \eta_{\widehat{NC}_{T1}}, \eta_{\widehat{NC}_{I1}}, \eta_{\widehat{NC}_{F1}} \right\}$$

$$\text{and } \widehat{NC}_2 = \left\{ \tilde{\zeta}_{\widehat{NC}_{T2}}, \tilde{\zeta}_{\widehat{NC}_{I2}}, \tilde{\zeta}_{\widehat{NC}_{F2}}, \eta_{\widehat{NC}_{T2}}, \eta_{\widehat{NC}_{I2}}, \eta_{\widehat{NC}_{F2}} \right\}$$

such that $NCM_1 = (T_1, P_1, NC_1)$ and $NCM_2 = (T_2, P_2, NC_2)$, respectively, and let $P_1 \cap P_2 = \phi$. The direct product of $\widehat{NC}_1$ and $\widehat{NC}_2$ is defined by

$$\widehat{NC}_1 \times \widehat{NC}_2 = \left\langle T_1 \times T_2, \widehat{NC}_1 \times \widehat{NC}_2, P_1 \times P_2, NC_2 \times NC_2 \right\rangle$$

and is defined as follows. For truthness

(a) $\begin{cases} \left(\tilde{\zeta}_{\widehat{NC}_{T1}} \times \tilde{\zeta}_{\widehat{NC}_{T2}} \right)(n_1, n_2) = \text{rmin} \left\{ \tilde{\zeta}_{\widehat{NC}_{T1}}(n_1), \tilde{\zeta}_{\widehat{NC}_{T2}}(n_2) \right\} \\ \left(\eta_{\widehat{NC}_{T1}} \circ \eta_{\widehat{NC}_{T2}} \right)(n_1, n_2) = \max \left\{ \eta_{\widehat{NC}_{T1}}(n_1), \eta_{\widehat{NC}_{T2}}(n_2) \right\} \end{cases}$ for all

$(n_1, n_2) \in T_1 \times T_2.$

(b) $\begin{cases} \left(\tilde{\zeta}_{NC_{T1}} \times \tilde{\zeta}_{NC_{T2}} \right)((m_1, m_2), p_1, (n_1, n_2)) = \tilde{\zeta}_{NC_{T1}}(m_1, p_1, n_1) \\ \left(\eta_{NC_{T1}} \times \eta_{NC_{T2}} \right)((m_1, m_2), p_1, (n_1, n_2)) = \eta_{NC_{T1}}(m_1, p_1, n_1) \end{cases}$ for all

(m_1, m_2) and $(n_1, n_2) \in T_1 \times T_2$ and $p_1 \in P_1.$

(c) $\begin{cases} \left(\tilde{\zeta}_{NC_{T1}} \times \tilde{\zeta}_{NC_{T2}} \right)((m_1, m_2), p_2, (n_1, n_2)) = \tilde{\zeta}_{NC_{T2}}(m_2, p_2, n_2) \\ \left(\eta_{NC_{T1}} \times \eta_{NC_{T2}} \right)((m_1, m_2), p_2, (n_1, n_2)) = \eta_{NC_{T2}}(m_2, p_2, n_2) \end{cases}$ for all

(m_1, m_2) and $(n_1, n_2) \in T_1 \times T_2$ and $p_2 \in P_2.$

(d) $\begin{cases} \left(\tilde{\zeta}_{NC_{T1}} \times \tilde{\zeta}_{NC_{T2}} \right)((m_1, m_2), (p_1, p_2), (n_1, n_2)) = r \min \left\{ \tilde{\zeta}_{NC_{T1}}(m_1, p_1, n_1), \tilde{\zeta}_{NC_{T2}}(m_2, p_2, n_2) \right\} \\ \left(\eta_{NC_{T1}} \times \eta_{NC_{T2}} \right)((m_1, m_2), (p_1, p_2), (n_1, n_2)) = \max \left\{ \eta_{NC_{T1}}(m_1, p_1, n_1), \eta_{NC_{T2}}(m_2, p_2, n_2) \right\} \end{cases}$

for all (m_1, m_2) and $(n_1, n_2) \in T_1 \times T_2$ and $(p_1, p_2) \in P_1 \times P_2$. The conditions for indeterminacy is same as truthness.

And for falsity,

(a) $\begin{cases} \left(\tilde{\zeta}_{\widehat{NC}_{F1}} \times \tilde{\zeta}_{\widehat{NC}_{F2}} \right)(n_1, n_2) = r \max \left\{ \tilde{\zeta}_{\widehat{NC}_{F1}}(n_1), \tilde{\zeta}_{\widehat{NC}_{F2}}(n_2) \right\} \\ \left(\eta_{\widehat{NC}_{F1}} \circ \eta_{\widehat{NC}_{F2}} \right)(n_1, n_2) = \min \left\{ \eta_{\widehat{NC}_{F1}}(n_1), \eta_{\widehat{NC}_{F2}}(n_2) \right\} \end{cases}$ for all

$(n_1, n_2) \in T_1 \times T_2.$

(b) $\begin{cases} \left(\tilde{\zeta}_{NC_{F1}} \times \tilde{\zeta}_{NC_{F2}} \right)((m_1, m_2), p_1, (n_1, n_2)) = \tilde{\zeta}_{NC_{F1}}(m_1, p_1, n_1) \\ \left(\eta_{NC_{F1}} \times \eta_{NC_{F2}} \right)((m_1, m_2), p_1, (n_1, n_2)) = \eta_{NC_{F1}}(m_1, p_1, n_1) \end{cases}$ for all

(m_1, m_2) and $(n_1, n_2) \in T_1 \times T_2$ and $p_1 \in P_1.$

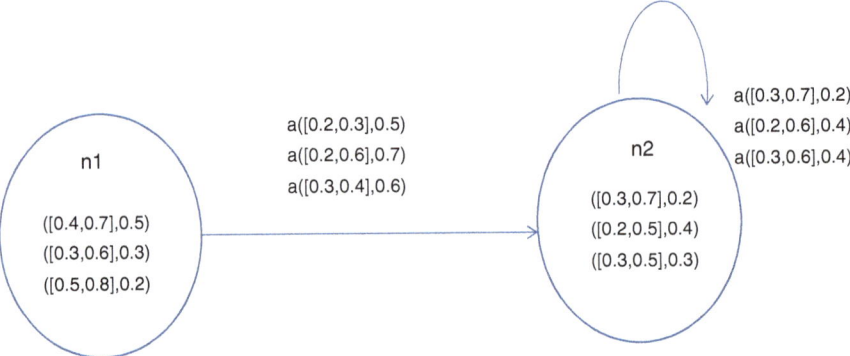

Fig. 10 Neutrosophic cubic subsystems $\widehat{NC}_1$

(c) $\begin{cases} \left(\zeta_{NCF1} \times \tilde{\zeta}_{NCF2}\right)((m_1, m_2), p_2, (n_1, n_2)) = \tilde{\zeta}_{NCF2}(m_2, p_2, n_2) \\ \left(\eta_{NCF1} \times \eta_{NCF2}\right)((m_1, m_2), p_2, (n_1, n_2)) = \eta_{NCF2}(m_2, p_2, n_2) \end{cases}$ for all

(m_1, m_2) and $(n_1, n_2) \in T_1 \times T_2$ and $p_2 \in P_2$.

(d) $\begin{cases} \left(\tilde{\zeta}_{NCF1} \times \tilde{\zeta}_{NCF2}\right)((m_1, m_2), (p_1, p_2), (n_1, n_2)) = r\max\left\{\tilde{\zeta}_{NCF1}(m_1, p_1, n_1), \tilde{\zeta}_{NCF2}(m_2, p_2, n_2)\right\} \\ \left(\eta_{NCF1} \times \eta_{NCF2}\right)((m_1, m_2), (p_1, p_2), (n_1, n_2)) = \min\left\{\eta_{NCF1}(m_1, p_1, n_1), \eta_{NCF2}(m_2, p_2, n_2)\right\} \end{cases}$

for all (m_1, m_2) and $(n_1, n_2) \in T_1 \times T_2$ and $(p_1, p_2) \in P_1 \times P_2$.

Example Let $\widehat{NC}_1$ and $\widehat{NC}_2$ be the two neutrosophic cubic subsystems of neutrosophic cubic $FSMs$. As

$$\widehat{NC}_1 = \left\{\tilde{\zeta}_{\widehat{NC}_{T1}}, \tilde{\zeta}_{\widehat{NC}_{I1}}, \tilde{\zeta}_{\widehat{NC}_{F1}}, \eta_{\widehat{NC}_{T1}}, \eta_{\widehat{NC}_{I1}}, \eta_{\widehat{NC}_{F1}}\right\}$$

and $$\widehat{NC}_2 = \left\{\tilde{\zeta}_{\widehat{NC}_{T2}}, \tilde{\zeta}_{\widehat{NC}_{I2}}, \tilde{\zeta}_{\widehat{NC}_{F2}}, \eta_{\widehat{NC}_{T2}}, \eta_{\widehat{NC}_{I2}}, \eta_{\widehat{NC}_{F2}}\right\}$$

such that $NCM_1 = (T_1, P_1, NC_1)$ and $NCM_2 = (T_2, P_2, NC_2)$, respectively, as shown in Figs. 10 and 11;

Then their corresponding direct product $\widehat{NC}_1 \times \widehat{NC}_2$ is shown in Fig. 12

Clearly, $\widehat{NC}_1 \times \widehat{NC}_2$ is a neutrosophic cubic subsystem of $NCM_1 \times NCM_2$.

Proposition 5.4 *The direct product of two neutrosophic cubic subsystems is a neutrosophic cubic subsystem.*

Proof The proof is similar as that the proof of the Proposition (5.2).

Definition 5.5 Let $\widehat{NC}_1$ and $\widehat{NC}_2$ be the two neutrosophic cubic subsystems of neutrosophic cubic $FSMs$. As

$$\widehat{NC}_1 = \left\{\tilde{\zeta}_{\widehat{NC}_{T1}}, \tilde{\zeta}_{\widehat{NC}_{I1}}, \tilde{\zeta}_{\widehat{NC}_{F1}}, \eta_{\widehat{NC}_{T1}}, \eta_{\widehat{NC}_{I1}}, \eta_{\widehat{NC}_{F1}}\right\}$$

and $$\widehat{NC}_2 = \left\{\tilde{\zeta}_{\widehat{NC}_{T2}}, \tilde{\zeta}_{\widehat{NC}_{I2}}, \tilde{\zeta}_{\widehat{NC}_{F2}}, \eta_{\widehat{NC}_{T2}}, \eta_{\widehat{NC}_{I2}}, \eta_{\widehat{NC}_{F2}}\right\}$$

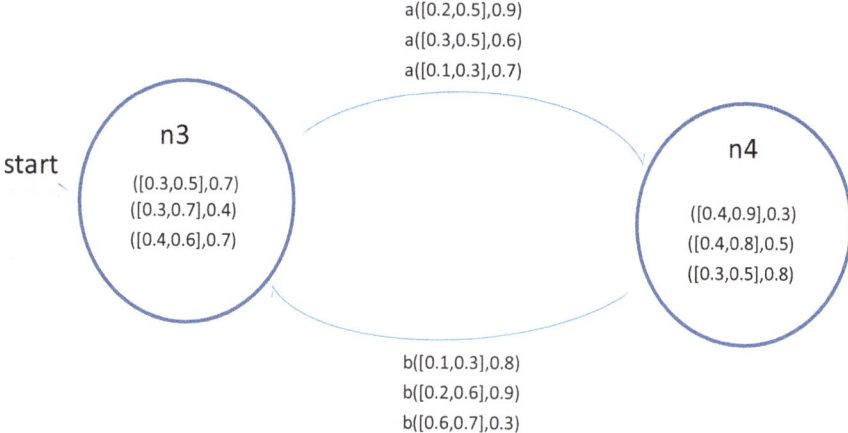

Fig. 11 Neutrosophic cubic subsystems $\widehat{NC}_2$

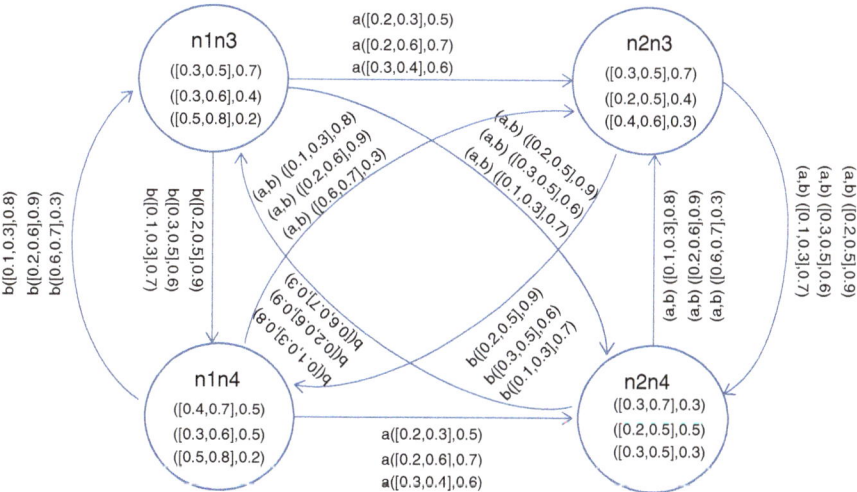

Fig. 12 Direct product $\widehat{NC}_1 \times \widehat{NC}_2$

such that $NCM_1 = (T_1, P_1, NC_1)$ and $NCM_2 = (T_2, P_2, NC_2)$, respectively. The restricted direct product of $\widehat{NC}_1$ and $\widehat{NC}_2$ is denoted by

$$\widehat{NC}_1 \wedge \widehat{NC}_2 = \langle T_1 \times T_1, \widehat{NC}_1 \wedge \widehat{NC}_2, P, NC_1 \wedge NC_2 \rangle$$

and is defined as follows. For truthness

(a) $\begin{cases} \left(\tilde{\zeta}_{\widehat{NC}_{T1}} \wedge \tilde{\zeta}_{\widehat{NC}_{T2}} \right) (n_1, n_2) = r \min \left\{ \tilde{\zeta}_{\widehat{NC}_{T1}} (n_1), \tilde{\zeta}_{\widehat{NC}_{T2}} (n_2) \right\} \\ \left(\eta_{\widehat{NC}_{T1}} \wedge \eta_{\widehat{NC}_{T2}} \right) (n_1, n_2) = \max \left\{ \eta_{\widehat{NC}_{T1}} (n_1), \eta_{\widehat{NC}_{T2}} (n_2) \right\} \end{cases}$ for   all

$(n_1, n_2) \in T_1 \times T_2$.

$$(b) \begin{cases} \left(\tilde{\zeta}_{NCT1} \wedge \tilde{\zeta}_{NCT2}\right)((m_1, m_2), p, (n_1, n_2)) = r\min\left\{\tilde{\zeta}_{NCT1}(m_1, p, n_1), \tilde{\zeta}_{NCT2}(m_2, p, n_2)\right\} \\ \left(\eta_{NCT1} \wedge \eta_{NCT2}\right)((m_1, m_2), p, (n_1, n_2)) = \max\left\{\eta_{NCT1}(m_1, p, n_1), \eta_{NCT2}(m_2, p, n_2)\right\} \end{cases}$$

for all (m_1, m_2) and $(n_1, n_2) \in T_1 \times T_2$ and $p \in P$.

Same for indeterminacy.

For falsity, we have

$$(a) \begin{cases} \left(\tilde{\zeta}_{\widehat{NC}T1} \vee \tilde{\zeta}_{\widehat{NC}T2}\right)(n_1, n_2) = r\max\left\{\tilde{\zeta}_{\widehat{NC}T1}(n_1), \tilde{\zeta}_{\widehat{NC}T2}(n_2)\right\} \\ \left(\eta_{\widehat{NC}T1} \vee \eta_{\widehat{NC}T2}\right)(n_1, n_2) = \min\left\{\eta_{\widehat{NC}T1}(n_1), \eta_{\widehat{NC}T2}(n_2)\right\} \end{cases}$$ for all $(n_1, n_2) \in T_1 \times$

T_2.

$$(b) \begin{cases} \left(\tilde{\zeta}_{NCT1} \vee \tilde{\zeta}_{NCT2}\right)((m_1, m_2), p, (n_1, n_2)) = r\max\left\{\tilde{\zeta}_{NCT1}(m_1, p, n_1), \tilde{\zeta}_{NCT2}(m_2, p, n_2)\right\} \\ \left(\eta_{NCT1} \vee \eta_{NCT2}\right)((m_1, m_2), p, (n_1, n_2)) = \min\left\{\eta_{NCT1}(m_1, p, n_1), \eta_{NCT2}(m_2, p, n_2)\right\} \end{cases}$$

for all (m_1, m_2) and $(n_1, n_2) \in T_1 \times T_2$ and $p \in P$.

Proposition 5.6 *The restricted direct product of two neutrosophic cubic subsystems is a neutrosophic cubic subsystem.*

Proof The proof is similar as that the proof of the proposition (5.2).

Definition 5.7 Let $\widehat{NC}_1$ and $\widehat{NC}_2$ be the two neutrosophic cubic subsystems of neutrosophic cubic FSMs. As

$$\widehat{NC}_1 = \left\{\tilde{\zeta}_{\widehat{NC}T1}, \tilde{\zeta}_{\widehat{NC}I1}, \tilde{\zeta}_{\widehat{NC}F1}, \eta_{\widehat{NC}T1}, \eta_{\widehat{NC}I1}, \eta_{\widehat{NC}F1}\right\}$$

$$\text{and } \widehat{NC}_2 = \left\{\tilde{\zeta}_{\widehat{NC}T2}, \tilde{\zeta}_{\widehat{NC}I2}, \tilde{\zeta}_{\widehat{NC}F2}, \eta_{\widehat{NC}T2}, \eta_{\widehat{NC}I2}, \eta_{\widehat{NC}F2}\right\}$$

such that $NCM_1 = (T_1, P_1, NC_1)$ and $NCM_2 = (T_2, P_2, NC_2)$, respectively. Then the $R-$union of $\widehat{NC}_1$ and $\widehat{NC}_2$ is defined by, for truthness

$$(a) \left(\tilde{\zeta}_{\widehat{NC}T1} \cup_R \tilde{\zeta}_{\widehat{NC}T2}\right)(m) = \begin{cases} \tilde{\zeta}_{\widehat{NC}T1}(m) & if\, m \in T_1 - T_2 \\ \tilde{\zeta}_{\widehat{NC}T2}(m) & if\, m \in T_2 - T_1 \\ r\max\left\{\tilde{\zeta}_{\widehat{NC}T1}(m), \tilde{\zeta}_{\widehat{NC}T2}(m)\right\} & if\, m \in T_1 \cap T_2 \end{cases}$$

$$(b) \left(\eta_{\widehat{NC}T1} \cup_R \eta_{\widehat{NC}T2}\right)(m) = \begin{cases} \eta_{\widehat{NC}T1}(m) & if\, m \in T_1 - T_2 \\ \eta_{\widehat{NC}T2}(m) & if\, m \in T_2 - T_1 \\ \min\left\{\eta_{\widehat{NC}T1}(m), \eta_{\widehat{NC}T2}(m)\right\} & if\, m \in T_1 \cap T_2 \end{cases}$$

$$(c) \left(\tilde{\zeta}_{NCT1} \cup_R \tilde{\zeta}_{NCT2}\right)(m, p, n) = \begin{cases} \tilde{\zeta}_{\widehat{NC}T1}(m, p, n) & if\, m, n \in T_1 - T_2 \\ \tilde{\zeta}_{\widehat{NC}T2}(m, p, n) & if\, m, n \in T_2 - T_1 \\ r\max\left\{\left(\tilde{\zeta}_{\widehat{NC}T1}(m, p, n), \tilde{\zeta}_{\widehat{NC}T2}(m, p, n)\right)\right\} & if\, m, n \in T_1 \cap T_2 \\ \qquad\qquad for all\, p \in P \end{cases}$$

$$(d)\ \left(\eta_{NC_{T1}} \cup_R \eta_{NC_{T2}}\right)(m, p, n) = \begin{cases} \eta_{NC_{T1}}(m, p, n) & if\, m, n \in T_1 - T_2 \\ \eta_{NC_{T2}}(m, p, n) & if\, m, n \in T_2 - T_1 \\ \min\left\{\eta_{NC_{T1}}\left((m, p, n), \eta_{NC_{T2}}(m, p, n)\right)\right\} & if\, m, n \in T_1 \cap T_2 \\ \quad for\, all\, p \in P \end{cases}$$

For indeterminacy same as above for truthness. And for falsity

$$(a)\ \left(\tilde{\zeta}_{\widehat{NC}_{F1}} \cup_R \tilde{\zeta}_{\widehat{NC}_{F2}}\right)(m) = \begin{cases} \tilde{\zeta}_{\widehat{NC}_{F1}}(m) & if\, m \in T_1 - T_2 \\ \tilde{\zeta}_{\widehat{NC}_{F2}}(m) & if\, m \in T_2 - T_1 \\ r\min\left\{\tilde{\zeta}_{\widehat{NC}_{F1}}(m), \tilde{\zeta}_{\widehat{NC}_{F2}}(m)\right\} & if\, m \in T_1 \cap T_2 \end{cases}$$

$$(b)\ \left(\eta_{\widehat{NC}_{F1}} \cup_R \eta_{\widehat{NC}_{F2}}\right)(m) = \begin{cases} \eta_{\widehat{NC}_{F1}}(m) & if\, m \in T_1 - T_2 \\ \eta_{\widehat{NC}_{F2}}(m) & if\, m \in T_2 - T_1 \\ \max\left\{\eta_{\widehat{NC}_{I1}}(m), \eta_{\widehat{NC}_{I2}}(m)\right\} & if\, m \in T_1 \cap T_2 \end{cases}$$

$$(c)\ \left(\tilde{\zeta}_{NC_{F1}} \cup_R \tilde{\zeta}_{NC_{F2}}\right)(m, p, n) = \begin{cases} \tilde{\zeta}_{\widehat{NC}_{F1}}(m, p, n) & if\, m, n \in T_1 - T_2 \\ \tilde{\zeta}_{\widehat{NC}_{F2}}(m, p, n) & if\, m, n \in T_2 - T_1 \\ r\min\left\{\left(\tilde{\zeta}_{\widehat{NC}_{F1}}(m, p, n), \tilde{\zeta}_{\widehat{NC}_{F2}}(m, p, n)\right)\right\} & if\, m, n \in T_1 \cap T_2 \\ \quad for\, all\, p \in P \end{cases}$$

$$(d)\ \left(\eta_{NC_{F1}} \cup_R \eta_{NC_{F2}}\right)(m, p, n) = \begin{cases} \eta_{NC_{F1}}(m, p, n) & if\, m, n \in T_1 - T_2 \\ \eta_{NC_{F2}}(m, p, n) & if\, m, n \in T_2 - T_1 \\ \max\left\{\eta_{NC_{F1}}\left((m, p, n), \eta_{NC_{F2}}(m, p, n)\right)\right\} & if\, m, n \in T_1 \cap T_2 \\ \quad for\, all\, p \in P \end{cases}$$

$P-$union, $R-$intersection, and $P-$intersection can be defined in a similar way.

Example Let $\widehat{NC}_1$ and $\widehat{NC}_2$ be the two neutrosophic cubic subsystems of neutrosophic cubic FSMs. As

$$\widehat{NC}_1 = \left\{\tilde{\zeta}_{\widehat{NC}_{T1}}, \tilde{\zeta}_{\widehat{NC}_{I1}}, \tilde{\zeta}_{\widehat{NC}_{F1}}, \eta_{\widehat{NC}_{T1}}, \eta_{\widehat{NC}_{I1}}, \eta_{\widehat{NC}_{F1}}\right\}$$

$$\text{and } \widehat{NC}_2 = \left\{\tilde{\zeta}_{\widehat{NC}_{T2}}, \tilde{\zeta}_{\widehat{NC}_{I2}}, \tilde{\zeta}_{\widehat{NC}_{F2}}, \eta_{\widehat{NC}_{T2}}, \eta_{\widehat{NC}_{I2}}, \eta_{\widehat{NC}_{F2}}\right\}$$

such that $NCM_1 = (T_1, P_1, NC_1)$ and $NCM_2 = (T_2, P_2, NC_2)$, respectively, as shown in Figs. 13 and 14:

Then their corresponding $R-$union $\widehat{NC}_1 \cup_R \widehat{NC}_2$ is shown in Fig. 15;

Clearly $\widehat{NC}_1 \cup_R \widehat{NC}_2$ is a neutrosophic cubic subsystem of $NCM_1 \cup_R NCM_2$.

Proposition 5.8 *The R−union (respectively, P-union, R-intersection, and P-intersection) of two neutrosophic cubic subsystems is a neutrosophic cubic subsystem.*

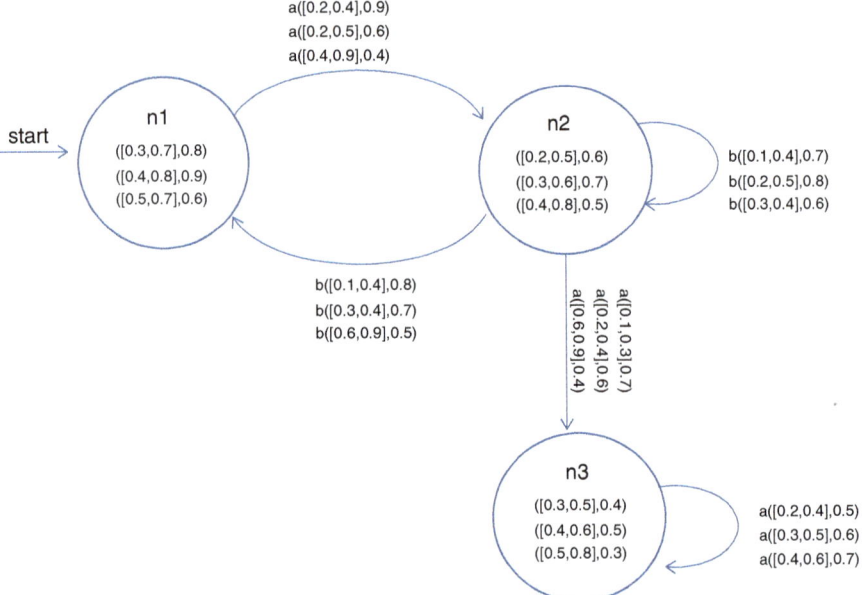

Fig. 13 Neutrosophic cubic subsystems $\widehat{NC}_1$

Proof For truthness. The condition (a) and (b) of definition (5.7) are obvious, therefore, we need only condition (c) and (d). Let $m, n \in T_1 \cap T_2$ and $p \in P$. Then

$$r \min\left\{\left(\tilde{\zeta}_{\widehat{NC}_{T1}} \cup_R \tilde{\zeta}_{\widehat{NC}_{T2}}\right)(m), \left(\tilde{\zeta}_{NC_{T1}} \cup_R \tilde{\zeta}_{NC_{T2}}\right)(m, a, n)\right\}$$

$$= r \min\left\{r \max\left\{\tilde{\zeta}_{\widehat{NC}_{T1}}(m), \tilde{\zeta}_{\widehat{NC}_{T2}}(m)\right\}, r \max\left\{\tilde{\zeta}_{NC_{T1}}(m, a, n), \tilde{\zeta}_{NC_{T2}}(m, a, n)\right\}\right\}$$

$$= r \max\left\{r \min\left\{\tilde{\zeta}_{\widehat{NC}_{T1}}(m), \tilde{\zeta}_{NC_{T1}}(m, a, n)\right\}, r \min\left\{\tilde{\zeta}_{\widehat{NC}_{T2}}(m), \tilde{\zeta}_{NC_{T2}}(m, a, n)\right\}\right\}$$

$$\preceq r \max\left\{\tilde{\zeta}_{\widehat{NC}_{T1}}(m), \tilde{\zeta}_{\widehat{NC}_{T2}}(m)\right\} = \left(\tilde{\zeta}_{\widehat{NC}_{T1}} \cup_R \tilde{\zeta}_{\widehat{NC}_{T2}}\right)(m),$$

and

$$\max\left\{\left(\eta_{\widehat{NC}_{T1}} \cup_R \eta_{\widehat{NC}_{T2}}\right)(m), \left(\eta_{NC_{T1}} \cup_R \eta_{NC_{T2}}\right)(m, a, n)\right\}$$

$$= \max\left\{\min\left\{\eta_{\widehat{NC}_{T1}}(m), \eta_{\widehat{NC}_{T2}}(m)\right\}, \min\left\{\eta_{NC_{T1}}(m, a, n), \eta_{NC_{T2}}(m, a, n)\right\}\right\}$$

$$= \min\left\{\max\left\{\eta_{\widehat{NC}_{T1}}(m), \eta_{NC_{T1}}(m, a, n)\right\}, \max\left\{\eta_{\widehat{NC}_{T2}}(m), \eta_{NC_{T2}}(m, a, n)\right\}\right\}$$

$$\preceq \min\left\{\eta_{\widehat{NC}_{T1}}(m), \eta_{\widehat{NC}_{T2}}(m)\right\} = \left(\eta_{\widehat{NC}_{T1}} \cup_R \eta_{\widehat{NC}_{T2}}\right)(m).$$

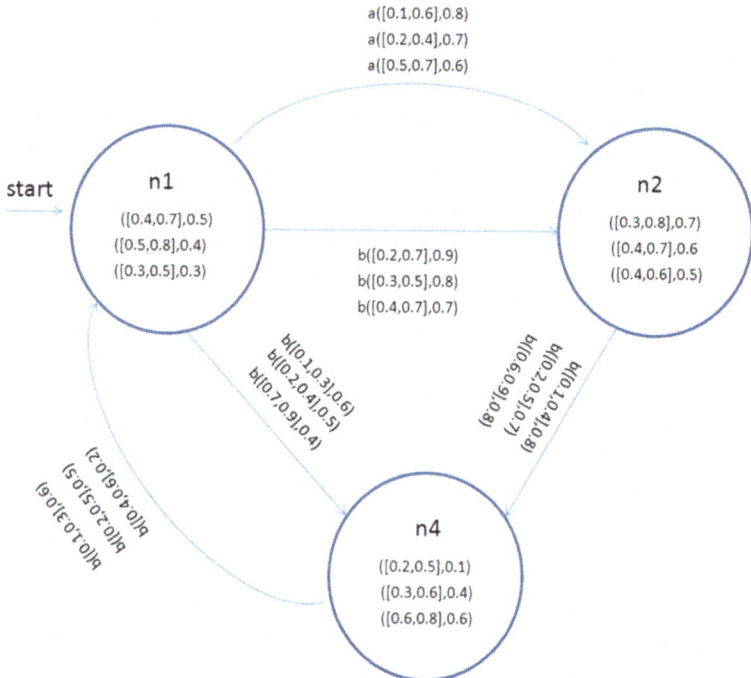

a([0.1,0.6],0.8)
a([0.2,0.4],0.7)
a([0.5,0.7],0.6)

Fig. 14 Neutrosophic cubic subsystems $\widehat{NC}_2$

Thus the $R-$union of two neutrosophic cubic subsystems is a neutrosophic cubic subsystem. Similarly, the prove of the $P-$union, $R-$intersection and $P-$intersection of two neutrosophic cubic subsystems is a neutrosophic cubic subsystem. The indeterminacy case is straightforward. And for falsity, the condition (a) and (b) of definition (5.7) are obvious, therefore, we need only condition (c) and (d) Let $m, n \in T_1 \cap T_2$ and $p \in P$. Then

$$r \max \left\{ \left(\widetilde{\zeta}_{\widehat{NC}_{F1}} \cup_R \widetilde{\zeta}_{\widehat{NC}_{F2}} \right)(m), \left(\widetilde{\zeta}_{NC_{F1}} \cup_R \widetilde{\zeta}_{NC_{F2}} \right)(m, a, n) \right\}$$

$$= r \max \left\{ r \min \left\{ \widetilde{\zeta}_{\widehat{NC}_{F1}}(m), \widetilde{\zeta}_{\widehat{NC}_{F2}}(m) \right\}, r \min \left\{ \widetilde{\zeta}_{NC_{F1}}(m, a, n), \widetilde{\zeta}_{NC_{F2}}(m, a, n) \right\} \right\}$$

$$= r \min \left\{ r \max \left\{ \widetilde{\zeta}_{\widehat{NC}_{F1}}(m), \widetilde{\zeta}_{NC_{F1}}(m, a, n) \right\}, r \max \left\{ \widetilde{\zeta}_{\widehat{NC}_{F2}}(m), \widetilde{\zeta}_{NC_{F2}}(m, a, n) \right\} \right\}$$

$$\geq r \min \left\{ \widetilde{\zeta}_{\widehat{NC}_{F1}}(m), \widetilde{\zeta}_{\widehat{NC}_{F2}}(m) \right\} = \left(\widetilde{\zeta}_{\widehat{NC}_{F1}} \cup_R \widetilde{\zeta}_{\widehat{NC}_{F2}} \right)(m),$$

and

$$\min \left\{ \left(\eta_{\widehat{NC}_{F1}} \cup_R \eta_{\widehat{NC}_{F2}} \right)(m), \left(\eta_{NC_{F1}} \cup_R \eta_{NC_{F2}} \right)(m, a, n) \right\}$$

$$= \min \left\{ \max \left\{ \eta_{\widehat{NC}_{F1}}(m), \eta_{\widehat{NC}_{F2}}(m) \right\}, \max \left\{ \eta_{NC_{F1}}(m, a, n), \eta_{NC_{F2}}(m, a, n) \right\} \right\}$$

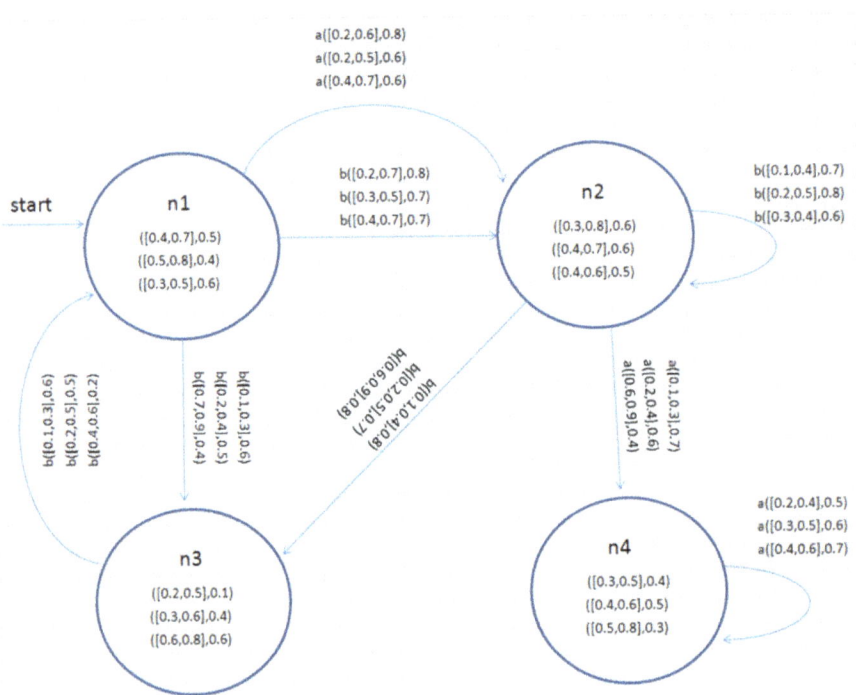

a([0.2,0.6],0.8)
a([0.2,0.5],0.6)
a([0.4,0.7],0.6)

start

n1

([0.4,0.7],0.5)
([0.5,0.8],0.4)
([0.3,0.5],0.6)

b([0.2,0.7],0.8)
b([0.3,0.5],0.7)
b([0.4,0.7],0.7)

n2

([0.3,0.8],0.6)
([0.4,0.7],0.6)
([0.4,0.6],0.5)

b([0.1,0.4],0.7)
b([0.2,0.5],0.8)
b([0.3,0.4],0.6)

b([0.1,0.3],0.6)
b([0.2,0.5],0.5)
b([0.4,0.6],0.2)

b([0.7,0.9],0.4)
b([0.2,0.4],0.5)
b([0.1,0.3],0.6)

b([0.6,0.9],0.8)
b([0.2,0.5],0.7)
b([0.1,0.4],0.8)

a([0.6,0.9],0.4)
a([0.2,0.4],0.6)
a([0.1,0.3],0.7)

n3

([0.2,0.5],0.1)
([0.3,0.6],0.4)
([0.6,0.8],0.6)

n4

([0.3,0.5],0.4)
([0.4,0.6],0.5)
([0.5,0.8],0.3)

a([0.2,0.4],0.5)
a([0.3,0.5],0.6)
a([0.4,0.6],0.7)

Fig. 15 R-union of two subsystems $\widehat{NC}_1$ and $\widehat{NC}_2$

$$= \max\left\{\min\left\{\eta_{\widehat{NC}_{F1}}(m), \eta_{NC_{F1}}(m,a,n)\right\}, \min\left\{\eta_{\widehat{NC}_{F2}}(m), \eta_{NC_{F2}}(m,a,n)\right\}\right\}$$

$$\preceq \max\left\{\eta_{\widehat{NC}_{F1}}(m), \eta_{\widehat{NC}_{F2}}(m)\right\} = \left(\eta_{\widehat{NC}_{F1}} \cup_R \eta_{\widehat{NC}_{F2}}\right)(m).$$

Thus the $R-$union of two neutrosophic cubic subsystems is a neutrosophic cubic subsystem. Similarly, the prove of $P-$union, $R-$intersection and $P-$intersection of two neutrosophic cubic subsystems is a neutrosophic cubic subsystem.

6 Applications

Neutrosophic cubic FSM is a competent index having an unlimited range of appliance in Mathematics. Neutrosophic cubic FSM are extra broad and active way in our daily life. Here in this section we see the usefulness of our proposed model to applying on the different kind of students.

Example Let us have the number of non-empty finite set of states PS_1, FS_2, GS_3, ES_4. These four states are the students of different kinds, that is, poor student PS_1, fair student FS_2, good student GS_3, excellent student ES_4. We give them degree of membership in the form of NCS (neutrosophic cubic set) , the truthful membership indicates books of the pupil, indeterminate membership shows uniform of the students and untruthful membership expresses money,

(1) Poor student $PS_1 = \left(\left(\tilde{\zeta}_{TS_1}, \eta_{TS_1}\right), \left(\tilde{\zeta}_{IS_1}, \eta_{IS_1}\right), \left(\tilde{\zeta}_{FS_1}, \eta_{FS_1}\right)\right)$,

(2) Fair student $FS_2 = \left(\left(\tilde{\zeta}_{TS_2}, \eta_{TS_2}\right), \left(\tilde{\zeta}_{IS_2}, \eta_{IS_2}\right), \left(\tilde{\zeta}_{FS_2}, \eta_{FS_2}\right)\right)$,

(3) Good student $GS_3 = \left(\left(\tilde{\zeta}_{TS_3}, \eta_{TS_3}\right), \left(\tilde{\zeta}_{IS_3}, \eta_{IS_3}\right), \left(\tilde{\zeta}_{FS_3}, \eta_{FS_3}\right)\right)$,

(4) Excellent student $ES_4 = \left(\left(\tilde{\zeta}_{TS_4}, \eta_{TS_4}\right), \left(\tilde{\zeta}_{IS_4}, \eta_{IS_4}\right), \left(\tilde{\zeta}_{FS_4}, \eta_{FS_4}\right)\right)$. Where interval valued membership indicates future prediction and single value express present time,

T	$\tilde{\zeta}_{\hat{N}_{T,I,F}}$	$\eta_{\hat{N}_{T,I,F}}$
PS_1	[0.3, 0.4], [0.2, 0.3], [0.4, 0.5]	0.5, 0.4, 0.6
FS_2	[0.5, 0.6], [0.4, 0.6], [0.3, 0.5]	0.4, 0.3, 0.2
GS_3	[0.6, 0.7], [0.5, 0.6], [0.2, 0.4]	0.5, 0.3, 0.2
ES_4	[0.7, 0.8], [0.5, 0.7], [0.1, 0.3]	0.6, 0.5, 0.4

These are the set of states and input symbols in NCFSM. Now of input symbols now we give the set of transitions for completing our model that is in the form of NCS, and we apply them in two sense one is to providing them a good facilities and second one is to provide them input a bad facilities. As the truthful membership indicates institutions to pupil, indeterminate degree of membership expresses practical work, and untruthful degree of membership shows environmental effect upon the pupils, now we express the membership to our transitions in the form of NCSs. Let $T = \{PS_1, FS_2, GS_3, ES_4\}$ and $P = \{G_F, B_F\}$ and let $NC = \{\tilde{\zeta}_{NT}, \tilde{\zeta}_{NI}, \tilde{\zeta}_{NF}, \eta_{NT}, \eta_{NI}, \eta_{NF}\}$ be a neutrosophic cubic subset in $T \times P \times T$.

$T \times P \times T$	$\tilde{\zeta}_{\hat{N}_{T,I,F}}$	$\eta_{\hat{N}_{T,I,F}}$
PS_1, G_F, FS_2	[0.4, 0.5], [0.3, 0.4], [0.2, 0.3]	0.5, 0.3, 0.2
FS_2, G_F, GS_3	[0.5, 0.7], [0.4, 0.5], [0.3, 0.7]	0.6, 0.5, 0.1
GS_3, G_F, ES_4	[0.3, 0.7], [0.4, 0.6], [0.1, 0.2]	0.8, 0.7, 0.3
ES_4, G_F, ES_4	[0.4, 0.7], [0.3, 0.6], [0.2, 0.3]	0.5, 0.4, 0.2
ES_4, B_F, GS_3	[0.2, 0.5], [0.3, 0.6], [0.4, 0.7]	0.4, 0.2, 0.1
GS_3, B_F, FS_2	[0.3, 0.4], [0.2, 0.5], [0.4, 0.7]	0.3, 0.2, 0.5
FS_2, B_F, PS_1	[0.2, 0.3], [0.1, 0.2], [0.5, 0.6]	0.6, 0.5, 0.4

The set of states, inputs, and NCSs which are used in our system are shown in Fig. 16

And also it shows that S1 is initial or starting state and S4 is its final or terminating state and also we see that our next coming state depends upon the previous state and the input value that we give to our FSM. This NCFSM also obeys the property of NC subsystems of Neutrosophic cubic FSM.

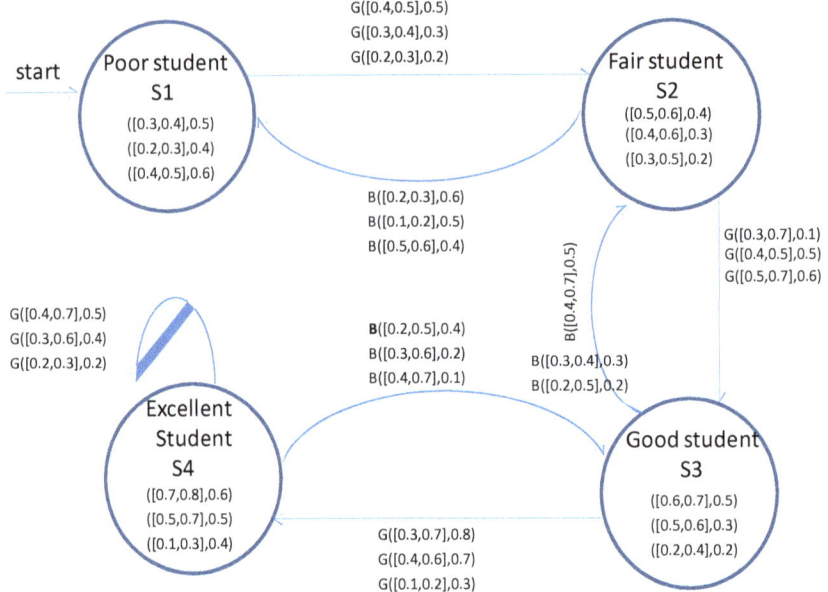

Fig. 16 The set of states, inputs, and NCSs for our system

7 Comparison Analysis

In the year 1994 D. S. Malik, J. N. Mordeson, and M. K. Sen [25] discuss the concept
of (fuzzy finite state machine) FFSM, submachine of FFSM [24], product of FFSM
[26] and also considered some properties related to these FFSM. In FFSM they
used FS (fuzzy subset) that gives only the truthful membership, therefore, this is not
enough for handling uncertainty. As the intutionistic fuzzy set is the abstraction
of FS they give us more knowledge as compared to FS, therefore, in 2005 Jun
[13] used the (intutionistic fuzzy set) IFS in the structure of FSMs, submachines of
intutionistic FFSM. In 2018 N. Yaqoob and N. Abughazalah [1] used CS (cubic sets)
in the structure of FSMs because CS is more informative implement to solve the
vagueness as that of FS and IFS. In this paper we suggest the idea of (neutrosophic
cubic set) NCS in the structure of FSMs. As the NCS is the most generalized form
of FS, IFS, and CS, this gives us more and more knowledge about problem and a
good tool to determine the complications in our routine life, therefore, in FSMs we
observed that this is the best choice to get finest answer and has ability to capture
the vagueness in a better way.

8 Conclusion

Abstraction of the old approach is the crucial purpose of research. So in this chapter by using the clue of (neutrosophic cubic set) NCS we prescribe the idea of (neutrosophic cubic finite state machine) NCFSM, we display that the cartesian composition, direct product and union of two subsystems of neutrosophic cubic finite state machines is a subsystems of NCFSM. Numerous examples have been set up on each case. We considered situations for subsystems of NCFSM to be for both an internal neutrosophic cubic subsystem of neutrosophic cubic FSM and an external neutrosophic cubic subsystem of neutrosophic cubic FSM.

References

1. Abughazalah, N. & Yaqoob, N. (2018). Applications of cubic structures to the subsystems of finite state machines. *Symmetry, 10*(11), 598.
2. Ali, M., Deli, I., & Smarandache, F. (2016). The theory of neutrosophic cubic sets and their application in pattern recognition. *Journal of Intelligent and Fuzzy systems, 30*(4), 1957–1963.
3. Atanassov, K. T. (1986). Intutionistic fuzzy sets. *Fuzzy sets and Systems, 20*(1), 87–96.
4. Gau, W. L. & Buehrer, D. I. (1993). Vague sets and intutionistic fuzzy sets. *Fuzzy sets and Systems, 23*(2), 610–614.
5. Gulistan, M. & Hassan, N. (2019). A generalized approach towards soft expert sets via neutrosophic cubic sets with applications in games. *Symmetry, 11*(2), 289.
6. Gulistan, M., Wahab, H. A., Smarandache, F., Khan, S., & Shah, S. I. A. (2018). Some linguistic neutrosophic cubic mean operators and entropy with applications in a corporation to choose an area supervisor. *Symmetry, 10*(10), 1–30.
7. Gulistan, M., Yaqoob, N., Rashid, Z., Smarandache, F., & Wahab, H. A. (2018). A study on neutrosophic cubic graphs with real life applications in industries. *Symmetry, 10*(6), 1–22.
8. Gulistan, M., Khan, M., Kadry, S., & Alhazaymeh, K. (2019). Neutrosophic cubic Einstein hybrid geometric aggregation operators with application in prioritization using multiple attribute decision-making method. *Mathematics 7*(4), 346.
9. Gulistan, M., Ali, M., Azhar, M., Rho, S., & Kadry, S. (2019). Novel neutrosophic cubic graphs structures with application in decision making problems. *IEEE Access, 7*, 94757–94778.
10. Gulistan, M., Mohammad, M., Karaaslan, F., Kadry, S., Khan, S., & Wahab, H.A. (2019). Neutrosophic cubic Heronian mean operators with applications in multiple attribute group decision-making using cosine similarity functions. *International Journal of Distributed Sensor Networks, 15*(9), 1550147719877613.
11. Gulistan, M., Beg, I., & Yaqoob, N. (2019). Decision making problems under the environment of neutrosophic cubic soft matrices. *Journal of Intelligent and Fuzzy Systems 36*(1), 295–307.
12. Hashim, R.M., Gulistan, M., I. Rehman, N., & Hassan, A. M. (2020). Nasruddin, neutrosophic bipolar fuzzy set and its application in medicines preparations. *Neutrosophic Sets and Systems 31*(1), 7.
13. Jun, Y.B. (2005). Intutionistic fuzzy finite state machine. *Journal of Applied Mathematics and Computing, 17*(1–2), 109–120.
14. Jun, Y.B. (2006). Intutionistic fuzzy finite switchboard state machine. *Journal of Applied Mathematics and Computing, 20*(1–2), 315–325.
15. Jun, Y. B. (2014). Improved correlation coefficients of single valued neutrosophic sets and interval neutrosophic sets for multiple attributes decision. *Journal of Intelligent and Fuzzy systems, 27*(5), 2453–2462.

16. Jun, Y.B. & Kavikumar, J. (2011). Bipolar fuzzy finite state machine. *Bullet Malaysian Mathematical Science Society, 34*(1), 181–188.
17. Jun, Y. B., Kim, C. S., & Yang, K. O. (2012). Cubic sets. *Annals of Fuzzy Mathematics and Informatics, 4*(1), 83–98.
18. Jun, Y. B., Smarandache, F., & Kim, C. S. (2016). Neutrosophic cubic sets. *New Mathematics and Natural Computation, 12*(4), 1–14.
19. Khan, Z., Gulistan, M., Chammam, W., Kadry, S., & Nam, Y. (2020). A new dispersion control chart for handling the neutrosophic data. *IEEE Access, 8*, 96006–96015.
20. Khan, M., Gulistan, M., Ali, M., & Chammam, W. (2020). The generalized neutrosophic cubic aggregation operators and their application to multi-expert decision-making method. *Symmetry, 12*(4), 496.
21. Khan, M., Gulistan, M., Hassan, N., & Nasruddin, A. M. (2020). Air pollution model using neutrosophic cubic Einstein averaging operators. *Neutrosophic Sets and Systems, 32*(1), 24.
22. Khan, Z., Gulistan, M., Hashim, R., Yaqoob, N., & Chammam, W. (2020). Design of S-control chart for neutrosophic data: An application to manufacturing industry. *Journal of Intelligent and Fuzzy Systems, 38*(4), 4743–4751.
23. Kumbhojkar, H. V. & Chaudhari, S. R. (1994). *On Covering of Products of Fuzzy Finite State Machine, 125*(4), 215–222.
24. Malik, D. S., Mordeson, J. N., & Sen, M. K. (1994). Submachines of fuzzy finite state machine. *Fuzzy Mathematics, 2*(4), 781–792.
25. Malik, D. S., Mordeson, J. N., & Sen, M. K. (1994). Fuzzy finite state machine. *Advances in Fuzzy Finite State Machines, 2*(4), 87–98.
26. Malik, D. S., Mordeson, J. N., & Sen, M. K. (1997). Products of fuzzy finite state machine. *Fuzzy sets and Systems, 92*(1), 95–102.
27. Sato, Y., & Kuroki, N. (2002). Fuzzy finite switchboard state machines. *Journal of Fuzzy Mathematics, 10*(4), 863–874.
28. Smarandache, F. (1999). A unifying field in logics. In: *Neutrosophy: Neutrosophic probability, set and logics* (pp. 1–141). Rehoboth: American Research Press.
29. Turksen, I. (1986). Interval valued fuzzy sets. *Fuzzy Sets and Systems, 20*(2), 191–210.
30. Wang, Y. (2014). Single valued neutrosophic cross-entropy for multicriteria decision making problems. *Applied Mathematical Modelling, 38*(3), 1170–1175.
31. Zadeh, L. A. (1965). Fuzzy sets. *Information and control, 8*(3), 338–353.
32. Zhang, W. H. (1998). Bipolar fuzzy sets. IEEE international conference on Fuzzy systems proceedings. *IEEE World Congress on Computational Intelligence, 1*, 835–840.
33. Zhan, J., Khan, M., Gulistan, M., & Ali, A. (2017). Applications of neutrosophic cubic sets in multi-criteria decision-making. *International Journal for Uncertainty Quantification, 7*(5), 377–394.

Analyzing Shortest Path Problem via Single-Valued Triangular Neutrosophic Numbers: A Case Study

Gözde Koca, Ezgi Demir, Özgür İcan, and Çağlar Karamaşa

1 Introduction

In an increasingly competitive environment, businesses are working harder to survive and to make a difference against their competitors. In such an environment, successful management of projects is extremely important for businesses. Scheduling of projects is one of the most important parts of project management. This schedule is the distribution of pioneering relationships and all activities related to each other over time. The project schedule determines all the time variables of the project process and helps project managers to use the activities to be carried out by the project purpose in the right direction and to ensure coordination.

Many factors make the solution difficult in project scheduling. One of these factors is uncertainties that cannot be completely removed. Increasing product variety and shortening of the products' lifetime make planning and therefore scheduling in uncertainty environment necessary. Studies on project scheduling are mostly

G. Koca (✉)
Faculty of Business and Administrative Sciences, Bilecik Şeyh Edebali University, Bilecik, Turkey
e-mail: gozde.koca@bilecik.edu.tr

E. Demir
Faculty of Business and Administrative Sciences, Piri Reis University, İstanbul, Turkey
e-mail: edemir@pirireis.edu.tr

Ö. İcan
Department of International Trade and Logistics, Ondokuz Mayıs University, Samsun, Turkey
e-mail: ozgur.ican@omu.edu.tr

Ç. Karamaşa
Department of Business Administration, Anadolu University, Eskişehir, Turkey
e-mail: ckaramasa@anadolu.edu.tr

© The Author(s), under exclusive license to Springer Nature Switzerland AG 2021
F. Smarandache, M. Abdel-Basset (eds.), *Neutrosophic Operational Research*,
https://doi.org/10.1007/978-3-030-57197-9_25

deterministic structures. It is very difficult to overcome uncertainty in deterministic structures. Therefore, to overcome the uncertainty in project scheduling problems, the fuzzy set theory or neutrosophic set theory is used. Within the scope of this study, the shortest path problem that deals with the activity times with neutrosophic numbers instead of deterministic numbers is used.

Broumi et al. [1–7] developed the concepts of neutrosophic graphs, interval-valued neutrosophic graphs, bipolar neutrosophic graphs, and single-valued neutrosophic graphs as generalization of fuzzy and intuitionistic fuzzy graphs by allowing the truth, indeterminacy, and falsity membership functions to the edges and vertices. Another version of neutrosophic graphs including refined literal indeterminacy is presented by Smarandache [8]. Shortest path problem can be defined as finding the minimum length between any pair of nodes (or vertices) in graph theory. Edge lengths (time, cost, etc.) between different nodes are represented by crisp numbers in classical shortest path problems. Apart from that edge lengths are defined by single-valued neutrosophic numbers under neutrosophic environment.

The main purpose of this study is to apply an algorithm developed by Broumi et al. [9, 10] for solving the shortest path problem in a real-world construction case study where each edge length is represented by single-valued triangular neutrosophic numbers.

The rest of the paper can be arranged in the following way. A brief literature review related to shortest path problem is presented in Sect. 2. The basic concepts of the neutrosophic sets and single-valued triangular neutrosophic sets are explained in Sect. 3. A network terminology is presented in Sect. 4. An algorithm for finding the shortest path and distance in single-valued triangular neutrosophic graph is given in Sect. 5. A real case study as a part of city hospital project including single-valued triangular neutrosophic numbers is given and solved by mentioned algorithm in Sect. 6. Concluding remarks and future recommendations are presented in the last section.

2   Literature Review

One of the most important problems encountered in the business world from the past to the present is the problem of bringing a series of activities together in accordance with a purpose and finishing them as soon as possible. This search, which resulted in the emergence of the project concept and project management discipline, has turned into a project-based business model, which is a highly popular approach today. So much so that, businesses can innovate with the help of projects; For this purpose, projects are gaining importance in all areas regardless of the sector. The methods and techniques based on certain principles in the management of these projects thus remain popular. Studies in the field of project management have accelerated after the Second World War. As a result of the great developments in the defense industry especially after the war, the necessity of managing the projects carried out regarding the weapon systems in the best way has emerged [11]. In the following process, for managers, project management has become a very interesting approach not only in the defense industry but also in other sectors.

One of the most important areas of project management is the scheduling of the project [12]. The purpose of scheduling projects is determining the completion time of the project, determining which activities are critical in order to ensure that the project ends on time, determining which activities can be delayed for how long without delaying the completion time of the project when necessary, determining when the activities will start and when they will end, how much money should be spent at any time of the project, and determining whether it will be worth spending extra to speed up some activities [13, 14]. The better the project is scheduled, the better the efficiency and effectiveness of the decisions, management, and practitioners will make. Three approaches are generally used to schedule the activities in the projects: Gantt scheme, CPM (critical path method) and PERT (program evaluation and review technique) [12–16]. The common aim of these approaches is to minimize the total duration of the project. Apart from these, PEP—project evaluation procedure, LESS—less cost estimating and scheduling technique, GERT—graphical evaluation and review technique, and PDM—priority diagram precedence diagramming method are the approaches that can be used in this field and especially in modeling business processes [17, 18]. The project scheduling discussed in this study was solved as the shortest path problem using neutrosophic numbers. There are many shortest path problems that are analyzed via various input data such as fuzzy sets, intuitionistic fuzzy sets, and neutrosophic sets, which are summarized in Table 1.

Table 1 Studies related to shortest path problems

Author(s)	Environment
Ngoor & Jabarulla [19]	Intuitionistic fuzzy sets
Okada & Soper [20]	Fuzzy sets
Kumar & Kaur [21]	Fuzzy sets
Majumdar & Pal [22]	Intuitionistic fuzzy sets
Kumar & Kaur [23]	Fuzzy sets
Yadav & Biswas [24]	Fuzzy sets
Jayagowri & Geetha Ramani [25]	Trapezoidal intuitionistic fuzzy sets
Hernandes, Teresa Lamata, Verdegay, & Kami [26]	Fuzzy sets
Kumar, Bajaj, & Gandotra [27]	Interval-valued intuitionistic trapezoidal fuzzy sets
Broumi, Bakali, Talea, & Smarandache [10]	Single-valued neutrosophic sets
Broumi, Bakali, Talea, Smarandache, & Ali [28]	Bipolar neutrosophic sets

3   Neutrosophic Sets

Smarandache [29] introduced the concept of neutrosophic sets (NS) having the degree of truth, indeterminacy, and falsity membership functions in which all of them are totally independent. Let U be a universe of discourse and x U. The neutrosophic set (NS) N can be expressed by a truth membership function $T_N^{(x)}$, an indeterminacy membership function $I_N^{(x)}$, and a falsity membership function $F_N^{(x)}$, and is represented as $N = \{<x : T_N(x), I_N(x), F_N(x)>, x \in U\}$. Also the functions of $T_N^{(x)}$, $I_N^{(x)}$, and $F_N^{(x)}$ are real standard or real nonstandard subsets of $[0^-, 1^+]$, and can be presented as T, I, F: $U \to [0^-, 1^+]$. There is not any restriction on the sum of the functions of $T_N^{(x)}$, $I_N^{(x)}$ and $F_N^{(x)}$, so $0^- \leq \sup T_N(x) + \sup I_N(x) + \sup F_N(x) \leq 3^+$. The complement of a NS N is represented by N^C and described as follows:

$$T_N^C(x) = 1^+ \ominus T_N^{(x)} \tag{1}$$

$$I_N^C(x) = 1^+ \ominus I_N^{(x)} \tag{2}$$

$$F_N^C(x) = 1^+ \ominus F_N^{(x)} \quad \text{for all } x \in U \tag{3}$$

A NS, N is contained in other NS P in other words, $N \subseteq P$ if and only if $\inf T_N(x) \leq \inf T_P(x)$, $\sup T_N(x) \leq \sup T_P(x)$, $\inf I_N(x) \geq \inf I_P(x)$, $\sup I_N(x) \geq \sup I_P(x)$, $\inf F_N(x) \geq \inf F_P(x)$, $\sup F_N(x) \geq \sup F_P(x)$ for all $x \in U$ [30].

3.1   Single-Valued Neutrosophic Sets (SVNS)

Wang [31] developed the term of "single-valued neutrosophic set (SVNS) which is a case of NS in order to deal with indeterminate, inconsistent, and incomplete information. They handle the interval [0,1] instead of $[0^-, 1^+]$ in order to better apply in real-world problems. Let U be a universe of discourse and x U. A single-valued neutrosophic set B in U is described by a truth membership function $T_B^{(x)}$, an indeterminacy membership function $I_B^{(x)}$, and a falsity membership function $F_B^{(x)}$. When U is continuous, an SVNS B is depicted as follows: $B = \int_x \frac{<T_B(x), I_B(x), F_B(x)>}{x} : x \in U$. When U is discrete, an SVNS B can be represented as $B = \sum_{i=1}^n \frac{<T_B(x_i), I_B(x_i), F_B(x_i)>}{x_i} : x_i \in U$ [24]. The functions of $T_B^{(x)}$, $I_B^{(x)}$, and $F_B^{(x)}$ are real standard subsets of [0,1], that is $T_B^{(x)} \to [0,1]$, $I_B^{(x)} \to [0,1]$ and $F_B^{(x)} \to [0,1]$. Also the sum of $T_B^{(x)}$, $I_B^{(x)}$, and $F_B^{(x)}$ [0,3] that $0 \leq T_B^{(x)} + I_B^{(x)} + F_B^{(x)} \leq 3$ [30]. Let a single-valued neutrosophic triangular number $\tilde{b} = \left\langle (b_1, b_2, b_3) ; \alpha_{\tilde{b}}, \theta_{\tilde{b}}, \beta_{\tilde{b}} \right\rangle$ is a special neutrosophic set on R. Additionally $\alpha_{\underset{b}{\sim}}, \theta_{\underset{b}{\sim}}, \beta_{\underset{b}{\sim}} \in [0, 1]$ and $b_1, b_2, b_3 \in R$ where $b_1 \leq b_2 \leq b_3$. Truth, indeterminacy, and falsity membership functions of this number can be computed as follows [32]:

$$
T_{\underset{b}{\sim}}(x) = \begin{cases} \alpha_{\underset{b}{\sim}}\left(\frac{x-b_1}{b_2-b_1}\right) & (b_1 \leq x \leq b_2) \\ \alpha_{\underset{b}{\sim}} & (x = b_2) \\ \alpha_{\underset{b}{\sim}}\left(\frac{b_3-x}{b_3-b_2}\right) & (b_2 < x \leq b_3) \\ 0 & \text{otherwise} \end{cases} \tag{4}
$$

$$
I_{\underset{b}{\sim}}(x) = \begin{cases} \left(\frac{b_2-x+\theta_{\underset{b}{\sim}}(x-b_1)}{b_2-b_1}\right) & (b_1 \leq x \leq b_2) \\ \theta_{\underset{b}{\sim}} & (x = b_2) \\ \left(\frac{x-b_2+\theta_{\underset{b}{\sim}}(b_3-x)}{b_3-b_2}\right) & (b_2 < x \leq b_3) \\ 1 & \text{otherwise} \end{cases} \tag{5}
$$

$$
F_{\underset{b}{\sim}}(x) = \begin{cases} \left(\frac{b_2-x+\beta_{\underset{b}{\sim}}(x-b_1)}{b_2-b_1}\right) & (b_1 \leq x \leq b_2) \\ \beta_{\underset{b}{\sim}} & (x = b_2) \\ \left(\frac{x-b_2+\beta_{\underset{b}{\sim}}(b_3-x)}{b_3-b_2}\right) & (b_2 < x \leq b_3) \\ 1 & \text{otherwise} \end{cases} \tag{6}
$$

According to Eqs. (4)–(6) $\alpha_{\underset{b}{\sim}}, \theta_{\underset{b}{\sim}},$ *and* $\beta_{\underset{b}{\sim}}$ denote maximum truth membership, minimum indeterminacy membership, and minimum falsity membership degrees, respectively. Suppose $\widetilde{b} = \left\langle (b_1, b_2, b_3) ; \alpha_{\underset{b}{\sim}}, \theta_{\underset{b}{\sim}}, \beta_{\underset{b}{\sim}} \right\rangle$ and $\widetilde{c} = \left\langle (c_1, c_2, c_3) ; \alpha_{\underset{c}{\sim}}, \theta_{\underset{c}{\sim}}, \beta_{\underset{c}{\sim}} \right\rangle$ as two single-valued triangular neutrosophic numbers and $\lambda \neq 0$ as a real number. Considering abovementioned conditions, an addition of two single-valued triangular neutrosophic numbers are denoted as follows [32]:

$$
\widetilde{b} + \widetilde{c} = \left\langle (b_1 + c_1, b_2 + c_2, b_3 + c_3) ; \alpha_{\underset{b}{\sim}} \wedge \alpha_{\underset{c}{\sim}}, \theta_{\underset{b}{\sim}} \vee \theta_{\underset{c}{\sim}}, \beta_{\underset{b}{\sim}} \vee \beta_{\underset{c}{\sim}} \right\rangle \tag{7}
$$

Subtraction of two single-valued triangular neutrosophic numbers are defined as Eq. (8):

$$
\widetilde{b} - \widetilde{c} = \left\langle (b_1 - c_3, b_2 - c_2, b_3 - c_1) ; \alpha_{\underset{b}{\sim}} \wedge \alpha_{\underset{c}{\sim}}, \theta_{\underset{b}{\sim}} \vee \theta_{\underset{c}{\sim}}, \beta_{\underset{b}{\sim}} \vee \beta_{\underset{c}{\sim}} \right\rangle \tag{8}
$$

Inverse of a single-valued triangular neutrosophic number $\left(\tilde{b} \neq 0 \right)$ can be denoted as follows:

$$\tilde{b}^{-1} = \left\langle \left(\frac{1}{b_3}, \frac{1}{b_2}, \frac{1}{b_1} \right) ; \alpha_{\tilde{b}}, \theta_{\tilde{b}}, \beta_{\tilde{b}} \right\rangle \tag{9}$$

Multiplication of a single-valued triangular neutrosophic number by a constant value is represented as follows:

$$\lambda \tilde{b} = \begin{cases} \left\langle (\lambda b_1, \lambda b_2, \lambda b_3) ; \alpha_{\tilde{b}}, \theta_{\tilde{b}}, \beta_{\tilde{b}} \right\rangle & \text{if } (\lambda > 0) \\ \left\langle (\lambda b_3, \lambda b_2, \lambda b_1) ; \alpha_{\tilde{b}}, \theta_{\tilde{b}}, \beta_{\tilde{b}} \right\rangle & \text{if } (\lambda < 0) \end{cases} \tag{10}$$

Division of a single-valued triangular neutrosophic number by a constant value are denoted as Eq. (11):

$$\frac{\tilde{b}}{\lambda} = \begin{cases} \left\langle \left(\frac{b_1}{\lambda}, \frac{b_2}{\lambda}, \frac{b_3}{\lambda} \right) ; \alpha_{\tilde{b}}, \theta_{\tilde{b}}, \beta_{\tilde{b}} \right\rangle & \text{if } (\lambda > 0) \\ \left\langle \left(\frac{b_3}{\lambda}, \frac{b_2}{\lambda}, \frac{b_1}{\lambda} \right) ; \alpha_{\tilde{b}}, \theta_{\tilde{b}}, \beta_{\tilde{b}} \right\rangle & \text{if } (\lambda < 0) \end{cases} \tag{11}$$

Multiplication of two single valued triangular neutrosophic numbers can be seen as follows:

$$\tilde{b}\tilde{c} = \begin{cases} \left\langle (b_1 c_1, b_2 c_2, b_3 c_3) ; \alpha_{\tilde{b}} \wedge \alpha_{\tilde{c}}, \theta_{\tilde{b}} \vee \theta_{\tilde{c}}, \beta_{\tilde{b}} \vee \beta_{\tilde{c}} \right\rangle & \text{if } (b_3 > 0, c_3 > 0) \\ \left\langle (b_1 c_3, b_2 c_2, b_3 c_1) ; \alpha_{\tilde{b}} \wedge \alpha_{\tilde{c}}, \theta_{\tilde{b}} \vee \theta_{\tilde{c}}, \beta_{\tilde{b}} \vee \beta_{\tilde{c}} \right\rangle & \text{if } (b_3 < 0, c_3 > 0) \\ \left\langle (b_3 c_3, b_2 c_2, b_1 c_1) ; \alpha_{\tilde{b}} \wedge \alpha_{\tilde{c}}, \theta_{\tilde{b}} \vee \theta_{\tilde{c}}, \beta_{\tilde{b}} \vee \beta_{\tilde{c}} \right\rangle & \text{if } (b_3 < 0, c_3 < 0) \end{cases} \tag{12}$$

Division of two single-valued triangular neutrosophic numbers can be denoted as Eq. (13):

$$\frac{\tilde{b}}{\tilde{c}} = \begin{cases} \left\langle \left(\frac{b_1}{c_3}, \frac{b_2}{c_2}, \frac{b_3}{c_1} \right) ; \alpha_{\tilde{b}} \wedge \alpha_{\tilde{c}}, \theta_{\tilde{b}} \vee \theta_{\tilde{c}}, \beta_{\tilde{b}} \vee \beta_{\tilde{c}} \right\rangle & \text{if } (b_3 > 0, c_3 > 0) \\ \left\langle \left(\frac{b_3}{c_3}, \frac{b_2}{c_2}, \frac{b_1}{c_1} \right) ; \alpha_{\tilde{b}} \wedge \alpha_{\tilde{c}}, \theta_{\tilde{b}} \vee \theta_{\tilde{c}}, \beta_{\tilde{b}} \vee \beta_{\tilde{c}} \right\rangle & \text{if } (b_3 < 0, c_3 > 0) \\ \left\langle \left(\frac{b_3}{c_1}, \frac{b_2}{c_2}, \frac{b_1}{c_3} \right) ; \alpha_{\tilde{b}} \wedge \alpha_{\tilde{c}}, \theta_{\tilde{b}} \vee \theta_{\tilde{c}}, \beta_{\tilde{b}} \vee \beta_{\tilde{c}} \right\rangle & \text{if } (b_3 < 0, c_3 < 0) \end{cases} \tag{13}$$

Score function $s\left(\tilde{b} \right)$ for a single-valued triangular neutrosophic number $\tilde{b} = \left\langle (b_1, b_2, b_3) ; \alpha_{\tilde{b}}, \theta_{\tilde{b}}, \beta_{\tilde{b}} \right\rangle$ can be found as follows [33]:

$$s\left(\widetilde{b}\right) = \frac{1}{16}[b_1 + b_2 + b_3] X \left(2 + \alpha_{\widetilde{b}} - \theta_{\widetilde{b}} - \beta_{\widetilde{b}}\right) \tag{14}$$

4 Project Network Terminology

Assume a directed network R(S,T) comprising a finite set of nodes $S = \{1, 2, \ldots, n\}$ and a set of n directed edges $T \subseteq SxS$. Each edge is shown by an ordered pair (i,j) where i,j $\in S$ and $i \neq j$. Two nodes can be specified as source and destination nodes and denoted by e and f, respectively. A path $P_{ij} = \{i = i_1, (i_1, i_2), i_2, \ldots, i_{l-1}, (i_{l-1}, i_l), i_l = j\}$ consists of alternating edges and nodes. The existence of one path P_{ei} in R(S,T) is considered for every i $\in S - \{e\}$. d_{ij} indicates single-valued triangular neutrosophic number related with the edge (i,j) like to the length necessary to traverse (i,j) from i to j. Lengths can be described as time, cost, distance, etc., in real-world problems. Neutrosophic distance along the path P is written by d(P) as follows:

$$d(P) = \sum_{i,j \in P} d_{ij} \tag{15}$$

Additionally node i can be considered as a predecessor of node j if a) node i is directly connected to node j and b) there is path direction that connects node i and j from i to j.

5 Algorithm for Analyzing Single-Valued Triangular Neutrosophic Path Problem

The main steps of an algorithm developed by Broumi et al. [9, 10] for the purpose of finding the shortest path between source and destination node in single-valued triangular neutrosophic graph can be summarized as follows.

Step 1. Consider $\widetilde{d_1} = < (0, 0, 0) ; 0, 1, 1 >$ and label the source node (i.e., node 1) as $\left[\widetilde{d_1} = < (0, 0, 0) ; 0, 1, 1 >, -\right]$. The label shows that the node has no predecessor.

Step 2. $\widetilde{d_j}$ can be found as $\widetilde{d_j} = \min\left\{\widetilde{d_i} \oplus \widetilde{d_{ij}}\right\}$; $j = 2, 3, \ldots, n$.

Step 3. If minimum occurs corresponding to the unique value of I, i.e., $i = r$, then label node j as $\left[\widetilde{d_j}, r\right]$. If minimum occurs corresponding to more than one values of I, then it can be said that there are more than one single-valued triangular neutrosophic path between source node and j. In other words, we need to choose any value of i in order to obtain single-valued triangular neutrosophic distance $\tilde{d}_j$ along the path.

Step 4. The destination node (i.e., node n) can be labelled as $\left[\tilde{d}_n, l \right]$. In that case, the single-valued triangular neutrosophic shortest distance between source and destination node is denoted as $\tilde{d}_n$.

Step 5. In order to find the single-valued triangular neutrosophic shortest distance between source and destination node due to labelling destination node as $\left[\tilde{d}_n, l \right]$, it needs to check the label of node l. By assuming it $\left[\tilde{d}_l, p \right]$ the label of node p is checked and this procedure repeats until obtaining node l.

Step 6. The single-valued triangular neutrosophic shortest path can be achieved by combining all nodes acquired at step 5.

Consider $\tilde{B}_i; i = 1, 2, \ldots, n$ as a set of single-valued triangular neutrosophic numbers, if $s\left(\tilde{B}_f \right) < s\left(\tilde{B}_i \right)$, for all i, the single valued triangular neutrosophic is the minimum of $\tilde{B}_f$.

6   Analysis

In order to show the applicability of the algorithm, a real case study as a part of city hospital consisting of 24 activities where each arc length is represented as single-valued triangular neutrosophic numbers is analyzed. Project-related information including activity explanation, predecessor(s), and durations denoted by single-valued triangular neutrosophic numbers (consisting T,I,F values) is shown as Table 2.

After that project network diagram is drawn in terms of Activity on Arc (AOA) consisting of 17 nodes via Graphviz program and shown as Fig. 1.

Then project network diagram where each arc length is represented by single-valued triangular neutrosophic numbers can be depicted as Fig. 2.

Calculations for the purpose of finding the shortest path problem related to single-valued triangular neutrosophic numbers can be explained as follows.

Since destination node is node 17, it can be stated that n = 17. Consider $\tilde{d}_1 =< (0, 0, 0); 0, 1, 1 >$ and the source node (i.e., node 1) can be labeled as $[<(0,0,0);0,1,1>, -]$ the value of $\tilde{d}_j$; j = 2,3,4 ... 17 can be acquired as follows:

Iteration 1. Only node 1 is the predecessor of node 2, so putting i = 1 and j = 2 according to the second step of the algorithm, the value of $\tilde{d}_2$ can be computed as follows:

Table 2 Project-related information

Activity codes	Activity explanation	Predecessor(s)	Single-valued triangular neutrosophic durations (consisting of T,I,F values)
A	Excavation, bracing, adjustment	–	(150,178,210); (0.6,0.25,0.15)
B	Groundwork	A	(40,44,52); (0.5,0.3,0.2)
C	3 basement until +1,00 grade	B	(112,128,143); (0.6,0.4,0.3)
D	Tower layers	C	(176,199,242); (0.6,0.2,0.4)
E	Facade work	D	(244,289,342); (0.55,0.2,0.4)
F	Electric work	B	(388,437,501); (0.5,0.4,0.2)
G	Mechanical operations	B	(388,437,501); (0.5,0.4,0.2)
H	Isolation work (over plinth)	E	(38,44,52); (0.65,0.2,0.3)
	Building wall	E	(162,176,223); (0.7,0.5,0.1)
J	Infrastructure works	I	(100,123,155); (0.5,0.4,0.2)
K	Plaster works	I	(146,160,184); (0.55,0.4,0.2)
L	Ceramic and marble related works	K	(180,202,245); (0.6,0.2,0.4)
M	Cement finish	K	(141,158,177); (0.7,0.5,0.1)
N	Ceiling floor	M	(165,188,221); (0.7,0.55,0.15)
O	Steel work (plinth)	D	(103,117,134); (0.75,0.35,0.1)
P	Steel work (roof)	O	(50,57,67); (0.65,0.3,0.2)
Q	Isolation work (roof)	P	(73,80,96); (0.6,0.1,0.3)
R	Paint	P	(128,159,223); (0.5,0.2,0.4)
S	Parquet and baseboard	L	(52,59,73); (0.7,0.25,0.15)
T	Lift works	D	(320,352,396); (0.4,0.2,0.5)
U	Pool	G	(75,86,100); (0.7,0.1,0.2)
V	Landscape	Q	(78,86,103); (0.6,0.1,0.4)
X	Built-in and movable furniture	V	(40,45,55); (0.75,0.15,0.2)
Y	Control and acceptance	X	(15,30,60); (0.7,0.3,0.2)

$$\tilde{d}_2 = \min \left\{ \tilde{d}_1 \oplus \tilde{d}_{12} \right\}$$

$$= \min \left\{ < (0, 0, 0) ; 0, 1, 1 > \oplus < (150, 178, 210) ; 0.6, 0.25, 0.15 > \right\}$$

$$\tilde{d}_2 = < (150,178,210); 0.6,0.25,0.15 >.$$

Since minimum occurs related to $i = 1$, node 2 can be labelled as $[<(150,178,210); 0.6,0.25,0.15>, 1]$.

Fig. 1 Project network
diagram

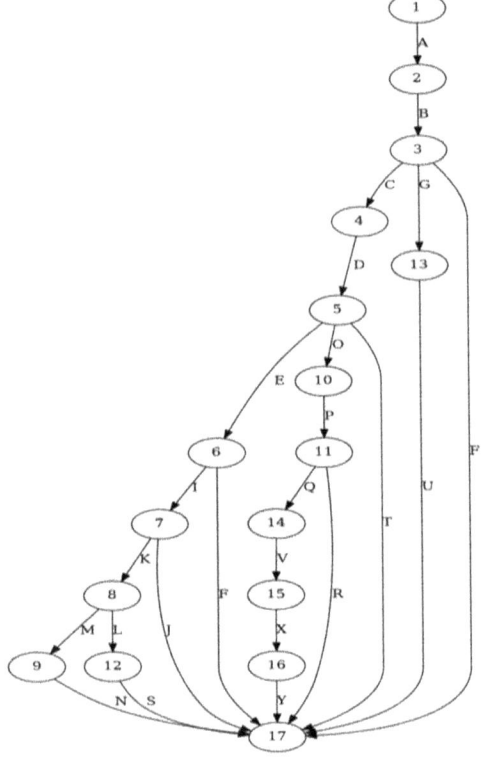

Iteration 2. Only node 2 is the predecessor of node 3, so putting i = 2 and j = 3 according to the second step of the algorithm, the value of $\tilde{d_3}$ can be computed as follows:

$$\tilde{d_3} = \min\left\{\tilde{d_2} \oplus \tilde{d_{23}}\right\}$$

$$= \min\{< (150, 178, 210)\,;\,0.6, 0.25, 0.15 > \oplus < (40, 44, 52)\,;\,0.5, 0.3, 0.2 >\}$$

$\tilde{d_3} = < (190,222,262);0.8,0.075,0.03>.$

Since minimum occurs related to i = 2, node 3 can be labelled as [<(190,222,262); 0.8,0.075,0.03>, 2].

Iteration 3. Only node 3 is the predecessor of node 4, so putting i = 3 and j = 4 according to the second step of the algorithm, the value of $\tilde{d_4}$ can be computed as follows:

$$\tilde{d_4} = \min\left\{\tilde{d_3} \oplus \tilde{d_{34}}\right\} = \min\{< (190, 222, 262)\,;\,0.8, 0.075, 0.03 >$$
$$\oplus < (112, 128, 143)\,;\,0.6, 0.4, 0.3 >\}$$

Fig. 2 Project network
diagram where each arc
length is represented by
single-valued triangular
neutrosophic numbers

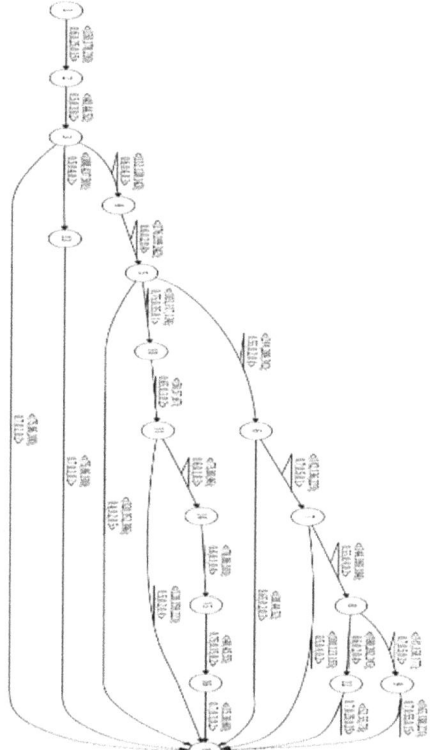

$\tilde{d_4} = \;<(302,350,405);0.92,0.03,0.009>$.

Since minimum occurs related to i $= 3$, node 4 can be labelled as
$[<(302,350,405); 0.92,0.03,0.009>, 3]$.

Iteration 4. Only node 4 is the predecessor of node 5, so putting i $= 4$ and j $= 5$
according to the second step of the algorithm, the value of $\tilde{d_5}$ can be computed as
follows:

$$\tilde{d_5} = \min\left\{\tilde{d_4} \oplus \tilde{d_{45}}\right\} = \min\{< (302, 350, 405) \; ; 0.92, 0.03, 0.009 >$$
$$\oplus < (176, 199, 242) \; ; 0.6, 0.2, 0.4 >\}$$

$\tilde{d_5} = \;<(478,549,647);0.968,0.106,0.0036>$.

Since minimum occurs related to i $= 4$, node 5 can be labelled as
$[<(478,549,647); 0.968,0.106,0.0036>, 4]$.

Iteration 5. Only node 5 is the predecessor of node 6, so putting i $= 5$ and j $= 6$
according to the second step of the algorithm, the value of $\tilde{d_6}$ can be computed as
follows:

$$\tilde{d_6} = \min \left\{ \tilde{d_5} \oplus \tilde{d_{56}} \right\} = \min \{ < (478, 549, 647) \, ; 0.968, 0.106, 0.0036 >$$
$$\oplus < (244, 289, 342) \, ; 0.55, 0.2, 0.4 > \}$$

$\tilde{d_6} = < (722,838,989);0.9856,0.0012,0.00144 >$.

Since minimum occurs related to i = 4, node 5 can be labelled as [<(722,838,989); 0.9856,0.0012,0.00144>, 5].

Other iterations from 6 to 15 can be made as similar way.

Iteration 6. The predecessor nodes of node 17 are node 3, 5, 6, 7, 9, 11, 12, 13, and 16 so putting i = 3, 5, 6, 7, 9, 11, 12, 13, 16 and j = 17 according to the second step of the algorithm, the value of $\tilde{d_{17}}$ can be computed as follows:

$$\tilde{d_{17}} = \min \left\{ \tilde{d_3} \oplus \tilde{d_{317}}, \tilde{d_5} \oplus \tilde{d_{517}}, \tilde{d_6} \oplus \tilde{d_{617}}, \tilde{d_7} \oplus \tilde{d_{717}}, \tilde{d_9} \oplus \tilde{d_{917}}, \tilde{d_{11}} \right.$$
$$\left. \oplus \tilde{d_{1117}}, \tilde{d_{12}} \oplus \tilde{d_{1217}}, \tilde{d_{13}} \oplus \tilde{d_{1317}}, \tilde{d_{16}} \oplus \tilde{d_{1617}} \right\}$$

$\tilde{d_{17}} = \min(<(190,222,262);0.8,0.075,0.03>\oplus<(388,437,501);0.5,0.4,0.2>,$
<(478,549,647);0.968,0.006,0.0036> $\oplus$ <(320,352,396);0.4,0.2,0.5>, <(722,838,989);0.9856,0.0012,0.00144> $\oplus$ <(38,44,52);0.65,0.2,0.3>,

<(844,1014,1212);0.99568,0.0006,0.000144>$\oplus$ <(100,123,155);0.5,0.4,0.2>,
<(1171,1332,1573);0.9994168,0.00012,0.0000288>$\oplus$<(165,188,221);0.7,0.55, 0.15>,

<(631,723,848);0.9972,0.00063,0.000072>$\oplus$<(128,159,223);0.5,0.2,0.4>,
<(1210,1376,1641);0.9992224,0.000048,0.00001152>$\oplus$<(52,59,73);0.7,0.25, 0.15>,

<(578,659,763);0.9,0.03,0.006>$\oplus$<(75,86,100);0.7,0.1,0.2>,
<(822,934,1102);0.999888,0.00000945,0.000001728>$\oplus$<(15,30,60);0.7,0.3, 0.2>).

$\tilde{d_{17}} = \min$ (<(578,659,763);0.9,0.03,0.006>, <(798,901,1043);0.9808, 0.0012,0.0018>,
<(760,882,1041);0.9952,0.00024,0.000432>,<(984,1137,1367);0.99784, 0.00024,0.0000288>,<(1336,1520,1794);0.99982504,0.000066,0.000000432>, <(759,882,1071);
0.9986,0.000126,0.0000288>,<(1262,1435,1714);0.99976672,0.000012, 0.000001728>,<(653,745,863);0.97,0.003,0.0012>,<(837,964,1162);0.9999664, 0.000002835,0.0000003456>).

S(<(578,659,763);0.9,0.03,0.006>) = 358.
S(<(798,901,1043);0.9808,0.0012,0.0018>) = 510.
S(<(760,882,1041);0.9952,0.00024,0.000432>) = 502.
S(<(984,1137,1367);0.99784,0.00024,0.0000288>) = 653.
S(<(1336,1520,1794);0.99982504,0.000066,0.000000432>) = 872.
S(<(759,882,1071);0.9986,0.000126,0.0000288>) = 508.

S(<(1262,1435,1714);0.99976672,0.000012,0.000001728>) = 827.
S(<(653,745,863);0.97,0.003,0.0012>) = 419.
S(<(837,964,1162);0.9999664,0.000002835,0.0000003456>) = 556.

Since S(<(578,659,763);0.9,0.03,0.006>) is smaller than other values $\widetilde{d}_{17}$ = min   (<(578,659,763);0.9,0.03,0.006>,<(798,901,1043);0.9808,0.0012,0.0018>, <(760,882,1041);0.9952,0.00024,0.000432>,<(984,1137,1367);0.99784, 0.00024,0.0000288>,<(1336,1520,1794);0.99982504,0.000066,0.000000432>, <(759,882,1071);

0.9986,0.000126,0.0000288>,<(1262,1435,1714);0.99976672,0.000012, 0.000001728>,<(653,745,863);0.97,0.003,0.0012>,<(837,964,1162);0.9999664, 0.000002835,0.0000003456>) = <(578,659,763);0.9,0.03,0.006>.

Since minimum occurs related to i = 3, node 17 can be labelled as [<(578,659,763); 0.9,0.03,0.006>, 3].

After applying iterations, a project network consisting of single-valued triangular neutrosophic shortest distance of each node from node 1 can be shown as Fig. 3.

The single-valued triangular neutrosophic shortest distance between nodes 1 and node 17 can be written as <(578,659,763);0.9,0.03,0.006 > because of obtaining node 17 as the destination node of the project network. According to Fig. 3, it can

Fig. 3 Project network consisting of single-valued triangular neutrosophic shortest distance of each node from node 1

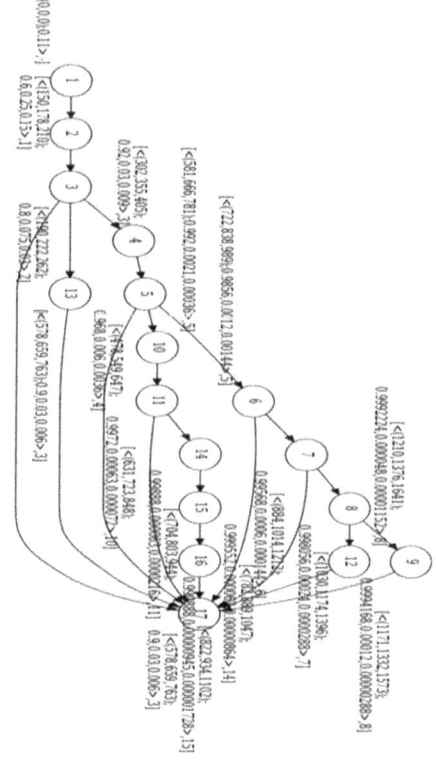

be said that the single-valued triangular neutrosophic shortest path from nodes 1 to 17 can be written as $1 \rightarrow 2 \rightarrow 3 \rightarrow 17$ and total duration computed via score function is 358 days.

7   Conclusions

In this study an algorithm developed by Broumi et al. [9, 10] is applied for solving the shortest path problem consisting of single-valued triangular neutrosophic numbers on a real case study as a part of city hospital. The outputs of this research show the applicability of this algorithm on real-world problems. For further researches a new algorithm can be designed for analyzing shortest path problems consisting of single-valued triangular neutrosophic numbers and the efficiency of them can be measured too. Additionally shortest path problem can be found for plithogenic, spherical, or other fuzzy set- based environments.

References

1. Broumi, S., Bakali, A., Talea, M., & Smarandache, F. (2016). Isolated single valued neutrosophic graphs. *Neutrosophic Sets and Systems, 11*, 74–78.
2. Broumi, S., Talea, M., Smarandache, F., & Bakali, A. (2016). *Single Valued Neutrosophic Graphs: Degree, Order and Size, IEEE International Conference on Fuzzy Systems (FUZZ)* (pp. 2445–2451).
3. Broumi, S., Talea, M., Bakali, A., & Smarandache, F. (2016a). Single valued Neutrosophic graphs. *Journal of New Theory, N, 10*, 86–101.
4. Broumi, S., Talea, M., Bakali, A., & Smarandache, F. (2016b). On bipolar single valued Neutrosophic graphs. *Journal of New Theory, N11*, 84–102.
5. Broumi, S., Talea, M., Bakali, A., & Smarandache, F. (2016c). Interval valued Neutrosophic graphs. *Critical Review, 12*, 5–33.
6. Broumi, S., Smarandache, F., Talea, M., & Bakali, A. (2016d). An introduction to bipolar single valued Neutrosophic graph theory. *Applied Mechanics and Materials, 841*, 184–191.
7. Broumi, S., Smarandache, F., Talea, M., & Bakali, A. (2016f). *Decision-Making Method Based On the Interval Valued Neutrosophic Graph, Future technologie* (pp. 1–8). IEEE.
8. F. Smarandache," Refined Literal Indeterminacy and the Multiplication Law of Sub-Indeterminacies," Neutrosophic Sets and Systems, 9, 2015, pp.58.63.
9. Broumi, S., Bakali, A., Talea, M., & Smarandache, F. (2016h). *Shortest path problem under triangular fuzzy Neutrosophic information, IEEE 10th international conference on software, Knowledge, Information Management & Applications (SKIMA)* (pp. 169–174).
10. Broumi, S., Bakali, A., Talea, M., & Smarandache, F. (2016i). *Computation of Shortest Path Problem in a Network with SV-Trapezoidal Neutrosophic Numbers, International Conference on Advanced Mechatronic Systems* (pp. 417–422).
11. Çubukçu, R. (2008). *Minimizing Time And Cost Risks Using Scheduling in Project Management, Unpublished Doctoral Thesis.* Adana: Çukurova University, Institute of Science.
12. Soltani, A., & Haji, R. (2007). A project scheduling method based on fuzzy theory. *Journal of Industrial and Systems Engineering, 1*(1), 70–80.
13. Kolaylıoğlu, Ö. (2006). *Project management and project management in civil engineering, unpublished Master's thesis.* Izmir: Dokuz Eylul University Institute of Social Sciences.

14. Kurt, Ö. (2006). *Project planning and programming technics; An application in construction sector, Unpublished Master's Thesis*. Antalya: Akdeniz University Institute of Social Sciences.
15. Kökçam, H. A., & Engin, O. (2010). Solving the fuzzy project scheduling problems with meta-heuristic methods. *Journal of Engineering and Natural Sciences, Sigma, 28*, 86–101.
16. Sönmez, E. (2007). *Why is the project management?, Unpublished Master's Thesis*. Istanbul: Mimar Sinan Fine Arts University Institute of Science.
17. Durucasu, H., İcan, Ö., Yeşilaydın, G., Gülcan, B., & Karamaşa, Ç. (2015). Project scheduling by means of fuzzy CPM method: An implementation in construction sector. *Ege Academic Review, 15*(4), 449–466.
18. Spinner, M. (1997). *Project Management, Principles and Practices*. New Jersey: Prentice-Hall.
19. A. Ngoor, M. M. Jabarulla, Multiple labeling Approach For Finding shortest Path with Intuitionistic Fuzzy Arc Length, International Journal of Scientific and Engineering Research,V3,Issue 11,pp.102-106,2012.
20. Okada, S., & Soper, T. (2000). A shortest path problem on a network with fuzzy arc lengths. *Fuzzy Sets and Systems, 109*, 129–140.
21. Kumar, A., & Kaur, M. (2011a). Solution of fuzzy maximal flow problems using fuzzy linear programming. *World Academy of Science and Technology., 87*, 28–31.
22. Majumdar, S., & Pal, A. (November 2013). Shortest path problem on intuitionistic fuzzy network. *Annals of Pure and Applied Mathematics, 5*(1), 26–36.
23. Kumar, A., & Kaur, M. (2011b). A new algorithm for solving shortest path problem on a network with imprecise edge weight. *Applications and Applied Mathematics, 6*(2), 602–619.
24. Yadav, A. K., & Biswas, B. R. (2009). On searching fuzzy shortest path in a network. *International Journal of Recent Trends in Engineering, 2*(3), 16–18.
25. Jayagowri, P., & Geetha Ramani, G. (2014). Using Trapezoidal Intuitionistic Fuzzy Number to Find Optimized Path in a Network. *Advances in Fuzzy Systems*, 1–6.
26. Hernandes, F., Teresa Lamata, M., Verdegay, J. L., & Kami, A. Y. (2007). The shortest path problem on networks with fuzzy parameters. *Fuzzy Sets and Systems, 158*, 1561–1570.
27. Kumar, G., Bajaj, R. K., & Gandotra. (2015). Algorithm for shortest path problem in a network with interval valued intuitionistic trapezoidal fuzzy number. *Procedia Computer Science, 70*, 123–129.
28. Broumi, S., Bakali, A., Talea, M., Smarandache, F., & Ali, M. (2017). Shortest path problem under bipolar Neutrosophic setting. *Applied Mechanics and Materials, 859*, 59–66.
29. Smarandache, F. A unifying field in logics Neutrosophy: Neutrosophic probability, set and logic., 1998. *Rehoboth: American Research Press*.
30. Biswas, P., Pramanik, S., & Giri, B. (2016). *Some distance measures of single valued Neutrosophic hesitant fuzzy Sets and their applications to multiple attribute decision making. F. Smarandache, & S. Pramanik içinde, in new trends in Neutrosophic theory and applications* (pp. 27–34). Brussels: Pons Publishing House.
31. Wang, H., Smarandache, F., Zhang, Y., & Sunderraman, R. (2010). Single Valued Neutrosophic Sets. *Multispace and Multistructure*, 410–413.
32. Abdel-Basset, M., Mohamed, M., Zhou, Y., & Hezam, I. (2017). Multi-Criteria Group Decision Making Based on Neutrosophic Analytic Hierarchy Process. *Journal of Intelligent & Fuzzy Systems*, 4055–4066.
33. Stanujkic, D., Zavadskas, E., Smarandache, F., Brauers, W., & Karabasevic, D. (2017). A Neutrosophic Extension of the Multimoora Method. *Informatica*, 181–192.

TrNN- EDAS Strategy for MADM with Entropy Weight Under Trapezoidal Neutrosophic Number Environment

Rama Mallick and Surapati Pramanik

1 Introduction

Multi-attribute decision-making (MADM) is useful in dealing with different aspects of our daily life. It is a process of selecting the best alternative from a set of available alternatives subject to predefined conflicting attributes. Since decision-making environments are different, powerful strategies have been developed and extended in the literature to encounter MADM problems.

MADM in crisp environment has been deeply studied ([1–4]; Yoon & Hwang, 1990; [5, 6]). In non-statistical uncertain environment, fuzzy set theory [7] has been widely employed ([8]; Ghorabaee et al., 2017; [9–11]).

To deal with uncertainty, indeterminacy, and falsity involved in worldly MADM, Smarandache [12] defined neutrosophic set (NS). Wang, Smarandache, Zhang, and Sunderraman [13] grounded a single-valued NS (SVNS). After that, researchers showed interest in neutrosophic hybrid sets (NHSs). NHSs have been proposed by combining NS and existing sets such as neutrosophic soft set (Maji, 2013), rough neutrosophic set (Broumi, Smarandache, & Dhar, 2014), complex neutrosophic set [14], neutrosophic soft expert set [15], tri-complex rough neutrosophic [16], neutrosophic bipolar set [17], neutrosophic cubic set [18], rough neutrosophic bipolar set [19], rough neutrosophic hyper-complex set [20], neutrosophic hesitant fuzzy set [21], neutrosophic refined set [22], etc. NSs and NHSs have been deeply employed in MADM such as SVNS environment [23–38], interval neutrosophic set (INS) environment [39–42], rough neutrosophic set (RNS) environment [16, 43–45],

R. Mallick
Department of Mathematics, Umeschandra College, Kolkata, West Bengal, India

S. Pramanik (✉)
Department of Mathematics, Nandalal Ghosh B.T. College, Kolkata, West Bengal, India

F. Smarandache, M. Abdel-Basset (eds.), *Neutrosophic Operational Research*,
https://doi.org/10.1007/978-3-030-57197-9_26

575

bipolar neutrosophic set (BNS) environment [40, 41, 46–50], neutrosophic hesitant fuzzy set (NHFS) environment [28, 29, 51–53], neutrosophic soft set environment [54–56], neutrosophic refined set (NRS) environment [57, 58]; Mondal, Pramanik, & Giri), neutrosophic number (NN) environment [59, 60], neutrosophic cubic set (NCS) environment [49, 61–64], and triangular neutrosophic set (TNS) environment [38, 53, 65, 66].

Ye [67] defined trapezoidal neutrosophic set. Ye [67] defined Trapezoidal Neutrosophic Number (TrNN) and presented some basic operations of TrNNs, score function for TrNNs, and two aggregation operators, namely, trapezoidal neutrosophic weighted arithmetic averaging (TrNWAA) operator and trapezoidal neutrosophic weighted geometric averaging (TrNWGA) operator. Liang, Wang, and Zhang [68] proposed score function, accuracy function, and certainty function of TrNNSs. In 2018, [69] developed VIseKriterijumska Optimizacija I Kompromisno Resenje (VIKOR) under TrNN environment. Biswas, Pramanik, and Giri [27] extended technique for order preference by similarity to ideal solution (TOPSIS) in TrNN environment. Pramanik and Mallick [70] developed TOmada de Decisao Interativa Multicriterio (TODIM) strategy under TrNN environment.

Keshavarz-Ghorabaee, Zavadskas, Olfat, and Turskis [71] proposed the evaluation based on distance from average solution (EDAS) strategy. EDAS ranks the alternatives according to the positive distance and negative distance between each alternative and average alternative. EDAS is a very effective MADM strategy and it is easy to solve MADM problems. Keshavarz-Ghorabaee, Zavadskas, Amiri, and Turskis [72] also developed an EDAS strategy for fuzzy multi-criteria decision-making (MCDM) to solve supplier selection problem. Kahraman et al. [73] extended EDAS strategy under interval-valued intuitionistic fuzzy environment. Li, Wang, and Wang [74] developed an EDAS strategy for linguistic neutrosophic multi-criteria group decision-making (MCGDM). Han and Wei [70] extended EDAS strategy for MCDM of multivalued NS.

To date, EDAS-based MADM strategy does not appear under TrNN environment in the current literature. This motivates us to fill the research gap and develop a novel EDAS-based MADM strategy under TrNN environment, which we call TrNN-EDAS strategy.

The organization of the paper is as follows: We describe some elementary definitions and operations of SVNS, TrNNs in Sect. 2. In Sect. 3, we present the weight of the attribute using entropy. Section 4 is devoted to developing TrNN-EDAS strategy under TrNN environment. In Sect. 5, an MADM problem is solved to show the feasibility and applicability of the developed EDAS strategy. In Sect. 6, we present the comparative study of TrNWAA and TrNWGA operator. Section 7 concludes the paper by stating the future scope of the research.

2 Preliminary

Definition 2.1 Assume that L is a fuzzy set [7] L in a universal set V' which is defined by

$$L = \left\{ < v, \kappa_L(v) > \big| v \in V' \right\} \tag{1}$$

where $\kappa_L(v) : V' \to [0,1]$is known as membership function of L and value of $\kappa_L(v)$ is called the degree of membership.

Definition 2.2 A trapezoidal fuzzy number (TrFN) [75] γ is defined by its membership function as follows:

$$\kappa_\gamma(v) = \begin{cases} \frac{z - \eta_{11}}{\eta_{12} - \eta_{11}}, & \eta_{11} \leq z < \eta_{12} \\ 1, & \eta_{12} \leq z \leq \eta_{13} \\ \frac{\eta_{14} - z}{\eta_{14} - \eta_{13}}, & \eta_{13} < z \leq \eta_{14} \\ 0, & \text{otherwise} \end{cases} \tag{2}$$

The TrFN γ is denoted by $\gamma = (\eta_{11}, \eta_{12}, \eta_{13}, \eta_{14})$ where $\eta_{11}, \eta_{12}, \eta_{13}, \eta_{14}$ are real numbers and $\eta_{11} \leq \eta_{12} \leq \eta_{13} \leq \eta_{14}$.

Definition 2.3 Assume that V' is a universal set and v' is an element in V'. An SVNS ([13]) M' in V' is defined by

$$M' = \left\{ v', < C_{M'}(v'), D_{M'}(v'), E_{M'}(v') > \big| v' \in V' \right\} \tag{3}$$

where $C_{M'}(v') : V' \to [0,1], D_{M'}(v') : V' \to [0,1]$ and $E_{M'}(v') : V' \to [0,1]$with $0 \leq C_{M'}(v') + D_{M'}(v') + E_{M'}(v') \leq 3$ for all$v' \in V'$. The functions $C_{M'}(v'), D_{M'}(v')$and $E_{M'}(v')$ are the truth membership, the indeterminacy membership, and the falsity membership functions, respectively.

Definition 2.4 Assume that β be a single-valued TrNN (SVTrNN) [67]. Its three independent membership functions are defined as:

$$C_\beta(z') = \begin{cases} \frac{(z' - m_{11})c_\beta}{(m_{12} - m_{11})}, & m_{11} \leq z' < m_{12} \\ c_\beta, & m_{12} \leq z' \leq m_{13} \\ \frac{(m_{14} - z')c_\beta}{(m_{14} - m_{13})}, & m_{13} < z' \leq m_{14} \\ 0, & \text{otherwise.} \end{cases} \tag{4}$$

$$D_\gamma(z') = \begin{cases} \frac{(m_{12} - z') + (z' - m_{11})d_\gamma}{(m_{12} - m_{11})}, & m_{11} \leq z' < m_{12} \\ d_\gamma, & m_{12} \leq z' \leq m_{13} \\ \frac{z' - m_{13} + (m_{14} - z')d_\gamma}{m_{14} - m_{13}}, & m_{13} < z' \leq m_{14} \\ 0, & \text{otherwise.} \end{cases} \tag{5}$$

$$E_\gamma \left(z' \right) = \begin{cases} \frac{m_{12}-z'+(z'-m_{11})e_\gamma}{m_{12}-m_{11}}, & m_{11} \le z' < m_{12} \\ e_\gamma, & m_{12} \le z' < m_{13} \\ \frac{z'-m_{13}+(m_{14}-z')e_\gamma}{m_{14}-m_{13}}, & m_{13} < z' < m_{14} \\ 0, & \text{otherwise.} \end{cases} \tag{6}$$

where $0 \le C_\beta(z') \le 1$, $0 \le D_\beta(z') \le 1$ and $0 \le E_\beta(z') \le 1$; and $0 \le C_\beta(z') + D_\beta(z') + E_\beta(z') \le 3$; $m_{11}, m_{12}, m_{13}, m_{14} \in R$.

Then SVTrNN β is presented as $\beta = ([m_{11}, m_{12}, m_{13}, m_{14}]; c_\gamma, d_\gamma, e_\gamma)$.

Definition 2.5 Assume that $m_1 = ([m_{11}, m_{12}m_{13}, m_{14}]; c_1, d_1, e_1)$ and $m_2 = ([m_{21}, m_{22}, m_{23}, m_{24}]; c_2, d_2, e_2)$ are any two TrNNs and $\zeta \ge 0$. Then we have the following operations [68]:

1. $m_1 \oplus m_2 = ([m_{11} + m_{21}, m_{12} + m_{22}, m_{13} + m_{23}, m_{14} + m_{24}]; c_1 + c_2 - c_1 c_2, d_1 d_2, e_1 e_2)$
2. $m_1 \otimes m_2 = ([m_{11} m_{21}, m_{12}m_{22}, m_{13}m_{23}, m_{14}m_{24}]; c_1 c_2, d_1 + d_2 - d_1 d_2, e_1 + e_2 - e_1 e_2)$
3. $\zeta m_1 = ([\zeta m_{11}, \zeta m_{12}, \zeta m_{13}, \zeta m_{14}]; 1 - (1 - c_1)^\zeta, (d_1)^\zeta, (e_1)^\zeta)$
4. $(m_1)^\zeta = ([(m_{11})^\zeta, (m_{12})^\zeta, (m_{13})^\zeta, (m_{14})^\zeta]; (c_1)^\zeta, 1 - (1 - d_1)^\zeta, 1 - (1 - e_1)^\zeta)$

Definition 2.6 Suppose $T = [m_{11}, m_{12}, m_{13}, m_{14}]$ is a trapezoidal fuzzy number on R, and $m_{11} \le m_{12} \le m_{13} \le m_{14}$; then, the center of gravity (COG) [68] of T defined by:

$$COG(T) = \begin{cases} a & \text{if } m_{11} = m_{12} = m_{13} = m_{14} \\ \frac{1}{3}\left[m_{11} + m_{12} + m_{13} + m_{14} - \frac{m_{13}m_{14}-m_{11}m_{12}}{m_{13}+m_{14}-m_{11}-m_{12}}\right] & \text{otherwise} \end{cases} \tag{7}$$

Definition 2.7 Let $m = \,<(m_{11}, m_{12}, m_{13}, m_{14}); C_m, D_m, E_m>$ be TrNN. Then the score function $S(m)$, accuracy function $Ac(m)$, and certainty function $E(m)$ [68] of TrNN are defined by

$$S(m) = COG(T) \times \frac{(2 + C_m - D_m - E_m)}{3} \tag{8}$$

$$Ac(m) = COG(T) \times (C_m - E_m) \tag{9}$$

$$E(m) = COG(T) \times C_m \tag{10}$$

Definition 2.8 Comparison of two TrNNs:

Assume that $m_1 = ([m_{11}, m_{12}m_{13}, m_{14}]; c_1, d_1, e_1)$ and $m_2 = ([m_{21}, m_{22}, m_{23}, m_{24}]; c_2, d_2, e_2)$ are two TrNNs in R. The comparison between TrNNs is presented as follows:

1. If $Sc(m_1) > Sc(m_2)$, then $m_1 > m_2$.
2. If $Sc(m_1) = Sc(m_2)$ and $Ac(m_1) > Ac(m_2)$, then $m_1 > m_2$.
3. If $Sc(m_1) = Sc(m_2)$ and $Ac(m_1) < Ac(m_2)$, then $m_1 < m_2$ and.
4. If $Sc(m_1) = Sc(m_2)$ and $Ac(m_1) = Ac(m_2)$, and $E(m_1) > E(m_2)$, $m_1 > m_2$ and when $E(m_1) < E(m_2)$, then $m_1 < m_2$ and $m_1 > m_2$ when $E(m_1) = E(m_2)$.

Definition 2.9 Assume that $m_j = \big([\, m_{h1}, m_{h2}, m_{h3}, m_{h4}]; c_{\tilde{t}_h}, d_{\tilde{t}_h}, e_{\tilde{t}_h}\big)$ $(h=1, 2, \ldots, p)$ is a TrNN, then TrNWAA operator and TrNWGA operator [67] are defined as follows:

$$TrNWAA\left(m_1, m_2, \ldots, m_p\right) = \sum_{h=1}^{p} w_h m_{hi}$$

$$= \left\langle \begin{pmatrix} \sum_{h \in \Omega_{max}} w_h m_{h1}, & \sum_{h \in \Omega_{max}} w_h m_{h2}, & \sum_{h \in \Omega_{max}} w_h m_{h3}, & \sum_{h \in \Omega_{max}} w_h m_{h4} \end{pmatrix}; \\ 1 - \prod_{h \in \Omega_{max}} (1 - c_h)^{w_h}, \quad \prod_{h \in \Omega_{max}} (d_h)^{w_h}, \quad \prod_{h \in \Omega_{max}} (e_h)^{w_h} \right\rangle \tag{11}$$

$$TrNWGA\left(m_1, m_2, \ldots, m_p\right) = \prod_{h=1}^{p} w_h m_{hi}$$

$$= \left\langle \begin{pmatrix} \prod_{h \in \Omega_{min}} m_{h1}^{w_h}, & \prod_{h \in \Omega_{min}} m_{h2}^{w_h} & \prod_{h \in \Omega_{min}} m_{h3}^{w_h} & \prod_{h \in \Omega_{min}} m_{h4}^{w_h} \end{pmatrix}; \\ \prod_{h \in \Omega_{min}} c_{n_h}^{w_h}, 1 - \prod_{h \in \Omega_{min}} \left(1 - d_{n_h}\right)^{w_h}, 1 - \prod_{h \in \Omega_{min}} \left(1 - e_{n_h}\right)^{w_h} \right\rangle \tag{12}$$

where w_h is the weight of $m_h (h = 1, 2, \ldots, p)$ such that $w_h > 0$ and $\sum_{h=1}^{p} w_h = 1$.

Definition 2.10 Assume that $m_1 = ([m_{11}, m_{12} m_{13}, m_{14}]; c_1, d_1, e_1)$ and $m_2 = ([m_{21}, m_{22}, m_{23}, m_{24}]; c_2, d_2, e_2)$ are any two TrNNs. The normalized Hamming distance ([27]) between m_1 and m_2 is defined as follows:

$$d(m_1, m_2)$$

$$= \frac{1}{12} \begin{pmatrix} |m_{11} (2+c_1-d_1-e_1) - m_{21} (2+c_2-d_2-e_2)| + |m_{12} (2+c_1-d_1-e_1) - m_{22} (2+c_2-d_2-e_2)| \\ + |m_{13} (2+c_1-d_1-e_1) - m_{23} (2+c_2-d_2-e_2)| + |m_{14} (2+c_1-d_1-e_1) - m_{24} (2+c_2-d_2-e_2)| \end{pmatrix} \tag{13}$$

3   Calculating Weight Using Entropy Strategy

Weight measure is an important part of MADM problem, and it is directly related with distance measure of two rating values. Here we utilize Hamming distance to measure the distance. The whole weight calculating process using entropy [76] is presented as follows.

The value H_j is the expression of deviation of the rating value which is calculated using average rating values:

$$H_p = \left(d\left(t_{1p}, t_p^+\right), d\left(t_{2p}, t_p^+\right), \ldots, d\left(t_{rp}, t_p^+\right)\right) \tag{14}$$

Here, t_p^+ is average rating value which is calculated using the following equation:

$$
t_p^+ = \left\langle \left(\frac{1}{r}\sum_{j=1}^r t_{j1}, \frac{1}{r}\sum_{j=1}^r t_{j2}, \frac{1}{r}\sum_{j=1}^r t_{j3}, \frac{1}{r}\sum_{j=1}^r t_{j4} \right); \right.
$$
$$
\left. 1 - \prod_{j=1}^r \left(1 - c_{j1}\right)^{\frac{1}{r}}, \prod_{j=1}^r \left(d_{j1}\right)^{\frac{1}{r}}, \prod_{j=1}^r \left(e_{j1}\right)^{\frac{1}{r}} \right\rangle \tag{15}
$$

$d(t_{1p}, t_p^+)$ is calculated by using Eq. (13).

The corresponding normalized distance vector is presented as follows.

$$H_P^t = \left[\frac{d\left(t_{jP}, t_P^+\right)}{\max_j d\left(t_{jP}, t_P^+\right)}, j = 1, 2, \cdots, r \right] \tag{16}$$

The entropy measure of the j-th attribute C_j for p available alternatives is obtained from:

$$r_p = -\frac{1}{\ln(m)}\sum_{p=1}^r \left[\frac{H_p'}{\sum\limits_{p=1}^r H_p'} \ln\left(\frac{H_p'}{\sum\limits_{p=1}^r H_p'} \right) \right], j = 1, 2, \ldots, r \tag{17}$$

We obtain the normalized weight of the j-th attribute as follows.

$$w_p = \frac{1 - r_p}{\sum\limits_{p=1}^n \left(1 - r_p\right)} \tag{18}$$

$$w = (w_1, w_2, \ldots, w_r), 0 \le w_r \le 1, j = 1, 2, \ldots, r.$$

4 TrNN-EDAS Strategy Under TrNN Environment

Assume that an MADM problem comprising y alternatives $T = \{T_1, T_2, \ldots, T_y\}$, and z decision attributes $U = \{U_1, U_2, \ldots, U_z\}$. The attribute value of the alternative T_i over the criteria U_j is presented in the form $m_{ij} = \left(\left[m_{ij}^1, m_{ij}^2, m_{ij}^3, m_{ij}^4\right]; c_{ij}, d_{ij}, e_{ij}\right), 0 \le c_{ij} \le 1, 0 \le d_{ij} \le 1, 0 \le e_{ij} \le 1, 0 \le c_{ij} + d_{ij} + e_{ij} \le 3$, and the decision matrix is constructed.

The steps of the EDAS under TrNNs environment are as follows:

Step-1: Standardize decision matrix:

Assume that $Q = (m_{rs})_{y \times z}$ is a neutrosophic decision matrix, where $m_{rs} = \left(\left[m_{rs}^1, m_{rs}^2, m_{rs}^3, m_{rs}^4\right]; c_{m_{rs}}, d_{m_{rs}}, e_{m_{rs}}\right)$ is the rating value of alternative T_r with respect to attribute U_s. The decision matrix $(m_{rs})_{y \times z}$ is standardized [54] to obtain the standardized decision matrix $X^* = (x_{rs})_{y \times z}$, where the component x_{rs}^k of the entry $x_{rs} = \left(\left[x_{rs}^1, x_{rs}^2, x_{rs}^3, x_{rs}^4\right]; c_{x_{rs}}, d_{x_{rs}}, e_{x_{rs}}\right)$ in the matrix S is considered as follows:

1. For benefit type attributes:

$$x_{rs} = \left(\left[\frac{m_{rs}^1}{k_s^+}, \frac{m_{rs}^2}{k_s^+}, \frac{m_{rs}^3}{k_s^+}, \frac{m_{rs}^4}{k_s^+}\right]; c_{x_{rs}}, d_{x_{rs}}, e_{x_{rs}}\right)\right\} \tag{19}$$

2. For cost type attributes;

$$x_{rs} = \left(\left[\frac{k_s^-}{m_{rs}^4}, \frac{k_s^-}{m_{rs}^3}, \frac{k_s^-}{m_{rs}^2}, \frac{k_s^-}{m_{rs}^1}\right]; c_{x_{rs}}, d_{x_{rs}}, e_{x_{rs}}\right) \tag{20}$$

Here $k_s^+ = \max\left\{m_{rs}^4 \,\middle|\, r = 1, 2, \ldots, y\right\}$ and $k_s^- = \min\left\{b_{rs}^1 \,\middle|\, r = 1, 2, \ldots, y\right\}$ for $s = 1, 2, \ldots, z$.

Then, the obtained standardized decision matrix is presented as follows:

$$X^* = (x_{rs})_{g \times h} = \begin{pmatrix} x_{11} & x_{12} & & x_{1z} \\ x_{21} & x_{22} & \cdots & x_{2z} \\ \vdots & & \ddots & \vdots \\ x_{y1} & x_{y2} & \cdots & x_{yz} \end{pmatrix} \tag{21}$$

Step-2: Using Eq. (18), calculate the weight vector of the attribute $w = \{w_1, w_2, \ldots, w_z\}$ where $w_i \in [0, 1]$, $\sum_{i=1}^{z} w_i = 1$.

Step-3: In this step we apply TrNWAA [67] operator which is presented as follows:

$$
TrNWAA\left(m_1, m_2, \ldots, m_p\right) = \sum_{h=1}^{p} w_h m_{hi}
$$

$$
= \left\langle \left(\begin{array}{cc} \sum\limits_{h \in \Omega_{\max}} w_h m_{h1}, & \sum\limits_{h \in \Omega_{\max}} w_h m_{h2}, & \sum\limits_{h \in \Omega_{\max}} w_h m_{h3}, & \sum\limits_{h \in \Omega_{\max}} w_h m_{h4} \end{array} \right); \\ 1 - \prod\limits_{h \in \Omega_{\max}} (1 - c_h)^{w_h}, \quad \prod\limits_{h \in \Omega_{\max}} (d_h)^{w_h}, \quad \prod\limits_{h \in \Omega_{\max}} (e_h)^{w_h} \end{array} \right\rangle \tag{22}
$$

Step-4: Using Eqs. (8, 9 and 10), we calculate score function, accuracy function, and certainty function value of standardized matrix and aggregated matrix.

Step-5: The positive distance from average (PDA) and the negative distance from average (NDA) matrices are calculated according to lower and upper values of matrix as shown:

$$
PDA_{rs} = \frac{\max\left(0, s\left(p_{rs}\right) - s\left(AV_s\right)\right)}{s\left(AV_s\right)} \tag{23}
$$

$$
NDA_{rs} = \frac{\max\left(0, s\left(AV_s\right) - s\left(p_{rs}\right)\right)}{s\left(AV_s\right)} \tag{24}
$$

Here, PDA_{rs} and NDA_{rs} reflect the positive and negative distance of rth alternative from average solution in terms of sth criterion for the lower level of decision matrix, respectively.

Remark 1 If $s(AV_s) = s(p_{rs})$ and $a(AV_s) \neq a(p_{rs})$, then replace score function with accuracy function in Eq. (23 and 24); if $s(AV_s) = s(p_{rs})$ and $a(AV_s) \neq a(p_{rs})$, then replace score function with certainty function in Eq. (23 and 24).

Step-6: Calculate weighted summation of the PD and ND from average matrix:

$$
SP_r = \sum_{s=1}^{z} v_s PDA_{rs} \tag{25}
$$

and

$$
SN_r = \sum_{s=1}^{z} v_s NDA_{rs} \tag{26}
$$

Step-7: Identify the normalized value of SP_i and SN_i for all alternatives as follows:

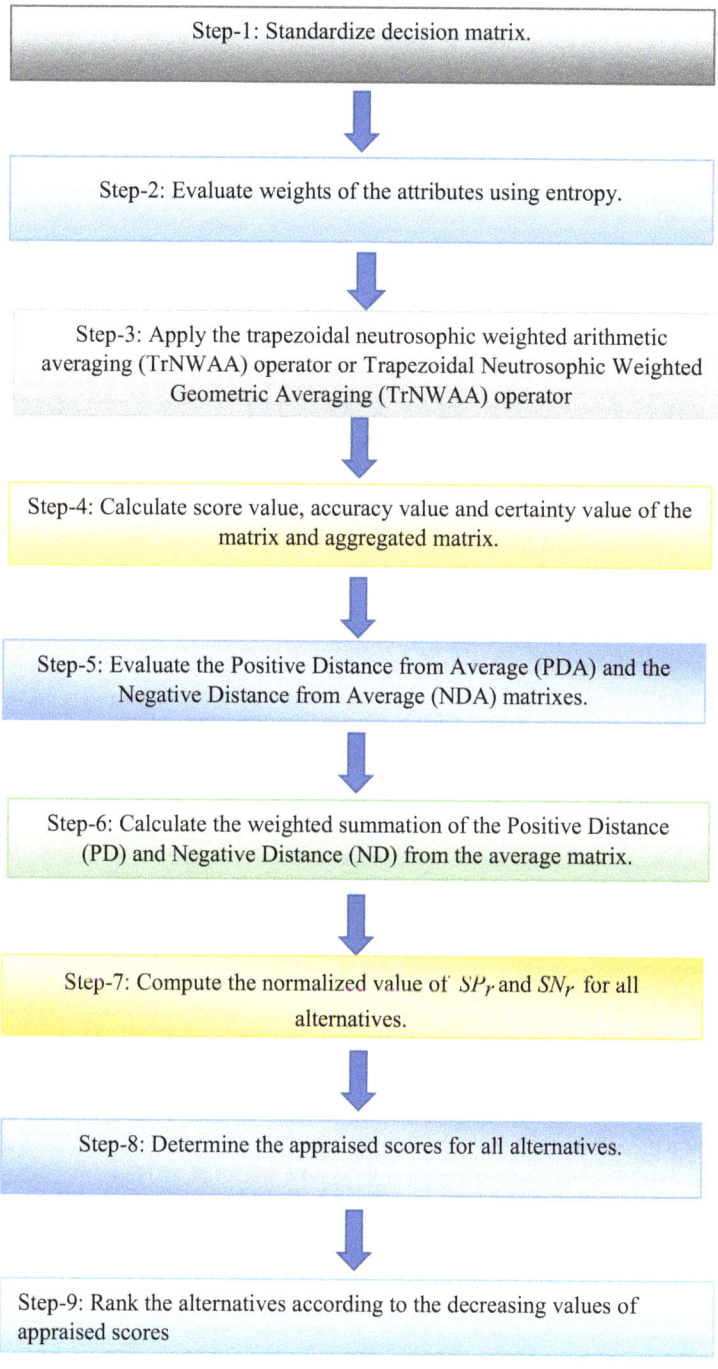

Fig. 1 EDAS strategy under TrNN environment

$$NSP_r = \frac{SP_r}{\max(SP_r)} \tag{27}$$

and

$$NSN_r = 1 - \frac{SN_r}{\max(SN_r)} \tag{28}$$

Step-8: Find the appraised score (AS) for all alternatives as follows:

$$AS_r = \frac{1}{2}(NSP_r + NSN_r) \tag{29}$$

where $0 \leq AS_r \leq 1$.

Step-9: Rank the alternatives according to the decreasing values of AS_r. The alterative with the highest AS_r reflects the best alternative (choice) (Fig. 1).

5 Numerical Example

We employ developed EDAS strategy to help an investor company to select the best country to invest. We consider an MADM problem adapted from [77].

An Indian businesses conglomerate company whose main business is producing and selling steel, energy, minerals, port, and infrastructure and cement. Recently, the overseas investment department decides to select a pool of alternatives from several foreign countries based on preliminary surveys. After thorough investigation, four countries (alternatives) are taken into consideration, i.e., T_1, T_2, T_3 and T_4.

Based on the experience of the department personnel, five factors are considered:

1. U_1; resources (suitability of the minerals and their exploration)
2. U_2; politics and policy (corruption and political risk)
3. U_3; economy (development vitality)
4. U_4 infrastructure (railway and highway facilities)
5. U_5; environment (such as energy).

Based on the surveys on the four countries and his/her knowledge and experience, expert provides his/her own evaluation as shown in Table 1. Here all the attributes are benefit type attributes.

Table 1 Weighted summation value of PD and ND

	$r = 1$	$r = 2$	$r = 3$	$r = 4$
SP_r	0.1112	0.1152	0.0944	0.1048
SN_r	0.1316	0.0693	0.1091	0.1216

The decision matrix is presented as follows:

$$
\begin{pmatrix}
 & U_1 & U_2 & U_3 \\
T_1 & ([0.15, 0.32, 0.35, 0.42]; 0.5, 0.2, 0.3) & ([0.53, 0.61, 0.73, 0.75]; 0.7, 0.3, 0.4) & ([0.32, 0.44, 0.52, 0.63]; 0.7, 0.3, 0.4) \\
T_2 & ([0.25, 0.33, 0.55, 0.63]; 0.6, 0.3, 0.3) & ([0.62, 0.64, 0.77, 0.82]; 0.8, 0.3, 0.2) & ([0.44, 0.51, 0.63, 0.7]; 0.6, 0.2, 0.2) \\
T_3 & ([0.32, 0.41, 0.52, 0.54]; 0.6, 0.4, 0.2) & ([0.41, 0.63, 0.74, 0.82]; 0.6, 0.4, 0.2) & ([0.32, 0.41, 0.52, 0.63]; 0.5, 0.3, 0.3) \\
T_4 & ([0.22, 0.34, 0.44, 0.5]; 0.6, 0.2, 0.1) & ([0.54, 0.61, 0.74, 0.81]; 0.8, 0.3, 0.2) & ([0.25, 0.32, 0.55, 0.64]; 0.6, 0.4, 0.2) \\
 & U_4 & U_5 & \\
 & ([0.45, 0.52, 0.71, 0.82]; 0.6, 0.2, 0.3) & ([0.42, 0.58, 0.61, 0.74]; 0.8, 0.5, 0.3) & \\
 & ([0.33, 0.42, 0.53, 0.64]; 0.7, 0.3, 0.4) & ([0.62, 0.73, 0.84, 0.91]; 0.6, 0.3, 0.2) & \\
 & ([0.42, 0.53, 0.62, 0.71]; 0.8, 0.4, 0.3) & ([0.53, 0.62, 0.74, 0.83]; 0.8, 0.2, 0.3) & \\
 & ([0.37, 0.44, 0.51, 0.72]; 0.6, 0.3, 0.2) & ([0.45, 0.64, 0.71, 0.87]; 0.6, 0.2, 0.3) &
\end{pmatrix}
$$

Step-1: Since all the attributes are benefit type, we do not need to standardize decision matrix.

Step-2: Calculate weights of the attributes using entropy.

Calculating average rating value using Eq. (15), we obtain

$$t_1^+ = ([0.235, 0.35, 0.465, 0.5225]; 0.5770, 0.2632, 0.2060)$$
$$t_2^+ = ([0.525, 0.6225, 0.745, 0.8]; 0.6278, 0.3224, 0.2378)$$
$$t_3^+ = ([0.3325, 0.42, 0.555, 0.65]; 0.6064, 0.2913, 0.2632)$$
$$t_4^+ = ([0.3925, 0.4775, 0.5925, 0.7225]; 0.6870, 0.2913, 0.2913)$$
$$t_5^+ = ([0.505, 0.6425, 0.725, 0.8375]; 0.7172, 0.2783, 0.2711)$$

Evaluate normalized distance vector using Eq. (16):

$$
\begin{pmatrix}
 & H_1 & H_2 & H_3 & H_4 \\
U_1 & 1 & 0.4325 & 0.3692 & 0.1624 \\
U_2 & 0.3281 & 1 & 0.3718 & 0.6513 \\
U_3 & 0.2329 & 1 & 0.4442 & 0.4970 \\
U_4 & 0.8592 & 1 & 0.3275 & 0.4146 \\
U_5 & 1 & 0.5403 & 0.3248 & 0.2472
\end{pmatrix}
$$

Using Eq. (17), we obtain entropy measures:

$$e_1 = 0.8636, e_2 = 0.9274, e_3 = 0.9077, e_4 = 0.9284, e_5 = 0.8957$$

Using Eq. (18), we obtain the weight vector as: $w = (0.2858, 0.1521, 0.1934, 0.1500, 0.2186)$.

Step-3: Using TrNWAA operator in Eq. (22), we obtain

$$\tilde{T}_1 = ([0.3447, 0.4741, 0.5515, 0.6407] \, ; 0.6682, 0.2811, 0.3314)$$
$$\tilde{T}_2 = ([0.4359, 0.5129, 0.6593, 0.7351] \, ; 0.6552, 0.2774, 0.2492)$$
$$\tilde{T}_3 = ([0.3946, 0.5073, 0.6165, 0.6888] \, ; 0.6765, 0.3252, 0.2512)$$
$$\tilde{T}_4 = ([0.3472, 0.4577, 0.5764, 0.6880] \, ; 0.6400, 0.2585, 0.1793)$$

Step-4: Using Eqs. (8), (9) and (10), we obtain the score value, accuracy value and certainty value of decision matrix as follows:

$$Sc\,(T_r) = \begin{pmatrix} 0.2022 & 0.4357 & 0.3180 & 0.4352 & 0.3903 \\ 0.2933 & 0.5466 & 0.418 & 0.3205 & 0.5414 \\ 0.2970 & 0.4288 & 0.2982 & 0.3984 & 0.5213 \\ 0.2857 & 0.5175 & 0.2936 & 0.3624 & 0.466 \end{pmatrix}$$

$$Ac\,(T_r) = \begin{pmatrix} 0.0607 & 0.1960 & 0.1431 & 0.1878 & 0.2927 \\ 0.132 & 0.4278 & 0.228 & 0.1442 & 0.3094 \\ 0.1782 & 0.2573 & 0.0942 & 0.2846 & 0.34 \\ 0.1863 & 0.405 & 0.1762 & 0.2071 & 0.1997 \end{pmatrix}$$

$$Ce\,(T_r) = \begin{pmatrix} 0.1517 & 0.4575 & 0.3339 & 0.3756 & 0.4683 \\ 0.264 & 0.5704 & 0.342 & 0.3365 & 0.4641 \\ 0.2673 & 0.3860 & 0.2354 & 0.4553 & 0.544 \\ 0.2236 & 0.54 & 0.2642 & 0.3107 & 0.3994 \end{pmatrix}$$

Similarly, we calculate the same of aggregate decision matrix:

$$Sc\left(\tilde{T}_r\right) = \begin{pmatrix} 0.3432 \\ 0.4156 \\ 0.3852 \\ 0.3798 \end{pmatrix}, \ Ac\left(\tilde{T}_r\right) = \begin{pmatrix} 0.1687 \\ 0.2378 \\ 0.2340 \\ 0.2385 \end{pmatrix}, \ Ce\left(\tilde{T}_r\right) = \begin{pmatrix} 0.3346 \\ 0.3838 \\ 0.3722 \\ 0.3311 \end{pmatrix}$$

Step-5: Evaluate positive distance from average (PDA) and the negative distance from average (NDA) each of the alternative and average alternative w.r.t each of the attribute using Eq. (23) and (24) we get,

$$PDA_{rs} = \begin{pmatrix} 0 & 0.2695 & 0 & 0.2681 & 0.1372 \\ 0 & 0.3152 & 0.0058 & 0 & 0.3027 \\ 0 & 0.1132 & 0 & 0 & 0.3533 \\ 0 & 0.3626 & 0 & 0 & 0.2270 \end{pmatrix}$$

$$NDA_{rs} = \begin{pmatrix} 0.4108 & 0 & 0.0734 & 0 & 0 \\ 0.1223 & 0 & 0 & 0.2288 & 0 \\ 0.2290 & 0 & 0.2258 & 0 & 0 \\ 0.2478 & 0 & 0.2270 & 0.0458 & 0 \end{pmatrix}$$

Step-6: In this step, we calculate weighted summation value of PDA and NDA of each alternative; using Eq. (21 and 22), we get:

Table 2 Normalize value of SP_r and SN_r

	$r=1$	$r=2$	$r=3$	$r=4$
NSP_r	0.9653	1	0.8194	0.9097
NSN_r	0	0.4734	0.171	0.076

Table 3 Comparative study between TrNWAA operator and TrNWGA operator

Operator	Alternative rank
TrNWAA operator	$A_2 > A_3 > A_2 > A_1$
TrNWGA operator	$A_1 > A_2 > A_4 > A_3$

Step-7: Using Eq. (23 and 24), we calculate normalized value of SP_r and SN_r. We get (Table 2):

Step-8: Using (29), the appraised score of each of the alternative w. r. t each of the attribute is calculated

$$AS_1 = 0.4826, AS_2 = 0.7367, AS_3 = 0.4952, AS_4 = 0.4928$$

Step-9: Ranking of the four alternative is

$$A_2 > A_3 > A_2 > A_1$$

The best alternative is A_2.

6 Comparative Study

Here we compare EDAS strategy for trapezoidal neutrosophic arithmetic average (TrNWAA) operator with trapezoidal neutrosophic geometric average (TrNWGA) operator. The result is shown in Table 3. Ranking order differs for TrNWAA operator and TrNWGA operator. Also, we observe that best alternative is different for the two operators. After applying TrNWAA operator, we see the best alternative is A_2 and by applying TrNWGA operator the best alternative is A_1.

7 Conclusion

In this chapter, we develop an MADM -EDAS strategy in TrNN environment. Here we briefly describe definition, operation law, aggregate operator, and distance operator for TrNN. In this chapter, we use entropy to calculate the weight of the attribute. In this strategy, TrNWAA operator is used to calculate the average value of the alternative under each of the attribute. Next, we calculate PD and ND between each of the alternative and the average alternative. A numerical example is solved to show the practicality and effectiveness of proposed strategy. A comparative analysis of TrNWAA operator and TrNWGA operator is presented. In the future, the application of the proposed strategy can be investigated in MAGDM problems. In future different directions of research are generated from this study.

Various types of problems can be solved using the EDAS strategy including teacher selection [78], renewable energy selection [79], and data mining [80] in TrNN environment.

References

1. Hwang, C. L., & Yoon, K. (1981). *Multiple attribute decision making methods and applications, a state-of-the-art survey*. New York: Springer-Verlag.
2. Karni, R., Sanchez, P., & Tummala, V. M. R. (1990). A comparative study of multiattribute decision making methodologies. *Theory and Decision, 29*, 203–222.
3. Zeleny, M. (2012). *Multiple criteria decision making Kyoto 1975 (Vol. 123)*. Springer Science & Business Media.
4. Zeleny, M. (2011). Multiple criteria decision making (MCDM): From paradigm lost to paradigm regained? *Journal of Multi-Criteria Decision Analysis., 18*(1–2), 77–89. https://doi.org/10.1002/mcda.473.
5. Weber, M., & Borcherding, K. (1993). Behavioral influences on weight judgments in multiattribute decision making. *European Journal of Operational Research, 67*(1), 1–12.
6. Zanakis, S. H., Solomon, A., Wishart, N., & Dublish, S. (1998). Multi-attribute decision making: A simulation comparison of select methods. *European Journal of Operational Research, 107*(3), 507–529.
7. Zadeh, L. A. (1965). Fuzzy sets. *Information and Control, 8*(3), 338–353.
8. Zavadskas, E. K., Turskis, Z., & Kildienė, S. (2014). State of art surveys of overviews on MCDM/MADM methods. *Technological and Economic Development of Economy, 20*(1), 165–179.
9. Aouam, T., Chang, S. I., & Lee, E. S. (2003). Fuzzy MADM: An outranking method. *European Journal of Operational Research, 145*(2), 317–3
10. Li, D. F., & Wan, S. P. (2014). Fuzzy heterogeneous multiattribute decision making method for outsourcing provider selection. *Expert Systems with Applications, 41*(6), 3047–3059.
11. Mousavi, S. M., Vahdani, B., Tavakkoli-Moghaddam, R., Ebrahimnejad, S., & Amiri, M. (2013). A multi-stage decision-making process for multiple attributes analysis under an interval-valued fuzzy.
12. Smarandache, F. (1998). A unifying field of logics. In *Neutrosophy: Neutrosophic probability, set and logic*. Rehoboth: American Research Press.
13. Wang, H., Smarandache, F., Zhang, Y., & Sunderraman, R. (2010). Single valued neutrosophic sets. *Multi-space and Multi-structure, 4*, 410–413.
14. Ali, M., & Smarandache, F. (2017). Complex neutrosophic set. *Neural Computing and Applications, 28*(7), 1817–1834.
15. Şahin, M., Alkhazaleh, S., & Uluçay, V. (2015). Neutrosophic soft expert sets. *Applied Mathematics, 6*, 116–127.
16. Mondal, K., & Pramanik, S. (2015e). Tri-complex rough neutrosophic similarity measure and its application in multi-attribute decision making. *Critical Review, 11*, 26–40.
17. Deli, I., Ali, M., & Smarandache, F. (2015). Bipolar neutrosophic sets and their application based on multi-criteria decision making problems. In *2015 International Conference on Advanced Mechatronic Systems (ICAMechS) (pp. 249-254)*. IEEE.
18. Ali, M., Deli, I., & Smarandache, F. (2016). The theory of neutrosophic cubic sets and their applications in pattern recognition. *Journal of Intelligent & Fuzzy Systems, 30*(4), 1957–1963.

19. Pramanik, S., & Mondal, S. (2016). Rough neutrosophic bipolar set. *Global Journal of Engineering Science and Research Management, 3*(6), 71–81.

20. Mondal, K., Pramanik, S., & Smarandache, F. (2016c). Rough neutrosophic hyper-complex set and its application to multi attribute decision making. *Critical Review, 13*, 111–124.

21. Ye, J. (2016). Multiple-attribute decision-making method under a single-valued neutrosophic hesitant fuzzy environment. *Journal of Intelligent Systems, 24*(1), 23–36.

22. Smarandache, F. (2013). N-valued refined neutrosophic logic and its applications in physics. *Progress in Physics, 4*, 43–146.

23. Biswas, P., Pramanik, S., & Giri, B. C. (2014a). Entropy based grey relational analysis method for multi-attribute decision making under single valued neutrosophic assessments. *Neutrosophic Sets and Systems, 2*, 102–110. https://doi.org/10.5281/zenodo.571363.

24. Biswas, P., Pramanik, S., & Giri, B. C. (2014b). A new methodology for neutrosophic multi-attribute decision-making with unknown weight information. *Neutrosophic Sets and Systems, 3*, 42–50. https://doi.org/10.5281/zenodo.571212.

25. Biswas, P., Pramanik, S., & Giri, B. C. (2016). TOPSIS method for multi-attribute group decision making under single-valued neutrosophic environment. *Neural Computing and Applications, 27*(3), 727–737. https://doi.org/10.1007/s00521-015-1891-2.

26. Biswas, P., Pramanik, S., & Giri, B. C. (2018a). Neutrosophic TOPSIS with group decision making. In I. Otay (Ed.), *Kahraman C* (pp. 543–585). Studies in Fuzziness and Soft Computing: Fuzzy Multicriteria Decision Making Using Neutrosophic Sets. https://doi.org/10.1007/978-3-030-00045-5_21.

27. Biswas, P., Pramanik, S., & Giri, B. C. (2018b). TOPSIS strategy for multi attribute decision making with trapezoidal numbers. *Neutrosophic Sets System, 19*, 29–39. https://doi.org/10.5281/zenodo.123533584.

28. Biswas, P., Pramanik, S., & Giri, B. C. (2019a). Non-linear programming approach for single-valued neutrosophic TOPSIS method. *New Mathematics and Natural Computation.* https://doi.org/10.1142/S1793005719500169.

29. Biswas, P., Pramanik, S., & Giri, B. C. (2019b). NH-MADM strategy in neutrosophic hesitant fuzzy set environment based on extended GRA. *Informatica, 30*(2), 213–242.

30. Mondal, K., & Pramanik, S. (2015a). Neutrosophic decision making model of school choice. *Neutrosophic Sets and Systems, 7*, 62–68. https://doi.org/10.5281/zenodo.571507.

31. Mondal, K., & Pramanik, S. (2015d). Neutrosophic tangent similarity measure and its application to multiple attribute decision making. *Neutrosophic Sets and Systems, 9*, 80–87. https://doi.org/10.5281/zenodo.571578.

32. Mondal, K., Pramanik, S., & Giri, B. C. (2018a). Single valued neutrosophic hyperbolic sine similarity measure based MADM strategy. *Neutrosophic Sets and Systems, 20*, 3–11. https://doi.org/10.5281/zenodo.1235383.

33. Mondal, K., Pramanik, S., & Giri, B. C. (2018b). Hybrid binary logarithm similarity measure for MAGDM problems under SVNS assessments. *Neutrosophic Sets and Systems, 20*, 12–25. https://doi.org/10.5281/zenodo.1235365.

34. Şahin, R., & Küçük, A. (2015). Subsethood measure for single valued neutrosophic sets. *Journal of Intelligent & Fuzzy Systems, 29*(2), 525–530.

35. Şahin, R., & Liu, P. (2016). Maximizing deviation method for neutrosophic multiple attribute decision making with incomplete weight information. *Neural Computing and Applications, 27*(7), 2017–2029.

36. Smarandache, F., & Pramanik, S. (Eds.). (2016). *New trends in neutrosophic theory and applications.* Brussels: Pons Editions.

37. Smarandache, F., & Pramanik, S. (Eds.). (2018). *New trends in neutrosophic theory and applications, Vol.2.* Brussels: Pons Editions.

38. Ye, J. (2014). Improved correlation coefficients of single valued neutrosophic sets and interval neutrosophic sets for multiple attribute decision making. *Journal of Intelligent & Fuzzy Systems, 27*(5), 2453–2462.

39. Dalapati, S., Pramanik, S., Alam, S., Smarandache, F., & Roy, T. K. (2017). IN-cross entropy based MAGDM strategy under interval neutrosophic set environment. *Neutrosophic Sets and Systems, 18*, 43–57. https://doi.org/10.5281/zenodo.1175162.

40. Dey, P. P., Pramanik, S., & Giri, B. C. (2016a). Extended projection-based models for solving multiple attribute decision making problems with interval valued neutrosophic information. In F. Smarandache & S. Pramanik (Eds.), *New trends in neutrosophic theory and applications* (pp. 127–140). Brussels: Pons Editions.
41. Dey, P. P., Pramanik, S., & Giri, B. C. (2016b). TOPSIS for solving multi-attribute decision making problems under bi-polar neutrosophic environment. In F. Smarandache & S. Pramanik (Eds.), *New trends in neutrosophic theory and applications* (pp. 65–77). Brussels: Pons Editions.
42. Pramanik, S., Dey, P. P., & Giri, B. C. (2015). An extended grey relational analysis-based interval neutrosophic multi-attribute decision making for weaver selection. *Journal of New Theory, 9*, 82–93. ISSN: 2149-1402.
43. Mondal, K., & Pramanik, S. (2015b). Rough neutrosophic multi-attribute decision-making based on grey relational analysis. *Neutrosophic Sets and Systems, 7*, 8–17. https://doi.org/10.5281/zenodo.571603.
44. Mondal, K., & Pramanik, S. (2015c). Rough neutrosophic multi-attribute decision-making based on rough accuracy score function. *Neutrosophic Sets and Systems, 8*, 14–21. https://doi.org/10.5281/zenodo.571604.
45. Mondal, K., Pramanik, S., & Giri, B. C. (2019). Rough neutrosophic aggregation operators for multi-criteria decision-making. In C. Kahraman & I. Otay (Eds.), *Fuzzy multicriteria decision making using neutrosophic sets, studies in fuzziness and soft computing* (p. 369). https://doi.org/10.1007/978-3-030-00045-5_5.
46. Abdel-Basset, M., Gamal, A., Son, L. H., & Smarandache, F. (2020). A bipolar neutrosophic multi criteria decision making framework for professional selection. *Applied Sciences, 10*(4), 1202. https://doi.org/10.3390/app10041202.
47. Dey, P. P., Pramanik, S., Giri, B. C., & Smarandache, F. (2017). Bipolar neutrosophic projection based models for solving multi-attribute decision-making problems. *Neutrosophic Sets and Systems, 15*, 70–79. https://doi.org/10.5281/zenodo.570936.
48. Pramanik, S., Dalapati, S., Alam, S., & Roy, T. K. (2018a). VIKOR based MAGDM strategy under bipolar neutrosophic set environment. *Neutrosophic Sets and Systems, 19*, 57–69. https://doi.org/10.5281/zenodo.1235341.
49. Pramanik, S., Dalapati, S., Alam, S., & Roy, T. K. (2018b). NC-VIKOR based MAGDM strategy under neutrosophic cubic set environment. *Neutrosophic Sets and Systems, 20*, 95–108. https://doi.org/10.5281/zenodo.1235367.
50. Pramanik, S., Dey, P. P., Smarandache, F., & Ye, J. (2018). Cross entropy measures of bipolar and interval bipolar neutrosophic sets and their application for multi-attribute decision-making. *Axioms, 7*(21), 1–25. https://doi.org/10.3390/axioms7020021.
51. Biswas, P., Pramanik, S., & Giri, B. C. (2016b). Some distance measures of single valued neutrosophic hesitant fuzzy sets and their applications to multiple attribute decision making. In F. Smarandache & S. Pramanik (Eds.), *New trends in neutrosophic theory and applications* (pp. 55–63). Pons Editions: Brussels.
52. Biswas, P., Pramanik, S., & Giri, B. C. (2016c). GRA method of multiple attribute decision making with single valued neutrosophic hesitant fuzzy set information. In F. Smarandache & S. Pramanik (Eds.), *New trends in neutrosophic theory and applications* (pp. 55–63). Brussels: Pons Editions.
53. Biswas, P., Pramanik, S., & Giri, B. C. (2016d). Aggregation of triangular fuzzy neutrosophic set information and its application to multi-attribute decision making. *Neutrosophic Sets and Systems, 12*, 20–40. https://doi.org/10.5281/zenodo.571125.
54. Pramanik, S., & Dalapati, S. (2016). GRA based multi criteria decision making in generalized neutrosophic soft set environment. *Global Journal of Engineering Science and Research Management, 3*(5), 153–169. https://doi.org/10.5281/zenodo.53753.
55. Pramanik, S., Dey, P. P., & Giri, B. C. (2016a). Neutrosophic soft multi-attribute group decision making based on grey relational analysis method. *Journal of New Results in Science, 10*, 25–37. https://doi.org/10.5281/zenodo.34869.

56. Pramanik, S., Dey, P. P., & Giri, B. C. (2016b). Neutrosophic soft multi-attribute decision making based on grey relational projection method. *Neutrosophic Sets and Systems, 11*, 98–106. https://doi.org/10.5281/zenodo.571576.

57. Mondal, K., & Pramanik, S. (2015f). Neutrosophic refined similarity measure based on tangent function and its application to multi attribute decision making. *Journal of New Theory, 8*, 41–50. https://doi.org/10.5281/zenodo.23176.

58. Mondal, K., & Pramanik, S. (2015g). Neutrosophic refined similarity measure based on cotangent function and its application to multi-attribute decision making. *Global Journal of Advanced Research, 2*(2), 486–494.

59. Mondal, K., Pramanik, S., & Smarandache, F. (2018). NN-harmonic mean aggregation operators-based MCGDM strategy in a neutrosophic number environment. *Axioms, 7*(12). https://doi.org/10.3390/axioms7010012.

60. Pramanik, S., Roy, R., & Roy, T. K. (2016). Teacher selection strategy based on bidirectional projection measure in neutrosophic number environment. In F. Smarandache, M. A. Basset, & V. Chang (Eds), Neutrosophic operational research *Volume II*, 29-53, Brussels: Pons asbl.

61. Dalapati, S., & Pramanik, S. (2018). A revisit to NC-VIKOR based MAGDM strategy in neutrosophic cubic set environment. *Neutrosophic Sets and Systems, 21*, 131–141. https://doi.org/10.5281/zenodo.1408665.

62. Pramanik, S., Dalapati, S., Alam, S., & Roy, T. K. (2017). NC-TODIM-based MAGDM under a neutrosophic cubic set environment. *Information, 8*(4), 149. https://doi.org/10.3390/info8040149.

63. Pramanik, S., Dalapati, S., Alam, S., Roy, T. K., & Smarandache, F. (2017). Neutrosophic cubic MCGDM method based on similarity measure. *Neutrosophic Sets and Systems, 16*, 44–56. https://doi.org/10.5281/zenodo.831934.

64. Pramanik, S., Dey, P. P., & Smarandache, F. (2017). An extended TOPSIS for multi-attribute decision making problems with neutrosophic cubic information. *Neutrosophic Sets and Systems, 17*, 20–28. https://doi.org/10.5281/zenodo.1012217.

65. Abdel-Baset, M., Chang, V., Gamal, A., & Smarandache, F. (2019). An integrated neutrosophic ANP and VIKOR method for achieving sustainable supplier selection: A case study in importing field. *Computers in Industry, 106*, 94–110.

66. Abdel-Basset, M., Mohamed, M., Hussien, A. N., & Sangaiah, A. K. (2018). A novel group decision-making model based on triangular neutrosophic numbers. *Soft Computing, 22*(20), 6629–6643.

67. Ye, J. (2017). Some weighted aggregation operator of trapezoidal neutrosophic number and their multiple attribute decision making method. *Informatica, 28*(2), 387–402.

68. Liang, R. X., Wang, J. Q., & Zhang, H. Y. (2018). A multi-criteria decision-making method based on single valued trapezoidal neutrosophic preference relation with complete weight information. *Neural Computing Application, 30*(11), 3383–3398. https://doi.org/10.1007/s00521017-2925-8.

69. Pramanik, S., & Mallick, R. (2018). VIKOR based MAGDM strategy with trapezoidal neutrosophic number. *Neutrosophic Sets System, 22*, 118–130. https://doi.org/10.5281/zenodo.2160840.

70. Pramanik, S., & Mallick, R. (2019). TODIM strategy for multi-attribute group decision making in trapezoidal neutrosophic number environment. *Complex and Intelligent Systems, 5*(4), 379–389.

71. Keshavarz-Ghorabaee, M., Zavadskas, E. K., Olfat, L., & Turskis, Z. (2015). Multi-criteria inventory classification using a new method of evaluation based on distance from average solution (EDAS). *Information, 26*(3), 435–451.

72. Keshavarz-Ghorabaee, M., Zavadskas, E. K., Amiri, M., & Turskis, Z. (2016). Extended EDAS method for fuzzy multi-criteria decision-making: An application to supplier selection. *International Journal of Computer Communication and Control, 11*(3), 358–371.

73. Kahraman, C., Keshavarz-Ghorabaee, M., Zavadskas, E. K., Cevikonar, S., Yazdani, M., & Oztaysi, B. (2017). Intuitionistic fuzzy EDAS method: An application to solid waste disposal site selection. *Journal of Environment Engineering and Landscape Management, 25*(01), 1–12.

74. Li, Y., Wang, J. Q., & Wang, T. (2019). A linguistic neutrosophic multi-criteria group decision making approach with EDAS method. *The Arabian Journal for Science and Engineering, 44*(3), 2737–2749.

75. Dubois, D., & Prade, H. (1983). Ranking fuzzy numbers in the setting of possibility theory. *Information Sciences, 30*(3), 183–224.

76. Chen, Y., & Li, B. (2011). Dynamic multi-attribute decision making model based on triangular intuitionistic fuzzy numbers. *Scientia Iranica, 18*(2), 268–274.

77. Uluc, V., Deli, I., & Sahin, M. (2018). Trapezoidal fuzzy multi-number and its application to multi-criteria decision-making problems. *Neutral computing and application, 30*, 1469–1478.

78. Pramanik, S., & Mukhopadhyaya, D. (2011). Grey relational analysis based intuitionistic fuzzy multi criteria group decision-making approach for teacher selection in higher education. *International Journal of Computer Applications, 34*(10), 21–29. https://doi.org/10.5120/4138-5985.

79. San, J., & Cristobal, R. (2011). Multi-criteria decision-making in the selection of renewable energy project in Spain: The VIKOR method. *Renewable Energy, 36*, 1927–1934.

80. Mondal, K., Pramanik, S., & Smarandache, F. (2016b). Role of neutrosophic logic in data mining. In F. Smarandache & S. Pramanik (Eds.), *New trends in neutrosophic theory and applications. Pons Editions* (pp. 15–23). Brussels.

Neutrosophic Fuzzy Goal Programming Algorithm for Multi-level Multiobjective Linear Programming Problems

Firoz Ahmad and Florentin Smarandache

1 Introduction

Most often, the mathematical programming problems consist of only one decision maker who takes the decisions all alone. Apart from that, many decision-making problems involve hierarchical decision structures, each with independent, and most often contradictory in nature. Such decision-making scenarios are termed as decentralized planning problems. Thus, the hierarchical decision-making texture of the problem is formulated as multi-level programming problems (MLPPs). If there are only two decision makers, then it becomes bi-level programming problems, tri-level for three decision makers, and so on. The fundamental concepts behind the MLPPs optimization techniques are that the leader-level decision maker defines his/her goals/target and then seeks the optimal solution from each subordinate level of the organization that has calculated individually. The follower-level decisions are then submitted and satisfied by the leader-level in view of overall benefit of the organizations. There may be more than one linear objective function that are to be optimized by different levels in MLPPs, then such kind of decentralized decision-making problems are termed as multi-level multiobjective linear programming problems (ML-MOLPPs).

There are several research works available in the literature that contribute to the domain of multi-level multiobjective linear programming problems. Based on fuzzy set theory, [1, 6, 19, 20, 22] presented fuzzy programming and fuzzy goal programming approaches to bi-level decision-making problems. Furthermore,

F. Ahmad (✉)
Department of Statistics and Operations Research, Aligarh Muslim University, Aligarh, India

F. Smarandache
Department of Mathematics and Science, University of New Mexico, Gallup, NM, USA

F. Smarandache, M. Abdel-Basset (eds.), *Neutrosophic Operational Research*,
https://doi.org/10.1007/978-3-030-57197-9_27

[8, 10, 12, 14, 15] suggested the fuzzy-based solution procedure for ML-MOLPPs. Later on, intuitionistic fuzzy set theory [7] is also introduced to solve the ML-MOLPPs under intuitionistic fuzzy environment. Recently, [9, 21] discussed the intuitionistic fuzzy techniques to solve the ML-MOLPPs by considering the membership as well as non-membership functions of all objectives at each level. Furthermore, the extension and generalization of fuzzy and intuitionistic fuzzy sets are presented and named as neutrosophic set. First, [18] proposed the neutrosophic set, and later on it was extensively used in the field of mathematical programming problems and their optimization techniques. Only few research work is available that captures the neutrosophic decision set theory in ML-MOLPPs. Only two research articles are cited below that have contributed to neutrosophic ML-MOLPPs domain. Maiti et al. [11] investigated neutrosophic goal programming strategy for ML-MOLPPs with neutrosophic parameters. Pramanik and Dey [13] also suggested a goal programming technique for neutrosophic ML-MOLPPs where the parameters have been taken as triangular or trapezoidal neutrosophic numbers. Thus, this chapter provides more emphasis toward the neutrosophic ML-MOLPPs research area and laid down a concrete base for neutrosophic ML-MOLPPs optimization domain.

In this chapter, the neutrosophic fuzzy goal programming (NFGP) algorithm is introduced to solve the multi-level multiobjective linear programming problems. Two different NFGP procedures based on neutrosophic fuzzy decision set are presented for ML-MOLPPs. To formulate any of these two proposed NFGP models of the ML-MOLPPs, the neutrosophic fuzzy goals of the objectives are determined by finding individual optimal solutions. Marginal evaluations of each objective functions are then depicted by the associated membership functions under neutrosophic environment. These membership functions are converted into neutrosophic fuzzy flexible membership goals by means of incorporating over- and under-deviational variables and assigning highest truth membership value (unity), indeterminacy value (half), and a falsity value (zero) as aspiration levels to each of them. To determine the membership functions of the decision variable vectors monitored by any level decision maker, the optimal solution of the corresponding MOLPP is separately solved. A marginal relaxation of the decisions is prescribed to avoid decision deadlock.

The first proposed NFGP solution algorithm provides an extension of the work presented by [1, 8, 16] under neutrosophic environment, which deals with bi-level linear single-objective programming problems. It also extends the work of [14] by introducing the NFGP algorithm to multi-level programming problems with a multiple linear objective at each level. The final fuzzy model groups the membership functions for the defined neutrosophic fuzzy goals of the objective functions and the decision variable vectors at all levels, which are determined separately for each level except the follower level of the multi-level problem.

The second proposed NFFGP algorithm may be seen as a method for solving multi-level multiobjective programming problems. First, it develops the NFGP model of the leader-level problem to obtain a satisfactory solution to the leader-level decision maker's problem. A marginal relaxation of the leader-level decision maker's decisions is taken into account to avoid a decision deadlock. These decisions of the leader-level decision makers are depicted by membership functions of neutrosophic fuzzy set theory and transferred to the second-level DM (SLDM) as an additional constraint. Then, the SLDM modeled its NFGP model that considers the neutrosophic fuzzy membership goals of the objectives and decision variable vectors of the leader-level decision makers. Afterward, the achieved solution is passed to the third-level DM (TLDM) who seeks the solution in a similar fashion. The same process is carried out until the follower level reaches. Thus, this procedure may be assumed as an extension of the fuzzy mathematical programming algorithm of [16, 17] under the neutrosophic environment.

The remaining part of the chapter is summarized as follows: in Sect. 2, the preliminaries regarding neutrosophic set have been discussed, while Sect. 3 discusses the formulations of multi-level multiobjective programming problems. The proposed neutrosophic fuzzy goal algorithm is developed in Sect. 4, whereas in Sect. 5, a numerical example is presented to show the applicability and validity of the proposed approaches. Finally, conclusions and future scope are discussed based on the present work in Sect. 6.

2 Preliminaries

Some basic preliminaries regarding neutrosophic set are presented in the following section.

Definition 1 ([4]) Let Y be a universe discourse such that $y \in Y$, then a neutrosophic set A in Y is defined by three membership functions namely, truth $\mu_A(y)$, indeterminacy $\lambda_A(y)$, and a falsity $\nu_A(y)$ and is denoted by the following form:

$$A = \{< y, \ \mu_A(y), \ \lambda_A(y), \ \nu_A(y) > \ | \ y \in Y\},$$

where $\mu_A(y)$, $\lambda_A(y)$ and $\nu_A(y)$ are real standard or non-standard subsets belong to $]0^-, 1^+[$, also given as, $\mu_A(y) : Y \rightarrow]0^-, 1^+[$, $\lambda_A(y) : Y \rightarrow]0^-, 1^+[$, and $\nu_A(y) : Y \rightarrow]0^-, 1^+[$. There is no restriction on the sum of $\mu_A(y)$, $\lambda_A(y)$ and $\nu_A(y)$, so we have

$$0^- \leq \sup \mu_A(y) + \lambda_A(y) + \sup \nu_A(y) \leq 3^+.$$

Definition 2 ([4]) A single-valued neutrosophic set A over universe of discourse Y is defined as

$$A = \{< y, \ \mu_A(y), \ \lambda_A(y), \ \nu_A(y) > \ | \ y \in Y\},$$

where $\mu_A(y)$, $\lambda_A(y)$, and $\nu_A(y) \in [0, 1]$ and $0 \leq \mu_A(y) + \lambda_A(y) + \nu_A(y) \leq 3$ for each $y \in Y$.

Definition 3 ([4]) The complement of a single valued neutrosophic set A is represented as $c(A)$ and defined by $\mu_{c(A)}(y) = \nu_A(y)$, $\lambda_{c(A)}(y) = 1 - \nu_A(y)$ and $\nu_{c(A)}(y) = \mu_A(y)$, respectively.

Definition 4 ([4]) Let A and B be the two single-valued neutrosophic sets, then the union of A and B is also a single-valued neutrosophic set C, that is, $C = (A \cup B)$, whose truth $\mu_C(y)$, indeterminacy $\lambda_C(y)$, and falsity $\nu_C(y)$ membership functions are given by

$$\mu_C(y) = \max\,(\mu_A(y),\ \mu_B(y))$$
$$\lambda_C(y) = \max\,(\lambda_A(y),\ \lambda_B(y))$$
$$\nu_C(y) = \min\,(\nu_A(y),\ \nu_B(y)) \text{ for each } y \in Y.$$

Definition 5 ([4]) Let A and B be the two single-valued neutrosophic sets, then the intersection of A and B is also a single-valued neutrosophic set C, that is, $C = (A \cap B)$, whose truth $\mu_C(y)$, indeterminacy $\lambda_C(y)$, and falsity $\nu_C(y)$ membership functions are given by

$$\mu_C(y) = \min\,(\mu_A(y),\ \mu_B(y))$$
$$\lambda_C(y) = \min\,(\lambda_A(y),\ \lambda_B(y))$$
$$\nu_C(y) = \max\,(\nu_A(y),\ \nu_B(y)) \text{ for each } y \in Y.$$

Definition 6 A solution set $Y^* \in S$ is said to be an efficient solution to the MLMOPPs if and only if there does not exist any other $Y \in S$ such that $O_{ij} \geq O_{ij}^*$ for all $i = 1, 2, \ldots, t;\ \ j = 1, 2, \ldots, m_t$, respectively.

Definition 7 For any ML-MOPPs, an efficient solution selected by the decision makers is the best compromise optimal solution which is chosen on the basis of decision makers' explicit and implicit criteria.

3   Description of ML-MOLPPs

Assume that a t- level multiobjective programming problem with minimization-type objective functions at different level. Consider that DM_i represents the i-th level decision maker and controls over the decision variable $\mathbf{y}_i = (y_{i1},\ y_{i2},\ \ldots,\ y_{in_i}) \in R^{n_i}$ for all $i = 1,\ 2,\ \ldots,\ t$, where $\mathbf{y} = (\mathbf{y}_1,\ \mathbf{y}_2,\ \ldots,\ \mathbf{y}_t) \in R^n$ such that $n = n_1 + 2 + \ldots + n_t$. Furthermore, we assume that

$$O_i(\mathbf{y}) = O_i(\mathbf{y}_1,\ \mathbf{y}_2,\ \ldots,\ \mathbf{y}_t) : R^{n_1} \times R^{n_2} \times \cdots \times R^{n_t} \rightarrow R^{m_i},\ \forall\ i = 1,\ 2,\ \ldots,\ t \tag{1}$$

represents the vector-set of a well-defined linear objective function to the i-th decision makers, $i = 1,\ 2,\ \ldots,\ t$. The equivalent mathematical expressions for the ML-MOLPPS with minimization-type objectives can be stated as follows:

[1st level]

$$\underset{\mathbf{y}_1}{\text{Min}}\ O_1(\mathbf{y}) = \underset{\mathbf{y}_1}{\text{Min}}\ \left(o_{11}(\mathbf{y}),\ o_{12}(\mathbf{y}),\ \ldots,\ o_{1m_1}(\mathbf{y})\right)$$

where $\mathbf{y}_2$, $\mathbf{y}_3$, ..., $\mathbf{y}_t$ solves

[2nd level]

$$\underset{\mathbf{y}_2}{\text{Min}}\ O_2(\mathbf{y}) = \underset{\mathbf{y}_2}{\text{Min}}\ \left(o_{21}(\mathbf{y}),\ o_{22}(\mathbf{y}),\ \ldots,\ o_{2m_2}(\mathbf{y})\right) \tag{2}$$

...

where $\mathbf{y}_t$ solves

[t-th level]

$$\underset{\mathbf{y}_t}{\text{Min}}\ O_t(\mathbf{y}) = \underset{\mathbf{y}_t}{\text{Min}}\ \left(o_{t1}(\mathbf{y}),\ o_{t2}(\mathbf{y}),\ \ldots,\ o_{tm_t}(\mathbf{y})\right)$$

subject to

$$\mathbf{y} \in S = \left\{ \mathbf{y} \in R^n | G_1 y_1 + G_1 y_1 + \cdots + G_t y_t (\le \ or\ =\ or \ge)\ \mathbf{q},\ \mathbf{y} \ge 0, \right.$$
$$\left. \mathbf{q} \in R^m \right\} \ne \phi, \tag{3}$$

where

$$o_{ij}(\mathbf{y}) = c_1^{ij} \mathbf{y}_1 + c_2^{ij} \mathbf{y}_2 + \cdots + c_t^{ij} \mathbf{y}_t, \quad i = 1,\ 2,\ \ldots,\ t,\quad j = 1,\ 2,\ \ldots,\ m_i$$
$$= c_{11}^{ij} \mathbf{y}_{11} + c_{12}^{ij} \mathbf{y}_{12} + \cdots + c_{1n_1}^{ij} \mathbf{y}_{1n_1} + c_{21}^{ij} \mathbf{y}_{21} + c_{22}^{ij} \mathbf{y}_{22} + \cdots$$
$$+ c_{2n_2}^{ij} \mathbf{y}_{2n_2} + \cdots c_{t1}^{ij} \mathbf{y}_{t1} + c_{t2}^{ij} \mathbf{y}_{t2} + \cdots + c_{tn_t}^{ij} \mathbf{y}_{tn_t}$$

$$\tag{4}$$

such that S is the multi-level convex constraints in feasible decision set under multi-level multiobjective programming problems. The notation m_i, $i = 1, 2, \ldots, t$ denotes the number of objective function under i-th decision maker, m is the number of constraints, $c_k^{ij} = \left(c_{k1}^{ij},\ c_{k2}^{ij},\ \cdots,\ c_{kn_k}^{ij}\right)$, $k = 1, 2, \ldots, t$, $c_{kn_k}^{ij}$ are constants, and the coefficient matrices of size $m \times n_i$ are depicted as G_i, $\forall\ i = 1, 2, \ldots, t$.

4 Proposed Neutrosophic Fuzzy Goal Programming Techniques

In the past few decades, it has been observed that the situation may arise in real-life decision-making problems where the indeterminacy or neutral thoughts about element into the feasible set exist. Indeterminacy/neutral thoughts are the region of the negligence of a proposition's value and lie between truth and a falsity degree. Therefore, the further generalization of fuzzy set (FS) [20] and intuitionistic fuzzy set (IFS) [7] is presented by introducing a new member into the feasible decision set. First, [18] investigated the neutrosophic set (NS) which comprises three membership functions, namely truth (degree of belongingness), indeterminacy (degree of belongingness up to some extent), and a falsity (degree of non-belongingness) functions of the element into the neutrosophic decision set (see [2, 3, 5]).

In ML-MOLPPs, if an imprecise aspiration level under neutrosophic environment is assigned to each of the objectives at each level of the ML-MOLPPs, then such neutrosophic objectives are termed as neutrosophic goals and dealt with neutrosophic decision-making techniques. Hence, the marginal evaluation of each neutrosophic goals is characterized through three different membership functions, namely truth, indeterminacy, and a falsity membership functions by defining the tolerance limits for attainment of their respective aspiration levels.

4.1 Characterization of Different Membership Functions Under Neutrosophic Environment

In multi-level decision-making problems, each DM intends to minimize its own objectives in each level over the same feasible region depicted by the system of constraints; hence, the individual optimal solutions are obtained by them and can be regarded as the aspiration levels of their associated neutrosophic goals.

Assume that $\mathbf{y}^{ij} = (y_1^{ij}, y_2^{ij}, \ldots, y_t^{ij})$ and $o_{ij}^{\min}$, $i = 1, 2, \ldots, t$, $j = 1, 2, \ldots, m_i$ be the best individual optimal solutions of each DMs at each level, respectively. Furthermore, consider that $l_{ij} \geq o_{ij}^{\min}$ denotes the aspiration level assigned to the ij-th objective $o_{ij}(\mathbf{y})$ (where ij means that when $i = t$ for t-th level decision makers then $j = 1, 2, \ldots, m_i$). Moreover, also consider that $\mathbf{y}^{i*} = (y_1^{i*}, y_2^{i*}, \ldots, y_t^{i*})$, $i = 1, 2, \ldots, t - 1$, be the optimal solutions for the t-th level decision makers of ML-MOLPPs. Consequently, the neutrosophic goals of each objective function at each level and the vector-set of neutrosophic goals for the decision variables monitored by t-th level decision makers can be stated as follows:

$$o_{ij}(\mathbf{y}) \lesssim l_{ij}, \quad i = 1, 2, \ldots, t, \quad j = 1, 2, \ldots, m_i \text{ and}$$

$$\mathbf{y}_i \cong \mathbf{y}_i^{i*}, \quad i = 1, 2, \ldots, t - 1,$$

where $\lesssim$ and $\cong$ represent the degree of neutrosophy of the aspiration levels.

One can note that the solutions $\mathbf{y}^{ij} = (y_1^{ij}, y_2^{ij}, \ldots, y_t^{ij})$; $i = 1, 2, \ldots, t$, $j = 1, 2, \ldots, m_i$ are probably different due to the conflicting nature of the objective functions at each level for all the decision makers. Therefore, it can be obvious to consider that the values of $o_{gm}(y_1^{gm}, y_2^{gm}, \ldots, y_t^{gm}) \geq$

$o_{ij}^{\min}$, $g = 1, 2, \ldots, t$, $m = 1, 2, \ldots, m_i$, and $\forall \ ij \neq gm$ with all values greater than $o_{gm}^{u} = \max [o_{ij} (\mathbf{y}_1^{gm}, \mathbf{y}_2^{gm}, \ldots, \mathbf{y}_t^{gm}), i = 1, 2, \ldots, t, j = 1, 2, \ldots, m_i$ and $ij \neq gm]$ are absolutely unacceptable to the objective function $o_{gm}(\mathbf{y}) = o_{gm}(\mathbf{y}_1, \mathbf{y}_2, \ldots, \mathbf{y}_t)$. As a result, $o_{gm}(\mathbf{y})$ can be taken as the upper tolerance limit $u_{gm}(\mathbf{y})$ of the neutrosophic goal to the objective functions. The upper and lower bounds for ij-th objective function under the neutrosophic environment can be obtained as follows:

$$U_{ij}^{\mu} = u_{ij}, \quad L_{ij}^{\mu} = l_{ij} \qquad \qquad \text{for truth membership}$$

$$U_{ij}^{\lambda} = L_{ij}^{\mu} + a_{ij}, \quad L_{ij}^{\lambda} = L_{ij}^{\mu} \qquad \text{for indeterminacy membership}$$

$$U_{ij}^{\nu} = U_{ij}^{\mu}, \quad L_{ij}^{\nu} = L_{ij}^{\mu} + b_{ij} \qquad \text{for falsity membership,}$$

where a_{ij} and $b_{ij} \in (0, 1)$ are predetermined real numbers.

Thus, the different membership functions, namely truth $\mu_{o_{ij}}(o_{ij}(\mathbf{y}))$, indeterminacy $\lambda_{o_{ij}}(o_{ij}(\mathbf{y}))$, and a falsity $\nu_{o_{ij}}(o_{ij}(\mathbf{y}))$ membership functions for the ij-th neutrosophic goals can be stated as follows:

$$\mu_{o_{ij}}(o_{ij}(\mathbf{y})) = \begin{cases} 1 & \text{if } o_{ij}(\mathbf{y}) \leq L_{ij}^{\mu} \\ 1 - \frac{o_{ij}(\mathbf{y}) - L_{ij}^{\mu}}{U_{ij}^{\mu} - L_{ij}^{\mu}} & \text{if } L_{ij}^{\mu} \leq o_{ij}(\mathbf{y}) \leq U_{ij}^{\mu} \\ 0 & \text{if } o_{ij}(\mathbf{y}) \geq U_{ij}^{\mu} \end{cases} \tag{5}$$

$$\lambda_{o_{ij}}(o_{ij}(\mathbf{y})) = \begin{cases} 1 & \text{if } o_{ij}(\mathbf{y}) \leq L_{ij}^{\lambda} \\ 1 - \frac{o_{ij}(\mathbf{y}) - L_{ij}^{\lambda}}{U_{ij}^{\lambda} - L_{ij}^{\lambda}} & \text{if } L_{k}^{l} \leq o_{ij}(\mathbf{y}) \leq U_{ij}^{\lambda} \\ 0 & \text{if } o_{ij}(\mathbf{y}) \geq U_{ij}^{\lambda} \end{cases} \tag{6}$$

$$\nu_{o_{ij}}(o_{ij}(\mathbf{y})) = \begin{cases} 1 & \text{if } o_{ij}(\mathbf{y}) \geq U_{ij}^{\nu} \\ 1 - \frac{U_{ij}^{\nu} \ o_{ij}(\mathbf{y})}{U_{ij}^{\nu} - L_{ij}^{\nu}} & \text{if } L_{ij}^{\nu} \leq o_{ij}(\mathbf{y}) \leq U_{ij}^{\nu} \\ 0 & \text{if } o_{ij}(\mathbf{y}) \leq L_{ij}^{\nu}. \end{cases} \tag{7}$$

To construct the different membership functions for the decision variables monitored by i-th decision makers, first, the optimal solution for the t-th level MOLPPs, $\mathbf{y}^{i*} = (\mathbf{y}_1^{i*}, \mathbf{y}_2^{i*}, \ldots, \mathbf{y}_t^{i*})$, $i = 1, 2, \ldots, t - 1$, should be carried out by using any appropriate method for MOLPPs optimization techniques.

Suppose that $T_k^{i\alpha}$ and $T_k^{i\beta}$, $i = 1, 2, \ldots, t - 1$, $k = 1, 2, \ldots, n_i$ be the maximum negative and positive tolerance limits on the decision variables imposed by the i-th level decision makers. Usually, the tolerances T_{ik}^{-} and T_{ik}^{+} may not be equal. The upper and lower bounds for ik-th decision variable vectors under the neutrosophic environment can be stated as follows:

$$\mu_{y_{ik}}^U = y_{ik}^* + T_k^{i\beta}, \quad \mu_{y_{ik}}^L = y_{ik}^* - T_k^{i\alpha} \qquad \text{for truth membership}$$

$$\lambda_{y_{ik}}^U = \mu_{y_{ik}}^L + a_{ik}, \quad \lambda_{y_{ik}}^L = \mu_{y_{ik}}^L \qquad \text{for indeterminacy membership}$$

$$v_{y_{ik}}^U = \mu_{y_{ik}}^U, \quad v_{y_{ik}}^L = \mu_{y_{ik}}^L + b_{ik} \qquad \text{for falsity membership,}$$

where a_{ik} and $b_{ik} \in (0, 1)$ are predetermined real numbers.

For each of the n_i components of the decision variable vector $\mathbf{y}_{ik}^* = (\mathbf{y}_{i1}^*, \mathbf{y}_{i2}^*, \ldots, \mathbf{y}_{in_i}^*)$ controlled by the leader $(t - 1)$-th level decision makers, the different linear membership functions under neutrosophic environment such as truth $\mu_{y_{ik}}(y_{ik})$, indeterminacy $\lambda_{y_{ik}}(y_{ik})$, and a falsity $v_{y_{ik}}(y_{ik})$ can be furnished as follows:

$$\mu_{y_{ik}}(y_{ik}) = \begin{cases} \frac{y_{ik}-\mu_{y_{ik}}^L}{y_{ik}^*-\mu_{y_{ik}}^L} & \text{if } \mu_{y_{ik}}^L \leq y_{ik} \leq y_{ik}^* \\ \frac{\mu_{y_{ik}}^U - y_{ik}}{\mu_{y_{ik}}^U - y_{ik}^*} & \text{if } y_{ik}^* \leq y_{ik} \leq \mu_{y_{ik}}^U \\ 0 & \text{otherwise.} \end{cases} \tag{8}$$

$$\lambda_{y_{ik}}(y_{ik}) = \begin{cases} \frac{y_{ik}-\lambda_{y_{ik}}^L}{y_{ik}^*-\lambda_{y_{ik}}^L} & \text{if } \lambda_{y_{ik}}^L \leq y_{ik} \leq y_{ik}^* \\ \frac{\lambda_{y_{ik}}^U - y_{ik}}{\lambda_{y_{ik}}^U - y_{ik}^*} & \text{if } y_{ik}^* \leq y_{ik} \leq \lambda_{y_{ik}}^U \\ 0 & \text{otherwise.} \end{cases} \tag{9}$$

$$v_{y_{ik}}(y_{ik}) = \begin{cases} \frac{y_{ik}-v_{y_{ik}}^L}{y_{ik}^*-v_{y_{ik}}^L} & \text{if } v_{y_{ik}}^L \leq y_{ik} \leq y_{ik}^* \\ \frac{v_{y_{ik}}^U - y_{ik}}{v_{y_{ik}}^U - y_{ik}^*} & \text{if } y_{ik}^* \leq y_{ik} \leq v_{y_{ik}}^U \\ 1 & \text{otherwise.} \end{cases} \tag{10}$$

Also, it should be noted that the range of y_{ik} may be shifted according to the decision makers' choices.

In a neutrosophic decision environment, the neutrosophic goals comprising the decision makers' objective functions at different level and the neutrosophic goals of the decision variable vectors are monitored by leader $(t - 1)$-th level decision makers. The attainment degrees to their aspiration levels to the extent possible are virtually achieved by the possible achievement of their respective memberships, namely truth, indeterminacy, and a falsity membership functions to their utmost degrees. Obviously, this aspect of neutrosophic fuzzy programming approach enables a neutrosophic fuzzy goal programming technique as a justified approach for solving the leader t-th level MOLPPs and consequently ML-MOLPPs.

4.2 Neutrosophic Fuzzy Goal Programming Approach

In neutrosophic programming approaches, the neutrosophic membership degrees can be transformed into neutrosophic membership goals according to their respective maximum degrees of attainment. The highest degree of truth membership function that can be achieved is unity (1), the indeterminacy membership function is neutral and independent with the highest attainment degree half (0.5), and the falsity membership function can be achieved with the highest attainment degree zero (0). Now, the transformed membership goals under the neutrosophic environment can be expressed as follows:

$$
\left.\begin{aligned}
\mu_{o_{ij}}(o_{ij}(\mathbf{y})) + d_{ij\mu}^- - d_{ij\mu}^+ &= 1, \\
\lambda_{o_{ij}}(o_{ij}(\mathbf{y})) + d_{ij\lambda}^- - d_{ij\lambda}^+ &= 0.5, \\
\nu_{o_{ij}}(o_{ij}(\mathbf{y})) + d_{ij\nu}^- - d_{ij\nu}^+ &= 0,
\end{aligned}\right\} \forall\ i = 1, 2, \ldots, t, \quad j = 1, 2, \ldots, m_i
$$

(11)

$$
\left.\begin{aligned}
\mu_{y_{ik}}(y_{ik}) + d_{ik\mu}^- - d_{ik\mu}^+ &= 1, \\
\lambda_{y_{ik}}(y_{ik}) + d_{ik\lambda}^- - d_{ik\lambda}^+ &= 0.5, \\
\nu_{y_{ik}}(y_{ik}) + d_{ik\nu}^- - d_{ik\nu}^+ &= 0,
\end{aligned}\right\} \forall\ i = 1, 2, \ldots, t-1, \quad k = 1, 2, \ldots, n_i
$$

(12)

or equivalently represented as follows:

$$
\left.\begin{aligned}
1 - \frac{o_{ij}(\mathbf{y}) - L_{ij}^\mu}{U_{ij}^\mu - L_{ij}^\mu} + d_{ij\mu}^- - d_{ij\mu}^+ &= 1, \\
1 - \frac{o_{ij}(\mathbf{y}) - L_{ij}^\lambda}{U_{ij}^\lambda - L_{ij}^\lambda} + d_{ij\lambda}^- - d_{ij\lambda}^+ &= 0.5, \\
1 - \frac{U_{ij}^\nu - o_{ij}(\mathbf{y})}{U_{ij}^\nu - L_{ij}^\nu} + d_{ij\nu}^- - d_{ij\nu}^+ &= 0,
\end{aligned}\right\} \forall\ i = 1, 2, \ldots, t, \quad j = 1, 2, \ldots, m_i
$$

(13)

$$
\left.\begin{aligned}
\frac{y_{ik} - \mu_{y_{ik}}^L}{y_{ik}^* - \mu_{y_{ik}}^L} + d_{ik\mu}^{\alpha-} - d_{ik\mu}^{\alpha+} &= 1, \\
\frac{\mu_{y_{ik}}^U - y_{ik}}{\mu_{y_{ik}}^U - y_{ik}^*} + d_{ik\mu}^{\beta-} - d_{ik\mu}^{\beta+} &= 1, \\
\frac{y_{ik} - \lambda_{y_{ik}}^L}{y_{ik}^* - \lambda_{y_{ik}}^L} + d_{ik\lambda}^{\alpha-} - d_{ik\lambda}^{\alpha+} &= 0.5, \\
\frac{\lambda_{y_{ik}}^U - y_{ik}}{\lambda_{y_{ik}}^U - y_{ik}^*} + d_{ik\lambda}^{\beta-} - d_{ik\lambda}^{\beta+} &= 0.5, \\
\frac{y_{ik} - \nu_{y_{ik}}^L}{y_{ik}^* - \nu_{y_{ik}}^L} + d_{ik\nu}^{\alpha-} - d_{ik\nu}^{\alpha+} &= 0, \\
\frac{\nu_{y_{ik}}^U - y_{ik}}{\nu_{y_{ik}}^U - y_{ik}^*} + d_{ik\nu}^{\beta-} - d_{ik\nu}^{\beta+} &= 0,
\end{aligned}\right\} \forall\ i = 1, 2, \ldots, t-1, \quad k = 1, 2, \ldots, n_i,
$$

(14)

where $d_{ik}^- = (d_{ik}^{\alpha-}, d_{ik}^{\beta-}), d_{ik}^+ = (d_{ik}^{\alpha+}, d_{ik}^{\beta+}); d_{ij}^-, d_{ij}^+, d_{ik}^{\alpha-}, d_{ik}^{\beta-}, d_{ik}^{\alpha+}, d_{ik}^{\beta+} \geq 0$; and $d_{ik}^{\beta-} \times d_{ik}^{\beta+} = 0, \forall\ i = 1, 2, \ldots, t-1, \ k = 1, 2, \ldots, n_i$ are the over and under deviations for truth, indeterminacy, and a falsity membership goals from their respective aspiration levels under neutrosophic environment.

In goal programming strategy, the over- and/or under-deviational variable vectors are considered in the objective function to minimize them and solely depend on the nature of objective function that is being optimized. In the proposed neutrosophic goal programming technique, the over-deviational variables for neutrosophic goals of each objective function, d_{ij}^+ $\forall$ $i = 1, 2, \ldots, t$, $j = 1, 2, \ldots, m_i$ and the over and under-deviational variables for the neutrosophic fuzzy goals of the decision variable vectors, $d_{ik}^{\alpha+}$, $d_{ik}^{\alpha-}$, $d_{ik}^{\beta+}$ and $d_{ik}^{\beta-}$ $\forall$ $i = 1, 2, \ldots, (t-1)$, $k = 1, 2, \ldots, n_i$ are needed to be minimized to attain the neutrosophic fuzzy goals.

4.3 Neutrosophic Fuzzy Goal Programming Approach for ML-MOLPPs

The proposed neutrosophic fuzzy goal programming (NFGP) algorithm for solving the multi-level multiobjective linear programming problems (ML-MOLPPs) is presented, and the two new algorithms are suggested under neutrosophic environment.

4.3.1 The First NFGP Algorithm for ML-MOLPPs

First of all, the first NFGP algorithm proposed in this chapter groups over the different membership functions for the prescribed neutrosophic fuzzy goals of the objective functions at each levels; it also groups the different membership functions of the neutrosophic fuzzy goals of the decision variable vector of the t-th leader-level problems that are optimized individually under neutrosophic environment. Thus, by assuming the goal attainment problems at the same preference level, the equivalent proposed neutrosophic fuzzy multi-level multiobjective linear goal programming model of the ML-MOLPPs under neutrosophic environment can be expressed as follows:

$$
\text{Min } F = \sum_{j=1}^{m_1} w_{1j\mu}^+ d_{1j\mu}^+ + \sum_{j=1}^{m_2} w_{2j\mu}^+ d_{2j\mu}^+ + \cdots + \sum_{j=1}^{m_t} w_{tj\mu}^+ d_{tj\mu}^+
$$

$$
+ \sum_{j=1}^{m_1} w_{1j\lambda}^+ d_{1j\lambda}^+ + \sum_{j=1}^{m_2} w_{2j\lambda}^+ d_{2j\lambda}^+ + \cdots + \sum_{j=1}^{m_t} w_{tj\lambda}^+ d_{tj\lambda}^+
$$

$$
- \sum_{j=1}^{m_1} w_{1j\nu}^+ d_{1j\nu}^- - \sum_{j=1}^{m_2} w_{2j\nu}^+ d_{2j\nu}^- - \cdots - \sum_{j=1}^{m_t} w_{tj\nu}^+ d_{tj\nu}^-
$$

$$
+ \sum_{k=1}^{n_1} \left(w_{1k.}^\alpha (d_{1k.}^{\alpha-} + d_{1k.}^{\alpha+}) + w_{1k.}^\beta (d_{1k.}^{\beta-} + d_{1k.}^{\beta+}) \right)
$$

$$+ \sum_{k=1}^{n_2} \left(w_{2k.}^{\alpha}.(d_{2k.}^{\alpha-} + d_{2k.}^{\alpha+}) + w_{2k.}^{\beta}.(d_{2k.}^{\beta-} + d_{2k.}^{\beta+}) \right) \quad \cdots \quad \cdots \quad \cdots$$

$$+ \sum_{k=1}^{n_{t-1}} \left(w_{t-1k.}^{\alpha}.(d_{t-1k.}^{\alpha-} + d_{t-1k.}^{\alpha+}) + w_{t-1k.}^{\beta}.(d_{t-1k.}^{\beta-} + d_{t-1k.}^{\beta+}) \right)$$

subject to

$$\mu_{o_{1j}}(o_{1j}(\mathbf{y})) + d_{1j\mu}^- - d_{1j\mu}^+ = 1, \quad j = 1, 2, \ldots, n_1$$

$$\mu_{o_{2j}}(o_{2j}(\mathbf{y})) + d_{2j\mu}^- - d_{2j\mu}^+ = 1, \quad j = 1, 2, \ldots, n_2$$

$$\cdots$$

$$\mu_{o_{tj}}(o_{tj}(\mathbf{y})) + d_{tj\mu}^- - d_{tj\mu}^+ = 1, \quad j = 1, 2, \ldots, n_t$$

$$\lambda_{o_{1j}}(o_{1j}(\mathbf{y})) + d_{1j\lambda}^- - d_{1j\lambda}^+ = 0.5, \quad j = 1, 2, \ldots, n_1$$

$$\lambda_{o_{2j}}(o_{2j}(\mathbf{y})) + d_{2j\lambda}^- - d_{2j\lambda}^+ = 0.5, \quad j = 1, 2, \ldots, n_2$$

$$\cdots$$

$$\lambda_{o_{tj}}(o_{tj}(\mathbf{y})) + d_{tj\lambda}^- - d_{tj\lambda}^+ = 0.5, \quad j = 1, 2, \ldots, n_t$$

$$v_{o_{1j}}(o_{1j}(\mathbf{y})) + d_{1jv}^- - d_{1jv}^+ = 0, \quad j = 1, 2, \ldots, n_1$$

$$v_{o_{2j}}(o_{2j}(\mathbf{y})) + d_{2jv}^- - d_{2jv}^+ = 0, \quad j = 1, 2, \ldots, n_2$$

$$\cdots$$

$$v_{o_{tj}}(o_{tj}(\mathbf{y})) + d_{tjv}^- - d_{tjv}^+ = 0, \quad j = 1, 2, \ldots, n_t$$

$$\mu_{y_{1k}}(y_{1k}) + d_{1k\mu}^- - d_{1k\mu}^+ = 1, \quad k = 1, 2, \ldots, n_1$$

$$\mu_{y_{2k}}(y_{2k}) + d_{2k\mu}^- - d_{2k\mu}^+ = 1, \quad k = 1, 2, \ldots, n_2$$

$$\cdots$$

$$\mu_{y_{t-1k}}(y_{t-1k}) + d_{t-1k\mu}^- - d_{t-1k\mu}^+ = 1, \quad k = 1, 2, \ldots, n_{t-1}$$

$$\lambda_{y_{1k}}(y_{1k}) + d_{1k\lambda}^- - d_{1k\lambda}^+ = 0.5, \quad k = 1, 2, \ldots, n_1$$

$$\lambda_{y_{2k}}(y_{2k}) + d_{2k\lambda}^- - d_{2k\lambda}^+ = 0.5, \quad k = 1, 2, \ldots, n_2$$

$$\cdots$$

$$\lambda_{y_{t-1k}}(y_{t-1k}) + d_{t-1k\lambda}^- - d_{t-1k\lambda}^+ = 0.5, \quad k = 1, 2, \ldots, n_{t-1}$$

$$v_{y_{1k}}(y_{1k}) + d_{1kv}^- - d_{1kv}^+ = 0, \quad k = 1, 2, \ldots, n_1$$

$$v_{y_{2k}}(y_{2k}) + d_{2kv}^- - d_{2kv}^+ = 0, \quad k = 1, 2, \ldots, n_2$$

$$\cdots$$

$$v_{y_{t-1k}}(y_{t-1k}) + d^-_{t-1kv} - d^+_{t-1kv} = 0, \quad k = 1, 2, \ldots, n_{t-1}$$

$$G_1 y_1 + G_1 y_1 + \cdots + G_t y_t (\leq \text{ or } = \text{ or } \geq) \mathbf{q}, \ \mathbf{y} \geq 0$$

$$d^-_{ij\cdot}, \ d^+_{ij\cdot} \geq 0 \text{ and } d^-_{ij\cdot} \times d^+_{ij\cdot} = 0, \ \forall \ i = 1, 2, \ldots, t, \ j = 1, 2, \ldots, m_i$$

$$d^-_{ik\cdot}, \ d^+_{ik\cdot} \geq 0 \text{ and } d^-_{ik\cdot} \times d^+_{ik\cdot} = 0, \ \forall \ i = 1, 2, \ldots, t-1, \ k = 1, 2, \ldots, n_i$$

$$(15)$$

Now the above model in Eq. (15) can be represented as follows:

$$\text{Min F} = \sum_{j=1}^{m_1} w^+_{1j\mu} d^+_{1j\mu} + \sum_{j=1}^{m_2} w^+_{2j\mu} d^+_{2j\mu} + \cdots + \sum_{j=1}^{m_t} w^+_{tj\mu} d^+_{tj\mu}$$

$$+ \sum_{j=1}^{m_1} w^+_{1j\lambda} d^+_{1j\lambda} + \sum_{j=1}^{m_2} w^+_{2j\lambda} d^+_{2j\lambda} + \cdots + \sum_{j=1}^{m_t} w^+_{tj\lambda} d^+_{tj\lambda}$$

$$- \sum_{j=1}^{m_1} w^+_{1j\nu} d^-_{1j\nu} - \sum_{j=1}^{m_2} w^+_{2j\nu} d^-_{2j\nu} - \cdots - \sum_{j=1}^{m_t} w^+_{tj\nu} d^-_{tj\nu}$$

$$+ \sum_{k=1}^{n_1} \left(w^\alpha_{1k\cdot}(d^{\alpha-}_{1k\cdot} + d^{\alpha+}_{1k\cdot}) + w^\beta_{1k\cdot}(d^{\beta-}_{1k\cdot} + d^{\beta+}_{1k\cdot}) \right)$$

$$+ \sum_{k=1}^{n_2} \left(w^\alpha_{2k\cdot}(d^{\alpha-}_{2k\cdot} + d^{\alpha+}_{2k\cdot}) + w^\beta_{2k\cdot}(d^{\beta-}_{2k\cdot} + d^{\beta+}_{2k\cdot}) \right) \quad \cdots \quad \cdots \quad \cdots$$

$$+ \sum_{k=1}^{n_{t-1}} \left(w^\alpha_{t-1k\cdot}(d^{\alpha-}_{t-1k\cdot} + d^{\alpha+}_{t-1k\cdot}) + w^\beta_{t-1k\cdot}(d^{\beta-}_{t-1k\cdot} + d^{\beta+}_{t-1k\cdot}) \right)$$

subject to

$$1 - \frac{o_{ij}(\mathbf{y}) - L^\mu_{ij}}{U^\mu_{ij} - L^\mu_{ij}} + d^-_{tj\mu} - d^+_{tj\mu} = 1, \ i = 1, 2, \ldots, t, \ j = 1, 2, \ldots, m_i$$

$$1 - \frac{o_{ij}(\mathbf{y}) - L^\lambda_{ij}}{U^\lambda_{ij} - L^\lambda_{ij}} + d^-_{tj\lambda} - d^+_{tj\lambda} = 0.5, \ i = 1, 2, \ldots, t, \ j = 1, 2, \ldots, m_i$$

$$1 - \frac{U^\nu_{ij} - o_{ij}(\mathbf{y})}{U^\nu_{ij} - L^\nu_{ij}} + d^-_{tj\nu} - d^+_{tj\nu} = 0, \ i = 1, 2, \ldots, t, \ j = 1, 2, \ldots, m_i$$

$$\frac{y_{ik} - \mu^L_{y_{ik}}}{y^*_{ik} - \mu^L_{y_{ik}}} + d^-_{1k\mu} - d^+_{1k\mu} = 1, \ i = 1, 2, \ldots, t-1, \ k = 1, 2, \ldots, n_i$$

$$\frac{\mu_{y_{ik}}^U - y_{ik}}{\mu_{y_{ik}}^U - y_{ik}^*} + d_{1k\mu}^- - d_{1k\mu}^+ = 1, \quad i = 1, 2, \ldots, t-1, \quad k = 1, 2, \ldots, n_i$$

$$\frac{y_{ik} - \lambda_{y_{ik}}^L}{y_{ik}^* - \lambda_{y_{ik}}^L} + d_{1k\lambda}^- - d_{1k\lambda}^+ = 0.5, \quad i = 1, 2, \ldots, t-1, \quad k = 1, 2, \ldots, n_i$$

$$\frac{\lambda_{y_{ik}}^U - y_{ik}}{\lambda_{y_{ik}}^U - y_{ik}^*} + d_{1k\lambda}^- - d_{1k\lambda}^+ = 0.5, \quad i = 1, 2, \ldots, t-1, \quad k = 1, 2, \ldots, n_i$$

$$\frac{y_{ik} - \nu_{y_{ik}}^L}{y_{ik}^* - \nu_{y_{ik}}^L} + d_{1k\nu}^- - d_{1k\nu}^+ = 0, \quad i = 1, 2, \ldots, t-1, \quad k = 1, 2, \ldots, n_i$$

$$\frac{\nu_{y_{ik}}^U - y_{ik}}{\nu_{y_{ik}}^U - y_{ik}^*} + d_{1k\nu}^- - d_{1k\nu}^+ = 0, \quad i = 1, 2, \ldots, t-1, \quad k = 1, 2, \ldots, n_i$$

$$G_1 y_1 + G_1 y_1 + \cdots + G_t y_t (\leq \ or \ = \ or \ \geq) \mathbf{q}, \ \mathbf{y} \geq 0$$

$$d_{ij.}^-, \ d_{ij.}^+ \geq 0 \text{ and } d_{ij.}^- \times d_{ij.}^+ = 0, \ \forall \ i = 1, 2, \ldots, t, \ j = 1, 2, \ldots, m_i$$

$$d_{ik.}^{\alpha-}, \ d_{ik.}^{\alpha+} \geq 0 \text{ and } d_{ik.}^{\alpha-} \times d_{ik.}^{\alpha+} = 0, \ \forall \ i = 1, 2, \ldots, t-1, \ k = 1, 2, \ldots, n_i$$

$$d_{ik.}^{\beta-}, \ d_{ik.}^{\beta+} \geq 0 \text{ and } d_{ik.}^{\beta-} \times d_{ik.}^{\beta+} = 0, \ \forall \ i = 1, 2, \ldots, t-1, \ k = 1, 2, \ldots, n_i,$$

$$(16)$$

where F represents the neutrosophic achievement function comprising the weighted over-deviational variables $d_{ij.}^+$, $\forall \ i = 1, 2, \ldots, t$, $j = 1, 2, \ldots, m_i$ of the neutrosophic goals l_{ij} and the under-deviational and over-deviational variables $d_{ik.}^{\alpha-}$, $d_{ik.}^{\alpha+}$, $d_{ik.}^{\beta-}$ and $d_{ik.}^{\beta+}$, $\forall \ i = 1, 2, \ldots, t-1$, $k = 1, 2, \ldots, n_i$ for the neutrosophic goals of all the decision variable vectors for the leader $t-1$-th levels. The corresponding weights $w_{ij.}^+$, $w_{ik.}^\alpha$ and $w_{ik.}^\beta$ depict the relative importance of attaining the aspired levels of the respective neutrosophic goals under the given constraints in the hierarchical decision-making scenarios.

To assign the different relative importance of the neutrosophic goals adequately, we have suggested the weighting scheme with the aid of u_{ij} and l_{ij}. The weighting scheme to each weight $w_{ij.}^+$, $w_{ik.}^\alpha$, and $w_{ik.}^\beta$ has been stated as follows:

$$w_{ij.}^+ = \frac{1}{u_{ij} - l_{ij}}, \quad \forall \ i = 1, 2, \ldots, t, \ j = 1, 2, \ldots, m_i \tag{17}$$

$$w_{ik.}^\alpha = \frac{1}{T_k^{i\alpha}} \text{ and } w_{ik.}^\beta = \frac{1}{T_k^{i\beta}}, \quad \forall \ i = 1, 2, \ldots, t-1, \ k = 1, 2, \ldots, n_i. \tag{18}$$

The NFGP model (16) gives the most satisfactory solution for the decision makers at all levels by attaining the aspired level of different neutrosophic membership goals at utmost possible in neutrosophic decision environment. The solution method is quite simple and demonstrated with the help of numerical examples in Sect. 5.

The step-wise solution procedure for the first NFGP algorithm for solving ML-MOLPPs can be stated as follows:

1. Solve each objectives individually for all levels under given constraints in order to find the maximum and minimum values of each objectives at all levels.
2. Depict the goals and upper tolerance limits—u_{ij}, l_{ij} ; $\forall\, i = 1, 2, \ldots, t,\; j = 1, 2, \ldots, m_i$—for each objectives at all levels.
3. Calculate the weights, $w_{ij\cdot}^{+} = \dfrac{1}{u_{ij} - l_{ij}}$, $\quad \forall\; i = 1, 2, \ldots, t,\;\; j = 1, 2, \ldots, m_i$ and set $g = 1$.
4. Evaluate the different membership functions $\mu_{o_{gj}}(o_{gj}(\mathbf{y}))$, $\lambda_{o_{gj}}(o_{gj}(\mathbf{y}))$ and $\nu_{o_{gj}}(o_{gj}(\mathbf{y}))$, $j = 1, 2, \ldots, m_g$ for each objective function under neutrosophic environment.
5. Develop the model given in Eq. (22) for the g-th level MOLPPs.
6. Obtain the value of $\mathbf{y}^{g*} = (\mathbf{y}_1^{g*},\, \mathbf{y}_2^{g*},\, \ldots,\, \mathbf{y}_t^{g*})$ by solving model given in Eq. (22).
7. Impose the maximum negative and positive tolerance limits on the decision variable vectors $\mathbf{y}_g^{g*} = (\mathbf{y}_{g1}^{g*},\, \mathbf{y}_{g2}^{g*},\, \ldots,\, \mathbf{y}_{gn_g}^{g*})$, $T_k^{g\alpha}$ and $T_k^{g\beta}$; $k = 1, 2, \ldots, n_g$.
8. Calculate the weights $w_{gk\cdot}^{\alpha} = \dfrac{1}{T_k^{g\alpha}}$ and $w_{gk\cdot}^{\beta} = \dfrac{1}{T_k^{g\beta}}$, $k = 1, 2, \ldots, n_g$.
9. Evaluate the different membership functions $\mu_{y_{gk}}(y_{gk})$, $\lambda_{y_{gk}}(y_{gk})$ and $\nu_{y_{gk}}(y_{gk})$ for the decision variable vectors $\mathbf{y}_g^{g*} = (\mathbf{y}_{g1}^{g*},\, \mathbf{y}_{g2}^{g*},\, \ldots,\, \mathbf{y}_{gn_g}^{g*})$ given in Eq. (12).
10. If $g > t - 1$, then proceed to step 11, otherwise go to step 4.
11. Depict the different membership functions $\mu_{o_{tj}}(o_{tj}(\mathbf{y}))$, $\lambda_{o_{tj}}(o_{tj}(\mathbf{y}))$ and $\nu_{o_{tj}}(o_{tj}(\mathbf{y}))$, $j = 1, 2, \ldots, m_t$ for each objective function at the p-th level under neutrosophic environment.
12. Calculate the weights, $w_{tj\cdot}^{+} = \dfrac{1}{u_{tj} - l_{tj}}$, $\forall\; j = 1, 2, \ldots, m_t$.
13. Formulate the model given in Eq. (16) under neutrosophic environment and solve it to get the satisfactory solution of the ML-MOLPPs.

4.3.2   The Second NFGP Algorithm for ML-MOLPPs

In the first NFGP algorithm, the final model contains the different membership functions for the neutrosophic goals of the decision variable vectors monitored by $t - 1$ levels, which separately solves for the i-th level MOLPPs. The second NFGP algorithm solves t MOLPPs that considers the decisions of the leader levels. After initialization steps 1 to 3 in first algorithm, the solution methods initiate with MOLPP of the first decision maker obtaining the compromise solution. A marginal evaluation of the first decision maker's decisions is taken into account to get rid of the decision deadlock. Hence, decisions of the first decision maker are depicted by the different membership functions under neutrosophic environment and sent to the second decision maker as additional auxiliary constraints. Afterward, the second decision maker considers the neutrosophic membership goals of the objectives as well as decision variable vectors of the first decision maker. After that, the achieved solution is passed to the third decision maker who tries to find out the optimal solution in a similar fashion. The processes of finding the optimal solution are repeated until the follower level is reached and consequently, the process is terminated.

The step-wise solution procedure for the second NFGP algorithm for solving ML-MOLPPs can be stated as follows:

1. Follow the same procedure from steps 1 to 9 as discussed in the first NFGP algorithm.
2. Formulate the model given in Eq. (16) for the ML-MOLPPs with $t = g$ under neutrosophic environment.
3. Solve the model given in Eq. (16) to get $\mathbf{y}^{g*} = (\mathbf{y}_1^{g*}, \mathbf{y}_2^{g*}, \ldots, \mathbf{y}_t^{g*})$.
4. Establish $g = g + 1$.
5. If $g > t$, then terminating with a satisfactory solution results $\mathbf{y}^{g*} = (\mathbf{y}_1^{g*}, \mathbf{y}_2^{g*}, \ldots, \mathbf{y}_t^{g*})$ to the ML-MOLPPs, otherwise proceed to step 7 of the first NFGP algorithm.

According to the solution priority, the second NFGP algorithm can be used to obtain the direct solution of the ML-MOLPPs to decisions of the first-level decision maker. After that, it directs the solutions to the decisions of second-level decision maker by preserving the solution closer to the decisions of first-level decision maker. Thus, the process goes on until the last level of the ML-MOLPPs preserving the solution closer to the decision of the leader levels.

5 Numerical Illustrations

The following numerical example consisting of tri-level multiobjective linear programming problems is presented to show the validity and applicability of the proposed NFGP optimization algorithms.

[1st level]

$$\underset{y_1}{\text{Min}}\ O_1(\mathbf{y}) = \underset{y_1}{\text{Min}}\ (o_{11}(\mathbf{y}) = y_1 - y_2 - 4y_3,\ o_{12}(\mathbf{y}) = -y_1 + 3y_2 - 4y_3),$$

where $\mathbf{y}_2$ and $\mathbf{y}_3$ solve

[2nd level]

$$\underset{y_2}{\text{Min}}\ O_2(\mathbf{y}) = \underset{y_2}{\text{Min}}\ (o_{21}(\mathbf{y}) = 2y_1 - y_2 + 2y_3,\ o_{22}(\mathbf{y}) = 2y_1 + y_2 - 3y_3,$$

$$o_{23}(\mathbf{y}) = 3y_1 - y_2 + y_3),$$

where $\mathbf{y}_3$ solves

Table 1 Individual minimum and maximum values for each objectives

	1st level		2nd level			3rd level	
	o_{11}	o_{12}	o_{21}	o_{22}	o_{23}	o_{31}	o_{32}
$\min_S o_{ij}$	-2.5	-3.5	-1	-1	-1	-0.5	0
$\max_S o_{ij}$	1	3	4	2	5	8.5	2

[3rd level]

$$\underset{y_3}{\text{Min }} O_3(\mathbf{y}) = \underset{y_3}{\text{Min }} (o_{31}(\mathbf{y}) = 7y_1 + 3y_2 - 4y_3, \; o_{32}(\mathbf{y}) = y_1 + y_3)$$

subject to

$$y_1 + y_2 + y_3 \leq 3, \quad y_1 + y_2 - y_3 \leq 1,$$

$$y_1 + y_2 + y_3 \geq 1, \quad -y_1 + y_2 + y_3 \leq 1,$$

$$y_3 \leq 0.5, \quad y_1, \; y_2, \; y_3 \geq 0.$$

The individual minimum and maximum values of each objective function for all the three levels of MOLPP under the given constraints S is furnished in Table 1. To apply the proposed NFGP algorithms, the aspiration levels and leader tolerance limits to the objective functions may be taken as the minimum and maximum individual optimal solutions.

The first NFGP algorithm can be elaborated through the solution method of the second NFGP algorithm. Thus, the following is the proposed first NFGP algorithm to tri-level multiobjective linear programming problem with the step-wise solution procedures.

First − level decision maker's NFGP model :

$$\text{Min } F_1 = 0.286d_{11\mu}^{+} + 0.154d_{12\mu}^{+} + 0.286d_{11\lambda}^{+} + 0.154d_{12\lambda}^{+} - 0.286d_{11\nu}^{-} - 0.154d_{12\nu}^{-}$$

subject to

$$-0.286y_1 + 0.286y_2 + 1.143y_3 + d_{11\mu}^{-} - d_{11\mu}^{+} = 0.714$$

$$-0.286y_1 + 0.286y_2 + 1.143y_3 + d_{11\lambda}^{-} - d_{11\lambda}^{+} = 0.143$$

$$-0.286y_1 + 0.286y_2 + 1.143y_3 + d_{11\nu}^{-} - d_{11\nu}^{+} = 0.03$$

$$0.154y_1 - 0.154y_2 + 0.62y_3 + d_{12\mu}^{-} - d_{12\mu}^{+} = 0.54$$

$$0.154y_1 - 0.154y_2 + 0.62y_3 + d_{12\lambda}^{-} - d_{12\lambda}^{+} = 0.21$$

$$0.154y_1 - 0.154y_2 + 0.62y_3 + d_{12\nu}^{-} - d_{12\nu}^{+} = 0.07$$

$$y_1 + y_2 + y_3 \leq 3, \quad y_1 + y_2 - y_3 \leq 1,$$
$$y_1 + y_2 + y_3 \geq 1, \quad -y_1 + y_2 + y_3 \leq 1,$$
$$y_3 \leq 0.5, \quad y_1, \ y_2, \ y_3 \geq 0, \tag{19}$$
$$d_{ij.}^-, \ d_{ij.}^+ \geq 0 \ \text{and} \ d_{ij.}^- \times d_{ij.}^+ = 0, \ \forall \ i = 1, \ j = 1, \ 2.$$

With the help of optimizing software, the optimal solution of the problem given in Eq. (19) is $\mathbf{y}^{1*} = (0.5, 0, 0.5)$. Assume that the first-level decision maker assigns $y_1^{1*} = 0.5$ along with the negative and positive tolerances $T_1^{1\alpha} = T_1^{1\beta} = 0.5$ and with the weights $w_{11.}^\alpha = w_{11.}^\beta = \dfrac{1}{0.5} = 2$, respectively.

Second − level decision maker's NFGP model :

$$\text{Min } F_1 = 0.286d_{11\mu}^+ + 0.154d_{12\mu}^+ + 0.286d_{11\lambda}^+ + 0.154d_{12\lambda}^+ - 0.286d_{11\nu}^-$$
$$- 0.154d_{12\nu}^- + 0.2d_{21\mu}^+ + 0.33d_{22\mu}^+ + 0.167d_{23\mu}^+ + 0.2d_{21\lambda}^+ + 0.33d_{22\lambda}^+$$
$$+ 0.167d_{23\lambda}^+ - 0.2d_{21\nu}^- - 0.33d_{22\nu}^- - 0.167d_{23\nu}^- + 2\left[d_{11.}^{-\alpha} + d_{11.}^{+\alpha} + d_{11.}^{-\beta} + d_{11.}^{+\beta}\right]$$

subject to

$$-0.4y_1 + 0.2y_2 - 0.4y_3 + d_{21\mu}^- - d_{21\mu}^+ = 0.2$$
$$-0.4y_1 + 0.2y_2 - 0.4y_3 + d_{21\lambda}^- - d_{21\lambda}^+ = 0.13$$
$$-0.4y_1 + 0.2y_2 - 0.4y_3 + d_{21\nu}^- - d_{21\nu}^+ = 0.04$$
$$-0.667y_1 - 0.33y_2 + y_3 + d_{22\mu}^- - d_{22\mu}^+ = 0.33$$
$$-0.667y_1 - 0.33y_2 + y_3 + d_{22\lambda}^- - d_{22\lambda}^+ = 0.18$$
$$-0.667y_1 - 0.33y_2 + y_3 + d_{22\nu}^- - d_{22\nu}^+ = 0.02$$
$$-0.5y_1 + 0.167y_2 - 0.167y_3 + d_{23\mu}^- - d_{23\mu}^+ = 0.17$$
$$-0.5y_1 + 0.167y_2 - 0.167y_3 + d_{23\lambda}^- - d_{23\lambda}^+ = 0.09$$
$$-0.5y_1 + 0.167y_2 - 0.167y_3 + d_{23\nu}^- - d_{23\nu}^+ = 0.01$$
$$2y_1 + d_{11\mu}^{-\alpha} - d_{11\mu}^{+\alpha} = 1, \quad 2y_1 + d_{11\mu}^{-\beta} - d_{11\mu}^{+\beta} = 1,$$
$$2y_1 + d_{11\lambda}^{-\alpha} - d_{11\lambda}^{+\alpha} = 0.5, \quad 2y_1 + d_{11\lambda}^{-\beta} - d_{11\lambda}^{+\beta} = 0.5,$$
$$2y_1 + d_{11\nu}^{-\alpha} - d_{11\nu}^{+\alpha} = 0, \quad 2y_1 + d_{11\nu}^{-\beta} - d_{11\nu}^{+\beta} = 0,$$

constraints (19)

$$d_{ik.}^{\alpha-}, \ d_{ik.}^{\alpha+} \geq 0 \text{ and } d_{ik.}^{\alpha-} \times d_{ik.}^{\alpha+} = 0, \ \forall \ i = 1, \ 2 \ k = 1. \tag{20}$$

$$d_{ik.}^{\beta-}, \ d_{ik.}^{\beta+} \geq 0 \text{ and } d_{ik.}^{\beta-} \times d_{ik.}^{\beta+} = 0, \ \forall \ i = 1, \ 2 \ k = 1.$$

The optimal solution for the second-level NFGP model in Eq. (20) is obtained as $\mathbf{y}^{2*} = (0.5, 0, 0.5)$, $(0.5, 0.998, 0.5)$, $(0.5, 0.5, 0)$. Suppose that second-level decision maker finalizes $y_1^{2*} = 0.998$ along with the negative and positive tolerances $T_1^{2\alpha} = 0.75$, and $T_1^{2\beta} = 0.25$ and with weights $w_{21.}^{\alpha} = \dfrac{1}{0.75} = 1.333$, and $w_{21.}^{\beta} = \dfrac{1}{0.25} = 4$, respectively.

Third − level decision maker's NFGP model :

$$\text{Min } F_1 = 0.286d_{11\mu}^+ + 0.154d_{12\mu}^+ + 0.286d_{11\lambda}^+ + 0.154d_{12\lambda}^+ - 0.286d_{11\nu}^-$$

$$- 0.154d_{12\nu}^- + 0.2d_{21\mu}^+ + 0.33d_{22\mu}^+ + 0.167d_{23\mu}^+ + 0.2d_{21\lambda}^+ + 0.33d_{22\lambda}^+$$

$$+ 0.167d_{23\lambda}^+ - 0.2d_{21\nu}^- - 0.33d_{22\nu}^- - 0.167d_{23\nu}^- + 2\left[d_{11.}^{-\alpha} + d_{11.}^{+\alpha} + d_{11.}^{-\beta} + d_{11.}^{+\beta}\right]$$

$$+ 1.33(d_{21.}^{-\alpha} + d_{21.}^{+\alpha}) + 4(d_{21.}^{-\beta} + d_{21.}^{+\beta})$$

subject to

$$- 0.78y_1 + 0.33y_2 + 0.44y_3 + d_{31\mu}^- - d_{31\mu}^+ = 0.06$$

$$- 0.78y_1 + 0.33y_2 + 0.44y_3 + d_{31\lambda}^- - d_{31\lambda}^+ =$$

$$- 0.78y_1 + 0.33y_2 + 0.44y_3 + d_{31\nu}^- - d_{31\nu}^+ =$$

$$- 0.5y_1 - 0.5y_3 + d_{32\mu}^- - d_{32\mu}^+ = 0$$

$$- 0.5y_1 - 0.5y_3 + d_{32\lambda}^- - d_{32\lambda}^+ = 0$$

$$- 0.5y_1 - 0.5y_3 + d_{32\nu}^- - d_{32\nu}^+ = 0$$

$$1.33y_2 + d_{21\mu}^{-\alpha} - d_{21mu}^{+\alpha} = 1.33, \quad 4y_1 + d_{21\mu}^{-\beta} - d_{21\mu}^{+\beta} = 3.99,$$

$$1.33y_2 + d_{21\lambda}^{-\alpha} - d_{21\lambda}^{+\alpha} = 0.94, \quad 4y_1 + d_{21\lambda}^{-\beta} - d_{21\lambda}^{+\beta} = 2.35,$$

$$1.33y_2 + d_{21\nu}^{-\alpha} - d_{21\nu}^{+\alpha} = 0.35, \quad 4y_1 + d_{21\nu}^{-\beta} - d_{21\nu}^{+\beta} = 1.86,$$

constraints (20)

$$d_{ik.}^{\alpha-}, \ d_{ik.}^{\alpha+} \geq 0 \text{ and } d_{ik.}^{\alpha-} \times d_{ik.}^{\alpha+} = 0, \ \forall \ i = 1, \ 2, \ 3 \ k = 1, \ 2.$$

$$d_{ik.}^{\beta-}, \ d_{ik.}^{\beta+} \geq 0 \text{ and } d_{ik.}^{\beta-} \times d_{ik.}^{\beta+} = 0, \ \forall \ i = 1, \ 2, \ 3 \ k = 1, \ 2.$$

$$\tag{21}$$

Table 2 Comparison of optimal solutions and satisfactory degrees of the given example

Proposed NFGP algorithm	Baky approach	Abo-Sinna approach	Shih approach
$(o_{11}, \mu_{11}) = (-2.499, 0.999)$	$(-2.498, 0.99)$	$(-2.21, 0.92)$	$(-2.21, 0.92)$
$(o_{12}, \mu_{12}) = (0.4941, 0.399)$	$(0.494, 0.39)$	$(-0.569, 0.55)$	$(-0.569, 0.55)$
$(o_{21}, \mu_{21}) = (1.002, 0.59)$	$(1.002, 0.6)$	$(1.88, 0.56)$	$(1.88, 0.56)$
$(o_{22}, \mu_{22}) = (0.498, 0.50)$	$(0.498, 0.5)$	$(-0.09, 0.7)$	$(-0.09, 0.7)$
$(o_{23}, \mu_{23}) = (1.002, 0.67)$	$(1.002, 0.67)$	$(1.09, 0.65)$	$(1.09, 0.65)$
$(o_{31}, \mu_{31}) = (4.491, 0.47)$	$(4.493, 0.45)$	$(2.62, 0.65)$	$(2.62, 0.65)$
$(o_{32}, \mu_{32}) = (1, 0.50)$	$(1, 0.50)$	$(0.899, 0.55)$	$(0.899, 0.55)$
$\mathbf{y}* = (0.5, 0.9975, 0.5)$	$(0.5, 0.998, 0.5)$	$(0.339, 0.61, 0.5)$	$(0.339, 0.61, 0.5)$

Table 3 Theoretical comparison of proposed NFGP algorithms with others

Proposed NFGP approach	Other approaches
Proposed approach considers the indeterminacy degree in decision-making process.	Abo-Sinna [1], Baky [8], and Shih et al. [16] cannot deal with indeterminacy in decision-making processes.
The overall satisfactory degree is achieved by attaining the neutrosophic fuzzy goals.	In [1, 8, 16] approaches, satisfactory degree is achieved by attaining the fuzzy goals.
It characterizes neutrosophic membership functions for both objectives and decision variables under neutrosophic environment.	Abo-Sinna [1], Baky [8], and Shih et al. [16] do not cover this aspects.
Additional predetermined parameters in indeterminacy and falsity degrees make the decisions more flexible according to decision makers' choices.	This facility is not provided in Abo-Sinna [1], Baky [8], and Shih et al. [16]

The final optimal solution for the ML-MOLPPs given in Eq. (21) is obtained as $\mathbf{y}^{3*} = (0.5, 0.9975, 0.5)$ with the different objectives values $o_{11} = -2.499$, $o_{12} = 0.4941$, $o_{21} = 1.002$, $o_{22} = 0.498$, $o_{23} = 1.002$, $o_{31} = 4.491$, and $o_{31} = 1$, along with membership functions $\mu_{11} = 0.999$, $\mu_{12} = 0.399$, $\mu_{21} = 0.590$, $\mu_{22} = 0.50$, $\mu_{23} = 0.67$, $\mu_{31} = 0.47$, and $\mu_{31} = 0.50$, respectively. A comparative study is performed among the proposed NFGP algorithm and presented in the Table 2. Other approaches reveal that the solution results are very close to [8], whereas [1, 16] give the same solution results for the presented numerical examples. Furthermore, the theoretical contributions in the domain of ML-MOLPPs are also summarized in Table 3.

6   Conclusions

This chapter proposes two different neutrosophic fuzzy goal programming algorithms for the solutions of ML-MOLPPs. The neutrosophic goal programming model is constructed to minimize the group tolerance of satisfactory degree of all the decision makers and to attain the highest degree for truth (unity), indeterminacy (half), and a falsity (zero) of each kind of the defined membership functions' goals to the utmost possible by minimizing their respective deviational variables and so that obtain the optimal solution for all decision makers. The primary advantages of the proposed two different neutrosophic fuzzy goal programming algorithms that the chances of refusing the solution repeatedly by the leader-level decision maker and reevaluation of the problem again and again by restating the defined membership functions required to reach the optimal solution would not arise.

The first NFGP algorithm considers the different membership functions for the defined neutrosophic goals of the objective functions at all levels as well as the different membership functions for the neutrosophic goals for the decision variable vectors at each level except the follower level of the ML-MOLPPs. The second NFGP algorithm solves the MOLPPs of the ML-MOLPPs by taking into account the decisions of the MOLPPs for the leader level only. A numerical example is presented to show the validity and applicability of the proposed NFGP algorithms with the fact that the degree of indeterminacy may arise in the hierarchical decision-making processes and can be overcome by utilizing the proposed algorithms. In future, it can be applied to real-life applications such as transportation, assignment, vendor selection, inventory control, supply chain, etc. and problems in multi-level decision-making scenarios.

Appendix

The NFGP approach to solve the single-level MOLPPs is presented to facilitate the achievement function $\mathbf{y}^{g*} = (\mathbf{y}_1^{g*}, \mathbf{y}_2^{g*}, \ldots, \mathbf{y}_t^{g*})$, $g = 1, 2, \cdots, t-1$. By using the same notations and symbols of this chapter, the NFGP model can be formulated for any g-th level MOLPPs and can be stated as follows:

$$\text{Min } F = \sum_{j=i}^{m_g} w_{gj\mu}^+ d_{gj\mu}^+ + w_{gj\lambda}^+ d_{gj\lambda}^+ - w_{gj\nu}^+ d_{gj\nu}^-$$

subject to

$$c_1^{gj} \mathbf{y}_1 + c_2^{gj} \mathbf{y}_2 + \cdots + c_t^{gj} \mathbf{y}_t + d_{gj\mu}^- - d_{gj\mu}^+ = 1, \quad j = 1, 2, \ldots, m_g$$

$$c_1^{gj} \mathbf{y}_1 + c_2^{gj} \mathbf{y}_2 + \cdots + c_t^{gj} \mathbf{y}_t + d_{gj\lambda}^- - d_{gj\lambda}^+ = 0.5, \quad j = 1, 2, \ldots, m_g \tag{22}$$

$$c_1^{gj} \mathbf{y}_1 + c_2^{gj} \mathbf{y}_2 + \cdots + c_t^{gj} \mathbf{y}_t + d_{gj\nu}^- - d_{gj\nu}^+ = 0, \quad j = 1, 2, \ldots, m_g$$

$$G_1 y_1 + G_1 y_1 + \cdots + G_t y_t (\le \text{ or } = \text{ or } \ge) \mathbf{q}, \ \mathbf{y} \ge 0$$

$$d_{gj\cdot}^-, d_{gj\cdot}^+ \ge 0 \text{ and } d_{gj\cdot}^- \times d_{gj\cdot}^+ = 0, \quad j = 1, 2, \ldots, m_g$$

References

1. Abo-Sinna, M. A. (2001). A bi-level non-linear multi-objective decision making under fuzziness. *Opsearch, 38*(5), 484–495.
2. Ahmad, F., & Adhami, A. Y. (2019). Neutrosophic programming approach to multiobjective nonlinear transportation problem with fuzzy parameters. *International Journal of Management Science and Engineering Management, 14*(3), 218–229.
3. Ahmad, F., Adhami, A. Y., & Smarandache, F. (2018). Single valued neutrosophic hesitant fuzzy computational algorithm for multiobjective nonlinear optimization problem. *Neutrosophic Sets and Systems, 22*. https://doi.org/10.5281/zenodo.2160357.
4. Ahmad, F., Adhami, A. Y., & Smarandache, F. (2019). Neutrosophic optimization model and computational algorithm for optimal shale gas water management under uncertainty. *Symmetry, 11*(4). ISSN 2073-8994. https://doi.org/10.3390/sym11040544. http://www.mdpi.com/2073-8994/11/4/544.
5. Ahmad, F., Adhami, A. Y., & Smarandache, F. (2020). Modified neutrosophic fuzzy optimization model for optimal closed-loop supply chain management under uncertainty. In Smarandache, F. & Abdel-Basset, M. (Eds.), *Optimization Theory Based on Neutrosophic and Plithogenic Sets* (pp. 343–403). Academic Press. ISBN 978-0-12-819670-0. https://doi.org/10.1016/B978-0-12-819670-0.00015-9. http://www.sciencedirect.com/science/article/pii/B9780128196700000159.
6. Arora, S., & Gupta, R. (2009). Interactive fuzzy goal programming approach for bilevel programming problem. *European Journal of Operational Research, 194*(2), 368–376.
7. Atanassov, K. T. (1986). Intuitionistic fuzzy sets. *Fuzzy Sets and Systems, 20*(1), 87–96. ISSN 01650114. https://doi.org/10.1016/S0165-0114(86)80034-3.
8. Baky, I. A. (2010). Solving multi-level multi-objective linear programming problems through fuzzy goal programming approach. *Applied Mathematical Modelling, 34*(9), 2377–2387.
9. Huang, C., Fang, D., & Wan, Z. (2015). An interactive intuitionistic fuzzy method for multilevel linear programming problems. *Wuhan University Journal of Natural Sciences, 20*(2), 113–118.
10. Lachhwani, K., & Poonia, M. P. (2012). Mathematical solution of multilevel fractional programming problem with fuzzy goal programming approach. *Journal of Industrial Engineering International, 8*(1), 16.
11. Maiti, I., Mandal, T., & Pramanik, S. (2019). Neutrosophic goal programming strategy for multi-level multi-objective linear programming problem. *Journal of Ambient Intelligence and Humanized Computing*, 1–12.
12. Osman, M., Abo-Sinna, M. A., Amer, A. H., & Emam, O. (2004). A multi-level non-linear multi-objective decision-making under fuzziness. *Applied Mathematics and Computation, 153*(1), 239–252.
13. Pramanik, S., & Dey, P. P. (2019). Multi-level linear programming problem with neutrosophic numbers: A goal programming strategy. *Neutrosophic Sets and Systems, 29*, 242–254.
14. Pramanik, S., & Roy, T. K. (2007). Fuzzy goal programming approach to multilevel programming problems. *European Journal of Operational Research, 176*(2), 1151–1166.
15. Sakawa, M., Nishizaki, I., & Uemura, Y. (1998). Interactive fuzzy programming for multilevel linear programming problems. *Computers and Mathematics with Applications, 36*(2), 71–86.
16. Shih, H.-S., Lai, Y.-J., & Lee, E. S. (1996). Fuzzy approach for multi-level programming problems. *Computers and Operations Research, 23*(1), 73–91.
17. Sinha, S. (2003). Fuzzy mathematical programming applied to multi-level programming problems. *Computers and Operations Research, 30*(9), 1259–1268.
18. Smarandache, F. (1999). A unifying field in logics: Neutrosophic logic. In *Philosophy* (pp. 1–141). American Research Press, UK.
19. Wan, Z., Wang, G., & Hou, K. (2008). An interactive fuzzy decision making method for a class of bilevel programming. In *2008 Fifth International Conference on Fuzzy Systems and Knowledge Discovery* (Vol. 1, pp. 559–564). IEEE.

20. Zadeh, L. (1965). Fuzzy sets. *Information and Control, 8*(3), 338–353. ISSN 00199958. https://doi.org/s10.1016/S0019-9958(65)90241-X.
21. Zhao, X., Zheng, Y., & Wan, Z. (2017). Interactive intuitionistic fuzzy methods for multilevel programming problems. *Expert Systems with Applications, 72*, 258–268.
22. Zheng, Y., Liu, J., & Wan, Z. (2014). Interactive fuzzy decision making method for solving bilevel programming problem. *Applied Mathematical Modelling, 38*(13), 3136–3141.

Multi-Attribute Neutrosophic Decision-Making in Dosimetric Assessment of Radiotherapy Imaging Techniques

R. Binu and Paul Isaac

1 Introduction

Neutrosophic set was described by Smarandache [13] as a generalization of fuzzy set and intuitionist fuzzy set with each element x of the universal set X characterized by three components consisting of truth, indeterminacy, and falsity membership values of x in X. Neutrosophic set theory offers a detailed understanding of the theoretical and mathematical paradigm under which speculative and ambiguous conceptual events can be handled by hierarchical representation of the Truth, Indeterminacy, and Falsehood elements. The main objective of the neutrosophic set is to narrow the gap between the vague, ambiguous, and imprecise real-world situations. The neutrosophic set has proliferated due to its wide range of conceptual specializations and application fields. It can be used to construct a mathematical model for real science and engineering problems. The consolidation of the neutrosophic set hypothesis with classical theories is a growing trend in mathematical research. Neutrosophic algebraic structures and its properties provide us with a solid mathematical foundation to clarify connected scientific ideas in designing, information mining, economic science, control system, and decision-making [1, 11].

In classical multi-attribute decision-making models (MADM), the weight of each attribute and the threshold of alternatives with crisp numbers are naturally considered. Moreover, the decision maker needs to test the parameters in the real-world problems by using linguistic hedges due to the lack of knowledge

R. Binu (✉)
Rajagiri School of Engineering and Technology, Kakkanad, India

P. Isaac
Bharata Mata College, Thrikkakara, India

and imprecise data. Bellman and Zadeh [2] first explored problems of decision-making in fuzzy environment. Pramanik and Surapati [8] researched MADM under a single valued neutrosophic set for modeling MADM in vague and imprecise data scenario. Jun Ye [20] has proposed a decision-making technique using the weighted correlation coefficient or the weighted cosine similarity measure of a single valued neutrosophic set. In this method data on alternatives with parameters are measured by the degree of three different level membership values, such as truth, indeterminacy, and falsehood, which provide a three-dimensional assessment of the information available.

P. Majumdar and S.K. Samanta [7] launched the notion of distance and similarity measure between two single valued neutrosophic sets and the idea of neutrosophic entropy measure. Said Broumi and Irfan Deli [4, 5] presented weighted similarity measure and matching function in decision-making of uncertain data with parameters. Biswas [3] implemented the entropy-based approach of gray relational analysis using hamming distance for MADM under a single valued neutrosophic set. These techniques are widely used in decision-making of imprecise and uncertain data, including image thresholding, precise contour estimation radiotherapy treatment and in stock exchange.

We consider the clinical application of the neutrosophic set in radio therapy treatment. An external machine [Linear Accelerator] focuses the external radiation beam on the treatment area using high-energy X-rays. The main aim of radiotherapy is to provide a specified dose of radiation to a tumor area as accurately as possible while limiting the spread of the dose to surrounding normal tissues. The degree of success of radiation therapy of tumor cells depends on localization of tumor position in treatment planning and tracing of tumor motion in treatment delivery. This can be achieved by verification of treatment field position through X-ray imaging procedures at the start of treatment and at regular intervals during the course of therapy. This helps to visualize the organ to be treated and to make sure there is minimization of setup errors and delivery of radiation to the patient is accurate. There are several strategies for examining set-up errors due to patient movement defects, inadequate positioning of shielding blocks, skin mark change in relation to internal anatomy and incorrect beam alignment during treatment. This has resulted in precision of treatment and exact location of the tumor position. Unnecessary exposure to patients during the delivery phase of radiation therapy for imaging technique is also a concern because of its potential to trigger secondary tumor formation and additional expose of radiation dose [6, 12]. The paper's aim is to develop a decision-making model from the available radiotherapy imaging techniques to optimize additional radiation exposure and gross tumor volume localization during treatment. The subjective and objective evaluation of the weight of criteria and the weighted similarity measure between alternatives and the ideal point leads to a better result and differentiate from other existing techniques.

The remaining portion of this paper is organized as follows. Section 2 discusses briefly the neutrosophic set operations and prerequisite for the comprehension of next sections. Section 3 provides the mathematical modeling framework for decision-making in detail with calculation of the decision matrix and the weight of the criteria. Section 4 constructs the application of the neutrosophic set in medical science and explains the process of finding the best radiotherapy imaging technique to minimize the radiation exposure. The findings and description of the proposed work relating to the comparative analysis and view of experts are also briefed in Sect. 5. Finally Sect. 6 presents a valid summary and future work of the proposed research.

2 Background

We recall some of the preliminary definitions and results in this section that are essential for a clearer and better interpretation of the coming sections.

Definition 2.1 ([16]) A *neutrosophic set* A of the universal set X is defined as $A = \{(x, t_A(x), i_A(x), f_A(x)) : x \in X\}$ where $t_A, i_A, f_A : X \rightarrow (^-0, 1^+)$. The three components t_A, i_A, *and* f_A represent membership value (Percentage of truth), indeterminacy (Percentage of indeterminacy), and non-membership value (Percentage of falsity), respectively. These components are functions of non-standard unit interval $(^-0, 1^+)$. □

Remark 2.1 ([14, 19]) If $t_A, i_A, f_A : X \rightarrow [0, 1]$, then A is known as Single Valued Neutrosophic Set (SVNS). □

Remark 2.2 In this thesis, we consider only SVNS. For simplicity SVNS will be called neutrosophic set. □

Remark 2.3 U^X denotes the set of all neutrosophic subsets of X or neutrosophic power set of X. □

Definition 2.2 ([16, 17]) Let A and B be two neutrosophic sets of X. Then A is contained in B, denoted as $A \subseteq B$ if and only if $A(x) \leqslant B(x) \forall x \in X$, this means that $t_A(x) \leq t_B(x)$, $i_A(x) \leq i_B(x)$, $f_A(x) \geq f_B(x)$, $\forall x \in X$. □

Definition 2.3 ([15, 16]) The complement of a neutrosophic set $A = \{x, t_A(x), i_A(x), f_A(x) : x \in X\}$ of X is denoted and defined as $A^c = \{x, f_A(x), 1 - i_A(x), t_A(x) : x \in X\}$. □

Definition 2.4 ([16]) Let $A, B \in U^X \forall x \in X$. Then

1. The union C of A and B is denoted by $C = A \cup B$ and defined as $C(x) = A(x) \vee B(x)$ where $C(x) = \{x, t_C(x), i_C(x), f_C(x) : x \in X\}$ is given by

$$t_C(x) = t_A(x) \vee t_B(x)$$
$$i_C(x) = i_A(x) \vee i_B(x)$$
$$f_C(x) = f_A(x) \wedge f_B(x).$$

2. The intersection C of A and B is denoted by $C = A \cap B$ and is defined as $C(x) = A(x) \wedge B(x)$ where $C(x) = \{x, t_C(x), i_C(x), f_C(x) : x \in X\}$ is given

by

$$t_C(x) = t_A(x) \wedge t_B(x)$$
$$i_C(x) = i_A(x) \wedge i_B(x)$$
$$f_C(x) = f_A(x) \vee f_B(x).$$

□

Definition 2.5 ([21]) A similarity measure between two neutrosophic sets A and B is a function defined as $S : X \times X \rightarrow [0, 1]$ which satisfies the following properties:

1. $S(A, B) \in [0, 1]$
2. $S(A, B) = 1 \Leftrightarrow A = B$
3. $S(A, B) = S(B, A)$
4. $A \subset B \subset C \Longrightarrow S(A, C) \leq S(A, B) \wedge S(B, C)$.

□

Definition 2.6 ([7]) The Entropy E of a neutrosophic set A of X defined as a function $E : X \rightarrow [0, 1]$ which satisfies the following properties:

1. $E(A) = 0$ if A is a crisp set
2. $E(A) = 1$ if $(t_A(x), i_A(x), f_A(x)) = (0.5, 0.5, 0.5)$ $\forall x \in X$
3. $E(A) = E(A^c)$
4. $E(A) \geq E(B)$ if A more uncertain than B
   i.e. $t_A(x) + f_A(x) \leq t_B(x) + f_B(x)$ and $|i_A(x) - i_{A^c}(x)| \leq |i_B(x) - i_{B^c}(x)|$.

□

Definition 2.7 ([7]) Let $A = \{x_i, t_A(x_i), i_A(x_i), f_A(x_i) : x_i \in X\}$ be a neutrosophic set defined on a universal set $X = \{x_1, x_2, \ldots, x_n\}$. The entropy measure E of a neutrosophic set A can be defined as

$$E(A) = 1 - \frac{1}{n} \sum_{i=1}^{n} (t_A(x_i) + f_A(x_i)) |i_A(x_i) - i_{A^c}(x_i)|.$$

□

Definition 2.8 ([10]) The Analytic Hierarchy Process (AHP) is a theory of measurement through pairwise comparisons and relies on the judgements of experts to derive priority scales. The comparisons are made using a scale of absolute judgements that represents, how much more, one element dominates another with respect to a given attribute. □

3 Decision-Making Procedure Using Subjective and Objective Evaluation in Neutrosophic Environment

The decision-making model is a scientific process based on the gathering and analysis of objective and structured information. The model helps decision makers to consider the situation, organize the information, and evaluate it, and then take action. Neutrosophic set and weighted similarity measure is used to model a real-world decision-making problem containing imprecise and vague data. The subjective and objective evaluation of the weights of criteria and decision making model in neutrosophic environment is developed through the analytical hierarchy process (AHP).

3.1 Basic Concepts

Definition 3.1 Let $A = \{x_i, t_A(x_i), i_A(x_i), f_A(x_i) : x_i \in X\}$ and $B = \{x_i, t_B(x_i), i_B(x_i), f_B(x_i) : x_i \in X\}$ be two neutrosophic sets defined on a universal set $X = \{x_1, x_2, \ldots, x_n\}$ where the components of A are not a scalar multiple of corresponding components of B at each point x_i. The orthogonal distance between two neutrosophic sets A and B can be denoted and defined as

$$d^{\perp}(A, B) = \sum_{i=1}^{i=n} \frac{\sqrt{(T_{AB}(x_i))^2 + (I_{AB}(x_i))^2 + (F_{AB}(x_i))^2}}{\max(|A(x_i)|, |B(x_i)|)},$$

where

$$T_{AB}(x_i) = [t_A(x_i)i_B(x_i) - i_A(x_i)t_B(x_i)]$$

$$I_{AB}(x_i) = [i_A(x_i)f_B(x_i) - f_A(x_i)i_B(x_i)]$$

$$F_{AB}(x_i) = [f_A(x_i)t_B(x_i) - t_A(x_i)f_B(x_i)].$$

Definition 3.2 The normalized orthogonal distance $d^{\perp}$ between $A, B \in U^X$ where $A \neq rB$ and r is a scalar, satisfies the following axioms:

1. $d^{\perp}(A, B) \geq 0$
2. $A = B \Rightarrow d^{\perp}(A, B) = 0$
3. $d^{\perp}(A, B) = d^{\perp}(B, A)$
4. $d^{\perp}(A, C) \leq d^{\perp}(A, B) + d^{\perp}(B, C)$ where C is any third neutrosophic set.

$\square$

Theorem 3.1 *The function* $S^\perp(A, B) = \frac{1}{1+d^\perp(A,B)}$ *defined between two neutrosophic sets A and B using normalized orthogonal distance is a similarity measure.*

□

Proof It is clear from the Definitions 2.5 and 3.2. □

Definition 3.3 Let $A = \{x_i, t_A(x_i), i_A(x_i), f_A(x_i) : x_i \in X\}$ and $B = \{x_i, t_B(x_i), i_B(x_i), f_B(x_i) : x_i \in X\}$ be two neutrosophic sets defined on a universal set $X = \{x_1, x_2, \ldots, x_n\}$. Let $w_i \in [0, 1]$ be the weight of each element x_i $(i = 1, 2, \ldots, n)$ with the property $\sum_{i=1}^{i=n} w_j = 1$. The weighted similarity measure using normalized orthogonal distance can be defined as follows:

$$WS^\perp(A, B) = \frac{1}{1 + \sum_{i=1}^{i=n} w_i \frac{\sqrt{(T_{AB}(x_i))^2 + (I_{AB}(x_i))^2 + (F_{AB}(x_i))^2}}{\max(|A(x_i)|, |B(x_i)|)}}.$$

Proposition 3.1 *The weighted similarity measure using normalized orthogonal distance satisfies the following properties:*

1. $0 \leq WS^\perp(A, B) \leq 1$
2. $WS^\perp(A, B) = WS^T(B, A)$
3. $WS^\perp(A, B) = 1$ *if* $A = B$.

□

3.2 Mathematical Model of Decision Problem

3.2.1 Alternatives and Criteria

Alternatives should represent varying solutions to the problem or different goals across objectives which create a set of innovative policy or management options tailored to achieve the goals.

1. Identify the alternative—Let $A = \{A_1, A_2, \ldots, A_m\}$ be a set of alternatives.
2. Select the criteria through which each alternative is evaluated—Let $C = \{C_1, C_2, \ldots, C_n\}$ be the set of parameters or criteria in relation to objects in A.

3.2.2 Construction of Decision Matrix

The characteristic of the alternatives A_i where $i = 1, 2, \ldots, m$ on criterion C_j where $j = 1, 2, \ldots, n$ is denoted by the single valued neutrosophic set defined on C_j as

$$A_i = \{C_j, t_{A_i}(C_j), i_{A_i}(C_j), f_{A_i}(C_j)\}.$$

Table 1 Pairwise
comparison scale

Verbal judgement	Numeric value
Extremely important	9 or 8
Very strongly more important	7 or 6
Strongly more important	5 or 4
Moderately more important	3 or 2
Equally important	1

A decision matrix $D = (\alpha_{ij})_{m \times n} = [a_{ij} \ b_{ij} \ c_{ij}]$ where $i = 1, 2, \ldots, m$ and $j = 1, 2, \ldots, n$ can be constructed from the above equation which is the evaluation of an alternative A_i with respect to a criterion C_j by the expert or decision maker. This step derives a pairwise comparison of each criterion, using a numerical comparison scale developed by Saaty [9] in analytic hierarchy process. Numeric values are derived from the subjective preference of individuals and the consistency ratio is calculated to avoid inconsistency in judgement (Table 1).

3.2.3 Calculation of the Criteria Weights

The weights of each criterion are calculated according to two parameters called subjective and objective weights. It is termed model synthesis derivation. The subjective and objective weight calculation corresponding to the criteria gives the decision-making procedure additional features. The subjective and objective weights are then combined to provide a hybrid model for the calculation of criteria weight.

Subjective Weight Calculation Subjective weight is the derivation of overall criteria priorities from the local priorities that represent the desired alternative for each criterion. Here the instructional tool is Analytic Hierarchy Process. Let $w_j(S)$ be the subjective weight of criterion C_j where $j = 1, 2, \ldots, n$ fixed by the decision maker such that, each $w_j(S) \in [0, 1]$ and $\sum_{j=1}^{j=n} w_j(S) = 1$.

Objective Weight Calculation Entropy is a significant element in the processing of uncertain information. The entropy measure can be used to evaluate the weights of attributes when it is imprecise and totally unknown for decision makers.

Definition 3.4 [7, 18] The entropy E_j of the criteria C_j where $j = 1, 2, \ldots n$ is defined as

$$E_j = \frac{1}{n} \sum_{i=1}^{i=n} (a_{ij} + c_{ij}) |b_{ij} - (b_{ij})^C| \tag{1}$$

and the entropy weight $w_j(E)$ of jth criteria C_j where $j = 1, 2, \ldots n$ is defined as

$$w_j(E) = \frac{1 - E_j}{\sum_{j=1}^{j=n}(1 - E_j)} \tag{2}$$

with $w_j(E) > 0$ and $\sum_{j=1}^{j=n} w_j(E) = 1$. $\Box$

Definition 3.5 ([3]) The combined weight w_j of subjective and objective weight can be defined as follows:

$$w_j = \frac{w_j(S)w_j(E)}{\sum_{j=1}^{j=n} w_j(S)w_j(E)} \tag{3}$$

with $w_j > 0$ and $\sum_{j=1}^{j=n} w_j = 1$. $\Box$

The objective weight and subjective weight are aggregated by non-linear weighted method addressing the limitation of considering only the effect of either subjective or objective factors in neutrosophic information.

3.2.4 Classification of Criteria and Ideal Point

Definition 3.6 ([21]) In the decision-making procedure, criteria are classified into two, according to their nature

1. Benefit criteria:- Maximum operator is used for identifying ideal alternative in benefit criteria
2. Cost criteria:-Minimum operator is used for identifying ideal alternative in cost criteria

$\Box$

In multi-criteria decision-making neutrosophic environment, the concept of ideal point has been used to identify the best attribute in the decision set.

Definition 3.7 In the multi-attribute decision-making process, the ideal point α_j $(j = 1, 2, \ldots, n)$ can be denoted and defined as follows:
For benefit criteria

$$\alpha_j = \langle \max_i a_{ij}, \min_i b_{ij}, \min_i c_{ij}, \rangle = \langle a_j, b_j, c_j \rangle.$$

For a cost criterion

$$\alpha_j = \langle \min_i a_{ij}, \max_i b_{ij}, \max_i c_{ij}, \rangle = \langle a_j, b_j, c_j \rangle.$$

$\Box$

Definition 3.8 The weighted similarity measure (W_{ij}) between an alternative $A_i, 0 < i \leq m$ and the ideal point $\alpha_j, \ 0 < j \leq n$ can be defined as

$$W_{ij}^{\perp}(A_i, \alpha_j) = \frac{1}{1 + \sum_{j=1}^{j=n} w_j d^{\perp}(A_i, \alpha_j)} = [W_{ij}]_{m \times n},$$

where

$$d^{\perp}(A_i, \alpha_j) = \frac{\sqrt{(T_{A_i \alpha_j})^2 + (I_{A_i \alpha_j})^2 + (F_{A_i \alpha_j})^2}}{\max(|A_i|, |\alpha_j|)}$$

$$T_{A_i \alpha_j} = a_{ij} b_j - b_{ij} a_j, \quad I_{A_i \alpha_j} = b_{ij} c_j - c_{ij} b_j, \quad F_{A_i, \alpha_j} = c_{ij} a_j - a_{ij} c_j$$

$$|A_i| = \sqrt{(a_{ij})^2 + (b_{ij})^2 + (c_{ij})^2}, \ |\alpha_j| = \sqrt{(a_j)^2 + (b_j)^2 + (c_j)^2}.$$

3.2.5 Ranking Order of Alternatives

The ranking order of all alternatives can be determined using the relation

$$A_i^* = \sum_{j=1}^{i=n} |(W_{ij} - s_{ij})|,$$

where

$$s_{ij} = \frac{2}{3} + \frac{a_{ij}}{3} - \frac{b_{ij}}{3} - \frac{c_{ij}}{3} \qquad (4)$$

s_{ij} represents test function of decision matrix $D = (\alpha_{lj})_{m \times n} = [a_{ij} \ b_{ij} \ c_{ij}]$ and we can arrive at a final decision using the above result in neutrosophic environment.

4 Decision-Making Problem in Medical Science

During radiotherapy there are several imaging methods for the precision of diagnosis, tumor volume localization and dosimetric analysis. The principal object of this analysis is to identify optimal dosimetric evaluation of imaging techniques of radiotherapy in neutrosophic scenario. The decision is based on the available imaging techniques (alternatives) evaluated on various dose evaluations (criteria) in radiotherapy. Imprecise and inconsistent data are obtained due to the setup errors in radio therapy due to patient movement, skin mark shift relative to internal anatomy, and incorrect beam alignment. For this purpose, we consider neutrosophic data with certain parameters as input data.

Table 2 Dosimetric assessment of exposure in radiotherapy and medical parameters

Alternatives	Imaging techniques	Criteria	Medical parameters
A_1	Portal imaging techniques	C_1	Dosimetry (Benefit Criteria)
A_2	C-True imaging	C_2	Prognosis (Benefit Criteria)
A_3	Cone beam computerized tomography	C_3	Environmental impact (Cost Criteria)
A_4	X-ray film method		

4.1 Data Set

The first stage of this frame work is to define available alternatives as a data set. Alternatives are represented by a set $X = \{A1, A2, A3, A4\}$ with four elements that are the available methods of dosimetric measurement in radio therapy imaging techniques for exposure accuracy. The oncologist must take a decision according to three criteria or parameters $E = \{C_1, C_2, C_3\}$. The table below gives a detailed overview of the four alternatives and three criteria we take for experimentation (Table 2).

4.1.1 Alternatives (Imaging Techniques)

1. Portal Imaging Techniques (A_1):- Portal imaging is the use of a therapeutic X-ray beam to form an image of the area being irradiated to study the set-up errors. Portal imaging has resulted in improved precision of treatment and can provide both geometrical and dosimetric details.
2. C -True Imaging (A_2):- C-True imaging technology from Tomo Therapy in partnered daily online images are quantifiable representations of patient anatomy originating from a mega-voltage X-ray source enabling patient set-up and dose-guided radiation therapy.
3. Cone beam computerized tomography (CBCT) (A_3):- Three-dimensional imaging technique using Kilovolt (KV)-X rays.
4. X- ray film Method (A_4) :-Using X-ray films instead of digital detectors for verification

4.1.2 Criteria (Medical Parameters)

The first stage of this frame work is to define available alternatives as a data set. Alternatives are represented by a set $X = \{A_1, A_2, A_3, A_4\}$ with four elements that are the available methods of dosimetric measurement in radio therapy imaging techniques for exposure accuracy. The oncologist must take a decision according to three criteria or parameters $E = \{C_1, C_2, C_3\}$.

1. Dosimetry (C_1):-The process or method of measuring the dosage of ionizing radiation by an object, usually the human body (benefit criteria).
2. Prognosis (C_2):-Forecasting of the probable course and outcome of a disease, especially the chance of recovery (benefit criteria).
3. Environmental impact (C_3):-It incorporate patient movement defects, inadequate positioning of shielding blocks, skin mark change in relation to internal anatomy, and incorrect beam alignment during treatment (cost criteria).

All alternatives are evaluated under the available criteria and the neutrosophic decision matrix D is constructed as follows using analytic hierarchy process [9] and dosimetric assessment.

$$
D = \begin{bmatrix} 0.45 \ 0.25 \ 0.35 & 0.50 \ 0.20 \ 0.30 & 0.80 \ 0.25 \ 0.45 \\ 0.65 \ 0.15 \ 0.25 & 0.65 \ 0.15 \ 0.25 & 0.45 \ 0.40 \ 0.45 \\ 0.45 \ 0.25 \ 0.35 & 0.55 \ 0.25 \ 0.35 & 0.45 \ 0.30 \ 0.80 \\ 0.75 \ 0.50 \ 0.15 & 0.65 \ 0.15 \ 0.20 & 0.65 \ 0.35 \ 0.85 \end{bmatrix} = (\alpha_{ij})_{4 \times 3}.
$$

4.1.3 Calculation of Subjective and Objective Weights

- Subjective Weight:-The weight $w = (0.30, 0.33, 0.37)$ of the criteria is given by experts in decision-making using analytic hierarchical process [9].
- Objective weight:-Entropy measure is used to calculate objective weight. From the Eqs. (1) and (2), the objective weight $w_j(E) = (0.45, 0.15, 0.40)$ using the matrix D.
- The combined weight w from the Eq. (3) after the subjective and objective evaluation on alternatives and criteria is $(0.40, 0.15, 0.45)$.

4.2 Evaluation

4.2.1 Ideal Point

In this phase we calculate an ideal point, a weighted measure of similarity. The ideal point α_j $(j = 1, 2, 3)$ from the above neutrosophic decision matrix D can be defined as follows using the Definition 3.7. α_j is constructed from the evaluation of alternatives A_i $(i = 1, 2, 3 \ and \ 4)$ with respect to the criteria C_j by the decision maker or experts.

$$
\alpha_j = \begin{bmatrix} \alpha_1 \\ \alpha_2 \\ \alpha_3 \end{bmatrix} = \begin{bmatrix} 0.75 \ 0.05 \ 0.15 \\ 0.65 \ 0.15 \ 0.20 \\ 0.45 \ 0.40 \ 0.85 \end{bmatrix}.
$$

4.2.2 Test Function of Decision Matrix

The test function $s = [s_{ij}]_{4 \times 3}$ of a decision matrix $D = [\alpha_{ij}]_{4 \times 3}$ is calculated as follows using the Eq. (4).

$$s = [s_{ij}]_{4 \times 3} = \begin{bmatrix} 0.625 & 0.670 & 0.703 \\ 0.753 & 0.753 & 0.538 \\ 0.601 & 0.654 & 0.455 \\ 0.703 & 0.769 & 0.489 \end{bmatrix}.$$

The following tables provide the experiment and observation data.

5 Results and Discussions

The ranking order of alternatives can be calculated from Table 3 by the parameter Ai^*. Choose the most elevated position (rank hierarchy) alternative as output and the most favorable alternative considering the above-mentioned criteria (Table 4).

Table 3 Proposed decision procedure between the alternatives (A_i) and criteria (C_j)

Alternative	Ideal point	$d^{\perp}(A_i, \alpha_j)$	$W_{ij}(A_i, \alpha_j)$	$\|W_{ij} - s_{ij}\|$	A_i^*
A_1	α_1	0.5211	0.8275	0.2025	0.6192
	α_2	0.0836	0.9876	0.3176	
	α_3	0.5483	0.8021	0.0991	
A_2	α_1	0.6635	0.7897	0.0367	0.6869
	α_2	0.1721	0.9748	0.2218	
	α_3	0.0772	0.9664	0.4284	
A_3	α_1	0.4116	0.8586	0.2576	0.6294
	α_2	0.0016	0.9997	0.3457	
	α_3	0.0813	0.7160	0.0261	
A_4	α_1	0.0000	1.0000	0.2970	0.6552
	α_2	0.0000	1.0000	0.2310	
	α_3	0.2030	0.91625	0.1427	

Table 4 Ranking order of alternatives A_i

Attributes	A_1	A_2	A_3	A_4
Rank	4	1	3	2

5.1 Comparative Study

Comparative research plays a vital role in the decision-making process. Through analyzing a comparative study, experts will identify potential benefits and predict future developments with accuracy. Here we consider three available weighted similarity measures (WSM) in literature [21] with proposed similarity measure for comparative study and identify the significant benefits with the same alternatives and criteria.

Table 5 Comparative study demonstration between the alternatives (A_i) and criteria (C_j)

Method	Alternatives	Ranking order
Cosine similarity	A_1	2
	A_2	1
	A_3	3
	A_4	4
Entropy method	A_1	1
	A_2	2
	A_3	4
	A_4	3
Matching function method	A_1	2
	A_2	1
	A_3	4
	A_4	3

Table 5 indicates that there is a difference in the order of the ranking. Not all strategies yielded the same results. To eliminate the variation and identify the best alternative, pair wise evaluation of alternatives, consistency ratio of criteria and clinical observations were applied by experts.

5.2 Experts Comments

Experts use statistical evidence and a direct interview on the above problem in decision taking to deduce the following conclusions.

1. Gross tumor volume coincides with planned and clinical target volume in C-True imaging which prevents additional exposed dose to normal structures and identify proper dosimetric analysis.
2. The risk factor for cone beam computerized tomography and X-ray film method is high because it affects the underlying anatomical structures (unsafe for patients due to additional exposure).
3. C-True imaging is the only imaging technique in which dosimetric analysis is conducted by anatomy of a patient position and isodose curves.

5.3   Inference and Results

The following points are listed in order to summarize the inference and discussions on the proposed decision-making problem.

- The best choice of alternative is A_2 (C-True imaging), considering given parameters dosimetry, prognosis, and environmental impact.
- The single valued neutrosophic set is used to analyze the data in all possible forms.
- Evaluation is based on subjective and objective weight of criteria.
- This approach reduces the data processing time and computational complexity.
- This method is very simple and effective.
- In the proposed decision-making model, the medium of instruction is the analytical hierarchy system, which involves a pair wise evaluation of alternatives and criteria priorities with data consistency ratio.

So it can be concluded that C -True imaging technique is the best dosimetric measurement of exposure in imaging techniques for radiotherapy (considering parameters and alternatives) with minimization of exposure dose and treatment precision.

6   Conclusion

The proposed multi-attribute decision-making technique for describing imprecise and ambiguous data is one of the generalized versions of the classical theories in decision-making. In this paper we consider weighted similarity measure as a parametrization tool and derive neutrosophic data ranking relationship. The best imaging technique in radiotherapy is analyzed taking into account all the alternatives and criteria as an application of single valued neutrosophic set in medical science. The procedure proposed in this paper for decision-making is convenient and simple to adopt for practical purposes. The decision-making method using weighted similarity measure can be applied to different fields such as engineering and medicine, as well as other highly complex circumstances of decision-making. We propose extending in our future research the decision-making methodology for medical diagnosis, data mining, and forecasting theory.

References

1. Ali, M. et al. (2014). Generalization of Neutrosophic Rings and Neutrosophic Fields. Infinite Study.
2. Bellman, R. E. & Lotfi A. Z. (1970). Decision-making in a fuzzy environment. *Management Science, 17*(4), B-141.

3. Biswas, P., Surapati P., & Bibhas C. G. (2014). Entropy based grey relational analysis method for multi-attribute decision making under single valued neutrosophic assessments. *Neutrosophic Sets and Systems, 2,* 102–110.
4. Broumi, S. & Florentin, S. (2013). Several similarity measures of neutrosophic sets. Infinite Study.
5. Broumi, S., Irfan, D., & Florentin, S. (2014). Distance and similarity measures of interval neutrosophic soft sets. In *Critical review, center for mathematics of uncertainty* (vol. 8, pp. 14–31). USA: Creighton University.
6. Gunderson, L. L. & Joel, E. T. (2015). Clinical radiation oncology. In *Elsevier health sciences.*
7. Majumdar, P. & Syamal Kumar, S. (2014). On similarity and entropy of neutrosophic sets. *Journal of Intelligent and Fuzzy Systems, 26*(3), 1245–1252.
8. Pramanik, S. et al. (2017). Multi criteria decision making using correlation coefficient under rough neutrosophic environment. Infinite Study.
9. Saaty, T. L. (1988). What is the analytic hierarchy process? In *Mathematical models for decision support* (pp. 109–121). Berlin: Springer.
10. Saaty, T. L. (2008). Decision making with the analytic hierarchy process. *International Journal of Services Sciences, 1*(1), 83–98.
11. Schumann, A. & Florentin, S. (2007). *Neutrality and many-valued logics.* Infinite Study.
12. Small, W. & Eric, D. D. (2012). Leibel and Phillips textbook of Radiation oncology. *JAMA, 307*(1), 93–93.
13. Smarandache, F. (1998). *Neutrosophy: neutrosophic probability, set, and logic: Analytic synthesis and synthetic analysis* (p. 105). Philosophy: American Research Press.
14. Smarandache, F. (1999). *A unifying field in logics: Neutrosophic Logic* (pp. 1–141). Philosophy: American Research Press.
15. Smarandache, F. (2003). *Proceedings of the first international conference on neutrosophy, neutrosophic logic, neutrosophic set, neutrosophic porbability and statistics.* Infinite Study. www.Gallup.Unm.Edu/$\sim$Smarandache/NeutrosophicProceedings.Pdf
16. Smarandache, F. (2005). Neutrosophic set-a generalization of the intuitionistic fuzzy set. *International Journal of Pure and Applied Mathematics, 24*(3), 287.
17. Smarandache, F. & Mumtaz, A., (eds.) (2015). *Neutrosophic sets and systems, book series* (vol. 9). Infinite Study.
18. Wang, J. Q. & Zhong, Z. (2009). Multi-criteria decision-making method with incomplete certain information based on intuitionistic fuzzy number. *Control and Decision 24*(2), 226–230.
19. Wang, H. et al. (2010). Single valued neutrosophic sets. Infinite study.
20. Ye, J. (2013). Multicriteria decision-making method using the correlation coefficient under single-valued neutrosophic environment. *International Journal of General Systems, 42*(4), 386–394.
21. Ye, J. (2014). Vector similarity measures of simplified neutrosophic sets and their application in multicriteria decision making. Infinite Study.

Index

Lightning Source UK Ltd.
Milton Keynes UK
UKHW020624150922
408905UK00002B/22